Brain Asymmetry and Ne

David W. Harrison

Brain Asymmetry and Neural Systems

Foundations in Clinical Neuroscience and Neuropsychology

 Springer

David W. Harrison
Behavioral Neuroscience Laboratory
Williams Hall
Virginia Polytechnic Institute
Blacksburg
Virginia 24061-0436
USA

ISBN 978-3-319-35911-3   ISBN 978-3-319-13069-9 (eBook)
DOI 10.1007/978-3-319-13069-9

Springer Cham Heidelberg New York Dordrecht London
© Springer International Publishing Switzerland 2015
Softcover reprint of the hardcover 1st edition 2015

The artwork on the cover was adapted from artwork entitled "The Good Earth" by Dr. Joseph Germana.

Printed on acid-free paper

Publishing is part of Springer Science+Business Media (www.springer.com)

The weight and the essence of the world rest squarely on the shoulders of the right brain as it watches over all else, including its literal and loquacious brother.
D. W. Harrison

To my friend and mentor Bill James

To my Children Nathan Andrew and Kendrick Connor

To the love of my life Patti Kelly Harrison

Preface

The modern behavioral neuroscience perspective extends directly across the centuries of time from early writings of Hippocrates stating that "From the brain and the brain alone arise our pleasures, joys, laughter and jests, as well as our sorrows, pains and grief." This evidenced-based philosophy, unlike some, finds direct support from science and is foundational for the discussions throughout the book. The initial discussions, though, will lend weight to other perspectives ultimately derived from one's belief system.

The subsequent writings are influenced more directly by the teachings of my patients, former and current students, and great thinkers within the field, including Alexander Luria, Roger Sperry, Kenneth Heilman, and many others. I have had many fine teachers over the years. Foremost among these have been the individuals that I have met recovering from one or another brain disorder and the caregivers, family members, and therapists actively engaged in recovery and the rehabilitation process with them. I have seen grace in their eyes and in their demeanor, where life is a passion and worth the process even in the struggle for recovery and with loss of functions.

There is clearly something in life that is greater than the pieces of us that fall subject to injury or disease. Each of these individuals has provided bits of the puzzle necessary to better understand the working brain and how to facilitate recovery and compensation for residual loss of the fragments of our behavior, emotion, and cognition. Many of the tools in our rehabilitation chest came directly from these teachers. Many individuals contributed to these writings through these teachers. Many other individuals contributed to these writings through their own scientific inquiry and theoretical developments.

Luria provided a conceptual framework, which is fundamental to modern neuropsychology and neuroscience, through his functional cerebral systems perspective and the appreciation of his basic functional units of the brain. Roger Sperry contributed much to the appreciation of two separate but integrated brains differing in their processing specialization, view of the world, and often oppositional take on it all. *They* clearly appear to see, hear, and feel the world differently as might two distinct individuals with oppositional bias in perception and expression, one against the other, but communicating and seeking balance and harmony through the interfaces or interconnections of the corpus callosum. Kenneth Heilman clearly advanced these topics in his theoretical accountings of emotion, spatial analysis, movement, memory, and so many other constructs.

Following a discussion of the mind body issue and the issue of localization of function, the basic functional units of the brain will be discussed, including the arousal systems, sensory perceptual or attentional systems, and the motor or intentional systems. This will be followed by discussion of functional neural systems theory. This will be followed by a discussion of the basic brain syndromes common to left sided or right sided brain pathology leading up to a brief account of neuropsychopathology for the clinician and clinical neuroscientist. The final sections of the book provide for specific topics on probable differences in specialization of each brain. Some of these topics evolve directly from our laboratory research findings and those of former students in the Behavioral Neuroscience Laboratory who each have contributed significantly to our literatures on cerebral lateralization of functions.

It has been a great pleasure to work with these fine folks and to come to know them a bit. I apologize in advance for any and all of the misconceptions or ill conceived adventures in the book. The errors are entirely mine. But some of the revelations of past and present science are expressed in the book from the perspective of the brains involved, I think.

Blacksburg, Virginia   David W. Harrison
Date

Acknowledgments

The artwork on the cover was adapted from a painting by Dr. Joseph Germana entitled "The Good Earth."

Contents

Part VII Brain Pathology

About the Author

David W. Harrison, Ph.D., is a clinical neuropsychologist, Director of the Behavioral Neuroscience Laboratory, and an Associate Professor within the Department of Psychology at Virginia Tech. He is also director of Neuropsychological & Counseling Services at Columbia Lewis Gale Medical Center in Roanoke, Virginia. Dr. Harrison received his Bachelor of Science in psychology/biology from the University of New Mexico and a Master of Science degree in biopsychology from the University of Georgia. He received his Ph.D. in biological psychology/ neuropsychology from the University of Georgia and went on to complete a clinical post-doctoral fellowship in neuropsychology, geriatrics, and behavioral medicine. He is a Charter Member of the American Psychological Society and a Fellow of the National Academy of Neuropsychology. He has authored/co-authored over 130 peer-reviewed publications, including books, book chapters and journal articles with his students in the fields of clinical neuropsychology and neuroscience.

Behavioral Neuroscience Laboratory
Virginia Polytechnic Institute
Blacksburg, VA 24062

Part I
Historical Foundations

Chapter 1
Introduction (Ignotum Per Ignotius: A Motto)

For any display or constriction in the range of behavior from among the vast reper-toire of possibilities available to the human organism and within its many settings or environmental contexts, there exists an equally large possibility of attributions and language descriptors to explain its origin and its meaning or lack thereof. In psychology, for example, the many theoretical positions within this one field have lead to a redundancy and exclusivity of languages, which separate our contributions and divide our scientific energies. Ultimately, this may hinder our progress in sci-ence and there is a need for a universal underlying construct and perhaps language to integrate and coordinate our accumulation of knowledge and progress toward a collective understanding. The fundamental basis for this integration may be found in neuroscience, and arguably languages, which are developed without a foundation in neuroanatomical bedrock, must ultimately undergo some degree of modification or transformation. This is the overriding perspective presented within this book.

We have the functional and anatomical equivalent of two separate brains (Sperry 1966, 1982; see also Springer and Deutsch 1998) in our head just as we have two kidneys, two lungs, and two of each of the nerves that interface with our body and with our world. The two brains may function in unison as a concerted system through communication, which occurs across the corpus callosum and other com-missures or tracts (cables henceforth). There are commonalities, which are shared by each brain as they each go about analyzing and controlling their side of the world and their side of the body. The left and the right side of the body and each side of our external world are largely evaluated and controlled by the opposing brain. Thus, the left side of the body and the left side of our external sensory world (vision, sound, touch, temperature, position senses) are controlled and understood largely by the right brain and vice versa.

The physical energies within the environment, which we are capable of converting into neural energies, undergo transduction and conversion through our sensory receptors and, subsequently, higher-order associative analysis perhaps as a function of our previous experiences with these energy bandwidths and patterns. This process may ultimately lead to the comprehension or understanding of the

© Springer International Publishing Switzerland 2015
D. W. Harrison, *Brain Asymmetry and Neural Systems,*
DOI 10.1007/978-3-319-13069-9_1

event(s). The sensory modalities are representative of our left and right hemibody and our left and right hemispace (the external world). Our first task is to understand the aspects of sensation and muscle control that are similar in comparisons of the left and right brains. Each cerebral hemisphere shares a fundamental structural and functional layout with the other brain with the relative location of the sensory and motor representations of our body and our world dispersed to specific regions. However, beyond these simpler comparisons the left and right brains become highly specialized and differentiated in their view of the world, their methods of communication, and their emotional identities.

Though researchers and clinicians have often naively expected symmetry of function, wherein two brains that look alike should behave and process information in a like fashion, instead the evidence exposes two separate but communicating brains that often do not agree (Gazzaniga et al. 1962; Sperry 1966, 1982; Gazzaniga 1998, 2000; see also Springer and Deutsch 1998). Indeed, these two entities are perhaps better viewed in the light of their differences. They do not see, hear, or feel the world in the same way. They do not process information in the same fashion. Rather than being two like structures charged with the left and right sides of things (our body and our world), they are much like two distinct representatives of vastly different cultures tied together by an interconnecting cable, the corpus callosum, wherein each cerebral hemisphere contributes in its own way to aspects of perception, comprehension, socialization, thought, and our emotional survival.

Each of these brains exists, at once collectively, as part of the overall functional entity we identify as self. Yet, they are clearly, at times, oppositional and at odds in the battle for a dominant position over the processing, comprehension, and perception of our world and ultimately in the regulation of our responsive nature or expression, which others may use too for their own attributions and views toward us. This oppositional nature was apparent to early scientist-clinicians, and our knowledge of this relationship ultimately evolved toward basic theoretical positions on the functional interactions between the cerebral hemispheres. This was expressed in various accounts of the cerebral "balance theory," where the vastly different processing styles and emotional tendencies of each brain were ultimately communicated across the corpus callosum with inhibition or downregulation of the homologous region of the adjacent brain. The logical, rationale, verbal, and linguistic analysis of the left brain might be inadequate to counter the processing bias of its neighbor residing at the other side of the skull. This oppositional neighbor residing at the right side of our skull is more intently perceptive of threat or spatial arrays conveying altered meaning to that conveyed in logic. A practical example is evident in the demonstration that patients with left-sided cerebral vascular accidents or strokes and with diminution of their speech–language systems are improved in their detection or analysis of lying or dissimilation. Patients with left cerebral lesions to Wernicke's area were now more capable in the detection of lying from others (Etcoff et al. 2000).

Intimate to the balance theory is the appreciation of changing metabolic states or activation and inactivation of specific functional neural units, which may at one time yield mild left frontal stress, for example, with social approach and linguistic speech demands and perhaps even with some associated happiness and enjoyment of the communicative person with whom we have interacted. The inhibition may be of homologous regions of the brain more prone to socially avoidant behavior, negative reflection on past social events, and negative emotional expressions, including anger, sadness, or social apprehension (fear). One might easily imagine stressors or cerebral activation conditions wherein the right hemisphere regions prone to the oppositional tendency of social avoidance, negative views toward others, and negative affective expressions would inhibit brain systems more prone to desire and to intend to positive social interactions, linguistic speech, and positive affect.

Thus, the interesting thing about these two oppositional brains attempting to achieve balance at any one point in time or another is that the balance is necessarily dynamic rather than static. In a normal brain, we may respond to our environmental stressors through metabolic activation of systems specialized to process that particular stressor. But, metabolic activation may be costly wherein we achieve heightened efficiency through a concurrent decrement in metabolism of the oppositional brain system. An analogy here might be of two antagonistic muscles charged with moving the skeleton in opposite directions (e.g., extending or flexing an arm). Incremental stress on the flexor muscle will result in increased metabolic rate to resist the stretch on that muscle concurrent with inhibition of the extensor muscle. A simplified accounting of this has been expressed in "quadrant theory" (Shenal et al. 2003; Foster et al. 2008; Foster et al. 2008), where inhibition runs not only across the cerebral hemispheres via the corpus callosum but also from the anterior to the posterior regions of each brain via the longitudinal tracts. Denny-Brown (1956) proposed a reciprocally balanced relationship between the frontal lobes and the posterior regions of the brain that is characterized by mutual inhibition. The dorsal pathways provide associative strength between visual and somatosensory analyses, and the ventral pathway provides associative strength between visual and auditory analyses. Regulatory control or inhibition over these associations arises from the frontal lobes and specifically over affective associations or kindling responses at the amygdaloid bodies, via the uncinate tracts.

Science clearly favors the role of the frontal lobe, within each cerebral hemisphere, as one of regulatory control over the regions in the back of the brain where sensory information is processed and comprehended. This leaves open the ready interpretation of angry outbursts and/or panic-like states as resultant from inadequate regulatory control by the right frontal region, which was expressed in "capacity theory" as follows (Carmona et al. 2009; Williamson and Harrison 2003; Foster et al. 2008; Mitchell and Harrison 2010; see also Klineburger and Harrison 2013). Again by inference from analogy, a muscle once stretched, with a corresponding resistance to stretch through heightened metabolic activation, may eventually reach an intolerable level beyond the capacity of the tissue to resist the stressor demands placed upon it. With every effort mounted to throw the resources of this muscle against the stressor, alas the capacity to challenge may be exceeded. In such a state of stress, one may see an abruptly diminished oppositional response with an immediate release of tone or regulatory control. In the case of the right frontal lobe's regulatory control over negative emotional dynamics, the response may now appear unexpectedly robust or reactively labile and unbridled. Such events might be recorded in the clinical research setting with panic attack and with rage, (for example, Foster and Harrison, 2002a; however, see Feinstein et al. 2013), with emotional release or "chills" on exposure to an especially provocative musical piece (Klineburger and Harrison 2013), and within the language systems with the onset of expressive speech deficits or stuttering (Foster and Harrison 2001).

In the following sections, we gain more familiarity with the arrangement and processing specializations of each brain region. We begin with a discussion of the three basic functional units of the brain (Luria 1973). Understanding these basic divisions provides a substantial basis for the understanding of any brain disorder. In addition, they provide a conceptual basis for understanding how information flows through the brain as with interactive functional cerebral systems contributing to the flow of information from the sensation of the event, to the comprehension and cross-modal analysis of our previous experiences with similar sensory events, and ultimately to the frontal lobes for expression or regulatory control (e.g., Denny-Brown 1956; Tucker and Derryberry 1992). We follow the visual pathways from the eyes to the occipital lobes and begin to appreciate the nature of problems which may develop secondary to dysfunction within this system. We discuss the somatosensory projections and ultimately the higher cortical processing or comprehension of our body parts and the basis for gesturing, postural changes, and eventually coordinated movement with sequential rule-regulated posturing underlying coordinated and complex movement. Auditory and vestibular systems are followed to higher brain regions, as are olfactory and gustatory systems. Through each of these accountings, the novice will initially appreciate the likeness or similarities between the two brains.

The student is always encouraged to place these seemingly symmetrical biological designs into the context of the identities, needs, and realities of the left and the right brains. In doing so, it is hoped that the reader may gain a somewhat improved understanding of brain pathology and functional change resulting from stroke, cancer or neoplasm, and perhaps traumatic brain injuries. And, within this discussion, perhaps, we develop new insights, allowing each of us alternative attributions toward our family members or others with brain disorders. This might, for example, include the negative attributions many family members or therapists may feel, and even express, toward an individual with a left frontal lobe syndrome of the amotivational and apathetic type (see Scott and Schoenberg 2011; see also Grossman 2002). To understand that activation of these left frontal systems results in increased happiness or positive affect, increased energy or "get up and go," increased social approach behavior/desire, and optimistic anticipation of future events and that deactivation through stroke, for example, may leave us with diminished desire or motivation and energy and decreased optimistic anticipation for future events, and may free us from the common but incorrect attributions that this person is "not trying" or that this person "wants me to do everything for them." Once we have improved attributions and understandings of brain disorders, we may at least hope for improved interventions. For many family members, this may reduce the emotional components, which are part of the loss or grieving we experience from acquired brain pathology in a previously independent and thriving family member.

Chapter 2
Mind–Body Issue

Neuropsychology and neuroscience, more broadly, is a relatively new area of study with no clinical neuropsychology textbooks available until the 1970s, with the notable exception of Alexander Romanovich Luria's initial publication of his text in 1966. However, there were many relevant resources even within the earliest writings in science. Indeed, the Egyptians described brain lesions some 5000 years ago and even provided insight into head injuries with a case study of "contra coup" damage to the brain. The early Greeks, including Hippocrates (460–379 BC; see Fig. 2.1), Aristotle (384–322 BC; see Fig. 2.2), and Galen (130–200 AD; see Fig. 2.3) each contributed their own accounting of the functions of the brain, which allow for some insights into their philosophical views on the mind–body issue. Hippocrates was clearly ahead of his time when he localized movement to the contralateral brain in his study of "the sacred disease" we know as epilepsy. Aristotle, having touched a brain without producing "feelings," related these functions to the heart. But Hippocrates stated that "from the brain and the brain alone arise our pleasures, joys, laughter and jests, as well as our sorrows, pains and grief." In this prescient publication, he established a basis for scientific inquiry that continues with the strongest vigor through the present day, locating the vast territory for scientific exploration squarely inside of the skull.

Notable among the great thinkers was Galen, later known as "the great physician" who provided an anatomical basis for human functions with his ventricular theory. Following his in-depth study of human brains, including those of the Roman gladiators that he studied, he reached the conclusion that inspired air interacted with the fluids within the ventricles to produce "pneuma." Intentional actions resulted with the flow of pneuma through the nerves out to the muscles resulting in volitional acts or behaviors. Thus, intentional action, involving movements of the musculoskeletal system, was conveyed from the ventricles within the brain down through the nerves and into the muscles, resulting in their enlargement upon exertion. It might even seem reasonable then that a skeletal muscle (e.g., bicep) appears to inflate with movement at the limb following in a predictable fashion. This theory

© Springer International Publishing Switzerland 2015
D. W. Harrison, *Brain Asymmetry and Neural Systems,*
DOI 10.1007/978-3-319-13069-9_2

Fig. 2.1 A likeness of Hippocrates, the "father of medicine." Originally published in *Historical Aspects of Hydatidosis of the Central Nervous System,* Turgut and Mehmet 2017, p. 7

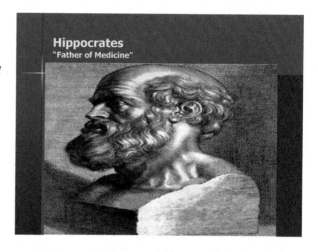

Fig. 2.2 A likeness of Aristotle who wrote "mind is the form that substance body takes." Originally published in *History of Regenerative Medicine,* Steinhoff and Gustav 2013, p. 4

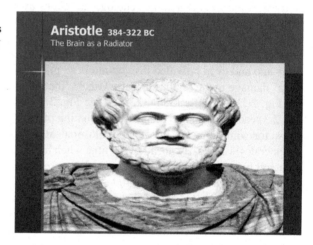

was eventually disproven by the selective cutting of a peripheral nerve without the predicted flow of pneuma from the nerves and out of the brain's ventricular system. In addition to the brain, little was known of the function of the spinal cord prior to Galen's anatomical studies. Galen carried out dissection and vivisection of animals providing details of the structure of the vertebral column, spinal cord, and nerve roots (see Pearce 2008). Despite different mechanisms of action, these early writers appear to ascribe aspects of personality, thought, and behavioral functions exclusively to the human body! Minimal, if any, reference is provided through these propositions of entities beyond the body or the need for an interface with a spiritual realm or divine entity located outside and beyond the bodily tissues fixed within a physical state of existence.

Fig. 2.3 A likeness of Galen, referred to during the dark ages as "the great physician." Originally published in *Concept of Aging as a Result of Slow Programmed Poisoning of an Organism with Mitochondrial Reactive Oxygen Species,* Skulachev, Vladimir P.; Bogachev et al. 2013, p. 318

Galen 130-200 AD
Cerebrum – Receives Sensory Imprints
Cerebellum – Commands the Muscles
Ventricles – Holds Fluids

Free-Will Construct

The overriding question or concern, which had spurred the mind–body controversy, was the issue of "free will" or what was referred to in earlier literatures in science as "volition." When laypeople are asked the question of their belief in *free will,* many endorse the construct. This belief defies much of the scientific perspective that we may eventually understand the human brain with the ability to predict behavior, thought, and emotions based on the understanding of these anatomical structures responding to, and interacting with, environmental events. The belief is that, no matter how advanced our scientific understandings, I will always be able to defy the predictions by implementing free will. Even as well-educated scientists, we may view the science of behavior with derision beyond its more basic and elementary efforts to explain intentional actions. For many, this lofty perch allows for a distinct individuality or identity, free of any physical or anatomical constraint, consistent with the core components of their belief system or philosophical view of the world. And for many, this is the fundamental basis for distinguishing human animals from the others, that the latter might be viewed as lower in the overall scheme of things.

Philosophical Doctrines

Throughout our history, there have been three primary philosophical doctrines relevant to the mind–body issue. The first philosophical doctrine for consideration is that of monism. The monistic doctrine holds that mind and body are the same thing and that there is no basis for attributing human functions to imagined or contrived entities beyond our bodies and their interface with the sensory array conveying information from our world or environment. La Mettrie, writing about 1750, with the publication of his manuscript "L'homme, la machine" provides a classic example of one advocating this perspective and also the dangers associated with this viewpoint. La Mettrie held that the machine was ripe for inquiry into its workings and the contributions of its parts.

Throughout history, scientists have attempted to understand human brain function based on the technology prevalent at the time. For example, the cognitive psychologist of the 1980s up until about 1995 developed elaborate theories of brain function using computer processing models and analogies. Let me say clearly, at this point, that the human brain is not a computer. But, this was viewed as a meaningful way to develop scientific language and to derive scientific predictions that might be tested experimentally in the laboratory. La Mettrie (Fig. 2.4) was impressed with the hydraulic technologies of his day with movement initiated, even in a previously fixed statue in the town square, through hydraulic force. It may help the reader to appreciate that volition or free will was often thought of specifically as the movements that we engage in or the volitional act of doing things. La Mettrie reasoned that if a statue can engage in volitional activities, and if we can understand these activities through understanding the machine, then man might be studied directly through scientific inquiry. The bottom line for La Mettrie was that the human body was a machine and that we could learn of its functions through the understanding of its mechanisms. The human body, though, was "the temple of God" and not subject to study from the dictates of the church. La Mettrie's view was insulting to many, including the Catholic Church, and to some was considered the view of a heretic.

Add to this the history of the torture of scientists, by the church, and the risks become more apparent. Many suffered under the merciless dictates of those representing the church. It was in 1997 when these actions of torture, imprisonment, and sometimes worse were publicly acknowledged through an apology by the pope. Many in science felt that it was too little and too late. But, regardless of your viewpoint, we can all acknowledge the conflict and the history of political and religious influence, which is most intimate to this controversy. We can also acknowledge the failings of science and the gradual evolution of ethical standards over time.

Fig. 2.4 Julien Offrey de La Mettrie, who argued that man could be studied like a machine. Copyrighted by Springer Science + Business Media, LLC

La Mettrie 1750
"L'homme, la machine"

The second philosophical doctrine is that of the dualist, which acknowledges the body and bodily functions but also the "soul" as a determinant of human activities. For our purposes here, "mind" is synonymous with "soul" and, as such, the mind–body issue is essentially an issue of whether or not our philosophical doctrine demands an accounting of a second entity (the soul). In modern times, for example, you may hear expressions of human functions either being "psychological" in origin or "organic." There is no basis for this distinction in science. Thus, the distinction is specifically identifying that individual's philosophical viewpoint more publicly as a dualist or one with a dualistic philosophy. Those "psychological" functions, then, are ascribed to something beyond the organism or involving "mental" activities that we cannot see or relate to brain functions. Void of these terms and ill-defined constructs, such as "mind," "mental," "unconscious," and the like, we may establish language derived specifically from functional aspects of a given brain location and within a functional brain system. These more functional terms having a basis in anatomy, include the aphasias, aprosodias, agnosias, alexias, agraphias, apraxias, and others.

Rene Descartes (1596–1650; see Fig. 2.5) was a politically astute gentleman. One might think him different from the monist La Mettrie for his involvement with the Catholic Church, while contributing substantially to the expanding willingness of the church to allow for the scientific study or inquiry into human functions. His influence, early on with the church, facilitated our movement out of "the dark ages" where the church substantially dictated and prevented (via torture when necessary) the activities of those who would do science. Descartes argued that much of human behavior is similar to that of "lower animals" wherein automatic or reflexive behaviors might be the subject of research without offense to the human body as "the temple of God." He reasoned, though, that other behaviors required conscious decision or free will, at which point the soul would interact with the body allowing for these high level or distinctly human processes. The church was responsive to such reasoning and Descartes cemented his place in science as "the father of physiological psychology."

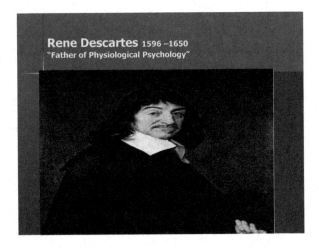

Fig. 2.5 The "father of physiological psychology," Rene Descartes. Originally published in *A Short History of Mechanics,* Allen and David H 2013, p. 13

Rene Descartes 1596 –1650
"Father of Physiological Psychology"

Descartes also reasoned that the soul would need a perfect place to interact with the body. After some inspection, he settled on the head due to its spherical shape perhaps. Many did not agree at that time, attributing aspects of personality, and the like, to the heart. Descartes also needed a location within the brain for the soul to interact for free-will processes. He selected the pineal body, partly due to its seemingly perfect shape, a sphere, and partly due to its apparent, but not actual, uniqueness as a unitary structure in the brain. This same structure is the focus of much research in modern neuroscience on the effects of ambient lighting in circadian entrainment, and it is known for its rich resources of melatonin associated with serotonin, one of our principle neurotransmitters involved in depression and violent-prone behavior. Of interest is the finding that levels of serotonin differ significantly in comparisons of those successful in committing suicide when compared with those who tried but were unsuccessful, with lowered serotonin levels in the former group (e.g., Stanley and Mann 1983; Fergusson et al. 2005).

The important thing for the reader is to know your own philosophical viewpoint or belief for it will determine what you make of the scientific findings. Indeed, research findings indicate that your philosophical view determines that which you are able to see. So, in knowing your viewpoint, you may be better able to appreciate your perceptual biases, whatever they may be. You will tend to see that which you have learned to see. The effect of learning on perceptual bias is well established in our literatures. A classical example, within the news media and within psychology, was the beating of Rodney King. The nation witnessed much of the same "objective evidence" presented within the courtroom where an African American man was repeatedly beaten by police officers in California. But, the perception of what was being viewed differed as a function of the learning history and perceptual biases of the observers. Two people might look at exactly the same evidence and perceive with vastly different interpretations, based on their experiences.

One man experienced something somewhat similar as an undergraduate after volunteering in a variety of settings, including an adult day care facility with the elderly and an inpatient psychiatric ward. In the latter setting, he met many wonderful folks dealing with one or another difficulty from a psychiatric perspective. One had gelastic lability with auditory hallucination as he carried on elaborate conversations with nonexistent people. His speech consisted of fluent but meaningless speech, which was often incoherent. He suspected that this man's "schizophrenic" diagnosis more accurately reflected a left temporal lobe or thalamic brain disorder. Another was a young woman who had self-admitted to the psychiatric facility with a history of suicidal ideation and behavior. After several months, the student was accepted into the University of Georgia's doctoral program. Prior to leaving, he visited the ward once more to say goodbye and that he would be moving away from the area.

He had very little money as a student and felt lucky to have a basement apartment in downtown Albuquerque (near "Old Town Albuquerque") with a partial dirt floor. He also had a Volkswagen bus that was a world of fun to drive even though he could not afford a battery for it! It required only a small 6-V battery, but that was an unnecessary expense at the time. Albuquerque rests on the foothills of the Sandia Mountains on a descending slope into the valley below. Thus, he was always parked on a hill, needing only to let the brake off to begin to roll forward, at which point he would "pop the clutch" to engage the engine. It was hard to get dates with this vehicle, though. At least those were his attributions for his state of affairs! His apartment door was something out of an Alfred Hitchcock thriller with glass panes defending him from the homeless, alcoholics, and drug users who rested against his basement windows.

It was some 2 weeks later when, in the middle of the night, he awoke with pervasive apprehension, a sense of urgent desperation, and the stark realization that something was terribly wrong. He had been awakened by the sounds of her woodensoled "clog" shoes coming down his basement stairs and ultimately as she struggled through the glass pane of his door next to the door knob and lock. It was at that very moment that he learned something about himself with deep insight and still without adequate comprehension. He found himself possibly on death's doorstep with only one objective as she was breaking into his apartment. He was determined to put his pants on! Accomplishing this seemingly irrelevant task, he met his assailant and struggled with her in self-defense. As her hand went into her bag, he was certain that she had a gun. None was ever found. As they struggled still up the stairs, they literally flopped onto the grounds at the back of the large house above his apartment.

Albuquerque has a long history by American standards, including its role in fighting tuberculosis (TB) as many of those afflicted by the disease relocated to the southwest for the dry and arid conditions, thought to be good for these patients. TB has an affinity for high levels of oxygen and is often found disproportionately in the upper lobes of the lungs. Therefore, higher altitudes with reduced oxygen was hoped to be beneficial in treating the disease. The city has many small Quonset huts, otherwise known as "TB huts" built for some of these patients and now used for student housing! I doubt that this information is revealed in the student's lease agreement. Regardless, there were two of these Quonset huts located in the backyard of the dwelling, which rested above his basement apartment. He had not met the coeds residing there and, as they were aroused by the battle, they saw what they had learned to see. One commonly held stereotype applied to the perceptual analysis of a man and woman engaged in fisticuffing is that "men do bad things." Indeed, many will stand vigorously behind such stereotypes, until confronted with conflicting data. Classic examples might be provided by the Iraq War, where we learned of torture and rape of prisoners held under American command (Armed Forces Press Release 2006). The woman commanding all Iraq detention facilities, Brigadier General Janis Karpinski, was reprimanded for dereliction of duty and then later demoted in rank. Specialist Lynndie England (Fig. 2.6) was sentenced to a prison term for these acts. In actuality, women are capable of and indeed do harmful things, like men.

The women from the Quonset huts saw what they had learned to see and swiftly came to the aid of his assailant, swinging their purses onto him. At this point, he clearly had his best stuff out with three women on him and nothing but his physical ability and athleticism to overcome these attacks. Actually, he informed them that this was a psychiatric patient and asked them to please call for assistance, at which point they broke off and assisted him through the point of relief by the officers. For

Fig. 2.6 Torture and rape at Abu Ghraib prison in Iraq. *Wikipedia, the Free Encyclopedia*. Wikimedia Foundation Inc., Web. 07 Oct. 2011. <http://en.wikipedia.org/wiki/Baghdad_Central_Prison>

the young man, it was a learning experience. He went with this woman through the admission process and tried to assist with the health-care professions for readmission to the psychiatric ward. She indicated to the young man her anger and despair as she felt that he had abandoned her like so many others had done before. But the point here is that we see what we have learned to see. The young man saw a gun that did not exist. The coeds initially saw a man attacking a woman, instead of a more valid perception.

For science, the mind–body issue developed into an entirely new issue of localizing behavior, cognition, and emotions to functional brain systems. Specifically, the methods of science allow only for the study of physical events, which may be seen, heard, or felt, but which are ultimately measurable or quantifiable in some respect. We have all heard that "a mind is a terrible thing to waste." But no one has, at this point, ever seen a mind. No one has ever held one in their hand. No one has ever measured it in its breadth or depth or circumference. No one has ever recorded its activity in any meaningful way. This does not mean that the soul or mind does not exist. But, it comes down to your belief or philosophy. Many neuroscientists are dualists or interactionists and many are monistic in their thinking. One such Nobel laureate and neuroscientist had spent his many years in science studying the neuron and dendritic fields. Despite these monistic activities derived from the scientific method, this gentleman expressed his philosophical views as a dualist. The methods of science may be flawed and inadequate in this respect for many of those reading this book.

Chapter 3
Localization Issue

For the neuroscientist, what had initially been an issue of mind or body evolved into a controversy among those with evidence of the unique contributions of each brain area to a given function and those impressed by the function of the brain as a whole, where brain mass was often viewed as more important than brain location in determining function or functional loss with its removal. This became a controversy among the radical viewpoints of the localizationist and the equipotentialist. The controversy among these two opposing schools of thought hinged on several major historical events, including the discovery of electricity with Reverend Abraham Bennett first demonstrating the production of electric charges by the contact of two dissimilar metals in 1789 (see Sanford 1915). This discovery was followed shortly with the scientific insight that Luigi Galvani (1791; see Fig. 3.1) shared in his discovery of the principle of ionic flow, where muscle movement resulted from these processes. Moreover, Galvani concluded that the muscle contractions resulted from electrical stimulation and that the source of the electricity must be in the tissues of the animal's body, since the metal only served for conduction of the current. This seemed even more probable, since it was known that certain eels and fish were capable of producing violent electrical shocks. Galvani concluded that all animals were capable of producing electricity from their body tissues.

By 1849, Emil Du Bois-Reymond (1848; see Fig. 3.2) had used a galvanometer to measure the amplitude of a nerve impulse; followed the next year by Hermann von Helmholtz's (1850a, b, c; see Fig. 3.3) measurement of the velocity of a neural impulse (about 50 m/s.). Emil Du Bois-Reymond and his friend Hermann von Helmholtz had been students of Johannes Peter Muller. It was Johannes Muller (1838; see Sinclair 1955) who first clearly formulated the bases for the separation of the sensory modalities, such that the same stimulus can elicit different sensory modalities as a function of the anatomical pathway that it activates. Moreover, he

© Springer International Publishing Switzerland 2015
D. W. Harrison, *Brain Asymmetry and Neural Systems*,
DOI 10.1007/978-3-319-13069-9_3

Fig. 3.1 Luigi Galvani discovered the principle of ionic flow, and that muscle movement resulted from these processes

related the anatomical basis for his *doctrine of specific nerve energies* not only to the nerve pathway which was activated but also to the terminal processing at the other end of the pathway within the brain. For example, mechanical pressure may activate visual sensations if applied to nerve pathways or brain areas devoted to visual processing. Conversely, he believed that the peripheral terminations of nerves were specialized to react preferentially to certain stimuli, a function which he referred to as *specific irritability*. This specificity in the irritability of a sensory pathway identifies the *adequate stimulus* for each modality where electromagnetic energy or light is the adequate stimulus for vision, whereas the retinal receptors may also activate to a pressure stimulus (e.g., rubbing the eyes) and again producing a visual response (light). These principles are well established in modern neuroscience and remain central to the area of psychophysics and the study of sensation and perception.

Measurement would become more and more in vogue with the eventual and continuing discoveries of regional maps within the brain. A rough estimate might indicate that about 40–60 % of the brain has been mapped at this writing. But, these efforts account for much of the modern research in neuroscience using the ever more sophisticated imagery equipment (see Raichle 1998). This process of discovery and survey of regional areas began with Woolsey (see Fig. 3.4) and many still

Fig. 3.2 Emil Du Bois-Reymond measured the amplitude of the neural impulse. He and his friend Hermann von Helmholtz had been students of Johannes Peter Muller

refer to the functional distributions across the cortex as "Woolsey's maps." In his initial studies, Woolsey focused on the somatosensory cortex, followed shortly by the extension of his work to auditory and visual areas. It was known that regions of the cochlea responded selectively to different tone frequencies, but little was known about the auditory cortex. Woolsey selectively stimulated localized regions in the cochlea with projection fibers exiting the cochlea in the auditory nerve. In doing so with great technical savvy, he and his colleagues mapped the patterns of evoked responses on the auditory cortex of the cat and monkey. This was the first clear demonstration of the tonotopic or cochleotopic organization of the auditory cortex. Early in the 1940s, Woolsey discovered the existence of a second somatosensory receiving area in the cortex of the cat, dog, and monkey and subsequently discovered secondary auditory and visual areas. He continued his cortical mapping work with humans and made comparative discoveries where the cortical map of the body, for example, would correspond to the relative separation of the functions of the face and hand for that species (e.g., Woolsey et al. 1979, 1942; Woolsey 1952).

Fig. 3.3 Hermann von
Helmholtz measured
the velocity of a neural
impulse. *Wikipedia, the Free
Encyclopedia*

Fig. 3.4 Clinton N. Woolsey
(1904–1993). He mapped
cortical responses to stimuli
across species using barbiturates to depress background
activity. The mapping
activities he initiated continue
today. Originally published in
*The Historical Development
of Ideas About the Auditory
Cortex,* Winer et al. 2011,
p. 21

Fig. 3.5 Hitzig, along with Fritsch, used electrical stimulation of the brain to produce movement

With electrical apparatus available in the laboratory, researchers began to stimulate specific brain regions and by 1870 Fritsch and Hitzig (see Fig. 3.5), using galvanic stimulation (direct current, DC), had localized motor functions across the motor cortex in the frontal lobe in the dog. Shortly after Fritsch and Hitzig's initial discovery of the motor cortex in the dog brain, Ferrier (1876, 1886) went on to study the precentral gyrus in monkeys; establishing the motor map on the cortex in the primate brain with the leg located dorsally and the mouth located ventrally. He also evoked eye movements and functions over areas that would become known as the frontal eye fields (dorsolateral frontal) and the parietal eye fields (lateral intraparietal area; e.g., Suzuki and Gottlieb 2013). Importantly, Ferrier used alternating current (AC) rather than DC stimulation, which could be maintained for a longer duration allowing for movements to unfold over time. Thus, twitches in muscle became seemingly coordinated, purposeful, and intentional movements. For example, stimulation of the motor cortex area representing the hand and the mouth yielded this account of the actions: "…brings the hand up to the mouth, and at the same time the angle of the mouth is retracted and elevated" in a manner resembling a feeding movement (1874, p. 418; see Graziano 2009). Dorsal stimulation over the cortical representation of the leg had the following effect. "…just such as when a monkey scratches its abdomen, with its hind leg."

The first direct stimulation of the human brain was performed by Bartholow (1874; see Fig. 3.6). He applied the techniques used by Fritsch and Hitzig to a

Fig. 3.6 Robert Bartholow applied a small electrical current to different sections of Ms. Rafferty's exposed brain, causing her body parts to move

30-year-old woman named Mary Rafferty with an open ulcer on her scalp, extending across the bilateral parietal regions. This opening had been produced from the friction of wearing a whale bone in her wig. This provided a 2-in. diameter access through which the pulsation of the brain could be clearly viewed and where insulated needles were inserted providing faradic stimulation on closing the circuit. This resulted in contralateral movements with "distinct muscular contractions in the right arm and leg. The arm was thrown out, the fingers extended, and the leg was projected forwards. The muscles of the neck were thrown into action and the head was strongly deflected to the right." Also, the low-voltage electrical current that he applied to the brain did not seem to cause her any pain. However, when Bartholow applied a larger amount of current she developed convulsions, went into a coma, and eventually died 3 days later following a major seizure. Bartholow continued on professionally and by 1893 had achieved the title of Professor Emeritus at Jefferson Medical College in Philadelphia, despite some criticism of the experiment by the American Medical Association.

Application of a low-voltage AC on the order of 3 V at 60 Hz is optimal for this purpose. Stimulation at the base of the left frontal lobe will yield movements especially of the right hemiface and with speech output more probable. Above this area, stimulation will result in the movement of the right arm and/or hand. The right leg may be activated by stimulation at the dorsal most part of the motor cortex (see Fig. 3.7). Hitzig, working with humans, stimulated the dorsolateral frontal lobe region and discovered the "frontal eye fields" where activation of the left frontal eye

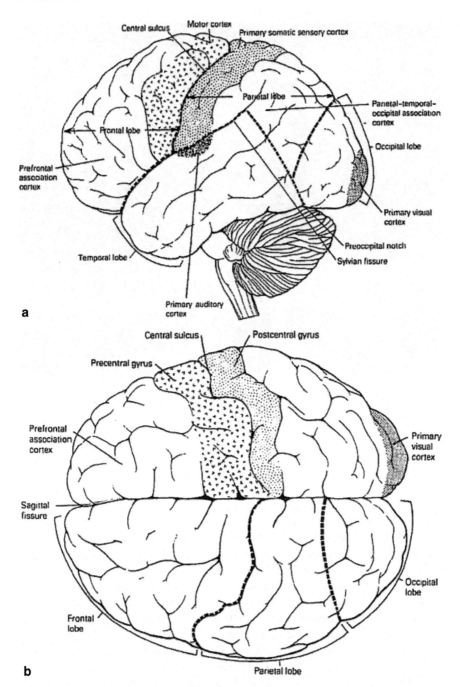

Fig. 3.7 The topographical representation of the primary sensory and motor projection areas. Originally published in *Brain Structures, Transmitters, and Analyzing Strategies,* Başar and Erol 2011, p. 56

Fig. 3.8 Penfield, along with his friend Jasper, mapped much of the functional locations of the cerebral cortex as epileptic patients underwent commissurotomy. Copyright Osler Library, McGill University

field results in rightward directed eye movements and where right frontal eye field activation results in leftward directed eye movements. By comparison here, Exner's hand area was found to be close or proximal to the frontal eye field, allowing for the contralateral regulatory control of coordinated hand movements. Note that these are intentional or volitional acts previously ascribed to "freewill."

By 1950, Penfield, a student of Sherrington's (see Fig. 3.8), was working with epileptic patients undergoing commissurotomy or section of the corpus callosum, which connects the two cerebral hemispheres. High-voltage electrical spikes, as may occur with epileptiform activity or seizure, have been found to produce mirror lesions at the homologous brain region, with sectioning performed as a preventive technique (Van Wagenen and Herren 1940). But the brain has no pain receptors in it! Thus, with local anesthetic to the scalp, skull, and meninges, surgery may be initiated on the brain with the patient alert and awake. In this setting, the neuropsychologist may evaluate the function of various brain regions (e.g., Gazzaniga et al. 1962, 1967; see Gazzaniga 2012). Like Fritz and Hitzig before him, Penfield was able to make many functional comparisons with stimulation of a specific cortical area yielding corresponding sensory or motor functions (Penfield 1967; Penfield and Rasmussen 1950). Penfield along with his friend Jasper mapped much of the functional locations of the cerebral cortex as epileptic patients underwent commissurotomy. It should be clear, at this point, that these are localizationist findings with unique and specific functions being mapped onto specific cortical regions.

The behaviorist B. F. Skinner defined reinforcement in operant conditioning as a stimulus which follows a behavior in a consequential way so as to alter the probability of the behavior and for the establishment of response habits (see Olds 1973). In 1953, James Olds and Peter Milner (Olds and Milner 1954; Milner 1991) observed a relationship between electrical stimulation to the septal area of the brain and the behavioral probability of returning to the region of the test apparatus where the stimulation had been received. Social psychologist James Olds had just arrived at Donald Hebb's laboratory for a postdoctoral fellowship, with little to no experience in brain stimulation techniques in rats. He was fortunate to have the help of surgically skilled graduate student Peter Milner, as he attempted to arouse a rat by electrically stimulating an area deep in the midbrain, known at the time as the "ascending reticular activating system." Although he failed to demonstrate these predicted effects, he did notice that one particular rat kept returning to the corner of the table where it had been stimulated before. This "brilliant mistake" was followed by Olds' ingenious design of an apparatus that allowed the rat to press a bar to stimulate its own brain. To the amazement of all, the rat pressed the bar over and over, and to the exclusion of all other activities. Olds and Milner reported that: "…the control exercised over the animal's behavior is extreme, possibly exceeding that exercised by any other reward previously used in animal experimentation" (Olds and Milner 1954; see also Fig. 3.9).

Based on these observations, stimulation of the nucleus accumbens and medial forebrain bundle was thought to have exceptional reinforcement value. They would show that rats could be trained to execute selected behaviors, including lever pressing responses, in order to receive the brain stimulation. This research would be fundamental to the discovery of the neural bases of reward and reinforcement and the demonstration of these effects across species including humans (Heath 1963; Moan and Heath 1972; see Knutson and Gibbs 2007; Narcisse et al. 2011). Robert Heath, for example, noted that rats would become exhausted and brave significant hardships in order to stimulate the nucleus accumbens, a dopaminergic structure most commonly associated with our reward circuits and pleasure. His human subjects likened the stimulation to sexual orgasm

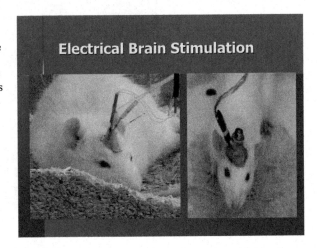

Fig. 3.9 Rat configured for electrical stimulation of the deep brain structures that are involved in Parkinson's disease and remediation. Image provided courtesy of Doctors Anja Hiller, Susanne Löfler, Andreas Moser, and Ulrich Hofmann

and eventually the structures would be associated with extreme pleasure in response to opiate drugs like heroin or morphine (see Moallem et al. 2012). Apparently, his research was financed by the Central Intelligence Agency and by the US military as the behavioral control was exceptional and provocative (see Robert Heath at Wireheading: http://www.wireheading.com/robert-heath.html).

Brain mapping research was then designed to determine the location of reward-relevant neurons and pathways affected by stimulation with divergent behavioral characteristics. Relevant locations included sites along the medial forebrain bundle, the lateral and posterior hypothalamus, and the ventral tegmentum (Wise 1989; 1996). The medial forebrain bundle consists of ascending dopaminergic fibers and it was found to relay information from the ventral tegmentum to the nucleus accumbens relating reward and addiction to the dopamine system (Wise et al. 1989). Modern techniques, using noninvasive electrical and magnetic stimulation methodology often extend the findings of earlier researcher. These laboratory manipulations were often completed through painstaking and deliberate research and often with morally questionable procedures, including those used by Robert Heath (Heath 1963; Moan and Heath 1972). Clinical findings would complement these laboratory studies where brain pathology or injury would provide much relevant information on localization of function within specific brain areas and neuronal systems.

Fig. 3.10 Paul Broca donated his own brain to science. It has since spent most of its time on a shelf in Paris

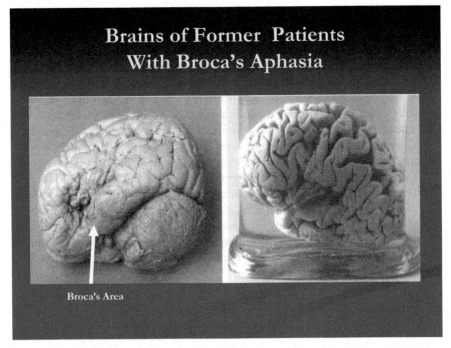

Fig. 3.11 Two brains of patients with lesions in "Broca's area"

The single most important contribution to our understanding and localization of brain function across the cerebral hemispheres may have come not from a large study with many participants, but instead from two single case studies by Paul Broca (1861: see Fig. 3.10). Broca was working with his patient, Leborgne, who was most remarkable for his expressive speech deficits, being severely nonfluent. The expressed speech consisted largely of perseverative phonemes or slow, labored, effortful speech. Broca followed this unfortunate gentleman to autopsy, where a distinct lesion from syphilis presented at the inferior posterior frontal lobe within the left brain (Fig. 3.11). Another of his patients, Lelong, exhibited reduced productive speech largely restricted to five words, "yes," "no," "three," "always," and "lelo" (a mispronunciation of his own name). Following a replication of his findings on autopsy of a similar lesion to that of Leborgne, Broca came out with perhaps the most remarkable conclusion in the history of our science. He stated, "We speak with the left brain."

By 1864, Broca had become convinced of the importance of the left hemisphere in speech, as the loss of speech, which he initially called aphemia (aphasia) was the result of pathology in this region. Specifically, he stated:

I have been struck with the fact that in my first aphemics the lesion always lay not only in the same part of the brain but always the same side—the left. Since then, from many postmortems, the lesion sas always left sided. One has also seen many aphemics alive, most of them hemiplegic, and always Hemiplegic on the right side. Furthermore, one has seen at autopsy lesions on the right side in patients in patients who had shown no aphemia. It seems from all of this that the faculty of articulate language is localized in the left hemisphere, or at least that it depends chiefly upon that hemisphere. (Broca1864; see also Springer and Deutsch 1998)

Eventually, Paul Broca donated his own brain to science. Despite this gesture and potentially the opportunity for science, it has largely spent time on a shelf in Paris.

Subsequent to Broca's declaration, the German neurologist Carl Wernicke (1874; see Fig. 3.12) noted that his patient also had a speech disorder. However, the patient's speech was fluent, while being largely devoid of meaningful content. This speech disorder corresponds to a lesion at the posterior, superior temporal gyrus within the left hemisphere, a location now referred to as "Wernicke's area" responsible for the production of meaningful, coherent, or comprehensible speech (Fig. 3.13). Such a patient may have normal expression or fluency in propositional speech, producing something like a "word salad" of content or meaning. Indeed, with dysfunction surrounding this area, we may see hyperfluency with heightened verbal output. Clinical psychology students might easily spend hours with such patients with a failure to acquire any meaningful assessment data and with the student not aware that they were working with a speech disorder. Others might ascribe high

Fig. 3.12 The German neurologist Carl Wernicke recognized the role of the left posterior superior temporal lobe in meaningful speech

Fig. 3.13 Wernicke's aphasia patient with fluent but meaningless speech. Originally published in *The Neuropathology of Disease*, Conn and Michael 2008, p. 753

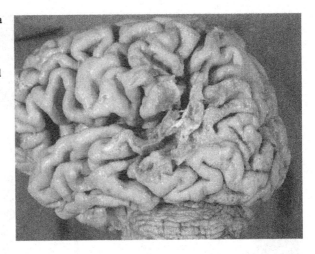

levels of intelligence to such an individual, being unable to follow the verbiage in the face of prolific output. Carl Wernicke recognized the role of the left posterior superior temporal lobe in speech comprehension and in the production of meaningful speech.

Later still, researchers (Heilman et al. 1975; Tucker et al. 1977; Ross 1981; Ross and Mesulam 1979) would appreciate the role of the right cerebral hemisphere in receptive and expressive speech with the description of expressive and receptive dysprosodia, corresponding to lesion location in the right frontal and right temporal regions, respectively. The processing of the affective or nonpropositional components of speech eventually was ascribed primarily to the right brain with logical linguistic or propositional speech within the left brain and with the frontal lobes involved in the expression of speech and the temporal lobes for the comprehension of speech components. Later, and on a very modest scale, Brian Shenal and Paul Foster (Shenal et al. 2003; Foster et al. 2008a, b) introduced quadrant theory (see Fig. 3.14), where expressive as opposed to receptive and affective as opposed to literal components would be derived from both a balance between the cerebral hemispheres, mediated by the interhemispheric communications across the corpus callosum, and an intrahemispheric regulatory role by the frontal lobes via the longitudinal tracts. This would provide a theoretical foundation, along with capacity theory, for some of the fundamental perspectives which will be discussed later in these writings.

Many technological and methodological advances developed throughout the history of neuroscience through the painstaking research and intellectual craftsmanship of the founding fathers of neuroscience and neuropsychology. In more modern times, the development of functional magnetic resonance imaging provided for elaboration of these earlier discoveries and a broadened appreciation of the interactions of disparate brain regions and systems. Functional magnetic resonance imagery, which offered temporal resolution on the order of seconds and spatial resolution on the order of millimeters, provided sometimes stunning images of brain activa-

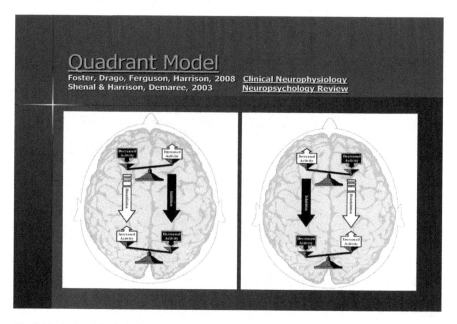

Fig. 3.14 A visual depiction of the predictions of quadrant theory based on the heavily intercon-nected cerebral hemispheres via the corpus callosum and the extensive interconnections between the anterior and posterior cerebral regions via the longitudinal tract

tion under behavioral, cognitive, and affective challenges. This technology became sufficiently available and spurred many to enter the field and ultimately to pro-vide their own unique contributions and discoveries. The first functional magnetic imagery study was published in 1992 (Kwong et al. 1992; see Mather et al. 2013) and many others were presented that year at scientific conferences (for review, see Frahm et al. 1993).

Initial validation of this instrumentation would come from demonstrating the basic localization constructs, including the well-established localization of basic sensory and motor regions. Beyond this, much has been accomplished and much re-mains to be discovered through the beneficial contributions of these newer technol-ogies and those that have been inspired to pick up the neuroscience studies. Mather et al. (2013) note the subsequent accomplishments using functional magnetic reso-nance imaging, including mapping the human retinotopic cortex (Engel et al. 1994), demonstrating the neural correlates of visual motion aftereffects in humans (Tootell et al. 1995), and demonstrating attentional modulation of basic perceptual process-ing (O'Craven et al. 1997). For many in the field, these modern projects, like those before them, provide unequivocal evidence for localization of function to specific regions of the brain. Moreover, the evidence of activation to functional variants in the sensory or motor arrays continues in modern efforts to develop functional maps of more complex variants in behavior, cognition, and emotion and the stimulus properties to which they respond.

The derivations of the strict localizationist view were challenged early on and from multiple fronts. Hughlings Jackson (1874) is often referred to as "the original skeptic." Jackson noted that complex mental processes are organized and reorganized within different *levels* of the brain and that they were not localized to one single and discrete area! He introduced now a well-established view of the hierarchical organization of the brain from brainstem on up to the cortex. This would become clearer within the emerging literature on speech processing, for example, where thalamic or subcortical dysphasia might present with a waxing and waning of symptoms (the arousal component), but where the localized cortical representations would remain essentially intact.

Marie-Jean-Pierre Flourens (see Pearce 2009; see Fig. 3.15) used experimental ablation techniques to study brain function. He would eventually derive a most influential conclusion from his work with pigeons. For this French physiologist, it was clear that the brain operates as a "concerted integrated system." Thus, the specific cortical zones identified by Broca and Wernicke could no longer be viewed in isolation. Through Jackson's work we could more readily appreciate the vertical and hierarchical organization of the brain, wherein higher levels, and by inference,

Fig. 3.15 Flourens describes the brain as a "concerted integrated system"

more recently evolved brain regions, would process the function with higher-level complexity.

Flourens was able to demonstrate convincingly for the first time that the main divisions of the brain were specialized for largely different functions. Removal of the cerebral hemispheres abolished all perceptions, motricity, and judgment, whereas the removal of the cerebellum altered equilibrium and motor coordination. In contrast, destruction of the brainstem (medulla oblongata) caused death. These experiments led Flourens to the conclusion that the cerebral hemispheres are responsible for higher cognitive functions; that the cerebellum regulates and integrates movements; and that the medulla controls vital functions, such as respiration, blood pressure, or circulation and related life-preserving activities. Possibly due to the relatively primitive brain structures and the small size of his experimental subjects, he was unable to localize memory and cognition, supporting his belief in diffusely organized brain functions. This view was partially vindicated by the recent appreciation of cellular networking and spreading activation among astrocytes (see Koob 2009). In Flourens' theoretical framework, different functions were ascribed to particular regions of the brain, but a finer appreciation of the localization of functions was lacking. Flourens' understanding of the brain provides for an integration of processing across systems in concert with one another and for any given behavior. This concept is fundamental to modern functional neural systems theory and seems

Fig. 3.16 Karl Lashley's search for the engram was featured in *Brain Mechanisms & Intelligence* in 1929. Originally published in *The Birth of a New Science,* Simpkins et al. 2013, p. 18

consistent with advances in understanding the role of glial cells in neuronal activity and interconnectivity, perhaps.

Notable among those challenging the localizationist viewpoint was the West Virginia-born Karl Lashley (see Fig. 3.16). Lashley was a student of the behaviorist John B. Watson who had argued that learning was the result of the formation of an association between a behavior and its consequences. Lashley set out to find the anatomical location of this association, which he referred to as the "engram." Within the laboratory, he trained rats on a variety of maze learning tasks, with acquisition followed by a surgical resection or ablation of various brain regions. Following the expenditure of substantial effort in this regard, Lashley eventually conceded that he was not able to find the engram as memory appeared to be a most elusive construct, at least anatomically. Instead, Lashley appreciated that the amount of brain tissue was more important than the location of the tissue removed. Lashley proposed his *mass action principle* wherein the brain acts as a whole or concerted unit, his *principle of equipotentiality* wherein brain tissue is equal regardless of its location, and his *principle of vicarious function* wherein one area of the brain may take over the function of another.

Lashley's work was criticized on statistical and methodological grounds. Roger Thomas (personal communication) noted, for example, that a rat learning a maze through exploration down one and/or another arm of the maze would necessarily be using parts of the brain processing olfactory, gustatory, visual, thigmotactic, vestibular, proprioceptive, tactile, pressure, temperature, and auditory events. Said more directly, the entire brain would be involved in the acquisition of these memories. Thus, the methods were inappropriate or inadequate for differentiating localization of brain functions. Modern statistical analyses may not have been performed by Lashley as these were not standard in the field until sometime after Lashley had reached his conclusions from his research.

One of the many benefits of being a doctoral student at the University of Georgia in my time period was the opportunity to take what may have been one of the better history of psychology courses in the nation offered by Lelon Peacock, a tall, burly, and imposing figure of a man who always seemed to be in a long white laboratory coat with a large Cuban cigar in his hand and another, at the ready, in his coat pocket. History in this course was learned by the student's exposure to quotations and sayings from the historical figures in the field rather than a presentation of names, dates, and events. In this manner, the student might come to know the people that established neuroscience through their own thoughts and expressions, perhaps.

Lelon describes his early days working as a young neuroscientist at the renowned Yerkes Primate Center in Atlanta, Georgia. Apparently, he had managed to establish a rat colony in a back room of this center devoted to primate research and without approval or official sanctions. Imagine his astonishment when 1 day Lashley arrived to inspect the laboratory and remarked immediately upon entering the facilities, "I smell rats!" Though it was hard for me to imagine this robust figure of a man intimidated and fearful, he was just that. But, all was well when Lashley took a deeper breath through his nostrils, savored the experience, and slowly exhaled stating "I love rats."

Fig. 3.17 Alexander R. Luria's 1973 publication of *The Working Brain* continues to influence modern neuropsychology and neuroscience

Regardless, Lashley's impact on the field was large and many still express his conclusions in their views of functional brain anatomy. More important than the accuracy of Lashley's conclusions was the impact that his derivations had on neuroscience, where there was now an improved appreciation of the varied and concerted role that even diffuse brain areas may play in any given behavior. Karl Lashley's search for the engram was featured in 1929 in *Brain Mechanisms & Intelligence*.

Indeed, this localization controversy within science was argued from the extreme positions of the radical localizationist and the radical equipotentialist. The echoes of the extreme viewpoints continue to exist today in various forms. But, as is often the case in science when two radically different viewpoints are expressed with such enthusiasm, both sides are correct to some extent. Progress was made with the integration of these two extreme viewpoints into a coherent theory of brain function. This essentially was the birth of *functional cerebral systems theory* and the Russian neuropsychologist Alexander Romanovich Luria (1973; see Fig. 3.17) said it best.

Part II
Functional Neural Systems Theory

Chapter 4
Basic Functional Brain Units

One of the early advances in communication among scientists in the USA and those in the former Soviet Union occurred between the prominent American neuropsychologist Karl Pribram and his Soviet correspondent Alexander Luria. Indeed, during the Cold War Luria and Karl were collaborating in California. I understand that Luria was a bit enthusiastic with his camera and especially as they travelled near some of the California military bases. All the while, these two correspondents and scientific colleagues were followed by the Central Intelligence Agency (CIA) and Komitet gosudarstvennoy bezopasnosti (KGB) as they travelled within the USA (personal communication) stirring up a good bit of excitement. Luria's work was a major influence on the creation of the field of neuropsychology. He published two case studies shortly before his death. In one he described a Russian journalist with synesthesia and a seemingly unlimited memory. This case was presented in a book *The Mind of a Mnemonist* (1987). In another of Luria's well-known books, *The Man with a Shattered World,* he presented the insightful account of Zasetsky, a man who suffered a traumatic brain injury (1987). These case studies illustrated Luria's methods, which combined with his clinical remediational approaches.

The eminent Russian neuropsychologist Alexander Romanovich Luria (1973) provided a most important conceptual advance for the understanding of brain function as he integrated the radical localizationist view of specialized brain "centers" with the radical equipotentialist view of disparate brain areas contributing to all aspects of behavior, cognition, and emotion. Luria provided a basis for the integration of these conceptually disparate views through a *functional cerebral systems* perspective, wherein he proposed three primary functional units specialized for distinctly different functions, but which are connected such that each functional unit contributes to the other and to all behavior, cognition, and emotion. The first unit is involved in maintaining tone or arousal level. The second unit is responsible for the reception, analysis, and comprehension of sensory input and the third unit for the regulatory control of all other units and for higher level "executive functions" involving planning, sequencing, and organization of behavioral, cognitive, and emotional *expressions*. Alexander R. Luria's 1973 publication of *The Working*

© Springer International Publishing Switzerland 2015
D. W. Harrison, *Brain Asymmetry and Neural Systems,*
DOI 10.1007/978-3-319-13069-9_4

Brain and the second edition of his *Higher Cortical Functions in Man* in 1980, continue to influence modern neuropsychology and neuroscience.

First Functional Unit

The first functional unit is responsible for general arousal level or cortical tone and consists largely of the brainstem reticular activating system (RAS; see Fig. 4.1). Sensory information from each modality projects to the RAS, affecting the overall arousal state of each brain. The RAS diffusely projects to each cerebral hemisphere recruiting or activating the higher-level brain systems.

Although the brainstem contains much of our sleep–wake system relevant to our progression through high arousal (rapid eye movement, REM) and low arousal sleep stages (non-REM), damage to brainstem structures is most often associated with diminished arousal level or wakefulness of one and/or the other brain. Thus, even small strokes or lesions here may, at least initially, appear more devastating as the entire brain may be unable to maintain an alert and aroused state sufficient to benefit from therapeutic rehabilitation efforts, for example. A general characteristic of damage to Luria's first functional unit includes diminished activation or arousal and lethargy and often with the appearance of a "waxing and waning" of abilities or performance. So, within session or across time samples the performance being evaluated may range from severely impaired, during low arousal states, to essentially normal during heightened arousal states.

The prognosis for such a patient is typically derived from the better performances. Indeed, brainstem lesions often surprise the team of therapists as the patient may present as more severely impaired and with more generalized deficits due to an overriding deficiency in arousal. The arousal construct is the most universal function in neuroscience and psychology, wherein optimal performance of any task is contingent on optimal arousal level. Either over arousal or under arousal may lead to performance deficits in an inverted-U fashion, where performance varies as a function of arousal level. Though this is a basic tenet for arousal theory in general, the phenomenon has been referred to as the Yerkes–Dodson law (see Fig. 4.2; Yerkes and Dodson 1908).

Fig. 4.1 Luria's first functional unit or brainstem reticular formation receives sensory input and, through recruitment, alters the arousal of the cerebral hemisphere

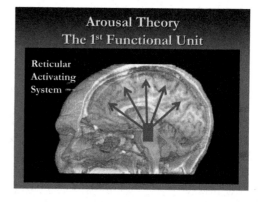

Fig. 4.2 The Yerkes Dodson Law states that performance varies as a function of arousal level. The performance of any function may be maximized by the achievement of an "optimal point" of arousal

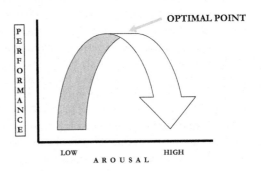

Second Functional Unit

Luria's second functional unit consists of the back half of each cerebral hemisphere. Each sensory modality projects to the back of each brain such that the initial reception of the sensory event, irrespective of sensory modality, occurs in this unit. Vision, audition, touch, pressure, pain, itch, warm, cold, olfaction, and gustation project to this unit charged with the sensory reception, analysis, and comprehension of this information. A family member with a stroke or with damage within the posterior brain regions will, by definition, have problems with the reception, analysis, and/or comprehension of sensory information. Such a person may be deficient in *comprehending* speech. Since sound projects to the temporal lobe and since the left hemisphere is specialized for linguistic or propositional speech processing, the critical location for this disturbance lies in the left temporal lobe in an area known as the superior temporal gyrus or "Wernicke's area." However, the same is true for the *reception* of linguistic speech sounds corresponding with the anterior superior temporal gyrus.

Many lay people and again many health care professionals may attribute the auditory receptive deficit derived from dysfunction in this area to the patient being "hard of hearing." Although the neural transduction and reception of sound is initiated in the inner ear, the cable projects to the primary auditory cortex for sound recognition. Thus, the attribution, while seemingly valid, is not accurate. This is more clearly recognizable with the demonstration that specialized auditory events processed by the intact right cerebral hemisphere are detected without the need for repetition and/or elevated volume. Moreover, as has been demonstrated for the visual projections from the retina to the occipital lobe, peripheral or inner ear damage may yield anatomical alterations at the other end of the cable as the auditory nerves and tracts find their way to higher cortical brain systems.

In a subsequent section, we follow the flow of information from the primary cortex on to the association areas eventually culminating in the highest level of analysis in the tertiary association areas. In these areas, the comprehension of each sensory modality is then integrated with other sensory systems and where a cross modal comprehension of the sensory event may now occur. So, for example, the somatosensory reception of touch, pressure, and warmth at one's left hand occurs initially in the primary somatosensory cortex of the right brain, followed by comprehension of the stimulus form (e.g., Mazzola et al. 2012) and associations with existing memories of the feelings experienced with holding this hand in the past. But, as the level of comprehension extends further into the tertiary association cortex, there is increasing potential to experience the greater depth of feelings, including visual and auditory associations of this person (e.g., wife or husband) and, more richly, even the emotions previously experienced from holding this precious hand in the past.

The tertiary association cortex is distinguished from the primary projection cortices not only in its cross-modal sensory analysis and integration capabilities but also by its abundant access to the corpus callosum and connections to homologous analytical regions in the opposite brain. This allows for the ultimate comprehension, at the human level, of the sensory array and the potential knowledge, associations, and previous experiences acquired by both brains, each providing a specialized perceptual bias and processing style in the overall analysis. In such a system, the written word might be initially converted from visual analyzers in the occipital cortex into cross-modal linguistic associations within the left hemisphere and now ultimately with an emotional response to the gist or provocative nature of the written word in the context of activation within the right cerebral hemisphere.

Despite the emotional processing advantage of the right brain, though, the left brain maintains a positive emotional bias. This has been demonstrated across both verbal or linguistic content and nonverbal spatial analysis of faces or tone of voice. Even the common QWERTY keyboard can be used to demonstrate the emotional processing bias as a function of positional location of the keys within the left or the right side of the keyboard (Jasmin and Casanto 2012). In this example, the researchers investigated whether differences in the way words are spelled, some with more letters on the right side of the keyboard and others with more letters on the left, influence their evaluations of the emotional valence of the words. Across three experiments, the relationship was supported between emotional valence and QWERTY key position across English, Spanish, and Dutch languages. Words containing more right-sided letters were rated as more positive in valence than words with more left-sided letters.

Third Functional Unit

Information processing may eventually move toward the third functional unit (the frontal lobe), via an enormous bundle of cables known collectively as the longitudinal tract. Information flows in a rostral direction toward the frontal lobe(s) and this follows the reception, analysis, and comprehension of sensory events and the subsequent cross-modal integration and analysis of those events within and across the second functional unit(s). It is within the third functional unit that we organize, plan, and sequence our own actions and responses to our sensory environmental array. Several robust and mature literatures exist on this unit reflecting the frontal lobe involvement in executive functions, regulatory control, effortful control, self-regulation, intentional behavior and cognition, and many other functions discussed elsewhere in this book (see also Stuss and Knight 2013). The activities of the third functional unit ultimately may lead to overt or covert actions or behavioral displays that allow for modification of our social, emotional, and/or environmental situations.

Ultimately, the activities of this unit culminate in expressive actions involving muscles, glandular functions, and organ functions throughout the entirety of the body. For example, at the end of the processes involved in the preparation of a desire or intent to respond and the organization of the motor sequences necessary for completion of the task, information flow moves toward the motor cortex where descending fibers of upper motor neurons differentially regulate the reflexive functions of the lower motor neurons which innervate muscle tissue at the motor unit. The final motor pathway originating at the primary motor cortex is now ready to send signals to distant effectors, including skeletal muscles involved in the movement of our body parts. But beyond this practical role in the regulation of motor intent and muscular responses, the descending pathways from the frontal lobe will regulate aspects of cardiovascular, cardiopulmonary, hormonal, and glandular functions. The loss or diminution of frontal lobe regulatory control may result in disinhibition or the release of reflexive responses of innervated muscles (e.g., Futagi et al. 2012) and glands (Cannon and Britton 1925), altering cardiovascular, cardiopulmonary, hormonal, and glandular functions.

The frontal lobes are responsible for those activities, which the clinical neuroscientist refers to collectively as "executive functions." This includes the organization, planning, sequencing, generativity, and regulatory control over behavior, cognition, emotion, and more broadly all bodily functions (see Stuss and Knight 2013). The left frontal lobe appears to be responsible for executive control over expressed propositional speech. Expressive speech problems are a common outcome of reduced metabolic activity in this region such that in milder cases one's speech output may be sparse and other's attributions may be that the individual is "shy" or just not very talkative. More pathological involvement of these regions might be apparent with expressive speech deficits, perhaps generally disorganized with tangential and often irrelevant speech expressions. But, the involvement of Broca's area specifically in the inferior, posterior frontal lobe is characterized by phonemic paraphasic errors, perseverative phonemic expressions, and may include articulation errors as with stuttering (see Duffau 2012). This system appears to provide not only for the inhibition of reflexive responses but also for the smooth transition and coordination among synchronized reflexes. From this vantage point, the frontal lobes may provide for flexibility within one or the other of these systems rather than for rigidity and inflexibility in response to environmental challenge or confrontational demands. Moreover, this construct of flexibility in coordinating responses is linked to more adaptive responding and the successful navigation of bodily responses through rough and demanding waters posed by a demanding ecosystem. Friedman and Thayer (1998) provide a relevant discussion of the role of autonomic flexibility in cardiovascular viability, for example.

The right frontal lobe appears to be responsible for executive control over negative emotional expression with dysfunction here underlying nonpropositional speech expression such as the volume or loudness of speech and the prosodic or emotional conveyance of speech through tone of voice. Right frontal lobe deficits are predictive of one or another reactive or exaggerated emotional expression within the negative affective valences, including anger, sadness, or fear. For example, damage within the right basal ganglia or lentiform bodies is predictive of emotional lability with poorly regulated control over crying behavior. This relationship was appreciated early on and demonstrated by Constantini in 1910 (see Oettinger 1913; see also Parvizi et al. 2009; see also Woodworth and Sherrington 1904) where lentiform lesions accounted for spastic crying and spastic laughter in the majority of cases and where redundant lesions accounted for much of the remaining sample. Discussion of a sad theme might easily decompensate a person afflicted with this brain disorder with the onset of crying behavior and quite out of character for the person by history. The chief of cardiology from a major medical center was confronted with the challenges of altered function in these pathways secondary to a glioma. The family noted that the man that they had known would never cry and that now, quite out of character, he was remarkably tenderhearted and emotionally fragile.

By extension of this discussion, damage or incapacity within the right orbitofrontal region may result in reduced regulatory control over negative emotions (e.g., Agustín-Pavón et al. 2012) and reactive expression of anger or aggressive behaviors (e.g., Fulwiler et al. 2012). Reduced regulatory control from this region may be evident in an expressive dysprosodia with an avalanche of speech volume, uncharacteristic shouting, or loud abrasive features. Moreover, the loss of regulatory control over anger and fear, for example, may reflect a concurrent loss of control over the cardiovascular system with corresponding reactivity in systolic blood pressure, heart rate (Emerson and Harrison 1990; Demaree et al. 2000; Herridge et al. 2004; Williamson and Harrison 2003; Everhart et al. 2008; Shenal and Harrison 2004; Rhodes et al. 2013; see also Foster et al. 2008), glucose (Walters & Harrison 2013a, b), skin conductance (Herridge et al. 1997), facial expression (Rhodes et al. 2013), and temporal lobe activation on the quantitative electroencephalogram (Mitchell and Harrison 2010).

Interestingly, the orbitofrontal region is seen in a lateral view of the brain proximal to the motor cortex. The motor cortex provides for directed efferent motor control with a topographical representation of the body wherein facial motor control is located inferior to the other body regions and at the basal region of the brain. This topographical anatomy provides for proximal brain areas in control of the face and in control of anger (right orbitofrontal). Our language provides the clue to this where *loosing face* occurs with the loss of emotional control or of a stable self-regulated state. The functional neural anatomy for the face and for anger control are likely consistent with Kinsbourne's (Kinsbourne and Hicks 1978) concept of *functional cerebral space* and where concurrent tasking demands may result in interference or a loss of control over the secondary task performed by that brain region. The dual-task challenge in this case might be the maintenance of regulatory control for anger and sympathetic control of the heart, while concurrently maintaining a stable, well-regulated facial expression. Herath and colleagues (Herath et al. 2001) used functional magnetic imagery during the performance of dual reaction time tasks. Performance of the dual task was found to activate cortical regions in excess of those activated by the performance of component single tasks. Moreover, these investigators reported that the dual-task interference was specifically associated with increased activity in a cortical field located within the right inferior frontal gyrus. This area has been previously implicated in emotion regulation and hostility, lending to the notion that areas of the brain that regulate emotion in particular, may be subject to the burden of dual-task interference effects.

The tasking demands might simply involve contracting facial muscles involved in the emotion. Matthew Herridge (Herridge et al. 1997) did investigate this as did others (Ekman et al. 1983; Kraft and Pressman 2012). Matthew found that contracting the corrugator muscle elevated sympathetic tone using a skin conductance measure. Hostile, violence-prone individuals maintained an elevated skin conductance at the left hemibody with reduced habituation at the left hemibody to repeated contractions of this muscle. Low hostile individuals evidenced a reduced sympathetic tone and a reduced rate of habituation at the right hemibody. Earlier on, George Demakis (Demakis et al. 1994) had demonstrated elevated skin conductance responses at the right hemibody with positive affective displays of spastic laughter. Hostiles maintain elevated cardiovascular measures of sympathetic tone as indicated by heart rate, blood pressure, and skin conductance measurements. The results support the role of the right orbitofrontal region in the regulatory control over anger (see also Agustín-Pavón et al. 2012; see also Fulwiler et al. 2012) and the dual concurrent management demands for regulating facial expressions and sympathetic drive. Moreover, the results support the conclusion of diminished resources at this region in hostile, violence-prone men. The evidence indicates diminished frontal capacity in hostile men. Facial dystonia was evident in the electromyogram recordings in this group and especially at the left hemiface (i.e., relatively diminished right frontal capacity; Rhodes et al. 2013).

The influence of facial postures, depicting one or another affective valence, on cardiovascular function and stress responses was also investigated in a project where the participants' positive facial expressions were covertly manipulated. Kraft and Pressman (2012) asked the participants to complete two different stressful tasks while holding chopsticks in their mouths producing a Duchenne smile, a standard smile, or a neutral expression. One group was made aware of the manipulation by asking them to smile, while covert manipulation of the comparison group was made using the chopsticks. Interestingly, both the covert and the posed facial configuration depicting a smile reliably lowered heart rates during stress recovery in comparisons with the neutral facial configuration.

The basis for the next experiment seems obvious. Although this experiment has not yet been conducted, our historical language is rich in the use of the warning statement to *stay out of my face!* The sensory approach to the face would seem to be proximal to these emotional systems not only through the representation of the face in the somatosensory cortex at the postcentral gyrus but also in the processing of the visual menace reflex, where an abrupt approach to the face results in a characteristic defensive reaction with an eye blink and potentially contact guard. The experiment might monitor negative affective behavioral, cognitive, and physiological responses (e.g., skin conductance) resulting from approach or physical manipulation of the foot, the leg, the hand, the arm, or the face. This functional anatomy controlling the face may indeed have relatively direct access to systems involved in regulating anger and cardiovascular/cardiopulmonary functions. Manipulation proximal to the face task or challenge the functional capacity of this anatomy to maintain control over the secondary functions presumed to be served by this brain region. More specifically, intrusion with proximity to the face should yield sympathetic drive with increments in blood pressure and defensive posturing. It should concurrently task the regulatory control over anger. With an individual lacking adequate capacity at this region, the outcome might be predictable with anger behavior and the heightened perception of external threat or provocation. A word to the wise—if you do this experiment, you had better stay out of arm's length or face the predictable consequences.

The final motor pathway, emanating from the motor cortex, ultimately expresses the desires and intentions established by more rostral (premotor) and subcortical (basal ganglia) frontal lobe structures. Directly rostral to the motor cortex we find the premotor cortex, which fires before the motor cortex (see Rizzolatti et al. 1996; see also Fogassi and Semone 2013) and prepares the motor actions and sequential self-regulation of our movements. More rostral still we find brain regions capable of processing increasingly abstract concepts and levels of analysis reflective of more recent evolutionary accomplishments. This includes the prefrontal cortex and eventually the far frontal regions and the frontal poles. The functional anatomy of the prefrontal cortex makes it extraordinarily powerful, yet the capacity demands for high-level flexibility and regulatory control may leave it vulnerable to the interference effects inherent in dual or multiple concurrent task demands. It provides for the management and processing of the highest levels of stress. Nonetheless, it suffers capacity limitations which, when exceeded, release more primitive brain systems (e.g., the limbic system) more evolutionarily adept at reactive responses and less well attuned to complex social rules and regulations.

The more complex the processing demands, the more susceptible the brain region may be to cellular exhaustion due to high metabolic demands. The prefrontal cortex is largest in humans where it has shown the greatest expansion in size in the most rostral regions approximating the frontal poles (Öngür and Price 2000; Öngür et al. 2003). This rostral expansion of the prefrontal cortex underlies the increasing capacity for complex and abstract thought as well as the capacity for regulatory control underlying complex working memory (Hasher and Zacks 1988), executive functions, and the challenges of complex social demands. The rostro-caudal model (Christoff and Gabrielli 2000) acknowledges that the rostro-caudal axis of the prefrontal cortex supports a control hierarchy whereby the posterior-to-anterior convexities of this region mediate progressively abstract, higher-order regulatory control functions. This implies that relatively complex, flexible, and integrative processing computations are subserved by the most anterior prefrontal regions, while relatively more fundamental, simple, and fixed processing computations are subserved by more posterior regions of the prefrontal cortex.

Several functional models exist for the role of the prefrontal cortex (e.g., see Miller and Cohen 2001; see also Koechlin and Summerfield 2007). But, a common proposal offers that the rostral and caudal prefrontal cortex can be distinguished on the basis of processing domain-general, versus domain-specific, representations (Christoff and Gabrieli 2000; Fuster 2004; Courtney 2004). A hierarchical version of this perspective proposes that domain-specific posterior frontal regions can be modulated by the more domain-general rules in anterior prefrontal regions (e.g., Sakai and Passingham 2001, 2006). Evidence supporting these models suggests that the anterior regions handle the most abstract, nebulous, and complex tasks, while the posterior regions handle simpler, lesser complex tasks lower in the functional hierarchy. This distinction is consistent with Luria's organization of primary, secondary, and tertiary cortical divisions and with the frontal lobes progression from primary cortical regions eventually to tertiary association cortex associated with increasing complexity of executive, regulatory control, and motor processing.

Chapter 5
Functional Cortical Levels Within Units 2 and 3

Primary Projection Areas and Secondary Association Areas

Luria further divides functional units 2 and 3 based on the level of reception, analysis, and comprehension (functional unit 2) and the level of organization and preparation of the motor response (functional unit 3) resulting from higher-order anatomical processors and the spread of information across surrounding cortical zones. More specifically, sensory information from our eyes and ears and from each of the other sensory modalities arrives at its own primary projection cortex in the part of the brain specialized for the processing of that type of sensory information. Vision projects onto the calcarine or striate cortex located at the medial occipital lobe (see Fig.5.1; Brodmann's area 17). Audition projects onto Heschl's gyrus at the anterior superior temporal lobe represented by Brodmann's area 41. Somatic senses including touch, pressure, cold, warm, pain, and itch project onto the anterior parietal lobe at the somatosensory cortex represented by Brodmann's areas 3, 1, 2. The German anatomist Korbinian Brodmann published his anatomical maps of human and nonhuman primates based on the cytoarchitectural features of the neurons he observed in the cerebral cortex using the Nissl stain. These maps are potentially very useful for communication purposes and for readily locating a cortical region.

The brain is arranged in a seemingly logical fashion with tracts, bundles, or "cables" carrying different types of sensory information to different and specific areas of the brain. Thus, there is a topographical representation of each sensory modality within each brain. Moreover, the spatial location and intensity of a sensory event is coded logically and in a highly topographical fashion, somewhat like the wiring in a computer or in the man-made machines with which we are more familiar. The topographical representation of the primary projection areas of several of the basic sensory modalities are depicted in Fig. 5.1. Neuronal projections carry visual information from the eyes to the primary projection area for vision located in the back of the brain (occipital lobe). Neuronal projections from the inner ear carry acoustic information and vestibular information largely

© Springer International Publishing Switzerland 2015
D. W. Harrison, *Brain Asymmetry and Neural Systems*,
DOI 10.1007/978-3-319-13069-9_5

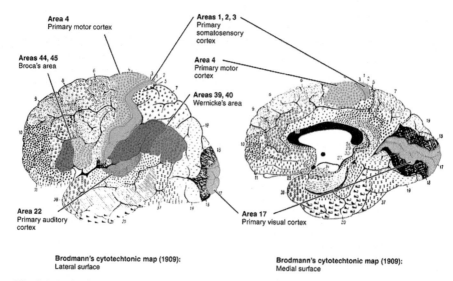

Fig. 5.1 Brain shown facing the left side and with a midsagittal section (*midline*) viewed on the right side of the figure: *Pink* is Brodmann area 17 (primary visual cortex). *Anatomy & Physiology,* Connexions website. http://cnx.org/content/col11496/1.6/, June 19, 2013

to the temporal lobes. Other cables carry somatosensory information (touch, pressure, cold, warm, pain, and itch; see Han et al. 2012; see also Fields and Sutherland 2013) from all areas of our body onto the primary projection area in the parietal lobe (somatosensory cortex), and so forth.

The reception of sensory information at the primary projection cortex starts a process of elaboration and increasingly complex levels of analysis moving onto the secondary association cortex and ultimately onto tertiary association cortex. Each primary projection area is surrounded by the secondary association area for that sensory modality. As information moves from the primary projection cortex to the secondary association cortex, it is processed at a more complex and higher level of analysis. For example, the reception of tactile stimuli occurs initially at the primary projection cortex. However, more complex tactile analysis in the secondary association cortex may allow for the recognition, by touch alone, of a pencil or key or other object on the corresponding body part. This somatosensory analysis provides for *stereognosis* or tactile object recognition through three-dimensional touch processed through the cutaneous receptors in the skin. Yet, at this level of analysis, it does not provide for a visual or auditory representation of the object. Nor does this level of processing provide for access to the speech systems sufficient for verbally naming the object. These related processes result from the flow of information out and onto these other functional neural systems with some levels of analysis being proximal and some more distally located from this system.

For audition, the primary projection area would allow for the detection of sound largely arriving from the contralateral hemispace. For the left brain, this would be the sound originating initially through the right ear with delayed arrival time for the left ear based on its placement further away from the location of the sound. Once detected by the primary projection cortex, the secondary association area for audition would allow for the analysis of the sound as a linguistic or propositional speech sound (left brain) or perhaps as a musical melodic or affective intonation; should the processing be performed by the right brain's auditory analyzer? Yet, this level of analysis does not provide for a visual or somatosensory representation of the object. The auditory primary projection areas are instead proximal to, or nested within, the speech systems. This cortex is near the somatosensory projections for the tongue and face providing gestural associations for lingual praxis or oral motor movements for the production of speech. Other associations are more distal to this brain region where access, like travelling from one town to another town, may require an axonal highway. This would certainly be the case if auditory verbal processing within the left brain was to produce emotional associations through connections with the right brain across the corpus callosum. Again, these related processes result from the flow of information out and onto these other functional neural systems with some levels of analysis being proximal and some more distally located from this system.

The vestibular projections provide initial detection of vectional force toward the left or toward the right as we change body positions. This information originates from the semicircular canals in the inner ear, which remind me of a carpenter's level. Any change in position within one of three distinct planes of movement (three separate semicircular canals within these planes) will provide the stimulus detected at the primary projection cortex. With higher-order analysis by the surrounding secondary association area, the left temporal region (via activation) may result in the perception of the body moving toward the right hemispace. Activation of the homologous region at the right brain may result in a sensation of spinning or moving toward the left hemispace (see Carmona et al. 2009). Again, the level of analysis initially within this modality may acquire intermodal associative strength resulting from associative activation within the tertiary association cortices.

A lesion within the primary projection area, as might result from a stroke, can have a specific impact on a sensory system based on its topographical location. For example, diminished activation of the somatosensory projection area may yield numbness or a sensation of being cold in the corresponding body area represented at the location of the stroke. A more severe loss of function here may result in an anesthesia. We might be able, then, to operate or perform what would previously have been a painful manipulation of this tissue with the patient unable to feel at that body location. Similarly, a stroke within the vestibular analyzer may alter position sense and, too often, results in dangerous falls and/or altered postural placement. Lesion of the left temporal region may result in a corresponding release or relative activation of the right temporal lobe and a sense of spinning toward the left. Any effort by the patient to compensate for these previously trustworthy signals might be catastrophic and result in additional injury. Indeed, the prognosis for many of our patients is shortened needlessly to about a 5–6-year window after a fall resulting in a broken hip (Maggi et al. 2010) and/or head injury and the iatrogenic effects of general anesthesia and surgical repair. Although the reduction in life expectancy was similar for both genders in the Maggi et al., investigation, the proportion of the years of life lost was higher in men, suggesting a worse impact of hip fracture on survival in men, even after consideration of the higher mortality rate in the general male population.

The third functional unit or frontal lobe is specialized eventually for motor output or the expression of our responses in our bodily movements. The complex organizational demands of planning and sequencing movement and the development of the intention and desire to perform the movement are processed by our secondary association areas in this unit. The primary projection area or motor cortex sends cables down to all parts of our body to effect movements or to manage the strength in the muscles represented topographically at the motor cortex. A lesion or even metabolic deactivation of the primary projection area for this unit may immediately alter strength in the corresponding musculature. So, with a more severe stroke in this area or within the cables projecting down from this area, we may see *paresis* (muscle weakness) or even *plegia* (paralysis). The reader might recall, though, that the overriding function of the frontal lobes has been to *regulate or inhibit* other brain areas and other parts of our body (e.g., Denny-Brown 1956; Tucker and Derryberry 1992). This regulatory control, expressed at the level of the primary motor projection cortex, is over muscle reflexes (e.g., Futagi et al. 2012).

With this thought, the brain disorder that initially leaves our family member with paresis or plegia of the muscles innervated by these descending pathways often is expressed, following recovery, as hyperreflexia, increased muscle synergy, spasticity, or dystonia. One of the primary goals of rehabilitation therapies is to minimize this outcome through intervention efforts designed to range or to extend the limb, if there is hyperreflexia of an upper extremity (arm or hand). With abnormal or increased flexor tone, exertional therapies might be implemented, wherein the exercise is specifically for the antagonistic muscle group. Extensor exercise might be more appropriate with hyperreflexia/synergy of a flexor muscle group. Alternatively, frontal lobe lesions or abnormalities often result in lower extremity extensor synergy. In milder cases, we might inspect the shoe of a patient for the associated heightened wear under the toe as suggestive of extensor tone. Here at the lower extremities, the oppositional or antagonistic muscle group to be exercised might be that promoting flexor synergy. Unchecked these features of upper motor neuron (primary motor cortex) pathology progress over time and, all too frequently, result in chronic pain syndromes and fixation or dislocation of the skeletal joint or pivot point for that limb. Developmental variants exist in the form of individual differences, which may also be useful in the analysis of variations in the functional capacity of these brain regions.

The secondary motor association cortex (premotor cortex), when activated with electrical stimulation techniques by the neuroscientist, yields evidence for activation of the desire or the intention to perform the movement. Indeed, over the years that I have worked teaching our doctoral students in neuropsychology diagnostically relevant clues, I have shared with them the diagnostic relevance of their own and others attributions toward the patient. Particularly relevant here, is the *fundamental attribution* toward others of *intent*. Indeed, the frontal lobes provide the anatomical basis underlying intentional activities or desire to perform. At one level, this might be seen as an aspect of memory, wherein maintaining your intention to the task affects your memory function, but where this system is not basic to the learning or consolidation of information, per se. Forgetting what you intended to do might be another example of inadequate processing within this system. The memory is there for what to do and how to do it—if you could just have a prompt or cue!

Laboratory evidence shows that the secondary association cortex in the frontal lobe activates a few milliseconds prior to activation of the primary motor cortex. The secondary association area functions to organize, plan, and sequence the muscle movements, which are subsequently and directly controlled by the primary motor cortex (see Joseph 2000). For example, Exner's writing area overlaps the premotor cortex. This area activates prior to and during intentional hand movements with the premotor cortex contributing to the organizational planning and programming of hand movements. Relatedly, the frontal eye fields overlap premotor cortex, which activates prior to and during directional eye movements (e.g., Suzuki and Gottlieb 2013). Broca's area too provides for the preparatory organizational planning of oral–laryngeal and lingual movements directly effected by the motor cortex (Foerster 1936; Fox 1995). Broca's area also becomes active prior to vocalization and during subvocalization as indicated by functional imaging research. Activation of these premotor areas and of their connections with the basal ganglia does not directly activate movement but instead the "impulse" or desire to initiate the act (e.g., Penfield and Boldrey 1937).

Stroke or pathological alteration to the secondary association area may result in clumsy or poorly coordinated complex movements of the corresponding body part, along with diminished utilization of that body part for purposeful actions. This might be apparent in manual dexterity deficits and poorly planned or sequenced movements, for example. Imagine, if you will, the level of coordination of finger movements achieved by a concert pianist from years of practice. Localized dysfunction within the premotor cortex may affect the coordinated, integrated, and rapidly alternating movements providing for the musical sounds from the piano. When we follow someone over time with damage to these brain regions, we often see a progression of functional pathology resulting from a differential learning history for the contralateral limb affected by the stroke and the ipsilateral body parts unaffected by the stroke. With experience, the discoordinated limb is used less, whereas the unaffected limb is used more, potentially aggravating the dysfunctional limb movements and their utility in contributing to our activities of daily living (Taub and Morris 2001).

Therapeutic efforts here focus on requiring the patient to use the affected limb, encouraging and promoting experiential learning or response-modified compensation for improved performance. *Constraint-induced movement therapy* (see Miltner et al. 1999) was initially demonstrated with nonhuman primates resulting in improved outcomes over time (however, see Bowden et al. 2013). In 1998, Liepert et al. (1998) were the first to report changes in the brain in response to constraint-induced movement therapy applied to human stroke patients. Using transcranial magnetic stimulation, they found an increase in the number of scalp locations that produced a motor-evoked potential in the paretic hand. Physically restricting the unaffected limb is designed to force intentional interactive use of the dysfunctional body part. Both the active use and the restricted use of body parts alter the underlying anatomy within the brain systems supporting that function. Also, neuroimaging studies show that this therapy promotes increased blood flow to the somatosensory and motor cortices (see Wittenburg and Schaechter 2009). Active use or exertional activity promotes improved efficiency in the neuronal organization and preparation for the response. Moreover, these same regions, along with the underlying basal ganglia, have been associated with the "impulse" or the desire to use that body part (e.g., Penfield and Boldrey 1937) as with directed intention to initiate a response with the body regions represented along the premotor and eventually the motor homunculus. Related and proximal systems provide for the anticipated consequence of the action following completion. Interestingly, these systems also contribute to "theory-of-mind" components (e.g., Carrington and Bailey 2009; Haxby et al. 2002; see Shallice and Cooper 2011; see also Cabeza et al. 2012), allowing for the prediction of other's behaviors and probable consequences of the act(s).

A more extreme example may be proffered by intervention efforts with children suffering from spasticity of the motor systems with unilateral cerebral palsy. Evidence supports the contention that constraint-induced movement therapy is beneficial in promoting motor capacity and implementation use of the affected upper limb in children with unilateral cerebral palsy (e.g., Sakzewski et al. 2011; Geerdink et al. 2013). Much of the current emphasis in this research area relates to the developmental constraints and emergence of frontal lobe regulatory control as a developmental and age-related process. Differences are found related to intervention characteristics, like duration of treatment, in children of different ages. For example, Geerdink et al. (2013) conducted a randomized control trial intervention with constraint-induced movement therapy. These authors conclude that their most cardinal finding was for the relevance of age on the speed of dexterity gain with the affected extremity during the period of intervention. They note that, compared to older children, children younger than 5 years of age had a 2.3 times greater chance of reaching maximum performance on their measures of motor dexterity within the 6-week period of intensive training. Moreover, the gains for the younger children were more rapid and more robust within the first 3–4 weeks of intervention, after which they approached stabilization of progress. They contrast these findings with those of the children 5 years of age and older with slower and more extended improvement of functional motor control. They note developmental differences in the regulatory control of these reflexes (e.g., grasp reflex) as relevant in determining the optimal intervention approaches and the optimal duration of treatment.

Tertiary Association Areas

In the discussions above, the basic functional features and anatomical locations for the primary projection areas and the secondary association areas were addressed. For the second functional unit of the brain charged with the reception, analysis, and comprehension of sensory input, the primary projection area and the secondary association area for each sensory modality receive, recognize, and perform elaborative associations *within that particular sensory modality*. This makes a bit more sense as the secondary association area surrounds the primary projection area anatomically and with the understanding that the flow of information is from the simple to the more complex association cortices. In and among the brain areas representing each sensory modality and surrounding the secondary association areas are the most complex and intelligent cellular constellations in the brain, which make up the tertiary association areas. Glial and neuronal cells in these areas have access to dendritic fields or anatomical extensions within and across the various sensory regions. Thus, for the first time in our discussion of functional brain anatomy, we are in regions charged with cross-modal analysis and the integration of our sensations.

Perceptually, this is a most important accomplishment as information known and recognized within one sensory system (e.g., vision) may be associated with the level of understanding and familiarity for this stimulus across the remaining sensory modalities. For example, a visual grapheme recognized by the left occipital cortex, through the primary and secondary association cortex, may now be converted into the corresponding sound via associations with the superior temporal gyrus (Wernicke's area). By extension, concurrent access onto the somatosensory cortices may provide associations or representations for the configuration of the body parts necessary to produce speech sounds in naming the object. In this fashion, the visual stimulus, which was initially identified by the visual association cortex (e.g., pencil), may be converted into a somatosensory and gestural event for relevant body parts. The tertiary association cortex provides a substantial and higher-order depth of understanding as far-field, but relevant, sensory experiences and associations may be brought to bear to maximize our comprehension of the event initially presented only within the visual modality (this example).

The complexity of analysis attributed to the tertiary association cortices is beyond that of other brain areas. As described thus far, these areas provide for the integration of information from multiple sensory modalities. The tertiary association areas, though, are also rather unique in their access to the major highways, which extend from the back of the brain to the frontal lobe (the longitudinal tract) and to those extending to the homologous regions within the opposite brain (the corpus callosum). The longitudinal pathways (intrahemispheric) convey the highest level of comprehension of the sensory array to the third functional unit (frontal lobe) to be acted upon through the acquired rule-regulated organization and preparation of our expressions or responses. The crossing (interhemispheric) pathways to the other brain provide for integration of processing approaches of the left and the right cerebral hemisphere. But, these cables have been shown to involve inhibitory control. So, communication across the corpus callosum at least provides an anatomical basis to achieve balance between the two cerebral hemispheres. However, priming or relative activation of either of the tertiary zones of the left or the right brain may serve to achieve a dominant perspective toward the information as, for example, a positive and optimistic view of the left hemisphere, which might be suppressed by activation of a more negatively valenced or deeply emotional perception by the right brain.

Regulatory control, by the frontal lobe of each brain, will partially be achieved through inhibition along the longitudinal tract. For example, with activation of the tertiary zone within the right hemisphere, we may achieve heightened negative emotionality such as fear or anger. The right frontal lobe would provide for regulatory control, in a normal brain, to prevent the emergence of a panic attack through the suppression of fear or the emergence of violence-prone behavior, through the suppression of anger. Additional information on these processes is available through manuscripts detailing these theoretical and neuroanatomical positions (e.g., see Thompson-Schill et al. 2005; see also Carmona et al. 2009; Shenal et al. 2003; Foster et al. 2008a, b; Walters et al. 2010; Mitchell and Harrison 2010). However, the literature on frontal lobe regulatory control is perhaps the more robust derivation of psychological science (see Thompson-Schill et al. 2005). Very much in line with Flourens' evidence for a hierarchical organization within the brain, each frontal lobe may function in a "top-down fashion" where descending frontal lobe tracts provide for regulatory control or inhibition over the first functional unit or the arousal systems within the brain stem (e.g., the mesencephalic reticular activating system).

For the frontal lobes or the third functional unit, once again the tertiary association areas surround the secondary association cortex, providing for the most complex aspects of human behavior, cognition, and emotion. Indeed, these far frontal areas have been shown to appreciate, and to provide for, the intention to highly abstract social rule-related behaviors conveying social and emotional intelligence. For example, damage within the basal or far frontal regions of the right frontal lobe often results in some degree of social anarchy or impaired intention to social proprieties and pragmatics within that culture (e.g., Anderson et al. 1999; Damasio 1994; see also Keenan et al. 2003; Tompkins 2012). By way of example, one distinguished business man was brought in to the neuropsychologist's office by his spouse with the complaint "this is not the man that I married." This corporate CEO and founder wore an expensive suit coat and a gold Rolex watch. But, he had fecal material under his nails and was picking his nose across the desk from me. Similarly, one might appreciate and be able to state the social rules for entering an office, but fail to carry out these rule-regulated behaviors with the self-attribution that "these rules do not apply to me." Another man was referred by the courts for inappropriate heterosocial behavior, groping young women at a local high school. He described his trip to the office where he yelled expletives at strangers on the road side, while knowing that the social and cultural rules dictate more regulated behavior or self-restraint.

Chapter 6
Functional Cerebral Systems Theory: An Integrated Brain

Wernicke–Geschwind Model

The challenge now is to appreciate the concerted interactions of multiple brain regions in performing one or another functional activity. The Wernicke–Geschwind model (see Bear et al. 2007; see also Anderson et al. 1999) is intimate to this discussion as the model provides a fundamental framework for the flow of information interactively across specialized brain areas. Carl Wernicke proposed an early model of language, which was elaborated upon and expressed again by Norman Geschwind, with the two now sharing credit for its inception. This language-based system might be used to understand the underlying processes in reading and, therefore, a reading disability of one or of another form.

Initially, the retinal projections of the written word travel via a thalamic relay (the lateral geniculate nucleus) to the occipital cortex with specialization within the left cerebrum for the recognition of the visual grapheme. For comprehension, the visual analysis of the graphical lexical image travels to the angular gyrus (tertiary association area) where cross-modal comprehension and conversion into the auditory sounds of the symbols are conveyed. A more dorsal pathway, through the angular gyrus, provides for the fundamental praxicons or gestural associations through integration with the somatosensory association areas within the parietal lobe. But, the model focuses on the ventral pathway through Wernicke's area in the superior posterior temporal gyrus, where the acoustic association and comprehension of the word sound may now be conveyed, via the arcuate fasciculus, toward Broca's area in the inferior premotor frontal cortex. Broca's area allows for the planning, organization, and sequencing of motor movements necessary for reading out loud, although activation will occur here to some extent in silent reading or

© Springer International Publishing Switzerland 2015
D. W. Harrison, *Brain Asymmetry and Neural Systems*,
DOI 10.1007/978-3-319-13069-9_6

covert speech (e.g., see Hickok and Rogalsky 2011). From Broca's area, the information flows to the final motor pathway within the motor cortex and eventually affecting movement through the contraction of skeletal muscles necessary to produce or express the speech sounds. Again, the pathway dorsal to Broca's area, which projects onto Exner's hand area, allows for the sequencing of hand movements necessary for writing what had been read in the form of graphical expressions and sequential postures/gestures of the hand. Ultimately, the well-sequenced series of movements underlies ideational praxis within the broader context of movement and movement disorders.

For listening to and understanding spoken words, linguistic or propositional sounds project onto the primary auditory cortex (Heschl's gyrus). A higher-order analysis and comprehension occur within the secondary association area at the posterior superior temporal gyrus known as Wernicke's area. Repetition of speech sounds requires the loop through the arcuate fasciculus toward the expressive speech region within the left frontal lobe. Although not expressed within the Wernicke–Geschwind model, many additional systems contribute to this basic functional cerebral system, including the frontal eye fields (e.g., Suzuki and Gottlieb 2013) for directional eye movements across the page of written words and Exner's hand area for the organization of writing movements or graphics, within the premotor cortex. Also relevant are right cerebral systems, where affective intonation and the emotional gist of the linguistic material might be understood. The right cerebral systems allow for the addition of prosody or melody in the comprehension of the text and, if reading aloud, in the expression of the written word. Dorsal pathways within the right cerebral hemisphere also allow for appreciation of spatial relationships and the conveyance of postural tone, for example. But, again, these analyses extend well beyond the early and more specifically targeted Wernicke–Geschwind account.

The seven components of the Wernicke–Geschwind model are depicted in Fig. 6.1. The workings of this model are depicted in Fig. 6.2. The components include the primary auditory cortex, the primary visual cortex, Wernicke's area, the angular gyrus, the arcuate fasciculus, Broca's area, and the motor cortex. The workings of the model were reassessed in 1999 (see Anderson et al. 1999). Subsequent to its inception, the model found support not only from lesion research but also from displays of metabolic activity resulting from various language activities using positron emission tomography (e.g., Petersen et al. 1988). Moreover, the model receives support from electroencephalogram and functional magnetic resonance methodologies. For example, one project (Schönwiesner et al. 2007) used event-related functional magnetic resonance imaging (MRI) and electroencephalography (EEG) in a parametric experimental design to determine the cortical areas in individual brains that participate in the detection and processing of acoustic changes. The results suggest that automatic change processing consists of at least three stages: initial detection in the primary auditory cortex, detailed analysis in the posterior superior temporal gyrus and planum temporal

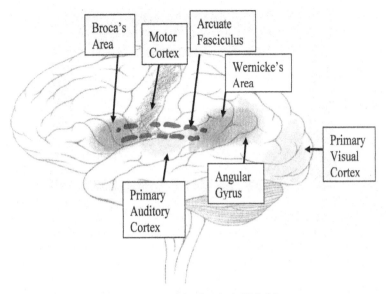

Fig. 6.1 The seven components of the Wernicke-Geschwind Model

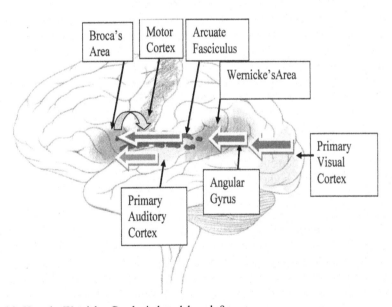

Fig. 6.2 How the Wernicke–Geschwind model works?

(see also Hugdahl 2012), and judgment of sufficient novelty for the allocation of attentional resources in the mid-ventrolateral prefrontal cortex. The scans from the Petersen project show metabolic activation during the generation, hearing, seeing, and expression of words (see Fig. 6.3).

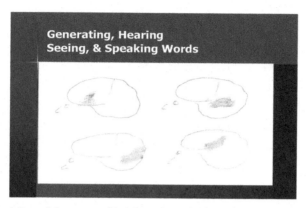

Fig. 6.3 From left to right, areas of heightened metabolic activation during the generation, hearing, seeing, and expression of words (see actual PET images at *Nature,* Vol. 331, No. 6157, pp. 585–589, 18 February 1988@ *Macmillan Magazines Ltd.* 1988). *PET* positron emission tomography

Marsel Mesulam (1990, 2000) proposed an alternative model based on the earlier version, wherein a hierarchy of neuronal networks process information by level of complexity. Performing simple over rehearsed language processes such as repetition of the days of the week might require activation primarily of the motor and premotor areas for language. More demanding linguistic tasks requiring heightened semantic and phonological analysis would require initial activation of the association areas. Mesulam's model provides for a spread of activation from unimodal language areas onto two additional sites for integration. These systems include the temporal region, providing access to remote memory or associations with this information, and the brain systems involved in emotional associations or the provision of meaning to the material content. Mesulam's model, in no way, abandons the established functional contributions of Wernicke's and Broca's area, but it instead elaborates on the broader contributions of previous experience with the content and the emotional associations that they convey.

These progressive, higher-order analytical epicenters are not unlike what we have discussed earlier on within Luria's functional cerebral systems with a flow of information from primary projection area, to the secondary association cortex, on to the tertiary association areas for complex and multimodal analysis/comprehension. These convergence zones, achieved through the formation of cell assemblies or groups of interconnected glial and neuronal cells, are collectively strengthened through their concurrent activation as proposed by Donald Olding Hebb in what is known as Hebb's law or, more formally, cell assembly theory. Hebbian theory (1949, 1955, 1959) describes a basic mechanism for synaptic plasticity, where synaptic efficacy arises from *repeated* and *persistent* stimulation of the cell and states:

> Let us assume that the persistence or repetition of a reverberatory activity (or "trace") tends to induce lasting cellular changes that add to its stability.... When an axon of cell *A* is near enough to excite a cell *B* and repeatedly or persistently takes part in firing it, some growth process or metabolic change takes place in one or both cells such that *A*'s efficiency, as one of the cells firing *B,* is increased.

The general idea is an old one, that any two cells or systems of cells that are repeatedly active at the same time will tend to become "associated", so that activity in one facilitates activity in the other. (Hebb 1949, p. 70). More succinctly put "cells that fire together, wire together."

Although Hebbian theory focused on neuronal activity and structures, it seems to be clear that these cellular assemblies are complex and relate substantially to more broadly defined cellular groups or cellular assemblies, including astrocytes, and ultimately their import and direction over neuronal pathways. It is most clearly the case, that the angular gyrus, also known as the left inferior parietal lobule, comprises one of these convergence zones. It is at this tertiary association area or location where multiple sensory modalities are integrated. The integration results from the expansive astrocytic and dendritic fields of the neurons within, and with access to, the frontal lobe, via the longitudinal tract and the opposite cerebral hemisphere, via the corpus callosum. This parietal occipital temporal region is, indeed, one of the most intelligent regions of our brain providing for the highest complexity of analysis, integration, and directional input for activity eventually conveyed by the frontal lobes and final motor pathways.

The associative tendencies for cellular assemblies to fire together more and more frequently through their repeated use may also provide at least one theoretical basis for creativity as defined here by loose or less common associations. For example, Paul Foster (Foster et al. 2012) has demonstrated a capacity relationship for the frontal lobe in word generation or verbal fluency. Diminished left frontal capacity may be evident in functionally diminished verbal fluency scores. However, it is also evident with the generation of words with less associative strength or common usage within the population. Redundant capacity may, by necessity, extend the associative demands to cellular assemblies less commonly activated in normal usage of the language.

This theoretical perspective on creativity awaits additional confirmation just as multiple pathways or neural and glial mechanisms should ultimately allow for novel or creative associative thought. But, some initial confirmation already exists where the authors (Foster et al. 2012) extend the conceptual analysis to dementia patients treated with acetylcholinesterase inhibitors (AChEI) in comparison with those not taking the medications. Patients taking AChEIs, which increase cortical ACh levels, differed from those not on these medications, possibly indicating reduced spreading activation with beneficial effects on the neural circuits involved. The effect of the drug(s) was to reduce the need for far-field associations found in individuals with reduced verbal fluency. The authors propose that the AChEIs reduce spreading activation and hence the activation of fewer associative nodes. Beyond this, the argument that although AChEIs enhance attention they also concomitantly reduce spreading activation with a potentially deleterious effect on memory is provocative. This research provides evidence of relevance to the distinction between frontal systems contributing to attentional processes and those posterior regions (hippocampal and perihippocampal regions) more commonly related to memory processes, per se.

Evidence from neural plasticity research after early deprivation or brain injury provides some evidence that language does not have an immutable neural architecture (see Blumstein and Amso 2013), but rather that the neural system is flexible and that it can change and reorganize in response to a number of factors. Functional magnetic imagery findings on the effects of early deprivation, for example, show that congenitally blind individuals recruit occipital areas in processing Braille, a language system used for tactile reading. This results in dramatic contrast with findings from sighted individuals where deactivation in these areas results from the performance of somatosensory tasks (Sadato 2005; Sadato et al. 1996; see Blumstein and Amso 2013). Blumstein and Amso (2013) note with interest that recruitment of occipital areas occurs in congenitally blind individuals during language processing (Bedny et al. 2011; Burton et al. 2002). Specifically, these authors argue that the visual cortex in congenitally blind individuals is recruited across a number of grammatical language components, including phonological, lexical semantic, and sentence processing. Moreover, the functional neural activation to language in this population is outside of the neural areas that are typically involved in these functions.

Part III
Three Functional Brain Units

Chapter 7
Arousal Syndromes: First Functional Unit Revisited

Arousal disorders involve alterations in the state or activation of the brain. These disorders may be generalized, affecting both cerebral hemispheres. But, very often, the disorder is localized to one or the other brain or to specific areas within the brain. In some cases, these may be visualized with functional magnetic resonance imaging (fMRI) or positron emission tomography (PET) scan technologies with behavior, affect, and cognition being the more sensitive indicators. We are more familiar with the arousal disorders, in the traditional sense, in individuals suffering from extreme arousal variants with coma or stupor. However, generalized arousal deficits have frequently been attributed to "mental retardation" and to other problems. Moreover, an alteration in arousal may be beneficial as Thomas Edison manipulated his sleep states to enhance creativity and to foster his inventions of such things as the light bulb, perhaps. By popular account (Maas et al. 1998), Edison would hold a steel ball or marble in his hand as he drifted off to sleep. But with the onset of dream sleep or rapid eye movement (REM) sleep, characterized by an abrupt onset of body paralysis, the ball would fall out of his hand and the noise, upon striking the floor, would awaken him. In this altered state, he would readdress his inventions and critical thinking for new inspirations and solutions.

Brain-Stem Sleep Mechanisms and Sleep Disorders

Each cerebral hemisphere sits on top of its brain stem consisting of the thalamic and hypothalamic bodies, the midbrain tectum (roof) and tegmentum (floor), the cerebellum, and the medulla. Luria's first functional unit, responsible for arousal or

© Springer International Publishing Switzerland 2015
D. W. Harrison, *Brain Asymmetry and Neural Systems*,
DOI 10.1007/978-3-319-13069-9_7

activation, is found here at the midbrain or mesencephalon in the diffusely projecting reticular activating system. The overriding commonality of a lesion to the brain
stem is an arousal disorder ranging from coma, to stuporous behavior, to mildly
lethargic states. This might be expressed with slurred and sluggish behavior as with
dysarthric speech. But as we begin to appreciate the complexities of the arousal
syndromes several disorders emerge. These are anatomically specific, which may
be useful in understanding the arousal syndromes and their varied features in the
clinical presentation of the patient. These are expressed as one or another sleep
disorder with each brain having its own sleep systems, more or less. This was appreciated early on as activation of the brain was found ipsilateral to the side of the
reticular activating system which was stimulated (Moruzzi and Magoun 1949).

Sleep has traditionally been defined by behavioral and psychophysiological
indices. Although the clinician and the scientist have been prone to rely on electrophysiological measures, a fairly adequate accounting of sleep may be had by
carefully observing behavioral indices. One of the better locations to appreciate
the behavioral indices of the sleep stages on a college campus is in the university
library. Here, you will find students actively involved in studying (i.e., sleeping)
with the stages of sleep visible for all to see in their public behaviors. Hopefully, this
does not include sleepwalking! These behavioral indices include the maintenance of
postural tone as opposed to body paralysis, myoclonic jerks or twitches, deep regular respiratory inhalation as opposed to variably irregular and shallow respiration,
REM, and overt indices of autonomic tone or emotional processing with sweating
and possibly pulse rate observed over a major artery.

The behavioral indices supporting a pleasant, quiescent, and restful state of low
arousal and with regular respiration, seemingly normal muscle tone, and normal dry
skin tone are those of non-REM sleep. Behaviorally, these non-REM sleep stages
differ largely in the depth of relaxation, lowered arousal level, and in their restorative features. The non-REM stages function as a "metabolic housecleaning." But,
the person participating in the deeper non-REM stages (3 and 4) may appreciate a
deeply restful and wonderfully restorative sleep period. These stages are accessed
acutely by those deprived and exhausted through the proverbial "power nap." In
this brief segue possibly lasting 20–30 min, the individual may awaken restored
as though they have just returned from a wonderful vacation without a care in the
world. If this is an exaggeration, then certainly the brief episode will feel restorative
on the order of a more extended sleep period.

In dramatic contrast, the behavioral features of REM sleep include REM, behavioral evidence in support of an "autonomic nervous system storm" with variable and
elevated respiratory rate, blood pressure, and heart rate, and with behavioral evidence
of paralysis accentuated by myoclonic jerks or twitches. The individual deprived of
REM sleep may have abrupt REM onset with a characteristic loss of motor tone
acutely signaling the onset of paralysis. I am reminded of students in a large introductory course with 700 classmates in two sections. These are very considerate folks
and they do their very best not to let their fatigue and sleep deprivation intrude into
the classroom setting. They carefully balance their head and trunk as they maintain
their posture at their desk. But, periodically, it will appear that some of the students
are pecking corn, like chickens in the field, as their head abruptly falls and they lose

control of muscle tone. Fortunately, this abrupt descent of the head is usually strong enough to wake them up. Otherwise, they would certainly wilt and fall, like a sack of flour, out of their chair and onto the floor if the body paralysis persisted.

Maintaining regularity in the daily schedule and consistency in activities may be helpful to establish stable circadian rhythms with a stable alternation across the stages of sleep throughout the sleep period. This will characteristically terminate in an REM period with awakening and the experience of being essentially "wide awake" and ready to get out of bed. This relates to the arousal state of REM sleep where the metabolic activity and the electrical activity of the brain are consistent with a highly aroused state and most like that of being awake. Indeed, the neural activation may be at a higher level than that which is present during wakefulness (e.g., see Riemann et al. 2012). College freshmen sometimes set their alarm back a few minutes during exams, which may be self destructive as they find themselves trying to awaken from deep non-REM sleep. They may be unable to get out of bed and be very sluggish if they are able to wake up, possibly missing the examination. This deep sluggishness and avid tendency to return to sleep is a feature of stage 4 sleep, in contrast to the high arousal and readiness to awaken from REM sleep.

The traditional method of determining sleep stage is derived from the electro-encephalogram (EEG) record. Most clinicians and neuroscientists use the international 10–20 system of electrode configuration. This system is a universally agreed upon placement configuration for the electrodes across the scalp (see Fig. 7.1). In the earlier days, the electrodes were more commonly attached to the scalp with conductive cream, which was somewhat like cement after it had hardened. But, the majority now uses electrode caps, where the 10–20 system is automatically config-ured with the cap placement and where conductive electrode gel is inserted through

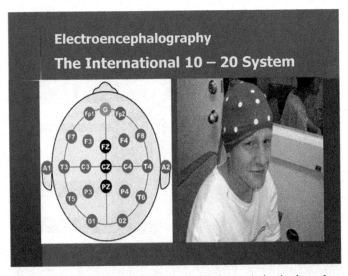

Fig. 7.1 The international 10–20 system of electrode placement. A simple electrode cap is shown on the right

the opening above the electrode and into the electrode cap. This placement system allows for communication among clinicians and neuroscientists as it provides standard locations for each electrode. My first EEG evaluation with a human was of a woman referred for a neuropsychological workup. Up to that time, I had performed EEG or evoked potential recordings on rats, dogs, cats, and other species. But, the first human evaluation was memorable.

This woman was a stripper at a nightclub. She was referred by her employer for "doin things I don't want done in my club." On additional inquiry, these were lewd sexual acts that fail to meet the social propriety standards of even the more shady side of this commercial enterprise. The behavioral complex included episodes of pelvic thrusting, which may result from temporal lobe seizures or from direct electrical stimulation of this region. But, the complaint centered on hypersexual behaviors, whereas the young woman evidently had no recall of these episodes. Sexual improprieties are more common in right frontal patients where rule-regulated behaviors may be appreciated at the level of social anarchy (e.g., Anderson et al. 1999; see also Damasio and Anderson 2003; Damasio et al. 2012; see also Hu and Harrison 2013b; Damasio 1994; see also Keenan et al. 2003). But, the amnesia for the event left me tendering an a priori hypothesis of right temporal lobe involvement. Figure 7.2 provides an example of the results from this assessment using the international 10–20 system.

In this comparison, the EEG record at electrode sites 6 and 7 featured low-voltage, high-frequency activity associated with an alert and aroused state. Similarly, cortical desynchrony was seen in comparisons at electrode sites 7 and 3 and electrode sites 8 and 5. In contrast, the activity over electrode sites 7 and 8 consisted of cortical synchrony with high-voltage, low-frequency activity within the delta bandwidth (1–4 Hz). Using the 10–20 system, I was able to localize a focal area of abnormal slowing over the right temporal lobe. The system allowed for the communication of

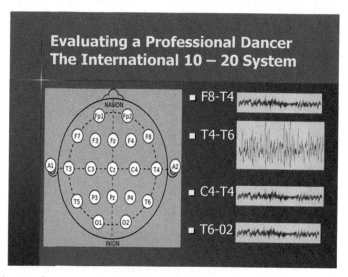

Fig. 7.2 A comparison among electrode sites using the international 10–20 system

the results to other clinicians and scientists as the locations are standard and replicable should she require a follow-up evaluation in another facility or at a later date.

Interestingly, the focal abnormality and the behavioral features resulted following exposure to visual oscillations with a flashing light. Rhythmic stimulation is present throughout the environmental surroundings. In a similar fashion, the brain consists of neural networks dominated by rhythmic activity (Adrian and Matthews 1934), measured as local fields potential on the EEG. It has long been theorized that these periodicities represent oscillations in neuronal excitability (e.g., Lindsley 1952) and neural entrainment to extraneous sensory conditions has been established (Mathewson et al. 2012). In this context, oscillating sensory events might be presented to the patient within visual, auditory, or somatosensory modalities based on the a priori hypothesis for the location of cerebral pathology derived from the interview and history. In this case, flashing lights worked. Of vocational relevance, she was dancing in front of a strobe light. Therapeutic interventions included vocational counseling and encouragement to avoid dancing in front of flashing lights. Ultimately though, sound oscillation with emotional musical melodies was sufficient in activating these right temporal lobe systems and inducing the event. The amnesia, which accompanied the event, was thought to be related to hippocampal and perihippocampal involvement and state-dependent learning phenomena (e.g., Milner 1965, 1968). Focal seizures of the right temporal lobe and amygdala have sometimes been referred to casually as "Madonna seizures" with pelvic thrusting and sometimes with penile erections in males with the episode.

In a basic sleep study, it might be reasonable to have some indication of the EEG record but without the complexities of a 120-electrode or even a 21-electrode configuration. The minimal configuration might consist of two EEG electrodes referenced to relatively neutral recording sites such as linked earlobes or a central electrode location over the vertex of the skull (e.g., central zero (CZ)). Alternatively, recordings from each of the active electrodes might be averaged, providing a reference voltage for each of the EEG bandwidths assessed. The basic sleep study in the laboratory might include other psychophysiological measures such as the electrooculogram (EOG) to measure muscle action potentials over the extraocular muscles (see Fig. 7.3). This might be useful to provide a clearer index of eye movements during sleep and, especially, during REM sleep or REM. An electromyogram is useful for the recording of muscle action potentials over some other skeletal or striate muscle involved in moving bone or intentional movement. This provides a unique comparison of muscle events controlling eye movement, in contrast to those muscles involved in volitional motor control or intention. During REM sleep, substantial activity over the extraocular muscles is expected but with diminished activity or paralysis over other skeletal muscles. These saccadic movements are mediated by the superior colliculus, which, in an awakened state, is under the regulatory control of the frontal eye fields (Holmes 1938; Guitton et al. 1985; Walker et al. 1998; see also Suzuki and Gottlieb 2013).

This configuration is sufficient for the basic sleep study. However, the study might include additional psychophysiological indices as appropriate for the specific question being answered. Respiration might be followed using a strain gauge or thermistor to record respiratory rate and volume during sleep. Skin conductance might

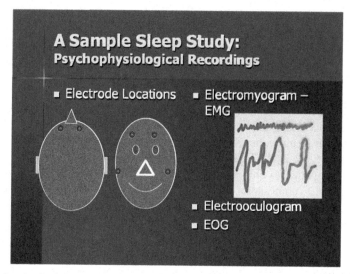

Fig. 7.3 Psychophysiological recordings for a basic sleep study using EEG, EOG, and EMG recordings. (*EEG* electroencephalogram, *EOG* electrooculogram, *EMG* electromyogram)

provide another index of sympathetic tone through changes in skin conductance with sweat. REM is also characterized by a significant increase in gastric contractions and acid flow into the stomach relative to non-REM sleep (see Pandi-Perumal et al. 2007). This might be accessed through the careful placement of a gastric tube into the stomach and recording of the electrogastrogram. Definitely, do not try this! The tube will end up in the stomach, the patient will aspirate, or other equally disastrous events may occur as this requires a good technique and good equipment. But, the gastric secretions with REM are recorded with volumetric measures, including a calibrated flask. These, and many other events, are all subject to assessment through various transducers, psychophysiological recording techniques, and measures.

Fig. 7.4 EEG activity varies across wakefulness, non-REM stages 1–4 and REM sleep. (Originally published in *Neurochemistry of Sleep,* Lajtha, Abel; Blaustein, Jeffrey D. 2007, p. 871). *EEG* electroencephalogram, *REM* rapid eye movement

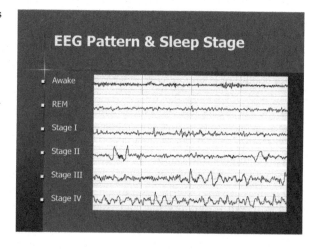

Figure 7.4 provides a visual comparison of the analogue signals and events that identify and define each sleep state. The primary change in EEG activity as we move from light stage 1 sleep to the deepest stage 4 sleep is an increase in voltage and a decrease in signal frequency recorded in cycles per second (i.e., hertz). This is best described as a shift from cortical desynchrony in the record to cortical synchrony as we move from stage 1 to stage 4 sleep or deep sleep. This transition to high-voltage, low-frequency cortical synchrony differs in every respect from the final stage of REM sleep represented by low-voltage, high-frequency activity or cortical desynchrony.

The stages of sleep require roughly a 90-min epoch for completion of a cycle in normal individuals. It seems to be relatively certain that this does not include university students. These typically sleep-deprived and highly stressed individuals, often transition to sleep readily in the laboratory with utter strangers around them and in a strange setting with wires attached to several parts of their anatomy. But, when they go to sleep, in defiance of the sleep literature on normal participants, they abruptly inter-REM sleep or directly interface with stage 4 sleep (anecdotal evidence only). And, they thank us when they leave! The stages of sleep and their defining features are presented for comparison in Fig. 7.5.

Stage W is actually not a stage of sleep but rather the period of drowsy wakefulness prior to transitioning into stage 1. This period is defined by alpha waves, a bandwidth ranging from 8 to 12 Hz (see Klimesch 2012 for review). Many have seen biofeedback apparatus for sale in magazines where the apparatus signals alpha activity. Alpha in adulthood is associated with a pleasant relaxed state of increased suggestibility. Alpha activity is found preferentially over the back of the brain and

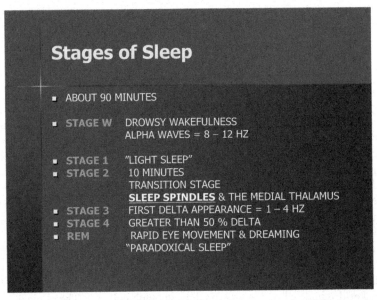

Fig. 7.5 The stages of sleep and their defining features on the EEG record. (*EEG* electroencephalogram)

especially over the visual cortex (Niedermeyer 1997; Feshchenko et al. 2001). Indeed, one validating technique for electroencephalographic recordings involves an eyes open, eyes closed or a light on, lights off manipulation, whereby alpha is increased with reduced visual activation in dim light and with alpha suppression with the onset of increased illumination (i.e., *the alpha suppression test*). A general rule in the industry features increased alpha (i.e., lower baseline cortical arousal) in healthier brains. In addition, alpha appears to be related to the availability of attentional resources for allocation with incremental cognitive load.

Stage 1 sleep is the first true sleep stage with a gradual slowing of the record and with a gradually incremental voltage. Individuals awakened from stage 1 sleep typically fail to appreciate that they have been sleeping. This might be grandpa on the sofa with the newspaper on his face and snoring. On awakening, he asserts that he was "not sleeping." Rather, he had just closed his eyes for a moment. The point here is that this is very light sleep and a stage not associated with restoration or deep relaxation, per se. Indeed, transitions into and out of stage 1 sleep may occur without awareness and with the continuation of ongoing behaviors and social interactions or conversation.

Stage 2 sleep is a critical sleep stage, both for the individual hoping to transition into deep restorative sleep stages 3 and 4 and for the diagnostician appreciating the sleep-related anatomy, wherein lesion localization may be a critical aspect of understanding and treating the various arousal syndromes. This "transition stage" of sleep conveys a variation from light sleep into deep sleep stages 3 and 4. It is defined on the EEG record by the occurrence of *sleep spindles* and these are displayed in Fig. 7.4. These high-voltage spindles reflect activation of the dorsomedial thalamus secondary to diminished inhibition of this structure by the ascending reticular formation (Dempsey and Morison 1942a, b; Dempsey and Morison 1943; see also Peter-Derex et al. 2012). Thus, with diminished arousal or activation by the reticular formation, inhibition over the dorsomedial thalamus is reduced, driving the cortical regions into high-voltage synchrony and eventually into delta activity with the progression into stage 3 and stage 4 sleep. Some of these functional relationships are discussed in Niedermeyer's *Electroencephalography: Basic Principles, Clinical Applications, and Related Fields (2011)*.

This thalamic system was discovered in 1942 by Dempsey and Morison (Dempsey and Morison 1942a, b, 1943) when low-frequency stimulation of the unit evoked widely distributed, high-voltage, cortical waves, characterized by long latencies and an initially progressive voltage increment. These recruitment responses produced waveforms which were thought to resemble spindle bursts produced by the administration of barbiturate anesthesia. Moreover, the evidence indicated that the neural pathways were the same and that the brain stem mediated the arousal response evident in the EEG (Lindsley et al. 1949; Moruzzi and Magoun 1949; see also Starzl and Magoun 1951). The significance of this neural pathway (reticular formation and the dorsomedial thalamus) was even more dramatic with Jasper's (Jasper and Droogleever-Fortuyn 1946; Jasper 1949; Hunter and Jasper 1949) hypothesis that the diffusely projecting thalamic system served as the subcortical pacemaker driving high-voltage cortical events with epilepsy. The clinical significance of this hypothesis was enormous and modulation of this system remains a primary intervention approach in seizure prophylaxis.

For the caregiver or therapist, a lesion of the dorsomedial thalamus may be evident with the patient complaining that he/she never sleeps, in contrast to others' accounts of the patient "sleeping all of the time." Indeed, the patient's presentation may feature a very light sleep disorder, wherein the individual may maintain posture in the chair and respond to questioning episodically, while transitioning into and out of light sleep. These transitions occur seemingly without awareness by the patient that they have been sleeping and even finishing a sentence that they had initiated prior to the previous bout of sleep. This sleep syndrome may result from cerebrovascular accidents or stroke involving the thalamoperforating artery, from anatomical or metabolic abnormalities or growths of one form or another or from pressure-related effects in proximity to the third ventricle and the flow of cerebrospinal fluid through the ventricular system and out to the cisternae.

Sleep stage 3 is traditionally defined by the first overt appearance of delta waves in the EEG record. These waveforms are high voltage and low frequency ranging from 1 to 4 Hz. Sleep stage 4 is traditionally defined by the preponderance of delta magnitude or minimally 50 % of the EEG record consisting of delta activity. From this point on, each of the stages 1 through 4 may be referred to as non-REM sleep in comparison with the features of REM sleep. Neurons with serotonin receptors are active during each stage of sleep until REM, so they appear to act as an REM inhibitor much of the time. When serotonin levels drop, the levels of the neurotransmitter acetylcholine rise in the brain. Serotonergic antidepressants may reduce REM (dream sleep); partly because increasing serotonin levels appear to inhibit the rise of acetylcholine (see Pagel 2008). Non-REM sleep appears to be potentiated by the indolamines, whereas REM sleep is potentiated by the cholinergic influences of the mesencephalic and pontine reticular formation and by the noradrenergic influences of the locus coeruleus (Jouvet 1972). However, the role of the raphe and serotonin is less clear than in the original theoretical accounts, since raphe cells begin to cease discharging during non-REM sleep.

When individuals are awakened from REM sleep, they typically confirm that they are indeed dreaming (see National Institute of Neurological Disorders and Stroke 2007; see also Riemann et al. 2012). This is the case even in many individuals who are adamant that they do not dream. With training, they may become more aware of their dreams and the content of the material processed during these episodes. REM sleep conveys the highest level of cortical arousal or desynchrony in the EEG and the arousal level in this stage of sleep may exceed that of wakefulness (e.g., see Riemann et al. 2012). This heightened cortical arousal easily flows toward the frontal lobes and with activation of the motor cortex for the intentional movement of our limbs. This might be the case with dream content where we are running or struggling or engaging in one or another physical activity. If we record over the motor cortex in the final motor pathway during this activity, we may see very remarkable action potential volleys, which would normally produce quite significant movements of our limbs.

Needless to say, this might be a problem for our safety in the bed and for that of our sleep partner. But, these action potentials continue down the motor pathways of the pyramidal tract and stop dead in the vicinity of the midbrain reticular formation. This area functions somewhat like an anatomical or physiological roadblock during REM (e.g., see Riemann et al. 2012) with body paralysis and with only the occasional action potential breaking through resulting in a spurious muscle twitch or myoclonus instead of a robust movement of the body. The resultant sensory and motor mismatch may yield illusory movements such as a sensation of falling with myoclonus or abrupt jerks perhaps waking us up or, even worse, awakening our partner.

This anatomical and physiological roadblock appears to function to restrict sensory input and to prevent overt motor activity during REM sleep as with a volitional movement. Some sensory information may make it past the blockade and these events may influence dream content. Generally, we only awaken to "significant" or relevant events. An example here might be the new mother waking to the sound of her baby crying. Men may awaken to these events also. But, there is a universal man agreement to never let the spouse know that we do! I am only kidding, of course. This stage of sleep leaves us relatively insensitive to environmental events. But this reduced sensitivity should not be confused with a diminished state of arousal. If the sensory event does provoke awakening from REM, we may be substantially aroused and relatively prone to awakening (see Riemann et al. 2012) and not returning directly to sleep. Thus, the overall effect of emotional stress may be to lighten sleep and, potentially, this may have survival value. Riemannet al. (2012) have also provided an improved understanding of chronic insomnia with evidence for generalized overarousal and the predominance of REM activity during sleep.

This functional disconnection provides some utility to the diagnostician and may provoke some confusion for the lay person. For the diagnostician, this is a sleep state in which we might evaluate basic reflexes, which would, otherwise, be inhibited to some extent by descending corticofugal fibers from the frontal lobes. One man had been poisoned over time by his wife using the chemical element arsenic. Arsenic has been called the Poison of Kings and the King of Poisons (Vahidnia et al. 2007) as it has often been used by the ruling class to murder one another and for its potency and ability to be discretely administered and over time. Eventually, this gentleman was seen by the neuropsychologist for the evaluation of functional neural systems with an advanced quadriparesis. Time had passed and he was in a new relationship. He indicated his basic concern about his "equipment" (when my sons were little they called it their "futures") and whether it still worked or not.

The assessment of impotency has traditionally involved the major distinction between *primary and secondary impotence*. With over-activation of the frontal lobes during sexual activities, possibly with perceived pressures to perform or due to cognitive stressors with the individual in an "observer role," the erectile reflexes may be inhibited. This is referred to as "secondary impotence" and therapy may focus on efforts to minimize the observer role and the cognitive preoccupation or performance anxiety. In contrast, primary impotence may imply faulty reflexive properties or abnormalities within the sexual apparatus lower down. A simple sleep evaluation will establish the basic issue of the integrity of the erectile functions as these are largely reflexive activities and the anatomical disconnect during REM sleep promotes hyperreflexia. Every 90 min or so, the individual enters REM sleep with the concurrent loss of descending regulatory inhibition. This promotes the increased probability of a reflexive engorgement of the penis. Many women have expressed concern that their partner was dreaming about another woman or that he was sexually unsatisfied only to discover that the erections during REM sleep were a natural and somewhat regular event.

Other disorders convey similarities with both non-REM or REM sleep components potentially providing diagnostic and treatment relevant information for the therapist or caregiver. Generally speaking, sleep disorders which involve movement of skeletal muscles other than the occasional jerk are relevant to the non-REM sleep periods, whereas high levels of arousal concurrent with body paralysis or illusory sensory and motor events are relevant to REM sleep (e.g., see Riemann et al. 2012). Narcolepsy provides a fair example of this with paralysis or "cataplexy" occurring with high arousal or excitement. One young woman was working in the elementary school system where she was known for her robust sense of humor. Early lesion studies provide for localization of pathological laughter and crying to the lentiform nuclei within the frontal lobes (Constantini 1910; see Oettinger 1913; Wilson 1923; McCullagh et al. 1999; Parvizi et al. 2001; Damasio et al. 2012; see also Parvizi et al. 2009). Her colleagues, students, and family members would bring her jokes as she was sure to find them funny and to laugh. But, the neuropsychological evaluation confirmed that she was labile for gelastic content with uncontrolled, deregulated, or hyperreflexic laughter. Moreover, these events were disabling as she would develop an acute onset of overwhelming sleepiness ("sleep attack") along with paralysis during these high-arousal events with positive affect. By the third provocative laughter episode in the office, she was immobilized!

Lesion studies provide for localization of laughter spasticity and crying spasticity in the majority of cases to the subcortical region of the lentiform nuclei within the frontal lobes (Wilson 1923). In a review of autopsy findings in 30 patients, pathological laughter and crying were never correlated with a single *cortical* lesion, whereas all 30 patients had damage within the internal capsule or lentiform area (Poeck 1985). Pathological laughter and crying are likely part of a larger corticobulbar system with a release of frontal cortical control over cerebellar pontine reflexes (Parvizi et al. 2001).

A case study is provided by a man with a long-standing developmental expressive dysphasia and stuttering. He described his educational experiences as the boy who could not read, along with having severe expressive speech deficits. The other children would laugh and make fun of him. But, he was physically robust and they paid the price for their antics. His gelastic seizures went undiagnosed until his mid-50s when he was seen for "problems on the job" and disability issues. His work as a welder was unfit by his supervisor's evaluation with "daydreaming" on the job next to a 2000- pound metal stamp that might easily amputate his fingers. Exploration of this vocational disability revealed an oscillating metal stamp and oscillating lights from his welding, which brought about absence seizures when these events were mimicked in the clinician's office. But, laughter had always decompensated him and he knew to avoid it quickly as he would lose consciousness and fall to the ground, again to the delight and pleasure of his classmates in school. Posed laughter in the office brought about nystagmus and myoclonus, whereas more provocative exposures left him light-headed and precipitously dropped his blood pressure.

Even though many societies and cultures appreciate laughter, it may be the expression of gelastic lability and brain pathology (Wilson 1923; see also McCullagh et al. 1999; Parvizi et al. 2001, 2009; Demakis et al. 1994) or incapacity just as a rage episode or recurrent crying spells might be (Wilson 1923; Ross and Rush 1981; McCullagh et al. 1999). Another man described his despair in trying to discipline his children at the dinner table where he would raise his voice and, in the excitement, find himself paralyzed on the floor and the children running back and forth over his abdomen. He also tried to explain his compromised role in his marital relationship as the sexual excitement was similarly disabling and immobilizing. Still another man would become excited at the bowling alley, sometimes collapsing with acute onset of paralysis as he approached the lane to discharge his bowling ball. Here again we appreciate the heightened arousal state concurrent with body paralysis.

Another man worked at the state penitentiary as a uniformed guard in charge of the cellblock and the prisoners residing therein. He was responsible for the security of the cellblock which housed convicted violent offenders. The cellmates were quick to appreciate his abrupt onset of sleep and paralysis with excitement, whereas he had been seemingly successful in hiding his narcolepsy from his employer. He self-referred for his neuropsychological workup after a near-successful attempt to escape from the prison under his watch. The goal had been to find him close to the bars so that he would fall close enough to the inmates for them to reach the keys and his weapon. This was easily implemented by fitful screaming and excitement from within the cellblock. He had been fortunate on more than one occasion. However, the culprits were improving their tactics and he was able to appreciate the risks involved.

The potential commercialization of animals susceptible to myoclonus or paralysis with high-arousal states includes the genetic propagation of fainting goats. Fainting goats are increasingly available, even on Craigslist, as these emotionally reactive animals evidence tonic immobility and paralysis upon excitement. These goats may be especially useful to some ranchers in coyote- or predator-infested areas as they will more readily fall victim to the attack with the onset of immobility to an abrupt startle stimulus. The abrupt incapacity of the goat would potentially spare the more expensive or valued livestock. Their presumptive role is to fall abruptly during the predator's assault with myoclonus providing the opportunity for the other valued animals to escape. In essence, the valued farm animals do not need to outrun the predator. They only need to outrun the fainting goat!

Generally, over the course of sleeping we may acquire our deepest stage 4 sleep early on and with ever-increasing episodes of REM sleep (e.g., see Riemann et al. 2012) as we progress toward the later part of the overall sleep period. Again, this may eventually result in circadian regularities where we may awaken from REM sleep and be alert and ready to get out of bed. This transition to ever more lengthy REM periods throughout the later part of the sleep period renders us ever more likely to awaken prematurely from an REM period and possibly during the early a.m. hours where we may be wide-awake and have difficulty returning to sleep. More serious disruptions may be characterized as sleep maintenance insomnia with the potent intrusion of high-arousal level REM and awakening with the processing of emotional stress. This may be more relevant as REM appears to be more probable and more potent following emotional events and where the emotional stress of the day may intrude into restful sleep at night (see National Institute of Neurological Disorders and Stroke: NINDS 2007).

Regardless, incremental time in REM is associated with awakening and feeling exhausted as though you have been working and solving problems all night rather than sleeping. These REM periods appear to be essential to process negative and also intense positive emotions and in their contribution to the mood state. One of the better predictors of depression is inadequate or altered sleep (e.g., see Riemann et al. 2012). REM may be suppressed through the promotion of serotonergic drive as, for example, with the administration of a selective serotonergic reuptake inhibitor (SSRI)antidepressant (e.g., Vas et al. 2013; see also Wilson and Argyropoulos 2005) or through increased exposure to bright light (see Dozier and Brown 2012), melatonin, and maybe to vitamin D (e.g., Partonen 1998; Holick 2007; see Parker and Brotchie 2011). Emotional processing might be advantageous in other situations where the potentiation of REM is achieved through administration of a noradrenergic agonist, cholinergic activation, or through arousing sensory input. This may include cholinergic afferents from the stomach with REM promotion following a "bedtime snack."

Figure 7.6 provides a summary of some of the features differentiating REM from non-REM sleep. Non-REM sleep allows for movement and postural support. This

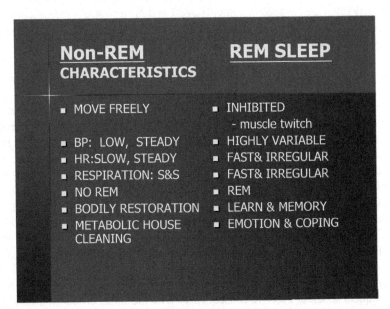

Fig. 7.6 Characteristics of non-REM and REM sleep. (*REM* rapid eye movement, *BP* blood pressure, *HR* heart rate)

may well include ambulation and even the engagement in complex activities as with somnambulism or sleepwalking. Blood pressure, heart rate, skin conductance, and related measures are relatively stable and reflect a "metabolic house cleaning" or activities which may be restorative and most restful. REM sleep, in contrast, is a stage of sleep with REM, body paralysis, and heightened arousal. Here, cardiovascular and respiratory functions are elevated and irregular. Rather than being associated with bodily restoration (non-REM sleep), REM sleep is more related to learning and emotional coping or emotional stress. This high-arousal sleep state is accompanied by an "autonomic nervous system storm" with many variants and stress-related activities (e.g., see Riemann et al. 2012). It is anything but restful! Stroke and heart attack may be promoted in this stage along with caloric utilization. In contrast, deep sleep is associated with the mobilization of growth hormones and may be important for caloric absorption.

Adult humans may expect to spend about 25 % of their sleep time in REM, whereas infants may spend as much as 50 % of their sleep time at this stage (see Pollak et al. 2010). Such developmental comparisons are crude, though, as the EEG record of an infant is sparse for higher-frequency bandwidths, including beta. Higher-arousal states in an infant, instead, are associated with alpha waves, a bandwidth in the middle of the arousal spectrum for adults (mean of 10 Hz; see Klimesch 2012). Students seem to require more REM sleep and, possibly, secondary to the demands for the consolidation of learning during this sleep state or, possibly, secondary to the heightened stress of this challenging environment and relocation away from family, friends, and their support networks. But, REM cycles with clinical disorders including the bipolar affect disorder (e.g., Duncan et al. 1979). It differs with schizophrenia (Kraepelin 1919; Chouinard et al. 2004) and with changes across the lifespan, where the elderly often are engaged in lighter (stage 1 and 2) sleep stages (see Siddiqui and D'Ambrosio 2012) and where they may perceive and complain of less sleep (e.g., Neikrug and Ancoli-Israel 2010). REM appears to be promoted by cognitive and emotional stress, whereas non-REM is promoted by physical exercise and relaxation.

Insufficient or disrupted REM sleep has been predictive of difficulties in carry-over or learning with impaired consolidation of the events and experiences during the day (Diekelmann and Born 2010; Martella et al. 2012). Social embarrassment or emotional coping failure may be minimized with consolidation of these events over the night and with the integrity of REM processing, whereas the same or similar coping failure may reoccur or the embarrassment may fail to diminish, with inadequate processing at these high levels of activation. REM deprivation has been associated with a failure to consolidate information from temporary memory storage into long-term memory (see Stickgold et al. 2001). REM deprivation in nonhuman animal research eliminates passive avoidance and spatial learning (Bridoux et al. 2012). REM deprivation eliminates active avoidance learning (Archer et al. 1984) acquired the previous day, whereas the animal is left with escape behaviors once exposed to the aversive events. An example here might be an animal in a one-way shuttle box (see Fig. 7.7). On day 1, the onset of a tone (CS) will predict a subsequent shock or aversive consequence. With the shock onset, the animal will escape to the opposite side of the shuttle box, which is safe and without shock. Many eutherian mammals learn this readily, such that they avoid the shock entirely

Fig. 7.7 An animal in a one-way shuttle box where a tone or other predictive stimulus (CS) signals an opportunity to avoid an aversive event by crossing over to the safe side of the two-way shuttle box

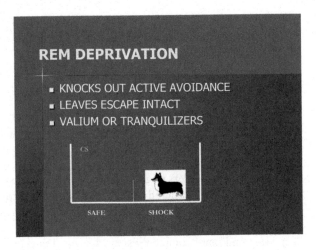

by jumping to the safe side with the onset of the CS. But, on return to their housing cage and with REM deprivation, the animal may fail the task on day 2 and be left to escape the shock after the fact. The end result is reflected in the impaired carryover of emotional learning with diminished coping behavior.

Valium or tranquilizers may also have this consequence, where the iatrogenic effects include REM deprivation while on these medications and where the consolidation of emotion or learning may be disrupted or minimized. Also, Valium (diazepam) may yield cumulative effects on REM deprivation and, eventually, with a rebound or heightened probability of the recurrence of REM (see Maisto et al. 2008). In the 1970s and 1980s, stomach ulcers were commonly ascribed to stress and aggravated by REM where stomach contractions and acid secretion into the stomach are prominent. The prescription of Valium during this time was of epidemic proportions. Valium might be used to minimize these effects by blocking REM sleep. However, with time and with cumulative deprivation effects, the patient would inevitably rebound with aggravation of the condition once again. On return to the physician, the prescription might be increased with more and more medication and a circular but worsening reliance on this medication concurrent with aggravation of the depression or dysphoric affective state.

Figure 7.8 shows an experiment on the effects of REM on anxiety consolidation or abatement (author unknown). College men were participants in an experiment

Fig. 7.8 Anxiety reduction as a function of group assignment where REM deprivation eliminated carryover or learning from day 1 to day 2. (*REM* rapid eye movement)

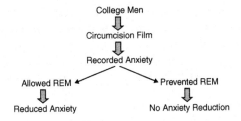

with exposure to a circumcision film. These macho men almost universally denied anxiety or apprehension, whereas the polygraph readings of their skin conductance responses and other psychophysiological indices of anxiety supported reliable increments in anxiety on exposure to this film. Subsequently, the men were randomly assigned to one of two groups with REM deprivation in one group. The comparison group was awakened an equal number of times but on a random sleep schedule. On day 2, both groups were again exposed to the circumcision film. The REM-deprived group's measures were comparable to day 1, while those of the group members allowed to dream that night were significantly reduced on the second exposure to the provocative film.

Several sleep disorders might be appreciated by the reader at this point. Insomnia has been related to a generally heightened state of central nervous system arousal (Riemann et al. 2012). Narcolepsy has strongly correlated features of REM sleep with high arousal and body paralysis. Somnambulism and enuresis are related to deep sleep and more commonly stage 4 sleep. Tricyclic antidepressants are sometimes prescribed for their noradrenergic influences in lightening sleep as these are often physically active deep sleepers. Night terrors are featured with deep stage 4 sleep as the individual is robustly capable of physical activity and may engage in overt activities and even aggressive behavior during deep sleep (e.g., Lyon 2009). These are more common in childhood and the parent will recognize them as being different from a more typical nightmare. With the nightmare, the parent may reassure the child upon awakening wide-awake from the bad dream. But, the apprehension from a night terror may be profound and often with an agitated and confused state. These might be promoted in an individual with mild cognitive impairments by the administration of a general anesthetic and where postoperative recovery is punctuated by an altered arousal state or encephalopathy.

The child experiencing a night terror, rather than awakening from a state of body paralysis, may be standing or mobile in posture or ambulation with negative emotional expressions. In our country's wars, many soldiers have been in combat settings with sleep deprivation and sometimes with amphetamines to promote survival. On return to their families or while on "rest and relaxation," some have injured their loved ones and even worse, during the night and with a rule out for the investigators of a night terror or stage 4 sleep event. These deep sleep phenomena are characterized by amnesia for the events, which may increase the soldier's vulnerability in the courtroom. Of course, these might present as iatrogenic episodes for some sleep-altering medications and with potentiation of the serotonergic or GABAergic pathways. Evidence indicates that Ambien (zolpidem), for example, may result in altered states and sometimes with amnesia for the episodes (Dolder and Nelson 2008). The US Congressman Patrick J. Kennedy stated that he was using Ambien and Phenergan when caught driving erratically at 3 a.m. "I simply do not remember getting out of bed, being pulled over by the police, or being cited for three driving infractions," Kennedy said (New York Times, May 5, 2006).

The apprehension experienced by an individual in a night terror is not readily resolved by reassurance or efforts to comfort them. Instead, the passage of time may be more important for the resolution of the symptoms. The parent may have faulty attributions of the child having a temper tantrum and being "spoiled." But, the adult may be at physical risk as others' attributions often are not appreciative of a sleep state. A Vietnam veteran and former prisoner of war was now residing in Florida. He had recurrent sleep terrors where he would be found in his backyard by his neighbors with an automatic weapon fully engaged in ambulation and apprehensive as he was awaiting dawn with the expectation that the enemy, rather than friendly neighbors, would be coming over the fence soon!

One young man suffered from epileptic seizures with altered states of awareness but with full mobility. He would elope in a deep sleep state and walk or run some distance in an apprehensive state. He was eventually shot by a local police officer on a highway, where the policeman noted that he had "resisted arrest." The man with the sleep disorder went to prison after he had recovered from his wounds. On release from prison, his mother initiated legal proceedings for what had happened to her son. Later, when asked her how her son was doing, she responded that he had been found shot to death with his body in a cave not far from her house. But, the point to be appreciated here is that the individual may be presumed awake during these events and the abilities to engage in motor activities are potentially intact through this non-REM and probable stage 4 sleep period.

Sleep apnea is a respiratory disorder, which was once euphemistically referred to as "Pickwickian syndrome" after the cartoon character that was always asleep on the job. This syndrome is ever more common in our society with obesity and obstructive apnea more and more prevalent. This is treated, in some cases, by placement of a continuous positive airway pressure (CPAP) apparatus at night to maintain positive pressure on the airway and to promote breathing. It may also be helped by postural changes to reduce the weight on the neck and chest areas possibly by propping up the head, shoulders, and trunk. This sleep disorder may often go undiagnosed with the patient experiencing suffocation like symptoms during sleep as there is breathing cessation, followed by a gasp for air. But, sleep apnea may result from brain dysfunction. Appreciation of the symptoms may be vital or result in interventions which affect the individual's health. This is more clearly the case with a newborn infant and possibly with an immature nervous system if the baby is sufficiently premature or if there was sufficient anoxia or other cerebral trauma during the delivery. The disorder is associated with sudden infant death syndrome where a seemingly healthy baby may die unexpectedly. Every now and then, a young couple appears in court and on the news with a deceased newborn and accusations that they may have suffocated their baby. Inadequate cerebral maturity or integrity, if severe, may yield a related respiratory disorder known as Cheyne–Stokes respiration.

This respiratory pattern consists of episodic cessation of respiration, followed by a gasp for air as the carbon dioxide receptors in the medulla may be activated triggering this reflexive gasp. This, in turn, is typically followed by a period of heightened respiratory rate until the blood is oxygen-saturated and the reflex resolves. Subsequently, the episode will reoccur and, if long standing, may promote anoxic events and encephalopathy. When identified during the neuropsychological evaluation, the physician and therapists are informed and encouraged to instrument cautions and possibly nocturnal monitors or an alarm apparatus, if appropriate. These events and dyspnea in general may be more common with frontal lobe disorders and especially diminished capacity within the right frontal lobe region. Here, dual task demands or stressors involving the concurrent tasking of the right frontal lobe may yield a cessation or disruption in the regularity and volume of inspiration and exhalation. But, Cheyne–Stokes patterns may be seen with brain stem and frontocerebellar involvement (Lee et al. 1976). High cervical lesions might also warrant monitoring for stable respiratory patterns in patients with cervical cord involvement.

It may also be relevant to mention the effects associated with general sleep deprivation as it appears to be increasingly common in modern societies (e.g., Finucane et al. 2011). Much concern has been expressed for the alarming increase in obesity and the metabolic syndrome, for example. Substantial evidence across controlled experimental laboratory investigations and large epidemiologic studies has exposed a close link between sleep disturbance and the pathogenesis of metabolic syndrome (see Tsang et al. 2013). Beccuti and Pannain (2011; see Tsang et al. 2013) reviewed this literature and concluded that the majority of studies support a relationship between short sleep duration and an increased incidence of obesity and metabolic diseases. These authors note that the prevalence of the metabolic syndrome has been increasing drastically over the past 30–40 years concurrent with evidence for diminished quantity and quality of sleep over the same time period (e.g., Finucane et al. 2011). Of interest here was a large-scale survey by the American Cancer Society (Kripke et al. 1979) conducted in the 1960s in the USA and compared with a "Sleep in America" poll conducted in 2008 by the National Sleep Foundation (National Sleep Foundation 2008). The comparison showed a reduction in average sleep duration from 8–9 to 6–7 h during workdays.

Functional Sleep Neuroanatomy

All areas of the brain are potentially relevant to the discussion of any function as the brain works as an integrated functional system (Luria 1980). For simplicity here, I will focus on the primary brain-stem structures involved in these processes as part of an oscillating sleep–wake system (Jouvet 1965; see also Riemann et al. 2012). In doing so, sleep or arousal syndromes might be exposed that may be presented by the individual with anatomical and pharmacological relevance. It is also hoped that this discussion may expose the therapist and caregiver to relevant components of the system, which may be altered by stroke or one or another form of brain pathology.

With an aroused and awake state, the mesencephalic reticular activating system is active and functions to maintain EEG arousal and cortical desynchrony (see Riemann et al. 2012). Lesions of the medial pontine reticular formation abolish REM sleep and injections of cholinergic agonists into this region, potentiating neural activity, and induce REM sleep (Jones 1993; cited in Kolb and Whishaw 2014). In opposition to the system activating REM sleep, slow-wave sleep characterized by cortical synchrony in the EEG record is promoted by activation of the raphe. The raphe is directly inhibitory over the reticular activating system, with raphe activation resulting in reticular formation deactivation and reductions in cortical arousal. The reticular formation, in turn, appears to be inhibitory over the medial thalamus. Thus, with the activation of the raphe, there is reduced inhibition of the reticular formation over the medial thalamus, allowing the thalamic structures to drive or recruit cortical synchrony in the form of high-voltage sleep spindles, signaling the onset of stage 2 sleep (e.g., see Peter-Derex et al. 2012). Moreover, the presence of sleep spindles in the EEG record signals a transition from light sleep into the deep sleep of stages 3 and 4.

This system promoting non-REM and the transition into the deeper sleep stages is directly inhibited by the locus coeruleus located dorsal to the raphe (e.g., see Riemann et al. 2012). As the locus coeruleus becomes active, the activity of the raphe is diminished, which lowers the inhibition over the reticular activating system. As the reticular activating system becomes active, there is a heightened inhibition over the medial thalamus and the development of desynchrony in the EEG associated with a highly aroused and possibly an awakened state. But the raphe is generally active to some extent. It might be pulling the individual toward relaxation and restful sleep. With depletion of the necessary transmitters or with lesions, which might result from a stroke or other malady, one or another component of this oscillating system might be less adequate in the promotion of its normal functions. Also, continuing controversy exists for the role of the raphe, where activation promotes non-REM sleep, but where the raphe may begin to decrease activation during non-REM sleep.

The role of the locus coeruleus is to increase or promote REM sleep and the associated body paralysis, which precludes the acting out of a dream state. Instead of overt and complex movements of limbs, only the occasional action potential makes its way to skeletal muscles producing the muscles twitches signaling REM activity. Interestingly, Jouvet (1972) observed remarkable behavior in cats subsequent to lesions of the locus coeruleus (subcoerulear nucleus). Specifically upon entering REM sleep, the cats failed to develop the atonia or general paralysis related to this sleep state. Instead, they stood up, looked around, and made movements of catching an imaginary mouse or running from an imaginary threat. Jouvet demonstrated that destruction of this critical brain region would release the motor system from paralysis, resulting in acting out the content of the dream.

Functional Sleep Psychopharmacology

Acetylcholine

The reticular activating system receives sensory input from each modality and promotes arousal or activation of each brain through a diffusely projecting network. The primary neurotransmitter involved in this activation or arousal influence is acetylcholine, a neurotransmitter involved in executive function at the basal forebrain (Whitehouse et al. 1982); learning at the hippocampus (e.g., Hasselmo 2006); found at all preganglionic fibers of the autonomic nervous system (Chang and Gaddum

1933; Feldberg and Gaddum 1934); and released from the postganglionic fibers of the parasympathetic branch of the autonomic nervous system. REM sleep is characterized by a unique pattern of neuronal activation involving a circumscribed group of cholinergic neurons in the dorsolateral tegmentum and the pontine tegmentum which activate thalamic structures triggering cortical desynchronization and a heightened state of cortical arousal (see Riemann et al. 2012). It has been unequivocally demonstrated that REM sleep represents the highest level of central nervous system arousal present during the sleep cycle (Maquet et al. 1996). Generally, larger lesions of the reticular formation promote a deeper and more pervasive "sleep" or comatose state. Stimulation here promotes arousal (Moruzzi and Magoun 1949). Physostigmine, which blocks acetylcholinesterase resulting in increased acetylcholine, awakens an animal if it is injected into the reticular activating system during REM sleep. Also, carbachol, a cholinergic mimic promotes the onset of REM sleep if injected into the reticular activating system during non-REM sleep (e.g., Yamuy et al. 1993).

More recently, Paul Foster (Foster et al. 2011, 2012) demonstrated the role of the acetylcholinesterase inhibitors often used to treat dementia, where they were shown to reduce spreading activation in frontal systems involved in intentional functions, including word fluency or generativity. Dementia patients in the group without acetylcholinesterase inhibiting medication showed significantly greater spreading activation as demonstrated through the generation of less common words on the verbal fluency test. Frontal capacity was improved as evident in reduced spreading activation in the acetylcholinesterase inhibitor group. Acetylcholine potentiation promoted more typical or common word usage and generation with reduced reliance on words with far-field associative strength.

For the sleep analysis, at this point, cholinergic activation appears to be important to REM sleep onset and maintenance, the consolidation of learning or memory, and the vagal influences promoting parasympathetic drive. The latter may promote acid secretion into the stomach and stomach contractions along with variable heart rate and blood pressure during this sleep stage (see National Institute of Neurological Disorders and Stroke: NINDS 2007). A stroke or damage to the upper pontine region and possibly with involvement of the superior cerebellar artery may diminish the arousal system in one or in both brains. But, the arousal disorder presenting as coma or stupor ranging to a hypoaroused state with dysarthria and slurred, sloppy behavior will be in the brain systems ipsilateral to the damaged reticular activating system. The functional lateralization of the sleep systems for each brain and, therefore, the potential to differentiate a left or right brain sleep disorder or alteration remains largely unappreciated within the neurosciences. But, lateralized lesions of the left or the right pontine region may alter the sleep for one or the other brain. Also, waxing and waning of symptoms may differentially occur in the processing specializations of the left brain with a left brain-stem lesion and in the processing specializations of the right brain with a right brain-stem lesion.

For the therapist or caregiver working with this patient, sensory stimulation techniques may be employed to promote heightened arousal. But, the sensory activationmight better be designed to activate the reticular activating system and not the raphe. The reticular formation may be activated by stimulation, including ranging the body, facial massage, illumination, and noise. However, the raphe activates to slow rhythmic stimulation like warmth on the face and gentle stroking of the skin. Again, these systems are oppositional and the interventions are more effective as they are more discretely tailored for one or another purpose. Also, if the arousal disorder is severe with coma or stupor, the physician may be maintaining the coma through barbiturates or drugs designed to suppress the activation of the reticular formation. This may be done for administrative and safety reasons with agitation a concomitant of, and predictor of, recovery from coma (Bogner et al. 2001; Bogner and Corrigan 1995). But, the reticular formation drives cortical desynchrony and recruitment where, with damaged cerebral systems, seizure provocation may be of concern and seizure threshold may be pharmacologically manipulated. So, the medical management efforts at maintaining coma may be based on concerns for seizure prophylaxis.

Two men presented to the neuropsychologist's office with complaints of episodic loss of consciousness that seemed to fall outside of the routine assessment of the arousal disorders. The first man was noticeably shorter than his spouse such that, when they hugged each other, his neck was compressed between her breasts. The couple complained that he would periodically pass out on the dance floor and sometimes when hugging his spouse. The couple provided a demonstration in the office, whereupon placement of his head between his wife's breasts did indeed result in altered consciousness and with diminished cerebral blood flow. This demonstration resulted in an angiogram with evidence for carotid artery occlusion and cerebral vascular disease and eventually an endarterectomy to restore blood flow. The second gentleman complained of losing consciousness during personal hygiene activities as he shaved his neck. Once again, a demonstration in the office confirmed his reports and again with occlusive arterial disease.

The Indolamines

The raphe is the principle serotonergic structure in the brain (e.g., see Lall et al. 2012). If the raphe is damaged with a lower pontine lesion, then the patient may experience insomnia for a few days with gradual but incomplete resolution of these features. For the therapist or family member, the stroke might be proximal to the distribution or branches of the inferior cerebellar artery. Many related functions may be altered due to the location of the lesion, including truncal ataxia or a high risk for falls if the cerebellum or frontocerebellar pathways are affected. The pontine area frequently affects visuomotor functions and several cranial nerves are nearby, which may alter some of the functions of the face or head region. The activation of the raphe appears to be promoted through agonistic interactions with other indolamine structures, including the area postrema at the back of the brain stem.

The *area postrema* was identified by Borison and Wang (1953; see Young 2012) as the locus of the chemoreceptor zone responsible for triggering vomiting. This structure is critically positioned and arrayed for the detection or sensation of blood-borne and cerebrospinal fluid-borne chemicals, including toxins. It is a highly specialized area of the brain in that it lacks a blood brain barrier and also lacks a cerebrospinal fluid brain barrier. It is arrayed with a relatively large complement of receptors for multiple neuroactive compounds capable of transducing these events into neural activity. The anatomy of this region also appears to promote leakage through the walls of the structure, exposing the anatomy to the chemicals carried within the blood and within the cerebrospinal fluid. The area postrema is relatively sensitive to circulating levels of Tryptophan and the hormone melatonin in the blood. Tryptophan, the precursor of serotonin and an essential amino acid, is only available in the diet (see Haleem 2012). With increments in Tryptophan, the area postrema activates serotonergic drive through the raphe, promoting deep sleep.

Also, serotonin injected outside of the blood brain barrier may result in incremental sleepiness through this interface just as the turkey dinner may increase circulating tryptophan and promote sleepiness and non-REM sleep. In summary, the indolamines appear to inhibit REM sleep and to activate or to promote slow-wave sleep. A large body of research, including animal and human studies, has confirmed the crucial role of the serotonin (5-HT) system in the regulation of anxiety-related behavior and traits (see Maron et al. 2012; see also Haleem 2012) and negative emotion more broadly defined. Currently, the most common class of effective antidepressants is the SSRIs, which act to selectively block the high-affinity reuptake of serotonin. Since tryptophan is only available in the diet, it seems likely that excessive dietary restriction and malnutrition decrease serotonin reserves by depletion of the necessary precursor for the production of this neurotransmitter. This raises concerns for sex differences if women are prone to dieting and if women have lowered overall levels of serotonin compared with men. Indeed, women are the recipients of approximately 78 % of the prescription SSRIs (Kessler 2003). In a review of the literature, Haleem (2012) concludes that "diet restriction-induced exaggerated feedback control over 5-HT synthesis and the smaller availability of tryptophan decreases serotonin neurotransmission at postsynaptic sites, leading to hyperactivity, depression, and behavioral impulsivity."

The Catecholamines

The substantia nigra in the ventral tegmentum consists of densely packed dopaminergic fibers and is part of the extrapyramidal motor system. A lesion here appears to promote lethargy and hypokinesis, whereas stimulation may yield the opposite effects with heightened energy and sometimes hyperkinesis. Much research and treatment efforts with the hyperkinetic disorders has been focused on this system of dopaminergic neurons. Dopamine is the precursor for the neurotransmitter norepinephrine, which is most densely concentrated in the locus coeruleus. The locus coeruleus is directly inhibitory on the raphe or serotonergic system, where activation results in the inhibition of slow-wave sleep and the promotion of REM or cortical arousal (Jouvet 1972; see also Samuels and Szabadi 2008; Andrews and Lavin 2006). Thus, the locus coeruleus is indirectly agonistic with the reticular formation through inhibition over the raphe. Norepinephrine is involved in mood and was manipulated with some success using the antidepressant medication *Wellbutrin* (bupropion), which promotes the catecholamines norepinephrine and dopamine via reuptake inhibition (see Stahl et al. 2004). A lesion to the locus coeruleus promotes hypersomnia and cortical synchrony as with deep sleep.

In summary, the proposed system (Hobson et al. 1975; Hobson et al. 1986) consists of an awakened state with the reticular formation active and maintaining EEG arousal and cortical desynchrony. Slow-wave sleep develops with activation of the raphe and through its inhibitory influences over the reticular activating system. Downregulation of the reticular formation provides for disinhibition of the dorsomedial thalamus. The dorsomedial thalamus becomes active driving the cortical neurons into a synchronous pattern with the onset of stage 2 sleep and the appearance of high-voltage sleep spindles (e.g., see Peter-Derex et al. 2012). The locus coeruleus is inhibitory over the raphe, resulting in the activation of the reticular formation. Thus, slow-wave sleep known as non-REM oscillates with periods of fast wave sleep known as REM and with the associated functional aspects, which have been discussed. Lu et al. (Lu et al. 2006) have discussed the mechanisms underlying the cyclical patterns of oscillating REM and non-REM sleep, proposing a switching "flip-flop" mechanism for REM initiation and for REM termination within the midbrain tegmentum. These neuroanatomical components contribute to the functional sleep system through their respective neurotransmitters. Among these are the indolamines which promote non-REM and the catecholamines which promote REM sleep patterns and sympathetic drive. The overall system oscillates (e.g., Mathewson et al. 2012) and these oscillations form much of the basis for the circadian rhythms and for health more broadly defined.

This was partially evident in a project collecting a large body of clinical data over three decades in a subgroup of depressed patients where sleep, temperature, hormone, and mood changes were found consistent with disturbances in circadian-related processes (McClung 2007). Circadian processes regulate daily physiological and biological rhythms approximating a 24-h cycle. Circadian-related abnormalities are present in virtually all subtypes of depression including seasonal affective disorder (SAD), major depressive disorder, and bipolar disorder (see Lall et al. 2012). In their review of the depression literature, Bunney and Potkin (2008) provide converging evidence from three areas of clinical research for circadian abnormalities in depression. These include physiological and behavioral abnormalities, the rapid (within hours) reversal of depression with circadian manipulations and the effects of antidepressants (e.g., SSRIs, MAOIs, tricyclics), and mood stabilizers (lithium and sodium valproate) on circadian-related mechanisms.

Under various research protocols, it has been revealed that REM sleep propensity is closely linked to the time of the core body temperature nadir (Dijk and Czeisler 1995; Czeisler et al. 1980). Based on this evidence, researchers (Kojima et al. 2013) have argued that the expression of REM sleep is most likely regulated by circadian processes whereas non-REM sleep characteristics are more dependent on prior waking duration (Dijk and Czeisler 1995). Circadian abnormalities are one of the prominent factors underlying sleep disturbance with depression. Approximately,

70–80 % of depressed patients complain of difficulty in falling asleep, staying asleep, or experiencing early morning awakening. A meta-analytic review of the literature (Baglioni et al. 2011) revealed a link between depression and insomnia. Moreover, recordings of electroencephalographic activity during sleep in depressed patients are suggestive of a dysfunction in several sleep parameters, including a more rapid onset of REM sleep and a reduction in slow-wave sleep (see Lam 2006). These findings support an interpretation of central nervous system overarousal (see Riemann et al. 2012) with many individuals suffering from depression.

Research on cerebral lateralization differences in circadian rhythms has not yet been completed. Oscillating patterns between the two cerebral hemispheres remain largely a topic for future scientific investigations. But, with the processing specializations of each brain, and with the evidence for significant cerebral asymmetry for the indolamines and the catecholamines (e.g., see Fitzgerald 2012), such research promises to shed light on many clinically relevant concerns, including depression and emotion more broadly defined. These investigations would be directly relevant to the mood disorders and potentially provide improved precision in the understanding and treatment of depression.

Chapter 8
Sensation and Perception: Second Functional Unit Revisited

The following sections explore the processes and pathways involved in the reception and comprehension of sensory information from our environmental surroundings eventually culminating in our perceptions, feelings, and emotional associations for these events. For each sensory modality, this process requires the conversion of physical energies from our world into neural energy and eventually the activation of specific functional cerebral systems for their comprehension, storage within memory, and operational or intentional activity derived and expressed in response to these events. The sense organs are not isolated and passive receptacles of energy information. Instead, the sensory systems originating with the receptor have been found to actively probe and to search the personal and the extrapersonal spaces to update the internal representations of the world and the contents of emotional and cognitive schema (Droogleever-Fortuyn 1979; Gitelman et al. 2002). This process involves interactions among the three principle functional units of the brain.

The ascending sensory projections begin with the transduction and propagation of sensory information originating at the receptor and traveling along specific highways in the central nervous system known as tracts, bundles or funiculi, or "cables" for the novice. These cables eventually arrive at their respective destinations within the brain, otherwise known as the primary projection area (cortex) for that (specific) sensory modality. By analogy, if you were headed from Chicago to Seattle, you would need to be on a specific route or highway. A different highway might take you to New Orleans. Glial cells are instrumental in the construction and repair of these highways, and, in many ways, these pathways provide for communication among interconnected clusters of glial cells (see Koob 2009; see also Becker and McDonald 2012).

Within your nervous system, visual information travels within the optic nerves to the thalamic relay (lateral geniculate nucleus) with second-order fibers forming the optic radiations projecting through the temporal lobes and onto the primary visual projection cortex. In a like manner, audition travels within the auditory cables to the thalamic relay (medial geniculate nucleus) and onto the auditory projection cortex.

© Springer International Publishing Switzerland 2015
D. W. Harrison, *Brain Asymmetry and Neural Systems,*
DOI 10.1007/978-3-319-13069-9_8

The visual projection pathway does not mix with the traffic projecting within the auditory system. Each system remains largely separate and exclusive in this anatomy. This mutually exclusive arrangement of the anatomy and physiology of the sensory systems is maintained up until the association cortices, where crossmodal integration is possible. Alternative mechanisms are available for the crossmodal influence of one sensory system on another sensory system via recruitment, facilitation, occlusion, and interference. Most notable among these influences is that of Luria's first functional unit discussed elsewhere in these writings under arousal theory and the reticular activating system.

Damage to any specific sensory pathway will result in a specific loss within that sensory modality. Depending on the location of the lesion within the cable at different geographical locations within the nervous system, other effects may be expected because of the proximity of other systems at that location. For example, damage to the dorsal spinal cord will very likely affect touch and pressure but not temperature and pain, which travel up a different cable within the lateral spinothalamic tracts. In contrast, at the level of the somatosensory projection cortex, in the parietal lobe, a lesion may very well affect the somatic senses more broadly to include touch, pressure, temperature, and pain. The geographical configuration of the nervous system, and especially of the brain, allows the diagnostician to validate a syndrome and, in many cases, to localize the damaged area. For the patient, the caregiver, and the therapists, lesion localization provides a basis for education and an understanding of the brain disorder and specific indications or contraindications for interventions. Ultimately, an improved understanding of the brain processes involved in sensation and perception should improve patient care and lighten the burden often drawn from misunderstandings of the human system and problems arising within it.

Visual Systems

Why does the world look the way that it does? This was a preliminary examination question for one woman as she was working toward the completion of her first doctoral degree. Part of the answer lies in what each brain is specialized for or prepared to see. Each brain, when activated, has a bias to see what it is prepared to see, and these preparations are fostered and facilitated through experience across the life span. One patient named "Cleopatra" provides an eloquent example, in that she had recurrent visual hallucinations or formesthesias at the right side consisting of "cute, funny, little frogs hoping around," whereas at the left side, she had recurrent hallucinations of "snakes" which scared her. She was attracted to those happy images at the right side and enjoyed watching them. Instead, at the left side, she was preoccupied with avoiding or escaping these provocative and menacing visual forms.

It is hard to imagine the interesting little frogs at her right in the absence of previous experiences with them through interactions with the environment and in a joyful or pleasant way. The preparation was in the architecture of her brain and specifically the positive bias and the location at the right visual field. These projections are connected early in development, and they appear to process what the left brain is better able to see, just as the right brain receives its visual projections from the left visual field. Similarly, the snakes that were accompanied by fear and avoidance behaviors might not have been seen unless she had learned to see them through interactive experiences within the environment. But, the right brain which sees from the left visual field was prepared for dealing with these threats and for mounting a fear response, which might save her life in some situations.

Another patient with negative associations and experiences with her former spouse was now "seeing him" as a devil trying to harm her. He was seen at the left with negative affective valence, sympathetic autonomic responses, and behavioral features consistent with an activated right brain. So, experience may be fundamental to the processing advantages of each brain. But, there appear to be limitations born from the compartmentalization or specialization of each brain for the content. The next challenge here is to begin with the physical energies that are transduced into neural events in vision and to follow them through the neural systems which process this information. These fundamental features will be useful in the discussion about brain disorders elsewhere in these readings.

Many individuals tend to think that they see all and that they know all that is present within their environmental surroundings. But, this is far from the actual case. In reality, the human retina is capable of the transduction of only a small fraction of the electromagnetic energy that we are exposed to through contact with our receptor cells in the retina. Other species may be able to see parts of the spectrum that we do not see, due to the limitations of our receptors and the adequate stimulus. The adequate stimulus for vision consists of electromagnetic wavelengths or light energy ranging in length from about 780 to about 360 nm (see Hood and Finkelstein 1986), where the longer visible wavelength is seen as red and where the shorter visible wavelength is seen as violet. Transduction of the visible spectrum ranging from longer to shorter wavelengths yields colors in the following order: red, orange, yellow, green, blue, indigo, and violet. Students in introductory courses on the visual system recall this arrangement using a mnemonic device like "Roy G. Biv," where the three-part name provides for the first letter of each color across the visible spectrum as follows: R for red, O for orange, Y for yellow, and so forth.

Light enters the eye through the cornea, traverses the aqueous humor, and travels through the pupil on its way into the vitreous humor and eventually onto the retinal surfaces. Pupil diameter is determined directly by the contractile structure of the iris, which consists mainly of smooth muscle. The iris serves to regulate the amount of light entering the eyecup. The iris is composed of two oppositional muscle groups (intrinsic muscles), a circular group known as the sphincter pupillae and a radial group known as the dilator pupillae. Contraction of the sphincter pupillae results in a decrement in the size of the pupil, whereas contraction of the dilator pupillae results in dilation of the pupil.

Lowenstein and Lowenfeld (1950a, b) demonstrated that the autonomic nervous system controls the size of the pupils with activation of the cholinergic parasympathetic nervous system constricting the pupil and the adrenergic sympathetic nervous system dilating the pupil. Moreover, the left and right brains differentially influence the parasympathetic and sympathetic branches of the autonomic nervous system (Oppenheimer et al. 1992; Oppenheimer and Cechetto 1990; Zamrini et al. 1990; Hoffman and Rasmussen 1953; Heilman et al. 1978; Foster et al. 2010; Andersson and Finset 1998). The tectoreticular fibers from the superior colliculus project to the mesencephalic reticular formation mediating hemispheric arousal. Ipsilateral fibers in this pathway are more abundant than contralateral fibers (Truex and Carpenter 1964) and stimulation of the colliculus induces a hemispheric arousal response (Jefferson 1958). Thus, each superior colliculus, via the tectoreticular system, appears to be able to activate or arouse the ipsilateral hemisphere.

Since ambient light entering each eye projects to the contralateral superior colliculus via the retinocollicular pathway (Rafal et al. 1990) and since this input alters the ipsilateral reticular arousal systems and the corresponding cerebral hemisphere, constrained ambient light to one eye may differentially impact sympathetic tone (Burtis et al. 2013). Also, activation of the sympathetic nervous system to novelty and to preferred sexual stimuli has been assessed using pupillometric recording techniques (Garrett et al. 1989), and others (Bar et al. 2005) have provided evidence for pupillometrics as a sensitive and reliable means of assessing autonomic nervous system functions.

Charles Darwin famously wrote that the eye caused him to doubt that random selection could create the intricacies of nature (see Masland 2012). In a prelude to reviewing the complex neural organization of the retina, Masland (2012) remarks that "Fortunately, Darwin did not know the structure of the retina: if he had, his slowly gestating treatise on evolution might never have been published at all." Masland further notes that although the retina forms a sheet of tissue only about 200 mm thick, that the neural networks within carry out image processing feats which were not yet imagined even a few years ago (see also Gollisch and Meister 2010).

Perhaps more than 60 distinct types of retinal neurons exist with specific roles in processing visual images. Masland (2012) describes three main stages in the arrangement of the mammalian retina. The first is charged with processing the outputs of the rod and cone photoreceptors into parallel streams of information. The second stage provides for access to retinal ganglion cells conveying the information out of the orbit and toward the thalamic relays. The third arrangement combines bipolar and amacrine cell activity to create diverse encodings of the visual world, which again are transmitted away from the orbit toward the brain. The author notes that perhaps 50 % of the ganglion cell response selectivities, which create diverse encodings of the visual world, remain to be discovered.

In addition to the bipolar cells and ganglion cells found within the retina are the horizontal cells. The role of the horizontal cell appears to be the provision of inhibitory feedback to rods and cones and also to bipolar cells (however, see Herrmann et al. 2011). Historically, the horizontal cells were thought to improve visual contrast via downregulation or inhibition of adjacent rods and cones. In this fashion, a darker or blackened surround might make a dim light appear to be brighter, for example. Masland (2012) notes that the horizontal cell functions to control gain or amplification of the visual signal. This would serve to hold or to maintain the signal input to retinal circuitry within its functional operating range. Thus, in the natural world, with great variation in the brightness of visual components within a scene, the horizontal cells function to stabilize and to adjust very bright and very dim objects for optimal viewing. This function is reminiscent of trying to photograph a very bright object in a dim room, where the camera might be saturated by the bright light. The inhibitory regulation of the horizontal cells, by analogy, allows for the viewing of the bright light at the same time as the dimmer ones.

Light enters the cornea, our first refractive lens, travels through the aqueous humor, traverses the lens or zonula, crosses the vitreous humor of the eyecup, and "falls" upon the back of the eye. Most engineers can follow this part of the story. But, they begin to pull their hair out with what happens next in the design of the visual system. If the engineer were to design this system, they might build something like a camera to produce clear crisp images on the retina. But, our retina is not designed in this fashion, nor is the image allowed to fall cleanly on the transducer surface. Instead, we have an *inverted retina* where light must cross several layers of tissue and fluids prior to its conversion to neural energy by the transducer.

To understand this, it may help to recount a simple story of a father convinced that his young son was infested with the devil or demonic spirits as he was difficult to parent and often somewhat high-spirited or oppositional and defiant. The father, feeling as though he was inadequate in raising the child, called the Catholic Church and, in his desperation, requested that an exorcism be performed on the boy. But, the church informed him that it would need proof in order to believe the father and certainly before investing in the exorcism. The boy's father had his proof one day when he photographed the child with a Kodak camera using an older-style flash. What was the proof? The boy had red eyes! The eyes were of the devil!

The story was fictional, of course. But, the question at this point is why did the camera reveal red eyes and what is the red stuff? The answer is blood that is crossing over the retina in addition to the blood vessels and supporting structures. Indeed, light now must go through the blood and blood vessels, the ganglion cells, the amacrine cells, and the horizontal cells before it finally interfaces with the visual receptors! For the engineer, this architecture defies logic and any hope of developing a clear, clean, crisp visual image of high contrast. But, the system works. The transduction of light energy into neural images is now performed by a duplex retina in humans, whereas this may not be the case in other species. The eagle, for example, has a preponderance of cones. The rat has a preponderance of rods. But, these species seem specialized for acuity (eagle) or sensitivity (rat), whereas we are afforded some combination of these with cones and rods dispersed disproportionately across our central and peripheral retina, respectively.

The rods are located largely in the periphery of the retina with specialization for *scotopic* (dim) lighting levels. The cones are more densely concentrated at the fovea or central retina for acuity in *photopic* (higher intensity) light conditions. Functionally, this might offer some advantage when viewing the stars in the night sky with your sweetheart under scotopic conditions. Here efforts to focus light on the periphery of the retina by looking slightly off center will increase the apparent brightness of the star. Instead, when looking at your beautiful gal across the distance in photopic conditions, try to focus the reflected light directly onto the fovea for the highest level of acuity. The horizontal cells in the retina work to improve the contrast of the visual image via lateral inhibition. Specifically, each cone will announce that it "sees something" concurrent with an effort to regulate the adjacent cellular activity via lateral inhibition. This process is somewhat like surround inhibition in the auditory system. It promotes acuity. It diminishes competing information as spurious or artifactual. Once understood, this functional anatomy was manipulated by changing the human factors engineering of our television and computer visual displays, using a dark background for the light emitter to improve contrast.

Some readers may recall the old cathode ray visual displays on the early television sets. My father purchased three of these at the hefty sum of five dollars when I was a young fellow. We were in the middle of the west Texas desert in Pecos, Texas best known for the historical figure Judge Roy Bean or the "Law West of the Pecos River." We had a substantial historical collection of "shoot-outs" at our old cowboy saloon downtown. I later learned that my interest in invertebrates might have been fostered by my childhood environment. One professor of invertebrate biology at the University of New Mexico described this desert area as the "invertebrate capital of the world." Tarantula migrations would periodically cover the roads in our town and also the highways and byways that we traveled. The professor would travel there to establish geographical scorpion plots (scorpions per square meter) using a black light (they glow a yellowish green). I learned to shoot accurately using the stinging scorpions on my bedroom wall and a flashlight held close to my Red Ryder BB gun. You cannot rid your home of these critters, and they easily wiggle over thresholds and architectural barriers. Whether it is true or not, I do not know, but according to legend, these little hellions are not just resilient. One was found alive at "ground zero" in White Sands, NM, after the explosion of the atom bomb. Recent engineering efforts to develop lightweight "skins" for jet aircraft to be resistant to scratching or etching of the surface have attempted to model after the scorpions' seemingly smooth exoskeleton, which has been found, instead, to be a multidimensional surface and ideal for surface protection.

I questioned my parents as I wondered who had selected our backyard for testing the atom bomb. This had direct implications for my childhood, since I loved to make and to eat snow ice cream. I remember my parents trying to explain to me that I could no longer eat snow ice cream as the snow was potentially contaminated with the radioactive fallout from "testing" the bomb. As for the five-dollar televisions, I was promoted to antennae specialist. My job was to hold the antennae, effectively extending the sets reception range and clarity of the picture of our two channels. The goal was to see Walter Cronkite on the evening news! It was not long before we would see President Kennedy's assassination on this set. Seeing these events along with Cronkite's dialogue may have influenced our perception of the world and the trust that we may have held in our government.

Working in Chester Karwoski's visual psychophysics laboratory at the University of Georgia recording from single cells within the eyecup, it was apparent that the eye is actually an extension of the brain, since complex visual analysis is initiated within the eyecup and prior to gaining access to the ganglion cells and their projections on toward the thalamic relays. Chet impressed on his students the varied specialization of the cells at this level. But, I was more impressed with the prover-bial "bug detector," a cellular arrangement providing for the detection of a moving spot across the retina. This might be useful for a reflexive tongue extension in the frog. But, I remain concerned of its role in my own eye.

Ultimately, the events detected and processed at the retinal level are projected on toward the thalamic relay (lateral geniculate nucleus) and eventually, via the optic radiations, onto the primary projection cortex also known as the calcarine or striate cortex in the occipital lobe. The pathway is characterized by a lamination of the cells or striations. But, normal development of the visual pathway is illu-mination dependent. Early perceptual researchers (Crair et al. 1998) attempted to restrict visual experience through rearing the newborn in dark lighting conditions. They quickly discovered that they were not studying perception subsequent to these environmental manipulations as some minimal level of ambient lighting is neces-sary just for normal cell growth throughout the visual system. The absence of light shortly after birth resulted in anatomical aberrations consisting of the loss of the normal striate or laminated appearance of these projections.

The brains are arranged in a seemingly logical fashion with the functional equivalent of "cables" carrying each sensory modality to its primary projection area within each cerebral hemisphere (see Figs. 3.7 and 8.2). In this fashion, information is carried from the left and the right retina of each eye back to the lateral geniculate nucleus in the thalamus, where second-order fibers eventually make up the optic radiations as they continue on toward the farthest points in the back of the brain pro-jecting onto the primary visual cortex in the occipital lobe (Figs. 8.1 and 8.2). In a practical sense, then, one may understand the basics of an underlying brain disorder

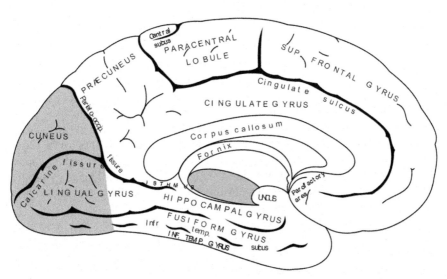

Fig. 8.1 Midsagittal view of the left brain showing the cuneus and lingual gyri above and below the striate/calcarine fissure. (This image is from *Gray's Anatomy* and is no longer subject to copyright)

Fig. 8.2 The visual pathways originating from two retinas within each eye as they continue on to the thalamus with second-order fibers traveling via the optic radiations to the primary visual cortex in the occipital lobe of each brain. (Originally published in *The Six Major Anatomic Decussations with Clinical Correlation*, Alpert, Jack N., 2012, p. 134)

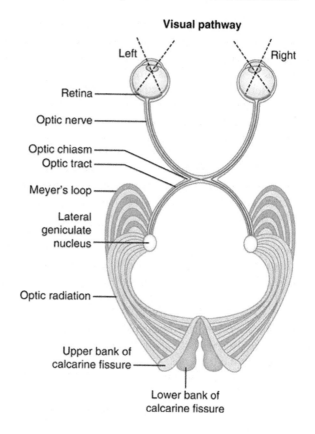

based on inspection of the locations of these cables and through appreciation of the location of the visual array found within each of these cables. For example, damage to the optic nerve of the left eye would likely affect both the left and the right visual fields of that eye and largely leave both the left and the right visual field functionally intact for both the left and the right brain. Why? Each eye has two retinas with one devoted to the left visual field (right retina) and one devoted to the right visual field (left retina). The left retina of each eye goes almost exclusively to the left brain, whereas the right retina of each eye goes almost exclusively to the right brain. This is the case, beyond 2° of visual angle (see Winkler 2012) as foveal vision is processed bilaterally at the visual projection cortices. In a similar fashion, upper visual quadrants project primarily to the lower occipital region or lingual gyrus, just as the lower visual quadrants project more dorsally through this system to the upper occipital or cuneus region (see Fig. 8.1).

One important aspect to appreciate, initially, in order to understand these pathways is that vision crosses and reverses vertically *as the image enters through the pupil* with left-sided images falling on the right retina of each eye and vice versa. It does not cross or reverse again! A stroke or damage within the left temporal lobe or left occipital lobe, though, may well cut the cable with a loss of visual images from the entire right visual field. As shown in Fig. 8.2, disturbance of the optic radiations within the left temporal lobe results in a loss of the right visual field (for each eye). Such a loss is referred to as *right homonymous hemianopsia* or loss of one entire visual half-field. A lesion at the homologous region, within the right temporal or occipital lobe, might instead result in left visual field blindness referred to as a *left homonymous hemianopsia*.

Four specific locations within the visual projections correspond with either left or right eye blindness (optic nerve damage) or left- or right-sided hemianopsia (right or left optic radiation damage, respectively). Damage within the lower parts of the optic radiations and/or occipital lobes within the left or right brain would yield a corresponding loss at the contralateral visual field for the upper quadrant. For the right brain, this would be identified as a *left upper quadrantanopsia*. An upper quadrantanopsia within either visual field heralds a lower lesion within the tracts or occipital lobe at the contralateral hemisphere. An additional site that is somewhat prone to tumor development consists of the pituitary gland located proximal to the optic chiasm (Fig. 8.3). Here pressure effects from an enlarging growth may impact the visual fibers originating from the nasal side retina of each eye and result in a corresponding loss of peripheral vision (i.e., *tunnel vision*).

Interestingly, a pituitary tumor might be expressed in the syndrome of "gigantism" with facilitation of growth hormones and oversized physical development. At the time of these writings, the largest known living man was a little over 8 ft tall and residing in Virginia (National Public Radio, August 2010). The awesome height of a giant conveys great strength and physical prowess. Instead, individuals with pituitary tumors resulting in gigantism often suffer many physical maladies, including joint pain. These problems may eventually be expressed in a shortened life span with multiple health-related problems. Arguably, the most famous case of gigantism is presented in the Bible where the young David is in the position of challenging Goliath, the Philistine champion. By this account, seemingly hopeless little David rises to the challenge and, in combat, slays the giant! If Goliath's physical size and descriptive features are accurate, then he very well may have been the product of a pituitary tumor located over the midline at the optic chiasm. Pituitary tumors are notorious for the resultant loss of the crossing retinal fibers, which convey peripheral vision from each eye. The hallmark of a pituitary tumor is "tunnel vision," which might have been a distinct advantage for young David in his approach to the Philistine (Bear et al. 2007). This would be most beneficial on approach from the giant's periphery rather than midline, as in a flanking maneuver. And, if I were little David, that is exactly what I would do.

Descartes (1649) projected a visual image through the eye of a bull and onto a wall, with the discovery that the transmission of light through the cornea and the lens results in an image on the retina, which is essentially upside down and backwards. Vertical components within the visual array are best demonstrated at the level of the optic radiations where information high in the visual array is located low within the optic radiations on their way through the temporal lobes and onto the visual primary projection cortex in the occipital lobe. In contrast, visual information within the lower visual quadrants projects through the upper levels of the optic

Fig. 8.3 (*Top*) Bihemispheric
reconstructions of the optic
nerve and tract (*red*) as well
as of the optic radiation
(*yellow*). (*Bottom*) Dissection of the optic radiation
into Meyer's loop (*yellow*),
central bundle (*green*), and
dorsal bundle (*blue*).
(*R* right, *L* left)

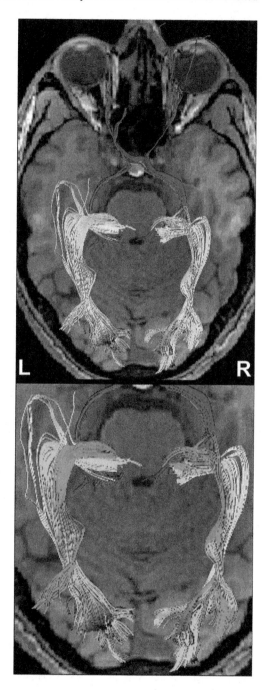

radiations, which approximate the upper extent of the temporal lobe and the transition into the lower levels of the parietal lobes. Practically speaking, pathology at the lower visual quadrants is a functional indication of a dorsal lesion location within the optic radiations (parietal proximity) and vice versa. These anatomical locations within the vertical visual array may be processed differently. For example, faces seen from below may be perceived as more positive and less negative, while faces seen from above appear more negative and less positive (Kappas et al. 1994). Also, safety issues relate to the location of the lower extremities within the lower visual quadrants in a patient at high risk for falls.

David Hunter Hubel and Torsten Nils Wiesel received the Nobel Prize in 1981, along with the neuropsychologist Roger Sperry who studied "split-brain" patients (see Berlucchi 2006). Hubel and Wiesel were lauded for their work on the visual system at the level of the occipital lobes. Hubel and Wiesel (1962) provided for detailed study of the visual system of the cat with the discovery of higher-order visual processing through these cortical regions culminating in their receipt of the Nobel Prize for their contributions. In this research, the cat's head might be fixed in position using a stereotaxic apparatus with visual stimuli of various forms presented onto a screen in front of the animal. Concurrent recordings of action potentials amplified and played through a speaker system were obtained from cells within the visual cortex in order to determine what aspects of the stimulus the occipital cells responded to, through the development of action potential volleys. Initially, they were frustrated as they seemed to elicit responses to the spot placed on an acetate sheet and moved across an overhead projector with the image displayed in front of the cat. But, they would discover that it was the edge of the acetate sheet, and not the spot, to which this cell was responding. Moreover, the orientation of the line was relevant to the elicitation of action potential responses from the cell. The seemingly spurious action potential volleys were actually valid responses of that cell to a line of a particular orientation!

This technique eventually resulted in the identification of simple cells, complex cells, and hypercomplex cells responding to the ever-increasing complexity of the visual stimulus (Hubel and Wiesel 1965). More specifically, these cells responded differently based on the orientation of a linear stimulus, the length and/or movement of the stimulus, and the angularity of the image. Subsequent to this research, concept cells were discovered to participate in the appreciation of concepts across sensory modalities and even the concept of familiarity. One concept cell for the number six might respond to the visual image "six," to the auditory sound "six," and to the graphic stimulus presented within the somatosensory modality from writing the number "six" on the subject's hand. Concept cells have been discovered across species and even for the concept of "the nest" for a mouse (Lin et al. 2007).

Quiroga et al. (2005) report one case where a unit responded only to three completely different images of the ex-president Bill Clinton. They report that another unit (from a different patient) responded only to images of the The Beatles, another one to cartoons from *The Simpsons* television series, and another one to pictures of the basketball player Michael Jordan. Subsequently, these researchers described a remarkable subset of medial temporal lobe neurons "that are selectively activated by strikingly different pictures of given individuals, landmarks, or objects and in some cases even by letter strings with their names." Using single-neuron recording techniques, Quiroga et al. (2013) describe experiments that led to the discovery of a neuron in the hippocampus of one patient who responded very preferentially (strongly) to "different photographs of actress Jennifer Aniston but not to dozens of other actors, celebrities, places, and animals." They describe another patient, where a neuron in the hippocampus "lit up at the sight of pictures of actress Halle Berry and even to her name written on the computer screen but responded to nothing else. Another neuron fired selectively to pictures of Oprah Winfrey and to her name written on the screen and spoken by a computer-synthesized voice. Yet another fired to pictures of Luke Skywalker and to his written and spoken name, and so on."

Once visual information arrives at the primary visual cortex in the occipital lobe, we begin to see our world in ever more complex perceptual ways. Information within each of the major sensory modalities (e.g., vision, audition, touch) moves, at a cortical level, from the primary sensory cortex to the secondary association area for that modality and then on to the tertiary association cortices. Research has demonstrated that the complexity of analysis increases in a manner corresponding with the flow of information across these regions. The left occipital region appears specialized for the processing of visual graphemes representing letters in our alphabet, which may be assembled in a serial processing fashion into words, sentences, and eventually paragraphs conveying semantic content. In rather stark contrast to this, the right occipital region appears specialized for the processing of visual patterns, including those patterns that may eventually be recognized as faces or places as higher level analysis continues within this modality (Tranel et al. 2009; Benton and Tranel 1993; see also Farah and Epstein 2012). Ultimately, the information may be integrated across other sensory modalities, providing a crossmodal analytical integration and maximal depth of understanding and familiarity with the stimulus array.

Damage within the occipital cortices may result in one or another *visual agnosia* or inability to recognize a particular type of image. Common visual agnosias include letter and number (Park et al. 2012), facial and line orientation (Tranel et al. 2009), and tool and hand (Bracci et al. 2012) recognition deficits, for example. That these problems are attributable to a brain disorder may be most strange and counterintuitive to a layperson comfortable in assigning "vision" to the "eyes" as this is what we tell our children, as we were also told. It is simply not true! The specific visual agnosia resulting from a brain disorder may be a function of the specific part of the visual cortex involved and, more notably, a function of the processing specializations of the left or the right cerebral hemisphere.

A common example would present with dyslexia from a left brain lesion, where visual pattern recognition or aspects of facial recognition remain relatively unaffected. Indeed, many laypeople along with many health-care professionals attribute visual agnosias to peripheral visual problems such as glaucoma or macular degeneration. Though the neural damage from glaucoma or other "peripheral" visual neuropathies may initially reside within the retina, second-order cell loss and cortical changes secondary to diminished activity in these pathways are now commonly established within the scientific community. Many of the clues that we have in

understanding a brain-related visual disturbance is in the specific type of visual image that cannot be processed and, if the lesion is within the projection pathways, the location of the visual defect within the visual fields. Interestingly, it appears as if the first indications of an evolving retinal defect may come from functional and anatomical changes at the occipital cortex (e.g., Van Buren 1963)!

Steeves et al. (2004) describe a woman with prior carbon monoxide poisoning 14 years earlier. She had developed a profound visual agnosia for objects with bilateral damage to the lateral occipital cortex, despite preservation of her abilities to recognize scenes of beaches, forests, deserts, cities, markets, and rooms. Functional neuroimaging showed activation with integrity for the parahippocampal regions near the ventromedial temporal lobe, suggesting that recognition and integration of the broader components of one or another scene does not depend upon recognition of the objects located within the scene. This separation of the elemental objects in a visual scene from the perception of the scene has supported arguments for the role of the parahippocampal regions in visual place recognition and the identity of this region within the ventral and medial stream as the *parahippocampal place area* (see Carlson 2013; see also Shinohara et al. 2012). Other relevant visual specializations appear at the visual association cortex, including the *extrastriate body area* for the visual recognition of body parts (Downing et al. 2001) and the *fusiform face area* (Cox et al. 2004; see Kanwisher and Dilks in press).

With sufficient damage to the occipital association cortices, the patient may develop a simultanagnosia with an inability to integrate the complexities of a visual scene. The patient might be seated next to the large window and marvelous view from the medical center overlooking the valley and the city of Salem, VA, for example. But among this vast visual array, this person may only be able to see individual elements at any given moment. Visually recognizing another aspect of the scene as an element within the surroundings will require separate visual analysis. Beyond this level of impairment, but not by much, is Anton's syndrome (Anton 1899) or cortical blindness presenting with an associated disorder of la belle indifference or a failure to appreciate the blindness. A Vietnam veteran and former prisoner of war had recurrent or episodic complaints identified as visual formesthesias where he stated "Gooks are throwing fire on me!" But, aside from these events with fearful apprehension or panic, he was cortically blind. He had no appreciation for this and neither did his staff. When asked if he could see, he said "of course I can." Often family members sit, initially in dismay, as the neuropsychologist tries to explain this to them. Otherwise said, many put much store in the self-assessment of the patient, and this is often inaccurate. This would be the *fundamental flaw of clinical inquiry,* where verbal self-report indices rule the day, potentially leaving the clinician as blind as the client/patient. By simple analogy here, using self-report, the patient might report "I feel cold...I am freezing," where the clinician might then attempt to warm the patient with a high fever, potentiating the fever to the point of damaging bodily systems or even with vital implications.

In addition, the percept of color and image is dynamic with rapid adaptation of these receptors altering the visual stimulus in the form of aftereffects or the persistence of the iconic trace. Normal function of the receptor is critically dependent on the visual motor apparatus, including the functions of the pons and the superior colliculi under the regulatory control of the frontal eye fields (Holmes 1938; Guitton et al. 1985; Walker et al. 1998; see also Suzuki and Gottlieb 2013). Moreover, the parietal eye fields appear to be organized in gaze-centered coordinates such that goal-related activity (e.g., reaching for an object) is remapped when the eyes move (Medendorp et al. 2003).

For example, a patient with a brain stem lesion involving the pons, cerebellum, or superior colliculus may have a reading problem without damage in the brain regions classically involved in lexical analysis or language processing. One avid reader expressed his frustration this way. "Every time I try to move my eyes to the next line of text my eyes go to the wrong place." The diagnosis for this gentleman might be descriptively stated as a "dyslexia without dysgraphia atypical" resulting from optic ataxia or visuomotor discoordination (dysmetria). This syndrome is also known within the neuroscience literature as brain stem dyslexia.

Persistence of the icon or visual trace is a critical issue in ongoing perception, and it is altered by arousal state or projections from the mesencephalic reticular activating system. But, the ongoing visual sensation and perception of events within our surroundings are critically dependent on refresh rate or the elimination of the previous image so that the subsequent image might be seen. The eyes are constantly in motion with barely detectable saccades. Researchers (Ditchburn and Ginsborg 1952; Riggs and Ratliff 1952; Yarbus 1961) asked what the function of these normally occurring reflexive eye movements are by using a curare derivative as a muscle relaxant to temporarily paralyze the extraocular muscles rendering the position of the eyes fixed on the visual stimulus. In doing so, they discovered that these movements facilitate the refresh rate and functions necessary to prevent adaptation and the sometimes illusory afterimages that they produce.

The frontal eye fields initiate intentional eye movements and inhibit reflexive saccades by the superior colliculus (Holmes 1938; Guitton et al. 1985; Walker et al. 1998; see also Suzuki and Gottlieb 2013). These intentional and reflexive eye movements serve as a mechanism for visual fixation, rapid eye movement, and the fast phase of optokinetic nystagmus. Apparently, the term saccade was coined in the 1880s by French ophthalmologist Émile Javal. He had used a mirror on one side of a page to observe eye movement during silent reading with the discovery that it involves a succession of discontinuous individual movements (Javal 1878). Visual saccades allow for quick movement of the fovea, with its high spatial resolution but limited spatial extent, around a large visual field to detect and process key environmental events (Yarbus 1961). Damage to this system at the level of the pons may result in visual nystagmus or irregular jerking, beating, or rolling movements of the eyes and perhaps nonconvergent gaze. Depending on the irregularity produced by the structural lesion, the patient may have a variety of visual complaints and often dizziness. If the frontal eye fields are damaged, there may be a diminished ability to direct the eye toward the side opposite the lesion and an ipsilesional increase in visual distractibility, through inappropriate saccades, to provocative stimuli within the setting. This functional anatomy underlies a variety of rapid eye movement disorders, whereas a locked or fixed visual gaze and sparse blink rate with masked faces is prominent with lesion of the basal ganglia or extrapyramidal motor system (Karson 1983; see also Bologna et al. 2012).

The visual array is rich in diversity and abundant in complexity, consisting of natural images and those created by human architecture, engineering, and scholarly activity. These include the subtle nuance of a facial expression, the stars at a great distance in the night sky, the divergent lines of buildings seeming to converge toward the horizon, and a seemingly infinite array of color, contrast, and perceptual delight. The amount of potentially relevant visual information requires perceptual selectivity and an equally abundant ability to ignore redundant information. The primate brain devotes perhaps 50% of the neocortex to visual processing regions (Van Essen et al. 1992). Yet, the processing must still be selective and narrow and especially if the visual information is worthy of a motoric response. Ignoring certain aspects of the visual array and selectively responding to other aspects is a fundamental attentional ability (Egeth and Yantis 1997; Proulx 2007, 2013). However, some researchers have asked questions at the other end of this spectrum, where vision is restricted or lost secondary to peripheral blindness.

The World Health Organization provides an estimate that over 39 million people suffer from blindness, whereas over 246 million are visually impaired ("Visual impairment and blindness," 2012; see also Proulx 2013). Many visual problems are acquired secondary to diabetes (see Proulx 2013) and others arise as age-related disorders, including glaucoma and macular degeneration. Efforts to compensate for visual deficits or to augment visual processing are increasingly focusing on the other sensory modalities for assistance. With these efforts, discoveries are being made on crossmodal influences or augmentation of one sensory modality at the cortical level via peripheral stimulation within another modality. The mechanisms of action remain unclear and speculative at this point in history. However, it appears to be established that visual cortical processing varies with input to other modalities, subsequent to diminished visual input. Multiple possibilities exist for this within functional neural systems theory, including the mechanisms of the diffusely projecting arousal systems originating within the first functional unit or brain stem reticular formation, and the altered regulatory demands placed upon frontal regions with the loss of one modality. Regardless, efforts to assist the blind individual or to better understand blindness provide the opportunity for scientific inquiry into the functional neural systems involved and the influence that single modality deprivation or augmentation may have on the adjacent sensory neural architecture.

The absence of visual acuity appears to enhance processing within the remaining sensory modalities (James 1950). Pasqualotto and Proulx (2012) note in their review of this literature that blind persons often have superior perceptual discrimination and localization within preserved sensory modalities, along with augmentation of verbal processing and memory capacity. Proulx (2013) notes that substantial evidence exists for the recruitment of visual cortex to support these enhanced abilities (see also Pasqualotto and Proulx 2012). Neuroimaging studies have demonstrated occipital activity in blind individuals during auditory, haptic, and olfactory stimulation (Amedi et al. 2003; Kupers and Ptito 2011; Rombaux et al. 2010). Proulx

(2013) asserts that "the enhanced abilities of blind individuals arise either by the recruitment of the otherwise dormant visual cortex by the remaining modalities or perhaps by the unmasking of visual cortical activity that normally supports such functions."

This argument, though, excludes alternative explanations, including the recruitment of the visual cortex by redundant visual pathways. The anatomical foundation for these effects is well established as ambient lighting is processed at multiple levels, including the superior colliculi and the pineal body. Moreover, activity within these systems responding to ambient light, outside of the visual acuity dimension, is influential in the recruitment of diffusely projecting arousal systems (Isaac and DeVito 1958) and in the alteration of sensitivity or thresholds within other sensory modalities. For example, auditory thresholds have long been known to vary with ambient light levels (Delay et al. 1978; Kallman and Isaac 1980). Also, relevant is the well-established effect of ambient light on the motor systems affecting locomotor activity (Isaac and Reed 1961; Isaac and Troelstrup 1969; Kallman and Isaac 1980) and response to psychopharmacological manipulations (Isaac and Troelstrup 1969; Lowther and Isaac 1976). Regardless, Proulx (2013) aptly points out that the brain has traditionally been viewed as a subdivided structure with each domain restricted to its separate location within the neural architecture. Instead, each region may better be appreciated by the sensory context which may be influential within and across these sensory modalities.

Perhaps the most notable redundant pathway for vision is found in the ganglionic projections of the retina to the superior colliculus. Perry and Cowey (1984) concluded from converging lines of evidence that not more than 10 %, and perhaps as few as 7 %, of the retinal ganglion cells project to the superior colliculus. These authors also found that the division of the crossed and uncrossed projections is asymmetric. They estimate, as did Pollack and Hickey (1979), that about 70 % of the retinal projections to the superior colliculus comes from the contralateral eye.

Somatosensory Systems

Imagine for a moment all of your feelings. Some of these are derived from movements of the gut, possibly expressed as a "gut feeling." Many of our feelings involve the detection of subtle reflexive responses throughout our body such as a tensing of the brow or contraction of our hand into a grasp posture. Some might involve the detection of a reflex posture as with arching of the back. Needless to say, many of us relate these "feelings" easily to emotion with some feelings conveying a defensive posture, discomfort, and negative emotion. Positive feelings abound as with gentle stroking of the skin or relaxation and with the digestion of our food in friendly company. The degree to which these feelings convey positive or negative emotion provides insight into the cerebral regions activated. Also, the degree to which we have access to language systems to describe the feelings provides a clue. Clinical and counseling psychologists spend much of their time helping the client verbalize feelings originating from the relatively nonverbal right cerebral hemisphere. Moreover, the brain regions providing feelings may be damaged as with a large right parietal lesion, leaving the patient bankrupt of feelings and their emotional associations. Another example might be provided from research on individuals with congenital agenesis of the corpus callosum effectively reducing connections between the cerebral hemispheres. Many of these individuals (about 40%) were identifiable as suffering from an autism spectrum disorder (Lau et al. 2013).

The body or *somatic senses* include the classic modalities of touch, pressure, temperature, and pain. Also, we might include aspects of kinesthesia, proprioception, and sensation from the gut. The psychophysicist would begin this discussion at the receptor where natural energy in various forms is converted into neural energy via the process of transduction. The one universal property of protoplasm might be irritability (Culbertson and Hyndman 1879, p. 305)! And, as such, the receptor is specialized for this purpose, being uniquely situated and designed for irritable responses to changes in energy within our external world.

Transduction or energy conversion results in a generator potential at the receptor. This occurs in response to the *adequate stimulus* (see Willis and Coggeshall 2004), as light might be the adequate stimulus for vision. The inference here is that the receptor is specialized for the detection of a specific bandwidth and intensity range of one form of energy emanating from our external world. But, the receptor is not entirely restricted to this stimulus event, as, for example, the retinal receptor might also respond to pressure as we rub our eyes, resulting in the perception of a flash of light, where none actually existed. Apparently, "eye-pushers' in ancient Greece argued that "deformation phosphenes" were evidence that the eye itself was generating light. This may have held sway until Kepler, Newton, and Descartes were able to appreciate that the pressure was activating the same tissue that was activated by the presence of light on the retina. Also noteworthy here, the morphology of the receptor itself is dynamic and undergoing a relatively continuous process of wear and tear and of regeneration.

The dynamic aspect of receptor morphology was apparent to early researchers who proposed specialized receptors for each cutaneous sensory modality, including touch, pressure, cold, warm, and pain. Some have said that progress results in science through contact with *reality* where our beliefs are faced with contradictory evidence forcing revision of our theory. This came from the staining of the human cornea revealing only one receptor type, the free nerve ending, which had been identified as "the pain receptor." Yet, the cornea, with an array of free nerve endings (Lele and Weddell 1959; see also Marfurt et al. 2010), is capable of the transduction of the energies of multiple sensory modalities, including touch, pressure, cold, warm, and pain. These modalities are each conveyed through skin senses with various classic receptor types associated with the adequate stimulus for that modality (see Willis and Coggeshall 2004).

The stimulus for touch consists of deformation gradients of the skin (Kenshalo and Nafe 1960; see also Willis and Coggeshall 2004). The relevant variables for this include the *rate of stimulation* and the presence of *motion,* with faster deformation gradients requiring less intensity to activate the receptor. Once the deforming process involving motion is discontinued, the stimulus for touch is no longer present. Hair follicles may function to amplify the mechanical deformation gradient with basket ending receptors wrapped around the root of the hair follicle. In hairless body areas, the adequate stimulus is, again, mechanical deformation. The deformation may involve either inward or outward movement of the skin with the direction of movement being inadequate as a discriminative stimulus. Instead, the inward or the outward gradient corresponds with touch and both movements are perceived as just that.

The continued mechanical deformation of the skin in the absence of movement is the stimulus for pressure (see Loomis and Lederman 1986). The relevant variable for pressure sensitivity is *area*. Touch is not pressure. Unlike touch receptors, pressure receptors do not stop responding at the maximum intensity of the stimulus, although adaptation may occur through a gradual reduction in receptor responses with continued or repeated stimulation. The distinction here is among the adequate stimulus characteristics and the basic property of all receptors to adapt. I would appreciate adaptation at the pain receptor, first thing in the mornings, when I would hop into the shower that my son had left on for me. Initially, I would indeed be in pain along with the sensation of the intense heat of my son's shower. I could not believe that he could have tolerated such pain. After a few seconds, though, I would be left with a very nice warm shower as pain receptors, in this case, adapt more rapidly than temperature receptors (see Willis and Coggeshall 2004).

Once the sensory event (*distal stimulus*) has been converted to the generator potential (*proximal stimulus*) in the form of neural energy, the information travels up the afferent projections of unipolar neurons, which have their cell bodies located within the dorsal root ganglia (see Willis and Coggeshall 2004) adjacent to the spinal cord. Up to this point, the information has been traveling through a peripheral nerve, which separates prior to entry at the spinal cord into the dorsal root carrying sensory afferents and the ventral root, which conveys the efferent fibers exiting the spinal cord on their way out to the nerve. One point here is that somatic nerves are mixed and typically carry sensory and motor fibers. Thus, a somatic nerve lesion typically results in mixed sensory and motor deficits, which may resolve as long as the cell body survives and the glial or Schwann cells are able to support regeneration of the cellular processes (axons and dendrites).

But, on entry into the nerve roots, and within the spinal cord, sensory information travels in a dorsal direction into the spinothalamic tracts. This was initially expressed as the Bell and Magendie law (Bell 1811; Magendie 1822a,b; Mueller 1831) as Charles Bell (see Fig. 8.4) and Francois Magendie (see Fig. 8.5) shared credit for this discovery. But, the student of neuroscience will do well to appreciate this basic distinction as sensory systems remain essentially at the back of the neural tube, with motor systems located at the front of the neural tube. Bell and Magendie described this for the spinal cord. But, the relationship holds for the brain with specialization of the frontal lobe for expression or intent and for the posterior brain regions for sensory reception, analysis, and comprehension, more broadly defined. Charles Bell (1774–1842) discovered that sensory roots are located dorsal in the cord, whereas motor roots exit the ventral side of the cord. Francois Magendie

Fig. 8.4 The English
medicine professor Charles
Bell

(1783–1855) shared the discovery with Bell on the location of sensory and motor roots at the ventral and dorsal cord, respectively. Bell's discoveries laid claim over the motor function, whereas the Frenchman Magendie's discoveries laid claim over the sensory function within the cord. With long-standing conflict between England and France, the contributions to the Bell–Magendie law became a matter of national pride, contributing in part to this outcome.

Touch and Pressure

Aristotle included touch as one of the five human senses, alongside vision, audition, olfaction, and gustation. It appears to be the first sensory modality to develop (Montagu 1971), and touch sensitivity is present in the human fetus after about 6 weeks of gestation in the womb and prior to the development of the eyes, ears, or any of the other sensory organs (Montagu 1986). Moreover, the organ providing for cutaneous touch sensation is the largest (perhaps extending in total area at over 1.6 m^2 in an adult) and the oldest of the sense organs (Field 2001; Montagu 1971; see Kosnar 2012). Kosner (2012) aptly points out that "…we are continuously using our skin to perceive vast amounts of inputs from our environment, such as informa-

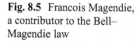

Fig. 8.5 Francois Magendie, a contributor to the Bell–Magendie law

tion about size, weight, texture, temperature, structure and other characteristics of objects that we touch. Moreover, we sense air flow in an environment, and localize and recognize objects that we grab. Additionally, we use the skin not only to receive information about the objects around us, and environment that we live in but we also use skin to socially interact with people around (us); to communicate feelings, and emotions."

In the spinal cord, *touch and pressure* travel together in the *dorsal funiculus,* whereas *temperature and pain* travel together in the *lateral funiculus*. Both the dorsal and the lateral funiculi are spinothalamic tracts, with the name reflecting origin at the spine and emergence at the thalamus where these fibers will synapse onto the ventral posterolateral (VPL) nucleus of the thalamus. From here they will be relayed on up to the parietal lobe, projecting onto the somatosensory cortex. The dorsal funiculus, carrying touch and pressure, maintains a topographical homunculus or body representation, which will eventually be expressed across the somatosensory cortex of the parietal lobe. But, the arrangement seems more practical in the spinal cord, where the axons first entering the lowest levels arise from nerves innervating the lower extremities (genitals, bladder, toes, foot, leg, and trunk). These fibers form the *fasciculus gracilis,* which remains near the midline of the dorsal funiculus all the way up the spinal cord and at the somatosensory cortex.

Eventually emerging alongside this "highway" or traffic lane carrying information for the lower part of our body will be fibers coming from higher up in our body, including the upper extremities (fingers, hand, and arm). These fibers form the *fasciculus cuneatus,* which remains laterally placed in the dorsal funiculus on its way up the spinal cord. These pathways, for the novice, may as well be named Highway 95 and Highway 81 as the traffic on each is largely determined by its point of origin across the upper or the lower parts of our body and, of course, the left and the right hemibody. The messenger, in this case, is simply traveling north, by analogy. But, among the more advanced issues relevant to neuroscience and clinical practice are the proximal and distal relationships established through this anatomy. These relationships provide a rich source for inquiry extending the effect of proximal anatomy to implications for proximal functions. For example, the right hemiface and the right hand are proximal within the cortical architecture and share many functions in linguistic and propositional communication.

Upon entering the spinal cord, the projections ascend to the level of the medulla at the base of the brain stem (see Fig. 8.6). Here they synapse onto the *nucleus gracilis* (lower body) or the *nucleus cuneatus* (upper body). The second-order fibers

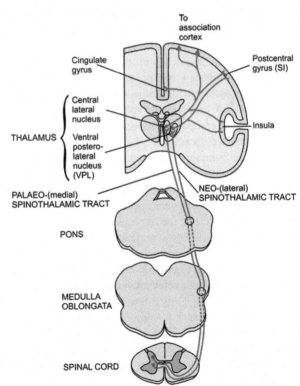

Fig. 8.6 The somatosensory projections from dorsal root, to dorsal funiculus, to the nucleus gracilis and nucleus cuneatus, up the medial lemniscus to the VPL (thalamus), to the somatosensory cortex. (*VPL* ventral posterolateral. Originally published in *Central Nervous Mechanisms of Muscle Pain: Ascending Pathways, Central Sensitization, and Pain-Modulating Systems,* Mense, Seigfried; Gerwin, Robert D., 2010, p. 118)

Fig. 8.7 The somatosensory
and motor strips with the
location of the body parts
represented across these
regions. (Originally published
in *Methods of Assessment of
Cortical Plasticity in Patients
Following Amputation,
Replantation, and Composite
Tissue Allograft Transplantation,* Siemionow, Maria Z.,
2011, p. 236)

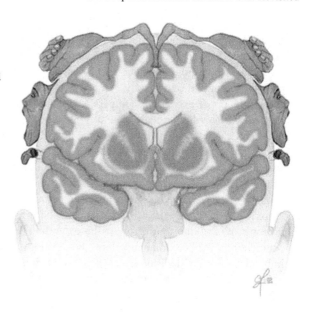

cross the midline joining a few redundant fibers (enough for tickle!) traveling up the
anterior spinothalamic tract where they will continue their ascent within the *medial
lemniscus* up to the thalamic relay nucleus, the VPL. Projections from the VPL
nucleus of the thalamus continue up to the postcentral gyrus, the primary projection
area for somesthesis (see Fig. 8.7). The medial lemniscus will see the merging of
touch and pressure, with temperature and pain, as they travel together now toward
the thalamus. A lesion in the spinal cord may discretely affect touch and pressure
or temperature and pain, whereas at the brain stem level, on up to the level of the
somatosensory cortex, these modalities are likely to be affected, collectively.

The small number of redundant fibers in the anterior spinothalamic tract were
discovered from working with patients suffering from tabes dorsalis or a wasting
away of the dorsal funiculus (see Wall 1970). As one might expect, these patients
lost the somatosensory awareness of their body parts, which is basic to proprioception at higher levels of the nervous system and specifically within the parietal lobe
association cortices. These patients might have to look at their legs, using vision to
compensate for the placement of the limbs, during ambulation. But, tickle remained
intact! Indeed, there were early concerns that tickle might actually be a separate cutaneous sense. But, the relevant variable for tickle was determined to be the number
of fibers responding. Only a few fibers were necessary for tickle, in contrast to the
large number necessary for proprioception and so forth. For the family member of
a patient with altered sensation from a dorsal spinal cord lesion, this redundant access to the somatosensory cortex in the brain provides a potential route to activate
this brain region, and normally, the route is activated by light brushing stimulation
or "tickle."

Although tickle did not turn out to be a separate somatosensory modality, the case for an "itch" appears to be quite different. This conclusion was evident with the discovery that itching is distinct from pain, touch, and temperature (Han et al. 2012; see also Fields and Sutherland 2013). Some sensory neurons with cutaneous receptors were found to have protein receptors on them which were identified as Mrgpr-A3. Chemicals known to create itching caused these neurons to generate electrical signals, whereas painful stimuli, such as hot water or capsaicin, did not. The researchers selectively killed the population of Mrgpr-A3 neurons in mice, resulting in animals that no longer scratched upon exposure to itch-producing substances or allergens. Instead, they maintained sensitivity to touch- and pain-producing stimuli. Subsequently, these investigators selectively engineered capsaicin-responsive receptors injected into Mrgpr-A3 neurons in a strain of mice lacking the capsaicin receptor. Subsequently, the itch-specific cells were found to evoke scratching behavior rather than pain behavior by injecting capsaicin into a focal location on the rodent's skin. Moreover, the authors appreciate that their discovery may allow for the development of new techniques that block Mrgpr-A3 receptors to silence the raging itch of poison ivy or eczema exclusively and without altering the additional cutaneous sensory modalities.

Temperature

The stimulus for temperature is *absolute temperature*. Threshold or sensitivity to temperature is typically relative to *physiological zero* or the temperature of the human body at roughly 32 °C. Temperatures above physiological zero will be perceived as warm and those below this reference will be perceived as cold. These mechanisms may underlie altered physiologic conditions as with fever, where the skin is hot to the touch, but where the patient may complain saying "I am freezing." The sensation of either cold or warm is relative to baseline temperature as was partially demonstrated with Ernst Heinrich Weber's (1846) classic three-bowl experiment. In this experiment, one hand was placed in a bowl of water at 20 °C (cold) and the other hand in a bowl of water at 40 °C (warm). Now if the hand from the first bowl is placed in a bowl of water at body temperature, the water will be perceived as warm. The hand from the warm bowl when placed in the bowl at body temperature will be perceived as cold. Diametrically opposite interpretations of the "neutral" temperature may be derived, perceptually, relative to the temperature of that part of the body at the time of emersion.

Warmth and cold, though, appear to be two separate sensory modalities (e.g., see Sinclair 1955; see also Willis Coggeshall 2004) as demonstrated by the larger thresholds for warmth than for cold, differing chronaxie values, and discrete responses to moderately intense temperature stimuli. The classical receptor types include the Ruffini cylinder for warmth and the Krause's end-bulb for cold. But, once again, either may be detected via a free nerve ending or even other receptor configurations. Also, Nafe and Wagoner (1936) emphasized vascular changes with temperature, arguing that temperature is a kinesthetic sensation with warmth resulting in dilation of the arterioles and with cold resulting in the constriction of the arterioles in the area affected. Since arteries, arterioles, and capillaries are innervated, the contractions and dilation of the arterial wall and lumen may provide for discriminative events for the perceptual analysis by the brain. The critical aspect is ultimately the anatomical system with a topographical arrangement of the pathways conveying each modality. This anatomical arrangement is most clearly distinct from the spinal cord projections on up to the thalamic relays and eventually to the primary cortical projections onto the somatosensory strip.

There appears to be no sensory receptor or modality specific to *burning heat*. Two clues to a better understanding of this were developed in the psychophysicist laboratory. One was the phenomenon referred to as paradoxical cold (Long 1977) where the exploration of the skin with a very warm stimulus (45–50 °C) would periodically be perceived as cold. Also providing insight here was the phenomenon of "psychological heat" (Ferrall and Dallenbach 1930), where alternating cold (12–15 °C) and warm (40–42 °C) grids simultaneously exposed to the surface of the skin would be perceived as "burning heat." These clues were integrated with the firing rate of the cold and warm receptors plotted as a probability distribution onto a graph. Here we could see cold receptors firing to cold stimuli and warm receptors firing to warm stimuli. But, hot stimuli yielded a recovery of the response probability of the cold receptors such that both cold and warm receptors responded *simultaneously* to convey burning heat.

Temperature travels along with pain in the lateral spinothalamic tract and eventually projects onto the VPL along with the other somatosensory modalities (see Willis and Coggeshall 2004). Temperature subsequently arrives at the primary projection area for somesthesis in the anterior parietal lobe. Interestingly, the shared pathway with pain deviates within the functional realm as direct electrical stimulation of the cortical projection area within temperature columns can elicit temperature sensations. This is not the case for pain (Penfield and Boldrey 1937, 1967; Mazzola et al. 2012a, b).

Pain

Pain remains the most elusive sensory modality, though champions exist here for each of several theories (see Mollet and Harrison 2006; Mitchell and Harrison 2010). More than any other modality, pain remains a significant challenge both for the neuroscientist and for the clinician. Many, and complex, pain syndromes exist well beyond the scope of these writings. And, all too often our understanding and our interventions for pain syndromes are inadequate. Indeed, treatment for pain, and especially surgical intervention, may be the precipitating event for the aggravation of the pain syndrome (e.g., see Perkins and Kehlet 2000). Moreover, much in the way of pain disorders within the patient population arise as iatrogenic disorders, secondary to other intervention efforts performed with hope that the patient would be helped.

The *threshold or sensitivity* for pain is relatively constant across the population but with wide disparity in the *tolerance* for pain (e.g., Johnson et al. 2012). Frontal lobotomies were thought to pacify violent-prone or emotionally reactive individuals. This expectation was based on research by Carlyle Jacobsen and John Fulton, at Yale in the early and mid-1930s (see Fenton 1999), where tolerance was increased for delayed gratification in two chimpanzees (Becky and Lucie) using the Wisconsin General Test Apparatus. Nonhuman primates in the apparatus could view the raisin kernel or food reward at which point the screen was lowered, and a delay was required, prior to gaining access to the treat. Monkeys generally tolerated this delay poorly with agitation and sometimes aggressive behavior. But, after the lobotomy, heightened tolerance was displayed and with increased passivity in behavior.

One doctoral student in psychology was working with this apparatus and "normal monkeys" but with very curious data and atypical delayed response deficits (see Malmo 1942). Like any good scientist might do, she looked behind the closed curtain. She expressed her dismay when she discovered that her alpha male was masturbating and that he was seemingly detached from the experimental contingencies! This might be the only case study to confirm the proverbial account relating masturbation to growing hair on the palmer surfaces of the hand, albeit with correlational data. And, as far as the other heralded relationship, the monkey was not blind. It was equally certain that he did not receive as many raisin kernels as the other, more experimentally compliant, monkeys. This side track was prompted by one of my colleagues who alternated with me in teaching a large introductory class at the University of Georgia with about 1600 students divided into two sections. She was horrified when she mentioned this mythical relationship in class only to discover that the man in the first row was visually impaired. Following this prompt, he stood up and showed the class the palms of his hands to clear his record, though her reputation remained somewhat tarnished, perhaps!

Egas Moniz, formerly the Portuguese foreign minister, received the Nobel Prize in 1949. He returned from exposure to these research findings and, within the year, implemented these procedures on humans with the frontal lobotomy or lobectomy. The first psychosurgery operations were conducted on 12 November 1935 on four patients, two with depression and two with schizophrenia, using alcohol injections into the depths of the frontal lobes. The technique was transitioned to the leukotome, and further refinements in the technique followed with time and practice. Altering the expressive or intentional frontal lobe systems often leads to erroneous conclusions or attributions by others toward the frontal lobe patient. This distinction should resonate with Moniz as a fundamental error in his use of the lobotomy procedures to control violent aggression. Moniz provides this, by example, as one of his lobotomized patients returned and shot him in the back, rendering him a quadriplegic. Many, including the clinical psychologists that I have worked with over the years, have mistakenly attributed social and emotional insensitivity to the frontal lobe patient who may present with masked faces or bland, flat affective expressions (Karson 1983; see also Bologna et al. 2012). But, this again may be a mistake.

Damaged or diminished frontal lobe regulatory control may well disinhibit the sensory systems received and comprehended within the posterior brain, the second functional unit (e.g., see Thompson-Schill et al. 2005). Indeed, sensitivity may well be increased with lowered thresholds (Woods et al. 2013). Depending on location of the lesion, the individual may be hypersensitive to the clothing touching their skin. They may be bothered by lights, sounds, smells, or events within other sensory modalities. The hostile violent-prone individual often expresses very high tolerance for pain. These features overlap with masculine gender role stress in "high macho" men. But, this tolerant male suffering pain to demonstrate machismo may be more sensitive to the pain than the normal woman, who readily avoids or terminates the pain stimulus with little tolerance and possibly better judgment! Hostile men reactively increase their blood pressure to painful events, whereas low hostile individuals maintain cardiovascular stability on exposure to the same stressors (e.g., Herridge et al. 2004). Incremental blood pressure provides a potential compensatory function as pain sensitivity has been reduced with elevated blood pressure in many projects (see Mollet and Harrison 2006).

By extension, pain sensitivity and sensitivity to emotional cues conveying negative affect appear to be relative specializations of the right cerebral hemisphere (see Pauli et al. 1999; see also Mitchell and Harrison 2010). For example, the right inferior frontal gyrus is implicated in pain processing, pain anticipation, and modulation, and abnormal gray matter volume has been identified in this region in chronic pain (Symonds et al. 2006; Moayedi et al. 2011; see also Kucyi et al. 2012). Consistent with a larger body of evidence, Kucyi et al. (2012) conclude that a disruption in the connection between the right temporoparietal junction and the inferior frontal gyrus may have a role in chronic pain. Although this conclusion is based substantially on intrahemispheric evidence within the right cerebral hemisphere, additional evidence further indicates that the right hemisphere may be inhibited via activation within left cerebral systems through the interhemispheric fibers of the corpus callosum. Processing specialization by the right cerebral hemisphere may diminish with attention to the verbiage processed by left cerebral systems. This conclusion was suggested, for example, by recent research indicating that, with lesion to Wernicke's area within the left hemisphere, patients were now more capable of detecting lying from others (Etcoff et al. 2000). The interpretation provides for a release of the right brain secondary to left hemispheric damage, in this case, resulting in improved emotional perception derived from the right cerebral systems analysis.

Temperature and pain from each side of the body travel up the lateral funiculus (spinothalamic tracts) to the brain. This anatomical distinction of pain and temperature traveling together and in a separate section of the cord, away from touch and pressure, may be diagnostically relevant in distinguishing a spinal cord injury from one at the level of the thalamic relays or that at the level of the primary projection cortex for somesthesis in the parietal lobe. Brain lesions within these pathways, including the thalamus and the parietal projections, will characteristically affect multiple body senses, whereas localized lesion of the spinal cord may impact either temperature and pain *or* touch and pressure. But the cable is one in the same, otherwise. Lesion or altered innervation at one end of this cable with loss of the cell body and cell death may, indeed, result in cellular loss through the projections and beyond the synapses in this pathway (see Chen et al. 2002).

Peripheral nerve lesions are predictive of hyperalgesia or increased sensitivity for pain and sometimes phantom pains from the area previously innervated by the nerve fibers (e.g., Perkins and Tracey 2000). A favorite example here comes from a visit to the dentist. This professional may work diligently to suppress the activation of the peripheral nerves innervating the roots of the teeth. But, infection or loss of the dual or cross-innervated dental structures can produce heightened sensitivity even in the presence of fewer fibers now carrying pain information to the brain (e.g., Djouhri et al. 2012). These findings may partially invalidate claims that pain results simply from overstimulation as pain follows denervation in this case. Moreover, conclusive evidence exists that neuropathic pain is not simply a symptom of disease but, instead, a consequence of disordered functioning of the nervous system (e.g., Scholz and Woolf 2002; Tsuda et al. 2012). Central lesions are more associated with intractable pain or even thalamic pain syndromes and often with an alteration in the quality of the pain with the patient describing excruciating pain complaints and even with recruitment into other modalities (e.g., nausea or vectional disturbance, auditory hyperacusis, and/or visual disturbances such as photophobia).

Higher Cortical Processing

The somatic senses of touch, pressure, cold, warm, and pain eventually arrive at the somatosensory projection cortex (Brodmann's areas 3, 1, 2) located at the anterior parietal lobe and immediately posterior to the central fissure, which separates them from the motor strip. Mountcastle, a graduate of Roanoke College in Virginia, was instrumental here for his discovery that the neuronal cells in the somatosensory cortex are arranged in vertical columns and that this arrangement directly relates to their collective function in processing sensory information. Published in a classic 1957 paper, this discovery is acclaimed as being foundational in the field of neuroscience. Subsequently in 1978, he proposed that all parts of the neocortex operate based on a common principle, with the cortical column being the fundamental unit of computation.

The sensory modalities located here in the somatosensory projection areas are arranged in a columnar fashion of cellular ensembles, which play a role in coding intensity. Also, there are horizontal associations which may play a role in contrast coding and lateral inhibition. The somatosensory cortex is arranged topographically from dorsal to ventral in the representation of the form of an oddly shaped upside down and backward little man or little woman, referred to as the somatosensory homunculus. Each brain, of course, has its own homunculus charged with the processing of cutaneous sensation from the contralateral hemibody. The odd shape of this homunculus is derived largely secondary to the amount of cortex devoted to various body parts. For example, the face, fingers, and hand are spaciously represented here related to the relative sensitivity of these body parts in contrast to the legs and the trunk or back. The proportion of fibers crossing to the contralateral body varies with the location for proximal and distal body parts and somewhat like that discussed for the motor cortex elsewhere in these writings.

Deactivation of the primary projection cortex through metabolic reduction or damage, as might result from a stroke, characteristically results in numbness or anesthesia for that part of the body represented on the homunculus. Interestingly, the patient with a lesion in this projection pathway onto the somatosensory cortex may complain of temperature dysesthesia, being cold at the affected body region. This, in and of itself, might not be so surprising, as we have many similar examples of lesion location and functional features located within the brain regions which have been lost. For example, a lesion to Wernicke's area results in impaired speech comprehension on the input side of things. But just as the speech output from the Wernicke's patient is altered to match the receptive deficits, the output of the patient with a somatosensory lesion is altered to match the receptive deficit. More specifically, the patient may express a sensation originating from the contralateral body of coldness concurrent with the actual reduction in temperature of that body part. The patient with a right parietal lobe lesion in the somatosensory cortex often complains of cold at the left hemibody. When the examiner holds the patient's left and right hand, the left hand *is* cold, relative to the homologous body part at the right side of the body.

It would be easy to miss the potential significance of this as the lesion originates in the second functional unit charged with the reception, analysis, and comprehension of sensory information. Yet, by inference, activation or deactivation of this tissue provides for alterations in blood flow or vasodynamic responses specific to that part of the body represented within the homunculus. Deactivation of the somatosensory cortex may reduce blood flow or engorgement of that part of the body and may be associated with edema or a buildup of fluid in that area. One might ask if therapies or interventions may be developed with activation of somatic cortex to improve blood flow to an extremity (e.g., with diabetes and peripheral neuropathies). It would not be surprising if these effects are state dependent as patients are often restricted to a wheelchair and somewhat inactive. But, the effects are often robust and lateralized in a contralesional fashion to the cerebral hemisphere affected by the pathology. It is well established that anesthesia in the surgical setting is accompanied by cold sensations and shivering (Abdelrahman 2012). But, the central basis for this at the level of the somatosensory cortex and its projections is not yet fully appreciated.

Blood flow throughout the body varies with brain activation during cognitive processing and also with habituation to the environmental context. If a person is carefully balanced on a tiltboard during a relaxed state and, subsequently, asked to engage in cognitive activities, blood will flow to the brain resulting in tilt of the board downward as the head becomes heavier from the dynamic shift in blood flow. Habituation also affects blood flow in an oppositional manner to the previous example, with a relative redistribution of the blood from the cephalic region to the distal extremities. But the habituation of the two brains affecting blood redistribution is not symmetrical (Harrison et al. 1989; Harrison 1990). Indeed, there appear to be asymmetries in the habituation process with a relative increase of blood flow to the distal left upper extremity in comparison with that at the distal right upper extremity. These findings indicate that habituation to one or another context is lateralized to the processing demands of that brain and that diametrically opposite laterality effects, with temperature changes at the cephalic and distal extremities, might be produced under contextual settings where one or the other brain is active (e.g., light or dark, linguistic or prosodic speech processing, and positive or negative affective events).

Manipulations performed to increase hand temperature through placement in warm water have sometimes been successful in treating pain from migraine, perhaps secondary to shifting blood flow away from the head to the distal extremity or reduced vasospasm (Harrison et al. 1989; see also Shaw 2012). These remain potential variants for diagnostic purposes and for interventions in individuals suffering from brain disorders with somatosensory involvement. But, to date, the efforts here have been minimal. Paul Foster (Foster et al. 2010) provided a creative approach to activation of the left and the right somatosensory cortices, while testing the theoretical proposition of the relatively prominent role of the left brain in parasympathetic drive, in contrast to the role of the right brain in sympathetic drive in his efforts to extend the autonomic nervous system to cerebral systems and frontal lobe regulatory control. Paul simply placed a vibrator in the subject's left or right hand to differentially activate the left or the right somatosensory cortex. Vibrotactile stimulation at the left hand elevated sympathetic tone as evidenced by increased heart rate and systolic blood pressure, whereas vibrotactile stimulation at the right hand lowered heart rate and systolic blood pressure.

Mathew Herridge had explored this relationship earlier (Herridge et al. 1997; see also Ekman et al. 1983) using posed facial contractions of the corrugator muscle (anger expression) and posed facial contractions of the zygomatic muscle (smile). Simply contracting the corrugator muscle resulted in increments in sympathetic tone recorded using skin conductance measures. In contrast, the contraction of the oppositional zygomatic muscle lowered skin conductance or sympathetic tone. Moreover, when these responses were compared in hostile violent-prone individuals and those low on hostility indices, the hostiles were slow to habituate skin conductance responses recorded from the left hand (right brain), whereas the low hostiles were slow to habituate skin conductance responses at the right hand (left brain). These and other projects (e.g., Rhodes et al. 2013) provide an initial indication that the somatosensory activation of the left or right brain may alter autonomic function, including that recorded using cardiovascular indices and that recorded using the more elegant skin conductance response measure.

Historically, the primary somatosensory cortices have been assessed using threshold measurement for light touch stimuli. This might be assessed more formally with von Frey hairs (1896). This assessment instrument consists of calibrated horse hairs varying in thickness and in the intensity of touch prior to the point of bending when applied to the surface of the skin. The hairs are typically attached to a wooden handle for administration purposes. But, these procedures are costly in time requirements and in the instrumentation as this is a precisely calibrated apparatus. So, the clinician commonly uses light touch or vibrotactile stimulation at the left or the right hemibody and then evaluates for relative cerebral weakness using *extinction* procedures. In this modality, the stimuli are *dichaptic* presentations, usually across homologous body parts (e.g., left and right index finger). But, the sensitivity of the dichaptic extinction procedure might be increased on concurrent face and finger stimulation from opposite sides of the body.

The role of primary somatosensory cortex in pain perception remains controversial, and this modality provides the great challenge for neuropsychological analysis and assessment. In the early twentieth century, Head and Holmes (1911) observed that patients with long-standing cortical lesions did not show deficits in pain perception. Similarly, Penfield and Boldrey (1937), in their work using electrical stimulation of patients' exposed cerebral cortices during epilepsy surgery, concluded that pain probably has little or no cortical representation. Pain stimuli activates the somatosensory cortex as seen with metabolic and with electrical measures. Metabolic changes might be recorded using positron emission tomography or functional magnetic resonance imagery (fMRI), whereas electrical changes are commonly evaluated in the laboratory using evoked potentials.

Although pain activates this region (e.g., Mazzola et al. 2012a), pain has not been elicited by direct stimulation of the somatosensory cortex. However, Mazzola et al. (2012b) reassessed this peculiar state of affairs by analyzing 4160 subjective and videotaped behavioral responses subsequent to cortical stimulation using intracerebral electrodes across all cortical regions. These were conducted over a 12-year period during presurgical evaluations of epilepsy in 164 consecutive patients. The authors report scarce pain responses to cortical stimulation (1.4%). These largely resulted from deep cortical stimulation concentrated in the medial parietal operculum and neighboring posterior insula, with pain thresholds showing a rostrocaudal decrement. Further, they note that this deep cortical region was largely inaccessible in Penfield's work (Penfield and Boldrey 1937) with intraoperative stimulation of the cortical surface after resection of the parietal operculum.

Importantly, no pain response was elicited to stimulation within any other cortical region, including the regions which are consistently activated by pain in the majority of functional imaging studies (i.e., the somatosensory projection and associations cortices, the cingulate gyri, the supplementary motor areas, and anterior frontal regions). This is in direct contrast to the other somatosensory events elicited by stimulating this cortex, where touch and pressure are perceived across the respective parts of the somatosensory homunculus.

Pain is often the more relevant consideration clinically with a lesion in the projection pathways. Most notorious among these pain disorders is the thalamic pain syndrome or intractable pain. The thalamic pain patient may be poorly understood with the pain fluctuating or variably presented over time (waxing and waning) and variably presented over location, sometimes appearing to the patient to be moving over their body parts from one period to the next. All of this is frequently evident to the caregiver, therapist, or diagnostician against a background of vague and cross-modal sensory complaints. Even the more informed caregiver or therapist may find that they are a bit suspicious of dissimulation or feigned symptoms.

Subsequent to the arrival of the somatosensory fibers onto the primary projection cortices for the reception or sensation of bodily events, the information moves to a higher level of somatic analysis onto the secondary association cortex. The secondary association cortex surrounds the primary projection area which, in turn, is surrounded by the tertiary association cortex. Information arriving at the secondary association cortex remains largely restricted to the somatic modalities. But, now there is increasing awareness of what the event represents in the form of a somatotopic map. This allows for stereognosis or the recognition of objects on the skin such as a pencil placed on the palmar surface of the hand or a pair of scissors, for example. This relates by comparison or by analogy to stereoscopic vision, providing for depth perception in the visual modality. Stereognosis provides this same depth of three-dimensional analyses through the somatosensory modalities. The inability to recognize an object through touch alone is identified by the diagnostician as tactile astereognosis. Tactile recognition is basic to gestural praxis or ideomotor praxis derived from proximal brain regions within the parietal lobes (e.g., Vingerhoets et al. 2012). Within this modality, an object is appreciated by its form through tactile stereognosis and the postural positioning necessary to manage the implement for fundamental tool use. This process is essential and foundational to the expression of basic postures necessary for combing our hair, brushing our teeth, walking, or any of the multiple activities of daily living and of tool or implement utilization.

As the therapist or caregiver may appreciate, the failure to produce the appropriate gesture or body posturing is fundamentally a safety issue. But, the deficit is somewhat analogous to the more ventral lesion in Wernicke's area, where speech comprehension is impaired. The failure to comprehend the somatic event, though, is a bit removed from the comprehension of the body posturing or gestures appropriate to use the object. It is also fundamental to gestures at each part of our body depending on the location of the parietal lesion higher (lower extremity confusion) or lower (facial, lingual, and upper extremity confusion) in its proximity to the somatosensory homunculus. So, object utilization may be affected and/or postural placements and gestures. This might include facial configurations expressing one or another emotional valence or tongue placements during swallowing or speaking. In the left parietal lobe, this may also present as body confusion and/or left–right confusion and ultimately the broader deficit complex which we identify as Gerstmann's syndrome (Gerstmann 1940, 1957; see also Cabeza et al. 2012).

The secondary association cortices are capable of further analyses within the somatic sensory modalities, as, for example, with graphesthesia or the recognition of a letter or tactile grapheme or a pattern drawn on the hand. The cerebral specialization is similar to that discussed elsewhere in these writings with lexical graphemes processed with some advantage by the left brain and with spatial patterns showing a right brain advantage. This may present also in gestural praxis with sign language derived from left cerebral systems (e.g., Damasio et al. 1986; Leonard et al. 2012; however, see Newman et al. 2002) in contrast to spatial directional and inferential gesturing derived from right cerebral systems. It is also notable that the detection of somatic sensations can be facilitated or inhibited by observing visual stimuli (e.g., Vandenbroucke et al. 2012) consistent with the interactive influences across sensory modalities at the level of the association cortex.

For example, Leonard et al. (2012) used magnetoencephalography restricted by individual cortical anatomy derived from magnetic resonance imaging to examine sensory processing and lexico-semantic integration in deaf individuals and speech in hearing individuals. These processes were found to activate a highly similar left frontotemporal network, which included association areas within the superior temporal region. Activation was found in deaf individuals during lexico-semantic processing and during speech in hearing individuals. The authors conclude that neural systems dedicated to processing high-level linguistic information are utilized for processing language regardless of modality or hearing status. Moreover, they assert that there was no evidence for rewiring of afferent connections from visual systems to auditory cortex.

Beyond the primary and secondary association cortex for each sensory modality, the information moves on to the highest level of perceptual analysis within the tertiary association cortex. Here the somatic event is subjected to crossmodal analysis and a deep level of comprehension associating the tactile event with the auditory and visual relationships occurring both in the present analysis and in previous experiences with this stimulus. A letter written on the right hand can be identified with its visual and auditory associations and with the posturing needed to produce the sound in speech or the image on paper through graphics. The high complexity of analysis and comprehension provides for access to the homologous brain region via the corpus callosum for oppositional analysis as, for example, with logical linguistic left cerebral analysis as opposed to right cerebral emotional associations.

With deep comprehension, the information may access the longitudinal tract on its way to the frontal lobe for intentional associations, desire or motivation, regulatory control or downregulation (e.g., see Barbas et al. 2013), and executive functions culminating in our expressions. But, oppositional activities are prominent in this pathway as well and where frontal lobe regulatory control might suppress either the spreading activation of fearful associations or linguistic associations and especially if we should act on the processed information via rule-regulated behaviors. For the right frontal lobe, this might be rule regulation based on social proprieties and social pragmatics (e.g., Bechara et al. 1994; Anderson et al. 1999; Damasio et al. 1994; see also Damasio and Anderson 2003; Damasio et al. 2012; Eslinger et al. 2004; see also Yeates et al. 2012; see also Hu and Harrison 2013b; see also Keenan et al. 2003; also Tompkins 2012). Similar social rules underlie the regulatory control over the frontal regions. Here, for example, talking a little may be appropriate, whereas talking too much may not be appropriate. For the right frontal lobe, some laughter might be appropriate, whereas too much laughter or laughter that is, in one dimension, too loud may convey an impropriety or insult within the social setting.

The right parietal region appears to be more important in the complex analysis of emotion as occurs through variants of a facial expression, tone of voice, or control issues relevant to spatial analyzers in these brain regions. But, the right posterior brain region is more clearly specialized for emotion with the perceptual analysis and comprehension of negative affect as with an angry or fearful facial configuration (e.g., Harrison et al. 1990; Harrison and Gorelczenko 1990; Indersmitten and Gur 2003; see also Gauthier et al. 2000). Gina Mitchell (Mollet and Harrison 2006) appreciated the shared functional cerebral systems involved in negative affect and pain. Gina proposed that much of these anatomical systems are shared with those processing pain and, therefore, the lateralization of pain was proposed to be a specialization of the right brain. Tolerance for pain, though, has been found altered with frontal lobotomized patients, whereas diminished sympathetic activation and diminished emotionality have been appreciated with right parietal lesions (Heilman and Bowers 1990; Heilman et al. 1978).

Pain exposure elicits an oppositional response system through the mobilization of endorphins or endogenous opioid polypeptides (see Willis and Coggeshall 2004). People suffering from chronic pain may have elevated levels of these substances, which function much like neurotransmitters. These are essentially endogenous opiates with structural properties resembling that of morphine or heroin (Moallem et al. 2012). Indeed, our affinity for these "painkillers" and tendency to develop addictions may be due to their structural similarity to our endogenous opiates. These compounds are mobilized in response to painful stress, including exercise, where the pain is often followed by oppositional analgesia and a feeling of well-being. This may be described by those engaging in regular exercise as a "runner's high," reflecting something of the high experienced from the exogenous opioid compounds (Boecker et al. 2008). Endogenous opiates work to some extent as "natural pain relievers" (Moallem et al. 2012) with stimulation of the central and periventricular gray matter.

Many have witnessed this system in action by viewing the predator chase scenes frequently recorded for television, where the predator initiates the attack and the prey engages in evasive and ever more strenuous exertion to elude capture. The run continues until, if captured, the prey awaits the final kill. As the camera zooms in for a close-up of the kill, we might expect a look of horror or extreme apprehension on the face of the prey. Instead, what we will see may be a rather placid, glassy-eyed victim with facial features conveying calm or resolution. If these features are valid indicators of pain, then the function of the chase may well be to induce analgesia through the mobilization of endorphins. This interpretation, if valid, would provide a mechanism to significantly modify the pain of a cold-blooded kill without the fight. It might also remind us that we may benefit from the hard fight if we find ourselves in peril at the hands of a predator.

W. David Crews III (Crews and Bonaventura 1992; Crews et al. 1999) manipulated this system with a young woman who had been maintained within an institutional setting using multiple point physical restraints. Physical restraint, even with consent, is generally frowned upon in an institutional setting. Moreover, the implementation of restraint is often used as a measure for evaluating the quality of the care setting and compliance with ethical and legal standards. The family, physicians, and administrators were faced with some difficulty in these decisions secondary to self-abusive behavior on freeing her limbs or even her head and neck from the restraints. If her head was freed, she would slam it against the table or wall to induce injury. If an arm was freed, she would beat herself and inflict injury. Following a review of her case, David decided to accentuate the pain that others had tried to manage through physical restraints. Naltrexone hydrochloride (Revia) was used to promote exquisite pain as an opioid antagonist. More specifically, naltrexone is purported to block opiate receptors (see Moallem et al. 2012). A heroin addict, for example, might go into an acute withdrawal with onset of exquisite pain upon administration of naltrexone.

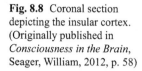

Fig. 8.8 Coronal section depicting the insular cortex. (Originally published in *Consciousness in the Brain*, Seager, William, 2012, p. 58)

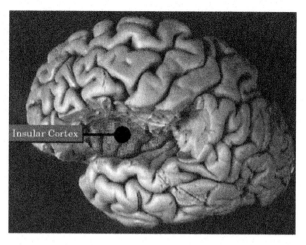

David used an ABAB single-subject experimental design to evaluate the treatment (B phase) in comparison with the baseline, no treatment condition (A phase). As difficult as it was, those involved and especially the family consented to freeing a restraint with the recording of baseline self-abusive behavior, followed by that recorded during treatment with naltrexone. Across each reversal, she showed a reliable reduction and eventually cessation of the self-injurious behavior. David followed this woman for an extended period with good outcome. Despite the success in this case, he continued to consider this a fortunate, if not a fortuitous, outcome as it might not have worked so well by any logical reason and due to the chronicity of the disorder.

These discussions of the somesthetic senses, including touch, pressure, cold, warm, and pain, would be incomplete without mention of the sensations arising from the gut. Although these projections contribute in small part to those conveyed through the somatosensory projections, they are largely processed at the insular cortex and especially the insular cortex within the temporal and parietal lobes (see Fig. 8.8). Contemporary neuroscience has identified the insula as the main cortical target for signals from the interoceptive system (e.g., Mesulam and Mufson 1982; Craig 2009; see also Damasio and Carvalho 2013) and functional neuroimaging studies consistently implicate the human insula in both interoceptive and emotional feelings (e.g., Stephan et al. 2003; see also Damasio and Carvalho 2013). Stimulation of the right temporal lobe and insular cortices has produced uterine and lower abdominal contractions (Penfield and Faulk 1955). Electrical stimulation of the right insular cortex can produce an increase in heart rate and blood pressure, implicating a role for the right insula in sympathetic cardiac arousal (Oppenheimer and Cechetto 1990; Oppenheimer et al. 1992; see also James et al. 2013). Moreover, there are direct sensory pathways from the bowel via the vagus nerve to the solitary nucleus of the medulla (see Paxinos et al. 2012), which is heavily connected to the amygdala. These can be activated during intestinal contractions and with hypermotility (Peppercorn and Herzog 1989, p. 1296).

Temporal lobe epilepsy may produce abdominal symptoms similar to those of irritable bowel syndrome. These might be distinguished from the latter condition by the presence of altered consciousness during some of the attacks, a tendency toward tiredness after an attack, and by an abnormal electroencephalogram (Zarling 1984, p. 687). Right posterior cerebral stroke patients are often present with constipation and eventually with these symptoms alternating with diarrhea. Although these brain regions are primarily involved in perception and sensory analysis, disturbance here may present as incomprehensible or meaningless stomach contractions, peristalsis, and/or abdominal contractions. In this context, like that with Wernicke's aphasia for meaningless speech, the disruption of meaningful perceptual analysis and impaired comprehension of gut movements appears to alter the meaningfulness of the expressions. By analogy, the word salad speech from a Wernicke's patient might be expressed as a gut or gastric mobility "salad."

One young woman had presented to a regional hospital to deliver her baby. She developed complications. A "code blue" was called in the postoperative recovery setting and with some anoxia as she was revived. She had left visual field deficits and deviation of the mandible toward the left along with a pervasive, agitated, and apprehensive state. With her recovery, many of her symptoms resolved. But, she continued to have recurrent panic episodes or fearful states and especially related to control issues. The episodes were characterized by abdominal contractions, sympathetic drive, apprehensive facial configurations, left-sided hyperreflexia, along with the emotional state of panic. The evaluation showed evidence for right frontal lobe regulatory control deficits with elevated activation over the right temporal lobe during the episode. The dramatic shift at the right frontal lobe in delta magnitude, depicting deactivation of this region, can be seen in Fig. 8.9 as she entered the panic state.

These results support the limited capacity interpretation discussed elsewhere (see Carmona et al. 2009; see also Klineburger and Harrison 2013), where this woman

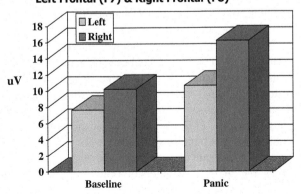

Fig. 8.9 Abrupt deactivation of the right frontal region evidenced by delta magnitude (in microvolts), recorded over the left and right frontal lobe at baseline and during a panic attack with abdominal contractions

Panic Attack
Frontal Lobe Delta Magnitude
Left Frontal (F7) & Right Frontal (F8)

Fig. 8.10 Abrupt and unbridled activation of the right temporal lobe evidenced by beta magnitude (in microvolts), recorded over the left and right temporal lobe at baseline and during a panic attack with abdominal contractions

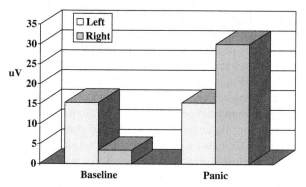

can be easily decompensated with a loss of regulatory control from the right frontal lobe secondary to control issues. The shift in regulatory capacity going into the panic state is much like abandoning the front line in military combat, where once turned the resources resisting the emotional panic are abandoned by this analogy. Upon provocation, this frontal region is shown to deactivate or abandon the resistance, leaving the arousal and fear systems within the right temporal and temporal parietal regions unbridled, unregulated, or uninhibited. The right frontal lobe provides inhibitory regulation over the right temporal lobe via longitudinal pathways, including the uncinate and the arcuate fasciculi. The uncinate provides for direct regulatory control over the amygdaloid bodies from the orbitofrontal region and the cingulate gyrus. The former is proximal to facial processing, where this woman would lose facial motor control presenting extreme facial apprehension. The loss of this regulation resulted in marked and clinically significant activation of the right temporal lobe as reflected in beta magnitude recorded over this region and also in comparison with that at the left temporal lobe (see Fig. 8.10).

With unbridled activation of the right temporal lobe, she presented a facial expression which conveyed horror along with strong abdominal contractions, profuse sweating, and rapid pulse rate. She was in a panic state. With these results in hand, efforts were made to inform her psychiatrist thinking that this would likely be justification for canceling the planned electroconvulsive shock treatments (ECT or "Edison's medicine") that had been scheduled for this young woman. With a background in psychoanalytic thinking, he responded "No it will be fine. I have seen this many times. She is simply reliving the birth experience." His interpretation was that the abdominal contractions resulted from "reliving the birth experience," discounting the findings of the evaluation that the right temporal lobe activation was unbridled with her brain disorder. But, when her condition worsened, precipitated by the additional insult of electroconvulsive shock placed directly over/across the right frontal region, the jury understood and summarily increased her award 200 % above what her attorney had requested. Relatedly, abdominal contractions with nausea and cardiovascular activity may be altered by activation of proximal parietal and temporal brain regions via vestibular stimulation (Penfield and Faulk 1955; see Carmona et al. 2008, 2009; see also Paxinos et al. 2012).

In the broader discussion of the somatic senses, there is substantial evidence that the right parietal area is dominant in regard to the processing of bodily sensations. Functional neuronal systems within the right cerebral hemisphere appear to be more sensitive or more responsive and to more effectively monitor events occurring on either half of the body, but particularly those events occurring at the left hemibody. In fact, this relationship was noted over 150 years ago by Weber. According to Weber (1834; see Joseph 2000), detection at the left half of the body exceeds that at the right in regard to most forms of tactual sensitivity. The left hand and the soles of the left foot, as well as the left shoulder, are more accurate in judging weight, have a more delicate sense of touch and temperature, such that "a greater sense of cold or of heat is aroused in the left hand" (p. 322). That is, the left hand judges warm substances to be warmer and cold material to be colder as compared to the right hand, even when both hands are simultaneously stimulated. Heightened cardiovascular responsivity was also found to vibrotactile stimulation at the left hand in comparison to that resulting from stimulation at the right hand (Foster et al. 2012).

Auditory Systems

The stimulus for audition consists of the compression and rarefaction of some conducting medium. This might be initiated through a hand clap with the hands coming rapidly together compressing air and with the rarefaction resulting as the hands disjoin and separate. Air is a reasonable conduction medium. But, it is not necessarily the better conductor as some species, like the whale, may communicate over a hundred miles using water as the medium. Many folks have anguished over the loss of whale and dolphin as they varied from their migratory pathways seemingly confused and disoriented and possibly even out of their element on the beach. It was the 1990s, perhaps, when we began to understand that these events might be related to experimentation with ultralow frequency sound by the navy and possibly from large engines or machinery emitting sound within this bandwidth (e.g., see Cox et al. 2006).

A general rule of thumb here is that higher frequencies require additional energy for the sound to cross a given distance (see Heller 2012). The bat, engaged in high-frequency echolocation, must expend sufficient energy to cross these near distances with sound emissions. But, the energy requirements for low-frequency sound transmission are reduced in comparison with higher frequencies. A pure tone consists of a stable or fixed sinusoidal waveform of a given frequency. Instead, most sounds are complex with multiple and overlapping frequencies and with variable durations. Humans are capable of sound transduction for frequencies ranging up to about 22 kHz. Similar to vision, the range of sound energy which humans are responsive to is small in comparison to the broad spectrum of sound energy within our environment. Once again, many species may be able to transduce and process higher-frequency sounds beyond the range for humans. Canines certainly may respond to a dog whistle that is of no relevance to people as it exceeds the human frequency range. These differences are also age related with children capable of hearing higher frequencies than detected by older age groups.

Needing a new set of earphones for the laboratory, the director approached a local sound system merchant for a comparison of their standard stock. The salesman was knowledgeable and showed her three basic apparati, which varied in price consistent with their specifications. The specifications on "the cheap one" were "not too good" as they accurately conveyed sound only up to 25 kHz. The "middle of the road" device was accurate to about 50 kHz, whereas the "the best one" was good to 250 kHz, an impressive engineering accomplishment, no doubt. The salesman seemed a bit confused that the customer purchased "the cheap one," as he was knowledgeable in the sales of these products to audiophiles. The laboratory director tried to explain to the salesman that she appreciated the specifications of the equipment, but that no human ear would be able to hear the difference as the frequency range is beyond 22 kHz! Indeed, much is spent on sound equipment for specifications exceeding those of the human sound transducers (receptors) and our auditory system more broadly defined. However, these relationships are complex and vary with the sound fields, oscillating frequencies, obstructions, and many other factors (e.g., see Heller 2012).

Sound frequencies are represented as oscillating waveforms, which eventually may be collected by the pinna or external ear. The canine pinna (variation noted) may be large in comparison to humans. This size difference is important for sensitivity as the Welsh corgi, for example, is substantially more sensitive to environmental noise, and in a college town, these are numerous throughout the night. An interactive science display that I took my children to provided the human equivalent of a bat's pinna. By placing your head into this apparatus, you might appreciate an acute improvement in your ability to hear the conversations of others within the building. Hunters have spent moderate sums of money on apparatus to improve their hearing in the forest to detect game. I suspect that large pinna might work as well as the electronic gear. But, this might look very strange in the forest! Native Americans would wear fox earlobes or pinna, which were thought to be decorative clothing by many European Americans. But, these were primarily for hunting purposes to improve sensitivity or sound collection as these stimuli were channeled into the middle ear.

The Eustachian tube extends upward close to the eardrum, allowing for the equalization of air pressures by swallowing or yawning or other techniques. This is important with acute change in altitude, traveling up and down the mountain ranges or possibly ascent and descent within a jet as we travel. This is often a problem with an infant or child in the plane where, with takeoff or landing, the child may become highly distressed with the pain and in need of pressure equalization between the inner ear and the ambient atmospheric pressure. But, Mom will relieve this with chewing gum or nursing opportunities dependent upon the age and capabilities of the child and the mother. Members of the Navajo tribe in New Mexico purportedly experience significantly more ear infections in early childhood development, secondary to the shape of the skull and a more horizontal placement of the Eustachian tube, allowing for infectious material to access the ear canals (Jaffe 1969). The typical anatomy of the auditory apparatus is shown in Fig. 8.11. Figure 8.12 shows the auditory pathways up to the primary projection cortex.

Sound intensity is quantified on a decibel scale, with the decibel being a unit of energy measurement somewhat like a meter is a measurement unit for distance. The human ear supports a broad range of sound intensities. Stevens and Davis (1938)

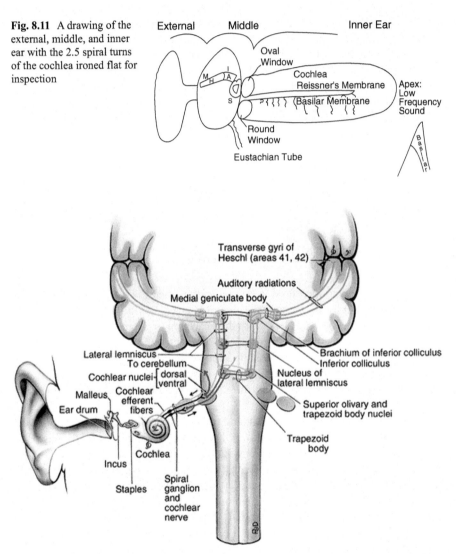

Fig. 8.11 A drawing of the external, middle, and inner ear with the 2.5 spiral turns of the cochlea ironed flat for inspection

Fig. 8.12 Anatomy of the auditory system from periphery to the auditory projection cortex. (Originally published in *Auditory and Vestibular Systems*, Strominger, Norman L.; Demarest, Robert J.; Laemle, Lois B.; 2012, p. 285)

calculated that the amplitude of vibration of the air molecules at the threshold of best human hearing (2000 Hz) was less than the diameter of a hydrogen molecule. Moreover, this system is capable of supporting sounds 1 million-million times more intense than the least intense stimulus that it is able to detect (see Heller 2012). With such a broad range, a log scale is used in the formula for the decibel. The formula requires a reference pressure against which the observed sound pressure may be measured. Since humans have their highest sensitivity (lowest threshold) for sound within the range of human speech sounds, and specifically at 2000 Hz (Stevens and Davis 1938; see also Heller 2012), this is often used as the fixed value in the computation. Based on this criterion, the reference pressure would be 0.002 dyn/cm², where a dyne refers to the force required to move 1 g in 1 cm/s. The dyne is a measure something like the erg, which is used in human factors research focusing on ergonomics or, for example, the force available at each of our major muscle groups for operating equipment or opening lids on a jar of peanut butter, perhaps.

The scientist or clinician might select another reference pressure for these purposes based upon one or another rationale. For example, in work with nonhuman primates, the neuropsychologist might consider an alternate reference based on that particular species' sound thresholds, which may be shifted a bit toward higher-frequency sounds (e.g., 8000 Hz in squirrel monkeys; see Harrison and Isaac 1984). Given this information, the researcher is able to measure sound intensity and the formula for this is as follows:

$$10 \log \frac{\text{Observed Pressure}}{\text{Reference Pressure}} = \text{Decibel (dB)}$$

The intensity of sound measured in the clinical or scientific use of a decibel meter is then represented by its numerical value at a given sound pressure level. These are selectable on the decibel meter with the provision of alternate scales (e.g., A, B, or C scales). The human ear supports sound to about 120 dB. In other words, there are only about 120 levels of loudness that can be detected, ranging from the lowest intensity level to the loudest. Smith (1997) describes this system as "…amazing; when listening to very weak sounds, the eardrum vibrates less than the diameter of a single molecule!"

In the human auditory system, pain and tissue destruction occur at about 120 dB and, yes, some folks install equipment in their cars to accomplish exactly this! But, more recently, it has become apparent that even moderate levels of sound (e.g., 80 dB) may damage the auditory system. This occurs on a college campus as many students, staff, and faculty wear sound systems as they move about the grounds. But, the user may think that she is listening to modest intensity sound, whereas the equipment may deliver spurious sounds well beyond the intensity range necessary for damage and without the user being aware of the variation in the volume secondary to the brevity of exposure.

According to a classical definition, the term loudness describes the "magnitude of an auditory sensation" (Fletcher and Munson 1933; see Röhl and Uppenkamp

2012). Röhl and Uppenkamp (2012) point out that the subjective rating of loudness is the perceptual correlate of sound intensity, which is known to depend on a number of other acoustical variables, such as frequency, spectral bandwidth, stimulus duration, temporal fluctuations, or monaural versus binaural stimulus presentation (Fletcher and Munson 1933; see also Heller 2012). Also relevant are nonauditory factors like context effects and personality traits like anxiety which can affect loudness (Stephens 1970). Röhl and Uppenkamp (2012) used fMRI to investigate the relationship of subjective loudness to activity within the auditory pathways. The authors conclude that the neural activity in the auditory cortex as measured by the blood oxygen level-dependent effect (fMRI) appears to be more a linear reflection of subjective loudness sensation rather than a display of physical sound pressure level, as measured using a sound-level meter.

The anatomy of the ear is traditionally divided into three units consisting of the external ear or pinna; the middle ear consisting of the tympanic membrane (eardrum) and three small bones known as the malleus, incus, and stapes (hammer, anvil, and stirrup) due to their shape or appearance; and the inner ear consisting of the auditory portion or cochlea (derived from the Greek word for snail) and the vestibular portion or vestibules, which have three semicircular canals designed for detecting positional movement. Collectively, this is the statoacoustic organ with vestibular and auditory portions within this anatomy. But, for simplicity, the discussion here will focus on the auditory component.

The sound oscillations collected by the external ear or pinna are channeled in toward the tympanic membrane or eardrum. This vibrating drum performs as a stiff diaphragm only up to about 8000 Hz, at which point it takes on the characteristics of a complex vibrator and often with overlapping frequencies of sound concurrently present across the drum. Rutherford's sound theory (1886) was partially based on the early telephones, where a speaker or diaphragm was able to replicate sounds via oscillations or vibration. But, again this works only up to about 8000 Hz, with speech frequencies generally below 4000 Hz (e.g., Lauter et al. 1985; Talavage et al. 2004).

The eardrum has an area of about 60 mm^2. This oscillating drum is connected via three small bones to the oval window of the inner ear. The oval window has an area of roughly 4 mm^2. Since pressure is equal to force divided by area, this difference in area increases the sound wave pressure by about 15 times as the oscillations are conveyed from the eardrum into the oval window (see Smith 1997). The three small bones of the middle ear (see Fig. 8.13) function to increase the force of the oscillations supported by the tympanic membrane, while decreasing the distance necessary for the oscillations (e.g., see Heller 2012). Thrust going into the oval window of the cochlea is roughly 25 times that at the tympanic membrane. This functions to increase the intensity of the stimulus, with the concurrent reduction in space requirements on moving into and up through the cochlea.

By World War II, the hazard of explosively intense sound was most apparent along with the hearing loss of our soldiers. This was appreciated in the artillery and tank divisions where the concussive force of a canon discharge was deafening and where permanent hearing loss was a hazard of combat requiring combat prepara-

Fig. 8.13 The malleus, incus, and stapes bones of the middle ear amplify force and decrease distance for sound oscillations going into the oval window. (Originally published in *Encyclopedia of Otolaryngology, Head and Neck Surgery,* Kountakis, Stilianos E., 2013)

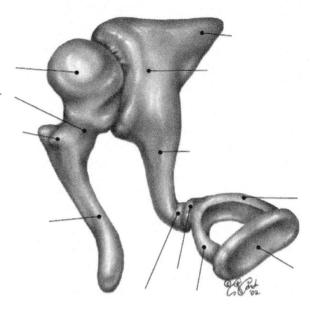

tions and military calculations. The efficiency of a soldier was lost with the onset of deafness and research was needed to reduce these effects. Damage to the olivo-cochlear reflex mechanisms following acoustic insult might render the soldier more susceptible to loud noise. Others tried to dampen the impact of intense sounds, and this might be partially accomplished through the manipulation of the olivocochlear reflex.

The olivocochlear reflex may prevent acoustic injury (Galambos 1956; Wie-derhold 1970; Maison and Liberman 2000) and chronic cochlear deefferentation may increase the vulnerability to permanent acoustic injury (Kujawa and Liberman 1997). The reflex allows the three small bones of the middle ear to shift out of phase as the result of an acoustic precipitant, such as a loud click. The malleus, incus, and stapes are arranged in series with one end at the tympanic membrane and the other end of the chain at the oval window. The phase shift now allows for the dissipation of energy at the middle ear rather than directly amplifying the intensity with transfer of the energy into the inner ear and onto the delicate basilar membrane containing our auditory receptors or hair cells. So, a tank driver might first hear a loud click prior to the discharge of the canon and now with less devastating impact on the receptive apparatus.

The inner ear consists of the auditory portion or cochlea. The cochlea is a dense structure arranged in a 2.5-turn spiral, looking much like the housing for a snail. The cochlea sits deep into the temporal bone, resulting in a structure that has, histori-cally, been difficult to access or to visualize its inner chambers and contents. The basilar membrane is arranged throughout this 2.5-turn spiral with the area of the membrane actually increasing as it ascends up into the upper part of the cochlea. Note the inversion here as space is becoming increasingly restricted in the upper

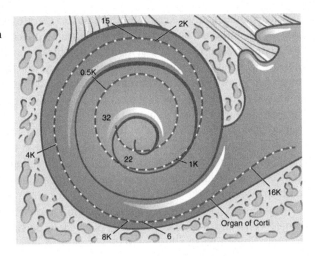

Fig. 8.14 Frequency map of the cochlear membranes from the base to the apex. (Originally published in *Physiology of Cochlea*, Kountakis, Stilianos E., 2013, p. 2156)

part of the cochlea, whereas the basilar membrane area is increasing in width (see Fig. 8.11; right lower corner). The anatomy looks like a fortress and conveys the appearance that sound must be very important to our survival to protect it this well and with such anatomical architectural feats. The receptors for sound transduction consist of hair cells. The basilar membrane at the base of the cochlea supports high-frequency sounds with the lower frequencies, including the speech range, located upward and into the apex of the cochlea (see Fig. 8.14). This also seems to defy engineering logic, as the size of the receptor hair cells supporting lower-frequency sounds are longer (see Fig. 8.11), whereas the size of the hair cells supporting higher-frequency sounds are shorter. To summarize here, as we move up into the apex of the cochlea, the basilar membrane becomes wider and the hair cells become longer!

Hermann von Helmholtz (1885) had posited early on that the hair cells would resonate at specific locations along the basilar membrane to specific frequencies, much like the strings on a harp or piano might do in response to resonant frequencies of sound. This theory became known as the "place–resonance theory of Helmholtz." The place part of this theory received some support with greater validity supported for high-frequency sounds. But, the resonance component fell into dispute and was ultimately untenable as the hair cells are attached only at one end to the membrane with the hair extending into the endolymph or fluid-filled cochlea. Eventually, Georg von Bekesy (1960), a member of the American Psychological Association, would receive the Nobel Prize in 1961 for uncovering the mechanisms of this apparatus, which had long been held a secret largely due to the protective anatomy surrounding the system. Von Bekesy used an elephant cochlea and found that specific frequencies of sound would produce an area of maximum deflection, much as Helmholtz had thought. This focal point for a pure sound frequency would result in *traveling waves* spreading across the adjacent basilar membrane, which function for surround inhibition. This is somewhat like the lateral inhibition discussed for vision as a function of the horizontal cells, and where inhibition improved contrast or clarity of the image.

The cochlea and the vestibules of the inner ear are innervated by one common nerve, the statoacoustic nerve. The dual features of the nerve are conveyed in the title with the acoustic part innervating the cochlea and with the static part innervating the vestibules or semicircular canals. This is the eighth cranial nerve and is more commonly known as the auditory nerve. The nerve exits the cochlea on its way to the brain stem, where the axonal fibers synapse on the cochlear nuclei and superior olivary bodies (auditory portion). From here, there are multiple auditory pathways with most traveling up the lateral lemniscus (recall that touch, pressure, temperature, and pain are found ascending together in the medial lemniscus) to the thalamus. Collectively, these are the auditory projections, and the information that they convey is ultimately analyzed for perceptual detail and acuity by the higher cortical systems.

But, there is a more primitive pathway, similar to that discussed for vision, which projects onto the inferior colliculus at the back or dorsal part of the brain stem (the tectum). Some fibers will travel to the tectum and synapse in the inferior colliculus (audition), whereas vision, outside of visual acuity or visual perception, projects to the superior colliculus of the tectum (e.g., Sedda and Scarpina 2012: see Fig. 8.15).

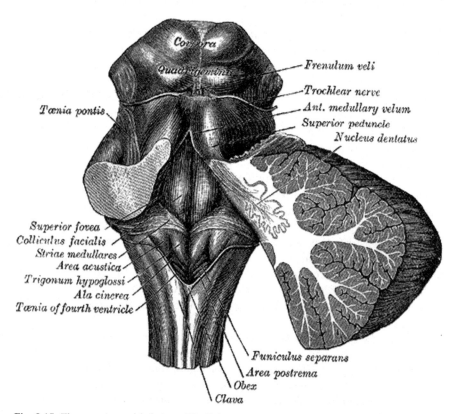

Fig. 8.15 The superior and inferior colliculi (corpora quadrigemina) viewed at the dorsal brain stem. (Originally published in *Gray's Anatomy of the Human Body*)

The inferior and superior colliculi appear as four bumps on the back ("the roof") of the brain stem (two at each side) with the general reference to these structures being the corpora quadrigemina. In this case, the terms inferior and superior are simply anatomical and not functional comparisons.

Within the cochlea, and throughout the auditory system, there exists a topographic representation of the frequencies of sounds (Lauter et al. 1985). This is found as proposed by the place component in the Helmholtz theory, in the cochlea, with low frequencies processed at the apex and with high frequencies at the base of the cochlea on the basilar membrane. This topographic relationship is found again at the medial geniculate nucleus of the thalamus and the inferior colliculus at the tectum. Eventually, the projections arise at the superior temporal gyrus and with ever more complex higher-level processing up these pathways. Topographical representation of sound exists again at the primary projection cortex in the temporal lobe, where a tonotopic map may be found (e.g., Talavage et al. 2004). But, a focal lesion does not produce a pure tone hearing loss as the projections overlay with divergent representations, at the cortical level, of the sounds processed at the base, middle, and apex of the cochlea.

Subthalamic auditory processing in mammals is mediated by bilaterally organized pathways that originate with the entry of the auditory nerves into the cochlear nuclei and culminate in the inferior colliculi (for review, see Oliver and Huerta 1992; Malmierca and Hackett 2010). Although inputs from both the left and right ears contribute to the pathway on each side, the representation in each inferior colliculus is dominated by the information about the contralateral sound field (Jenkins and Masterton 1982; Kelly and Kavanagh 1994; Delgutte et al. 1999; see Orton et al. 2012). Regions within the inferior colliculi at each side interact across the commissure of the inferior colliculus, a bundle of fibers that connects the two inferior colliculi and which provides both inhibitory and excitatory activation between these homologous structures at the tectum (Moore 1988; see Orton et al. 2012).

Individuals exposed to high-intensity or high-frequency sounds may suffer with "boiler room deafness" with the audiogram showing a notch with a loss of hearing at that bandwidth but often for the higher frequencies transduced near the base of the cochlea at the oval window. Many older patients complain of presbycusis or tinnitus with a constant high-frequency ringing sound from damage or loss of the hair cells along the basilar membrane (e.g., see Heller 2012). But, the laterality of this persistent noise is relevant as the reticular activating system of each brain stem, and the regulatory influences of each frontal lobe, may affect the perceived loudness, the laterality of origin, and the persistence of the sound. Moreover, the affect or emotional tolerance may vary with the brain involved in processing the sound from each ear. Ultimately, this may be detectable in anatomical alterations at the cortical level and, most commonly, attributions by the layperson, and even health professionals, that the patient is "hard of hearing."

The attribution that the patient is "hard of hearing" has traditionally been ascribed to loss or damage within the receptor apparatus or hair cells along the basilar membrane in the cochlea. Occasionally, the attributions might be that there is auditory nerve damage at the periphery. Once again the attribution that a peripheral cochlear defect in this system remains restricted to the peripheral end of the neural pathway appears to be incorrect. The loss of auditory receptors, the hair cells, may result in profound changes in the structure and function of the central auditory system. This has been demonstrated by a reorganization of the projection maps in the auditory cortex (see Sharma and Dorman 2012). These plastic changes higher up in the brain occur not only as a consequence of mechanical lesions at the cochlea or biochemical lesions of the hair cells by ototoxic drugs but also as a consequence of the loss of hair cells in connection with aging or noise exposure (see Syka 2002). The lesion or loss here may ultimately be conveyed throughout the pathway across the synapses and onto the cortical systems. So, a receptive hearing loss is a neuropsychological loss and eventually localizable at the other end of the system at the primary auditory projection cortex of the superior temporal gyrus (Heschl's gyrus) and the surrounding secondary and tertiary association cortices (see Sharma and Dorman 2012).

For some individuals suffering hearing loss, sensory prostheses are available in the form of electromechanical devices (see Cosetti and Waltzman 2011) that interface with the brain areas that normally process sensory information. The most extensively developed technology of this sort, which is available today, is the cochlear implant (see Gluck et al. 2014), though higher cortical implant technology is under development (e.g., Smith et al. 2013). The cochlear implant apparatus is designed to stimulate auditory nerves to produce auditory sensations in deaf individuals and typically to facilitate speech or auditory language processing demands. By using multiple electrodes or electrode configuration arrays (however, see Sharma and Dorman 2012) implanted in the cochlea, response patterns in the auditory nerve may be modified in ways that simulate naturally occurring sound patterns. Gluck et al. (2014) note in their review that the effectiveness of the intervention is improved in young children and in those adults having recently lost their hearing. Blamey et al. (2013) investigated the factors affecting the auditory performance of 2251 postlinguistically deaf adults using cochlear implants in a follow-up study and concluded again that patients with longer durations of severe to profound hearing loss were less likely to improve with cochlear implants than those with shorter duration experience.

Gluck et al. (2014) further note that the virtual speech sounds generated by cochlear implants do not replicate normal speech. Instead, it appears evident that the auditory patterns may provide for discriminative abilities for analytical interpretation to augment receptive communication efforts. Rapid learning or improvement in perceptual processing with the device may occur in the early months with more gradual improvement over the subsequent years with experience (Tajudeen et al. 2010). Neuroanatomical evidence relates these functional gains to neuroplasticity with cortical changes evident over time. For example, cochlear implantation in deaf cats produced substantial changes in the auditory cortex (e.g., Klinke et al. 1999; see Sharma and Dorman 2012; see also Gluck et al. 2014). Moreover, the investigation of neuroplasticity with sensory deprivation or augmentation has uncovered a remarkable capacity of the brain to be shaped by environmental input. In particular, a wealth of studies has documented striking effects of sensory deprivation in one modality on the neural arrangement and responsivity of the remaining modalities. These studies indicate that the brain remains substantially amenable to sensory alterations through a dynamic reorganization of cortical functions. It is generally agreed that multimodal brain areas show enhanced processing and recruitment of input to the remaining modalities in unimodally deprived animals and in blind and deaf humans (see Bavelier and Neville 2002), though the exact mechanisms for these changes remain unclear.

Gilley et al. (2006) examined cortical reorganization after congenital deafness by recording 64-channel electroencephalogram in normal-hearing, early-implanted, and late-implanted children while they listened passively to a phonemic speech sound "ba." Current density reconstructions using standardized low-resolution brain electromagnetic tomography (sLORETA) and dipole source analyses were performed using temporal components of the evoked response. In this project, auditory stimulation activated the superior temporal sulcus bilaterally and the right inferior temporal gyrus in normal-hearing children. Normal activation was observed in children implanted before age 3.5 years (i.e., within the sensitive period), along the contralateral superior temporal sulcus to the implanted ear and along the right inferior temporal gyrus, irregardless of the ear stimulated. In contrast to this encouraging finding for early developmental implants, children implanted after the end of the sensitive period (i.e., after age 7 years) produced more diffuse activity and lower-amplitude signals. Late-implanted children showed activation primarily of multimodal cortical processing regions and areas of visual cortex contralateral to the stimulated ear. The authors conclude (Gilley et al. 2006) that auditory stimulation activates similar neural networks in early-implanted children to that shown in normal-hearing children, a finding consistent with the improved prognosis for these implant patients. In contrast, primarily, multimodal cortical areas are activated in late-implanted children with a distributed network of brain areas and lower-amplitude responses, which occur along with reduced optimism for the intervention outcome (see Sharma and Dorman 2012).

Important features in the assessment of the auditory system are the sound frequencies lost, where an acute cortical lesion will, characteristically, not induce a pure tone hearing loss. Second, the lateralization of the sound loss is relevant as might be assessed using a dichotic listening test (see Pollmann 2010; Hugdahl et al. 2009; Hugdahl 2003, 2012) or electrophysiological recordings (e.g., Gilley et al. 2006). With dichotic presentations, the concurrent arrival of suprathreshold sound at each of the auditory cortices will yield extinction of the weaker cortical region. For example, the patient might detect sound and accurately locate the origin of the stimulus within left or right hemispace. But, with the concurrent arrival of sound to the left and the right ear, the sound might only be detected at the left ear, an indication of relative weakness or lesion of the left auditory cortex through contralateral inhibition by the dominant cortical system. Third, the nature or content of the auditory events not processed well, in comparison with those that are, will contribute to the confirmation of the lesion or dysfunctional brain region. For the receptive auditory cortices, a hearing loss for logical linguistic speech sounds with preservation for intonation, pitch, prosody, or melody is directive for further investigation of the left temporal lobe and diagnostic checks for a receptive dysphasia.

Once the receptive analysis at the primary auditory cortex is complete, the neural processing continues at the secondary association area for sound. Within the left temporal region, this overlays Wernicke's speech area, where the sequence of phonemic sounds is associated with meaningful memories of these sequentially analyzed sounds over time as with verbal language analysis and comprehension. The homologous region of the secondary auditory association cortex of the right hemisphere appears to provide comparative analysis and comprehension of the prosody, melody, or affective nuance of the sound (e.g., Bourguignon et al. 2012).

If incongruent linguistic speech sounds (e.g., pa, ba, ca, da, and so forth) are provided to each ear concurrently, there will typically be a right ear advantage, whereas auditory affect perception will more typically present with a left ear advantage (Kimura 1967; Ley and Bryden 1982; Hugdahl and Anderson 1987; Hugdahl 1988, 2012; Bryden and MacCrae 1989; Snyder et al. 1996; Demaree and Harrison 1997b; Mitchell and Harrison 2010; see also Pollmann 2010). Moreover, imaging data indicate that emotional dichotic listening tasks produce bilateral activation in the frontal, temporal, and parietal lobes (Jancke et al. 2001, 2003; Jancke and Shah 2002; see also Hugdahl 2012). Frontal lobe activation may result from vigilance to the stimuli (Jancke and Shah 2002), while temporal lobe activation results from the auditory events (e.g., Hugdahl 1995, 2012). Bilateral activation may result from callosal transfer of verbal information to the left hemisphere and emotional information to the right hemisphere (Jancke et al. 2001). Alternatively, bilateral activation may be a result of presentation of both positive and negative emotional tone. Perhaps consistent with this interpretation, Gadea et al. (2005) found that negative emotional induction produced an increase in identification of dichotic stimuli at the left ear and a decrease in identification of dichotic stimuli at the right ear.

Moreover, sound frequencies have long been related to emotional valence within an evolutionary context. The concept is based on empirical data, first pointed out by Collias (1960, p. 382; see Morton 1977), showing that natural selection has resulted in the structural convergence of many animal sounds used in "hostile" and "friendly" contexts. Morton (1977) states simply, "birds and mammals use harsh, relatively low-frequency sounds when hostile and higher-frequency, more pure tonelike sounds when frightened, appeasing, or approaching in a friendly manner. Thus, there appears to be a general relationship between the physical structures of sounds and the motivation underlying their use." Morton argues that proximity lessens the difficulties of communicating. However, in close proximity, the consequences of communication are more immediate, producing selective pressures favoring their existence.

The specialization of one or the other homologous auditory cortical regions is well established across species. This appears to be the case even though the anatomical redundancy is comparatively large for this modality with a relatively greater percentage of bilateral projections than what we find for vision and for the somatosensory modalities. Suffice it to say that the auditory modality has sufficient redundancy within each cerebral hemisphere and within the brain stem lateral lemniscus pathway to provide some resistance to a complete hemispatial loss of this modality. This may have evolutionary significance as the role of sound for defense and self-preservation is certainly large. But, for the diagnostician, this difference provides some utility in the evaluation of functional neural systems. An example might hold for the evaluation of neglect disorders where a homonymous hemianopsia and/or a hemianesthesia are/is more probable and with the utility of investigating sensory extinction for dichotic presentations of sound. Dichotic listening is one of the most frequently used auditory tests, and many (Hugdahl 1988, 2003, 2012; Pollmann 2010; Hugdahl et al. 2009) provide evidence in support for routine use of the dichotic method in the assessment of cerebral function and, specifically, at the superior temporal lobe and the planum temporale. The planum temporale is defined anatomically by the triangular surface on the supratemporal plane posterior to Heschl's gyrus (see Hickok and Saberi 2012).

For the family member or therapist working with the patient, the temporal lobe auditory defect usually is noticed, but may be poorly understood and with attributions which range to psychopathy, rather than to a distinct variety of auditory processing deficit. For example, dysfunction at the right temporal lobe may result in an auditory affect agnosia or inability to derive the emotion conveyed in others speech. This patient may be differentially sensitive to the literal speech conveyed and understood by the left brain. This has been demonstrated by Ley and Bryden (1982) and Bryden and MacRae (1989) with emotionally intoned verbal content and Blonder et al. (1991) with the use of incongruent literal and affective auditory content. For the normal brain, this incongruence may be "funny" and provoke laughter. Many comedians manipulate affect in this way to provoke similar reactions.

But, for the therapist or caregiver, there may be minimal appreciation of the amount of communication that is actually conveyed nonverbally through tone of voice. When this system is not working properly and when the patient must rely on the literal passage in the statement, it is critical that others attend to just that! This is indeed an emotional disorder wherein the auditory spatial analysis of the right temporal and parietal region is defunct. By homology, the caregiver or therapist may accentuate the prosodic and melodic aspects of sound for advantage in communication with a patient suffering from a left temporal auditory verbal processing deficit where the comprehension and the conveyance of logical, literal content is in disarray. Also, facial expressions and pantomime might make for a new and improved communication style. Agenesis of the corpus callosum connecting the two cerebral hemispheres accounts for additional varieties of speech processing pathology. For example, in one project using the Child Behavior Checklist, Badaruddin et al. (2007) accessed a data set ($n=733$) of individuals with a community diagnosis of agenesis of the corpus callosum. Evaluating a subset of high-functioning children, ages 6–11 years, they found that 39% produced clinically significant scores for social problems and that 48% produced clinically significant scores for attentional difficulties.

In training a new technician to administer a standardized memory battery, the technician was frustrated at the extended time required for the administration, which exceeded the typical guidelines. But, the clue was to come from the patient's right-sided gaze preference and preoccupation with right hemispace. This preoccupation presented with occasional smiles and friendly facial expressions toward the right hemispace, and the occasional chuckle. Upon sitting down with the patient, the neuropsychologist inquired as to what was over there and, since *they* were at the right, if he could hear their voices as in conversational speech. To the technician's chagrin, he replied that his deceased father, brother, and uncle were at his right and that he very much enjoyed listening to them as they talked. He considered them to be most positive as they were his "guardian angels." This was his priority in the setting, rather than the standardized testing presented to him by the technician. Any normative interpretation of the test results would be skewed by the focal point of the patient's intention and the concurrent overlay for auditory processing demands on the patient's overly active left cerebral hemisphere.

These are features of heightened or spurious activation of the association cortex within the left temporal lobe, whereas acute damage to the left temporal region may result in a receptive or Wernicke's aphasia and where damage at the right temporal region might convey auditory affect agnosia, dysmusia, and/or receptive dysprosodia. With these events, the patient may decompensate with the promotion of the auditory paracusia from the oscillating stimuli that this tissue is selectively responsive to and, more specifically, auditory linguistic speech sounds for the left superior temporal gyrus. Oscillating auditory spatial, melodic, and/or emotional sounds might be more provocative for the right superior temporal gyrus. So, the very nature of the testing materials appears to disable the patient and the technician invalidating the normative comparisons from the memory battery.

Another man referred for neuropsychological evaluation with a long-standing diagnosis of schizophrenia was undergoing quantitative encephalography in the neuropsychologist's office. He complained to the technician that the auditory hallucinations were very intrusive and that he listens to rock music to "make them go away." The technician asked him up front "what do they say?" He replied "I can't tell you because you will have me arrested." In fact, he provided more than one account of his history where he had been involuntarily confined with concerns about the content of the paracusia. He located the voices to the left side but acknowledged that they were trying to make him do very bad things. During the quiet recordings of the electroencephalogram, he abruptly stated "could you see it…I just heard them?" Indeed, we had recorded concurrent activation over the right temporal lobe with the hallucinated event. When the technician showed him the data and explained that it was his own voice that he had heard and that it related to the abnormal electrical activation in his brain, he said "it doesn't sound like me." The technician responded that "it would not sound like you even if it were a recording of your voice played over a speaker." He agreed, and this seemed to help him put the voices aside, to some extent, acknowledging the evidence of his right temporal lobe epilepsy rather than externally located devils at his left side attempting to control him.

Vestibular Systems: Dizziness and Falling

The vestibular system originates in the inner ear with five distinct end organs: three semicircular canals that are sensitive to angular acceleration of the head and two otolith organs that are sensitive to linear accelerations (e.g., see Carmona et al. 2009). The semicircular canals are arranged as a set of three mutually orthogonal sensors with each canal positioned at a right angle to the other two. This arrangement is similar to that at the corner of a room, where three sides are at right angles to one another. The relative position of each canal to that of the others provides sensory information necessary to determine the direction and amplitude of any head rotation. The canals are organized into functional pairs wherein both members of the pair lie in the same plane. Any rotation in that plane is excitatory to one of the members of the pair and inhibitory to the other. The otolith organs include the utricle and the saccule, which sense motion in the horizontal plane (utricle) and in the sagittal plane (saccule), where horizontal movements might include movements toward the front and the back or from the left to the right and where sagittal movements might be from supine to stand, and so forth.

The statoacoustic nerve leaves the vestibules and synapses at the brain stem vestibular nuclei and the superior olivary nucleus. Second-order fibers travel toward the thalamic relays onto the ventrolateral nucleus and the VPL. Third-order fibers then continue to the vestibular projection cortex at the temporal lobe and insular cortex and with projections to the somatosensory cortex via the radiations from the VPL. To a large extent, the vestibular reflex is a function of cerebellar, pontocerebellar, and frontocerebellar interactions, while the projection pathways continue to the temporal and parietal regions for reception and comprehension.

Aristotle set the stage for much of our modern perspective on the sensory modalities. Although he was aware of vestibular phenomena and recorded experiences involving dizziness, the vestibular modality was excluded from discussion of the classic sensory systems detailed in *De Anima* (Wade 1994, 2003). The vestibular modality has been difficult to localize, was the last of the basic sensory modalities to be discovered, and relatively little exists in our science to extend the analysis of this system to shared functional cerebral systems in psychology or in neuroscience more broadly defined. Progress was made with the verification of the vestibular end organs in the temporal bone by Prosper Meniere in 1861 (Baloh 2001). But, the effects of Aristotle's omission were long term and the contributions of this system to emotional processing, hemispatial neglect, and higher cerebral functions, in general, have only recently been explored (Carmona et al. 2009).

The vestibular system may be better known to many during childhood subsequent to activation of the apparatus through rotational behavior. Often several rotations toward the right will yield an aftereffect sensation of spinning in the oppositional direction and a tendency to fall in the direction of the spin as we can no longer trust the sensations in judging our position. This may be an exhilarating experience and profitable too for the carnival owner selling tickets to activate the vestibular apparatus along with the other modalities and beyond rationale comprehension. This came home to haunt me and my wonderful wife when my youngest child, then in middle school, was considering his options for the science fair. It might either be a project on one or the other fungus that he was constantly growing in every corner of the house or, perhaps, an inquiry into the vestibular system.

He broached the question in the following fashion with the clarity and reason of the young scientist. He asked boldly "Pop! What makes folks barf?" The rest is history. After spinning Mom 30 rotations in a chair, he found that the photographs of his mother were not admissible for public display in his science poster, since Mom was literally green and clearly nauseated. The picture was all but flattering, in this respect. But, his quantitative electroencephalogram results were spectacular, even to his father, as his mother's frontal lobe deactivated with delta activity prevalent in the record and especially at the oppositional frontal lobe. This loss of regulatory control by the frontal region (functional unit 3) over the vestibular projections in functional unit 2, if asymmetrical, may literally destabilize our position sense and result in illusory aftereffects. But, the phenomenon is again not restricted to the sensory modality, but rather with lateral asymmetry in activation of the temporal and parietal regions, activation of the autonomic nervous system, and emotional valence may co-occur within this modality (Carmona et al. 2009).

Joseph Carmona demonstrated that right temporal activation, negative affective valence, and dominate vectional complaints of spinning to the left result along with differential activation of sympathetic tone using skin conductance responses with increased sweating (Carmona et al. 2008). The carnival owner might well produce a vectional bias with oppositional activation of one or the other cerebral hemisphere through the direction of the spin. But, this system appears to be asymmetrically positioned with a right cerebral dominance for vestibular processing, sympathetic tone, and intense emotions. This asymmetry is potentially consistent with the specialization of the left brain for personal space and the right brain for peripersonal or extrapersonal space, wherein this system provides for analysis and comprehension of our body position in extrapersonal space. Also, this system includes projections to the temporal lobe and the insular cortex. This is relevant as electrical stimulation of the insular cortex or vestibular stress may yield movement of the gut and maybe to the point of precipitating emetic responses along with nausea.

In these respects, the vestibular system appears to be critical to survival in that over activation yields emetic responses to purge our digestive tract of potential toxins. But, this system also provides a distinct and usually trustworthy foundation to our sense of where we are in space. This is critical for our balance in ambulation and our ability to avoid falls. Indeed, with stroke or metabolic disturbances in the right vestibular projection cortex and the insular region, spatial delusions are to be expected. The primary clue for this is often risk for falls. Fall risk is a primary safety issue and a critical determinant for return to home or to a nursing facility (e.g., Ashburn et al. 2001). This can be demoralizing for the patient and for the family after spending a stressful stay in the medical center recovering from a stroke only to be discharged with a fall precipitating readmission to the hospital. Patients with vestibular dysfunction who report dizziness have a much greater risk of falling, and balance-related falls account for more than one half of accidental deaths in the elderly (Agrawal et al. 2009). If the fall involves a broken hip or fractured limb and the patient requires general anesthesia for fixation or repair of the fracture, then the overall prognosis is more guarded. Moreover, the patient's expected longevity may be diminished possibly to a 5- or 6-year window (Maggi et al. 2010).

Disturbance of this system within the right brain is predictive of more elaborate spatial delusions well beyond those of a vectional disturbance or dizziness. In stoke patients with delusion, the ischemic lesion is most often localized to the right temporal lobe and especially in the inferior distribution of the right middle cerebral artery (e.g., see Piechowski-Jozwiak and Bogousslavsky 2012). With extension of the damage into the right temporal parietal area, many may perceive very real relocation or movement of their body to other geographical locations. For the caregiver, this may seem delusional as in our historical view of being "crazy or mentally disturbed." But, these may be the attributions of the uninformed. The patient may very well appreciate movement through external space and even to another location. With right temporal lobe pathology and especially within the distribution of the right posterior cerebral artery, the patient may express negative attributions of external control or threat and often with the emotional overlay of great fear or apprehension along with an agitated state (e.g., Perez et al. 2011). The agitated state, as a safety management issue on a hospital or nursing care unit, may be managed pharmacologically through prescription haloperidol (Haldol) or lorazepam (Ativan). These interventions are not without cost to the patient as damage to the extrapyramidal motor or dopaminergic systems and tardive dyskinesia have been demonstrated with chronic use (e.g., Casey 1991). These brain regions play a role in panic states with recent evidence suggesting heightened sensitivity to carbon dioxide (CO_2) and aggravation of the fear or apprehension with respiratory irregularities or dyspnea.

The caregiver or therapist may play a role here in providing gentle reassurance and possibly an anchor for the patient within external space through reality therapies for location or place, providing frequent reminders and prompting. For the spatial reference, the patient may benefit from contact comfort therapies possibly through the provision of a comforting shawl, blanket, or familiar items from home. Something to cuddle might be helpful and especially if this a familiar spouse, significant others, or a family pet associated with safety and security. Olfactory stimuli consisting of positive and pleasant odors might be useful and even improve the attributions and treatment by the nursing staff working with the family member (e.g., see Applebaum et al. 2010) by activating positive emotional brain systems. Gentle reassurance does not include highly prosodic speech and loud speech volume. Although the patient may elicit this from the caregiver or staff, this may be a function of mirroring of the fearful or angry affect state of the patient. These interactions are known to activate the right temporal lobe and may precipitate agitation or decompensate the patient with dysfunction in this system. Motherese may be effective in some situations, though. Specifically, beneficial effects have been found when the fragile individual is within their own secure setting (e.g., their home) and when the motherese is reassuring and gently conveyed to this person (Bunce and Harrison 1991).

The vestibular system appears to be differentially lateralized to the right cerebral hemisphere (see Carmona et al. 2009; see also Chakor and Eklare 2012) specialized for extrapersonal spatial analysis by these functional anatomical systems. It may, therefore, differentially contribute to brain systems involved in negative emotion or in more intense affective perception and comprehension (see Carmona 2009, 2008). Like Wernicke's area for speech sound comprehension and the somatosensory association cortex for the comprehension of somesthetic events, dysfunction may not only alter comprehension but also it appears to yield meaningless expressions. Like the word "salad" speech in receptive dysphasia, the expressions in postural balance and control may be a "salad" of compensatory motor movements ill-conceived and without coherence for maintaining balance or postural control. Vestibular therapies may be useful in some cases not only for improved comprehension of balance but also for improved analysis and comprehension of extrapersonal space, including the spatial neglect disorders, which accompany lateralized brain lesions in the right temporal and parietal regions (see Heilman et al. 2003, 2012; Heilman and Gonzalez Rothi 2012b). These therapies might involve rotation or movement within and through one or more specific planes of section represented within the vestibules or semicircular canals of the inner ear.

A distinction might be useful here specifically in the location of the vectional disturbance arising from brain pathology. From functional neural systems theory, the frontal lobe plays a regulatory role over vestibular activation at the ipsilateral temporal and insular regions of the brain. A right frontal stroke may deregulate the right temporal and cerebellar regions, resulting in a perception of leftward vection. A left frontal stroke, in contrast, might disinhibit or release the left temporal and cerebellar region with activation here perceived as rightward vection. The direction of perceived vection has been supported by quantitative electroencephalogram analysis (see Carmona et al. 2008), whereas the direction of falls in patients with lateralized temporal or parietal activation did not reach statistical significance in the project. But, the sample size was small and additional participants are needed. In contrast to these lateralizing effects are ipsilesional vectional complaints with posterior lesions, where right cerebellar strokes may increase rightward vection complaints and behaviors and where left cerebellar strokes may increase leftward vection complaints and behaviors (e.g., see Troost 1980; see also Chakor and Eklare 2012).

The discussion thus far has addressed the oppositional systems of the left and right cerebral hemisphere and brain stem regions with a prevalent tendency of the left hemisphere to initiate activities toward and within right hemispace and a right hemispheric affinity for movement toward and within left hemispace. Quadrant theory (Foster et al. 2008a, b; Shenal et al. 2003; Walters and Harrison 2013a) and traditional functional systems theory (Luria 1973, 1980) provide for regulatory control by the frontal lobes for lateral movement. But, the frontocerebellar systems appear oppositional for vertical spatial analysis and vectional disturbance with anterior or posterior falls resulting from frontal or cerebellar pathology, respectively.

Frontal lesions were appreciated early on with the prevalent tendency of the Parkinson's patient to fall forward somewhat like a missile and with otherwise down going postures of the face, head, and trunk (e.g., Ashburn et al. 2001). Cerebellar patients with midline lesions may be prone to posterior falls and retropulsion (e.g., Heilman et al. 2012) and sometimes with an arching of the back or truncal regions. With these considerations, one may gain heightened respect for our typically automatic navigation systems that control the dimensional rotations of our body much like an airplane must adjust for yaw, pitch, and linear acceleration. Although these general considerations may be useful to the therapist or caregiver, careful analysis of the directional tendency for a risk of falls may be useful and ultimately promote safety. Physical and occupational therapists routinely provide such assessments and competent safety recommendations for others.

Three Chemical Senses?

Common Chemical Sense

The human body is capable of processing at least three chemical senses with the transduction of the adequate stimulus into neural energy. These include the "common chemical sense," which appears to be processed by free nerve endings within the oral membrane (e.g. Krasteva and Kummer 2012). These nerve endings are sensitive to mildly irritating vapors (e.g., pepper). The afferent pathway is through the fifth cranial nerve, the trigeminal nerve (see Willis and Coggeshall 2004). That this nerve is processing irritating vapors may make more sense in the broader scheme of things as this is one of the largest cranial nerves and it is loaded with pain fibers as reflected in the clinical pain syndrome "trigeminal neuralgia." Many are unaware of the 13th cranial nerve (Brookover 1913: Fuller and Burger 1990; see also Vilensky 2012; however, see Wyatt 2003), the "nervus terminalis" (also referred to as cranial nerve zero), as it was discovered after the other 12 cranial nerves and primary references beget additional references. The 13th cranial nerve (also called cranial nerve zero), in humans, is located anterior and medial to the first cranial nerve (the olfactory nerve). The fibers project from the nasal septum. Although the function of this nerve remains controversial, it appears to play a role in vasomotor control of the nasal septum and in processing aspects of taste and smell originating from this region. This nerve is purported to play a role in chemical communication in humans, including the detection of pheromones (see Meredith 2001).

This 13th cranial nerve (see Vilensky 2012) may be involved in these processes as was appreciated earlier on with work on the pit viper and Jacobson's organ. Venomous snakes appear in two primary configurations known as the front-fanged snake or "pit viper" and the back-fanged snake, such as the coral snake. I grew up with the former and would marvel, after a rattlesnake hunt, at how many steaks might be rendered from one very large diamondback rattler! The diamondback rattler is a most impressive creature in size and skill. The sidewinder rattlers are noted for their speed. But, do not underestimate the mobility and agility of an agitated diamondback rattler. Pit vipers essentially slap you with their jaw extended flat. The cartilaginous head allows for remarkable flexibility extending the mandible into a flat continuous plane with the maxillary region of the head. The slap will insert the needlelike fangs and collapse the venom sack filling the opening with a mixture of meat tenderizer, cardiovascular agents, and/or poison. The coral snake, in contrast, must grind the poison in with the aid of small teethlike protrusions.

The pit viper flicks the tongue out collecting vapors to be applied to Jacobson's organ. This serves the predator well after a hunt with the prey, injected but not deceased, finding its way across the landscape. The pit viper may track its prey using the chemical trail provided by the animal. If Jacobson's organ is removed, the snake may no longer be able to track the prey, supporting the notion that this organ is responding to vaporous chemical stimuli. The analogy here is that vaporous chemicals entering the nasal septum may provide information relevant to taste and smell originating from this region.

Olfaction

The remaining chemical senses include smell or olfaction and taste or gustation. In their manuscript entitled "The Color of Odor," Morrot et al. (2001) describe smell as a "peculiar sensory modality, the main function of which remains to be specified." They note that the peripheral components of the system have low substrate specificity, with a single receptor recognizing multiple odorants and with a single odorant recognized by multiple receptors (Malnic et al. 1999). The olfactory projections are largely ipsilateral (Powell et al. 1965; Price 1973) with fibers from the left olfactory mucosa and olfactory bulb conveying their information to the left cerebral hemisphere and vice versa. This may be aptly appreciated in that the left nostril differentially exposes the left brain to smell, whereas the right cerebrum draws largely from the olfactory experiences at the right nostril and its sensory array. In contrast, the retina of each eye contains the receptive fields for both cerebral hemispheres, where damage to the left occipital region will affect the vision in both eyes within the right visual half-field. Also by way of comparative contrast, the somatic projections arise predominantly from contralateral sensory arrays.

Moreover, the olfactory projections are distinct in that they avoid the thalamic relay structures common to the remaining sensory modalities. This neuroanatomical substrate in which olfaction reaches the olfactory cortex without a thalamic relay suggests to some (Sela et al. 2009) that the more typical or characteristic role of the thalamus in processing and relaying sensory information is performed instead at the olfactory bulb or olfactory cortex. Olfaction is the only sensory modality that projects directly into cortical regions without preliminary processing at the thalamic level (Powell et al. 1965; Price 1973). Morrot et al. note further that sensory transduction at olfactory receptors and the conduction of olfactory information along unmyelinated axons are the slowest within the nervous system. Olfactory receptor transduction and neuronal conduction allow for the detection of the odorant stimulus by about 400 ms, approximately ten times slower than visual detection (Herz and Engen 1996). Beyond these distinctions among the sensory modalities, it is well established that odors can modify behavior outside of language processing systems or "conscious awareness" (e.g., Epple and Herz 1999), evoke emotions (e.g., de Groot et al. 2012), and evoke past memories (e.g., Chu and Downes 2000).

Considerable interest exists on the laterality of olfactory processing relevant to positive and negative emotional valences and to approach and withdrawal models. Using fMRI, Bensafi et al. (2012) found left insula activation to an unpleasant odorant mixture, whereas the pleasant mixture induced activation of the right insula. Support for cerebral asymmetry in olfactory processing with a right hemispheric processing specialization for pleasant stimuli and a left hemispheric specialization for unpleasant stimuli has been proposed by others (Anderson et al. 2003). Also, Sela et al.(2009) found that right thalamic lesions altered olfactory hedonics by reducing the pleasantness of pleasant odors.

However, the findings of Sela et al. are apt to be relevant to the intensity and arousal dimensions of the emotional perception rather than specifically to valence. The former is well established within the literature as a function of right cerebral regions and especially those involving projections to the right temporal and parietal regions (e.g., Heilman et al. 1978; see also Heilman and Valenstein 2012). Also relevant here is that the project by Anderson et al. failed to control for sex differences in laterality with the majority of the participants (20 of 26) being female. Differential laterality is also well established as a function of sex or gender. However, sex differences in emotional responding confound the interpretation as well, with evidence of left amygdale activation in women, for example, on exposure to negatively valenced emotional provocation. Women are also reported to display higher complexity and differentiation in their articulation of emotional experiences (Barrett et al. 2000) and to score higher than men on self-report measures of empathy (e.g., Davis 1996; Baron-Cohen and Wheelwright 2004). These differences and others may be foundational for a better understanding of response differences, including nurturance and open verbal linguistic processing advantages for women, perhaps.

Olfactory stimuli must be soluble in water or fat and have a vapor pressure sufficient to arrive at the olfactory mucosa, a small patch found in the upper nasal cavity. For substances entering either nostril, the pathway continues on toward the throat where the oral cavity and the nasal cavities merge on their way toward the pharynx, uvula, and epiglottis. For the present discussion, the anatomy at the upper nasal cavity region is most relevant. Here we will find an extension of the nervous system with hair cells and bipolar neuronal cells extending from the olfactory bulbs at the base of the brain, down through the cribriform plate and into a mucous-covered patch collectively known as the olfactory mucosa. The unpleasant insight at this juncture is that the odors we smell are molecules that we are in direct physical contact with at the olfactory mucosa.

Olfactory dysfunction from head injury was described early on by Jackson (1864) in a case of posttraumatic anosmia in a middle-aged man after a fall from a horse. The relationship between head trauma and olfactory disturbance is well established. Moreover, converging evidence relates head trauma to olfactory deficits and to emotional processing deficits within the domain of sociopathy with reduced empathy. Nonetheless, olfactory sensitivity is likely the most neglected component in the assessment of head injury, whereas anosmia and/or olfactory agnosia are sensitive indicators of neural dysfunction or damage from such trauma (Deems et al. 1991; Neumann et al. 2012).

The anatomy of the olfactory apparatus is of considerable relevance to the assessment and diagnosis of closed head injury or traumatic brain injury. The soft delicate neuronal fibers emanating from the olfactory bulb at the base of the brain travel through a thin boney *cribriform plate* at the floor of the anterior cranial fossa. This apparatus is configured somewhat like a guillotine where physical forces of torsion and shear, with abrupt movement of the brain within the skull chamber, may cut the fibers descending into the olfactory mucosa.

Neumann et al. (2012) hypothesized a relationship between damage to this anatomy and emotional pathology because these neural substrates overlap with the ventral circuitry of the orbitofrontal cortex, which plays a critical role in anger regulation and affective responses, such as empathy. They provide a comparative study between participants with traumatic brain injury with dysosmia and those with normal olfaction (normosmia). The authors used a convenience sample of participants ($N = 106$) of adults with moderate to severe traumatic brain injuries who were tested for olfactory function. Participants averaged 11.5 years post injury. Using multiple self-report measures, they provide support for the hypothesized relationship, with olfactory deficits from traumatic brain injury predictive of emotional impairments and reduced empathy.

The olfactory apparatus is at risk with a head injury, as described above, and from other forms of pathology. Within the clinical setting, the assessment of smell can be performed with a collection of commonly available and familiar items often readily available in a hospital room and at the patient's bedside. A kit might be assembled from commercially available oils, including oil of peppermint, cinnamon, wintergreen, and many others. The "scratch and sniff" test may also be used and efforts have been extended to provide normative data on this instrument with some success (Doty 1995; Doty et al. 1984, 1989). Another variant used in the assessment of odor identification is the "Sniffin' sticks" test, which has been used to assess olfactory performance by the combined testing of odor identification, odor discrimination, and olfactory threshold (Hummel et al. 1997).

The detection of odor, though, is regularly confounded by the measurement technique and not insignificant among these is the requirement to sniff, which may announce the presence of an odor. This confound was present in the early efforts to quantify olfactory stimuli using the Zwaardemaker olfactometer (see Wenzel 1948) consisting of a metal or glass tube inserted at variable depths into a porous clay tube, which might be saturated with an odorous stimulus. The quantity or amount of exposure to the odor stimulus was purported to be a function of the depth of insertion into the tube (see Fig. 8.16). Shallow insertion would leave more of the surface area of the odor-saturated porous clay tube exposed, allowing the conveyance of increased intensity of the odor stimulus.

Fig. 8.16 The Zwaardemaker olfactometer and the Elsberg's blast injection apparatus for quantifying the exposure to an odorous stimulus

ZWAARDEMAKER OLFACTOMETER

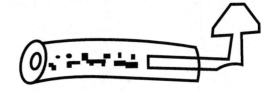

ELSBURG'S BLAST INJECTION

Alternatively, Elsberg's blast injection apparatus (Elsberg et al. 1936) might be used in an attempt to quantify the odor by the injection of air from a syringe through a flask containing the odorous stimulus and up to a stopper placed below the nostril. This apparatus eliminates the requirement that the subject or patient sniff the odor. But instead, this apparatus provides a blast of air which might be detected either through tactile cues or through sound to announce the presence of the odor stimulus (see Fig. 8.16). Although much progress has been made since the development of these early approaches to the quantification of smell in the laboratory using computerized administration and scoring techniques, the modern approaches are elaborations on these early themes with the sniff or injection potentially confounding the assessment. A modern olfactometer may be viewed in Fig. 8.17, where a bit of reminiscence may reveal remnants of Elsberg's blast injection apparatus.

Olfactory receptors consist of hair cells and bipolar cells covered by mucous located in a small patch in the upper nasal cavity. Receptors in each nostril are separated by the nasal septum. Stimulation of one olfactory bulb elicits activation in the other tract. However, this activation is relative and of smaller magnitude. This apparatus is a positive feedback system with tufted cells projecting back onto the glomeruli in the form of reverberatory neural circuits. Figure 8.18 and Fig. 8.19 provide for a visual inspection of this anatomy.

These glomeruli are important way stations for the transduction of olfactory stimuli with neural information headed toward the olfactory bulb. As such the glom-

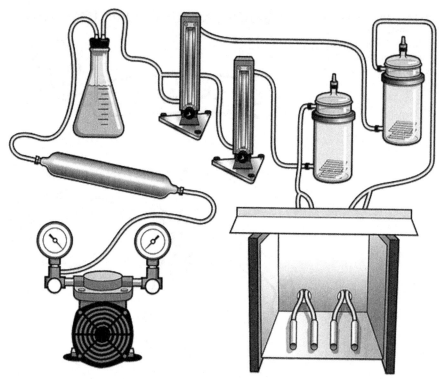

Fig. 8.17 A modern laboratory olfactometer. (Originally published by Walter, Abigail J., *Walking Response of the Mediterranean Pine Engraver,* Orthotomicus erosus, *to Novel Plant Odors in a Laboratory Olfactometer, Journal of Insect Behavior,* Volume 23, Issue 4, p. 255)

Fig. 8.18 A drawing of the glomeruli showing the tufted cells projecting back onto this apparatus

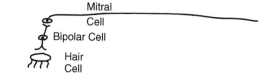

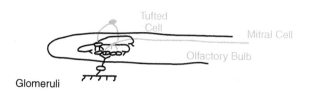

Fig. 8.19 The glomeruli and olfactory projections. (Originally published in *Gray's Anatomy of the Human Body*)

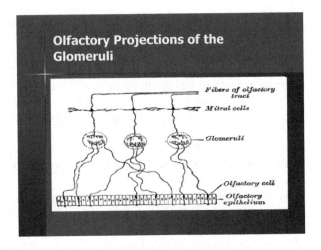

eruli may form the basic unit of the olfactory map of the olfactory bulb. The reverberatory circuits with tufted cells projecting back onto this apparatus provide for high absolute sensitivity. Substantial resources have been expended to elaborate on olfactory information to promote positive moods and attraction within the perfume industry. Similarly, much has been invested to minimize negative affective odors as, for example, with room deodorants and such. A long running advertisement for a room deodorant claimed that the product "killed germs." Two women were depicted in a room with a nasty yellow film in the air (germs). Each woman was spraying a room deodorant into the air. But, only the product used by one of the women removed the nasty yellow film in the air effectively killing the germs. I remember questioning what the active ingredient might be in the competitor's product. I was amazed to discover that the competitor's product contained formaldehyde, a known carcinogen. Formaldehyde is a gas, which is sold as a 40% solution and often "thinned" down even more for commercial use. It is used to treat furniture fabrics and in the construction industry. But, many will recall its use as an embalming agent in the morgue or at the funeral home. The use of this product might be less costly and more effective, if the consumer would simply spray a bit near the nose of each guest as they arrive for the party, rather than spraying it all over the house! But, of course, the hazard remains.

The Roman philosopher Lucretius (first century BCE) speculated that different odors may result from different shapes and sizes of the odor stimuli that activate the olfactory organ. Modern demonstrations of this theory include the cloning of olfactory receptor proteins by Linda B. Buck and Richard Axel and the subsequent pairing of odor molecules to specific receptor proteins. Each odor receptor appears to recognize only a particular molecular feature or class of odor molecules. Earlier on, the differential sensitivity of the system was poorly appreciated as the anatomy seemed more appropriate for sensitivity in the detection of odor than for discrimination among odors. This viewpoint was challenged with newer information supporting high differential sensitivity and substantially beyond the imagination of many neuroscientists within the field. Linda Buck and Richard Axel received the 2004 Nobel Prize for the discovery of an entire gene family (about 1000 genes) which codes for these olfactory receptors (Buck and Axel 1991). The genes were described as blueprints for a family of smell receptor proteins in the nose that work in different combinations so that the brain can identify a large array of odors—much like the letters of the alphabet are combined to form different words. These researchers provided evidence indicating that each odor-sensing cell in the nose possesses one type of odorant receptor and each receptor can detect a limited number of odorant substances. The identities of different odors are discretely recognized by the brain through sensory maps beyond the receptor level in parts of the brain known as the olfactory bulb and olfactory cortex.

Each odor that we perceive maps onto a pattern of activation across these receptors with pattern analysis and recognition capabilities further up the projection pathways in the olfactory bulb and olfactory projection and association cortices. Much of the earlier thinking on smell was based on a simple analogy to the rods in our retina, where sensitivity was high, but where differential sensitivity was low. With the improved appreciation of pattern analysis capabilities within the olfactory system, it became increasingly clear that we are indeed capable of distinguishing large numbers of olfactory stimuli and affording them unique aspects in memory. Indeed, olfactory processing may be one of our more powerful sensory modalities activating not only memories, which were seemingly long forgotten, but also strong emotional responses. The responses have been shown to be instrumental in the choice of our mate and even the continuity of our relationships with other people. Moreover, recent evidence (Crisinel and Spence 2012) relates specific aspects and qualities of music to the cortical representations of olfactory stimuli, potentially extending the associative analysis of this sensory modality to other brain regions.

Human olfactory capabilities may be miniscule in comparisons with other mammalian eukaryotes. This might be grossly apparent in the relative allocation of brain mass to the olfactory apparatus in humans in comparisons to other species. Figure 8.20 provides for a visual comparison of the relative size of the olfactory bulb at the base of the frontal lobe in the dog (*Canis familiaris*) and in the human brain. The canine olfactory bulb is roughly 40 times larger than that of humans, and the average canine possesses hundreds of millions of receptors for odors, compared with a few million for humans (see Dog Wikipedia 2005). Although this apparatus is robustly visible on gross inspection of the dog, careful inspection of the area at the base of the frontal lobe is required to visualize the olfactory bulb in the human brain.

The olfactory pathways are entirely central nervous system and very complex with substantial access to the limbic system structures which also project diffusely throughout the brain. Every traditional sensory modality, with the exception of olfaction, projects onto the thalamic relays prior to projecting on to the primary projection cortices. The primary olfactory projection cortices are identified with locations at the inferior medial temporal lobe and at the lateral olfactory gyrus of the insular cortex (Powell et al. 1965; Price 1973). Moreover, redundancy exists within this system at every level with other pathways providing olfactory information. If

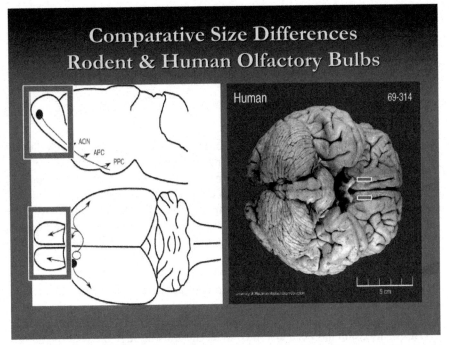

Fig. 8.20 The olfactory bulbs of the dog (*Canis familiaris*) and human. (The image at the *right* is adapted with permission from http://www.brains.rad.msu.edu, and http://brainmuseum.org, supported by the US National Science Foundation. The image at the *left* is modified from one copyrighted by Springer Science+Business Media, LLC)

a dog is conditioned to lift its leg to the presence of an odor and the olfactory tract is then severed, the dog will still respond to subsequent presentations of the odor with the leg lift response. Now if the trigeminal nerve is severed, the response may be eliminated. Thus, the trigeminal nerve is able to convey olfactory information at some level. Redundancy may also exist within higher brain regions in the combined processing of olfaction and other sensory systems, including music.

For example, Crisinel and Spence (2012) provide evidence that musical senses may be employed to help with the assessment of a smell. Subjects were asked to inhale 20 odors ranging from apple to violet and woodsmoke, which came from a wine-tasting instructional kit. Following a good sniff, the subjects were asked to process their way through 52 sounds of varying pitches, played by string or brass, by woodwind, or by piano to identify the best match for the odor. Sweet and sour smells were rated as higher pitched, smoky and woody ones as lower pitched. Vanilla too had elements of both piano and woodwind, whereas musk was strongly brass by association. Taste may also be linked through cortical associations as the authors have previously established that sweet and sour tastes, like odors, are linked to high pitch, while bitter tastes are associated with lower-pitched sounds. For example, pieces of toffee consumed concurrent with listening to musical sounds varied in flavor as a function of this relationship. Toffee eaten during low-pitched music was rated as bitterer than that consumed during the high-pitched rendition. By this manipulation and by others, perceptual processing in one sensory modality may be influenced by the ongoing processing demands within another.

Pheromones

Olfactory thresholds are relatively constant in men, with heightened variability in women (e.g., Dalton et al. 2002). Women may be less sensitive to odors during the menses, for example. Reduced sensitivity may result during pregnancy, but qualitative changes may outweigh the measureable differences in threshold or sensitivity. For example, Doty (1976) reviewed the initial studies on olfactory sensitivity during pregnancy. Although the results are often contradictory, he concluded that sensitivity during late pregnancy is lower than normal. He also concluded that anecdotal reports of pregnancy-related hypersensitivity, although they cannot be ruled out, have received little support in the literature (Doty 1986; see Gilbert and Wysocki 1991). Moreover, the less sensitive person may be more likely to be bothered or intolerant of the odor. This diminished tolerance is easily and often confused with sensitivity, whereas sensitivity is a measure of threshold.

This relationship has been disrupted in recent years through the widespread use of birth control pills, creating artificially elevated levels of estrogen throughout the cycle, rather than just at ovulation, when estrogen would normally peak. Research has indicated that women are attracted to men with pheromones, indicating a maximally dissimilar immune system when they are ovulating (Wedekind et al. 1995; Wedekind and Furi 1997; Elder 2001). The pill simulates pregnancy, potentially causing women to be attracted to men with similar immune systems. Divorce rates may indeed be elevated due to the artificial elevation of estrogen early on in the relationship and the subsequent cessation of estrogen supplements during the childbearing years. Following from this discussion, it might be argued that coming off of the pill during serious dating relationships might reduce the probability of divorce further down the road. But, this implication currently awaits confirmation through scientific investigation.

Derntl et al. (2013) investigated the impact of cycle phase and oral contraception using 20 women taking oral contraceptives and 40 women without oral contraception. Participants were healthy Caucasians who were further divided into follicular and luteal phase. These researchers assessed olfactory performance twice using the "Sniffin' sticks" battery along with intensity and pleasantness ratings of n-butanol. Women outperformed men in odor discrimination and odor identification. During the luteal phase, higher thresholds (reduced sensitivity) were found along with higher intensity ratings for the n-butanol. Also, improved olfactory performance correlated positively to duration of oral contraception. The authors conclude that odor performance is altered by menstrual phase and also the duration of use for oral contraception. These conclusions, of course, support the interpretation that olfactory processing can be modulated by hormonal changes.

Pheromones are subject to olfactory processing through multiple pathways, including the 13th cranial nerve (also called cranial nerve zero). Both men and women significantly increase their oxytocin blood levels in relation to birth or mothering pheromones. Oxytocin is generally considered to be the "parenting hormone" or "the cuddling hormone" essential for bonding at birth and eventually the development of close and trusting relationships. Current research avenues include the active and ongoing investigation of oxytocin in autism (Modahl et al. 1998; see also MacDonald and MacDonald 2010) and, with the development of the fusiform gyrus (Shultz 2005; Shultz et al. 2000), a brain region located within the temporal lobe important for facial recognition and familiarity (e.g., Kanwisher et al. 1997; see also Kanwisher and Dilks in press).

Dysfunction at the fusiform gyrus has been related to gaze aversion and diminished eye contact. Moreover, over activation of the posterior right temporal region and the ventral longitudinal pathways to the orbitofrontal region appears to promote fearful states and even left-sided formesthesias of devils or evildoers trying to coerce or to inflict harm upon this person. One possible variant of this neuroanatomical relationship has been recently expressed in the proposition that autism is a neural systems disorder with underconnectivity within the frontal–posterior pathways associated with the longitudinal tract (e.g., Just et al. in press). This might be reflected in poor downregulation over the posterior facial processing systems with altered or pathological activation and facial processing deficits. Within the right hemisphere, these systems appear specialized not only for faces but also negative emotion and primarily fearful apprehension. Also evident in research on orbitofrontal damage is the consequential loss of good decision making, including the selection of friends and the ability to establish strong relationships (e.g., see Bechara 2004).

Prolactin (PRL) or luteotropic hormone (LTH) appears to be the hormone responsible for milk letdown after the birth of the baby and one way by which babies recognize their mothers. A new mother might consider the benefits of leaving her nursing shirts out, when going out for the evening, so that the babysitter might swaddle the child in them to calm the child should it become upset in her absence. Plug-in air fresheners are now on the market containing the pheromones of lactating dogs or lactating cats to calm nervous pets during their owner's absence. We made use of these when our corgi became snappish as my sons would bring home the football or track-and-field team after a big event. These pheromones appear to help calm the dog and to offset some of the male hormones and smell of these players. Mothers can recognize their infant's blanket from among piles of identical wraps by smell alone! McClintock (McClintock 1971, 1998; Jacob and McClintock 2000) has demonstrated that pheromones, which we are generally unaware of, influence us in many ways, affecting our mood, our liking or disliking of even our mate, our sense of trust with another, and even the synchrony of the menstrual cycle of women sharing close quarters.

Chemosignals play a role in human communication, although the extent of these influences remains unknown. Research indicates, for example, that the exposure to sweat excreted by donors experiencing fear results in enhanced vigilance and caution in comparison with exposure to controls (e.g., donor sweat from playing sports;

Albrecht et al. 2011; Chen et al. 2006; Haegler et al. 2010; Zernecke et al. 2011). Moreover, these influential effects have been shown to occur largely outside of conscious awareness (Sobel et al. 1999; Lundström et al. 2008). The communication mechanisms have been clearly documented to include emotional mechanisms and neural systems potentially vital to survival and which were likely to have been operational prior to the establishment of higher cortical systems involved in verbal or linguistic communication, for example. The specifics of the chemosignal communication patterns or neural imprints which convey different emotions, though, appear to be less well established in our science than are the arguably newer linguistic patterns underlying language. Nevertheless, the evidence acquired using chemical analyses of stress-related odors reveals that male signals conveying fear are stronger than female signals, an effect complemented by the display of increased sensitivity to these signals in females (Wysocki et al. 2009; see also de Groot et al. 2012).

In one project, de Groot et al. (2012) investigated the role of chemosignals in the communication of emotions. Sweat was collected from men while they watched either a fear-inducing or a disgust-inducing movie. Subsequently, women were exposed to sweat samples during the performance of a visual search task. Exposure to the sweat obtained under differing emotional conditions was sufficient to induce a similar emotional response in the women. "Fear sweat" resulted in the production of fearful facial expressions in the women, whereas "disgust sweat" resulted in the production of disgusted facial expressions. Not only was the chemosignal sufficient in yielding a mirror emotional display in the women participants, it was also sufficient to alter the women's perceptions during the visual search task and to alter eye-scanning behaviors. These findings support the potential role of chemosignals in the communication and alteration of emotion in the recipient.

Gustation

The appreciation of basic taste modalities can be traced back at least to Aristotle who wrote of sweet, bitter, salt, and sour as succulent or harsh. Umami or savoriness was first described in 1908. Only recently was it recognized as the fifth basic taste as it might be evoked by some free amino acids such as monosodium glutamate (see Kawamura and Kare 1987). The stimulus for taste or gustation generally consists of water-soluble chemicals. The receptors are specialized epithelial cells located on the surface of the tongue, the pharynx, and the larynx. In children, the receptors may be dispersed over the inside surface of the cheek. The taste receptor may respond to more than one stimulus. The adequate stimulus consists of five primary taste qualities generally referred to as salt, sweet, sour, bitter, and umami. But, fat receptors may also be found on the surface of the tongue. The distribution of the taste modalities over the surface of the tongue has been a subject for some controversy. The surfaces of the tongue consist of a topographical arrangement with heighted probability of sweet receptors at the anterior or tip of the tongue. Sour receptors may be somewhat more prominent along the sides of the tongue and away from midline.

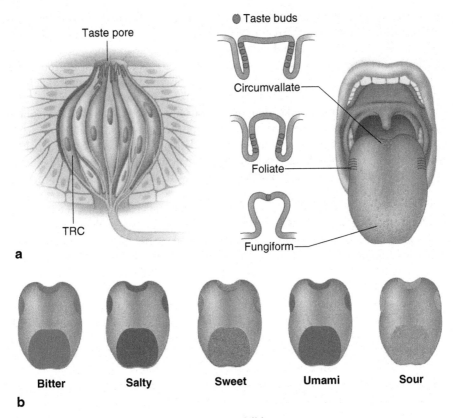

Fig. 8.21 Topographic distribution for the taste modalities

Bitter receptors may be more densely located at the back of the tongue. Salt receptors are prominently distributed across all but the center of the tongue. This leaves the center of the tongue relatively devoid of taste receptors (see Fig. 8.21). But the controversy has exposed a relative allocation for the tip and sides of the tongue, irrespective of flavor modality.

Although the adequate stimulus for taste consists of the chemicals corresponding with the respective taste modalities, taste may be elicited by electrical stimulation of the tongue. Electric taste was discovered by Sulzer in 1752 (see Bujas 1971), and this event was related to the discovery of the battery cell by Volta (see Sanford 1915). Electric taste has been widely used in gustatory testing and recordings from the surface of the tongue. This may be accomplished using the psychophysiological techniques of electrogustometry. With direct current (DC) stimulation at the tip of the tongue, the anode will taste sour whereas the cathode will taste soapy or alkaline (e.g., Grant et al. 1987).

This appears to be partially mediated by the chorda tympani nerve, since dissection of the chorda tympani during middle ear surgery may result in ageusia of the ipsilateral anterior tongue and an associated loss of "electric taste" at least within the intensity range typically used for inducing taste sensations (Pfaffman 1941; see Robins 1967). Alternating current (AC) may elicit bitter flavor at about 1000 Hz, whereas 50 Hz may correspond with sour. It is reasonable to expect that someday virtual reality devices will exist for the richness of various cuisines from around the globe! Taste augmentation research is well underway too, where electrical activation of the gustatory surfaces might be used to facilitate the intensity of taste like carbonation might promote the enjoyment of a soda pop. Facilitation or alteration of taste has been demonstrated with concurrent musical processing and flavors have been shown to correspond with musical stimulation (Crisinel and Spence 2012). Moreover, it is reasonable to expect the enjoyment or dislike of the food bolus to vary with multimodal processing activation and with relative activation of the left or of the right brain.

The basic taste modalities may eventually interact at the cortical level with other factors, including smell, texture, and temperature. Adaptation at the receptor level does occur for the taste modalities. The rates of adaptation differ among different salts and among different sugars. Any acid will adapt the tongue for all other acids. And, adaptation on one side of the tongue appears to affect sensory thresholds for taste on the other side of the tongue. Aftereffects are common with adaptation in other sensory modalities such as vision. These aftereffects provide a physiological basis for illusions of one form or another. One favorite visual illusion is presented below, where focused gaze on the green and yellow flag for 30 s results in adaptation and aftereffects on the color images (see Fig. 8.22). Aftereffects for taste may occur as aftertastes. For example, if a strong salt solution is applied to the surface of the tongue, water will now taste sour (e.g., Bartoshuk 1968). Variations in the concentration of salt on the tongue may yield the aftertastes of sweet, sour, salt, or tastelessness!

Developmental changes are remarkable for gustation with the threshold doubling roughly every 20 years (Cooper et al. 1959; see also Pavlidis et al. 2013). This made sense to one father as his assigned parenting responsibility (by the mother)

Fig. 8.22 Focus for 30 s on the *dot* in the *green* flag. Then shift your gaze to the *center* of the *white* flag

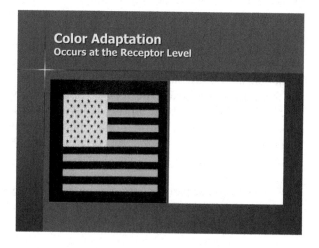

was to feed his boys "baby food." It was something like loading a mortar round. Upon insertion of the spoon with the bitter vegetable into the constantly mobile oral motor apparatus, the boy would spit it out as far as he could make it go! With age-related sensitivity differences, the bitter vegetables must have tasted something like battery acid might taste to an adult. He could not help but notice that Mom was happy to feed the desserts and that she never had this projectile problem! Along with decreased sensitivity over time, development ushers in a shift in preference from sweet to bitter foods (e.g., see Mennella and Ventura 2011). One young man tried to impress his young friends by serving them beer at about age 14. These were "real men" no doubt. But, everyone was glad that there was a lemon meringue pie to chase away the terribly bitter beer after each sip.

Pavlidis et al. (2013) evaluated fungiform papillae at the tip of the human tongue in 156 nonsmokers ranging in age from 10 to 80 years. Thresholds were found to be significantly elevated in older individuals, along with reduced and worsening vascular density at the tip of the tongue. Interestingly, significant differences were found in electrogustometry thresholds between the left and right side of the tongue and between men and women.

The psychophysicist studies the conversion of environmental energies into neural energies within the nervous system via transduction. Light presents in the form of electromagnetic energy and sound in the form of oscillation from the compression and rarefaction of some conducting medium such as air. If the energy within the environment cannot be converted into neural energy through the process of transduction, then it effectively does not exist perceptually. The nervous system is designed to process only the neural energies derived originally from these environmental interactions and, as such, it is arguably naïve as other species may be sensitive to energies and, therefore, knowledgeable of worldly aspects with which humans have no clue. The sensitivity or threshold for converting these energies is dynamic and may be altered by variations in the intensity or even the presence versus the absence of stimulus energies within another sensory modality.

The topographical areas initially responding to the various sensory events may well process and analyze the event independent of the language areas. Thus, individuals may be perceptually responsive to many events that they are unable to discuss or to talk about as the processing is underway outside of the speech or language regions. Prior to the development of neural systems theory, some might have ascribed these as "unconscious" events. But, alas, the term "unconscious" is no more a scientifically derived term than the word "soul." It is mystical and largely outside of scientific study. In contrast, we can demonstrate activation in neural circuits and systems that are outside of, and away from, language. This is better demonstrated with far-field processing at primary and secondary cortical areas and in right cerebral systems disconnected from the left brain's language areas.

For taste, the psychophysical property of sensitivity may be altered using temperature (Bartoshuk et al. 1982; Talavera et al. 2005) ranging from cold (about 17 °C) to warm (about 42 °C). The psychophysical properties of taste sensitivity vary with temperature. A young man was horrified to discover that, in his first hand-crafted love note, he had misspelled the salutation identifying his sweetheart to be as "Dear Sweaty!" He never fully recovered from this embarrassment and neither was the relationship able to survive. Eventually, he was off to college and living in a basement apartment with a partially dirt floor. It did have a small stove and he had made a proposal for a date, where he would cook. He did know how to make chili at that time, so he was set or at least he thought so, until he considered the beverage. He had wanted to serve wine but, alas, he could not afford a good wine and certainly not until he had a 6-V battery for his old 1961 Volkswagen bus. He had been pushing it down the road, jumping in, and "popping the clutch" to get it started. Albuquerque, NM, seemed to be a hospitable location for all of this as the basement was dry (between rains) and there were foothills of the Sandia Mountains to insure that gravity was on his side to start the bus. Honestly, he was living off of 19-cent boxes of macaroni and cheese cooked on a Bunsen burner.

After ruminating a bit over the wine issue, it became clear that he could either purchase a cheap bitter wine or a cheap sweet wine. But, he took heart in his reading of psychophysics where it was clear that, if he could not change the wine, he could change his sweetheart's sensitivity to the bitter or sweet taste. And, he could do this by serving the wine chilled to kill the bitter or at room temperature to kill the sweet. When sensitivity within a biological system is affected by temperature, it is tempting to infer that a metabolic process is represented by the functional change. Temperature has a strong influence on how we taste. For example, the perceived sweetness of diluted sugar solutions increases strongly with temperature (Bartoshuk et al. 1982). In addition, cooling or heating of the tongue by itself is sufficient to cause sensations of taste in about 50 % of humans (Cruz and Green 2000).

Taste signals are conveyed toward the central nervous system through multiple cranial nerves. The posterior one third of the tongue is innervated by the ninth cranial nerve, the glossopharyngeal nerve. The anterior two thirds of the tongue are innervated by the seventh cranial nerve, the facial nerve. The larynx and pharynx are innervated by the tenth cranial nerve, the vagus nerve. Afferent fibers project to the nucleus solitarius, with the second-order fibers traveling via the medial lemniscus in the brain stem to the arcuate nucleus of the thalamus. From here the pathway extends to the cortical projections within the temporal lobe and the insular region.

Chapter 9
Motor Functions: Third Functional Unit Revisited

Muscle Control: Strength, Coordination, and Endurance

The frontocerebellar systems and the dopaminergic contributions of the basal ganglia and substantia nigra are most directly under investigation with clinical and neuroscience inquiries involving strength, balance, coordination, and endurance. A classical distinction has been drawn between these two interactive neural systems providing the basis for normal movement and, when dysfunctional, the basis for many movement disorders. These two distinct, dynamically interactive, neural pathways are referred to as the pyramidal and the extrapyramidal motor systems. They overlap anatomically at the striate bodies, where the descending and myelinated pyramidal tract lays distinctly against the gray field of the basal ganglia (the caudate, putamen, and globus pallidus). The reference to the "striate bodies" originates from visual inspection of this brain region, where white stripes (myelinated pyramidal cell axons) lie against a gray field (the basal ganglia).

Pyramidal Motor System

We might easily take for granted the effortless ability we possess to move or to interact with objects using our body parts and appendages. If thirsty, we might reach for a glass of water and quench thirst. However, disruption of the motor pathways carrying our intentions out to the effector muscles in our arms and hands, for example, might result in this becoming a monumental or even impossible task. This is often the case with a spinal cord injury, with potentially devastating consequences for that person's quality of life accompanied by paralysis of the body regions distal to the cut. Technological advances have been made which may soon allow the brain to bypass some of these injuries using neural-interface systems.

© Springer International Publishing Switzerland 2015
D. W. Harrison, *Brain Asymmetry and Neural Systems*,
DOI 10.1007/978-3-319-13069-9_9

These brain–machine interfaces detect activation of neural systems at the motor cortex, for example, and provide for some degree of discriminative control over the muscles via direct stimulation techniques or over robotic arms, adaptive instrumentation, and assistive devices. In a clinical trial of "BrainGate," a neural interface enabled a patient with paralysis secondary to a spinal cord injury to use a computer cursor (Hochberg et al. 2006; see Jackson 2012). Hochberg and colleagues (2012) more recently reported that two patients suffering long-standing paralysis were able to reach and grasp with a robotic arm controlled using BrainGate. One of these individuals, paralyzed from a stroke 15 years prior, was reported to have been successful in drinking from a bottle using the robotic arm. Brain signals are accessed using thin silicon electrodes surgically inserted proximal to the motor cortex. Neurons in this area were responsive to the patient's efforts to imagine the use or movement of the robotic arm. Translational efforts had been performed earlier to convert these intentions, recorded via the electrodes, into three-dimensional movements of the robotic arm and hand.

There are two basic systems for the present discussion, which are generally referred to as the pyramidal and extrapyramidal motor systems. The first represents the motor projections from the primary motor cortex. This is the same area that Alexandria Luria (1973, 1980) referred to as the primary projection area from the third functional unit (the frontal lobe). This system is most notable for the giant pyramidal cells or Betz cells forming the pyramidal tract via the corticospinal projections from the motor cortex or precentral gyrus (see Fig. 9.1 and 9.2). These cells represent the *final motor pathway* exiting the frontal lobe to indirectly control striate or skeletal muscles located throughout the body. These descending cells, originating from the motor cortex, are also known as the upper motor neurons as distinguished from the lower motor neurons (alpha motor neurons) exiting the spinal cord and directly innervating skeletal muscle. This descending motor pathway is part of the frontocerebellar system (frontodentatorubrothalamic tract) involved in the production of *intentional and coordinated* movement.

The motor cortex consists of six layers with the pyramidal or Betz cell bodies located in layer 5 and with their dendrites extending up toward the surface and spreading primarily across layer 1 of the motor cortex. The overlapping dendritic field in layer 1 will yield activation of a complex muscle group through surface electrical stimulation, whereas deeper stimulation is associated with more specific activation of distinct muscle groups (see Rothwell et al. 1991). This arrangement of the motor cortex and the descending projections of the pyramidal cells are depicted in Fig. 9.3.

The upper motor neurons, descending from the motor cortex, form the pyramidal tract or *final motor pathway*. Fritz and Hitzig (1890) mapped the motor cortex with the facial region located ventrolateral and with the lower extremities located dorsomedial along this strip corresponding to a motor homunculus. John Hughlings Jackson (1958) appreciated the topography here even earlier, as his wife's epileptic seizure spread across this strip creating a *march* of muscle contraction across her body. These topographical changes, reflecting spreading activation, are now commonly referred to as a *Jacksonian march*. This topographical arrangement continues

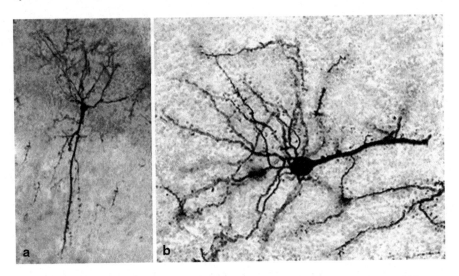

Fig. 9.1 Giant pyramidal cells like the corticospinal projections forming the pyramidal tract. (Originally published by Jia, He, *Dendritic morphology of neurons in medial prefrontal cortex* and hippocampus *in 2VO rats,* Neurological Sciences, Volume 33 Issue 5, p. 1065)

Fig. 9.2 Giant pyramidal cells like the corticospinal projections forming the pyramidal tract

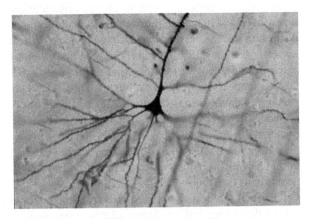

on down through the descending projections and eventually onto the specific alpha motor neurons exiting the spinal cord at the ventral root and out the peripheral somatic nerve to effect muscular contractions at the neuromuscular junction (synapse).

Lesion of the upper motor neuron may result in paresis or even plegia, initially. Muscle strength may be recorded using dynamometers designed for discrete muscle groups. The typical measurement is restricted to use of the hand dynamometer to measure strength and fatigue across repetitive trials (Dodrill 1978; Harrison and Pauly 1990; Crews et al. 1995). Loss of motor precision may be present, but this is due to weakness rather than discoordination or dysmetria. But, the loss of frontal lobe regulatory control with this lesion may result in a loss of inhibition over the lower motor neuron (e.g., Futagi et al. 2012). The upper motor neuronal lesion

Fig. 9.3 Descending projections from the motor cortex and the arrangement of the cell bodies and dendritic fields in layers 5 and 1, respectively

Pyramidal Cells: Motor Cortex

- **MOTOR CORTEX**

  6 Layers

- **LAYER 5**

  Pyramidal Cells of Betz

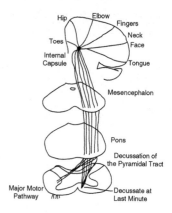

characteristically results in hyperreflexia or dystonia with antigravity synergy being prominent. In contrast, lesion of the lower motor neuron, the alpha motor neuron within the peripheral nerve, may result in denervation of the muscle and, if the alpha motor neuron cell body dies, muscle atrophy.

In advanced pathology, one might see fasciculations where muscle groups and the overlying skin may appear to crawl. History might associate these events with demons under the skin or possessions, perhaps. When the lower motor neuron is lesioned, but the cell body survives, then a slow and gradual regeneration of the axons may occur with reinnervation of the muscle. This process is thought to be largely under the control of astrocytes. Thus, glial cells (including astrocytes) are very much involved in this process and in control of regeneration more broadly defined; the development of scar tissues; and the removal of waste materials, including the process of phagocytosis. Glial cells, known to many as "mother cells," play an ongoing role in the construction and development of the neuronal highways throughout the nervous system and a neural protective role in the management of neurotoxins (see Koob 2009).

The final motor pathway descends from the motor cortex, initially forming one part of the corona radiate. Subsequently, it descends through the external capsule and travels through the basal ganglia, forming part of the striate bodies (white stripes from the myelinated axons travelling through the gray field of the basal ganglia). It emerges, densely configured now, into the internal capsule, continuing its descent through the brain-stem tegmentum, until it reaches the pyramidal decussation at the level of the medulla (see Fig. 9.4). The tract is easily visible at the brain-stem tegmentum as the cerebral peduncle, which occupies much of the tegmental space. Functional asymmetry exists in the decussation, which some have tried to relate to handedness or precision of motor control at the right hemibody.

At the level of the medulla, the descending fiber bundles, collectively known as the cerebral peduncles, cross abruptly at the pyramidal decussation. About 90 % of the fibers cross (Brodal 1981), whereas about 10 % of the fibers remain uncrossed and ipsilateral to the side of origin. The decussation of the pyramids is depicted in

Fig. 9.4 The descent of the pyramidal tract to the level of the pyramidal decussation at the medulla

The Pyramidal Tract Final Motor Pathway

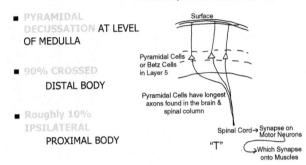

Precentral Gyrus = Motor Cortex; 6 Layers of Cortex

- PYRAMIDAL DECUSSATION AT LEVEL OF MEDULLA

- 90% CROSSED DISTAL BODY

- Roughly 10% IPSILATERAL PROXIMAL BODY

Surface

Pyramidal Cells or Betz Cells in Layer 5

Pyramidal Cells have longest axons found in the brain & spinal column

Spinal Cord → Synapse on Motor Neurons

"T"   ↳Which Synapse onto Muscles

Fig. 9.4. This nonhomotypic crossover differs from a homotypic crossover such as that formed by the corpus callosum. A homotypic crossover provides for input to homologous brain regions in the opposite hemisphere. This is integral to the balance theory discussed elsewhere in this document, where activation of one brain region inhibits the corresponding location in the opposite hemisphere. With the nonhomotypic crossover or decussation, cells are impacting other body regions and, in this case, muscles located on the opposite side of the body. A lesion in this pathway, above the medulla, may result in contralateral weakness and/or hyperreflexia, whereas below the medulla the lesion will have predominantly ipsilateral effects.

Although the majority of these fibers cross at the level of the medulla with the decussation of the pyramids, it remains relevant that roughly 10% do not cross (Brodal 1981; see Harrison 1991). This provides an anatomical basis for some redundancy in the motor system. This is more clearly the case for the proximal body regions, including the shoulder and face. These proximal body regions receive a disproportionate share of the ipsilateral projections. The descending corticospinal projections are known to have greater input to distal cervical motor neurons compared to proximal cervical motor neurons (Palmer and Ashby 1992; Porter and Lemon 1993; McKiernan et al. 1998; Turton and Lemon 1999). Quantitative data have shown the existence of greater strength deficits in the more distal versus the more proximal upper extremity regions in people with chronic hemiparesis (Colebatch and Gandevia 1989). This may play out during the recovery process after a stroke or lesion of the upper motor neurons, where the prognosis for recovery is substantially improved for the proximal or axial body regions (see review by Coupar et al. 2012). Moreover, the recovery of movement in the proximal body regions occurs first in this process and with more distal aspects of the extremities recovering later, if this is to be the case.

Lesion of the upper motor neuron, descending from the motor cortex, is characterized initially by weakness or paresis of the muscle group often followed by

Fig. 9.5 Electromyography recordings over the left and right facial region (masseter muscle) at baseline, following a cold pressor (CP) stressor, and in recovery in low- and high-hostile participants

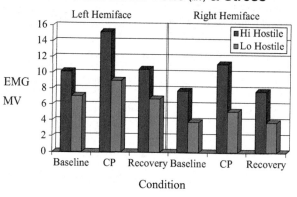

some spasticity or dystonia. These features are useful in the diagnosis of the brain disorder in the form of hard neurological signs corresponding to the lesion location. But, more often than not these present in milder versions within individuals presenting as essentially normal but with relevant motor features that are "soft signs" for mild brain dysfunction. Included here might be the stress-related disorders like that with social anxiety and depression (Everhart et al. 2002; Crews et al. 1995) or with affective disorders like hostility or the posttraumatic disorders (Rhodes et al. 2013; Demaree et al. 2002) or even with childhood learning disability (Huntzinger and Harrison 1992) and childhood depression (Emerson et al. 2005). For example, with diminished right frontal capacity in hostile, violence-prone individuals, increased left-hand flexor strength was found in comparison with other right-handed men. Moreover, facial dystonia was predicted in high hostile, violence-prone individuals based on this theoretical position. When electromyographic recordings were taken over the masseter muscle, facial dystonia was confirmed, which was significantly greater over the left hemiface (Rhodes et al. 2013; Herridge et al. 1997). However, bilateral facial dystonia was significant in these comparisons (see Fig. 9.5).

Extrapyramidal Motor System

The second system appreciated here is the extrapyramidal motor system, which consists of the basal ganglia, the substantia nigra, and the gamma motor neurons. The gamma motor neurons eventually exit the spinal cord in the peripheral somatic nerve. The extrapyramidal system is largely dopaminergic and pathology here is commonly associated with Parkinson's syndrome in clinical neuroscience research. There exists cerebral asymmetry for dopamine with lateralization to the left hemisphere (e.g., Glick et al. 1982). Onset of Parkinson's syndrome with resting tremor at the right hemibody is more strongly associated with the apathetic, amotivational features with behavioral slowing and diminished positive emotion or low self-esteem. David Gilley (Gilley et al. 1995) found depression to be more severe in Parkinson's dementia in comparison with cortical and subcortical vascular dementias. Depression is more strongly linked to dysfunction or deactivation at the head of the caudate within the left frontal lobe.

Other neurocognitive and emotional relationships have been described for activity within these systems, including relatively diminished size of the left medial frontal lobe at the cingulate gyrus (decreased capacity inferred), where pervasive worry, fearfulness in the face of uncertainty, shyness with strangers, and pervasive fatigue have been reported (Pujol et al. 2002). Also, Somerville et al. (2013) found these brain systems to be differentially involved in preoccupation with social evaluation. Child, adolescent, and adult participants viewed cues indicating that a recording camera was either off, warming up, or projecting their image to a peer during the acquisition of behavioral, autonomic, and neural-response (functional magnetic resonance imaging; fMRI) data. The belief that a peer was watching was sufficient to induce "self-conscious emotion" that was heightened in the adolescent group. fMRI findings supported the engagement of the medial prefrontal cortex (MPFC) and the striatum within the left hemisphere. The researchers conclude that these brain systems are critical to socioaffective processes in which social-evaluation contexts influence arousal. Regardless, the task at this point is to focus on the motor components of these extrapyramidal and largely dopaminergic structures.

My first experiment as an undergraduate under the tutelage of Gordon Hodge, involved the placement of a unilateral radiofrequency lesion in the pars compacta of the substantia nigra in rats (also see Hodge and Butcher 1980). This lesion resulted in unilateral hyperkinesis with rotational activity counts in the rats. Subsequently, I administered one or another of the dopaminergic agonists D-amphetamine (dextroamphetamine), Cylert (pemoline), or *Ritalin* (methylphenidate), resulting in an amelioration of the rotational behavior and rats that ran more normally rather than in a rotational fashion. Lesion within the extrapyramidal motor system, though, is more commonly associated with resting tremor and perhaps tardive dyskinesia in a more advanced state (e.g., Casey 1991). Choreiform movements, torticollis, tics, and oddly formed sequential movements are common with problems in this system. These appear, even to the patient, to be "intentional." And, they are, as intention or desire appears to be the overriding function of these frontal lobe, executive brain systems (see Damasio et al. 2012).

Just as the upper motor neurons of the pyramidal motor system eventually synapse onto the lower motor neurons or alpha motor neurons, the extrapyramidal motor system exerts its regulatory control through the lower gamma motor neuron. The gamma motor neuron regulates the tone or contraction of the intrafusal muscle fibers (see Willis and Coggeshall 2004), which run in parallel with the extrafusal muscle fibers. The extrafusal muscle fibers are innervated by the alpha motor neuron. It may help the reader now to know a bit more about these muscle types in order to appreciate the differential influences of these interdependent systems.

Three basic categories or muscle types include striate or skeletal muscle, cardiac muscle, and syncytial or smooth muscle. Cardiac and smooth muscle fibers do not require neural innervations to contract, the only thing that a muscle can do. Striate

Fig. 9.6 The nuclear bag-type intrafusal muscle fiber attached to bone at one end and to the extrafusal muscle fiber at the other end

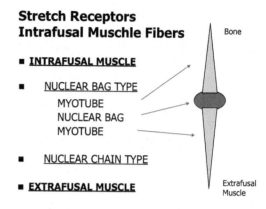

Stretch Receptors
Intrafusal Muschle Fibers

- **INTRAFUSAL MUSCLE**

  - NUCLEAR BAG TYPE
    MYOTUBE
    NUCLEAR BAG
    MYOTUBE

  - NUCLEAR CHAIN TYPE

- **EXTRAFUSAL MUSCLE**

Bone

Extrafusal
Muscle

muscles are arranged in an oppositional system in order to move bones. These oppositional muscles are antagonists, whereas muscles that work synergistically to move the skeleton in one direction are agonists. Extrafusal striate muscle fibers are connected at each end to bone. This arrangement allows for movement of the bone in one direction with contraction, whereas the antagonistic muscles move the bone in the opposite direction with contraction. This allows for flexion and extension of that body part or limb. In a normal system, there is a proportional balance between these muscle groups. Lesion of upper motor neurons alter this balance resulting in antigravity posturing with flexor synergy in the upper extremities and extensor synergy in the lower extremities.

Intrafusal muscle fibers function as stretch detectors for skeletal muscles. The anatomy of these muscle fibers differs from the extrafusal fibers, in that they are arranged in parallel with the skeletal muscle fibers via attachment at one end to bone and at the other end to the extrafusal muscle fibers. This arrangement is ideal for the detection of stretch in the skeletal muscles and for adjusting muscle tone through the extrapyramidal motor system. The extrapyramidal motor system allows for the adjustment or fine-tuning of the stretch receptors via contraction of the intrafusal fibers. The tighter the intrafusal muscle fiber, the more sensitive it is to stretch within the skeletal muscle. In a normal person, there is an alternation among the extensor and flexor muscles. The gamma motor neurons and the intrafusal muscle fibers provide for regulation of this alternating tone. If the system is sloppy or poorly regulated as may occur with damage to the extrapyramidal motor system, then tremor or the oscillating contractions of the flexor and extensor muscles may become more apparent. This, by definition, is a *resting tremor,* where the limb moves further in one direction or the other before the stretch receptors activate and trigger the oppositional muscle group. We might visualize this system in the lower extremities with excessive alcohol consumption, where the inebriated person leans too far forward prior to detecting and responding to the stretch with contraction of the oppositional muscle group, for example.

Figure 9.6 shows the basic anatomy of a nuclear bag-type intrafusal muscle fiber. Slight stretch appears to be detected first at the myotube ending with activation of the flower spray receptors. Stronger stretch will be detected by activation or irritation of the 1A annulospiral endings wrapped around the nuclear bag. But in conditions of extreme stretch, the Golgi tendon organ (GTO) located in the muscle tendon may activate to prevent the muscle from being torn from the bone. The response to activation of the flower-spray endings in the myotube and to activation of the 1A annulospiral endings in the nuclear bag is to trigger reflex contraction of that skeletal muscle to oppose the stretch. This occurs with reflex relaxation or inhibition of the antagonistic muscle group (see Willis and Coggeshall 2004). In contrast to this system, activation of the GTO normally signals an end to the attempt to increase resistance to mild or moderate stretch. Instead, with extreme stretch and with the muscle now at risk of tearing from the bone, the GTO results in reflex relaxation of the muscle giving way to the stressor and saving the muscle and tendon from damage. This is accomplished through synaptic contact onto Renshaw or internuncial neurons providing for reflex inhibition with gamma-aminobutyric acid (GABA).

These reflexes are largely under the involuntary control of the extrapyramidal motor system. With the exception of situations where this system has been damaged (e.g., Parkinson's disease) or perhaps deactivated through drugs or alcohol consumption, the extrapyramidal motor system provides for stable muscular responses to fine stretch. With lesion or deactivation, the extent of stretch required for activation is increased resulting in sloppy responses or tremor at rest. But, the reflexes protecting the muscle and tendon from strong injurious stretching and tearing may be overridden by intentional activation of efferent fiber pathways from the motor cortex. One all-American college athlete overrode these protective reflexes with his enormous bicep muscle now sitting unattached at one end and curled up on top of his shoulder. Fortunately, the team had access to an excellent sports injury physician, surgeon, and trainer. He was well enough to return to the field shortly with the insertion of a nylon screw to anchor the tendon back into the bone. Later, he would succeed in his dream to play professional football and to play it well.

The extrapyramidal motor system provides for variable sensitivity to muscle stretch, allowing for reflexive control over much of our posture. But, this variable sensitivity to muscle stretch is broad based as a component of all of our skeletal muscles. This means that variation in sensitivity may underlie a resting tremor in the hand, the face, or even the vocal motor apparatus. One less reflective graduate student failed to appreciate her neuroscience background and training after speaking with one of her faculty members. Her attributions toward this gentleman were that he was intimidated by her and that he was "anxious" in her presence. Instead, this gentleman had gracefully managed an essential resting tremor for several years. Regardless, her attributions provide a reasonable platform for a research investigation on components of anxiety or anxiety-related behaviors involving the extrapyramidal motor system, including the substantia nigra and the basal ganglia (caudate, putamen, and globus pallidus). The expression of facial affect and affective intonation in speech, as with a subcortical expressive dysprosodia or dysphasia, would be altered and unreliable without smooth transitions from one motor praxicon or postural position to another. Moreover, since these systems moderate the expression of the *intentional* motor system, the intent would appear to be more stilted or less dynamically responsive (e.g., masked faces in Parkinson's disease) and to lack flexibility from one affective transition to another.

A faulty basal ganglia and especially dysfunction within the head of the caudate nucleus, is reminiscent of a faulty transmission in an old car. Instead of making a smooth transition from one gear to the next, this transmission jerks with erratic movements that rendered the occupants of the car subject to the humor of any on-lookers! This is part of the neuropsychological evaluation of the extrapyramidal and the pyramidal motor systems using *passive range of motion* techniques. By holding the patient's bicep muscle or hamstring for the lower extremity, and carefully ranging the limb to and fro while the patient attempts to relax the limb, the examiner may be able to detect antigravity synergy from an upper motor neuron lesion with antigravity dystonia. With extrapyramidal motor dysfunction, though, the transition in smooth ranging of the extremity may be interrupted with notching or intermittent jerks.

In contrast, dysfunction of the frontocerebellar system may result in *tremor of intent* or dysmetria in the volitional movement of one or another extremity. This is more common with the cerebellar lesion, which may be accompanied by truncal ataxia and alterations in the processing of one or more cranial nerves due to the proximity of these brain-stem structures. The patient with cerebellar or frontocerebellar dysfunction may present as clumsy or discoordinated and this may become the primary rehabilitation goal as the discoordination may be a safety issue and increase the risk for falls. Cerebellar lesions at either the left or the right brain stem often mimic dysfunction within the ipsilateral frontal lobe (e.g., Heilman et al. 2012). This more commonly presents with organizational sequencing deficits and perseverative errors typically resulting from deactivation or lesion to the ipsilateral frontal lobe.

Jackson (1874, 1958) appreciated this relationship early on in his discussion of the organization and reorganization of function across each level of the nervous system. Schmahmann (2004) also appreciated this relationship in discussion of disorders of the cerebellum including ataxia, dysmetria of thought, and the "cerebellar cognitive affective syndrome." But, these systems are also involved in more basic functions, including the regulation of respiration and such. A patient with damage to the right frontocerebellar system may evidence dyspnea on exertion secondary to poorly organized respiratory reflexes and poorly regulated sympathetic nervous system activation. This patient may develop dyspnea secondary to a right frontal stressor as might be implemented through a negative affective challenge or even constructional or visuospatial challenges. Under more significant stress, the affective display may be unbridled or more substantially deregulated producing a vital state as with a panic or rage episode.

One example may be provided by a young man under his parents' care following two right frontal cerebrovascular accidents and a right frontal mass accompanied by dyspnea on exertion and/or on negative emotional stress. The physician had prescribed a continuous positive airway pressure (CPAP) apparatus, which seemed to aggravate the man with control issues and an inability to sleep with the machine over his face. The young man experienced panic symptoms, where this device might be helpful to many but where it appeared to be a bit beyond his ability to benefit from the intervention. The parents advocated for their son's right to a choice in the matter, whereas the physician referenced concerns for abuse and neglect with potential legal ramifications. These differing perspectives led ultimately to a conflict of action between the physician and the family, whereas they shared the same goals at a broader level.

Part IV
Clinical Syndromes

Chapter 10
Thalamic and Hypothalamic Syndromes

Thalamic Syndromes

Each sensory modality, with the exception of olfaction, travels from its receptors via tracts to the thalamic nuclei within each brain (Powell et al. 1965; Price 1973). From here, information is relayed on to the primary cortical projection area for each modality via the thalamic radiations. It follows that a thalamic stroke may present with vague, ill-defined sensory complaints that are initially somewhat hard to nail down and which may fluctuate or wax and wane over time. But the hallucinations derived from altered activation of the thalamic nuclei may be multimodal, corresponding with the very close proximity across systems with vision at the lateral geniculate (LG) nucleus, audition at the medial geniculate (MG) nucleus, somesthesis at the ventral posterior lateral (VPL) nucleus, and with motor fibers from the motor cortex projecting to the ventral lateral (VL) nucleus of the thalamus (see Fig. 10.1).

Olfactory information reaches the olfactory cortex without a thalamic relay. However, the thalamus may still play a significant role in olfaction. Sela and colleagues (Sela et al. 2009), for example, tested olfactory function in patients with unilateral focal thalamic lesions and in age-matched healthy controls. Thalamic lesions did not significantly influence olfactory detection but did significantly impair olfactory identification evident in perceptual measures, as well as in their sniffing patterns. Healthy participants modulated their sniffs in accordance with the content of the odor, whereas thalamic patients did not. Also, right thalamic lesions altered olfactory hedonics by reducing the intensity of pleasant odors. Of interest here was that this shift in pleasantness was apparent in the auditory control. The authors conclude that the thalamus plays a significant role in human olfaction even though it is outside of the projection pathways from the olfactory bulb to the cortex.

One patient with thalamic lesions (see Fig. 10.2) presented with a waxing and waning of symptoms with extreme pain (thalamic pain syndrome) and disturbance across sensory modalities at the left hemibody and left hemispace. The waxing and waning of her symptoms occurred daily, and often on an hourly basis. Following

© Springer International Publishing Switzerland 2015
D. W. Harrison, *Brain Asymmetry and Neural Systems,*
DOI 10.1007/978-3-319-13069-9_10

Fig. 10.1 Lateral view of the right thalamus with labeling of the thalamic nuclei. A transverse section allows for viewing of medial thalamic structures in this image

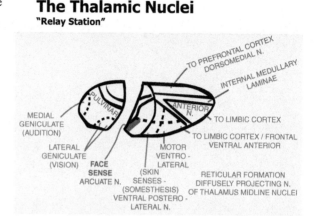

Fig. 10.2 Images depicting thalamic infarct in a patient with bilateral and multimodal formesthesias. (Reprinted by permission from *Cognitive Neuropsychiatry*, 12 (5), 422–436 (2007). http://www.tandfonline.com)

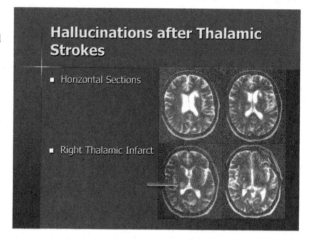

her thalamic stroke, she began experiencing cross-modal hallucinations, which differed in the emotional content at the left and right hemibody and hemispace. She provided vivid and consistent reports of the right-sided hallucinations as consisting of "college age boys in colorful Hawaiian shirts" who "are too happy and talk too much." She reported that these interesting and enjoyable boys are very "energetic." She called them "the he-haw boys," and reported that she could hear them talking and that she enjoyed listening to their conversations. She remarked on their funny hairstyles, describing them as "sort of kinky."

The left-sided hallucinations were described as "men in black religious clothing that make no noises." She called them "the eye drillers" and stated that "they look a hole right through you!" She provided vivid drawings of the hallucinations that accentuated the positive and negative associations she had with each hallucination (Fig. 10.3). Within the somatosensory modality, she reported pain and numbness at the left hemibody, especially the left hand and foot. She reported a cold dysesthesia

Fig. 10.3 The patient's own drawings of her formesthesias within the left and right hemispace. (Reprinted by permission from *Cognitive Neuropsychiatry*, 12(5), 422–436 (2007). http://www.tandfonline.com)

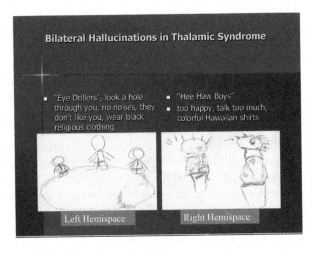

at the left hemibody, and this was accompanied by altered circulation at that extremity with her left-sided body temperature colder in comparison with that at the right (Mollet et al. 2007).

It is important to note that the term hallucination may no longer be a valid label as it was derived from an era substantially devoid of the scientific understanding of functional nervous systems. These events are commonly viewed by the layperson as reflecting "craziness" or "feeble mindedness" or "mental problems" or, even worse, a "nervous breakdown." With improved understanding of brain function, we can now appreciate the localization value of these events and the optimism that they provide us in demonstrating that the cells critical for processing forms within one or another sensory modality in the brain are alive. Indeed, they are likely overly active or disinhibited with recovery yielding improved dampening or inhibition of the activation and more regular oscillation patterns within these neural and glial networks.

It may be useful for the reader to appreciate that the primary technique used within neuroscience and neuropsychology to decompensate these brain regions and to elicit/induce the formesthesia (hallucination) is to activate that sensory system through oscillation. Rhythmic stimulation is present in many forms within the environment. In a similar fashion, the brain consists of neural networks dominated by rhythmic electrical activity (Adrian and Matthews 1934) recorded using an electroencephalogram. These periodicities represent oscillations in neuronal excitability (e.g., Lindsley 1952), and neural entrainment to extraneous sensory conditions has been established in normal brains (e.g., Mathewson et al. 2012). In the case of neuropathology, the neural networks may acquire altered sensitivity to oscillating events, and these may convey in the form of overactivation with hallucinations. For example, for a somatosensory formesthesia originating at the left hemibody (e.g., "snakes are biting my leg") oscillation of that body part might increase the probability of the occurrence of the perceptual event. For an auditory paracusia that consists of hearing voices within the right hemispace, rapid oscillation of auditory speech sounds stress the tissue proximal to Wernicke's area in the left brain. For the evocation of visual formesthesia, a flashing light or strobe at the contralateral visual field may be sufficient. Vestibular disturbances may be overtly activated by visuomotor and/or postural oscillations, for example.

One thalamic patient would decompensate with affective episodes where sad thoughts were followed by the sensation of the sounds of snakes at the left side and the sensation of snakes on his left hemibody. He would then see the snakes at the left and feel them biting with substantial fear or panic behavior and sympathetic nervous system tone. With environmental or contextual therapies concurrent with nutritional management and medication, he eventually made it a full year without an episode. At that point, he went for a follow-up medical evaluation out of town and stayed in a motel with a swimming pool. His wife described his decompensation with exposure to the oscillating shimmer of the water in the pool and her horror as he sank to the bottom of the pool. Fortunately, she was an athlete and a strong swimmer and was able to pull him back up to the surface where she restored his respiratory function and calmed his heart.

Another man with left temporal lobe seizure activity would decompensate to the therapist's oscillating logical linguistic or propositional speech during conversation. Entering the rehabilitation center's therapy rooms, from his quiet hospital room in the morning, would precipitate a receptive dysphasia with word salad and meaningless content or jargon concurrent with the onset of hearing voices within the right hemispace (auditory paracusia). Treatment was initiated to use minimal speech when possible, longer interspeech intervals, a quiet environment, and non-verbal methods of communication. The environmental context was relevant in his room with contraindications for auditory stressors emanating from the television and the hallway. Interestingly, individuals diagnosed with schizophrenia originating from left cerebral pathology may be recurrent visitors to social settings, which may decompensate them, as visible to others, through their conversations with individuals who are not present and perhaps with gelastic content or responses. Those experiencing hallucinations originating within right posterior cerebral systems may avoid people, if possible, and be somewhat more guarded or paranoid in their social responses, perhaps.

Thalamic Pain Syndrome

Individuals with thalamic lesions or pathology proximal to the thalamus and deeper limbic structures may present with vague and multimodal sensory complaints along with intolerable pain. Behavioral and clinical data indicate that pain is differentially perceived across the two sides of the body, and several neuroimaging studies have provided evidence for lateralization of pain to the right hemisphere (Symonds et al. 2006). Evidence for the lateralization of pain exists despite neuroimaging evidence for a widespread "pain matrix" distributed across the bilateral cerebral hemispheres. Thalamic pain syndrome appears more common with right thalamic involvement, possibly related to the general specialization of the right brain for pain and for negative emotion (see Mollet and Harrison 2006). The individual with this syndrome may experience exquisite or intractable pain precipitated by range of the left-sided extremity or from light brushing stimuli at the left hemibody. Thalamic features are generally confusing for the therapist and for the caregiver as they frequently will vary over time and even within session with a waxing and waning feature. The fluctuating course relates to brain-stem disorders more broadly defined where arousal systems may be dysfunctional with lowered-arousal and heightened-arousal fluctuations yielding less-than-optimal performance. The reticular formation projections through the intralaminar nuclei of the thalamus are part of this arousal system.

Research on thalamic pain syndromes has recently been extended to migraine with evidence for retinal projections to the pulvinar and centromedian nuclei of the thalamus (Noseda et al. 2010). Brain-mapping evidence was found for direct optic nerve pulvinar connections using diffusion magnetic resonance (MR) tractography (Maleki et al. 2012). The presence of this pathway has implications for photophobia, a somewhat common multimodal phenomena linking pain with ambient lighting level. The pulvinar receives trigeminal pain-sensitive neurons innervating vascular and dural structures, providing a link between sensory modalities critical for whole-body allodynia (Burstein et al. 2010). This term is relevant here in the cross-modal joining of pain to a stimulus which does not normally provoke or exacerbate pain (see Merskey and Bogduk 1994) and, in this example, potentially excruciating pain subsequent to incremental activation of the visual system with bright light. This visual pathway is not one that is directly involved in perceptual acuity, in contrast with the perceptual projections from the retina to the lateral geniculate nucleus of the thalamus and on to the occipital cortex.

For the therapist or caregiver, the symptoms of thalamic pain may occur against a background of vague multimodal complaints that vary over time. It may help to recall that the topographic arrangements of the sensory and motor projections are maintained as they traverse through the thalamic nuclei. This arrangement, with diverse sensory and motor systems in very close proximity, may result in confusing symptoms where the pain or other sensory complaint seems to move or to appear at different parts of the body over time. Moreover, the sensory system involved in the complaint may fluctuate from session to session or even within session. This inconsistency or uncertainty of location, along with the fluctuating intensity and modality of the complaint, may lead to erroneous conclusions that the disorder is feigned or "psychological."

Negative affective formesthesia and dysesthesia appear to be common with thalamic involvement within the right hemisphere and brain-stem systems. Qualitative complaints and extreme negative affect may present with thalamic pain, and these may occur without the presence of a peripheral pain origin. Treatment may defy the traditional approaches to pain management. The therapist or caregiver might appreciate the shared functional features of these right brain systems and work to manage them more broadly to promote improved adjustment in this person. This might include efforts to minimize exposure to those events which activate the right thalamic region and posterior cerebral systems. The patient may benefit from efforts to minimize exposure to provocative negative affect events. Efforts to minimize perceived external control within the setting may be helpful, along with the provision of options and choices in activities and in their care management.

The brain is sensitive to oscillating events, which are influential within the arousal systems and at higher cortical levels. Thus, the therapist and caregiver might implement cautions for oscillating events within each sensory modality and especially at the left hemibody/hemispace. This may include ranging the left hemibody or exertional therapies at the left side. The distinction among therapeutic interventions and effort to maximize the patient's functions are important here as therapy would involve developing improved tolerance to these activities. One woman with thalamic pain syndrome after a large right middle cerebral artery distribution cerebrovascular accident had episodic left-sided visceral pain and an expressed urgency for a bowel movement. The pain was also perceived to be emanating from the left upper extremity concurrent with her left hemiplegia. Therapists implemented a transcutaneous electrical nerve stimulation (TENS) unit for electrical activation of these muscles. However, the stimulation produced oscillating left upper extremity movements along with a profound upper gastrointestinal pain, which was perceived as an impending bowel movement. Subsequent to replicating this phenomenon, the therapists came to appreciate that the oscillating stimulation at the left hemibody required caution for seizure activation and potentially within the insular region receiving gastrointestinal projections. Perhaps relevant to this clinical case is evidence of right amygdala activation seen in an functional magnetic resonance imaging (fMRI) study in response to painful visceral (gastric) stimulation (Lu et al. 2004).

Thalamic pain more commonly presents at the left hemibody and sometimes with a qualitative description by the patient as sickening or intolerable. The emotional overlay may be dramatic and typically is consistent with the affective valences processed by the right brain, with anger, fear, and sadness depending on the functional systems affected (see Heilman and Valenstein 2012). The clinical features of the thalamic pain patient appear primitive. The tendency for the pain to travel or float across body regions combined with vague multimodal sensory complaints make this disorder difficult to diagnose and to treat. Common attributions toward these individuals, even by otherwise well-informed health-care professionals, include impressions that the disorder is "psychological" or hysterical in nature as with a conversion reaction or somatoform disorder. Appreciation of the anatomical proximity of pain to the other somatic senses may help as we appreciate that a light brushing stimulus on the body or efforts to range an extremity or the application of even mild temperature variance may yield recruitment in this system and aggravate or even decompensate the patient with a thalamic pain syndrome. This may be more pronounced if the dysfunction extends to the right amygdala, specialized for kindling and recruitment or the perpetual activation of the negative stimulus trace (e.g., Goddard and Douglas 1975).

It may be important to understand that the right brain is neither logical nor verbal in its fundamental mechanisms of perception and expression. To ask the question verbally or linguistically is to talk to the left brain, which may have no real clue to the disturbance (e.g., with lesion to the corpus callosum). Thus, the clues originating from the patient afflicted with these events are more characteristically nonverbal and may be expressed in directional gaze and through facial expressions as might denote fear or apprehension. Behavioral features, including agitation or psychiatric interventions with the administration of Haldol or Ativan, may be the initial clue to initiate a more thorough evaluation and to develop an improved approach with the individual suffering from these disorders.

Thalamic Sleep Syndrome: Stage 2 Sleep Disorder

Thalamic sleep disorder appears to be more commonly associated with the patient presenting, initially after stroke or injury, as hypoaroused and seemingly unable to awaken fully within the therapy or rehabilitation setting. The therapist and caregivers may describe the patient as "always sleeping," whereas the patient may fail to appreciate their sleep in the sense that it has been inadequate in providing the fulfillment of their needs or demands for rest and restoration. Thalamic sleep disorder may result from any injury or metabolic alteration, including changes within the dorsomedial thalamus. This might result from a stroke within the thalamoperforating artery, for example. The role of the dorsomedial thalamic region includes the promotion or generation of high-voltage sleep spindles (see Peter-Derex et al. 2012) within the electroencephalographic record. This defines stage 2 sleep, which provides for a transition from cortical desynchrony in high states of arousal to one eventually consisting of cortical synchrony with the occurrence and prominence of delta waves within the sleep record.

The individual with an acute/subacute thalamic syndrome may be unable to transition from light stage 1 sleep to the deeper and restorative episodes of sleep stages 3 and 4. An individual awakened from stage 1 sleep will typically deny that they had been asleep as this stage varies only somewhat from a state of drowsy wakefulness (stage W; see NINDS 2007; Niedermeyer 2011). The inference to be drawn here may be that the patient does not appreciate their sleep and that they are unable to process deep and restorative sleep stages necessary for a feeling of rest and well-being. One consideration here for the physician involved with this patient's care might be to avoid the demands that the patient be more awake or aroused for participation in therapies, which might involve consideration of central nervous system stimulants. Instead, this thalamic region is under the inhibitory control of the reticular activating system, which is regulated by the raphe. Promotion of the indolamines as with facilitation of the serotonergic systems through the raphe may promote inhibition of the reticular formation allowing the thalamic systems to drive cortical synchrony for deep sleep transition. This may seem a bit contraindicated with a chronically sleepy patient, whereas there are multiple approaches available in this oscillating sleep system. For the caregiver, the system might be promoted through tryptophan within the bloodstream from consumption of turkey, if this is appropriate from your discussions with your health-care professional. Regardless, additional research is needed to provide clarity to this area.

Hypothalamic Syndromes

The hypothalamus is a region of the brain involved in the mediation of an immense number of bodily functions. It is located at the base of the brain below the thalamic nuclei and proximal to the third ventricle. Lemaire et al. (2011) used in vivo diffusion tensor imaging tractography to explore the macroscopic white matter connectivity of the human hypothalamus. The researchers mapped the connectivity of six hypothalamic compartments in human participants according to hemisphere and region and concluded that the analyses, by subject, revealed tangible evidence for lateralization of the connections and a rightward hemispheric frontal connectivity of the preoptic, anteroventral, and lateral compartments. The authors report extensive white matter connectivity of the preoptic, anteroventral, lateral, and posterior compartments of the hypothalamus with the cortex and, in particular, with the right frontal lobe. Notably, the anteroventral compartment connects particularly with the prefrontal cortex, implicating this region for regulatory control over hypothalamic responses.

The pituitary gland or hypophysis extends out and down from the hypothalamus into the sella turcica (Turkish saddle)—a basal skull depression. The pituitary is richly vascularized providing for a most unique interface between the brain and the blood supply, where the blood–brain barrier is minimal allowing for communication to and from distal organs. In essence, the hypothalamus is a visceral efferent structure for distal influences across the body. Visceral regulatory control over this structure is largely derived from the medial prefrontal cortex (Öngür and Price 2000; Lemaire et al. 2011). It receives afferent information essential for temperature regulation and control of blood concentrations affected by diet with food and water intake. This system provides for the primary regulatory control of the hormone system regulated by descending fibers from the medial orbitofrontal region. Figure 10.4 provides a look at the location of the hypothalamus at the base of the brain.

The complexity of this system and the many hypothalamic syndromes are well beyond the scope of these writings (e.g., see Lemaire et al. 2011). But a few select

Fig. 10.4 An image of the diencephalon showing the location of the hypothalamus on the midline

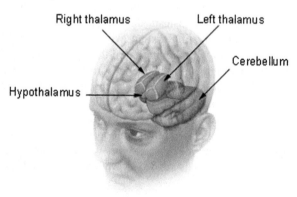

Diencephalon

Right thalamus   Left thalamus

Cerebellum

Hypothalamus

features are worth discussing here and foremost among these are the *ventromedial hypothalamic (VMH) syndrome* and the *lateral hypothalamic syndrome*. One of the primary areas of interest for eating or food consumption is the VMH region. Deactivation or lesion herein may result in the classic VMH syndrome featuring alteration of food intake and a loss of dietary regulatory control. VMH lesion may result in hyperphagia (Anand and Brobeck 1951; Anand et al. 1955) and the subsequent development of obesity. Students' reactions to rats that had undergone VMH lesions were remarkable. The level of obesity was most significant as the housing cage was now inadequate for these animals! From the localizationist perspective, the VMH was identified early on as the "satiety center" as the lesion seemed to render the animal insatiable. But, in humans with an insatiable eating syndrome where a VMH lesion was suspected, behavioral alterations may result which might, otherwise, be attributed to right frontal lobe dysfunction, including perseverative eating behavior. Frustrated nursing staff may attempt to implement behavioral therapy interventions more commonly using "response prevention techniques." These techniques have not been effective with the patients that I have served thus far. Figure 10.5 provides a look at the pituitary gland extending directly below the hypothalamus and its interface with the rich vascular supply surrounding the pituitary stalk.

In some cases, the staff member is adjacent to the patient at mealtime physically restricting the upper extremities and, if necessary, the head long enough for the patient to swallow their food, followed again by release and eating initiation. The eating behavior may convey left frontal release features with rapid, ballistic, eating behavior without termination (e.g., Ruch and Shenkin 1943). These are characteristics of inertia disorders resulting from right frontal lobe damage (e.g., Drewe 1975; see also Stuss and Benson 1986; see also Damasio et al. 2012). The need for additional research on the laterality of this midline structure or of its lateralized regulatory control by descending medial prefrontal cortical efferent projections appears obvious (Öngür and Price 2000). It is also of interest that these patients may be more responsive to external stimuli or events originating within extrapersonal space, which signal to initiate feeding behavior. This includes an apparent increase in response to food odors, the site of food, the sound of food being prepared or consumed, and

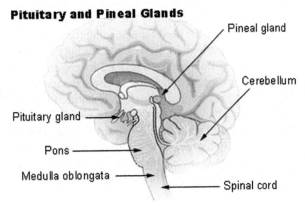

Fig. 10.5 The pituitary gland provides a rich vascular interface with distant organs. (*Wikipedia, the Free Encyclopedia. This work is in the public domain in the USA because it is a work of the US Federal Government under the terms of* Title 17, Chap. 1, Sect. 105 of the US Code)

even the time expressed by the clock on the wall. Classic research on obesity in social psychology revealed a heightened response to external cues (Schacter 1968; Wansink et al. 2007), which clearly warrants more research on these systems and the potential for lateralized frontal regulatory control for eating.

The medial prefrontal region effects regulatory control over visceral efferents, including the hypothalamic regions (Öngür and Price 2000). It is reasonable to predict that the regions within the right hemisphere activate to threatening stimuli and to those events requiring aggressive or predatory behaviors. The need for food and the activation of sympathetic responses to promote successful food gathering may be relevant here. In contrast, the left brain regions appear specialized for pleasant socialization, the consumption and digestion of food (e.g., Holland et al. 2011d), and quiescent states. From these existing lines of evidence, one might predict the right frontal mediation of food consumption, where diminished capacity would lead to ballistic, poorly regulated consumption behaviors, and perseverative eating well beyond the point of satiety in that setting. But, we have not yet conducted this research in my laboratory. David Cox and Kate Holland have provided initial confirmation of the digestive and quiescent processes after food consumption with evidence of left frontal lobe stress and activation using quantitative electroencephalography (Cox et al. 2008; Holland et al. 2011b), neuropsychological test performance (verbal fluency), and cardiovascular measures (Holland et al. 2011a, d, 2012).

The localizationist may appreciate the role of the lateral hypothalamic region in the initiation of feeding and drinking with the early attribution that this is the "feeding center." Lesion of the lateral hypothalamic nucleus may result in the cessation of food intake. The animal with a focal lesion here may present with *adipsia* (no drinking) and *aphagia* (no eating; Anand and Brobeck 1951; Anand et al. 1955), followed by a gradual recovery of water intake or the animal will not survive. The patient with dysfunction in this system may be identified initially with elevated sodium levels or hypernatremia. For the therapist or caregiver, appreciation of the depletion of water intake, and water intake behaviors, may very well be vital for this individual. In many health-care settings, the water or beverage is made available for consumption, whereas this may well be inadequate care for the maintenance and survival of the adipsic and/or anorexic patient.

One severely adipsic patient was closely followed up for over a year within a highly supportive head injury facility. Not only did the director and staff of this facility appreciate the need for managing the associated frontal lobe disorder in this gentleman with a high level of care needs but they also appreciated the requirement for monitoring fluid intake and fluid management. This severely adipsic patient was fitted with a water pack and solenoid device to deliver a small water bolus, on schedule, to his lips. The bolus was quantified by the length of the shaft from the solenoid, thus preventing aspiration or choking. Hypovolemia secondary to dehydration may be a right frontal stressor to elevate sympathetic drive for survival and to increase activity level along with consumption behaviors. Foundational support for this position is established with dehydration elevating angiotensin II, an index of sympathetic tone (e.g., Nazarali et al. 1987).

One of the more exciting areas for research on hypothalamic syndromes is the investigation of oxytocin, a neuropeptide produced by the hypothalamus. Oxytocin has been implicated in love, empathy, and trust. Psychologists were implementing developmental research techniques in the 1970s, where oxytocin was used to promote milk letdown for rat pups. The lactating dam might have milk letdown or not within the operant contingencies under investigation. With the additional decades of research behind us, researchers are now investigating oxytocin for its role in broader constructs that were not easily addressed in the past. Some cursory evidence exists in using commercial products available to pacify the family dog. Many have used oxytocin successfully, perhaps, as the family *dawg* was somewhat aggressive around the house, for example, with many adolescent males around. But, with oxytocin administered through a wall outlet, the dog may be more pleasant to be around and appear to be less defensive. Rather than defending the nest, he may seem to be absorbed in it and comfortable that all is well at his house. Of course, these are anthropomorphic musings at best and based on anecdotal evidence. Indeed, some evidence exists for oxytocin increasing defensive or aggressive behavior (Hahn-Holbrook et al. 2011).

Oxytocin is more commonly known for its role in childbirth, as its release allows for uterine contractions facilitating delivery. I remember the birth of my sons. They were clearly ugly and covered with stuff that I prefer not to describe here. But after many hours of labor and eventually nursing, these babies were beautiful! Oxytocin plays an important role in breast-feeding with milk letdown from the mammary glands (see Longo et al. 2012). Oxytocin release may be triggered by sounds of the baby's cry or even of other's babies. These reflexes again appear to be under frontal lobe regulatory control with elevations in anxiety or stress disrupting the nursing process and aggravating the mother's efforts often in a circular fashion. This initial socialization with contact comfort has a long history in psychology in the development of trust and the establishment of love or emotional bonds (e.g., Harlow 1962).

In a recent review of the effects of oxytocin on social interactions, the authors described it as "the peptide that binds," referring to its powerful role in the formation of relationships (MacDonald and MacDonald 2010). Not only does oxytocin increase trust in humans (Kosfeld et al. 2005) but it has also been linked to autism with lowered levels in autistic individuals (Modahl et al. 1998). Others (Hollander et al. 2003) have found reductions in repetitive behaviors in adults with autistic and Asperger's disorders with oxytocin infusion.

Moreover, Naber et al. (2012) provide a double-blind, placebo-controlled, within-subject experiment on the effects of parenting styles and hostility with intranasal administration of oxytocin to fathers of children with autism spectrum disorders. Fathers with their typically developing toddler ($n = 18$), and fathers of toddlers diagnosed with autism spectrum disorder ($n = 14$), were observed in two play sessions of 15 min each with an intervening period of 1 week. The authors conclude that, in all fathers, oxytocin elevated the quality of paternal sensitive play. They found that the fathers stimulated their child in a more optimal way, and they showed less hostility, with incremental gains in paternal sensitive play irrespective of the clinical status of their child. Still others have found reliable effects of oxytocin levels in moles, where monogamy was predicted by elevated numbers of oxytocin receptors and where sexual promiscuity was found with lowered levels (Inset and Shapiro 1992).

The hypothalamus plays an important role in sexual preferences. For example, Savic et al. (2005, 2006) used positron emission tomography to evaluate the hypothalamic response to common odors or pheromones. The scent of testosterone in male sweat and the scent of estrogen in female urine were differentially responded to as a function of sexual preference. Heterosexual men and homosexual women showed hypothalamic responses to estrogen, whereas the hypothalamus of homosexual men and heterosexual women both responded to testosterone. The hypothalamus of all four groups did not respond to the common odors, which instead produced a normal olfactory response in the brain.

Although sex differences are frequently ignored in neuroimaging and lesional studies, there is ample evidence for their existence and for their importance in neuropsychological research. For example, testosterone is another hormone involved in aggressive behavior. The hypothalamus stimulates testosterone production through gonadotropin-releasing hormone. This hormone, in turn, causes the production of two other hormones—follicle-stimulating hormone and luteinizing hormone—collectively known as gonadotropins. Luteinizing hormone is released into the bloodstream where it travels to the male testes and triggers the production of testosterone from cholesterol. High testosterone levels enhance attention to aggressive stimuli, downregulate the interaction between cognitive and emotional brain systems, and are associated with dominant aggressive behavior (Dabbs et al. 1995; Dabbs and Hargrove 1997). During moral decision making, individuals having high testosterone levels are more likely to make utilitarian decisions, especially when doing so implies acts of aggression and social cost (Carney and Mason 2010). Testosterone is also associated with diminished sensitivity to the affective signals that facilitate pursuit of empathic behaviors and choices (van Honk et al. 2005, 2010).

Chapter 11
Syndromes of the Left Brain

Dysphasia

One of my favorite activities in April and May each spring is to stand between the hives in my apiary and to watch the complex "dance of the honeybee." Although many ascribe "language" strictly to humans, this appears to be a generalization far outstretching the data. The 1973 Nobel laureate, Carl von Frisch, described this dance in his language hypothesis for this seemingly simple creature. This communication system, one bee to the others, appears to be effective in the shared localization of the nectar source. Also, the navigational and orienting skills of the relatively simple honeybee are apparent on exiting the hive where variable size flight loops and flight angles through the sunshine seem sufficient to ensure the accuracy of their path to the nectar source and their return home to the hive. In humans, we attribute similar functions in the first case to the left hemisphere, whereas navigation and orienting in extrapersonal space appear more clearly to be a capability of the right cerebral hemisphere (e.g., Heilman et al. 1995; Guariglia and Antonucci 1992; Foster et al. 2008; Committeri et al. 2007). "Ape language" is also a term used to describe the demonstrations of Washoe, the chimpanzee, and the bonobo Kanzi, to access human language through communication devices like signing and lexigrams (Gardner and Gardner 1969; Savage-Rumbaugh et al. 1993; see also Gannon 2010).

In humans, we have evidence of at least logical communication sequences being a function of the left cerebral hemisphere. Indeed, the most commonly used indication for a left-brain disorder is consistent with Broca's assertion (1861) that "We speak with the left brain." Speech disorders of left-brain origin commonly involve logical linguistic speech deficits. The basic distinction has been among the disorders derived from frontal lobe dysfunction and those derived from temporal lobe dysfunction. The distinction has focused on the fundamental constructs of fluency in expression and the comprehension of speech, where left frontal dysfunction is related to nonfluent or expressive dysphasia also referred to as Broca's aphasia. A disorder in the comprehension of speech is related to receptive dysphasia also

© Springer International Publishing Switzerland 2015
D. W. Harrison, *Brain Asymmetry and Neural Systems,*
DOI 10.1007/978-3-319-13069-9_11

Fig. 11.1 Broca's and Wernicke's area within the left cerebral hemisphere. Originally published in *Neuroanatomical Atlas of Key Presurgical and Cognitive Eloquent Cortex Regions*, Faro et al. 2012, p. 970

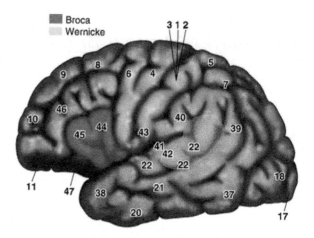

referred to as Wernicke's aphasia. Broca's area and Wernicke's area in the left hemisphere are depicted in Fig. 11.1.

Broca's area (the inferior frontal operculum) is largely premotor cortex involved in the organization, planning, and sequencing of motor movements that will then be effected by the motor cortex, the final motor pathway. In this inferior posterior frontal region, the upper motor neurons will effect movement with the oral motor apparatus involving the tongue, lips, mandible, and other functionally related muscle groups. Broca's dysphasia presents with nonfluency in expression often with articulation errors and phonemic paraphasic errors (e.g., see Hickok and Bellugi 2000; see Duffau 2012). Dysfunction in this area relates to a diminished intention or desire to speak and especially during confrontation conditions under the direct observation of another or with imposed time constraints to produce word output.

This reduction in speech initiation or intent to talk appears to be a function of the strong interactions of the premotor cortex with the underlying dopaminergic structures of the basal ganglia and especially the head of the caudate nucleus (Brunner et al. 1982; Draganski et al. 2008; Ford et al. 2013). In contrast, disorganization in the production of speech sounds (phonemic and perseverative paraphasic errors) corresponds more with dysfunction in processing at the premotor cortical region. President George W. Bush provided multiple public examples of halting, hesitant speech with perseverative and phonemic paraphasic errors and several videos are readily available online (e.g., George W. *Bush Stutters* Over and Over—YouTube 2011).

The typical speech evaluation addresses the basic constructs as identified in Fig. 11.2. The evaluation includes the assessment of speech fluency or generativity, the reception of speech sounds, the comprehension of speech, consolidation of verbal information or memory for speech passages, and speech repetition. The evaluation allows for a diagnosis of the frontal lobe speech deficits, which include the expressive and the subcortical expressive dysphasias. It allows for the identification of the temporal lobe speech deficits, which include the receptive and the subcortical receptive dysphasias. In addition, the transcortical speech disorders may be identified, including the transcortical expressive and the transcortical receptive

Fig. 11.2 Basic behavioral and anatomical components in the evaluation of dysphasia

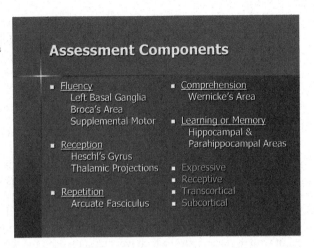

Assessment Components

- Fluency
  - Left Basal Ganglia
  - Broca's Area
  - Supplemental Motor
- Reception
  - Heschl's Gyrus
  - Thalamic Projections
- Repetition
  - Arcuate Fasciculus

- Comprehension
  - Wernicke's Area
- Learning or Memory
  - Hippocampal &
  - Parahippocampal Areas
- Expressive
- Receptive
- Transcortical
- Subcortical

dysphasias. Subcortical dysphasia may result following damage to the underlying left hemisphere structures, including the basal ganglia, hippocampus, and thalamic regions (e.g., Cox and Heilman 2011). The following paragraphs elucidate on the characteristic features of these propositional speech disorders.

An *expressive dysphasia* or nonfluent speech disorder with restricted word production under confrontation, with diminished desire or intent to speak or with sparse expression, and/or with articulation or stuttering errors may be indicative of dysfunction within Broca's area at the inferior-posterior frontal lobe. But, a subcortical lesion here in Broca's area may leave simple, over-rehearsed, casual conversational speech somewhat unaffected. The speech deficit may be more apparent under confrontation with demands to produce words under time constraints. This is performed using one or another verbal fluency test, such as the Controlled Oral Word Association Test (COWAT; Spreen and Benton 1977; see also Lezak et al. 2012). Moreover, the associative strength of the words produced provides evidence for the capacity of these brain regions with decreased capacity yielding spreading activation and less common word production and cognitive associations with expressive dysphasia (Foster et al. 2011).

Speech expression is typically effortful with halting, hesitant features and with meaningful content consisting of nouns and verbs. The *subcortical expressive dysphasia* results from a lesion impacting the left basal ganglia and especially the head of the caudate nucleus, whereas thalamic expressive dysphasia has been reported (Cox and Heilman 2011). These systems are intimate to initiation and, when damaged, the patient may require substantial prompts for speech initiation. These subcortical systems are critical to initiation, energy level, and desire to respond, and amotivational syndromes may result with the demise, deactivation, or decreased metabolic rate of this brain region (see Granacher 2008; Grossman 2002; Hu and Harrison 2013b). This is especially the case with medial frontal syndrome with damage proximal to the supplementary motor cortex. Greater pathology within this later region results in akinetic mutism (e.g., Freemon 1971; Fontaine et al. 2002;

see also Duffau 2012). The patient recovering from akinetic mutism or with milder syndromes shows hypokinesis and perhaps low volume, hoarse, whispered speech characteristics.

A somewhat atypical variant of pathology within this system might be appreciated in the history provided by a young man recently graduated from a university's finance program with employment opportunities at a major financial institution. Even though he had been remarkably successful in managing large capital investment funds and in his degree program, he had been painfully aware of an episodic inability to express his speech. This problem was evident to him, extending back at least to his early middle school years. Even in his sporting activities, he would have these episodes along with an inability to throw a ball with his right arm on the baseball field. One clue, which might signal an episode, was the onset of difficulties in carrying out movements of the face on demand (a buccofacial apraxia) and sometimes numbness and immobility of his tongue. He recalls much frustration and embarrassment with these episodes, which went undiagnosed for several years. Eventually, one of the doctoral students on the neuropsychology practicum team appreciated that the events might be reliably produced on the administration of verbal fluency tests under confrontation (e.g., the COWAT). The episodic nature of the events and the neural circuitry involved provided for a diagnosis of left temporal lobe seizure disorder with supporting evidence from the electroencephalogram record.

Some of the rough clinical guidelines from conversational speech, which may indicate fluency deficits, are depicted in Fig. 11.3. The left-sided column provides some potential indications for a nonfluent expressive dysphasia, whereas the right-sided column provides some indications for a fluent receptive or transcortical receptive dysphasia. Expressive dysphasia features phonemic paraphasic errors (e.g., see Duffau 2012) where the wrong phoneme is expressed but where the meaning of the statement remains intact. This would be the case by way of example, when the statement "I drove the car to the store" is expressed as "I dove the tar to the fore." The meaning remains intact and is conveyed to the listener despite the phonemic error(s). The expressive dysphasia may present with perseverative phonemes,

Fig. 11.3 The fluency construct in the evaluation of dysphasia

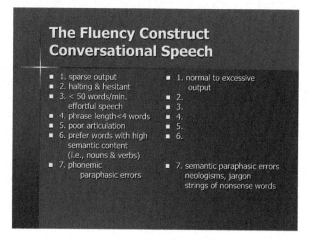

The Fluency Construct
Conversational Speech

- 1. sparse output
- 2. halting & hesitant
- 3. < 50 words/min. effortful speech
- 4. phrase length <4 words
- 5. poor articulation
- 6. prefer words with high semantic content (i.e., nouns & verbs)
- 7. phonemic paraphasic errors

- 1. normal to excessive output
- 2.
- 3.
- 4.
- 5.
- 6.
- 7. semantic paraphasic errors neologisms, jargon strings of nonsense words

words, or phrases. Broca's patient might state his name as "tan, tan, tan, tan, tan" with more severe perseverative features reflecting the extensive lesion identified at autopsy, which would impact these systems. One patient provides for an example as she perseverated on the statement "May I have this dance?" With each statement, she would generate increased sympathetic tone, blushing, and facial expressions of embarrassment. With increasing frustration, the perseverative statements seemed more common or probable, and to her dismay! Broca's patient Tan, perhaps with right hemispheric release, would appear frustrated, with the production of profane and vulgar expletives.

Another man injured his left frontal region with a midsagittal commercial bandsaw transection, followed immediately by a perpendicular transection proximal to Broca's area. He was expressionless in propositional speech and seemingly quite frustrated by his efforts to talk. He became fluent, though, only with anger where he would emit fitful and foul language with loud volume. His therapists likened his behavior to that of the historical figure Phineas Gage (see Damasio 1994). This man spent much of his time alone in a dimly lit room, whereas his family noted his prior appreciation for social activities, natural sunlight, and out-of-doors recreational activities.

One hallmark of left frontal dysfunction and an expressive dysphasia may be the patients' apparent awareness of and frustration with their disorder. This may relate to self-esteem and to subsequent word production in a negative fashion with social interactions providing for more negative encounters and decreasing both the enjoyment of the interaction and the probability for seeking additional social opportunities. For example, if normal individuals are brought into the laboratory and provided a series of failure experiences, the metabolic rate in these left frontal brain systems may diminish (see Davidson 2003), with heightened probability of the behavioral, emotional, and cognitive deficits previously discussed, including reductions in verbal fluency and negative self appraisal or self-depreciating remarks. Other common features which may be identified with an expressive dysphasia are depicted in Fig. 11.4. Of course, the extent and severity of dysfunction may vary over time both between individuals and within an individual.

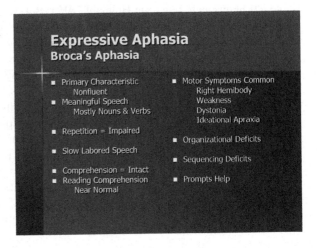

Fig. 11.4 Common features which may be identified with an expressive dysphasia

Expressive Aphasia
Broca's Aphasia

- Primary Characteristic
  Nonfluent
- Meaningful Speech
  Mostly Nouns & Verbs

- Repetition = Impaired

- Slow Labored Speech

- Comprehension = Intact
- Reading Comprehension
  Near Normal

- Motor Symptoms Common
  Right Hemibody
  Weakness
  Dystonia
  Ideational Apraxia

- Organizational Deficits

- Sequencing Deficits

- Prompts Help

Conceptually, the expressive dysphasia is an *ideational dyspraxia* if the dysfunctional regions involve the premotor cortex and its rich interconnections with the basal ganglia. Ideational dyspraxia of the oral motor facial regions is a disorder of Luria's third functional unit essential for the organization, planning, and sequencing of movement. Impairment in these systems may alter the normal expression of speech as an executive function with variation in the smooth transitions from one phoneme to the next and from one gestural posture of the oral motor unit to the next. Impaired cognitive fluency presents with poor transitions in the verbal basis of thought and with inflexibility or rigidity in these processes (e.g., see Gläscher et al. 2012). These same systems have been found to be essential in rule-regulated speech production underlying grammar, where agrammatical statements might be presented but where the meaning remains intact.

Just dorsal to Broca's area is Exner's hand area, consisting of the premotor region regulating, planning, and sequencing right hand movement or ideational praxis for the expression of graphics. A lesion in Exner's area may result in a motor or expressive dysgraphia (e.g., see Ardila 2012) and is often concurrent with an expressive dysphasia. This correlation may result from the anatomical proximity of Exner's area to Broca's area. Moreover, these areas jointly fall within the distribution of the left middle cerebral artery, where a stroke within the pre-Rolandic branch of this artery may impact both areas, as would a dam on a river affect the flow of the river down the stream. The specialization of Exner's area, within the left brain, for graphics and feeding movements at the right hand, is reminiscent of the fiddler crab with its tiny right-sided claw used for feeding and with the dramatic and powerful left-sided claw used for fighting and personal defense. Additional research on the laterality of Exner's area in motor graphics as opposed to support of a sheet of paper for writing purposes might be useful. These areas differ in laterality indices of emotion, including sign language or defensive and aggressive postural gestures and displays.

Receptive dysphasia involves impairments in the reception and comprehension of speech sounds. This disorder commonly results from dysfunction within the auditory projection area and Wernicke's area at the superior-posterior temporal region in the left hemisphere (e.g., see Hugdahl 2012). But, with more profound impairment such as the larger lesion or its extension more intrusively into Wernicke's area, the patients may be unaware of their speech deficits. They might even attribute any perceptual difficulty in communicating to others to flaws in the other person, rather than to a substantive deficit within their own language systems.

The diagnostician might initiate a word salad conversation with this person, providing jargon and incoherent meaningless speech to him or her in an interactive discourse. Therapists, family members, and health-care professionals are often surprised to see a patient who can handle very simple over-learned conversations respond in a like manner to totally nonsensical word salad productions as though they were meaningful. This word salad conversation might continue unabated other than for the time constraints on the assessment. Common features, which may be identified with a receptive dysphasia, are depicted in Fig. 11.5a. A lesion extending into the pathway dorsal to Wernicke's area may impact the parietal hand region, yielding

Fig. 11.5 **a** Common features which may be identified with a receptive dysphasia. **b** A social leader in her community with "word salad" speech and receptive dysphasia

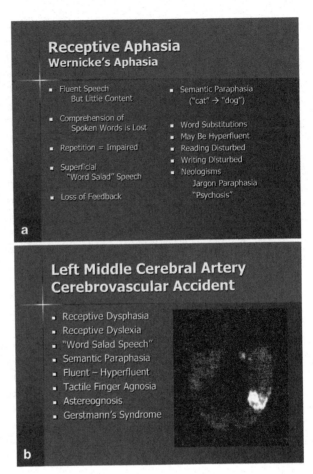

a receptive dysgraphia and/or ideomotor dyspraxia of the hand (e.g., Hermsdörfer et al. 2012; Vingerhoets et al. 2012; see Ardila 2012).

Figure 11.5b shows the effects of a stroke within the posterior branch of the left middle cerebral artery in a 72-year-old social activist. She has many visitors to her room and actively engages others in social interactions with fluent to hyperfluent speech, largely lacking meaning. She responds to "word salad" speech from others in a seemingly identical fashion to her response to meaningful speech. She initially has many visitors and eventually there is evidence of these and similar verbal stressors resulting in seizures accompanied by verbal hallucinations or paracusia. The evaluation reveals associated deficits consistent with Gerstmann's syndrome. She has receptive dysphasia, dyslexia with dysgraphia, dyscalculia, and finger agnosia. Her speech consists of anomia with circumlocution errors. The sensory examination reveals somatic deficits with tactile finger agnosia and astereognosis. Eventually, she had to be shielded from heavy social interactions with a maximum of one person in her room. She was able to benefit from the provision of minimal speech and

increased interspeech intervals to promote improved comprehension and to minimize seizure inductions. The hallucinations gradually resolved.

Just as the expressive dysphasia originates proximal to Exner's hand area and
may convey with a motor dysgraphia, the anatomical basis for a receptive dysphasia may overlay the parietal face and/or hand area. At the parietal facial region,
the effect may be an ideomotor lingual or facial dyspraxia with poor posturing or
placement within the oral motor apparatus and/or at the hand. The receptive dysgraphia in this scenario might convey impoverished content or meaning in the script
secondary to the temporal lobe disorder and gestural problems in holding the pen
properly with a parietal lobe dyspraxia. By analogy from the auditory systems of the
temporal lobe, the parietal dysgraphia might provide a "word salad" of postural or
gestural arrays with confusion for body parts and positions (e.g., see Ardila 2012).

The subcortical speech disorders allow for functional integrity of the cortical
speech regions, whereas these regions may fluctuate with variation in arousal or
activation level. Arousal mediation for these areas is affected with the subcortical
lesion as might occur with disruption of the thalamic projections of the reticular
activating system from the brainstem. Such a speech disorder (subcortical) may
fluctuate over time with a waxing and waning of the symptoms derived from the
cortical brain regions. A subcortical expressive dysphasia may vary over time with
integrity of the cortical region, such that one therapist may appreciate nonfluency
and another therapist might not see it. Alternatively, the dysphasia might be apparent at one moment and not be apparent at the next moment within the session.

Subcortical receptive dysphasia may follow from damage to the hippocampal
or perihippocampal regions of the left hemisphere. Such a lesion may impact the
primary memory system essential for the acquisition and consolidation of verbal
information, as with learning (e.g., Kelley et al. 1998; Fernaeus et al. 2013). These
subcortical systems, when lesioned, may result in a verbal learning disability or a
failure to consolidate speech. The therapist will appreciate this as the patient may
fail to show carryover of the verbally presented instructions and directions from one
therapy session to the next. These deficits may be confirmed through the assessment
of verbal learning and memory as with logical memory tests, auditory verbal learning tests, and related procedures. Performance on these tests might be compared
with nonverbal learning and memory tests. These assessments would best compare
the sensory systems and spatial processing advantages over auditory verbal processing skills. Although it may be difficult to rely on other than verbally mediated
therapeutic interventions, this patient may benefit more from the presentation of
information through the learning systems of the intact right brain.

Therapeutic recommendations might encourage lowered social confrontation
and verbal linguistic interactions and demands, along with more time involvement
in preferred independent activities. For some, this might be in the form of animal-
assisted interventions. For others, it might involve caring for a plant or participation
in independent music therapy interventions accompanied by fewer social demands.
Social and verbal stress may be elevated at the patient's level during the recovery
process. Routine levels of verbally interactive socialization may overwhelm this
individual early on in the recovery process. This discrepancy may be appreciated in

comparison with both their baseline exposure prior to hospitalization and baseline speech-processing abilities, which are now diminished.

The person suffering from a subcortical receptive dysphasia, if severe, may lack sufficient persistence of the auditory memory trace for the consolidation of verbal information. With verbal speech discourse in the normal brain, the hippocampal system provides for persistence (Bliss and Lomo 1973; Wang 2001) of the sounds for a duration sufficient to allow a response and sufficient to allow for learning or consolidation of the information. With impersistence of the verbal trace, the person may lack even the ability to recall what speech sounds were just received or even what speech sounds they had just produced. This is a significant disability and one which may be most disabling to the best intentions of therapists and institutional staff who have learned to rely on speech and maybe too much! The subcortical receptive dysphasia is essentially a learning disability with impaired carryover or acquisition for logical, linguistic speech.

A good exercise for any therapist in the making would be to function with the client or patient as though there were no speech or language tools available. The clinical psychologist might be asked to assess and treat behavioral and emotional disorders in nonhumans prior to being allowed to work with humans, as this would provide at least some exposure to nonlinguistic approaches and interactions. Clinical psychological science would clearly benefit from revisiting contextual therapies other than an increasing reliance on verbal therapies (e.g., acceptance and commitment therapy (ACT), rational emotive therapy (RET), and cognitive behavioral therapy more broadly defined). Along these lines, a clinical graduate student explained that bulimia was "caused by a parental overemphasis on food." A case study of a pet cat named "Bun Bun" was provided to the student as these verbally based attributions seemed somewhat less tenable and even a bit ridiculous. Bun Bun had developed an eating disorder due to which she would gorge herself followed by emesis, and this was recurrent and disturbing for her family. The doctoral student was asked, as a clinician, what she might do to help Bun Bun and what alternative causal attributions might be viable within a broadened diagnostic framework. At this point, this student was speechless!

Transcortical dysphasias result following lesion or deactivation of the brain regions proximal or adjacent to the primary cortical speech regions traditionally referred to as Broca's area and Wernicke's area (e.g., Joinlambert et al. 2012). The transcortical dysphasias are characterized by intact repetition with anatomical integrity of the arcuate fasciculus, which connects auditory receptive cortex and parts of Wernicke's area to Broca's area and eventually to the motor cortex. In general, the arcuate fasciculus, a large association tract connecting perisylvian areas of the frontal, parietal, and temporal lobes is larger at the ventral basal (frontotemporal) regions in the left brain (Good et al. 2001; Herve et al. 2006), possibly relevant to language functions. For comparative purposes, the uncinate fasciculus connecting the basal frontal and temporal regions is larger in the right brain, possibly related to emotional functions. These cortical areas, connected by a cable (the arcuate fasciculus), allow for echoes of received speech and the diagnostic speech indicator of *echolalia*. The novice might explain away an echolalia with attributions that the

Fig. 11.6 a Basic features
of a transcortical expressive dysphasia. **b** A patient
with transcortical dysphasia
following a hemorrhage
associated with a metastatic
deposit

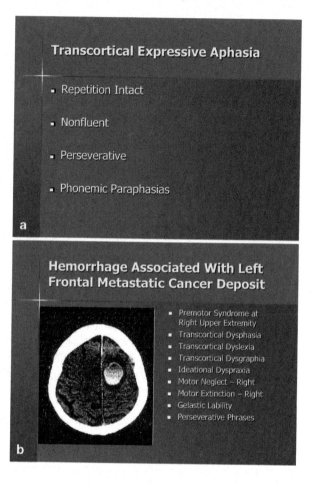

patient is merely seeking confirmation of another's speech or that they are "just
hard of hearing." But, the neuropsychologist is not afforded the luxury of explaining
away such behavior as this is a major diagnostic clue to the existence of a transcortical expressive, transcortical receptive, or mixed transcortical speech disorder.
Common features of a transcortical motor dysphasia are presented in Fig. 11.6a.

Figure 11.6b provides a case study of a 76-year-old man's status post a left frontal hemorrhage associated with a metastatic melanoma deposit. He presents the
behavioral manifestations of a premotor lesion for his upper right extremity with
a right-sided motor neglect and right-sided motor extinction during concurrent left-
hand use. He initiates movement with a preference for his nonpreferred left hand.
He is dysgraphic with perseverative and poorly organized writing. His laughter and
verbal phrases are perseverative with multiple phonemic paraphasic errors. His
rightward-directed gaze requires prompting and is impersistent, returning rapidly
to midline. Regardless, his speech is meaningful and others can understand what
he is saying.

The transcortical motor speech disorder may present as with the expressive dysphasia described above, but with integrity for repetition. Repetition may be quantified by the level of repetition that the patient is capable of, as with words, phrases, or sentence-level repetition. Similarly, the transcortical receptive dysphasia may look like receptive dysphasia, but with integrity for repetition. Mixed transcortical dysphasia may present with isolation of the speech loop allowing for echolalia, but with combined features of an expressive and a receptive dysphasia (e.g., Joinlambert et al. 2012). In contrast to the echolalia and intact repetition that present in the transcortical dysphasias, a lesion within the arcuate fasciculus may disconnect the temporal and frontal speech regions with a subsequent loss of speech repetition and a diagnosis of *conduction aphasia* (e.g., Anderson et al. 1999).

The production of speech by the individual with a receptive speech disorder is fluent, although a more posterior lesion surrounding Wernicke's area may result in hyperfluent speech. The later speech disorder is a *transcortical receptive dysphasia,* where the speech is often irrelevant. Many of these individuals might be described by the layperson as "ratchet jaws" to use more common language, where they show a tendency to "talk your ears off." The examiner may need to be intrusive and may invade the verbosity emanating from the patient in order to complete the assessment. Indeed, doctoral level students may easily spend an hour or more with such a patient and retreat from the session with very little in the way of meaningful case-related interview information. This is more common if the student is somewhat nonfluent, providing an unfortunate interaction in this social setting. The patient may be asked if he or she is talking more than usual after a left-sided far posterior stroke. Often, a far posterior lesion will allow the patient to appreciate this outcome and share dismay over the social implications. Common features of transcortical receptive dysphasia are depicted in Fig. 11.7. Interestingly, dysfunction of the homologous right cerebral region may result in a transcortical dysprosodia. This speech disorder is characterized by hyperfluency but not for logical linguistic speech. Instead, this disorder may result in hyperprosodia ranging from milder cases with too much

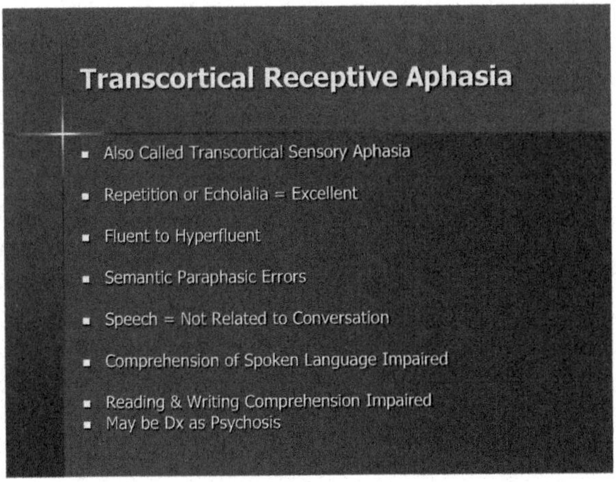

Fig. 11.7 Common features of a transcortical receptive dysphasia

Transcortical Receptive Aphasia

- Also Called Transcortical Sensory Aphasia
- Repetition or Echolalia = Excellent
- Fluent to Hyperfluent
- Semantic Paraphasic Errors
- Speech = Not Related to Conversation
- Comprehension of Spoken Language Impaired
- Reading & Writing Comprehension Impaired
- May be Dx as Psychosis

Fig. 11.8 Isolation of the speech areas with integrity of the arcuate fasciculus with a mixed transcortical dysphasia. Adapted with permission from http://www.brains.rad.msu.edu and http://brainmuseum.org, supported by the US National Science Foundation

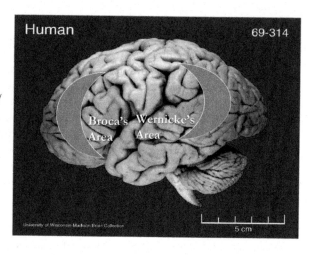

emotional tone in their speech to severe disorders with vocalizations, including moaning, grunting, and/or wailing behaviors (e.g., Harrison et al. 1990).

Mixed transcortical dysphasia may be, in essence, a functional and/or structural isolation and encroachment of the speech areas (see Fig. 11.8). This patient may feature an echolalia with impairments in both the expression and the reception of speech. The intact anatomy that is isolated may be a minute cortical slab from the anterior temporal and the inferior posterior frontal region in addition to the cable connecting them, the arcuate fasciculus. It is via the arcuate fasciculus that the speech trace moves from receptive speech areas to the expressive speech areas, providing for echolalia in speech.

Mixed transcortical dysphasia shows the combined features of expressive and receptive dysphasia. However, repetition remains intact with preservation of the arcuate fasciculus interfacing the proximal left frontal and temporal cortices. The nonfluent patient with impaired comprehension may echo the speech passages of others. Moreover, this person may be positively received by others as what goes in, may be what comes out. One man with mixed transcortical dysphasia was maintained near the nursing station for monitoring purposes and to ensure quality of care. The nurses would say to him "How are you today?" The patient would respond "How are you today?" The nurses would say "Fine and you?" The patient would respond "Fine and you?" Features of mixed transcortical dysphasia are presented in Figs. 11.9 and 11.10.

In one study of left-sided cortical border-zone infarcts, researchers (Joinlambert et al. 2012; see also Duffau 2012) found that the speech-related difficulties of their right-handed patients was initially that of a transcortical-mixed dysphasia; noting expressive speech disturbances and altered lexical and syntactic comprehension, while repetition was preserved. This initial presentation evolved over time toward a transcortical motor dysphasia, with reduced speech fluency, short sentence length, and phonemic paraphasia, with preservation of speech comprehension and repetition in patients with left anterior cortical border-zone infarcts within the middle

Fig. 11.9 Common features
of a mixed transcortical
dysphasia

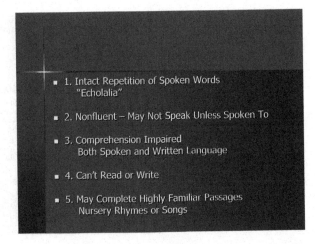

1. Intact Repetition of Spoken Words
   "Echolalia"

2. Nonfluent – May Not Speak Unless Spoken To

3. Comprehension Impaired
   Both Spoken and Written Language

4. Can't Read or Write

5. May Complete Highly Familiar Passages
   Nursery Rhymes or Songs

Fig. 11.10 Common features
of a mixed transcortical
dysphasia. (continued)

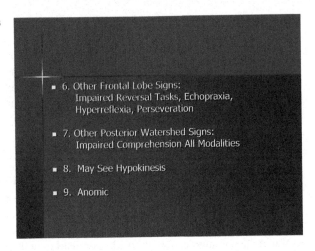

6. Other Frontal Lobe Signs:
   Impaired Reversal Tasks, Echopraxia,
   Hyperreflexia, Perseveration

7. Other Posterior Watershed Signs:
   Impaired Comprehension All Modalities

8. May See Hypokinesis

9. Anomic

cerebral artery distribution. Instead, the speech of the patients with left posterior cortical border-zone infarcts within the distribution of the middle cerebral artery evolved toward a transcortical sensory dysphasia with altered lexical and syntactic comprehension, preserved naming, and repetition.

Global dysphasia: If the dysphasia is severe, conveying deficits in reception, comprehension, repetition, and expression, then the diagnostic impression conveyed by the examiner might be a *global dysphasia*. The global dysphasic patient typically shows a broad spectrum of severe verbal linguistic deficits as depicted in Fig. 11.11. Nonetheless, the integrity of the patient's right brain may be a substantial resource for implementing caution in functional activities for safety. Moreover, this patient may well be capable of navigating through the geographical coordinates of the hospital grounds and eventually the home or neighborhood environment. Many of these individuals are eventually able to care for themselves at some level and

Fig. 11.11 Common features
of global aphasia

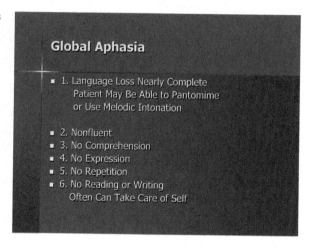

to obtain some significant degree of independence. This may not be the case with
a large lesion within the homologous regions of the right hemisphere. Following
the right-sided brain lesion, the basis for return to activities rests more upon the
patient's poor insight to deficits with anosagnosia and anosodiaphoria (e.g., Gerafi
2011; see Prigatano 2010; see also Pia et al. 2004). The left brain may be less able to
care for itself, maintain safety in daily functional activities, and find its way around
through extrapersonal space, just as the right brain may be less capable of social
engagement with verbal linguistic demands.

A local musician suffered a massive stroke within the distribution of the left
middle cerebral artery. With family permission and encouragement, his music was
played over the speaker system in the rehabilitation unit. He would sometimes join
in the melody and hum along. His first "speech" came from efforts to implement
melodic therapies and possibly the resources of the intact right cerebrum as he could
eventually sing a few words and especially if the words of the song actually were
useful in the communication of his likes or dislikes and his needs or wants to others.

Disorders of propositional speech may be accompanied by anomic features and
frequently with circumlocution in an effort to compensate for this disability. With
circumlocution, the person may be attempting to name a fork, for example, but
the expression might be "You know the thing you eat with." One man was in the
US Special Forces. He was unfortunate enough to be the only known survivor of
a helicopter unit, which had been shot down over a country where the president
had repeatedly stated that "We have no troops there!" Interestingly, the evidence
suggested that he had been married to a native woman while in a comatose state
with the woman subsequently signing his discharge papers to remove him from the
military (thus confirming the president's statement).

On his recovery, his record included completion of a standardized intelligence
test which severely penalized him for circumlocution errors with his anomia or
word-finding problems. The intelligence quotient was so low that there was active
discussion among his case managers of having him declared legally "incompetent"
(again), using the intelligence test measures as the basis for this action. However,

the test was readministered but with the provision of credit for circumlocution errors in speech that were, otherwise, correct in conveying the answer to the test items. With this procedural adjustment, his scores were clearly within the superior range. This was enough to eventually recover his losses and his mother was very instrumental in recovering his benefits. As for his new wife, he had many pictures of lovely ladies from the country in question, but no pictures of the woman that he had married, presumably while in the comatose state. She confessed on inquiry that she had been rewarded with citizenship for her actions! Remarkably, the couple grew ever closer together and the marriage prospered. She was with him through thick and thin, and he with her! When I left him after receiving a job offer in Virginia, he was busy writing his book or training manual for the Special Forces and still with circumlocution errors in the text. As for standardized test administration with brain-damaged individuals, many other problems exist (for examples see Satz 1993; see also Loring and Bauer 2010).

The cortical dysphasias are more frequently the result of a cerebrovascular accident or stroke involving the *middle cerebral artery* (e.g., Joinlambert et al. 2012). Occlusion or hemorrhage within the anterior branches of the middle cerebral artery may result in an expressive dysphasia, whereas occlusion or hemorrhage within the posterior branches of the middle cerebral artery may result in a receptive dysphasia. The subcortical expressive dysphasia may be derived from occlusion or hemorrhage within the distribution of the anterior cerebral artery and especially the arterial branches perforating the basal ganglia within the subcortical frontal lobe regions (Cox and Heilman 2011). Thalamic dysphasias more commonly result from occlusion or hemorrhage of the branches of the thalamoperforating artery within the left brain. Subcortical receptive dysphasia may involve the distribution of the posterior cerebral artery of the left brain, a distribution which supplies much of the hippocampal and perihippocampal regions essential for the consolidation, learning, or carryover of verbal information (e.g., Fernaeus et al. 2013).

The majority of these strokes produce infarction of the tissue with a loss of cells in the region secondary to cell death from inadequate perfusion of oxygen and nutrients supplied through the arterial blood supply and transfused across capillary beds. However, hemorrhagic infarctions occur with some frequency. With blood loss, there may be neural tissue loss directly on contact with the iron from the blood. The formation of a hematoma may result in potentially reversible brain dysfunction secondary to the space occupying features of this bolus. This may, in some cases, be extracted through neurosurgery or in others with resolution through natural processes. Regardless, some relative optimism for an improved prognosis may be derived from a hemorrhagic event if the bleeding stops, if there is natural resolution of the clot, and if the pressure effects and surrounding edema resolve with the resolute integrity of the adjacent brain tissue. Increased caution is advisable for the therapist and the caregiver in this situation as the patient may appear to be more severely afflicted than is actually the case. Time will tell the story along with the medical monitoring of the mass and careful decision making for intervention. Of course, pathology of multiple etiologies may affect these brain areas, and vascular etiology is only one among these.

One young woman had a severe expressive dysphasia after a hemorrhage and surgical evacuation of the hematoma over Broca's area. The severity of her speech disorder was overestimated though, as eventually the surgical sponges, which had been accidentally left by the neurosurgeon, were discovered and subsequently removed. Once discovered, the neurosurgeon was effectively able to correct the problem and the surgical team was able to scrutinize the surgical protocol and staffing needs. Seeing her again after the corrective surgery to remove the foreign material revealed a fluent young lady, proud of her recovery and with plans for marriage. This outcome was a far sight better than the original discharge plans to continue her recovery within a brain injury facility. She had received counseling to carefully consider the wedding plans and this had not been well received.

Dyslexia

Many literatures exist on the origins and subtypes of dyslexia. Some of these literatures are complex in their language and are deduced from theories other than brain theory. For simplicity in this respect, dyslexia will be discussed from two primary points of view, clearly based within the established architecture of functional neural systems theory. The first is the position of Hughlings-Jackson (1879) and A. R. Luria (1966, 1973, 1980) on the location of the lesion. If the reading problem resulted essentially from a visual deficit, then the diagnosis and treatment would be for one or another of the *visual agnosias* or anopsias. This might include visual letter recognition deficits subsequent to a left occipital lesion. In contrast, lesions within the language processing areas would result in one or another of the *aphasic alexias*. The second diagnostic system involves the basic distinction between dyslexia of subcortical origin and dyslexia of cortical origin. This second diagnostic system (e.g., Imtiaz et al. 2001; see also Ardila 2012) distinguishes among *dyslexia with dysgraphia* (cortical lesion) and *dyslexia without dysgraphia* (subcortical lesion). These two diagnostic systems are discussed below.

Luria's approach is parsimonious as the functions of the frontal lobe regions and the temporal lobe regions, involved in speech expression and reception, are conveyed with consistency in the nature of the reading disability. *Expressive dyslexia* conveys nonfluency with slow and effortful reading both out loud and in silent reading situations. Reading might involve sparse and/or effortful output, phonemic paraphasic errors, perseverative errors, and related frontal lobe impairments (see Duffau 2012). Lexicon is conveyed in a meaningful fashion by this patient and remembered (recognition) with carryover intact, but with deficits in output. Expressive dyslexia might be more frustrating for the patient and promotes diminished self-esteem on demands for lexical production under confrontational conditions. One example here might be the aggravation of expressive deficits when confronted with demands to read aloud in front of an audience or maybe in front of classmates. The expressive deficit may be aggravated with frustration, embarrassment, and lowered self-esteem with the additional social stress or confrontational demands.

Receptive dyslexia presents with fluent reading but with impaired comprehension. The conversion of the visual grapheme in the written text to the auditory association within Wernicke's area in the superior temporal gyrus would be faulty with semantic paraphasic errors, neologisms, and sometimes meaningless or word salad expressions. The lexical expressions would be fluent but meaningless and relatively empty of content. In contrast to the expressive dyslexic with frustration and embarrassment conveyed with the individual aware of their errors, the receptive dyslexics might not appreciate their deficit in reading or possibly consider the text to be problematic rather than their reading of it.

Subcortical expressive dyslexia might be appreciated with heightened initiation deficits or inertia, whereas the individual with *subcortical receptive dyslexia* might be more problematic in the area of learning or consolidation of the written word where lexical carryover might be impaired secondary to hippocampal and perihippocampal dysfunction. *Transcortical dyslexia* might be either transcortical expressive or transcortical receptive with the diagnosis hinging on the speech evaluation (e.g., Joinlambert et al. 2012). Auditory repetition or echolalia is intact and there are features otherwise suggestive of Broca's or Wernicke's type speech deficits. *Transcortical expressive dyslexia* frequently involves dysfunction within the left frontal eye field. This would be conveyed in a patient having difficulty with visual pursuit toward and within the right hemispace (e.g., Guitton et al. 1985; see also Suzuki and Gottlieb 2013).

One young boy had this problem and the teacher was trying an often-used approach, but one which was potentially inappropriate for this type of dyslexia. The teacher tried to improve reading by making the text larger. This required heightened integrity of the frontal eye fields as the visual pursuit demands were now exaggerated. Instead, the child benefitted from smaller text to maximize reading fluency and to minimize the demands for visual motor integrity and rightward gaze and pursuit. Therapy, though, is very often the opposite in the requirements for intervention techniques to that employed to maximize functional activities of daily living. Therapeutic interventions include visual motor exercises with the requirement to pursue a visual stimulus such as a pendulum from left to right without head movements. Intervention efforts with the boy were implemented away from the scrutiny of his peers and free of confrontation and negative evaluation by others.

Transcortical receptive dyslexia may arise from disconnection of the ventral pathways from the occipital lobe toward Wernicke's area. This directly impacts the conversion of the visual grapheme into the component sounds in transition to the auditory association cortex. The alternative diagnostic scheme may be useful to the extent that the lesion impacts the angular gyrus at the tertiary association cortex (parietal occipital temporal region). This syndrome is commonly diagnosed as dyslexia with dysgraphia, whereas the alternative subcortical lesion presents as dyslexia without dysgraphia (e.g., see Ardila 2012). More specifically, the distinction is between dyslexia of cortical and of subcortical origins. The latter reflects a classic disconnection syndrome. More specifically, the angular gyrus, essential for the conversion of a visually presented grapheme into auditory and somatosensory associations, is disconnected from the occipital cortex of each cerebral hemisphere.

These individuals might be able to write but unable to read written text and even their own writing samples. So, graphical repetition of the lexical image is impaired in contrast to integrity in writing to dictation through the auditory modality. This disconnect is, instead, between the visual lexicon projecting to the bilateral occipital lobes and the left angular gyrus providing for crossmodal conversion of the visual lexicon into graphical praxicons.

Vogel and colleagues (Vogel et al. 2012) further note that studies using functional neuroimaging methods have generally converged on a set of left hemisphere regions used for single-word reading (Jobard et al. 2003; Turkeltaub et al. 2002; Bolger et al. 2005; Vigneau et al. 2006), including a region near the fusiform cortex in the left occipitotemporal border. Indeed, some have referred to this region in the left hemisphere as the visual word form area (see McCandliss et al. 2003). Regions near the left supramarginal gyrus and angular gyrus have been functionally described as phonologic and/or semantic processors (Binder et al. 2005; Church et al. 2010; Graves et al. 2010). The frontal end of this neural architecture at the left inferior frontal gyrus has instead been related to phonological and/or articulatory processes with its direct proximity and influence over the motor cortex for the production of word sounds (e.g., Mechelli et al. 2003; Fiez et al. 1999; Booth et al. 2007). Vogel et al. (2012) also point out that many studies that do not require reading aloud (e.g., Cohen et al. 2003; Polk et al. 2002) show activity in these same regions.

Dysgraphia

Five years before Broca's influential paper, Marcé (1856) had described a number of cases, where spoken and written language disorders were derived from separable and not parallel systems. Similar observations were published by Ogle (1867), including one case of aphasia without agraphia. This case was taken by Ogle as evidence "that the faculty of speech and the faculty of writing are not subserved by one and the same portion of cerebral substance"(1867, p. 106). By similar convention and rationale as that used to identify and understand dyslexia, the theory and anatomical specificity applied to the speech systems involved in aphasia and dyslexia may provide a useful and theoretically consistent basis for understanding the common disorders of writing, the dysgraphias (e.g., see Ardila 2012).

Expressive dysgraphia may result from left frontal lobe disorders and especially of the premotor region. Prominent here is Exner's hand area (Exner 1881; Franck-Emmanuel et al. 2009) essential for the regulatory control, planning, and sequencing of right-hand movements resulting in written text. This area located in the middle frontal gyrus was identified as the "graphic motor image center." Problems here might be conveyed in nonfluent graphics with perseverative and/or phonemic errors in the writing sample. Though nonfluent, the writing would potentially be meaningful in content, which might be understood by the reader. *Receptive or transcortical receptive dysgraphia,* in contrast, might be fluent to the point of hypergraphia and sometimes with burgeoning volumes of meaningless ramblings and word salad content. Often, the patients express their misperception that this

Fig. 11.12 A writing sample from a patient with mixed agraphia, including features of hypergraphia and multiple perseverative errors

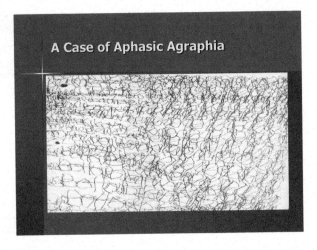

meaningless text, by any standard, is actually a very important expression of their thoughts and beliefs. *Mixed dysgraphia* based on this system might be conveyed with expressive organizational-sequencing deficits or inertia in addition to meaningless word salad content. Figs. 11.12 and 11.13 provide samples of two patients with aphasic agraphia.

The severity of dysfunction does vary from individual to individual and with variations in the pathology. This may present with odd features in some patients. Loss of the functional integrity of one or the other cerebral hemisphere may result in the release of the other cerebral hemisphere with its own efforts to complete activities normally processed by the damaged brain. Two examples are provided below. The first (see Fig. 11.14) provides a sample of the right brain's efforts to perform a graphical task, wherein the instructions are given with multiple prompts to "Write a sentence about a man and a car." This brain *writes* a picture of a man and a picture

Fig. 11.13 A writing sample from a patient with aphasic agraphia

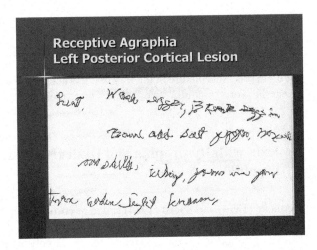

Fig. 11.14 An intact right brain attempting to perform a typically left brain task as it writes a sentence about a man and a car

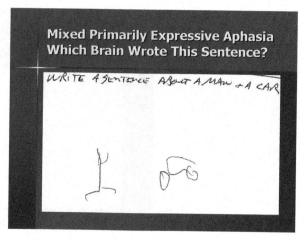

Fig. 11.15 An intact left brain attempting to perform a typically left brain task as it writes a sentence about a man and a car

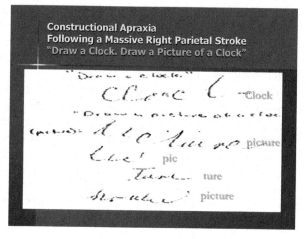

of a car. The second example (see Fig. 11.15) provides a sample of a left brain's attempt to "Draw a picture of a clock." With the initial graphics depicting the "clock", the patient is instructed "No, draw a picture of a clock." Upon these directions, the left brain *draws* its picture in the way that it is accustomed to doing things! It simply *draws* the word "picture," "pic," "ture," and "picture."

Dyslexia with Dysgraphia and Dyslexia Without Dysgraphia

Wernicke (1874) and Kleist (1934) spoke of cortical alexia or *alexia with agraphia* and subcortical alexia or *alexia without agraphia*. This diagnostic framework was intimately connected to the anatomical regions involved and is most consistent with

Fig. 11.16 Test results from a young man with a history of developmental dyslexia with dysgraphia since the first grade with a remote undiagnosed CVA and an acute CVA at age 20. *CVA* cerebrovascular accident

Dyslexia With Dysgraphia on Developmental Dyslexia With Dysgraphia

- R Grip = 17 kg.
- L Grip = 39 kg.
- R Anesthesia
- R Finger Agnosia
- R Visual Field Deficits
- Dyslexia
- Dysgraphia
- Dyscalculia
- Circumlocution Errors

the framework of the present writings. This framework, along with Luria's designation of the aphasic alexias, is useful as it forms the common basis for the majority of clinical and research accounts of these problems. Although the disorders are acquired with brain damage subsequent to the establishment of these language-processing skills, the same systems are thought to be involved in many developmental language disorders. One younger man with a long-standing diagnosis of developmental dyslexia with dysgraphia was eventually seen for a stroke at age 20. The location of the cerebrovascular accident identified it as *subcortical dyslexia with dysgraphia* with visual access to the angular gyrus intact for both hemispheres. Interestingly, his school-related deficits were apparent by the first grade of public education with his MRI providing some evidence of a remote cerebrovascular accident with an acute event overlaid on that anatomical region. Relevant findings from the evaluation are included in Fig. 11.16, including grip strength, speech deficits, and sensory findings.

The classic lexical pathways from the Wernicke–Geschwind Model are depicted in Fig. 11.17. Wernicke's model was revived by Norman Geschwind (1982) with reading requiring initial processing of the visual grapheme at the occipital cortex. From here, the visual grapheme would be processed by the angular gyrus consisting of tertiary association cortex. The ventral pathway converts the visual grapheme into associated sounds and comprehension within Wernicke's area. The dorsal pathway provides for conversion into somatosensory associations, including postures necessary for the graphical depiction of the text. Regardless, Wernicke's area provides access to the arcuate fasciculus to carry information to Broca's area for the organization of motor sequences or ideational praxis necessary for oral production in speech (reading out loud) or for motor graphics with sequential movements of the hand in writing (Exner's hand area).

From this model, the angular gyrus at the parietal-occipital-temporal region is essential for reading and for writing. Dejerine (1891) identified the anatomy of dyslexia with dysgraphia as a cortical lesion of the angular gyrus within the left cerebral hemisphere. Anything in the way of a structural defect or lesion in this

Classic Lexical Pathways

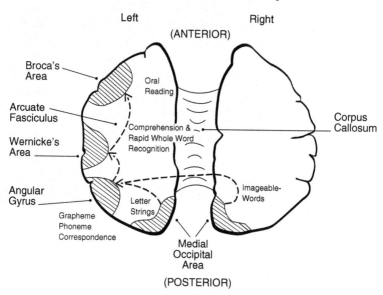

Fig. 11.17 The classic lexical pathways for visual input into the angular gyrus and on to Wernicke's area with transmission via the arcuate fasciculus to the frontal cortical processors at Broca's area and then to the motor cortex

region may result in *dyslexia with dysgraphia,* whereas disconnection of the access of the visual analyzers to this region results in *dyslexia without dysgraphia.* The former has sometimes been called "parietal alexia," although this is somewhat off base as the region is tertiary association cortex with multimodal sensory analysis and comprehension capabilities interfacing or juxtaposing the capabilities of the parietal, occipital, and temporal regions in the angular gyrus. The latter disorder of visual access to the angular gyrus is by definition a disconnection syndrome, though many others exist.

The clinical syndrome of dyslexia with dysgraphia is depicted in Fig. 11.18 below. This region (the angular gyrus) has a long history as the hotbed location for research on learning disabilities. This is exemplified in Josef Gerstmann's (Gerstmann 1940, 1957; see also Cabeza et al. 2012) work, identifying the features of a complex syndrome, which accompanies a lesion in the angular gyrus. Gerstmann's syndrome (Gerstmann 1940, 1957; see also Cabeza et al. 2012) includes a variety of learning disabilities or school-related deficits, but most commonly it is associated with dyslexia, dysgraphia, dyscalculia, ideomotor or gestural dyspraxia, anomia, left–right confusion, and finger agnosia. Fig. 11.19 shows the location of the angular gyrus where cortical dysfunction may result in the syndrome of dyslexia with dysgraphia.

Subcortical Gerstmann's syndrome with subcortical dyslexic dysgraphia often provide for variable or fluctuating features secondary to the functional integrity of the overlying cortical processors, elements of stimulus persistent in the oscillating

Fig. 11.18 Features associated with alexia with agraphia

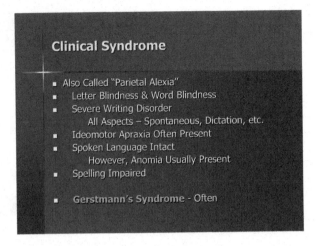

Fig. 11.19 The anatomy of dyslexia with dysgraphia with cortical dysfunction at the angular gyrus. Modified image: Adapted with permission from http://www.brains.rad.msu.edu and http://brainmuseum.org, supported by the US National Science Foundation

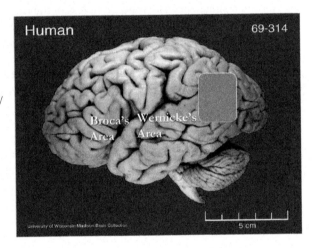

circuits for vision and audition, and acquisition or carryover issues with involvement of hippocampal and perihippocampal regions. Figure 11.20 provides a writing sample from a 45-year- old man with subcortical dysgraphia and Gerstmann's syndrome. The sample provides examples of the semantic difficulties associated with this lesion, which is also displayed in the figure.

History provides provocative accounts of the disconnection syndrome we know as dyslexia without dysgraphia. Valerius Maximus (30 AD) described a man who was struck on his head with an axe and lost his memory for letters, but had no other cognitive problems. Geronimo (Gerolimo) Mercuriale (1588) noted in astonishment "This man (after a seizure) would write but could not read what he had written." Subcortical dyslexia, otherwise referred to as dyslexia without dysgraphia is better thought of as a disconnection syndrome with severing of the access from the bilateral occipital lobes' visual analyzers to the angular gyrus. With this lesion, the cortical language systems remain intact and are sufficient for written language

Fig. 11.20 A writing sample and MRI section from a 45-year-old man with subcortical Gerstmann's and agraphic alexia. *MRI* magnetic resonance imaging

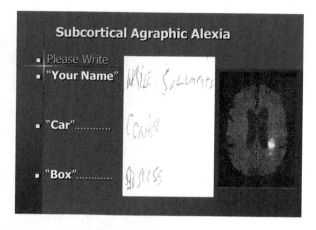

production, though lexical processing is impaired. Typically, this involves a single lesion, which simultaneously produces a loss of the right visual field with lesion of the left occipital region or of the optic radiations projecting to this area and with concurrent lesion of the splenium of the corpus callosum, such that the information from the left visual field can no longer directly access the left angular gyrus. Somatic and auditory access to language-processing areas may be intact and these pathways may provide for improved therapeutic intervention approaches. The lesion which was sufficient to produce dyslexia without dysgraphia in one 44-year-old man is depicted in Fig. 11.21.

Dyslexia without dysgraphia most commonly results from cerebrovascular infarct within the distribution of the left posterior cerebral artery (see Jacobson and Marcus 2011), whereas dyslexia with dysgraphia typically results following cerebrovascular infarct within the distribution of the posterior branches of the left middle cerebral artery. Both result from dysfunction within the left cerebral hemisphere and both reflect posterior brain disorders. However, dyslexia without dysgraphia, in

Fig. 11.21 MRI showing cerebrovascular accident within the distribution of the left posterior cerebral artery with damage to the left occipital region and to the splenium of the corpus callosum. *MRI* magnetic resonance imaging

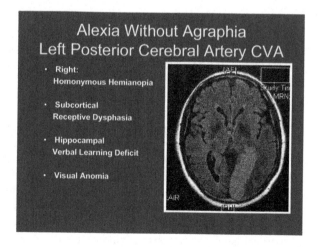

the classic literature, results from a subcortical lesion. This lesion effectively eliminates direct access of vision to the angular gyrus. It may lesion the optic radiations of the left temporal region travelling adjacent to the hippocampus and related structures. This lesion commonly will impact the hippocampus and perihippocampal regions, resulting in impaired learning or carryover (see Squire 1992, 2004; see also Iijima et al. 1996; Fernaeus et al. 2013) of verbally presented information within the auditory or visual systems.

The dorsal pathway with access to the somatosensory processors may or may not be affected. It is more probable that the dorsal pathway will be affected and this may show up behaviorally in features of a subcortical ideomotor or gestural dyspraxia. But the therapist and caregiver may appreciate that the posterior cerebral artery has primary access to brain regions involved in the somatosensory and proprioceptive awareness of the lower extremities, the bladder, and the genitalia. With a left posterior cerebral artery cerebrovascular accident, the right leg may well be unsafe and unreliable with the patient placing the limb in inappropriate postures for the task at hand. This may be apparent in a waxing and waning of safety in ambulation. The prognosis worsens secondary to the right visual field defect.

This lesion distribution characteristically results in a right homonymous hemianopsia or visual field defect within the right visual field. This finding is a staple or basic validating feature of the lesion distribution and its absence may bring into question the diagnostic impression. The lesion is subcortical to the angular gyrus and, as such, it may yield a subcortical Gerstmann's syndrome. This may wax and wane with intermittent deficits in left–right awareness, ideomotor praxis, and even in graphics. However, with the integrity of the cortical language systems, the therapists and the examiner may be able to demonstrate integrity of graphics relative to the reading impairment.

Atypical syndromes mimic these classic neuropsychological disorders. These may present with various locations or distributions of pathology. For example, *dyslexia without dysgraphia-atypical* might result from damage elsewhere in the brain and outside of the distribution of the posterior cerebral artery. A lesion within the pons or cerebellar region might result in *optic ataxia* where lexical processing is impaired with graphics intact, secondary to poorly coordinated eye movements or visual scanning. One gentleman was frustrated as he observed the input from his visual gaze and pursuit movements to be unreliable, where his gaze would fall from line to line and even within a line of graphical text. Also, bilateral optic nerve damage might alter lexical processing secondary to blindness with preservation of graphics, perhaps.

Dyspraxia

Within the functional neural system's theoretical framework, the movement disorders identified as one or another of the dyspraxias originate more broadly from dysfunction within Luria's second and third functional units. Damage to the second functional unit, consisting of the regions located toward the back of the brain which receive, analyze, and comprehend sensory information, may be sufficient to produce an *ideomotor or gestural dyspraxia* (e.g., Hermsdörfer et al. 2012; Vingerhoets et al. 2012). The third functional unit consists largely of the frontal lobe and regions of the brain responsible for the organizational sequencing and intentional basis for movement (see Duffau 2012). Damage within this unit may be sufficient to produce an *ideational dyspraxia* (Marcuse 1904; Pick 1905; Heilman 1973; see Heilman and Gonzales Rothi 2012).

The location of one or another form of the dyspraxias corresponds with the topographical representations of the body distributed along the homunculi across the somatosensory, motor, and premotor regions. Dorsal lesions more commonly impact the functional integrity of the lower body and lower extremities and ventral lesions more commonly impact the upper extremities and/or facial regions. The parietal lobe is a prominent contributor to normal praxis due to the strong fundamental role of the body senses or somatic senses in the appreciation of body postures or gestures. Alterations here may result in dyspraxia with a fundamental flaw in the somatosensory basis underlying the movement with erroneous placement of the body part or impaired praxicons.

Within Luria's second functional unit and within the parietal lobe of each cerebral hemisphere is the topographically arranged somatosensory cortex. The primary projection area for somesthesis is intimately connected with the premotor cortex via U-connectors and with the basal ganglia, including the head and the body of the caudate nucleus. This relationship provides for kinesthesia or the sensory bases for movement and gestural praxis. Disruption within this region may yield clumsy or discoordinated movement with the postural placement being incongruent with the planned and sequenced expression of that body part. The loss here is in the sensory basis of the movement, posture, or gesture and with an untrustworthy accounting of the expression or movement. Impaired placement or implementation of the body part may more directly be appreciated with anesthesia of the body region represented within the somatosensory cortex, where the loss of sensation is fundamental to the accuracy of placement of the body part relative to other body regions (proximal space) and relative to distal spatial coordinates within the environment (extrapersonal space). The former is more directly associated with specialization of the left cerebral hemisphere, whereas the later is more fundamental to the specialization of the right cerebral hemisphere (Chewning et al. 1998; Heilman et al. 1995; Jeerakathil and Kirk 1994; Jeong et al. 2006, Foster et al. 2008).

The relationship between the sensory processing specialization of the parietal regions, which underlies movement and movement disorders, increases in complexity with dysfunction of the secondary and tertiary association areas. The secondary association cortex within the parietal lobe provides for stereognosis and three-dimensional depth perception and comprehension within the somatosensory modalities. The functional disturbance arising from a lesion here will remain under the category of an ideomotor or gestural dyspraxia, but the comprehension of the bodily placement may be more meaningless or irrelevant to the planned movement underway. Here, the loss is not in the sensory receptive category with a pure loss of sensation (anesthesia) so much as it is a loss of the comprehension or meaning of the posture, gesture, or limb placement. This may occur without the patient's ability to appreciate the error, as this cortex appears to be specialized for that purpose. And, it has been compromised in doing so! This lesion will essentially produce body confusion or body part confusion and often this develops with some degree of compromise for left–right hemibody awareness. Dysfunction at the left parietal occipital region or tertiary association cortex may result in the characteristic components of Gerstmann's syndrome, including dyslexia, dysgraphia, left–right confusion, body confusion, and other maladies (Gerstmann 1940, 1957; see also Cabeza et al. 2012). Pathology at the angular gyrus and within the distribution of the angular artery branch of the middle cerebral artery may be expressed in these clinical features.

For the therapist or caregiver, it may help to appreciate that the fundamental basis of movement includes the awareness of the names of the body parts, where they are located on the body, and how to position them in meaningful ways. Therapy might initially establish a high level of accuracy in the identification of the body parts and perhaps with over rehearsal to the point of mastery. This may be followed by postural or positional therapies, where the patient replicates the placement of the limb or body part based upon either prompts or cues provided within extrapersonal and personal space to promote an integration of these spatial dimensions. The final stage of therapy might attempt the implementation of ideational components or the sequential processing and arrangement of movements underlying activities of daily living or functional activities defined by the individual's needs.

Proprioceptive deficits affecting postural placement of body parts or movements are common with parietal lobe involvement. This is more apparent with a right parietal lobe disorder and perhaps pathology within the secondary and tertiary association cortex at this region. The initial assessment includes the parietal drift test (Wyke 1966) where the arms are extended in front of the patient with the eyes closed. The patient is instructed to maintain the arms at that location. Right parietal lobe lesions may drift at the left arm into extrapersonal space and this is typically lateral drift. Lateral drift of the right arm appears to be less common, whereas that extremity might display a mild medial drift secondary to a parietal disorder. Frontal lobe deficits impinging on the motor projections may have some down drift of the contralateral arm secondary to weakness or a vertical neglect.

Also relevant to the tertiary association cortex or the dorsal pathways through the angular gyrus of the parietal occipital temporal region are the disconnection syndromes common to this region. This tertiary association cortex provides the neuroanatomical architecture for crossmodal sensory integration, analysis, and comprehension, where body posture or gesture is interfaced with visual feedback and analysis and with auditory feedback and analysis. Lesion within this region may disconnect one sensory modality's analytical capability from the needs of the brain region, analyzing and comprehending the other sensory modality. Here, the disturbance in ideomotor praxis may involve a loss of the visual contributions in matching postural gestures to the shape of the object (e.g., coffee cup) as it is recognized within the visual analyzers (e.g., Vingerhoets et al. 2012). Also, auditory contributions to the postural array may be lost or disturbed. These sensory disconnection syndromes may be assessed, for example, by asking the patient to identify, by touch alone, the object presented visually. Alternatively, the patient might be asked to recognize from among a sample of visually presented hand postures the one which corresponds with the current placement of their hand (out of sight).

Luria's third functional unit, consisting of the frontal lobe, provides the fundamental basis for organizational sequencing of movements or praxicons (e.g., see Duffau 2012). This is more clearly related to the premotor cortex and the associated functional deficits in motor praxis identified as the ideational dyspraxias. The specific body location for the expression of the ideational dyspraxia again relates to the locations along the homunculus of the premotor cortex, where lower-body structures are present dorsally and upper-body regions are present ventrally in this system. Ideational dyspraxias may present in the speech apparatus or vocal motor apparatus from dysfunction within Broca's area of the left hemisphere. The patient may produce sequential phonemic errors as with a phonemic paraphasia (see Duffau 2012). The sequencing and initiation of propositional speech sounds may be effortful and appear halting and hesitant, for example. Disarticulated movements with impaired sequencing may appear with an ideational dyspraxia at the hand from dysfunction within Exner's hand area or at any body region. This relationship with skeletal muscles is well established. In contrast, much research is needed to provide for independent investigations of the premotor control of smooth or syncytial muscles along with investigations of the regulatory control over cardiac muscle.

The left cerebral hemisphere appears specialized for precise motor movements and precisely sequenced praxicons. These well-regulated sequential actions may be produced under stress partially as a function of the capacity of the left premotor region for managing multiple concurrent task demands. In addition, research supports the role of the left brain in managing body regions and postures, whereas the right brain appears to be specialized for extrapersonal spatial analysis and activities. In this respect, ideational and ideomotor dyspraxias often signal a left-brain disorder, although these problems may originate with dysfunction of the right brain. In contrast, the positioning of the body part within extrapersonal space and the oppositional support to planned movements may be differentially a contribution of the right cerebral hemisphere.

Dysfunction approximating the motor cortex and the origins of the upper motor neurons may mimic dyspraxic disorders. However, the sloppy movement resulting from such a lesion results from muscle weakness or dystonia rather than sequential coordination. Also, the therapist and caregiver might appreciate that the frontocerebellar pathways provide for coordinated movements, with cerebellar dysfunction often presenting with movement problems that are somewhat similar to the ideational dyspraxia of frontal lobe origin. This was initially recognized by Jackson (1864; see also 1958). Cerebellar lesions result in ataxiaor dysmetria and feature brainstem symptoms originating from nearby cranial nerves. Historically, many tests of motor coordination have been referred to, collectively, as tests of frontocerebellar function with reference to the frontodentatorubrothalamic tracts. However, cerebellar disorders are not restricted to presentations of discoordinated intentional movements. Indeed, cerebellar dysfunction often mimics an ipsilateral frontal lobe impairment altering coordinated cognition and/or emotional expression (e.g., Heilman et al. 2012) associated with that region. Heilman et al. (2012) provide a case study of disordered affective communication or dysprosodia with cerebellar damage. Schmahmann (2004) discusses the greater breadth of cerebellar disorders involving features of ataxia, dysmetria of thought, and the "cerebellar cognitive affective syndrome."

Gerstmann's Syndrome

Josef Gerstmann (Gerstmann 1940, 1957; see also Cabeza et al. 2012) identified several features resulting from a left parietal lesion at the angular gyrus and this syndrome now bears his name. Gerstmann's syndrome includes a variety of school-related learning disabilities; but most commonly, it is associated with dyslexia, dysgraphia, dyscalculia, ideomotor or gestural dyspraxia (e.g., Hermsdörfer et al. 2012; see also Vingerhoets et al. 2012), anomia, leftright confusion, and finger agnosia. The features of Gerstmann's syndrome are identified in Fig. 11.22. Some refer to this same area as the inferior parietal lobule based upon its location and lobular appearance.

Dyslexia with dysgraphia is a feature of cortical dysfunction within the left angular gyrus or parietal occipital temporal region as discussed elsewhere in these writings. Left–right confusion is commonly conveyed with encroachment into the parietal lobe, where body confusion is common with damage to the association cortex around the somatosensory projection areas. The parietal lesion may result in body confusion and the loss of the representational postures or gestures which underlie an ideomotor or gestural dyspraxia (e.g., Hermsdörfer et al. 2012; see also Vingerhoets et al. 2012). The somatic basis for movement is disrupted with the dorsal pathways, whereas access to the language systems of Wernicke (1995) and Geschwind (1982) involves the ventral pathways. Both are relevant to anomia as

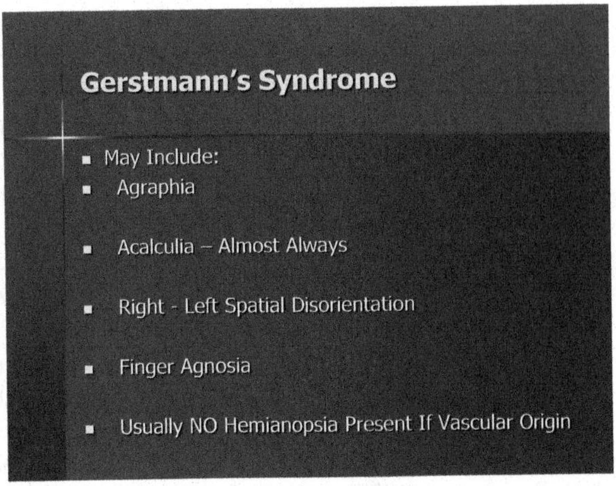

Fig. 11.22 Gerstmann's syndrome involves several features relevant to research on learning disabilities in school settings. This syndrome results from dysfunction within the angular gyrus and/or left parietal-occipital-temporal dysfunction

Gerstmann's Syndrome

- May Include:
- Agraphia

- Acalculia – Almost Always

- Right - Left Spatial Disorientation

- Finger Agnosia

- Usually NO Hemianopsia Present If Vascular Origin

somatic representations and auditory representations interface with the visual representations of the object for associative integration.

Additionally, Gerstmann's syndrome involves dyscalculia or impaired numerical processing of the operations involved in arithmetic computations. Salomon E. Henschen (Henschen 1925; see Kahn and Whitaker 1991) may have been the first individual to use the term "acalculia," the inability to perform the operations concerned with arithmetic. Prior to Henschen's work, Lewandowsky and Stadelmann (1908) published one of the first recognized cases of an acquired calculation disorder. Ultimately, Peritz (1918) argued for a "calculation center" in the left angular gyrus.

Neurological and neurophysiological findings suggest that the posterior parietal cortex is a critical component of the circuits that form the basis of numerical abilities in humans (see Roitman et al. 2012; see also Cabeza et al. 2012). Patients with parietal lesions are impaired in their ability to access the deep meaning of numbers. Patients with inferior parietal damage often have acalculia with performance deficits in arithmetic (2+4=?) or number bisection (what is between 3 and 5?) tasks, whereas the reading or writing of numerals and the ability to recite multiplication tables remain intact. Roitman and colleagues (Roitman et al. 2012) note in their review of functional imaging studies of neurologically intact humans that performing subtraction (Simon et al. 2002), number comparison (Pinel et al. 2001), and nonverbal magnitude comparison tasks (Fias et al. 2003) heightens activity in areas within the intraparietal sulcus. They conclude that clinical cases and imaging studies reviewed together support a critical role for parietal cortex in the mental manipulation of numerical quantities. Moreover, imaging studies have consistently found activation of parietal cortex in the processing of symbolic number, including number words or Arabic numerals (Dehaene and Cohen 1997; Pinel et al. 2001; Simon et al. 2002).

The posterior parietal region or intraparietal sulcus appears critical to the encoding of whole number estimates as well as proportions. In their review, Roitman and colleagues (Roitman et al. 2012) report that subsequent studies replicate the findings on the role of the intraparietal sulcus in processing nonsymbolic visual stimuli and extend these findings to proportions (Jacob and Nieder 2009). They report that once subjects had habituated to arrays of elements in which 50% of the totals (ranging from 4 to 32) were colored blue and the rest were red, infrequent probes with deviant stimuli composed of 60–90% red items dishabituated the BOLD signal, with greater recovery, as the distance between habituation and deviant proportion increased. Also, Piazza and colleagues (Piazza et al. 2004) tested whether representation of number in parietal cortex might extend to nonsymbolic stimuli. Participants passively viewed visual arrays of n elements. As subjects habituated to a standard number (16 or 32), deviant values, ranging from half to double the value of the standard number, were occasionally presented. Although participants were not explicitly required to discriminate the visual stimuli in any way, recovery of the BOLD signal along the intraparietal sulcus was proportional to the ratio of the standard and deviant stimuli. It is notable here that these authors report that the same region did not respond to changes in the shape of the elements nor did it depend on whether nonsymbolic (arrays of dots) or symbolic (Arabic numerals) stimuli were used (Piazza et al. 2007).

Fig. 11.23 A severe left parietal stroke with Gerstmann's
syndrome and other features

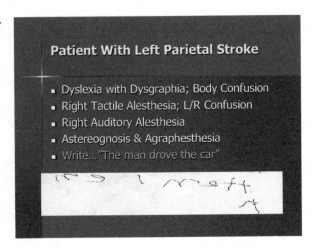

Cabeza et al. (2012) note in their review of the functions of the ventral parietal cortex that although activations of this region tend to be bilateral, there is an emphasis on the role of the left angular gyrus in numerical processing (Dehaene et al. 2003). They note that activation of the ventral parietal region is heightened for exact than for approximate calculations (Stanescu-Cosson et al. 2000) and for numerical answers stored in memory (Chochon et al. 1999). The latter includes the memory for answers to addition and multiplication problems. Moreover, this region demonstrates heightened activation when subjects solve previously trained multiplication problems or in other situations requiring numerical fact retrieval (Grabner et al. 2009a, 2009b).

Figure 11.23 presents a case of Gerstmann's syndrome with several of the identifying diagnostic findings but with additional symptoms derived from dysfunction within the surrounding cortical areas. A writing sample is provided along with the assessment results from this patient recovering from a left parietal stroke. This individual presents with the characteristic or expected deficits following damage to this brain region, including features of Gerstmann's syndrome. However, the stroke is severe such that the expected somatosensory association deficits, which might include finger agnosia and astereognosis, instead present with right-sided tactile allesthesia. With severe damage, the presentation of sensory information at the right side might be perceived by the patient as originating from the left side or left hemibody, reflecting the activation of the brain ipsilateral to the side of stimulation.

This patient's lesion is somewhat less localized as might be suggested by the auditory allesthesia and its implication of a severely damaged superior temporal gyrus as was the case with this individual. For the therapist or caregiver, the finding of allesthesia is a red flag for a more severe lesion than is typical in the recovery setting. It may be most useful in these cases to modify the expectations for recovery in rehabilitation settings to reflect the level of severity of the functional deficits. Otherwise, the patient may be viewed as failing to meet time deadlines and progress may well be underestimated, whereas the patient may indeed be making a good recovery and responding well to treatment. This mismatch between the

Fig. 11.24 A social leader in her community with "word salad " speech and receptive dysphasia

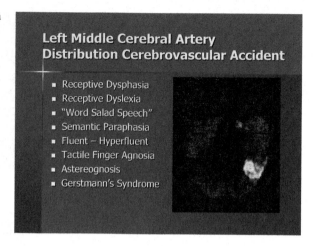

Left Middle Cerebral Artery Distribution Cerebrovascular Accident

- Receptive Dysphasia
- Receptive Dyslexia
- "Word Salad Speech"
- Semantic Paraphasia
- Fluent – Hyperfluent
- Tactile Finger Agnosia
- Astereognosis
- Gerstmann's Syndrome

therapist's expectations and the patients progress may result in ratio strain and has been known to lower the morale of the patient, the therapist, and family members, whereas modified expectations for progress might yield diametrically opposite attributions and outcomes.

The results of an evaluation of another woman, very actively involved in the social functions of her community, are depicted in Fig. 11.24. She suffered a stroke to the posterior branch of the left middle cerebral artery, resulting in Gerstmann's syndrome with receptive dyslexia and dysgraphia, receptive dysphasia, dyscalculia, tactile finger agnosia, and related features. Yet, she remained actively engaged in her social agenda from her room, despite being unable to distinguish "word salad" speech from normal language. She was verbally incompetent, while maintaining her enjoyment of talking, apparently a bit too much. The social stress and language processing demands of several people in her room were eventually shown to induce left temporal and parietal lobe seizures and auditory hallucinations or paracusia.

Left–Right and Body-Part Confusion

Much of the clinical investigation and treatment of right brain disorders has focused on the resulting disruption in the processing and awareness of extrapersonal space as with a left hemineglect syndrome (see Heilman and Gonzalez Rothi 2012b). In contrast, the patient with a left parietal brain disorder may suffer substantial confusion in the location and identification of body parts and the left or right side of the body (see Cabeza et al. 2012). This processing specialization difference between the cerebral hemispheres has been expressed within the theoretical literature through the conceptual appreciation of personal and peripersonal space (Heilman et al. 1995; Foster et al. 2008). Otherwise said, the left brain appears to be specialized for spatial analysis and comprehension of "personal space", including the basic ability to identify and localize the body and its parts, including the relational aspects of the left and right hemibody .

Indeed, this specialization for relational thought may contribute to seemingly unrelated constructs, including appreciation of mathematical and logical reasoning. By way of example, a linguistic relationship conveying increased or lessened proximity might be found within the family tree such as "her father is my uncle and your cousin, so she is your…, and so forth. The social and political impact of these relational thought disturbances remains largely undiscovered (however, see Tonkonogy and Puente 2009). In the present discussion of body part confusion, though, these abilities are fundamental to the dyspraxias or movement related disorders and mastering this component of left cerebral spatial analysis is a prerequisite for the accomplishment of basic gestures or postural placements. The patient with a left parietal lesion, presenting with body part confusion and/or left–right confusion might require interventions with over-rehearsal of these basic constructs to the point of mastery prior to the therapeutic interventions to master gestural placement and finally unfolding or developing sequential and coordinated body movements. Many individuals struggle with safety-related issues where, for example, the elbow might be identified as the patient's nose and where the ear might be confused with the hand. Again, the lesion distribution relative to the homunculus within the secondary and/or tertiary parietal association areas may determine the density of the confusion for specific bodily regions. For example, an inferior parietal lesion may alter facial gestures and/or produce facial confusion. The dorsal lesion near the midline may differentially influence the placement, localization, and comprehension of the lower extremity.

Chapter 12
Syndromes of the Right Brain

The right cerebral hemisphere appears specialized for survival with substantial evidence supporting its functionally preeminent role in threat detection, sympathetic drive, escape or avoidance behaviors, and awareness of locations and events within extrapersonal space. It was with these resources that I approached my apiary one fateful afternoon on a rainy day in June without my protective gear and equipment. It was cold and wet and I knew that the honeybees that I served cared not for either of these conditions or for me, perhaps. I opened the hive and instantaneously found myself automatically processing the more complex aspects of survival with nary a verbally based thought derived from my more logical analysis of these events. I was running now and with speed for shelter and for escape from the Pandora's box that I had so callously and naively opened but a moment ago. Surely, my heart rate, blood pressure, oxygen saturation, spatial analysis, and escape functions were released and in control of all that I was and all that I was going to do. By estimation, I outran tens of thousands of bees that day. I did not outrun 60 of them and they gifted me their lives through their suicidal stings. But these did not hurt on any measurement scale as much as the one bee that made it into the car with me and stung my lovely wife! I had no real difficulty with the beestings other than in my appearance like Rocky Balboa after his big fight. But with good conscience and reflection, I did call the local medical center for an emergency IQ test! The IQ test(s) though have been better indicators of the left hemisphere and especially based upon the rapid sequential processing and verbal linguistic analysis favored on these instruments. It would be hard to image the clinician even attempting the completion of an IQ test in an individual without substantial left cerebral functional capabilities. Instead, they would likely decline the consult as an "inappropriate referral."

The following sections provide for some elementary discussion of basic syndromes of the other hemisphere—the right brain.

© Springer International Publishing Switzerland 2015
D. W. Harrison, *Brain Asymmetry and Neural Systems,*
DOI 10.1007/978-3-319-13069-9_12

Limb Alienation Syndrome

Oliver Sacks (1985) described a patient who fell out of bed after pushing out what he perceived to be the severed leg of a cadaver, which had been hidden under his blanket by the staff. Patients may express their alienation in statements like, "I don't know whose hand that is, but they'd better get my ring off of it!" or, "This is a fake arm someone put on me. I sent my daughter to find my real one." One man, following a right middle cerebral artery distribution cerebrovascular accident, complained with disgust "why did the doctor take my arm off and put this thing on me?" The nature of the limb may take on delusional properties as with the university engineering professor that had been "hit by a Silkworm missile," which disconnected him from his wife, with the connection having been through the left arm. Alienation of the affected extremity may gradually resolve during recovery, first with the patient's recognition that the arm belongs to them and followed later by the effort to protect and care for the limb. Otherwise, the limb might easily be neglected and left in the spokes of the wheelchair while navigating the floor.

Bayne and Levy (2005) report on a potentially related scenario in which a Scottish surgeon by the name of Robert Smith was approached by a man asking that his apparently healthy lower left leg be amputated. The basis for the man's request for the amputation appears to be that his left foot was not part of him—it felt alien. By this account, Smith performed the amputation after a psychiatric consultation. Two and a half years later, the man stated that his life had been transformed for the better by the operation (Scott 2000). A second patient was also reported as having been satisfied with his amputation (Furth and Smith 2002). Apparently, additional amputations of otherwise healthy limbs were scheduled when the story broke in the media. Not surprisingly, there was a public outcry, and Smith's hospital administration instructed him to cease performing such operations. Although no hospital would encourage or condone healthy limb amputations, the authors suggest that the syndrome is fairly common with some organized communication among the afflicted. The authors extend their discussion without neuroscience evidence, to the possibilities for overlapping diagnostic categories here to include body dysmorphic disorder (Phillips 1996) and body image disturbance with anorexia nervosa (Garner 2002).

One woman had suffered a childhood traumatic brain injury which, by her family's account, had left her fitful and with foul language corresponding with intermittent panic attacks. She described limb alienation features for her left leg, along with episodes of rapid heart rate, fearful apprehension, and panic episodes. The left leg worsened in its developmental course and at age 60 the left hip was replaced, apparently without appreciation of the central origin of her symptoms. The leg was not fully incorporated into her concept of self and had never proved trustworthy along with multiple falls to the left side. The left hip replacement may have been doomed to failure, secondary to the right posterior pathology that left her with left-sided proprioceptive deficits, abduction of the left foot and leg, panic attacks, left limb alienation, and recurrent left-sided visual formesthesias. Figure 12.1 provides for an abbreviated list of her complaints and her drawing of the devil-like creature that she saw and heard at her left side concurrent with feelings of impending vital threat

Fig. 12.1 A patient with childhood-onset left lower extremity limb alienation syndrome with episodic panic attacks and left-sided hallucinations provides her drawing and a partial list of her complaints. *TBI* traumatic brain injury

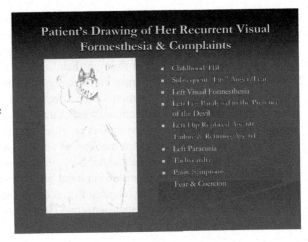

and coercion. She described a quiet, stealthy creature that would abruptly come at her and the voice of her father screaming warning sounds that indicated that "he wants to kill us!"

Limb alienation syndrome might reasonably be considered an amputation of the limb, wherein the central somatic and motor representations have been ablated with preservation of the limb itself. Within this framework, the visual or auditory projections and/or associations of the limb might be intact or the limb might be detected by its placement within the right hemispace or at the right hemibody. The syndrome implies some preservation of the negative affective analysis by right cerebral systems and these are typically reflective of disgust or anger with the attachment or foreign body, which is no longer integrated within the individual's self-concept. This limb may logically present obstacles for the afflicted person as it may become intertwined in the wheelchair spokes, provoking feedback from concerned therapists. With release of antigravity reflexes, the left hand might involuntarily grasp onto the chair or other objects making even routine activities of daily living more difficult. But, ultimately, the syndrome is rested within the negative affective processors of the right brain, where the limb is discordant with one's sense of self. This missing central element in the Gestalt or holistic perception of self may be fundamental to the emotional attributions that the contralesional limb should be removed or discarded. The therapist may work with the patient to reintegrate or assimilate the limb into self-concept, and this may be better done initially using the functions of the intact left cerebral hemisphere with logic, positive attributions, and placement within the right hemispace under the care of the right hemibody.

This emotional somatic disorder may be expressed with or without another somatic disorder more commonly lateralized to, but not exclusive to, the left cerebral hemisphere. Wolpert et al. (1998), for example, provide a case of "vanishing arm syndrome" that may be common to dense and focal lesions within the left parietal region. The resultant anesthesia for the body part represented within the lesioned area may be sufficient to remove tactile or somatic awareness of that body part.

The case described was of a 50-year-old woman (PJ) with a large left hemispheric cyst. She extinguished touch from the affected hand region when presented bilateral tactile stimulation at the homologous body region. In contrast, the visual fields were intact. She had astereognosis at the right hand, where she was unable to identify objects by touch and pressure alone. Periodically, she described that her limb would vanish without visual representation for the affected body part. Also, when the limb would "disappear" she might become aware of its position once again, if the limb is touched or if she could see it. The authors provide evidence on the temporal requirements for the limb to vanish as a function of somatic manipulation. Specifically, upon PJ placing her right arm behind her back, the limb would vanish as a function of the amount of weight placed within her hand. Heavier weights helped to maintain awareness of the limb, requiring longer time periods for the limb to vanish. Wolpert et al. (1998) demonstrate that somatosensation (feeling one's limb) is updated from visual (seeing the limb) as well as tactile inputs (something touching the limb).

Aprosodia

Aprosodia is a deficit in the comprehension, repetition, and/or expression of variations in tone of voice which are essential to convey both linguistic and emotional information (Monrad-Krohn 1947; Ross,1981; Ross and Mesulum 1979; Ross and Monnot 2008; Williamson et al. 2000; see also Leon and Rodriguez 2008). Affective aprosodia refers to a specific deficit in producing or comprehending the emotional or affective tones of voice. Aprosodia is most commonly associated with right hemisphere strokes; however, it may also result from other types of brain damage such as traumatic brain injury. Although research investigating hemispheric lateralization of prosody continues (e.g., Ross and Monnot 2008; Bourguignon et al. 2012), there is strong evidence that many aspects of affective prosody are directed by the right hemisphere. Disorders of emotional communication can have a significant impact on the quality of life for those affected and their families. However, there has been relatively little research regarding treatment for these disorders (e.g., Leon and Rodriguez 2008). The linguistic contributions of prosody vary across cultures and languages. However, the primary associations for prosodic variations in clinical neuropsychology have been related to the conveyance of emotional tone of voice and its related functional contributions to emotional intelligence. Disorders of emotional communication clearly have a significant impact on the quality of life for the individual and for their families. Yet, much remains to be done especially on the side of therapeutic interventions, which may improve the individual's relationships, and strengthen their overall prognosis (e.g., Leon and Rodriguez 2008).

Within the present theoretical framework, the aprosodias may be defined based on the origin of symptoms from within the second or third functional units, although prosodic variations may result from brain-stem dysfunction within the first functional unit. Receptive dysprosodia results from dysfunction within the right temporal lobe (Ross 1981; Ross and Monnot 2008). Assessment approaches include dichotic sound presentations with many variants described within the literature (see Pollmann 2010; Hugdahl et al. 2009; Hugdahl 2003, 2012). *Receptive dysprosodia* might be supported in the assessment with the finding of an auditory affect agnosia (Heilman et al. 1975). Testing includes the presentation of incongruent literal and affective prosodic words, phrases, and/or sentences (e.g., Emerson et al. 1999), where a literal or logical linguistic bias overrides the affective intonation. The assessment of a receptive dysprosodia may also be supported through documentation of a dysmusia, where tone and/or melody comprehension may be impaired. Cross validation of the syndrome might be useful through the assessment of hemispatial neglect and/or impairments in facial recognition, geographical confusion, constructional dyspraxia, or other syndromes resulting from right posterior cerebral pathology or malfunction (e.g., Fisher 1982; Tranel et al. 2009; Heilman et al. 2012b).

Expressive dysprosodia is distinguished in neuroanatomy largely based on cortical or subcortical origin within the right frontal lobe (e.g., Heilman et al. 2004). Expressive dysprosodia might present with poorly regulated speech volume and an avalanche in volume perceived as anger deregulation by others. This syndrome may be cross validated with anger regulation problems, facial dystonia especially at the left hemiface, left-sided motor and/or premotor deficits, motor constructional dyspraxia, and many other features discussed elsewhere in these writings. *Subcortical expressive dysprosodia* might be suggested by a staccato quality in speech and or elevated pitch. The latter might relate to hyperreflexia in the vocal motor apparatus and some patients show dyspnea especially when exposed to right frontal stressors or on exertion. The assessment of damage to the right frontal region yields many cross-validation features and many of these are affective in nature, including social improprieties, social pragmatics, social anarchy (e.g., Anderson et al. 1999; Damasio 1994; Eslinger et al. 2004; see also Yeates et al. 2012; see also Tompkins 2012), and anosognosia (e.g., Berti et al. 2005; see also Prigatano 2010). These features might be assessed if a dysprosodia is suspected not as a prerequisite for the diagnosis but based on proximity of the problem to these functional neural systems.

Transcortical dysprosodia is infrequently diagnosed or evaluated, whereas the clinical implications may be substantial. *Transcortical receptive dysprosodia* may present with hyperprosodia with emotional intonation in speech irrelevant to the context. This is homologous to the hyperfluent propositional speech presented with a transcortical dysphasia, whereas the hyperfluency in this case is for emotional sounds or prosodic variations in speech. In mild cases, the patient may appear overly emotional secondary to the elevated prosodic expressions. In extreme cases, the patient may be excessively vocalizing with moaning and wailing more common. This patient may be managed as though she were suffering from a pain disorder with medication and close nursing supervision, whereas this is a brain disorder with pathology within the right cerebral speech systems affecting prosodic variation or vocalization. Bright light therapy was successful, in one project, in reducing vocalizations with reliable reductions in prosody and corresponding increments in propositional speech expressions (Harrison et al. 1990). But, these interventions have not effectively eliminated the presenting complaint, and the clinical impact may be marginal. More often, the primary impact of this diagnosis is the realization by the staff and physicians that the hyperprosodic expressions are fundamental to these right brain speech systems with impaired comprehension of emotional variations in others' speech with *auditory affective agnosia*. The implication would be for interventions to focus on the literal conveyance in communication as the comprehension and meaning of prosody is impaired. The emotional expression in prosody may be meaningless or less relevant, similar to the word salad speech resulting from a lesion within the homologous brain region.

Although seldom documented, *echoprosodia* appears to be a common malady comparable to the frequency of echolalia among the clinical brain disorders. Echoprosodia may be evaluated through the presentation of a moan or groan with affective nuance and especially if these have been observed from the patient. The induction may be sufficient to drive the echo in many patients with transcortical dysprosodias. Regardless, confirmation of the integrity of prosodic repetition is fundamental to the diagnosis of a transcortical dysprosodia just as intact repetition of propositional speech is fundamental to the diagnosis of transcortical dysphasia (e.g., see Joinlambert et al. 2012).

Many caregivers and family members have had faulty attributions toward the hyperprosodic patient that the moaning and wailing are done intentionally to irritate them or for the purpose of obtaining pain medications. But, the syndrome is readily cross validated through the finding of congruent symptoms indicating dysfunction within proximal right cerebral systems or brain regions. Transcortical receptive dysprosodia involves brain systems where spatial confusion and delusions are common as well as fear disorders with formesthesias of negative affect within the left hemispace or at the left hemibody. In many cases, the patient may be at some increased risk for elopement with recovery of the lower extremities and increased mobility. This may result from right posterior circulation cerebrovascular accidents, anoxic encephalopathy where transcortical features are prominent, and from many other ailments. For example, many individuals with autism spectrum disorders present with unusual or odd-sounding prosody. Although this feature is a widely noted observation, it continues to be perceived as an under-researched area (McCann and Peppé 2003).

Hughlings-Jackson (1915) suggested the dominant role of the right hemisphere in emotional communication. Subsequently, Heilman et al. (1975) provided perhaps the first modern study showing right-lesioned patients to be impaired in the recognition of affective content in otherwise linguistically neutral passages. This emotional defect in speech was described as "auditory affect agnosia." Tucker et al. (1977) later provided evidence of impaired prosodic expression of neutral passages in patients suffering right cerebral damage. Ross and Mesulam (1979) also described two patients with significant psychosocial difficulties, unable to insert affective intonation into their speech, subsequent to right anterior supra-Sylvian infarctions. By 1981, Ross had provided a neuroanatomical model of the aprosodias, which is depicted in Fig. 12.2. This model is somewhat of a homologue for the Wernicke–Geschwind model of propositional speech designed for left cerebral systems. Although the model is used for the assessment of individuals with various acquired brain disorders, it has been applied to clinical populations with emotional disorders and even within the context of a developing child. For example, Carol

Fig. 12.2 A working model for syndromes of motor and sensory aprosodia (Ross 1981). (Adapted with permission from http://www.brains. rad.msu.edu, and http:// brainmuseum.org, supported by the US National Science Foundation)

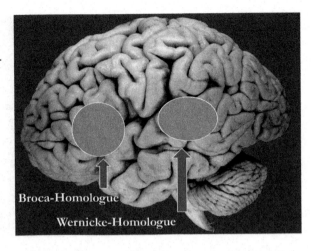

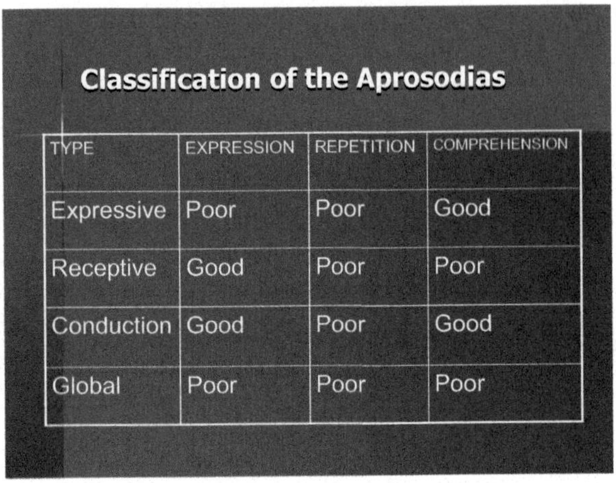

Fig. 12.3 A comparison of expression, repetition, and comprehension expectations for expressive, receptive, conduction, and global aprosodias

Classification of the Aprosodias

TYPE	EXPRESSION	REPETITION	COMPREHENSION
Expressive	Poor	Poor	Good
Receptive	Good	Poor	Poor
Conduction	Good	Poor	Good
Global	Poor	Poor	Poor

Emerson (Emerson et al. 1999; see also Emerson et al. 2005) investigated depressed and nondepressed school-aged boys on the processing of prosodic speech sounds conveying different affective valences (e.g., anger, sadness, happy, and neutral affective intonation). The depressed group was deficient in auditory affective analysis across each affective valence and with a negative emotional bias. Figures 12.3 and 12.4 provide for comparisons of some of the classification attributes of the various aprosodic disorders.

Amusia

Soon after Broca (1865) reported the first case of a linguistic speech disorder resulting from a lesion within the left hemisphere, Bouillaud (1865) presented a series of cases in which various musical abilities were lost secondary to brain insult. Notably,

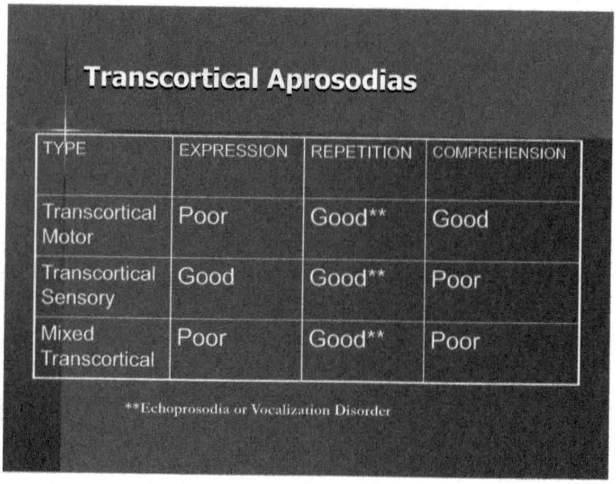

Fig. 12.4 A comparison of expectations for expression, repetition, and comprehension for the transcortical aprosodias

Transcortical Aprosodias

TYPE	EXPRESSION	REPETITION	COMPREHENSION
Transcortical Motor	Poor	Good**	Good
Transcortical Sensory	Good	Good**	Poor
Mixed Transcortical	Poor	Good**	Poor

**Echoprosodia or Vocalization Disorder

he also presented a patient with preserved musical processing, with the ability to compose and write music, while having lost the ability to speak or to write language. The former auditory spatial processing deficits are now better known within the literature on amusia, where reception and comprehension of musical sounds has been associated with auditory projection and association cortices within the temporal lobes and where expression and/or regulatory control has been related to the frontal lobes. Others (Calabrò et al. 2012) have reported musical hallucinations following a right temporal lobe ischemic stroke. Functional magnetic resonance imaging (fMRI) assessment revealed that although an activation pattern involving the primary auditory cortex and the temporal associative areas, bilaterally, was found in the patient and normal controls, the musical hallucinations were evident with significantly increased activation mostly located in the right temporal cortex (in the ischemic area).

Philosophers, musical theorists, and neuroscientists have appreciated a connection between emotions and tonality in speech. This seems to be commonsensical as the tones in music are like a variant of the tones in human speech, which indicate emotional content. Phonemic sounds and vowels may be elongated for a dramatic effect such that musical tones might be viewed simply as exaggerations of the normal verbal tonality. This simpler view attributed language lateralization to the left cerebral hemisphere and music lateralization to the right cerebral hemisphere. This widely held view that the cerebral hemispheres are specialized for dealing with entire categorical functions like language or music has generally lacked support. Instead, the research has generally supported a component analysis such that aspects of speech may be differentially lateralized, as may be the content of music. Nevertheless, the evidence for cerebral laterality of function is overwhelming and a fundamental facet to neuropsychology and to neuroscience.

The right hemisphere appears generally responsive to pitch, melody, and emotional intonation or the gist conveyed from an auditory spatial array, whereas the left hemisphere has frequently been described as an auditory sequential analyzer as with propositional speech comprehension or expression. There is also evidence suggesting that neural responses to pleasant music in general tend to be more lateralized to the left hemisphere, an effect consistent with the "valence lateralization model" (Altenmuller et al. 2002; Flores-Gutierrez et al. 2007). fMRI investigation of emotional valence and memory in musical processing (Green et al. 2012) found that subjective liking per se caused differential activation in the left hemisphere, of the anterior insula, the caudate nucleus, and the putamen. Furthermore, sequential rhythmic analysis, rather than melody, may be a left cerebral specialization (Gordon and Bogen 1974). Using magnetoencephalography, researchers (Kuchenbuch et al. 2012) have also found that long-term musical training affects the lateralization of processing of pitches toward faster processing in the left hemisphere when there is a strong rhythmic or temporal component to the tonal stimulation (see also Kung et al. 2013).

Music may serve as an amplifier for emotion challenging the frontal capacity for regulatory control over these systems. Philip Klineburger (Klineburger and Harrison 2013) has argued that peak emotional experiences may result more readily from the exposure to intense emotional music and especially in those individuals with minimal frontal resources afforded to self-regulation or regulatory control. Reduced capacity appears to foster musical creativity with more divergent associations and less common or typical organizational relationships in musical production. In situations providing concurrent emotional stressors or with inherent incapacity in these systems, the resources driving associative analysis and production may require broader associative resources. This prediction is intimate to the concept of shared functional cerebral systems (e.g., Hebb 1949, 1955, 1959; see also Kinsbourne and Hicks 1978) processing these temporal and spatial auditory events.

Auditory spatial analysis by the right hemisphere includes the comprehension of nonpropositional speech components or prosody (e.g., Ross and Monnot 2008; see also Bourguignon et al. 2012). The evaluation of a brain disorder within these systems overlaps in the assessment for a dysprosodia and for one or another dysmusia (e.g., receptive, expressive, and transcortical dysmusia). In a project evaluating music-processing ability in patients who had undergone unilateral temporal cortectomy for the treatment of intractable epilepsy (Liegeois-Chauvel et al. 1998), researchers found that a right temporal cortectomy impaired the use of contour and interval information in the discrimination of melody. Instead, a left temporal cortectomy impaired the use of interval information. The superior temporal gyrus was most critical for musical processing and a distinction was drawn with the anterior superior temporal gyrus more important to metric processing. In these comparisons, the posterior temporal gyrus of the right cerebral hemisphere was more crucial for processing pitch and temporal variation.

The left hemisphere's specialization has been described for rhythm or rapidly processed sequential sounds. Individuals preparing to tap out a rhythm of regular intervals activated the left frontal cortex, left parietal cortex, and the cerebellum (Tramo 2001). More difficult rhythms recruited activation across broader areas of the left brain and, if the rhythm demands activation of sympathetic drive, the right cerebral hemisphere may play an increased role, thereby shifting the dynamics with

heightened emotionality and elevated arousal. Others (Kung et al. 2013) have identified interacting cortical and basal ganglia networks underlying the process of finding and tapping to the musical beat.

Much remains to be accomplished for the standardized assessment of musical
processing abilities and for the incorporation of test battery data into the neuropsychological evaluation. Moreover, little has been done to provide for standard
assessment of musical recognition or recall, comprehension, or production although
these abilities remain a multifaceted feature of human abilities and aspects of music
appear to be processed by neural systems contributing to emotion and well-being
(see Klineburger and Harrison 2013). Indeed, music is intimate to many in their
day-to-day activities, whereas this has largely been lost by standardized approaches to the neuropsychological evaluation. However, some early contributions were
made in this area including, but not limited to: the Montreal Battery of Evaluation
of Amusia (Peretz et al. 2003), the Seashore Rhythm Test (see Strauss et al. 2006),
Denman's Tones and Melodies Test (Denman 1987), and the Rhythm Tapping Test
(see Luria 1973, 1980).

Spatial Processing Disorders

Disorders of extrapersonal space are common with right cerebral dysfunction (see
Heilman and Gonzalez Rothi 2012b). The assessment of spatial processing disorders includes the comparison of personal and extrapersonal spatial processing as
with left–right awareness and body awareness in contrast to geographical awareness, left hemispatial neglect syndrome, and constructional dyspraxia, for example.
Subsequently, a more refined comparison might be provided by analysis of these
spatial domains at and within each sensory modality (primary and secondary association cortices) and by analysis of multimodal processing. Careful attention is
necessary to identify the disconnection syndromes with impaired access from one
sensory modality to the other. An example here would be the inability to access
visual spatial analyzers via somatosensory systems.

Visual Spatial

In his Herbert Birch Memorial Lecture presentation (1992; see Tranel et al. 2009),
Arthur Benton credited Henry Hécaen and Oliver Zangwill for a critical shift in
thinking regarding hemispheric laterality and specifically, for transforming the
long-held doctrine of left hemisphere dominance to an account of hemispheric
asymmetry. By this juncture, the idea that we speak with the left hemisphere was
firmly established and with comparable importance the work by Hécaen and Zangwill had shown conclusively that right hemisphere damage leads to defects in visuospatial and visuoconstructional abilities (see Hécaen and Albert 1978; Zangwill
1960).

Neuropsychological research on visual–spatial processing has largely focused on a complex circuit with its primary anatomical foci, including the frontal eye fields for regulatory control, the parietal eye fields for sensory perceptual integration and analysis, and the superior colliculus for reflexive eye and hand or arm movements. In one project, for example, fMRI was used to image subjects while they visually searched for embedded targets (Gitelman et al. 2002). Visuomotor search activated the posterior parietal cortex and the frontal eye fields. Moreover, a greater number of activated voxels were recorded on the right brain, which the authors found to be consistent with "the known pattern of right hemispheric dominance for spatial attention." The superior colliculus showed prominent activation in situations demanding visual search versus eye movement, demonstrating, for the first time in humans, activation of this region specifically related to an exploratory attentional contingency. The researchers conclude that the search-dependent variance in the activity of the superior colliculus was significantly influenced by the activity in a network of cortical regions, including the right frontal eye fields and bilateral parietal and occipital cortices.

Visual spatial analysis within right cerebral systems appears to involve the processing and associative analysis of visual patterns beyond that of letters, the hands, or tool implements. Instead, the right cerebral hemisphere has long been associated with visual spatial analysis of faces and places, complex visual scenes, and constructional images or figures, and numbers. Recently, double dissociation of visual letter and number recognition was performed using fMRI (Park et al. 2012). Initially, the researchers replicated previous findings in experiment 1. Participants viewed strings of consonants and Arabic numerals with letters activating the left fusiform and inferior temporal gyri more than numbers. In contrast, numbers activated the right lateral occipital area more than letters. The authors argued that since the distinction between letters and numbers is culturally defined and relatively arbitrary, the double dissociation between the two categories of visuospatial stimuli provides evidence that a neural dissociation can emerge as a result of experience.

Experiment 2 was designed to test the hypothesis that lateralization of visual number recognition depends on lateralization of higher-order numerical processing. In this case, the participants performed addition, subtraction, and counting on arrays of nonsymbolic stimuli varying along the number dimension. This task produced neural activity in and around the intraparietal sulcus, a region long associated with relational thought and higher-order numerical processing. The ability to predict individual differences in the lateralization of number activity in visual cortex by individual differences in the lateralization of numerical processing in parietal cortex, suggests a functional relationship between the two regions. Moreover, the relationship appears consistent with associative relationships and higher-level processing as information moves from visual analyzers to cross-modal analysis within the tertiary association cortex.

In related visual perceptual research using fMRI, researchers (Bracci et al. 2012) tested whether the common role of hands and tools in object manipulation is also reflected in the distribution of response patterns to these categories in the visual cortex. The findings indicated that static pictures of hands and tools activate the left lateral occipitotemporal cortex. Activation of this region to tools selectively overlapped with responses to hands but not with responses to whole bodies, non-hand body parts, other objects, or visual motion. Moreover, the perception of object-directed actions performed by the hands or by tools recruits activation in the left frontoparietal cortex. Pattern analysis of responses at the left lateral occipitotemporal cortex suggests a high similarity between response patterns to hands and tools but not between hands or tools and other body parts. Finally, functional connectivity analysis show that this left hemisphere hand/tool region is selectively connected with regions in left intraparietal sulcus and left premotor cortex that have previously been implicated in hand/tool action-related processing. Taken together, these results suggest that action-related object properties shared by hands and tools are reflected in the organization of high-order visual cortex. The percept provides for associative analysis of tools and hands necessary for tool use with the frontal associations down the longitudinal tract to organize, plan, and implement the motor sequences and muscle movements.

Circles, squares, triangles, and other complex geometrical shapes and visual patterns may be recognized and perceptually integrated in these neural systems. Many examples exist in the development and deployment of assessment measures designed to be sensitive to nonverbal learning and memory with visual spatial patterns (e.g., Design Learning Test; Foster et al. 2009). The analysis and comprehension of visual spatial stimuli extend to some of the perceptual phenomena identified by the early Gestalt psychologists (Wertheimer 1923), including the laws of closure, proximity, and continuity in the processing of nonlinguistic shapes and forms. This might be followed by analysis of visual facial recognition, affective facial recognition, and eventually geographical visual analysis as with map-drawing activities. Constructional praxis is basic to this process much like ideomotor praxis is fundamental to hand/tool action (e.g., see Vingerhoets et al. 2012), and this may be assessed with complex (e.g., Rey Complex Figure) and simple figure copy tests. Alternatively, manipulations of building blocks or other geometrical shapes may be combined to construct three-dimensional designs, such as a house or a building of one form or another.

Visual–spatial memory assessed using constructional tests usually supports better performance on a figure-copying test (e.g., the Rey Complex Figure Copy Test) than when drawing from memory, where delayed recall has been related to a dissipation of the memory trace over time and to poorer performance. Regardless, the opposite effect may be found with frontal lobe impairments and where the patient may benefit from the provision of additional time or delayed recall conditions. This phenomenon may be more useful in individuals with organizational deficits, distractibility, or impaired performance under confrontative conditions, where additional time (delayed recall) may facilitate performance. In one project (Roth et al. 2013), consecutive patients in a neurocognitive disorders clinic were given a battery of clinical cognitive tests that included copying a figure of intersecting pentagons and then drawing the figure from memory. A subgroup of four patients with frontal dysfunction showed marked improvement in drawings at a delay compared to copying directly from an image, although it is generally evident from prior studies that most patients have declines in drawing performance at a delay. The unusual pattern of better performance following a delay, compared to an initial copy, occurred in patients with frontal lobe dysfunction.

Visuospatial processing may have long-standing utility in the evolutionary course of survival in situations of potential violent conflict, where the decision to fight, flee, or to attempt negotiation entails assessing many attributes contributing to the relative formidability of oneself and one's opponent. It is reasonable that, in these situations, cerebral laterality of systems processing and assessing issues of risk or potential for harm might be altered by the processing specializations arising from either cerebral hemisphere. Ultimately, the visual–spatial representations facilitate such assessments of multiple factors. Fessler and Holbrook (2013) note that "because physical size and strength are both phylogenetically ancient and ontogenetically recurrent contributors to the outcome of violent conflicts, these attributes provide plausible conceptual dimensions that may be used...to summarize the relative formidability of opposing parties." The authors also note that "the presence of allies is a vital factor in determining victory," leading them to hypothesize that men accompanied by male companions would therefore envision a solitary foe as physically smaller and less muscular than would men who were alone.

The predicted effect was document in two studies (see also Holbrook and Fessler 2013), one using naturally occurring variation in the presence of male companions and one employing experimental manipulation of this factor. Participants were shown an image of a man (a photo of Ali Beheshti, who was convicted of firebombing the home of the publisher of a novel about the Prophet Muhammad; Walker 2009) who was described as a "convicted terrorist" and asked to estimate his height, size, and muscularity using standardized six-point arrays. Although laterality was not assessed in this project, the findings are suggestive, given the negative and positive emotional bias (Tucker 1981; Tucker and Williamson 1984; Davidson and Fox 1982; Ehrlichman 1987; Silberman and Weingartner 1986; Davidson 1993; Ekman et al. 1990), socially avoidant and socially engaging (*BIS/BAS:* Gray 1982; McNaughton and Gray 2001; Davidson 1995; Harmon-Jones et al. 2003, 2004), negative reflective and optimistic anticipation of future events (Manuck et al. 2000), and submissive and dominant perspectives (Demaree et al. 2005) of the right and left cerebral hemispheres, respectively.

Auditory Spatial

Evaluation of auditory spatial analysis within right cerebral systems involves the use of dichotic presentations and perhaps incongruent linguistic and prosodic content with concurrent binaural sounds (see Pollmann 2010; Hugdahl et al. 2009; Hugdahl 2003, 2012). Ley and Bryden (1982) demonstrated a dissociation of right and left hemisphere analysis of auditory spatial variants with a right-ear (left brain) advantage for verbal linguistic sounds and a left-ear advantage for emotional sounds. Auditory affect agnosia might be acquired along with musical tone and melody-processing deficits after a right temporal lesion (see Peretz et al. 1994; Calabrò et al.

2012). Many family members, caregivers, and therapists rely on tone of voice along with facial expression cues to convey their concern and empathy for the patient. These resources may well be impoverished with right cerebral dysfunction with resultant confusion and frustration by these well-intentioned and supportive individuals. They may reasonably derive attributions that the patient is obstinate or insensitive where understanding of the underlying brain disorder is integral to improved attributions and care. Many auditory spatial variants exist, including the location of the origin of sounds within extrapersonal space. Moreover, changes in the relative location of sounds approaching or moving away from the body may provide critical data for analysis and for emotional associations (e.g., impending threat).

Neuronal mechanisms of auditory distance perception are poorly understood, as distance processing may be based on intensity-independent cues where we distinguish between soft-but-nearby and loud-but-distant sounds. Using fMRI measurements and computational analyses in a virtual reverberant environment, researchers (Kopčo et al. 2012) found that activations to sounds varying in distance, compared with activations to sounds varying in intensity only, were significantly increased in the planum temporale and posterior superior temporal gyrus, contralateral to the side of sound origin. These results, using fMRI, suggest that neurons in posterior auditory association cortices, in or near the areas processing other auditory spatial features, are sensitive to intensity-independent sound properties relevant for auditory distance perception.

Teshiba and colleagues (Teshiba et al. 2012) note that unlike the visual system, a direct mapping of extrapersonal space does not exist within human auditory cortex. They point out that "the prime candidates for attentional modulation include the frontoparietal network, which demonstrates right hemisphere lateralization across multiple attentional states." In this project, subjects completed an exogenous auditory orienting task while undergoing fMRI, providing evidence of hemispheric specialization in the localization of sounds within extrapersonal space. The authors conclude that the "where" component in auditory spatial processing involves the secondary auditory cortex within the right hemisphere when analyzing the auditory processing stream during both evoked (orienting task) and intrinsic (resting-state) activities.

Vestibular Spatial

Dizziness or vertigo, often accompanied by nausea, is among the more common patient complaints affecting approximately 20–30 % of the general population (see Della-Morte and Rundek 2012). Cerebrovascular accidents may be causal for these complaints accounting for 3–7 % among all causes of vertigo. The vertebrobasilar system supplies blood perfusion to the inner ear, brain stem, and cerebellum, and rupture or occlusion of this supply is frequently recognized from the accompanying symptoms of vertigo, nausea, and vomiting, along with nystagmus (see Chakor and Eklare 2012). However, the manifestations of disturbance within this system vary depending on the vascular areas affected and the underlying functional neuroanatomy. Also, these brain-stem structures process vestibular information and vectional forces as part of a larger system, including the sensory processing areas within the second functional unit (sensory analysis and comprehension) and the third functional unit or frontal lobes. They later appear to play a regulatory role over the lateralized vestibular systems with contralateral vectional complaints resulting from left or right frontal damage.

It is now well established that vestibular information plays an important role in spatial memory processes (e.g., Machado et al. 2012). One large mountain of a man and former star athlete was simply trying to walk after his stroke left him with a pervasive vectional disturbance. Subsequent to this stroke, he had a severe left hemineglect syndrome with concurrent leftward vection. This gentleman was literally spinning into space that did not exist for him in many ways! He learned to walk, though, by broadening his gait and through implementing thigmotactic cues as he had relative integrity of his left parietal lobe and somatosensory analysis from the right side of his body. Specifically, his left brain would run its right hand over the wall to his right side providing it with thigmotactic or somatosensory-based cues on his position both vertically and longitudinally as he navigated hallways within the building. In other settings, he would employ these techniques through placing his hand on his wife's shoulder as he walked with her.

Vestibular spatial analysis is assessed using one or another vestibular challenge, but this might better be done with caution and by those trained for this purpose or falls, nausea, and affective decompensation may follow. Generally, activation of the right posterior cerebral regions, and especially the right temporal and parietal region, lends itself to leftward vectional complaints and vice versa. This may result from deregulation with a frontal lobe disorder and reduced regulatory capacity and control of the vectional systems located toward the back of the brain (first and/or second functional unit). But relative activation of one or the other cerebral hemisphere may result from many and varied origins. Lateralized brain-stem lesions, more characteristically, result in ipsilateral vection complaints (see Brandt and Dieterich 1994) and, with extension of the lesion into the pons, visuomotor deficits, including nystagmus and/or nonconvergent gaze, which may aggravate these complaints. In this case, visuomotor exercises might be initiated and, again, with caution. Extension of the lesion toward the baroreceptor systems in the medulla may contribute to orthostatic hypotension along with the vectional disturbance.

A younger woman and exceptional public school teacher had managed a television with a persistent electrical short and a leaking school roof, above the television, for more than a year when fate brought these events together in a storm. She moved the children in her class to safety and, while standing in a puddle of water, tried to turn the television set off inadvertently placing her head against the defective television set. She was electrocuted with the current path, by assessment, traversing through the right temporal region and on down her left hemibody. Her chronic complaint from this accident was for a leftward vection with severe nausea and negative emotions, which correlated well with activation of the right temporal region on the quantitative electroencephalogram. Simply ranging her left arm would aggravate the symptoms with an emetic response and tachycardia with heightened sympathetic tone.

Patients recovering from vestibular insults face many obstacles with balance and gait disturbance and often with head and trunk tilt or postural misalignments. These postural and locomotor biases appear to result from changes in the spatial perception of self and the location of self within extrapersonal space. The importance of vestibular cues as a necessary foundation for the accurate representation of body orientation was demonstrated in patients with right or left vestibular neurotomy as a treatment for Menière's disease and healthy controls (Saj et al. 2012). The subjective location of straight-ahead was investigated using a method disentangling lateral shift and tilt components of error. In the horizontal plane, subjects were required to align a rod with their body midline. In the frontal plane, they were asked to align the rod with the midline of their head or trunk. The location of straight-ahead varied specifically according to the side of the lesion. The patients with left vestibular nerve lesions had a contralesional lateral shift of subjective straight-ahead along with an ipsilesional tilt of the head (less severe for the trunk). The evaluation of the right vestibular nerve lesion patients showed, by contrast, that the representation of the body midline was near accurate in both the horizontal and frontal planes and consistent with control subjects. Also of interest, only patients with left vestibular loss were concerned with these changes in perception of self-orientation in space, a finding potentially relevant to discussions of anosognosia/anosodiaphoria (see Prigatano 2010) and the shared functional role of spatial analysis with emotion or insight to one's deficits.

Somatic Spatial

Somatic spatial disorders of the right brain include proprioceptive deficits with poor appreciation of the location of the left hemibody within extrapersonal space. But this extends to spatial analyzers within the right parietal region specialized for more distal arrays with geographical coordinates (e.g., Benton et al. 1974). Right temporal lobe disorders may also yield delusional sensations of moving through space. This occurs with control issues or feeling as though the movement is forced or from external control or coercion. Thus, fear is common, as is anger, dependent on the proximity of the brain disorder to these cerebral systems.

Another common somatic spatial disorder frequently associated with left brain pathology is tactile astereognosis, where stereognosis is the ability to recognize an object by touch and pressure alone. Left brain damage is associated with bilateral tactile astereognosis, whereas right brain damage may be relatively restricted to left hand astereognosis, perhaps. However, a simpler take on this holds that when the hand is used for stereognostic discrimination, the integration of somatosensory patterns into spatial information about the objects takes place in the contralateral somatosensory hand area (Roland 1976). But, Cannon and Benton (1969) extended the understanding of tactile perception in their investigation of the perception of direction within this modality. In this project, although there was an equally severe deficit on the contralateral hand in the left and right brain-damaged groups, the patients with lesions of the right hemisphere showed a high incidence of defective performance on the ipsilateral hand, while the patients with lesions of the left hemisphere showed essentially normal performance in this regard. The authors concluded that the evidence supports the concept that the right hemisphere plays a more important role than the left in the mediation of behavior requiring the appreciation of spatial relations, extending this evidence to a test situation involving the tactile modality.

Tactile object perception may require multiple point physical contact, and restricted capacity of these brain regions may limit tactile form perception. Once recognized the tactile perception of the object might be conveyed to systems aware of the use of the object or tool or to systems specialized for the gesture or ideomotor praxis necessary to use the object (e.g., Vingerhoets et al. 2012). Other nearby systems may be useful in the incorporation of the object into language systems necessary for naming the object and discussing its utility within a logical linguistic context. Other neural systems provide for a visual representation of the tool or object (e.g., Bracci et al. 2012). Still others convey the sounds, which may accompany the object during its use or implementation.

Touch is the first sense to develop and it remains critical to the manipulation of tools and language processing through gestures and gestural praxis (e.g., see Schaefer et al. 2012). Research on mirror neuronal systems have demonstrated that the observation of another being touched or even being nearly touched is sufficient to activate somatosensory cortex (e.g., Keysers et al. 2010; see also Fogassi and Simone 2013). Much of this research indicates that activation of somatosensory cortex may not only be involved in the perception or experience of touch but also it might provide a somatic dimension to our perceptual understanding of another person's experiences related to being touched (e.g., Keysers et al. 2010; see also Schaefer et al. 2012). Schaefer, Heinze and Rotte (2012) used fMRI techniques to explore the responses of an observer in such settings. The authors conclude that the mirror-like responses recorded in this project reflect the peripersonal space of a seen body part. Moreover, the study shows that vicarious somatosensory responses are especially sensitive to touch seen in the peripersonal space of the other body, perhaps extending the analysisof comprehension of mirror neuron response complexes to spatial domain of the body being touched.

Olfactory Spatial

Less is known about the spatial aspects of olfactory processing. Nonetheless, this modality often provides our first clue to dangerous or desired stimuli within extrapersonal space and these events very well may precipitate withdrawal or approach behaviors as we hasten to escape a negative affective or putrid odor and to approach a preferred odor. Positive and pleasant olfactory stimuli may precipitate approach behaviors and proximity for the point of origin for these events. These spatial processing biases are potentially robust and this modality may well overwhelm the response system biases originating from the other sensory modalities (e.g., the smell from a glass of milk, which has gone sour). The olfactory modality, though minimized in the human nervous system, remains very powerful in the spatial processing domains of proximity/personal space and distal/extrapersonal space where the emotional attributions correspond with the cerebral hemisphere charged with processing these spatial domains.

Olfactory hallucinations by many patients, suffering right brain disorders, are described as being the "odors of hell." These accountings have included burning sulfur, rotting dead bodies, and vomit, whereas positive olfactory hallucinations are less common complaints as they may be pleasurable and enjoyed with pleasant and quiescent associations, perhaps. The brilliant 38-year old composer, George Gershwin, described these dreadful olfactory hallucinations several months before he died of a right temporal lobe glioblastoma multiforme (see Waxman 2010). Relevant to the conversation here, he had been referred to for psychotherapy as his physicians had thought these events to herald a "neurotic disorder." Syndrome analysis might have revealed a left upper quadrant anopsia due to passage of the optic radiations through this region, had he been seen for a neuropsychological evaluation.

Facial Recognition

Historically, visual facial recognition and visual line orientation recognition problems have been ascribed to dysfunction within the inferior posterior right cerebral hemisphere. Benton's Facial Recognition Test and Judgment of Line Orientation Test have been commonly used for the assessment of the posterior right cerebral region within the context of the neuropsychological examination. Using a lesion-deficit mapping technique, researchers (Tranel et al. 2009) provided additional evidence of this functional neuroanatomical relationship. The investigators found that, "consistent with clinical lore, the tests have good localization value that points to right inferior parietal and nearby temporal and occipital structures as being important for performance on these tests. These results support the clinical application of these tests as good measures of right hemisphere functioning, especially in the inferior parietal, occipitoparietal, and occipitotemporal sectors that have been associated with visuoperceptual discrimination and visuospatial judgment."

Several other facial recognition tests exist, including the Crews Test of Facial Memory, which provides for an assessment of immediate and delayed facial recognition. These are part of the larger battery of tests forming the Crews Tests of Neuropsychological Functioning (Crews 2012: see Cognicheck.com). W. David Crews III is the founder and chief executive officer (CEO) of CogniCheck, an online memory screening tool, which is capable of providing confidential neuropsychological assessment information and longitudinal trends for individuals to use with their health-care professional. It includes the assessment of verbal and nonverbal learning or memory with immediate and delayed test formats. Interestingly, the subtests evaluating facial memory were prepared and validated across age groups and across geographical regions within the USA. David's team took sample pictures of faces across the country. He offered financial incentives to a homeless man in one city, if he would clean up and get a haircut for the photograph. This man spread the word and soon there were many volunteers ready for their haircut, a shower, and their photography session. This good deed may have had a more lasting impact as more than one participant returned to let David know that he was now gainfully employed, with the haircut and bath being necessarily sufficient to acquire a job and to promote improved self-esteem. Indeed, this intervention appeared to be the booster shot that was needed to promote a return to work for some of these good people.

Following a right posterior cerebral artery stroke, one woman developed a facial agnosia and prosopagnosia. The former refers to the loss or inability to recognize faces, whereas the latter and more devastating disorder involves an inability to recognize familiar faces, including your own face in a mirror! Subsequent to this event, she would panic as her husband of 50 plus years would follow her at the grocery store. In her panic, the police would be called and the spouse would be nearly a wreck simply trying to acquire groceries. After her husband passed away and with the development of a multi-infarct dementia syndrome, there was some residual benefit from this stroke as she could be comforted and felt safe with her hand in another's hand using the somatic access for recognition of this surrogate gentleman. Facial recognition disorders may present in various ways with some unable to recognize faces and some seeing faces that are not there. A very nice video of two patients presenting with distinct types of visual facial agnosia was published under the title "Stranger in the Mirror" (NOVA 1993).

Much additional research is needed in the area of social anxiety, where some are gaze avoidant and perhaps with elevated affective intensity or negative emotional bias in social settings. In one project, hostile men also scoring high on anxiety measures were evaluated (e.g., Harrison and Gorelzenko 1990). These men viewed neutral affective faces as significantly more emotional where the neutral faces were responded to as angry faces. Moreover, this effect was restricted to faces viewed within the left visual field providing evidence of lateralization of the effect to the right cerebral hemisphere. In similar research, altered processing of sensory perceptual information by this group was found within the auditory (Demaree and Harrison 1997b), somatic (Herridge et al. 1997; Rhodes et al. 2013), and vestibular modalities (Carmona et al. 2008). Beyond this were demonstrations of motor differences (e.g., Demaree 2003; Rhodes, Hu and Harrison 2013) and premotor differences on measures of fluency (e.g., Williamson and Harrison 2003), which indicate that reduced frontal resources or capacity (see Klineburger and Harrison 2013) may underlie the heightened sensory perceptual intensity from systems found at the posterior end of the longitudinal tract.

It is reasonable that extremely shy or inhibited individuals who are slow to acclimate to new people may habituate slowly to facial novelty reflecting a social learning deficit (Blackford et al. 2012). fMRI was used to examine habituation to neutral faces in 39 young adults with either an extreme inhibited or extreme uninhibited temperament. The researchers focused on two brain regions involved in the response to novelty—the amygdala and the hippocampus. Habituation to neutral faces in the amygdala and hippocampus differed significantly by temperament group. Individuals with an uninhibited temperament demonstrated habituation in both the amygdala and hippocampus, as expected. In contrast, in individuals with an inhibited temperament, the amygdala and hippocampus failed to habituate across repeated presentations of faces. The failure of the amygdala and hippocampus to habituate to faces may mediate the behavioral differences seen in individuals with an inhibited temperament and social anxiety. In other research (Kranenburg 2012), the rate of habituation and parenting behavior of hostile men were significantly improved using nasal administration of oxytocin, a pheromone known to affect trust (Kosfeld et al. 2005), bonding, and nesting behaviors (e.g., MacDonald and MacDonald 2010; see also Longo et al. 2012).

Just as habituation reflects a learned reduction in responding with repeated or continuous exposure, the opposite may present in pathological states where the individual is preoccupied with otherwise irrelevant events. One of the better clues for the presence of visual hallucinations in those with autism or other maladies is visual preoccupation. This is more clearly the case with oscillating visual stimuli, and some will oscillate their own fingers or hands for this purpose. The oscillating visual stimulus will serve to lower the threshold for activation of one or the other occipital lobe and/or thalamic projections. With activation of these brain regions, the individual may see what and where that brain region is prepared to see! Neuropsychologists may agree that these events are seldom identified and diagnosed. Seldom are the parents or family members instructed on the relationship between sensory oscillation, sensory entrainment, and onset of hallucination. Moreover, the oddity of the behavior may lead to erroneous attributions. One example here is a young man with Down's syndrome who helps clean the exercise machines at a local weight club. With the oscillations in the runner's legs and the spinning of the stepper or treadmill, he becomes transfixed. Unfortunately, he frequently is at the runner's legs, and she may leave in disgust with negative attributions toward his behavior and intentions. His visual formesthesias are at the right hemispace accompanied by substantial interest and approach behaviors unwanted by the code. But, for another with visual formesthesias at the left these events may provoke fear and agitation and sometimes needless increments in psychoactive medication as contextual therapies may be more effective and without the iatrogenic effects.

Many of the stereotypical behaviors prevalent with the autism spectrum disorders are sufficient in this respect and may reflect oscillating induction of hallucinated events within one or another of the sensory modalities topographically located and processed within discrete but converging brain regions (see Behrendt 2003, 2012; Lutterveld and Ford 2012). Differential assessment and diagnosis of disorders affecting these functional neural systems are warranted and potentially with any repetitive or oscillating behavior or sensory event. Oscillation, of course, may occur within somatosensory, visual, auditory, vestibular, and other sensory modalities and often with multimodal or combined effects.

Facial processing is most notable for activation of the fusiform gyrus or *fusiform face area* (Cox et al. 2004), an area within the visual association cortex found at the inferior temporal region. Thomas et al. (2009), using diffusion tensor imaging, found that individuals with congenital prosopagnosia had decreased connectivity within the occipitoparietal cortex. Behrman et al. (2007) also found reduced size of the anterior fusiform gyrus in this population (see Carlson 2013). Carlson (2013) extends the discussion of complex figure or form processing within the ventral stream to the extrastriate body area located posterior to the fusiform face area and partially overlapping it. This region activates to photographs, silhouettes, and stick drawings representative of the human body or body parts and not to control stimuli, which included drawings or photographs of tools (see also Bracci et al. 2012), scrambled silhouettes, and scrambled stick drawings of human bodies (Downing et al. 2001). Also, Schwarzlose et al. (2005) showed that the fusiform facial area responded differentially to faces, whereas the extrastriate body area engaged to headless bodies and body parts. Reversible lesions or temporary disruption of the extrastriate body area from transcranial magnetic stimulation also was found to temporarily produce impaired recognition of photographs of body parts but not parts of faces or motorcycles (Urgesi et al. 2004).

Kanwisher and Dilks (in press) characterize the neural system involved in facial recognition as the ventral visual pathway, which extends from the occipital lobe into inferior and lateral regions of the temporal lobe. In their review of neuroimaging research on humans, they differentiate functional properties of the cortical regions within this system involved in visually perceiving people, places, and things. They conclude from this, now substantial body of work, that the ventral visual pathway is not homogeneous, but instead that it consists of a highly differentiated structure containing distinct functional regions. These regions include the fusiform face area responding selectively to faces (Kanwisher et al. 1997; McCarthy et al. 1997), the parahippocampal place area, which responds selectively to places (Epstein and Kanwisher 1998), the extrastriate body area, which responds selectively to bodies (Downing et al. 2001), the lateral occipital complex, which responds to object shape (Malach et al. 1995; Kanwisher et al. 1997) largely independent of object category, and the visual word form area, which responds selectively to both visually presented words and consonant strings (Baker et al. 2007; Cohen, Dehaene et al. 2000). The anatomical seat of each of these functional regions appears to be strong and perhaps reminiscent of the impact that Broca's findings had on the early localization controversy. Kanwisher and Dilks (in press) comment that, "Each of these regions is present in approximately the same location in virtually every healthy subject. These regions, and their cohorts (e.g., the occipital face area), constitute the fundamental machinery of high-level visual recognition in humans."

Ventral neuroanatomical connections running through the longitudinal tracts and interfacing the right fusiform gyrus with the orbitofrontal region and basal frontal lobe may eventually be understood for their relationships with negative affective facial perception with aversion. Right frontal incapacity especially within the orbitofrontal regions appears to alter and augment the intensity of a viewed face and eventually with the need for the perceiver to avert their gaze as a compensatory aspect for facial processing. Patients with dysfunction or lesion within this system often present with left visual formesthesias of intensely negative persons with a common attribution that these individuals "look a hole right through you!" This account may provide a clue for those afflicted with a milder disturbance of cerebral function. Specifically, there is inadequate dampening of the affective intensity perceived by the observer with the conveyance of an intense and negative facial expression. One possible variant of this neuroanatomical relationship has been recently expressed in the proposition that autism is a neural systems disorder with underconnectivity within the frontal–posterior pathways associated with the longitudinal tract (e.g., Just, Kellera, Malavea, Kanab, Varmac, in press). This might be reflected in poor downregulation over the posterior facial processing systems with altered or pathological activation and facial processing deficits. Within the right hemisphere, these systems appear specialized not only for faces but also for negative emotion and primarily fearful apprehension.

Right brain disorders within the posterior regions, including the occipital, temporal, and parietal areas, may present with an inability to recognize faces or facial agnosia. If the patient is unable to recognize familiar faces, including their own face in a mirror or in a picture, then the diagnosis is of prosopagnosia. One prosopagnosia case was followed the court system, since this gentleman acquired the brain disorder as the result of another driver's neglect (by jury verdict). This man was very soft-spoken and most thought him to be a very nice man, failing to appreciate his strong homicidal tendencies and anger following his brain injury. It was interesting to watch this man who could not even recognize his own face in a mirror, work with horses. He was a "horse whisperer" before most new what this term meant. His facial processing deficits and profound learning deficit for faces and for places were accompanied by intermittent spatial delusions, which he seemed determined to discuss at length. In these episodes, he is forced "out of my box and into a different place." But, this place is pleasant and peaceful and free of the troubles and tribulations of this world. The deep and abiding anger comes from being "forced back into this box against my will." The relationship of perceived external control within neural systems specialized for extrapersonal space and negative emotion is not lost in this example.

When he initially arrived at the head injury program, his front windshield was covered with sticky notes. He tried repeatedly to make a "groom box" for the university's horse stables, but could never finish it in one day. He would start over the next day, but never did finish it, despite his occupational background in residential and commercial construction as a contractor. He knew the regular staff at the head injury program by sight, but could not identify anyone with just a face photograph. Full-length recognition was basically intact; even with everyone wearing the same gym clothes. Even more remarkable, he could name every horse in the university stables (20 plus).

One woman was experiencing substantial paranoia after her right temporal lobe stroke with active left-sided visual formesthesias of the devil trying to do her harm. Since the stroke appeared to extend into the hippocampal region on the scans, the neuropsychological assessment included concerns for confabulation. Since the stroke was at the right cerebral hemisphere, the rule out on the diagnostics was for confabulation of faces and/or places. A lesion to the homologous region might, instead, have left components of confabulation for the content of speech or, more specifically, what others had said or done socially. When asked if she knew the examiner from before (this was their first meeting), she agreed that she knew him well. Indeed, she identified him as Michael Landon—a rather good-looking fellow from a television series entitled "Little House on the Prairie." Well, Mr. Landon played a strong religious and nearly angelic role in this series. With that in brain and with the good looks to boot, the examiner was feeling pretty good about himself! But, these attributions lost their validity as the right brain disorder revealed its affective nature. Indeed, this angel (Michael Landon—the neuropsychologist) was actually the devil and I had appeared to kill her and to forcefully drag her away to the "other place" against her will!

Facial stimuli may be poorly comprehended and gaze aversion may be related to pathology within these systems and within the fusiform gyrus (e.g., Pierce et al. 2001; Schultz, Gauthier et al. 2000; see also Blackford et al. 2012). This is an exciting area of current research interests on the autism spectrum disorders and social anxiety. Some evidence exists for mirror neuron dysfunction in individuals with autism (Dapretto 2006), although this remains controversial. The exclusion of the cuneus gyrus in the occipital region and its connections in favor of the ventral pathways may ultimately lack validity. But, the special role of the fusiform gyrus in face processing is commonly expressed to the point that it is frequently referred to as the fusiform "face" area. Moreover, the differing roles of the left and right fusiform face areas have been described, including the composite whole facial recognition by the right hemisphere and the recognition of discrete or sequentially processed facial components (e.g., mouth and eyes) by the left fusiform region (e.g., Rossion 2000; as cited by Hellige et al. 2010).

Some evidence indicates decreased neuronal mirroring and emotional empathy to faces in males (Schulte-Rüther et al. 2008; see also McClure 2000), whereas neuronal activation to tools or mechanical objects yields the opposite effect in research on sex differences. The data suggest that females recruit areas containing mirror neurons to a higher degree than males during processing in empathic face-to-face interactions. This may underlie facilitated emotional "contagion" in females. Based on these findings, some researchers are addressing the possibility of a masculinized brain in social anxiety disorders and/or the autism spectrum disorders with diminished interest in faces and focal preoccupation with inanimate objects or tools (see Baron-Cohen 2003; Montagne et al. 2005).

Behavioral studies suggest that women often perform better on emotional tasks than men. For example, a female advantage is described in the decoding of nonverbal emotional cues both in adults and children (see Hall 1978; Hall et al. 2000; see also McClure 2000). Consistently, studies of affective arousal and the expression of emotion (e.g., in response to the emotions of other people) demonstrate superior performance of women over men (see Brody and Hall 2000). Women are also reported to display higher complexity and differentiation in their articulation of emotional experiences (Barrett et al. 2000) and to score higher than men on self-report measures of empathy (e.g., Davis 1996; Baron-Cohen and Wheelwright 2004). It appears to be in good accordance with these findings that psychiatric disorders such as autism spectrum disease, conduct disorder, and antisocial personality disorder, which are often characterized by a lack of empathy, are far more common among males (Chakrabarti and Baron-Cohen 2006).

Cultural differences in facial processing and the neural architecture involved have also been uncovered. Chiao and Immordino-Yang (2013) state that "despite the apparent robustness of the organization of visual processing in the brain, culture appears to shape neural processing by influencing the process by which a visual stimulus is perceived, encoded and recognized, even within domain-specific neural regions along the ventral visual pathway." In their literature review, the authors conclude that "one of the hallmark findings from cultural psychology is the distinction between analytic and holistic processing." For example, East Asians have demonstrated a more holistic visual facial processing bias, attending to the central object and the surround, in contrast to a more analytic facial processing bias among Westerners. Westerners, instead, differentially allocate facial processing resources to the central object features over the surround (Nisbett and Miyamoto 2005; Nisbett et al. 2001; see Chiao and Immordino-Yang 2013).

The authors report that cultural neuroscientists have found that this analytic–holistic processing distinction affects neural responses even in domain-specific visual brain regions. For example, Goh and colleagues (Goh et al. 2007) compared elderly Westerners and elderly East Asians and found differences in visual processing in object-specific areas of the lateral occipital cortex, which they interpreted as reflecting lifelong cultural entrainment of analytic and holistic styles of perceiving the world. In addition, Goh et al. (2010) found that Westerners showed greater neural selectivity during face processing in the left fusiform gyrus face-processing area, whereas East Asians showed greater selectivity within this neural area within the right hemisphere. Chiao and Immordino-Yang (2013) conclude that this cultural difference in the lateralization of visual processing may reflect cultural differences in analytic–holistic processing style. This seems reasonable, from their analysis, as the right hemisphere is thought to process more holistically, whereas the left hemisphere is thought to process more analytically and sequentially. This difference in laterality and the functional neural systems involved are consistent with research findings in which Caucasians engage in more sequential processing of facial features such as the eyebrow and mouth. In contrast, East Asians devote relatively more attention to processing the eye region, a central facial feature that may allow for more holistic, simultaneous processing of peripheral facial features (Jack et al. 2011; see Chiao and Immordino-Yang 2013).

Place Recognition

Just as there is substantial evidence for the specialization of the fusiform region in facial recognition, there appears to be a high level of specialization in the perception and knowledge of places dependent upon the sensory modality involved in the process. Historical evidence has long been available from clinical lesion studies and the conclusions have found support in functional imaging work (see Kanwisher, & Dilks, in press). Visual pathway analysis, for example, provides evidence for the discrete functional processing and perceptual comprehension of people, places, and things. Among these is the parahippocampal place area, which responds selectively to places (Epstein and Kanwisher 1998; see also Shinohara et al. 2012) and the extrastriate body area, which responds selectively to bodies (Downing et al. 2001; see also Kanwisher and Dilks, in press).

Medial temporal lobe structures and especially the hippocampus have been implicated in spatial navigation. Early on, "place cells" were found in the rodent hippocampus (O'Keefe and Nadel 1978), and subsequently in humans (Ekstrom et al. 2003), which fired relative to position within the environment indicating that a cognitive map of the environment may be stored within the hippocampus. Hippocampal place cells interact with cells from the surrounding entorhinal cortex contributing to the perception of the spatial grid and localization within space (Hafting et al. 2005; Moser et al. 2008). Functional neuroimaging research using virtual reality tasks during navigation through simulated three-dimensional environments has consistently found lateralized activation at the right medial temporal and parietal regions (Iaria et al. 2007; Hartley et al. 2003; Grön et al. 2000; Maguire et al. 1998).

These findings are consistent with clinical research where neuropsychological testing of individuals with damage to the hippocampus and surrounding temporal and parietal lobe regions has demonstrated processing and memory deficits (e.g., Lee et al. 2005; Parslow et al. 2005; Astur et al. 2002; Nunn et al. 1999; Bohbot et al. 1998; Fernaeus et al. 2013). White et al. (2012) conclude from their review that "The hippocampus and surrounding medial-temporal lobe structures are thought to be important for long-term allocentric (world-centered) representations of the environment, whereas the parietal lobe has been proposed as providing short-term egocentric (self-centered) representations within the environment (Burgess 2008; Whitlock et al. 2008)."

The right brain seems especially capable of processing faces and places external to the body along with negative affective valences. In the patient with a new parietal lobe stroke, this may present rather robustly in the form of *spatial delusions* or confusion about places and their location within them. The patient may describe a robust sense of being put in places against his/her will with a prominent feeling of coercion or external control by others. The delusions may be restricted to relocation or placement in familiar places. However, the place is generally not pleasing as it conveys either a consistent theme with the delusion of fear or anger or destruction or alteration of a previously safe environ, such as the home or the hospital room. For some, the delusion is active and recurrent with the actual sensation of moving through space against their will along with the sensation of being out of control. But for others, the delusion is one of finding themselves in strange places or awakening to discover relocation to another place.

One patient provides an example after her large right middle cerebral artery distribution cerebrovascular accident with near-complete occlusion of the right internal carotid artery found on further assessment. She maintained strong linguistic conversational skills, but appeared a bit paranoid related to her delusions. She was finding herself periodically in her home with a man within her left hemispace that "looks like my husband...but he is not!" The patient's spouse is a contractor, and this devilish creature at the left had altered their home in some very bad ways, having plastered heavily over the doorways and such. With a large stroke at her right parietal region, she no longer recognized that her left arm was her own. We had an elaborate conversation addressing the question of "why did they stick that on me?" These components are typical of *limb alienation syndrome,* which should not be confused with *alien limb syndrome.* With the parietal lesion, the patient may be disgusted by their left-sided body parts and even want to throw them away, whereas alien limb syndrome usually presents from a disconnection of the two brains secondary to a lesion or encroachment upon the corpus callosum. Another aspect of the right parietal syndrome, which was present in this patient, was a dense proprioceptive torticollis with no real ability to appreciate the location of her head in space. Of course, this lost body part may be a fundamental feature underlying the limb alienation syndrome.

A doctor of engineering and CEO of his company shared his anger, combining spatial delusions and limb alienation syndrome, after a dense right parietal stroke. He noted that he "had been connected to his wife over there (gesturing to the left side of his body), when he had been hit by a Silkworm missile." He stated that, ever since this, he had not been able to use that side. He was adamant that the "thing" attached to his shoulder was not his arm! He went on to describe the subsequent events where someone or some group was moving him through space and placing him in strange places. He believed without a shadow of doubt, that he had been in California the previous night and that he was relocated to Virginia during the night. Another engineer described riding a "large bolt" around the hospital. He was terrified because he felt out of control, and there was no steering wheel on the bolt. Another man was adamant that his house was being moved without his permission and out of his control. He misperceived his hospital location to be at his home and demanded that family members and hospital staff "pack up the house." He was not referring to items within the house. He was referring to the house itself, as though it might be packed up and folded, like a tent.

Another nice woman frequently tendered in her conversations that nothing looked familiar any more. On inquiry, this complaint was largely restricted to faces and places. She had previously visited the post office with her spouse daily, and this place seemed more familiar to her, perhaps. She was unable to recognize the man that she had married and lived with through some 50 plus years and this resulted in police intervention as she would panic at the grocery store and try to escape from this "stranger." When the daughter received a desperate call from her that "there was a strange man in her house and Ken was nowhere to be found," she asked the woman to have the stranger pick up the other phone, where she recognized the voice as her husband's. But, he had to walk toward her while conversing to integrate the face and the place with the voice of her now familiar and cherished spouse. Writing an official-looking letter for him to carry to show to the police if they were to suspect him with her cries for help in public places, seemed to help this gentleman cope. The panic attacks that she had were somewhat similar to those of a middle-aged executive secretary who described episodes where she could not even recognize her home, her neighborhood, or even her spouse. But, she appreciated logically that each of these must be hers!

Dressing Apraxia

Among the many variations of movement disorders, right cerebral pathology is more predictive of a dressing apraxia (see Goldstein 2013; see also Tonkonogy and Puente 2009; see also Heilman and Valenstein 2012). Rather than the ideomotor or gestural components of impaired limb movement and differing from the ideational or executive sequencing deficits in limb use, dressing apraxia may be more akin to spatial confusion in using the limbs and clothing the body. After a large right middle cerebral artery distribution cerebrovascular accident, an accomplished professor and scholar maintained excellent logical linguistic speech and social skills. He failed to appreciate any deficits with plans to return immediately to his summer teaching schedule at the university. But, the occupational therapist described his efforts to simply put on his clothing. She stated that "this man gets lost in his clothes." She further noted that "he ties his shoe strings without having his foot in the shoe." The disability was significant as he intended to live alone and to return to work at the university level, which clearly would require additional time for recovery and rehabilitation efforts.

Early accounts of dressing apraxia were apparent in the stroke literature since 1910 (see Brown 1974). Subsequently, Brain (1941) labeled the disorder(s) dressing apraxia, attributing it to underlying deficits in visuospatial organization. Patients were thought to be unable to dress themselves due to an inability to visualize the spatial properties of clothing and to match the correct garment opening to the correct body part. This deficit was present without corresponding deficits in balance, strength, motivation, and cognition. Dressing apraxia may be confounded by a left hemiplegia or inability to use the left arm. However, the confusion extends beyond the motor modality to a more complex spatial disturbance and, indeed, one which may not be restricted to the visual modality, per se. It was apparent even in early research (MeFie et al. 1950) that a relationship or link exists between dressing apraxias and constructional apraxias in addition to associated disturbances with a left hemineglect syndrome, perhaps (see also Heilman and Gonzalez Rothi 2012b).

Constructional Disorders

By 1934, Kleist (see Ha et al. 2012) had defined constructional apraxia as an impaired ability to purposefully and accurately shape or assemble materials, or draw pictures, despite the absence of apraxia in isolated movements and certain aspects of performance. This definition distinguishes constructional apraxia from limb apraxia, although many include neglect or visuospatial difficulties as a part of the symptom complex. Regardless, the constructional apraxias currently include a potentially broad array of combinatory problems constructing a coherent and spatially meaningful structure or drawing (e.g., Roth et al. 2013).

As a young man, I was a foreman of a very small crew in the construction industry building residential houses in New Mexico. I was very fortunate to get to know some of the better folks on this planet and those that work with their hands. This was useful in paying my college expenses, with pay rates superior to working in other typical student labor positions. This would all come back to haunt me several years later when I was planning the purchase of a new sport utility vehicle (SUV) to handle the winter weather driving demands in the mountains of southwestern Virginia. My father-in-law was sitting on my front porch and stated that I would need a carport to cover the new car or it would be subject to the weather. Within 30 min, we had sketched out a double carport, figured the materials list, and gone to shop at our local lumberyard. We came directly back and began the footing and assembly process with my father-in-law working at one side, while I worked at the other. All was fine, until we met each other at the central beam and discovered that we had been working with different plans. Our walls were off by a good inch or so! This reminded me of an old saying within the construction industry that a good carpenter was not one that did a perfect job but one that could fix his or her mistakes. And we did! But we would long debate whose side of the structure was off.

Within the broader domain of combinatory disorders, where parts are assembled into a meaningful array, are the constructional disorders. These may be evaluated from our framework as originating within the second or third functional units, where the motor constructional dyspraxia may originate from right frontal dysfunction and where sensory constructional dyspraxia may derive from right posterior cerebral dysfunction. Spatial distortion or confusion may follow from dysfunction of the second functional unit, whereas organizational sequencing impairment may follow dysfunction within the third functional unit. Constructional disorders relate to multiple brain systems, and these include those originating from dysfunctional left cerebral systems. But, the prominent symptoms are usually of right brain origin. Motor constructional dyspraxia (see Fig. 12.5) features perseverative errors in design drawings and with a failure to implement a coordinated and well-organized strategy in the approach to the drawing or combinatory task. This lack of strategy may be costly (however, see Roth et al. 2013) in the final outcome where the spatial

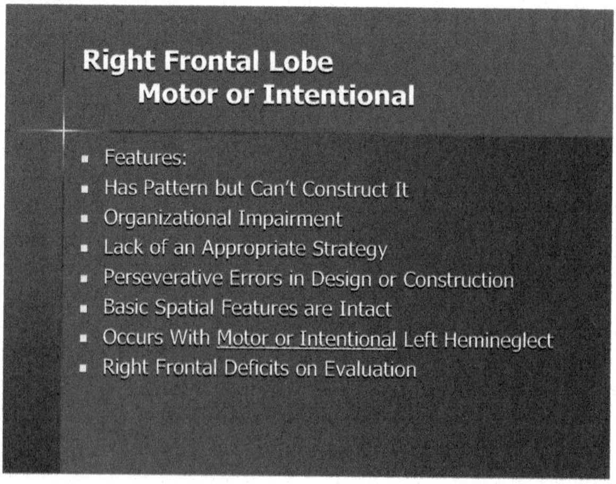

Fig. 12.5 Features of a motor constructional dyspraxia

**Right Frontal Lobe
Motor or Intentional**

- Features:
- Has Pattern but Can't Construct It
- Organizational Impairment
- Lack of an Appropriate Strategy
- Perseverative Errors in Design or Construction
- Basic Spatial Features are Intact
- Occurs With Motor or Intentional Left Hemineglect
- Right Frontal Deficits on Evaluation

features may be intact but the organization is impaired. Sensory constructional dyspraxia (see Fig. 12.6) involves spatial distortion and confabulation in the design, perhaps.

Vincent van Gogh may have revealed aspects and locations for his own brain disorder with some of his paintings characterized by what, otherwise, might be assessed as multiple perseverative marks or designs in the figure (e.g., The Starry Night 1889; Starry Night over the Rhone 1888; Wheat Field with Crows 1890). Moreover, his regulatory control over anger is, by some accounts, suspect and even the events leading up to his death as a result of a gunshot wound to the abdomen. These behavioral, cognitive, and affective features are at least suggestive of right frontal pathology to the neuropsychologist, which would require investigation of these systems on the assessment. In his case, we can appreciate the potential benefits of frontal lobe capacity limitations in his artwork, energy level, and emotional passion (see Naifeh and Smith 2011). The artistic style found in Vincent Van Gogh's famous self-portrait is not entirely unlike that seen in Fig. 12.7 with repetitive

Fig. 12.6 Features of a sensory constructional dyspraxia

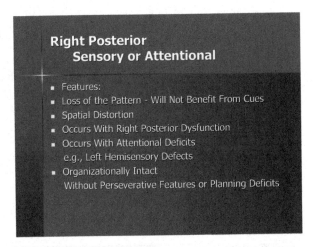

Fig. 12.7 Repetitive line drawing as an artistic style expression. Copyright permission from the artist Jessie Lee, 2014. See for comparisons, Vincent Van Gogh's self-portrait, oil on canvass from 1889.

strokes throughout the image, where multiple repetitions of each overlapping design broadens the creative expressions of the artist.

With severe right parietal lesions, the individual may be severely compromised not only in the understanding or comprehension of external space but often also in the ability to access spatial analyzers (tertiary association cortex) from one or more of the basic sensory cortices. For the visual access to these systems, the patient may lose the ability to visually image even familiar items, faces, or places. A long haul tractor-trailer driver expressed his well-rehearsed admiration for his 18-wheeler tractor-trailer rig. His wife shared that "He loves that truck a lot more than he loves me!" It seemed clear to everyone that he would probably agree. But, when asked to draw the beloved truck he stated "Oh you know…it is ole Model 109." When pressed again, and several times, he continued to state "it is ole Model 109." Eventually, he was able to "draw a picture" of his truck. But, instead of a drawing, the paper had "109" written on it. Now, the challenge is to guess, which brain drew this picture. The right parietal region, which is intimate to constructional praxis and combinatory functions within extrapersonal space, was severely compromised, and he was arguably without an internal representation of the image of his beloved rig.

Constructional dyspraxia resulting from left cerebral dysfunction is less common. Left frontal lobe pathology may yield sparsity in detail or multiple omission errors and this may overlap a more general amotivational and apathetic syndrome (see Scott and Schoenberg 2011; see also Grossman 2002; see also Duffau 2012). However, more complex presentations have been reported. Figure 12.8 provides a sample of the constructional difficulties experienced by a patient subsequent to a left-sided cerebrovascular accident. The syndrome included dyslexia and a modest right hemineglect syndrome. It may be noteworthy here, that he required multiple prompts to complete this drawing, was behaviorally slow, and featured an amotivational and apathetic frontal lobe syndrome.

Figure 12.9 shows a drawing sample from a Virginia engineer. He had spent much of his life mapping the Commonwealth of Virginia for commercial and gov-

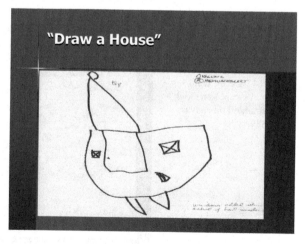

Fig. 12.8 Constructional dyspraxia (atypical) with left cerebral lesion and functional deficits including dyslexia and a right hemispatial neglect

Fig. 12.9 This is a map of Virginia drawn by an engineer after a right-sided cerebrovascular accident. The spatial relationships are substantially distorted with geographical confusion

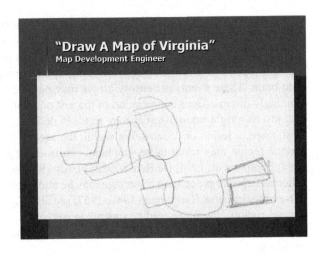

"Draw A Map of Virginia"
Map Development Engineer

ernmental purposes when he was asked to draw a picture of Virginia. He expected and appreciated no real problems in doing this! He expected no difficulties in any of his routine activities, including driving a car (e.g., anosognosia and anosodiaphoria). This combinatory or constructional disorder was diagnosed along with geographical confusion, facial agnosia, and a nonpropositional speech disorder (dysprosodia) subsequent to his large right-sided cerebrovascular accident.

Geographical Confusion

Impairment in geographic or topographic orientation may manifest in various ways with confusion for the spatial array of the principal streets in one's community, the route to be followed to arrive at a familiar destination, the directional location to the north, south, east, or west of major geographical locations within one's native country (e.g., California, Canada, Florida) or the ability to follow a map. These deficits were associated early on with the presence of parieto-occipital disease (see Benton et al. 1984). Many believed that bilateral parieto-occipital lesions were required for the occurrence of these disorders, since the confusion was attributed to a loss of visual memories. However, as early as 1895, Dunn (cited in Benton et al. 1984) rejected the hypothesis of a "loss of recollection of optical images." Instead, it was hypothesized to be a loss of a center for "the sense of location," in the right hemisphere. Based on Dunn's early proposition, subsequent research placed an increasing emphasis on the importance of disease of the right hemisphere in the production of these disturbances in geographic or topographic orientation.

The right cerebral hemisphere is particularly adept at processing spatial arrays, which might otherwise be identified as maps. Aside from the stereotypical map, one might use to find the way as we traverse geographical distances or locations, visual maps might include our friend's face with subtle redistributions of muscle tone providing sometimes intimate knowledge or insight into underlying emotions. Moreover, tone of voice or prosody provides an auditory spatial map, rich in meaning and nuance, which is followed within the analytical and perceptual systems of the right brain. These are samples of the many spatial processing advantages of the right brain. These events or sensory arrays may be perceived as a Gestalt or with seemingly instantaneous appreciation of the gist or meaning conveyed by the map. But, just as a right brain disorder may result in deficient or aberrant processing of these maps, a lesion or deactivation within the second functional unit at the right parietal region may result in geographical confusion (e.g., Paterson and Zangwill 1944; Benton et al. 1974; De Renzi 1982; Fisher 1982; see also Tranel et al. 2009). Whereas superior processing advantage may be allocated to the left hemisphere for left–right awareness (Gerstmann 1940, 1957; see also Cabeza et al. 2012), the right parietal region may more readily appreciate east–west coordinates. Disruption in vertical space or the three-dimensional array may further provide for north–south confusion.

Individual differences in navigational styles and tendencies provide clues for compensatory strategies within a rehabilitation setting or through the efforts of the caregiver to assist a patient recovering from a right-sided stroke. One couple was having a party at their house where the wife drew the map and reversed the east–west coordinates in the drawing. Interestingly, many of the women were able to find the house without difficulty and the men—well—not so good. In the couple's musings with the guests, it appeared as though the women had used verbal sequencing strategies to locate the house and that the somewhat bewildered men had used dead reckoning with a Gestalt view of the map. One neuropsychologist also recalls a colleague with a right brain disorder who uses linguistic strategies to find her way around at scientific conferences. The left cerebral hemisphere appears to be specialized for the awareness of body parts and positions or "personal space." The right cerebral hemisphere, by contrast, appears to be specialized for peripersonal space or the location of our body and of other objects or events within the external world (Heilman et al. 1995; Foster et al. 2008). Thus, geographical awareness corresponds with right brain integrity and geographical confusion may occur with dysfunction here (e.g., Paterson and Zangwill 1944; Benton et al. 1974, De Renzi 1982; Fisher 1982). In contrast, left–right confusion, body-part confusion, or gestural confusion (gestural dyspraxia) correspond with damage or deactivation in the homologous regions within the left brain. Assessment approaches differ for geographical processing disorders and may include asking the patient to describe a *route*, read a *map*, draw a map, follow a route, or point out geographical *landmarks* (e.g., Stark et al. 1996; see Dudchenko 2010).

Aguirre and D'Esposito (1999) proposed a taxonomic classification system for cases of topographical or geographical disorientation by distinguishing four categories of the syndrome. *Landmark agnosia* is characterized by the inability to use salient environmental features for orientation (Hécaen et al. 1980; Pallis 1955). This may be the consequence of damage to neural systems processing landmarks and possibly including the right ventral occipitotemporal association cortices at the fusiform, lingual, and parahippocampal gyri (Maguire et al. 1998; Maguire et al.

1998; Aguirre et al. 1996). *Egocentric disorientation* is proposed to describe disorders with the inability to represent the location of objects with respect to self even though they are able to identify objects within the scene or within the landscape (Levine et al. 1985; Stark et al. 1996). Damage to the right posterior parietal cortex appears to be responsible for this deficit. *Heading disorientation* with an inability to remember the direction to go within the route might also present as an intentional deficit subsequent to frontal pathology. However, some have argued for evidence presented (e.g., Takahashi et al. 1997; Cammalleri et al. 1996) subsequent to lesions within the retrosplenial cortex or posterior cingulate (see Turriziani et al. 2003; Cammalleri et al. 1996).

The final category of topographical disorientation described by Aguirre and D'Esposito is *anterograde disorientation,* possibly revealing the consolidation deficits attributable to right hippocampal and parahippocampal damage. The anterograde deficit refers to an impaired carryover or consolidation of new information. This disorder preserves the treasure chest of knowledge for routes, landmarks, and way finding learned prior to the onset of impairment. The deficit instead relates to an inability to acquire new representations of geographical information reflecting a learning disability (Epstein et al. 2001; Habib and Sirigu 1987). Right parahippocampal lesions were found most consistently in patients with anterograde topographical disorientation (Habib and Sirigu 1987; Pa, 1997). This might be evident even with early pathology with a temporal lobe dementia process and especially one with an affinity for cholinergic systems and the hippocampal/parahippocampal regions.

Neglect Disorders

Hemispatial neglect is characterized by defective detection of events and impaired exploratory or intentional actions in the contralesional hemispace. Research indicates the relevance of the specific frame of reference neglected within extrapersonal or peripersonal space and within egocentric or object-based frames of reference (e.g., Buxbaum et al. 2004; Heilman et al. 2012; see also Cabeza et al. 2012). The source of the neglect has been attributed more or less to the system affected, including sensory perceptual, motor intentional, and arousal-based components (Heilman and Gonzalez Rothi 2012b). This might ultimately involve an inability to disengage attention automatically from the intact region and direct it to the contralesional side (Corbetta and Shulman 2002; Corbetta et al. 2008). However, neglect is often accompanied by other spatial and nonspatial attentional deficits that affect both sides of space (Husain and Nachev 2007).

Substantial research has been conducted on the differential contributions of the left and right hemispheres in the allocation of attention, using both neurologically compromised and intact individuals. Overall, the evidence from normal, healthy individuals has revealed the importance of the posterior temporal parietal region of each hemisphere in the allocation of attention to contralateral hemispace, but the right hemisphere appears to be dominant (Adair et al. 2003; Barrett et al. 2000; Bowers and Heilman 1980; Heilman and Van Den Abell 1980). Moreover, whereas the left hemisphere has a proximal attentional bias, the right hemisphere has a distal bias for attentional allocation away from the body and into extrapersonal space (Chewning et al. 1998; Heilman et al. 1995; Jeerakathil and Kirk 1994; Jeong et al. 2006; see also Jewell and McCourt 2000). Furthermore, right hemisphere lesions (Heilman and Valenstein 1979; Ota et al. 2001), and right superior temporal gyrus/sulcus lesions in particular (Hillis et al. 2005; Watson et al. 1994), are associated with the neglect of left hemispace. In one project using voxel-based lesion mapping, specific relationships were found between lesions in the inferior parietal lobe and perceptual neglect, the dorsolateral prefrontal cortex and visuomotor neglect, and the temporal lobe and object-centered neglect (Verdon et al. 2010; see also Aimola et al. 2012).

Neglect disorders are divided into those involving either the left or the right hemispace and/or hemibody. Within each cerebral hemisphere, the neglect disorders are separated into the motor or *intentional neglect* syndromes derived from dysfunction within the frontal lobe and the sensory or *attentional neglect* syndromes derived from dysfunction within the posterior cerebral regions. The diagnosis of a sensory neglect requires some demonstration of the integrity of one or another sensory modality from the neglected hemispace. Thus, the information is received but is, in some respect, ignored or neglected. The motor neglect syndromes involve a failure to intend or to persist at or within the contralesional hemispace. A typical example might result from deactivation of the premotor cortex in the right frontal lobe, where there is functional integrity in the use of the left arm, since the motor strip is intact. But, the patient may appear to be hemiplegic at the left, failing to initiate movement at that extremity, and even with substantial encouragement or confrontation by the therapist. Nonetheless, the patient may initiate use of the affected extremity spontaneously and sometimes with the most curious appraisals by the therapist ranging to concerns for malingering of symptoms or noncompliance. The attribution by others that the functional disturbance is "intentional" again provides a clue for a frontal lobe disorder.

Riddoch (1935) first reported hemispatial neglect in the absence of central visual problems. Hemispatial neglect disorders are a common feature of brain dysfunction and may follow a stroke to one or the other cerebral hemisphere. However, the potent role of the right brain in processing extrapersonal or distal space and the relatively restricted role of the left brain in attending/intending to right hemispace provides a functional foundational base for the frequent association between neglect and disorders of the right brain. Right brain disorders not only accompany the vividly apparent left hemispatial neglect syndromes but often occur with affective features involving alienation toward the left hemibody. The latter is common in limb alienation syndrome. Limb alienation refers to the lack of integration of the left arm or leg into the concept of self-following right-sided brain damage. The individual may not only deny that the limb is their own but also they may want to discard it or throw it away as if someone or something had attached it to them without their permission.

Neglect involves a failure to respond or to orient to novel or meaningful stimuli presented contralateral to the dysfunctional cerebral hemisphere. The differential diagnosis includes the exclusion of pure sensory deficits resulting from pathology within the primary projection pathways from sensory receptor to the primary projection area for that sensory modality. For the somatosensory projections, the diagnosis of a left hemineglect syndrome might rule out a left hemianesthesia. For the visual pathways, the diagnosis of a left hemineglect syndrome might rule out a left homonymous hemianopsia. As such, the diagnostician has ultimately provided evidence of the arrival of sensory information from the contralateral hemispace onto the primary projection pathways adequate for the detection of that sensory event. The failure to appreciate the sensory event involves, to some extent, the higher-order processing or appreciation of that modality or multiple modalities originating from the contralateral hemispace. Karl Pribram (1991) has argued for a neural representation of extrapersonal spatial coordinates somewhat like a visual hologram. Disparity between this three-dimensional neural representation and extrapersonal spatial coordinates may be relevant to neglect syndrome. Moreover, perceptual disparity may provide a foundational discrepancy relevant to delusional disturbances involving spatial coordinates or movement through the three-dimensional array. Just as a simple mismatch between the activity in the motor cortex and the somatosensory cortex may produce an illusion of movement, delusional disturbances involving more complex associations may be derived from altered associations among neural systems.

From a functional cerebral systems perspective, the distinction to be made is among Luria's first, second, and third functional units for the brain contralateral to the neglected stimulus event. Neglect is more commonly diagnosed with right brain disorders. Common lesion sites include Luria's first functional unit consisting of the reticular activating systems and the arousal projections through the right thalamic region. Studies of both nonhuman animals and patients have revealed that damage to components of one brain's thalamic and mesencephalic reticular activating system can induce the ipsilateral attentional and intentional biases that are characteristic of the unilateral neglect syndrome (Watson, Heilman et al. 1974; Watson et al. 1981). Relative hypoarousal of the ipsilateral hemisphere was found with clinical lesions (Watson et al. 1974), resulting in the oppositional attentional and intentional biases of the intact and relatively hyperaroused hemisphere. Other studies demonstrated that patients treated with slow transcranial magnetic stimulation applied to the unlesioned hemisphere, reducing the treated hemisphere's activation, improved with a corresponding reduction in neglect behavior (Koch et al. 2008).

Left hemineglect may be robustly apparent after lesion of Luria's second functional unit involving the reception, analysis, and comprehension of sensory input. The right parietal lobe and the tertiary cortical region of the right parietal–temporal–occipital region provide a critical region for the development of neglect (Heilman et al. 1995; Heilman and Valenstein 1979; see also Heilman et al. 2012; see also Heilman and Gonzalez Rothi 2012b). Finally, dysfunction within Luria's third functional unit, the frontal lobe, may easily produce a motor neglect disorder and especially with pathology affecting the secondary and tertiary association cortices of the right frontal region. The fundamental distinction is made between the second and third functional units, where dysfunction of the former may produce a *sensory or attentional neglect syndrome* and where dysfunction of the frontal lobe may result in a *motor or intentional neglect syndrome*. Using this system, a *mixed neglect syndrome* would be one involving both the sensory perceptual systems within functional unit 2 and the motor intentional systems within functional unit 3. Subcortical/ thalamic neglect disorder may result from brain-stem diencephalic and/or mesencephalic origin(s).

The premotor or secondary association cortex within the frontal lobe of each brain intends largely toward the opposite hemispace and hemibody. Kenneth Heilman showed early on that damage to these brain regions results in a corresponding intentional or motor neglect. The patient with damage here may be able to feel the limb, see the limb, and appreciate that the limb belongs to them. However, the limb and the space within which the limb is used are not well intended to with left side impersistence in directional head and/or eye movements. This may result with a loss of inhibitory control over reflexive head and/or eye movements toward the ipsilesional side. In this scenario, the individual may provide the initial assessment clue with a right gaze bias or rightward rotation of the head and neck about the neuraxis. The postural displacement of the head and neck represents a motor torticollis, ultimately with asymmetry in cranial nerve functions (e.g., innervation of the sternocleidomastoid and trapezius muscles.

Just as a sensory neglect involves the demonstration of the integrity of the sensory projections, the demonstration of a motor neglect will involve the exclusion of a pure motor defect involving the final motor pathway and upper motor neurons. A left hemiplegia with an inability to use the left limb would not, by itself, be consistent with a motor or intentional neglect. Instead, the lesion might involve association cortex, wherein the desire or intent to that side may be altered. Involvement of the premotor dorsolateral frontal cortex is consistent with this premise, since the loss of the affected frontal eye field yields a relative dominance or release of the homologous region in the other brain and presents as a directional gaze bias ipsilateral to the lesion. For example, deactivation of the right frontal eye field would diminish intentional gaze toward the left hemispace and fundamentally increase the probability of rightward gaze bias or preference (e.g., Suzuki and Gottlieb 2013), sufficient for the diagnosis of a left motor or left intentional hemineglect syndrome. Another example may be the expression of a spasmodic torticollis, where rightward head rotation is derived from dystonia or spasticity subsequent to upper motor neuron pathology (Harrison et al. 1985).

Interestingly, the movement of the entire head toward left hemispace may require less functional tissue within the premotor area than does the movement of the eyes toward the left. For the therapist, the patient may benefit from implementing treatment initially requiring directed head movements to compensate for a left motor neglect. With recovery, the patient may then initiate more focused efforts to gaze and to pursue toward and within left hemispace. This patient may also be relatively distractible ipsilateral to the lesioned frontal lobe secondary to a release of brainstem reflexes at the level of the tectum or superior colliculus. The therapist may implement interventions to minimize distractibility ipsilateral to the lesioned frontal lobe using placement next to a wall or away from social distracters. The normal left frontal lobe, by way of this example, may be relatively dominant now and responsive to opportunities for social engagement or approach, propositional speech, and events within right hemispace, as this is what it appears prone to do.

Damage or deactivation of the left brain may produce contralateral neglect disorders. But, these are commonly milder in intensity at least for the processing of extrapersonal space. Instead, these deficits may be somewhat more intensely appreciated in the sensory or motor analysis of proximal space as with ideomotor or gestural praxis (e.g., Hermsdörfer et al. 2012; see also Vingerhoets et al. 2012) with involvement of Luria's second functional unit and with ideational praxis with involvement of Luria's third functional unit. But, the left brain does appear to contribute to the sensory and motor analysis of extrapersonal space to the right of midline. For the therapist or family member, the appreciation of a mild right hemineglect syndrome may very well be important and a failure to appreciate this disorder may affect the rehabilitation outcome if not the safety of this patient. The therapist might also appreciate that in some ways the left brain disorder may be a red flag for safety as a mild neglect of right hemispace may coexist with a more remarkable disability in ideomotor or ideational praxis. A right hemibody limb may be improperly placed or improperly sequenced concurrent with diminished attention or intention to the right side of the world and distal to the body. These neural systems have long been known for their shared role in functional tool use and in language or propositional speech (e.g., Vingerhoets et al. 2012; see also Bracci et al. 2012).

The individual with a left hemineglect syndrome will characteristically present with anosognosia and/or anosodiaphoria, failing not only to appreciate the left side of the world but also failing to appreciate that they have a problem. One man was notable in this regard as he was running for a public office at the time of his stroke. He would awaken in the morning, shave the right side of his face, dress the right side of his body, and attempt to navigate in his wheelchair. He would quickly lodge his wheelchair against the wall at his left using his right arm for mobility. He was frustrated that things were not working well as this wall did not exist and "Why doesn't my wheelchair work?" All the while, he was demanding to leave the hospital as he should not "be doing these silly therapies." He had "an election to win!" Another CEO of a major office supply product development corporation demanded to return to the golf course. I was invited to her country club for lunch one day, a couple of years later, and was somewhat astonished that she was indeed on the course. Even after a good recovery, she would swing at the golf ball and fall on the ground having been off to the right! Over and over again she fell down, while others expressed attributions of admiration for her attempting to overcome deficits that she really did not believe existed.

The assessment of the sensory basis of neglect might begin with the evaluation of the relative integrity of one or the other sensory modality using dual concurrent bilateral stimulation techniques (e.g., Hugdahl 1988, 2003, 2012; Hugdahl et al. 2009). These techniques are designed to assess for *extinction* within that sensory modality. Sensory extinction tests provide for dual concurrent stimulation to each cerebral hemisphere, where activation of the stronger or intact cerebral region in one brain may extinguish the activation of the homologous but weaker brain region. Extinction manifests itself through access to the corpus callosum, wherein each brain may use lateral inhibition across this commissure (homotypic crossover) to quiet the homologous region in the other cerebral hemisphere. In a normal brain, sensory input coming from each side of the world can be located within space using these techniques. Concurrent bilateral stimulation from the world may also be improved in contrast or contour using lateral inhibition within a neural system. Essentially, stronger activation within one cerebral hemisphere may inhibit the weaker activation of the other to reduce noise and to maximize the detection and perception of the stronger stimulus. But, if one or the other brain is relatively weaker in the visual, auditory, somatosensory, vestibular, or other modality, then concurrent input to that sensory modality projecting to the stronger brain will potentially eliminate the presence of the weaker stimulus activation in the homologous brain region.

For visual extinction, dual concurrent or stereoscopic visual stimulation is provided within the left and the right visual field (Anton 1899; Poppelreuter 1917). If the right occipital region is relatively weaker, then the stimulus originating from the left visual field may be extinguished by the dominant activation of the left occipital lobe and by stimulus presentation within the right visual field. Unilateral stimulus presentation at the left visual field or at the right visual field is detected normally, whereas the dual concurrent stimulation of both visual fields results in the extinction of the weaker one. For tactile stimulation, the assessment employs dichaptic techniques with dual concurrent tactile presentation at the left and the right hemibody (Loeb 1885; Oppenheim 1885). For audition, the assessment involves dichotic techniques, where concurrent bilateral stimulation at each ear may be used to assess for extinction (Bender 1952; Heilman 1970; Hugdahl 1987; Heilman and Valenstein 1972; Alden et al. 1997; Demaree and Harrison 1997b; see also Pollmann 2010; Hugdahl et al. 2009; Hugdahl 2003, 2012; see also Hugdahl and Westerhausen 2010).

Motor or intentional neglect may be assessed through the evaluation of motor impersistence, motor extinction, or through hemiakinesia. Duration of contralateral gaze or in the use of the contralesional extremities may be *impersistent* even with substantial prompting and encouragement. Motor extinction results with the use of the premotor region of the dominant frontal lobe where dominance is conveyed in relative activation or in the relative functional integrity of that region. Activation of the left frontal lobe may inhibit the homologous regions of the oppositional frontal lobe. This may be useful in normal brains where oppositional motor posturing may be fundamental to opening a jar or in ambulation, for example. But, with a premotor lesion any activation of the oppositional frontal lobe may potentially extinguish the response capability or activation of the weaker frontal lobe. For the therapist or family member of an individual with a motor neglect, maximizing the intention to the neglected side may be accomplished through minimizing the dual concurrent tasking demands at the opposite (ipsilesional) side of the body. Maximizing performance, though, differs from therapeutic intervention where bimanual task demands may effectively challenge the weaker frontal lobe and where persistence or intention to the task may be assessed using temporal measures, including the duration or persistence of the response. The patient may benefit from the therapist's caution to not demand more of this region than it is capable of and where response *shaping* using the *method of approximations* and hierarchical goal setting are preferred.

Such an individual may neglect near or far space, including that well beyond their reach. Standard pencil and paper tests may be sensitive to neglect and they are useful to the extent that they are sensitive. But, a patient passing such a test may still suffer significant neglect and ignore their left hemibody or even distant objects within the left side of the more expansive environmental surroundings. Many patients are unable to explain or to understand why they find themselves drifting on the road and onto the side road or highway median. The preponderance of research on hemineglect syndromes has focused on the neglected hemispace, whereas the science has itself neglected the potential alteration of the boundaries of the ipsilesional hemispace where the patient is often presumed to process the dimensional boundaries of that space well.

Many patients have been in error not only in defining the dimensional boundaries of the left hemispace but also of the right hemispace (ipsilesional space). This results in more complex spatial processing deficits than what may otherwise be appreciated. In some, an estimate of 90° to the left is near midline, whereas an estimate of 90° to the right is deviant and beyond the accurate mark. But, with bilateral cerebral dysfunction, the patient may have bilateral neglect disorders. Bilateral motor or intentional neglect disorders, for example, appear to be common with a frontal lobe dementia. Such an individual may be restricted largely to midline in lateral gaze and pursuit. Efforts to initiate and to persist may be restricted, bilaterally, other than intent to the midline. The therapist, caregiver, and the patient may maximize their efforts and interactions from an accurate assessment of the boundaries within which they work. If one or the other of these team members is out of the workable space of the other, then effort and time may be used inefficiently and recovery or rehabilitation progress may be delayed.

Pencil and paper measures of neglect typically evaluate neglect through the detection of rightward or leftward errors. The Line Bisection Test (Schenkenberg et al. 1980) provides an example. The patient is typically provided with linear stimuli and the challenge to bisect the line(s) at midline. Many variants of the test are available and often the samples are generated bedside for testing purposes. A sample is provided in Fig. 12.10. There are many variants of the Line Bisection Test including the sample provided, which was administered bedside for an initial screening of the patient's symptoms. Two separate attempts failed to accomplish the midline bisection goal and the patient failed to appreciate the errors with the overriding left hemineglect syndrome.

A performance sample using Luria's M&N task (Luria 1980; Christensen 1979) is presented in Fig. 12.11, where the patient neglects the left hemispace with each line of writing located increasingly within the ipsilesional side of space. This test requires the production of alternating M&Ns using cursive. The test is sensitive to perseverative errors, which may be seen in this figure. It may be noteworthy that this patient has lost cursive writing and is restricted to block printing. But, the left

Fig. 12.10 Administered bedside for screening purposes. The patient is asked to bisect the lines at the middle, whereas the patient is in error toward the right side across two administrations (I and II)

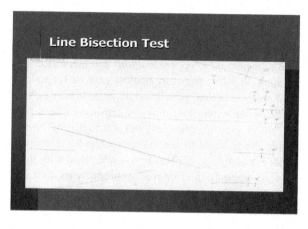

Fig. 12.11 A patient with a right parietal stroke attempts to complete Luria's M&Ns task with each line shifted more toward the right hemispace or ipsilesional side of the page

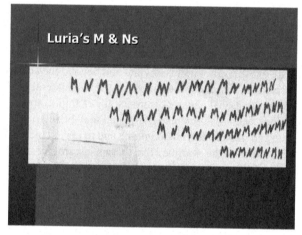

neglect is easily appreciated with a drift toward the right side of the page in the progression from one line to the next. Figure 12.12 provides a performance sample from a patient with right hemisphere damage attempting to copy a Greek cross. The patient has a right parietal lesion and a left hemineglect syndrome.

Figures 12.10 and 12.11 provide potential evidence for left hemispatial neglect in patients with right-sided brain injury. Figure 12.13 depicts the patient's efforts at Luria's circle, square, triangle task, where alternate sequencing of this three-design set is assessed for perseverative sequencing errors. However, after completion of the first line, this patient fails to return to the left side of the drawing. Figure 12.14 provides a sample of one person's efforts to copy a picture of a chair with noticeable perseverative errors in retracing lines and also a relative neglect of the left-sided components in the image. Figure 12.15 provides samples of a patient's improvement from a left hemineglect syndrome over a three-week period, following a right-sided cerebrovascular accident. This improvement is apparent in his efforts to draw

Fig. 12.12 A patient's attempt to copy the Greek cross

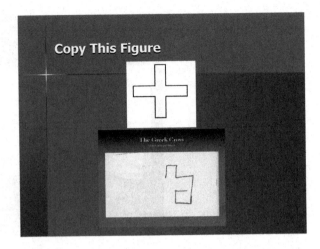

Fig. 12.13 A patient with left hemineglect attempts Luria's circle, square, triangle task with failure to return to the left side of the page

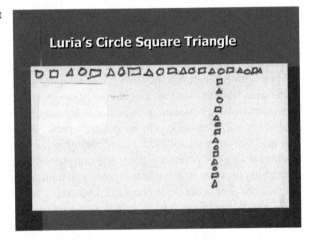

Fig. 12.14 A patient with left hemineglect syndrome attempts to copy the drawing of a chair

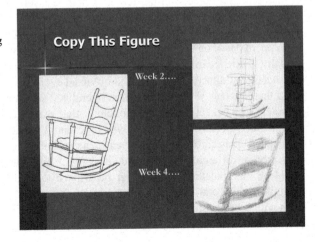

Fig. 12.15 Drawings of a house provide comparisons over a 3-week recovery period in a patient following a right-sided cerebrovascular accident

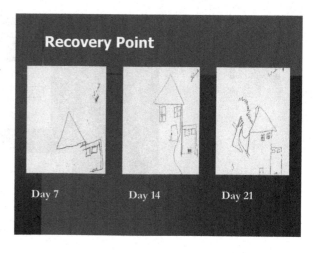

a house. Increasing complexity and detail is developed at the right side of the drawing along with extension of the figural components into the left side of the drawings.

Treatment of motor or intentional neglect may include exertional therapies wherein the head, eyes, trunk, and/or limb(s) are exercised toward and within the left hemispace. Treatment may involve developing ways to bring the patient's intentional (motor systems) or attentional (sensory) systems toward the left and in an incremental fashion based on the functional accomplishments or capabilities demonstrated (sensory perception) in the patient's performance of these activities. Again, this might be approached by the therapist using the method of successive approximations with ever-increasing goals and corresponding demands for reinforcement. This might be initiated just a few degrees past midline with gradual progression leftward as the previous goal is accomplished. Self-monitoring and therapist monitoring is encouraged to exceed the previous accomplishments over time. Throwing a nerf ball or colorful beach ball from the left may be useful. Playing a preferred game within the patient's left hemispace, music therapy provided within left hemispace, or manipulating objects on the left may help. Sensory or attentional neglect may benefit from efforts to promote novelty or saliency within the left hemispace. This might be accomplished through one or another modality or through multimodal enrichment based on the patient's relative strengths and weaknesses at each sensory modality. Saliency might be manipulated using intensity or duration dimensions for vibrotactile, auditory, vestibular, and/or visual events.

In their review of the literature on lateralized sensory interventions for neglect disorders, Schock et al. (2013) conclude that exogenous sensory events trigger stimulus-driven shifts of attention when presented within either the left or the right hemispace. This shift is evident even if the sensory cue and the target stimuli are presented in different sensory modalities (e.g., Eimer and Driver 2001). Salience might be manipulated within either hemispace to modify directed attention. For example, salience at the left hand might be manipulated using somatic modifications with vibrotactile cues. It might also be modified by placing a brightly colored (salient) glove over the left hand. It has also been demonstrated that modification of the emotional valence of lateralized events can influence attention-related processing (Eastwood et al. 2001; Fenske and Eastwood 2003). Schock et al. (2013) note that negatively valenced sensory events may serve as an attention "magnet." This may be useful with an individual suffering from neglect by narrowing the attention focus, which may facilitate enhanced processing of the negative event (Fenske and Eastwood 2003).

During a visual search task, negatively valenced stimuli are located with reduced reaction times and with fewer errors than positively valenced stimuli (Eastwood et al. 2001). Related evidence indicates that events with a negative emotional valence may even be sufficient to overcome neglect symptoms evident within the left visual field of patients with right hemisphere damage (Grabowska et al. 2011). However, Schock, et al. (2013) point out that showing negative emotionally arousing stimuli interferes with the processing of other less salient left visual field targets as a function of heightened competition for attentional resources (Hartikainen et al. 2007). These authors also identify the neural network, including the limbic regions such as the amygdale, insula, and anterior cingulated cortex, which are known to be relevant to salience detection and evaluation of threat (see Seeley et al. 2007; Menon and Uddin 2010).

Premotor lesions resulting in a motor neglect may present with integrity of motor function at the contralateral limb for spontaneous movements. But, the patient may not use it under confrontational demands. This may prompt considerable skepticism in others to infer that the patient is malingering or to inquire as to dissimilation being present. This is not the case as the integrity of the premotor regions and their interactions with the head of the caudate nucleus may be essential to the desire, intention, or preparation to use the affected limb. Also, with the addition of the learning history where the neglected limb conveys discoordination and functional ineptitude and where the unaffected limb functions well, the patient may differentially acquire a preference for the ipsilesional extremity. This may be counteracted through the use of limb restraint therapies where the intact limb is restrained to force the use of the affected extremity, at least for a part of the therapy day. Figure 12.16 provides a

Fig. 12.16 Premotor lesion at the right upper extremity in a man with hemorrhage associated with metastatic deposit at the left frontal region

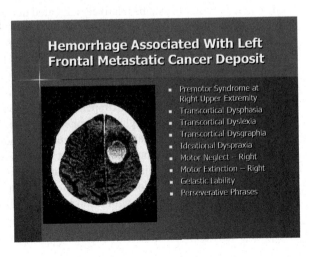

case study of a man with premotor deficits for his right upper extremity and a right motor neglect. His intervention efforts in rehabilitation included the introduction of constraint-induced movement therapy.

Constraint-induced movement therapy (see Miltner et al. 1999; Taub and Morris 2001) was initially demonstrated with nonhuman primates resulting in functional improvements in the affected limb over time. In 1998, Liepert et al. (1998) demonstrated positive brain-related changes in stroke patients inferred from motor-evoked potentials. Neuroimaging studies have also demonstrated improved blood flow to somatosensory and motor cortices (Wittenberg and Schaechter 2009) and structural changes in these cortical regions that are not apparent after other rehabilitation techniques (Gauthier et al. 2008; however, see Bowden et al. 2013). Patients receiving this intervention often recover substantially more function in their affected extremity than patients who are simply told to try to use the limb as often as possible (Taub et al. 2002; however, see Bowden et al. 2013). Active use promotes improved efficiency in the neuronal organization and preparation for the response. Moreover, these same regions have been associated with the "impulse" or the desire to use that body part (e.g., Penfield and Boldrey 1937) as with directed intention to initiate a response with the body regions represented along the premotor and eventually the motor homunculus. Vestibular therapies have also been effective in the treatment of neglect (e.g., Geminiani and Bottini 1992) as rotational or vectional stress may yield spatial shifts in the apparent location of the body and a positional shift into the contralateral hemispace to counteract the vection.

Paul Foster (Foster et al. 2008) demonstrated the vertical dimensions and the affective dimensions of space in his project asking normal participants at a major university medical center to place pegs labeled with an emotional valence on a pegboard. Significant lateral and vertical differences were found as a function of emotional valence raising the specter of neglect disorders to the emotional array and interfacing the emotion, such as depression with down-going features, to the spatial array. Positive affect may involve activation of neural systems elevating gaze vertically along with facial expressions and even prosody. It is reasonable to expect that elevation of gaze, facial musculature, and tone of voice may provide a release for neural activation of systems relatively activated during depression or dysphoric thoughts.

Others have demonstrated that the spatial aspects of left and right hemineglect disorders are, in fact, not balanced based on functional laterality specialization(s). This was evident earlier on with evidence that the right hemisphere may be better able to process extrapersonal space bilaterally, whereas the left hemisphere may be more restricted to processing extrapersonal space within the right hemispatial domain. Using patients with right hemisphere damage, researchers (Aiello et al. 2012) have recently shown that the imbalance transcends extrapersonal space and incorporates task specialization. They note that spatial reasoning has a relevant role in mathematics where it is widely assumed that in cultures with left-to-right reading, numbers are organized along the mental equivalent of a ruler. This consistent representation of a number line, with small magnitudes located to the left of larger ones was expressed in patients with right brain damage being able to disregard smaller numbers with a pathological rightward bias in mentally setting the mid-point of number intervals (Zorzi et al. 2002), i.e., reporting without calculation what number is halfway between two other numbers. The derived interpretation is that the enhanced attentional bias toward higher numbers at the right side of the mental number line represents spatial neglect for numbers at the left.

Contrary to this interpretation, Aiello and colleagues (Aiello et al. 2012) find that patients with right brain damage disregard smaller numbers both when these are mapped on the left side of the mental number line and on the right side of an imagined clock face. When viewed collectively, these findings indicate that the right hemisphere supports the representation of small numerical magnitudes independently from their mapping on the left or the right side of a spatial–mental layout. Moreover, the anatomical correlates identified through voxel-based lesion–symptom mapping and the mapping of lesion peaks on the diffusion tensor imaging-based reconstruction of white matter pathways showed that the rightward bias in the imagined clock face was correlated with lesions of high-level middle temporal visual areas that code stimuli in object-centered spatial coordinates. This includes stimuli that, like a clock face, have an inherent left and right side. These findings extend the discussion of neglect-related disorders to more complex conceptual elements that exist within the broader neural systems structure and which coalesce or converge among these neural structures.

The neural circuit involved in neglect disorders, includes the superior colliculus and this was evident originally in the phenomenon first described by Sprague (1966), where lesions of the superior colliculus in cats resulted in an unexpected recovery from deficits in spatial orienting caused by damage to other parts of the attention network. Following occipital or parietal cortex lesions on one side of the brain, cats exhibit visual neglect, with the tendency to ignore visual objects presented in the affected part of the visual field.

Suppression of activity in the superior colliculus on the opposite side of the brain was shown to relieve the symptoms of visual neglect. Relatedly, a clinical case was described in which the symptoms of visual neglect experienced after damage to the frontal cortex were relieved by additional and subsequent damage to the contralateral superior colliculus, providing evidence that the Sprague effect also extends to humans (Weddell 2004; see Krauzlis et al. 2013).

In their review of the functional anatomy of the superior colliculus, Krauzlis et al. (2013) conclude that these neurons respond to stimulus modalities in addition to vision, and that most exhibit activity patterns related to the planning and execution of orienting movements and covert attentional processes. Also, Kallman and Isaac (1980) demonstrated the role of the superior colliculus in processing ambient illumination underlying arousal propagation through the mesencephalic reticular formation. Burtis, Heilman, Mo, Wang, Lewis, Davilla, Ding, Porges, & Williamson (in press) may have manipulated this pathway using constrained left- and right-sided monocular viewing with evidence for lateralization of the sympathetic activation to illumination originating at the left eye in comparisons with that originating at the right eye. These conclusions were partially based on evidence that the retinal projections to the superior colliculi are largely contralateral and where left-sided visual input would differentially activate the contralateral right superior colliculus and right cerebral hemisphere. The ultimate potential for clinical intervention in neglect disorders, through manipulations of activity within the superior colliculus remains largely unexplored, despite evidence for its role in covert attention as well as arousal functions intimate to neglect disorders.

Emotional Disorders

Damasio and Carvalho (2013) note that "depression alone is the leading cause of disease in the United States and the leading cause of non-infectious disease worldwide." Opining on the neuropsychological bases underlying the emotional state, Damasio et al. (2013) state that "research on the neural basis of affective phenomena has established beyond doubt that the human insular cortices are involved in processing body feelings (such as pain, pleasure, and temperature) and feelings of emotions (Damasio et al. 2000; Kupers et al. 2000; Brooks, Nurmikko et al. 2002; Craig 2002). This vast and extensive cortical area includes regional sensory specializations (vision, audition, vestibular, touch, pressure, pain, taste, and olfaction) and cross-modal integrative areas, which clearly activate within the interpretive and perceptual analysis of one or of another emotional feeling. Beyond these findings, some have raised the question that the insular cortices are the "necessary and sufficient platform for feeling states in humans" and are, in effect, the exclusive source of these feelings or experiential states (Craig 2009, 2011; cited in Damasio et al. 2013; see also Damasio and Carvalho 2013). But, history may well be repeating itself in this strict analysis as, for example, in the earlier case of Hughlings Jackson (1874). Jackson argued against the strict localizations views of his time noting that complex mental processes are organized and reorganized within different *levels* of the brain, establishing the hierarchical organization of the brain from brain stem on up to the cortex. Jackson's arguments, alongside those of many other neuroscientists and clinicians ultimately lead to modern functional cerebral systems theory.

In their investigation of the proposal for exclusive localization, Damasio et al. (2013; see also Damasio and Carvalho 2013) report the findings of a patient whose insular cortices were destroyed bilaterally as a result of herpes simplex encephalitis. The authors conclude that "The fact that all aspects of feeling were intact indicates that the proposal is problematic. The signals used to assemble the neural substrates of feelings hail from different sectors of the body and are conveyed by neural and humoral pathways to complex and topographically organized nuclei of the brain-stem, prior to being conveyed again to cerebral cortices in the somatosensory, insu-lar, and cingulate regions." In this interpretation, the first neural substrate of feeling states is found within the subcortical brain-stem structures with elaborative analysis eventually relating feeling states to cognitive processes such as decision making and imagination. However, just as a bottom-up and top-down analysis receives support both in anatomy and in function(s), there is also undeniably robust connections between the frontal and posterior cortical regions (e.g., the longitudinal tract) and between homologous regions of the left and right brain via the corpus callosum. These robust interconnections provide for a functional neural systems theory of increased complexity. Moreover, additional pathways are relevant, including the interactive connections of cortical regions with the basal ganglia and corticocortical interconnections via "U-connections," for example.

The neuropsychological approach to emotional disorders, from a functional neural systems perspective, involves the determination of cerebral laterality differences in the specialization of emotional processes, where the left brain has a propensity for positive affect; social approach; optimistic anticipation of the future; an energetic and rapid sequential processing style; and logical linguistic speech. The right brain appears to have a propensity for negative affect, social avoidance, negative reflection on the past, an intuitive holistic appraisal of the gist, somatic concerns, and nonlinguistic prosodic speech. Secondly, this approach involves the determination of anterior (frontal) and posterior (parietal, temporal, and occipital) differences in the specialization of emotional processes where the frontal regions are involved in executive functions or regulatory control and expression and where the posterior brain regions are involved in the reception and comprehension of the emotional event. This functional analysis yields four distinct quadrants for the evaluation as discussed within quadrant theory (Foster et al. 2008; Shenal et al. 2003; Foster et al. 2008; Walters and Harrison 2013a). But, each quadrant can be subdivided into its components as with the primary projection cortices, the secondary and the tertiary association areas, and the subcortical regions and tracts.

Within the frontal systems charged with regulatory control and expression, the disorder might be evident in altered facial expression, altered speech expression (e.g., tone or volume of voice), emotional lability or deregulation, and/or altered muscular tone or hyperreflexia (e.g., dyspnea with anxiety). Other examples include inappropriate expressions of "rule-regulated behaviors" as with social improprieties, turn taking, or a failure to regulate or to produce emotional expressions congruent with the social context (e.g., Anderson et al. 1999; Eslinger et al. 2004; see also Yeates et al. 2012). Expressive deficits may also convey in the amotivational syndromes derived from left frontal pathology within the head of the caudate or medial frontal pathology (see Scott and Schoenberg 2011; see also Grossman 2002; see also Granacher 2008; Hu and Harrison 2013b). Within the posterior systems charged with the reception and comprehension of emotional events, the disorder might present within the auditory analyzers resulting in an inability to comprehend emotional tone or nuance within another's voice as with a receptive dysprosodia. It might present instead within the visual analyzers resulting in an inability to comprehend emotional tone or nuance within a facial expression. These, sometimes subtle cues, may be of primary importance in appreciating the gist of the communication with others where the literal conveyance may be restricted in emotional content. Activation of these analyzers may alter the intensity of emotional expressions indirectly as with anger or fear or result in an overassessment of the threat implied or imposed by another. A more extreme case may result from overactivation yielding

Fig. 12.17 Emotional disturbance in patient M. K. related
to right posterior cerebral
pathology from a brain tumor.
The constellation of clinical
symptoms is listed

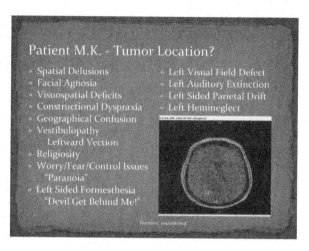

formesthesias accompanied by fear, apprehension, and paranoia as with an agitated
state.

Clinical imaging results from one such patient, afflicted with a brain tumor
within the right posterior cerebral region, are presented in Fig. 12.17. Patient M.K.
presented with spatial delusions, extreme apprehension reflective of paranoid content with preoccupation at the left hemispace. It was at the left hemispace that she
experienced visual formesthesias of a devil with implications for threatening approach and coercion. The religious relevance of these events was deeply meaningful
as she described them. She demanded assertively that the "devil get behind me!"
These negatively valenced emotional features were accompanied by features potentially validating right cerebral dysfunction, including left-sided sensory defects
and left-sided sensory extinction, left hemineglect syndrome, visuospatial deficits,
constructional dyspraxia, geographical confusion, facial agnosia, and vestibulopa-
thy. The latter may reflect right temporal parietal activity as conveyed with leftward
vectional complaints. Moreover, the emotional intensity and the affective valence
attached to these perceptual events reflect the processing advantages of the neural
systems involved and provide for a broader perspective in the assessment of her
brain pathology.

Chapter 13
Mixed Brain Syndromes

The majority of brain disorders may involve multiple locations or diffuse dysfunction attributable to more than one brain regions. This might be the case following a traumatic brain injury where physical events involving shearing forces and stretching of tissues are involved. These features may follow a seemingly localized cerebrovascular accident where vasodynamic influences, edema, and pressure effects are at play and potentially affecting far-field brain regions. The multitude of combined syndromes is well beyond the scope of these writings. However, one potential categorical example may be provided in the case of foreign-accent syndrome(s).

Foreign-Accent Syndrome

Prosody, rhythm, and timber of speech may all be altered in complex combinations to result in one or another distinct foreign accent. If diagnosed, the speech variant is called *foreign-accent syndrome* or alternatively alien-accent syndrome (e.g., Gurd et al. 2001). This disorder may compel changes in the social response of others and not always in a fashion that pleases the patient. The impact of a stroke, for example, may be sufficient to alter the prosody, pitch, and/or tempo of speech to the extent that the person may acquire a foreign accent that cannot be easily attributed to their culture or to their learning history. It may even simulate an accent from cultures, nationalities, or geographical regions which the individual dislikes.

One noteworthy patient was an Irish born woman who, at that time, was living in the hill country of western Virginia. Deeply disturbing to her was her loss of her native tongue and the acute acquisition, with her stroke, of a deeply English accent. Indeed, the therapists and others were attracted to her, by their own account, because they enjoyed hearing her accent. This was very disturbing to her as she related in astonishment "I am not an Englishman!" Indeed, she related the history

© Springer International Publishing Switzerland 2015
D. W. Harrison, *Brain Asymmetry and Neural Systems,*
DOI 10.1007/978-3-319-13069-9_13

of her mother country and what she viewed as the havoc imposed by the English invasion and domination of Ireland; even outlawing the use of the Irish language in Ireland (Gaelic). She continued with her accounting of the potato famine, where she suspected the English of taking what potatoes were produced in Ireland, while the Irish literally ate grass (emphasis added!). It may have relieved her distress just a bit when, in desperation, the neuropsychologist explained that "it all goes to show that an Englishman is just an Irishman with a brain disorder!"

Altered pitch and/or staccato speech may result from a deep right frontal lesion or sometimes from a lower brainstem lesion, usually right sided. By itself, the former is referred to as a subcortical expressive dysprosodia. Foreign-accent syndrome appears to result from a more complex array of lesions (e.g., Katz et al. 2012). No single brain location seems entirely adequate for these symptoms. More consistently, the lesion pattern includes the inferior frontal or parietal regions of the left hemisphere, possibly contributing to a lingual dyspraxia. But, altered pitch may be derived more directly from dysfunction within the right frontocerebellar system and cerebellar involvement has been found with this syndrome (Marien and Verhoeven 2007; Heilman et al. 2012; Katz, Garst et al. 2012). Alterations and variations in the accent might be derived from various combinations of lesions. Moreover, with a brainstem lesion, the arousal systems may be affected with a resultant waxing and waning of symptoms such that the acquired accent may come and go. The more common location for pathology may be at the inferior left frontal parietal region, where lingual dyspraxia and/or hyperreflexia may yield oddities in expression as described above. The quality of the accent is often British. But, Russian, Chinese, Irish, Eastern European, and other accents have been found. Pierre Marie (Marie 1907) may have been the first to describe this condition, followed by Pick (1919) in his early writings.

One woman, for example, was born and raised in southwestern Virginia near the Appalachian Mountains. After a new pontine infarct and with a history of a remote left inferior frontal–parietal cerebrovascular accident, she acquired a distinct German accent. The presence or absence of the foreign accent waxed and waned with fluctuation in her arousal state, such that with louder volume her expressed speech was clearly conveying a German accent whereas, with low volume speech, this variant might go undetected. These combined lesions are depicted in Fig. 13.1.

Moreno-Torres et al. (2012) describe foreign-accent syndrome as a condition at the mildest end of the spectrum of speech disorders. They note that the diagnosis does not hinge on the production of phonological errors, being instead associated with various alterations in the fine execution of speech sounds which cause the impression of foreignness. The researchers conducted a multimodal case study evaluation. The patient was a middle-aged bilingual woman who had chronic foreign-accent syndrome with segmental deficits, abnormal production of linguistic and emotional prosody, impaired verbal communication, and reduced motivation and social engagement. Magnetic resonance imaging was interpreted to show bilateral small lesions mainly affecting the left deep frontal operculum and dorsal anterior insula. The authors further note that diffusion tensor tractography suggested disrupted left deep frontal operculum–anterior insula connectivity. Metabolic activity measured

Fig. 13.1 German foreign-accent syndrome following combined lesion of left frontal parietal and left pontine regions

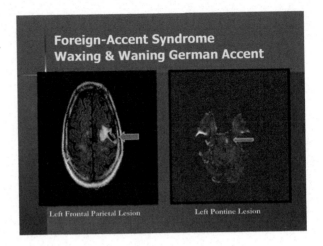

Foreign-Accent Syndrome
Waxing & Waning German Accent

Left Frontal Parietal Lesion   Left Pontine Lesion

with positron emission tomography was primarily decreased in Brodmann's areas 4,6,9,10,13,25,47, the basal ganglia, and anterior cerebellar vermis. This case study once again highlights multiple lesions with this syndrome, including the cerebellar region. The authors found compensatory activation of brain regions as a function of controlled attention and feedback, which they hope will provide a basis for eventual therapeutic approaches with individuals suffering from these rare disorders.

Chapter 14
Frontal Lobe Syndromes

The frontal lobes are anatomically defined by the central fissure at their posterior border as it is proximal to the somatosensory cortex of the parietal lobe. Anatomical landmarks exposed on the lateral surfaces of the frontal lobes include the superior, middle, and inferior frontal gyri (see Fig. 14.1). A gyrus is a cortical ridge surrounded by sulci or grooves produced with a folding of the surface of the brain to increase surface area. Prominent anatomical regions within the frontal lobes, which have been the subject of much research, include the motor cortex anterior to the central fissure, the premotor cortex anterior to the motor cortex, the prefrontal regions and frontal eye fields, the supplementary motor region, the orbitofrontal region, the cingulate gyrus (see Fig. 14.2), and the frontal poles. It is fair to say that the frontier of neuroscience and clinical research corresponds with the anterior-most regions of the frontal lobes: the frontal poles and the orbitofrontal and the basal frontal regions.

The frontal lobes have extensive connections to all lobes within the second functional unit and to the reticular activating systems within the first functional unit. It receives from these areas but also plays a prominent role in the regulatory control or inhibition of these other systems. Some of the regulatory influences are direct via the frontal lobes' connections to the posterior brain regions along the longitudinal tract. An example might include the orbitofrontal region and its influence over anger activation at the right amygdale via the uncinate fasciculus. The orbitofrontal region of the prefrontal cortex includes the rectus gyrus and orbital gyri, which constitute the inferior surface of the frontal lobes lying immediately above the orbital plates. Lesions of this region are often not restricted to the orbitofrontal cortex alone. Instead, clinical lesions more typically extend into neighboring cortical areas involving the ventromedial prefrontal region. The ventromedial prefrontal region includes the medial and the lateral orbitofrontal cortex, thus encompassing Brodmann's areas 25, 24, 32; medial aspects of 11, 12, and 10; and the white matter subjacent to all of these areas (see Bechara 2004). Bechara (2004) concludes from the literature reviewed that "patients with bilateral lesions of the ventromedial cortex develop severe impairments in personal and social decision-making, in spite of otherwise largely preserved intellectual abilities." Many difficulties arise following

© Springer International Publishing Switzerland 2015
D. W. Harrison, *Brain Asymmetry and Neural Systems*,
DOI 10.1007/978-3-319-13069-9_14

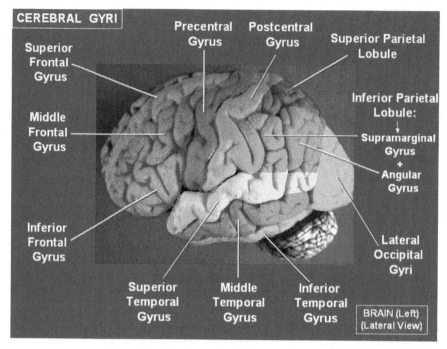

Fig. 14.1 The cortical gyri, including the location of the superior, middle, and inferior frontal gyri

Fig. 14.2 The anterior cingulate gyrus within the frontal lobe

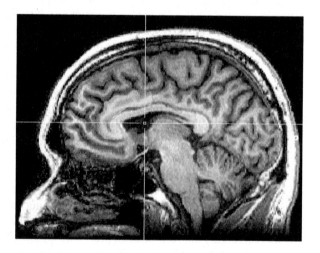

damage to these areas, including difficulties in planning their activities and difficulties in choosing friends, partners, and activities (Bechara et al. 2000, 2002).

Right-sided orbitofrontal lesions have long been associated with heightened reactive anger or rage behavior, whereas stimulation has been associated with pacification or diminished reactive anger behavior (e.g., Fulwiler et al. 2012; Agustín-Pavón et al. 2012). Alternatively, activation or stimulation of the right amygdale has been related to activation of anger (Everhart and Harrison 1996; Everhart et al. 1995; Demaree and Harrison 1997b; see also Blackford et al. 2012; see also Fulwiler et al. 2012), whereas a lesion here has been associated with pacification or anger reduction (Butter et al. 1970).

An important distinction in research on the neural mechanisms of emotion regulation involves the relatively limited duration of emotional states in comparison with emotional traits. Emotional traits are defined as the stable tendency to experience particular emotions in daily life. Fulwiler et al. (2012) note that neuroimaging investigations of the regulation of anger states point to the involvement of reciprocal changes in the prefrontal cortex and amygdala activity, but the neural substrate of trait anger has received less attention. Using resting-state functional magnetic resonance imaging, these researchers evaluated whether the variation in the strength of functional connectivity between the amygdala and the orbitofrontal cortex is associated with trait anger. Healthy men completed the Spielberger state-trait anger expression inventory. Correlational analysis for resting-state functional connectivity was carried out with the left and the right amygdala as separate seed regions. Anger measures were correlated to the right and the left amygdala on a voxel-by-voxel basis. Trait anger was inversely associated with amygdala–orbitofrontal connectivity. Anger control, the tendency to try to control expressions of anger, showed the opposite pattern of being positively correlated with amygdala–orbitofrontal connectivity. The authors argue that the results provide additional evidence for the role of this corticolimbic circuit as a neuroanatomical substrate underlying stable differences in anger regulation.

Jose Manuel Rodriguez Delgado (1969, 1977) was well known early on for his creative efforts with his invention which he called a *stimoceiver*. This apparatus allowed for full freedom in ambulatory movement, while providing electrical stimulation to one or another brain region. Using a remote control apparatus, which preceded our wireless garage door opener, he would provide electrical stimulation to specific areas of the brain and often concurrent with a receiver, which monitored electrical activity from specific brain areas using remote electroencephalographic recordings. Delgado was most notable for using stimulation techniques to alter emotion and to control behavior remotely at a distance from the subject.

Most remarkable within the media was his demonstration of the stimoceiver in Cordoba at a bull-breeding ranch. Delgado demonstrated not only his scientific acumen but also his tenacity as he stepped into the bull-fighting ring with a bull previously implanted with the remotely controlled stimoceiver. With dramatic flair and a bull snorting with dust rising below the nostrils, Delgado pressed a remote control button, which caused the bull to stop its charge and cease its aggression. The stimulation was directed to control the amygdale, whereas he described altered emotions beyond aggression with stimulation elsewhere and in other species, including primates. This inhibition of aggression was a replication of frontal lobe regulatory control over this structure using this implant technique. Critics argued that the stimulation did not alter the aggressive response, but instead forced the bull to turn toward the left.

Sensory Thresholds or Sensitivity

Redundant frontal control exists through its role in the regulation of the first functional unit or the reticular activating system. This *indirect* influence allows for arousal-related changes to affect the apparent intensity of sensory events projecting through the thalamic relay system on toward the primary projection cortices. Activation of the arousal-based brain-stem reticular formation alters sensory thresholds up to an optimal point of arousal. The descending projections from the frontal lobe allow for reduction of the arousal state and the indirect alteration of sensory thresholds. Sensory thresholds or sensitivity might be affected or manipulated by either diminished frontal lobe capacity or frontal lobe stress (Woods et al. 2013) and through the manipulation of ambient sensory conditions within other sensory modalities (e.g., Delay et al. 1978). One definition of frontal lobe stress is that originating from lateralized dual concurrent task demands. Examples might be provided from the laboratory such as verbal fluency challenge with incremental processing demands at the left frontal region (e.g., Wood et al. 2001; Baldo et al. 2006; Baldo et al. 2001; Foster et al. 2012) and figural fluency challenge with incremental processing demands at the right frontal region (e.g., Ruff et al. 1994, 1986; Williamson and Harrison 2003; Holland et al. 2012; Mitchell & Harrison 2010; see also Foster et al. 2005).

The individual with a frontal lobe syndrome may have reduced capacity within the descending corticofugal and posterior projecting corticocortical pathways, which indirectly dampen or diminish the intensity of sensory information within the brain. For example, patients with focal lesions of the dorsolateral prefrontal cortex exhibit abnormally increased auditory-evoked potential amplitude (Knight et al. 1989). Frontal incapacity or lesion might result in lowered somatosensory thresholds (heightened sensitivity) where even the clothing on their body is irritating accompanied by restlessness. Lowered auditory thresholds might be expressed by a patient bothered by noises easily tolerated by others. Sensitivity to lights might also be increased, where even previously subthreshold events might now be detected. With diminished frontal capacity, these perceptually more intense ambient or contextual environmental events may decompensate some and may promote related symptoms for others (e.g., ocular migraine with photophobia; skin sensitivity disorders with allergic-like responses). Moreover, Bechara (2004) states that lesion of the ventromedial (which includes the orbitofrontal) sector of the prefrontal cortex interferes with the normal processing of somatic signals underlying emotional values, while sparing most basic cognitive functions. Such damage leads to impairments in the decision-making process, which seriously compromise the quality of decisions in daily life. Bechara (2004) provides a review of the evidence in support of "The Somatic Marker Hypothesis," which provides a systems-level neuroanatomical and cognitive framework for decision making and which suggests that the process of decision making depends in many important ways on neural substrates that regulate not only emotions but also feelings. It is not surprising at this point, that much of the research on disturbed sensory gaiting mechanisms and schizophrenia has focused on disturbances in frontal lobe function (e.g., Egan et al. 2001).

A younger woman about 20 years old with a right frontal lobe syndrome expressed her symptoms with social improprieties, delayed response deficits (impulsive; e.g., Malmo 1942; see also Cai et al. 2012; see also Allman and Meck 2012), and heightened somatic sensitivity. The latter was expressed with great discomfort in her clothing. She had scarring on her skin from repetitive scratching and irritation and she painted her body with calamine lotion, while wearing only enough clothing to suffice. She had many therapists as her poor appreciation of social boundaries seemed unmanageable at times. She required much investment by community service organizations to help her along the way. Another woman with hypersensitivity to vaporous chemicals could hardly tolerate the medical center environment where she was to be evaluated. And another was overly sensitive to noise easily tolerated and habituated to by others. By way of a social example, descending regulatory control over subcortical and brain-stem structures might include the reward system within the pathways of the medial forebrain bundle and the nucleus accumbens (Öngür and Price 2000; Narcisse et al. 2011), where reduced capacity and/or concurrent frontal stressors within a social setting may alter the intrinsic intensity or reward value of being with other people. Much additional research is needed to develop effective strategies to supplement or support the basic neural systems involved in the down regulation or "self-regulation" of sensory systems.

Startle Responses, Orienting Responses, and Habituation

Early literatures on frontal lobe function were instrumental in exposing the "top-down" functional role of these brain regions in regulating or modulating arousal-related responses (e.g., Isaac and DeVito 1958) originating within the mesencephalic and diencephalic brain-stem regions (Dempsey and Morison 1942a, 1942b, 1943; Jasper and Droogleever-Fortuyn 1946; Hunter and Jasper 1949; Moruzzi and Magoun 1949). Diminished frontal capacity or lesion of the anterior cingulate gyrus may alter the startle response (e.g., Hazlett et al. 1998; Klineburger and Harrison 2012; Levenson et al. 2012). This is more often apparent in the exaggerated startle response to an acoustic event. A loud abrupt noise presented in a classroom setting may be sufficient to demonstrate individual differences in the reflexive reactions to the event. But, the alteration may be heightened with pathological processes in this region. This system may also be altered in hypoaroused states with minimal reactivity well below expectations drawn from a normal comparison.

Teuber (1964) suggested that the frontal lobes "anticipate" sensory events that result from behavior, thus preparing the sensory processing systems, within the second functional unit of the brain, for events that are about to occur. The expected results are compared with actual experience, and thus smooth regulation of activity results. Novel or salient events characteristically yield an orienting response and the anatomical underpinnings are lateralized through direct pathways from the cingulate gyrus to the amygdaloid bodies and onto the hippocampus, perihippocampal, and entorhinal cortices. Frontal lobe dysfunction frequently results in altered rates of habituation of the orienting response to redundant or irrelevant events. Since habituation may be considered a fundamental reduction in arousal with repeated or continuous exposure, dishabituation may reflect a hyperaroused or hypervigilant state just as overhabituation may be considered a state of diminished arousal (Harrison and Pavlik 1983; Harrison and Isaac 1984). Moreover, habituation may be lateralized to the dominant tendencies or actions of one or the other frontal region. For example, dishabituation secondary to a right orbitofrontal lesion more commonly presents with hyperkinesis and rapid internal clocking mechanisms with time urgency (e.g., Holland et al. 2009; see also Granacher 2008). In contrast, overhabituation secondary to a medial left frontal lesion more commonly presents with bradykinesia, behavioral slowing, and amotivational features (see Scott and Schoenberg 2011; see also Grossman 2002; Hu and Harrison 2013b; see also Granacher 2008).

Barbas et al. (2013) conclude that a cardinal function of the prefrontal cortex is selective attention, with the ability to use a stimulus when it is relevant to the task at hand. The authors note that it may be even more important to understand the mechanisms by which these neural circuits modulate and minimize the vast "roar" of irrelevant information impinging upon the sensory systems at the back of the brain. This was appreciated in classic research with individuals suffering even mild impairment in the prefrontal regions, where they were found to be distractible and poorly able to habituate to irrelevant or redundant environmental events. Barbas et al. (2013) aptly note that the neuroanatomical connections to the prefrontal regions arise preferentially from the sensory association cortices, providing multimodal influence and associative strength. Thus, one or another prefrontal region may be biased by the influence of a particular sensory modality. However, the prefrontal region is not segregated from the influence of the sensory events or broader sensory context arising from the other sensory modalities. Thus, selective attention within a sensory modality may be influenced by activity within the other sensory modalities. This arrangement is also consistent with the cross-modal influences of one or another sensory modality by activity in the other modalities within the first functional unit or brain-stem reticular formation (Luria 1966, 1973, 1980). For example, ambient lighting levels alter auditory thresholds (Delay et al. 1978; Kallman and Isaac 1980).

The neuropsychological evaluation of the startle and orienting responses and the rate of habituation to these events with repeated or continuous exposure may provide substantial insight as to the functional integrity of relevant frontal regions (e.g., Levenson et al. 2012). Startle and orienting responses may be assessed separately within each sensory modality. For example, startle may be assessed using repetitions of a visual menace stimulus. Acoustic startle might be evaluated with redundant abrupt and intense auditory events. Similarly, tactile startle might be assessed with redundant somatosensory probes or menace stimuli to the body. The intensity of the reactive startle response, the integrity of the orienting response toward and within each hemispace or hemibody, and the ability to habituate to redundant or irrelevant events with repeated or continuous exposure may enlighten the evaluation and diagnostic conclusions and provide for more appropriate interventions. Moreover, regression to more normal responses here might confirm progress over time and promote improvement in the prognosis.

One cold winter's week left the community under a severe blizzard and ice storm without electricity and water. Truck drivers were trying to haul water into town. But, many were using snow to activate the plumbing systems and for personal hygiene. During this severe weather event, many electrical linemen worked around the clock trying to return the electrical systems to the grid. One such lineman was working around the clock relying more and more heavily on stiff coffee and cigarettes to help maintain his arousal. Following several days of this, he ruptured his anterior communicating artery at the circle of Willis, which surrounds the thalamic brainstem region. The anterior communicating artery connects the origin of the anterior cerebral arteries within each brain. Surgical repair of the aneurysm left him with a hemorrhagic infarction within the distribution of the right anterior cerebral artery. Initially, bilateral frontal deficits were prominent with left supplementary motor deficits presenting as an akinetic mutism (e.g., Fontaine et al. 2002). But, with recovery of the left frontal region, the deficits associated with the right mesial frontal impairment were more apparent and he was increasingly hyperkinetic. He completed one evaluation where his examiner was walking in reverse at a fair clip, just to stay in contact with him. He had a severe dishabituation disorder and was intolerant to tactile probes or menace stimuli, where a tap on the shoulder produced an acute startle and elevation of the shoulders. This failure to habituate the tactile startle and the orienting response played out functionally as he was averse to his clothing and to physical contact with his spouse.

Reflex Regulation and Hyperreflexia

Luria (1973, 1980) referred to the frontal lobes as the third functional unit charged with the organization, planning, sequencing, and expression of behavior, emotion, and cognition (see also Stuss and Knight 2013). Moreover, this unit has been the subject of research primarily on executive functions and the regulatory control or effortful control of the first and second functional units involved in arousal or activation and in sensory reception and comprehension, respectively. For example, in highly impulsive individuals, regulatory structures including medial and lateral regions of the prefrontal cortex were isolated from subcortical structures associated with appetitive drive, whereas these brain areas clustered together within the same module in less impulsive individuals using functional magnetic resonance imaging (Davis et al. 2012; see also see Oades 1998; see also Emond et al. 2009; see also Granacher 2008; Helfinstein and Poldrack 2012; see also Barbas et al. 2013).

Consistent with this regulatory role, the frontal lobes directly control motor reflexes. The frontal lobes, though in many ways oppositional to each other (e.g., the expression of positive or negative affect), may be viewed as oppositional in their regulation of reflexes throughout the body. This may be seen in simple walking patterns, where activation of the left motor cortex may inhibit the dominant extensor reflex for the right leg and where concurrent contralateral inhibition of the right motor cortex via the corpus callosum may yield disinhibition over the extensor tone at the left leg, allowing for the support of our weight at that extremity. Crossed callosum control is restricted somewhat at the proximal muscles of each extremity (Brodal 1981; Harrison 1991). Reciprocal oscillations between the frontal lobes are thought to play a prominent role not only in simple walking activities and oppositional emotional expressions but also in the regulatory control of the cardiovascular and cardiopulmonary systems, through dynamic fluctuations in sympathetic and parasympathetic efferent activities.

The development of frontal lobe regulatory control is, to a great extent, progressive over time (see Munakata et al. 2012). Additional frontal capacity develops within the context of the age-related changes in childhood, adolescence, and even into early adulthood with evidence for active neural integration, especially of the far frontal regions, continuing perhaps into the early twenties. I see some validity in the euphemism that "all children are brain damaged!" At least we often expect more from older individuals in the areas of cognitive flexibility, sensitivity to the needs and concerns of others, decreased oppositional affect and behavior, respect for social and cultural rules, and, in general, regulatory control and effortful control. Moreover, there is substantial basis for this within the findings of developmental neuroscience (Paus et al. 1999; see also Goldberg 2001).

At birth, there exists a plethora of reflexes, which eventually come under the regulatory control of the frontal lobes and these are often assessed as developmental milestones. Prominent reflexes present at birth, and even prior to this, include the suck, the root, and the snout reflexes corresponding with the proximity of light-brushing tactile or pressure stimulation to the lips or the surrounding facial region. The Babinski reflex, related to antigravity synergy and down going toes on plantar stimulation, is also present and gradually resolves with improved frontal lobe inhibition. These early developmental reflexes may be expressed again following damage to the descending fibers of the frontal lobe (see Futagi et al. 2012). For example, the prominent feature of an upper motor neuron lesion might initially be paresis or plegia of the body part represented at the region of the motor homunculus lesioned. However, with time hyperreflexia may become one of the primary management issues with the development of antigravity posturing, typically with flexor synergy for the upper extremities and with extensor synergy for the lower extremities. But, hyperreflexia may present with altered pitch in speech through alterations in the regulatory control over the vocal motor apparatus. This may be a component of akinetic mutism resulting from mesial left frontal dysfunction with speech gradually recovering to a soft, hoarse, whisper. Alternatively, altered and typically elevated pitch may result from mesial right frontal dysfunction.

A young woman on the birth control YAZ concurrent with factor 4 clotting syndrome provides an example after her subcortical right frontal stroke. She made a remarkable recovery but with residual vocalization deficits characterized by elevated pitch and inadequate breath support. She struggled to vocalize or to express speech with elevated pitch. The rule out was for a hypophonia of various origins. However, inspection of the oral motor apparatus revealed a broad and open airway with elevation of the pharynx and with functional lingual praxis. Instead, at the onset of efforts to vocalize, the airway would close with hyperreflexic features, leaving her straining to provide sufficient respiratory support to express her speech and a concurrent elevation in pitch. These symptoms were aggravated by any additional impositions or restrictions on the airways, which would occur on efforts to talk on the phone while in a supine posture. More relevant vocationally, any sense of time urgency or negative affective stress would leave her gasping for air. So, in the broader sense, this subcortical expressive dysprosodia or speech disorder was also a dyspnea with hyperreflexic responses to stress, restricting the airway and the ability to breathe.

Although the layperson may think that when the physician taps the patellar tendon she/he is checking the knee, the assessment is more useful not only in determining the integrity of the upper motor neurons descending from the motor cortex but also for related functions. A peripheral nerve lesion more commonly results in diminished muscular activation and eventually with atrophy or muscle loss. Stress on the frontal lobe (see below) or diminished frontal lobe capacity as with upper motor neuron damage yields antigravity dystonia. This might be assessed through passive range of motion techniques, where the examiner holds the bicep muscle of the arm and passively ranges the patient's arm. The frontal lobe patient may show flexor synergy here and appear to be opposing the examiners' efforts to extend the arm. These "frontal release signs" present in the form of hyperreflexia or dystonia (e.g., Futagi et al. 2012).

It is relevant to mention here that one of the early descriptions of *oppositional-defiant disorder* was derived from the assessment of these reflexes. The frontal lobe includes the same brain systems which underlie emotional flexibility, as opposed to rigidity and inflexibility, in thought. Behavioral, cognitive, and emotional flexibility are likely components of social and emotional intelligence. This might be appreciated initially in fundamental social interactions, including eye contact and the preponderance of attention that we direct to the facial region and facial expressions. Facial dystonia has been demonstrated in emotional disorders of depression, anxiety, and hostility (Rhodes et al. 2013). This has been confirmed in systematic research extending the predictions for altered muscular tone to the upper extremities of adults (e.g., Crews et al. 1995; Harrison et al. 2002; Everhart et al. 2002) and of children (e.g., Emerson et al. 2005). The underlying postural difficulties in facial expression may render these individuals disabled in social settings and result in negative emotional reactions and dislike from others through initial, and almost instantaneous, observations of the individual's facial expressions.

Moreover, as frontal stress is increased, as for example within a social setting with other people present and with dual tasking or organizational demands, the facial region may become more dystonic and dyspraxic and the patient more debilitated. Indeed, this person may decompensate with reduced capacity in the regulation of anger. The regulation of anger, as with the inhibitory control of the orbitofrontal region over the amygdale via the uncinate fasciculus, might be better appreciated through the location of the critical orbitofrontal systems proximal to the frontal lobe regions regulating facial expressions. Clues from our language such as "stay out of my face!" likely have an emotional basis in this respect. The expression "lose face" may refer to a discounting of emotional integrity related to diminished facial motor control.

The expression "keep a stiff upper lip" may relate to facial motor control despite adversity and where this negative affective stress is somewhat specific to inferior, posterior frontal, and orbitofrontal functions (e.g., Agustín-Pavón et al. 2012; Fulwiler et al. 2012). Herath and colleagues (Herath et al. 2001) used functional magnetic imagery during the performance of dual-task demands. Performance of the dual task resulted in excess activation in comparisons with single-task performance. Dual-task interference was specifically associated with increased activity in a cortical field located within the right inferior frontal gyrus, an area that has been associated with emotional regulation. The results support the notion that areas of the brain that regulate emotion, in particular, may be subject to the burden of dual-task interference effects. Diminished capacity in these regions may yield deregulated facial expressions confirming your emotional reactions and potentially to those who might relish in them.

Ekman and colleagues (Ekman et al. 1983) demonstrated that simply contracting specific facial muscles was sufficient for the experience of the corresponding emotion. They demonstrated the reflexive release of the autonomic nervous system components consistent with the derived facial expression depicting either a positive or a negative emotional posture through the contraction of specific muscle groups in the face. In a subsequent experiment, Davidson et al. (1990) combined the measurement of observable facial behavior with simultaneous measures of brain electrical activity to assess hemispheric activation during the experience of happiness and disgust. Disgust was found to be associated with right-sided activation in the frontal and anterior temporal regions. These findings were in contrast to those recorded during the happy condition. Happiness was accompanied by left-sided activation in the anterior temporal region supporting valence-specific laterality of emotion.

When considered collectively, these studies provide evidence for the cerebral lateralization of emotion by positive and negative valence. They provide evidence that the corresponding control over the autonomic nervous system is functionally related to that involved in facial expressions conveyed directly through muscle contractions at discrete facial regions. These findings illustrate the utility of using facial behavior to verify the presence of emotion and the ability to alter volitional control over the autonomic nervous system through simple posturing of the facial regions. When the investigation of facial posturing was extended to those with high hostility and anger difficulties, evidence of facial dystonia was found with increased left hemifacial tone using electromyogram recordings (Rhodes et al. 2013). In addition, reduced habituation was found at the left hemibody using skin conductance measures following posed contractions of the corrugator muscle in high hostiles (Herridge et al. 1997).

The capacity of the frontal lobes to regulate or to inhibit reflexes elsewhere in the body may best be understood through the appreciation of two fundamental features of this functional neuroanatomy. First, there exists some basic capacity of these cells to do their job. This inherent *capacity* is one of the aspects of our nervous system that makes us somewhat unique from one person to the next. We differ, as individuals, in our capacity to regulate these reflexes. Moreover, this difference varies within us as a function of our developmental level or the progression across our life span (see Yeates et al. 2012), and as a function of the relative health and integrity of these cellular regions (e.g., the upper motor neurons and prefrontal regions).

Secondly, the capacity of these cells to function well and to regulate any given reflex is fundamentally related to the concurrent tasking demands placed on this anatomy. The frontal lobes are largely the systems that we are referring to when we discuss stress, and stress in the present context is derived from multiple concurrent tasking demands (see Williamson and Harrison 2003; Mitchell and Harrison 2010; Carmona et al. 2009). The extent to which the functional cerebral space (Kinsbourne and Hicks 1978) involved in the performance of two concurrent tasks is shared or overlaps will partially determine the resources available to complete the primary and the secondary tasks well or, alternatively, the extent of the interference and degradation that will occur in the performance of the secondary task. It may be helpful to first appreciate the narrow predictions of capacity theory and, subsequent to this, to think more broadly of the scope of bodily functions that must ultimately come under the regulatory control of these brain regions.

To begin with a simple example, the concurrent tasking demands might be defined as a rapid alternating movement such as tapping rate at the left or at the right index finger. These are largely controlled by the contralateral frontal lobe and, more specifically, upper motor neurons originating from the motor cortex. Their coordinated movements are partially a function of the premotor cortex surrounding this tissue. If the left frontal lobe is faced with a dual concurrent task demand or "stressor" as with reading out loud (Broca's area), then interference from the performance of the dual task may be costly at the right hand with relatively preserved tapping rate at the left hand (e.g., Kinsbourne and Cook 1971; Harrison 1991). Indeed, evidence indicates that the extent of interference to one or another extremity and to the proximal or distal aspects of the extremities, on such a dual task, corresponds with the percentage of fibers originating from the contralateral, as opposed to the ipsilateral, frontal lobe (Colebatch and Gandevia 1989; Harrison 1991).

It may help to digress a moment here in order to appreciate this anatomical distinction. Within the rehabilitation setting after a stroke, for example, the prognosis for recovery of the functional use of an affected extremity varies at the distal, as opposed to the proximal, location along that extremity. For example, Colebatch and Gandevia (1989) evaluated the strength of 12 muscle groups of the arm to determine the distribution of weakness derived from upper motor neuron damage. They evaluated three groups of subjects including 14 intact volunteers, 10 patients with unilateral arm paresis, and 6 patients with severe paralysis of the arm. The pattern of weakness was not the same in all patients at the contralesional side. Shoulder muscles were relatively spared while the wrist and finger flexors were relatively severely affected. Moreover, in hemiparetic and hemiplegic patients, the strength of muscles ipsilateral to the lesion was reduced compared with normal controls at least consistent with the evidence for heightened ipsilateral projections to the proximal extremity region.

However, this effect may not be apparent in the context of early post-stroke recovery (Beebe and Lang 2008). With a left frontal stroke affecting upper motor neurons, the prognosis is substantially better for recovery of some movement at the right shoulder (proximal regions) than for the distally located right index finger. This results at least partially from increasingly diminished ipsilateral control (from the intact right frontal lobe) as we move toward the distal regions of that extremity. Roughly 90% of the upper motor neurons cross to control the contralateral limbs. However, the proportion crossing is greatest for the right index finger, with gradually increasing ipsilateral control as we move across digits two through five, and then increasing still as we move toward the shoulder and facial regions (Brodal 1981).

One young man was referred by his occupational therapist with the goal of improved self-feeding after a motor vehicle accident with substantial trauma to the left frontal lobe. The left frontal region is especially critical for ideational praxis or the coordinated sequencing of bodily postures necessary for activities of daily living. But, faced with a dense right hemiplegia, the therapist was working with the left upper extremity ipsilateral to the brain damage. This man was severely discoordinated in the use of his left arm with self-feeding efforts being hazardous with the spoon sometimes hitting him in the face or even the eye. With appreciation of the functional anatomy, the relative discoordination was identified with the use of the *proximal* left arm and shoulder region in the feeding process. Again, these proximal body regions receive a larger proportion of ipsilateral fibers. In this case, the left arm was ipsilateral to a damaged frontal lobe. Through the temporary physical restriction of the proximal arm region, and through efforts to limit the feeding movements as far as possible to the distal region of the left arm, progress was made and the basic goal of self-feeding was accomplished. This prediction was derived as one of seeking movement control by the intact contralateral right frontal region using the digits of the left hand.

Evidence indicates that the frontal regions regulate the cardiovascular system, circulating blood glucose levels, oxygen saturation levels, the volume and affective intonation of speech, facial expressions conveying emotion to others, reward systems (e.g., social and appetitive), and many, other systems. Blood pressure and heart rate vary with the intensity of a frontal stressor or dual-tasking demands with heightened stress reactivity in individuals with minimal or reduced frontal capacity (Foster and Harrison 2004, 2006; Williamson and Harrison 2003; Foster et al. 2008). Functional outcomes in behavior, cognition, and emotion vary with the nature of the stressor, demanding resource allocation from regional cellular networks within the left or the right frontal lobe. The regulatory control over pain, anger, fear, and sympathetic mediation of the heart appears to involve shared or overlapping functional cerebral space (Mollet and Harrison 2006; Mitchell and Harrison 2010; Holland et al. 2012).

A dual-task challenge might include managing anger concurrent with pain management, with the demands of one reducing the capacity to regulate the other. Expressive prosody or the volume of our speech may be deregulated with concurrent stress demands related to perceived external control. Ultimately, if the right frontal lobe is diminished in capacity or if the dual concurrent stressor demands exceed this regions capacity, something must be sacrificed. This may diminish crossed inhibitory control over left ventricular cardiovascular reflexes with increments in blood pressure, heart rate, and sweating as with an anger management disorder or with tachycardia during a panic attack. Similar reactivity in glucose mobilization was evident in individuals with reduced right frontal capacity on exposure to pain at the left arm (Walters and Harrison 2013a).

An elderly man with a frontal dementia might serve as an example. His physical therapist was working on "sit-to-stand" techniques to promote improved functional independence so that this man might go home with his spouse. On efforts to stand, he developed significant flexor tone in the upper extremities, including a grasp reflex at the hands such that he lifted the chair with his movements. He lost control of expressive prosody with the onset of loud volume speech. He developed the onset of fearful facial expressions and complaints of fear of falling/panic and sharply elevated heart rate and blood pressure. Increasing demands for volitional organization, effortful control, and initiation of bodily movements appeared to have resulted in hyperreflexic posturing and deregulation of autonomic nervous system activity among the changes witnessed in the more general behavioral, cognitive, and emotional presentation.

Based on these examples, one may appreciate that the approaches we take in managing health problems are somewhat like the story of the blind men holding different parts of the elephant. The cardiologist will see a heart problem at the heart, whereas the neuropsychologist will appreciate that the problem is a function of the brain systems involved. This does not at all suggest that either professional is wrong. They may be simply looking at two ends of shared or overlapping neural systems. Long standing exposure to right frontal lobe stressors involving perceived threats, social defensiveness and attributions of external control by others, and eventually cynical hostility and a negative view toward others reflect a formula for cardiovascular disease (Siegel 1985; Smith et al. 2004; Smith and Pope 1990; Suls and Bunde 2005) and metabolic syndromes like diabetes mellitus (see Walters and Harrison 2013a, 2013b; see also Anthony et al. 2006). Furthermore, the metabolic syndrome and obesity appear to occur with alterations in the reward circuits and especially with increasing abdominal fat (Luo et al. 2013). Of great import to each of us as individuals are the inherent opportunities we have from appreciating these problems early in the disease process. It may provide a basis for our own interventions to improve our health, including exercise and vocational and work setting choices as these may be available and general stress management techniques. But, again these are features that may be present early on developmentally and they may be substantially advanced toward pathology by adolescence or early adulthood, in some cases (Raikkonen et al. 2003).

The frontal lobes are interfaced with distal body organs through the visceral efferent systems of the hypothalamic regions via the pituitary adrenal axis. Frontal lobe processing demands or stress, such as the response to a perceived abrupt environmental event, may trigger a startle response (e.g., Levenson et al. 2012) and the mobilization of adrenaline from the adrenal cortex on the kidney. The somewhat persistent activation, which follows, is familiar to many who have experienced a traumatic event (e.g., a motor vehicle accident) or the rush of excitement as a favorite team comes out onto the playing field. This blood-based interface allows for far-field effects on distal anatomy, such as the adrenal and cardiopulmonary systems.

But, it also provides a basis for prolonging the duration of the peripheral emotional responses. Through this mechanism of global and persistent activation, the brain may be freed from direct mediation of distal cardiovascular systems, allowing for improved efficiency under survival conditions or conditions of perceived threat. But, this global and persistent activation may instead be something of a nuisance if threat turns out to be bogus or spurious and as we await the return of our body to a stable state free from the activation of survival systems or sympathetic drive.

The implications for health are potentially substantial, as the brain may mobilize to any potentially vital threat, including hypoglycemic events (see Walters and Harrison 2013a, 2013b); dehydration (e.g., Nielsen et al. 2001); decreased oxygen saturation levels (Hu and Harrison 2013a; see also Homnick 2012); hyperthermia (e.g., Nybo 2012); negative affective threats for pain, anger, or fear (e.g., Mitchell and Harrison 2010; Holland et al. 2012; Demaree and Harrison 1997b); noise levels (Harrison and Kelly 1989); inadequate blood pressure and/or heart rate (e.g., Freeman 2006; Frysinger and Harper 1989); and many other events. This conclusion raises the specter of intervention to the appreciation of dynamic physiological events which develop across the day and those which occur over the life span.

The frontal lobes might be better known to the psychologist through research on stress. In the past, it was sufficient to think of stress as a general construct requiring mobilization of resources through the hypothalamic–pituitary–adrenal axis (Selye 1956). The problem with this approach was that the research findings were variable with some "stressors" not resulting in the expected adrenergic drive, as might be seen through increments in blood pressure and heart rate. John Williamson and others (e.g., Gray et al. 2012; Williamson and Harrison 2003; Williamson et al. 2013; Carmona et al. 2009; Holland et al. 2012; Mitchell and Harrison 2010) have proposed that stress be defined by the brain regions involved in the response to that stressor, through altered metabolic and/or functional activity corresponding to these brain regions. For example, John demonstrated that stressors corresponding to left frontal function may indeed lower blood pressure and heart rate, whereas right frontal stressors more readily correspond with increments in these measures and in sympathetic tone more broadly defined. Indeed, this fundamental assumption, though not universal in psychology, is foundational for neuropsychological investigations using methodologies such as the electroencephalogram and functional magnetic resonance imaging, where metabolic or electrical changes within specific brain regions result from a given stressor event.

Thus, with a verbal fluency stressor tasking left frontal capacity, we might lower blood pressure and heart rate, while a figural fluency stressor tasking right frontal capacity might increase blood pressure and heart rate in individuals with diminished regulatory control (see Lacey and Lacey 1958; Williamson and Harrison 2003; Mitchell and Harrison 2010). Diametrically opposite effects may be derived from stressors corresponding with activation demands from homologous frontal regions within the left cerebral hemisphere. Left frontal stress may include laughter or smiling and social approach, whereas right frontal stress may result from anger or contraction of the corrugator muscle, as with a frown and social avoidance. Contraction of the corrugator increases the experience of anger and elevates blood pressure and heart rate, while contraction of the zygomatic or smile muscle increases the experience of happy emotions and lowers blood pressure and heart rate (Ekman et al. 1983) and skin conductance (Herridge et al. 1997). Simply rehearsing or learning a list of positive affective words lowers systolic blood pressure, whereas rehearsing or learning a list of negative affective words increases systolic blood pressure (Snyder et al. 1998; Shenal and Harrison 2003). Even this relationship, though, is not perfect, as the lateralization of functions between the two brains is dynamic and efficient in this respect. And, the laterality of function has been found to differ as a matter of the specialization of one or the other brain for the task at hand (Gazzaniga et al. 1962; Sperry 1966, 1982; Gazzaniga 1998, 2000; see also Springer and Deutsch 1998).

It seems reasonable that the functional capacity of these frontal regions may be exceeded by the demands of extreme stress yielding an acute reactive response, which corresponds with the diminished activity of these regions and presents in a deregulated and potentially unbridled state. Paul Foster extended this to the regulatory demands placed on the frontal systems as a function of the intensity of a memory (Foster and Harrison 2002b). Significant correlations were found between the subjective intensity of angry memories and beta activation on quantitative electroencephalographic recordings at right frontal and temporal electrode sites. Control issues or negative affective threats may stress the right frontal lobe, whereas mildly pleasant affect and social approach may stress the left frontal lobe (Davidson 1995; Davidson and Fox 1982). When the capacity for regulatory control (e.g., frontal activation) is exceeded, though, a dynamic reduction in regulatory control may occur with disinhibition of oppositional systems within the neural network. This may be seen with a reactive panic or anger state (e.g., Kent et al. 2005; Everhart et al. 1995; Everhart and Harrison 1995, 1996; Demaree and Harrison 1997b, 1996) or the onset of crying and perhaps with gelastic lability secondary to frontal deregulation (Constantini 1910; cited in Oettinger 1913). Moreover, the regulatory control capacity for each negative affective valence originating with the second functional unit (e.g., sadness, anger, and fear) appears to be anatomically represented within separate frontal neuroanatomical systems.

Amotivational/Apathetic Syndrome and the Social Anarchist

Several prominent clinical literatures exist on frontal lobe functions, but none have been more influential than the popular case of Phineas Gage (see Fig. 14.3). Twenty-five-year-old Gage was struck by a tamping iron weighing 13 $\frac{1}{4}$ pounds after a work-related explosion. The iron was reported to have travelled some 80 ft. where it landed with blood and brain tissue largely from his left frontal lobe and head region (see Fig. 14.4). His case was influential upon the nineteenth century understanding of brain and particularly the debate on cerebral localization of emotions and behavior (Damasio et al. 1994; Damasio 2005). Moreover, this may have been

Fig. 14.3 A portrait of Gage with his "constant companion"—his inscribed tamping iron

Fig. 14.4 A depiction of the tamping iron and its projection through Gage's skull

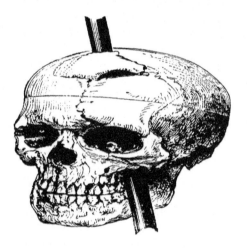

the first case indicating a relationship between frontal lobe damage and features previously attributed to "personality." Most characteristic in Damasio's reanalysis of Gage was the appreciation of the social impropriety and irritability following the event. Damasio's analysis also allows for a better conceptualization of the evolution and recovery from the brain injury over time, where social anarchy may be a characteristic of right frontal rather than left frontal pathology and perhaps subsequent to the inherent release of left cerebral systems.

The following presentations may be found in Damasio's writings (Damasio 2005). Prior to the accident, Gage was described by his employers and others as having been hard working, responsible, and "a great favorite" with the men in his charge, with his employers having regarded him as "the most efficient and capable foreman in their employ." But these same employers, after Gage's accident, "considered the change in his mind so marked that they could not give him his place again." Gage was described as having lost the equilibrium or balance between his intellectual faculties and animal propensities. Apparently, he was fitful and irreverent, indulging at times in the grossest profanity (unlike the former Gage). He was manifesting but little deference for his fellow workers, impatient of restraint or advice when it conflicts with his desires. He was, at times, pertinaciously obstinate, yet capricious and vacillating, devising many plans of future operations, which are no sooner arranged than they are abandoned in turn for others appearing more feasible. He was described as a child in his intellectual capacity and manifestations with the animal passions of a strong man. Previous to his injury, although untrained in the schools, he possessed a well-balanced mind, and was looked upon by those who knew him as a shrewd, smart businessman, very energetic and persistent in executing all his plans of operation. In this regard, his mind was radically changed so decidedly that his friends and acquaintances said he was "no longer Gage."

Some of the evidence on frontal lobe syndromes was derived from the unfortunate patients who were victims of the purposeful manipulation or extirpation of frontal lobe tissue through the frontal lobotomy or even the frontal lobectomy procedures (removal). The frontal lobotomy technique, early on, involved the insertion of a calibrated ice pick into the patient's frontal lobe and severing of the connections especially at the basal frontal or orbitofrontal regions. It defies imagination, but Moniz (1949), one of the pioneers in implementing these procedures, received the Nobel Prize for the dastardly deed. The use of the technique became especially common after World War II and many individuals who had earlier undergone this surgery have been followed over the years. The procedures were fairly easily performed in the office and, remarkably, without anesthesia. Indeed, one notable psychiatrist, Walter Freeman, had his son or others hold the patient down, while inserting the pick.

Freeman and Watts apparently performed America's first frontal lobotomy on Mrs. Hammatt on September 14, 1936. After she tried to change her mind and now determined not have the surgery, she was forcibly anesthetized. Freeman recorded her last words before the surgery as "Who is that man? What does he want here? What's he going to do to me? Tell him to go away. Oh, I don't want to see him,"

Fig. 14.5 Frontal lobe syndrome of the amotivational/
apathetic type (*left frontal*),
and the social impropriety
and social anarchy type (*right
frontal*)

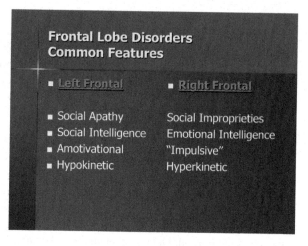

followed by a scream. Freeman apparently had no surgical training or certification
and often performed the procedure without anesthesia within some of the nation's
premier medical centers (Shorter 1997). Families might note, in dismay, the surgical
entry points on the forehead and above the eyes. So, the technique was perfected,
sliding the pick over the orbit under the eyelid and then punching the implement
through the basal skull and into the orbitofrontal brain region. The family or others
did not really need to know in some cases! So popular was the surgery, though, that
even the Kennedy family requested that it be performed on one of their girls.

The surgery was thought to be ideal for people who were too emotional. Indeed,
the expected results from surgery would be a blunting of affect with little or no effect on intelligence or memory. The clinical results would often support the claims
by the psychiatrist of improvement in emotionality without negative effects based
on standard pencil and paper or psychometric measures. Frontal lobe damage from
lobotomy would eventually be associated with personality changes and frequently
with great remorse or regret by family, friends, and/or caregivers of the patient.
For example, an individual suffering an acute frontal lobe injury may first draw
suspicion by their spouse with complaints like "This is not the man that I married."
Some of the descriptors of the personality changes with frontal lobe syndrome are
identified in Fig. 14.5 as they correspond with injury to the left or to the right frontal
region.

Frontal lobe syndrome of the amotivational/apathetic type corresponds with diminished energy level (see Scott and Schoenberg 2011; see also Grossman 2002).
Many with left frontal lobe pathology state it this way. They say "my get up and
go, got up and went!" Chronic fatigue is part of this presentation along with social
apathy (see Scott and Schoenberg 2011; see also Grossman 2002) and diminished
happiness and humor. In contrast, the frontal lobe syndrome of right orbitofrontal
origin is more commonly one of social or heterosocial impropriety, impaired social
pragmatics (e.g., turn taking: e.g., Anderson et al. 1999; see also Yeates et al. 2012;
see also Helfinstein and Poldrack 2012; see also Damasio et al. 2012; see also Hu

and Harrison 2013b), an apparent lowering of moral standards, and "impulsivity" (Davis et al. 2012; see also see Oades 1998; see also Emond et al. 2009; see also Granacher 2008; Helfinstein and Poldrack 2012). Moreover, significant progress has been made to differentiate the functional localization of complex value-based decision making within these frontal lobe circuits (e.g., Gläscher et al. 2012).

Very little research currently exists on social propriety and social pragmatics, though the subtle variations in the expression of these disorders have great impact and are generally dismissed under the rubric of emotional intelligence. One example is found in our reactions to the successful landing of a man on the moon. Neil Armstrong is oft remembered for his famous statement after landing on the moon and setting foot on lunar soil. He is quoted as saying "That's one small step for [a] man, one giant leap for mankind." Less is known of his statement purportedly made upon stepping off of the lunar surface and before he re-entered the lunar lander, where he is quoted as saying "Good luck Mr. Gorsky" (Mikkelson and David: urban legends). Many people at NASA apparently thought that it was a casual remark concerning some rival Soviet Cosmonaut. However, no Gorsky was found in either the Russian or the American space programs. On July 5 in Tampa Bay, Florida, after a speech by Armstrong, a reporter brought up the now 26-year-old question again. This time he explained the impropriety, since Mr. Gorsky was now deceased. When he was a kid, Neil was playing baseball with his brother in the backyard. His brother hit the ball, which landed beneath Mr. Gorsky's bedroom window. Leaning down to pick up the ball, he heard Mrs. Gorsky shouting at Mr. Gorsky, "Oral sex? Oral sex you want? You'll get oral sex when the kid next door walks on the moon!" Another Apollo astronaut, Jim Lovell, invested in his marriage from space when, on Christmas Day, his wife was gifted a mink coat from "the man in the moon" (see Lovell and Kluger 1994).

Intelligence and Memory

One of the greater concerns of aging in otherwise healthy people is that they may be developing memory problems and that this might be the first clue for an early dementia process, perhaps. W. David Crews III promoted a Pfizer-funded project for the provision of a memory screening outreach program for middle-aged and older adults (Crews et al. 2009). This project funded a clinic for the provision of 1000 free memory assessments throughout the Commonwealth of Virginia. Within a week, each of the available slots was fully booked by otherwise successful and active individuals, worried that they might be losing it, at least a bit.

Contrary to popular belief, frontal lobe disorders often defy traditional assessment approaches relying on memory measurement and the quantification of intelligence with standardized instruments. Importantly, these often-serious brain disorders, which convey significant vocational, social, and emotional disability, may not be detected using paper and pencil measures. For example, Damasio and Anderson (2003) state:

...impairments of cognitive abilities measured by standard intelligence tests are not striking in patients with frontal lobe lesions, even when bilateral. And yet, frontal lobe patients often behave in a most unintelligent way. Many of the real life problems encountered by patients with frontal lobe lesions are in social situations, raising the possibility of a selective defect in some aspect of "social intelligence." pp. 415

Relatedly, Bear et al. (2007) in their text entitled *Neuroscience: Exploring the Brain*, state:

While frontal lobotomy can be performed with little decrease in IQ or loss of memory, it does have other profound effects. The changes...are a blunting of emotional responses and a loss of the emotional components of thoughts. In addition, lobotomized patients often developed "inappropriate behavior" or an apparent lowering of moral standards. (These patients) had considerable difficulty planning and working toward goals. pp. 578

Damasio and Anderson (2003) state that frontal lobotomy:

...may result in major dissociations between well-preserved memory capacity (as demonstrated by normal performances on standardized memory tests) and severely impaired utilization of those abilities in real-life situations. pp. 418

An example of the potency of this issue within the clinical setting may be provided through discussion of a man who sustained traumatic brain injury resulting from a pedestrian motor vehicle accident. The lesions were substantially consistent with those who might be induced by frontal lobotomy with an encephalomalacia (softening of the brain) at the basal forebrain region and with relative left frontal lobe pathology (see Fig. 14.6). This minister, previously active in his parish, was unable to complete his routine duties and responsibilities despite excellent and, indeed, superior performance on the many pencil and paper measures used to assess him across multiple neuropsychological evaluations. Some of these results are depicted below (see Figs. 14.7–14.9).

It is difficult, not only for the layperson but also for the head injury professional, to appreciate the resulting disability from such a brain disorder in the presence of preserved memory function and intelligence, at least as assessed on pencil and paper measures. Frontal lobe function is intimate to the constructs of desire or intent. More specifically, in the training of graduate students in clinical neuropsychology, one of the tasks for the student is to appreciate that when they have personal attributions toward another of intent or desire, they are essentially referring to frontal lobe processes. To say, "he did that intentionally" may be a de facto statement of frontal lobe function. Thus, the attribution of blame (perhaps) is altered with a better understanding of brain function, where dysfunctional frontal lobe regions may alter desire, motivation, or the intention to initiate and to complete an act. For the healthy left frontal lobe, these intentional acts involve substantial energy, the desire or intent to engage socially, and the intent to engage in positive affective and verbally interactive activities. These appear to be diminished with the amotivational syndrome arising from dysfunctional left frontal lobe origin and medial frontal pathology with reduced activation of social reward systems and dopaminergic pathways (see Scott and Schoenberg 2011; see also Grossman 2002; see also Granacher 2008; Hu and Harrison 2013b).

Fig. 14.6 Midsagittal brain scan depicting the frontal lobe encephalomalacia subsequent to a traumatic brain injury

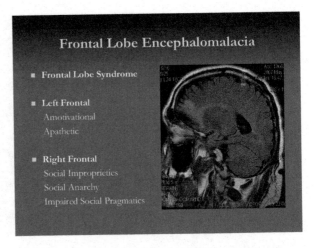

Fig. 14.7 Performance results on the WAIS-III following frontal lobe injury

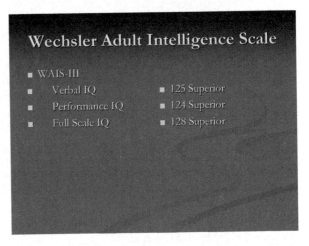

Fig. 14.8 Wechsler Memory Scale-III performance following frontal lobe injury

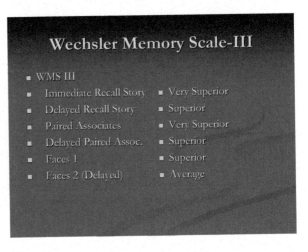

Fig. 14.9 Wechsler Abbreviated Scale of Intelligence (WASI) performance following frontal lobe injury

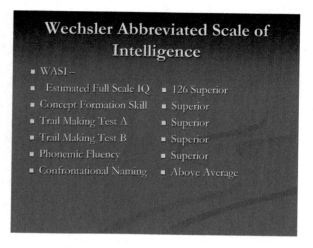

While the patient with left frontal pathology may present with diminished energy and behavioral slowing or inertia, heightened energy or "impulsivity" may result from right frontal pathology and, by inference, the release of left frontal systems (e.g., decreased caution). Both are described in the scientific literatures as "delayed response deficits" (Malmo 1942; see also Cai et al. 2012; see also Allman and Meck 2012). However, with left frontal pathology, the patient may initiate only following inordinate delays with behavioral slowing or bradykinesia, whereas the patient with right frontal pathology may initiate substantially prior to implementing the organizational planning or caution for successful task completion. This may include hypoactivity syndromes derived from medial frontal lobe or left frontal lobe pathology and hyperactive or "impulsive" features with pathology within the homologous regions of the right frontal lobe presupplementary motor or orbitofrontal cortex (see Oades 1998; see also Emond et al. 2009; see also Granacher 2008; Helfinstein and Poldrack 2012; see also Cai et al. 2012).

Inertia and Perseveration

Fuster (1980) proposed that the prefrontal cortex plays a role in the temporal structuring of behavior, synthesizing cognitive and motor acts into purposeful sequences. Stuss and Benson (1986) argued for a hierarchical role of the frontal lobes, as part of a functional cerebral system for the regulation of behavior. This occurs through the modulation of the processing of sensory information, performed within the second functional unit or the posterior areas of the brain, in contrast to the frontal lobe. Two frontal lobe counterparts were appreciated. The first involves the ability to sequence, change set, and integrate information. The second involves modulating drive, motivation, and will. The former are more strongly dependent on the integrity and capacity of the dorsolateral and orbitofrontal regions. The latter are related more to medial frontal lobe structures. Related functions of the frontal lobes were described in discussions of executive functions to include anticipation, goal selection, preplanning, and monitoring or vigilance.

Regardless, disorders of initiation, either with inadequate delay (impulsive type) or with behavioral slowing (amotivational type: see Scott and Schoenberg 2011; see also Grossman 2002) are considered frontal lobe disorders of *inertia,* presenting either with difficulty in the initiation of a response or in terminating a response. Either frontal lobe may demonstrate a disturbance in inertia, which may be assessed using a variety of procedures, including "go–no-go tasks" for example. But, the nature of the inertia differs as a function of the specialization of the left and the right brain. Indeed, inertia disorders of left frontal origin approximating Broca's area may present with inertia in phonemic enunciation, perhaps presenting slow, effortful speech, but often presenting with perseverative phonemes or phonemic paraphasic errors (see Duffau 2012).

Inertia resulting from right frontal pathology might present, more characteristically, with perseverative figural components either in an alternating design drawing (e.g., Ruff et al. 1994; Foster et al. 2005) or extending beyond the boundaries of the figure. For example, the drawing of a circle might continue with multiple perseverative loops and failure to terminate the drawing at its origin. Figure 14.10 presents the figural copy of a drawing of a flower arrangement by a patient recovering from a right frontal stroke. A careful look at the drawing reveals the tangential aspects of the patient's thoughts as the leaves become bird heads and wings and as the birds eventually fly away into the right side of the drawing. This is not a tangential linguistic feature but instead this tangent varies the gist or visual perceptual meaning of the figure. Figure 14.11 shows the performance of a patient with a right frontal stroke on the Draw a Clock Test and on Luria's ramparts (see Christensen 1979). The perseverative features are prominent with the failure to shift from one design

Fig. 14.10 A patient with a right frontal stroke attempts to copy a drawing with multiple perseverative errors and difficulty terminating each figural component

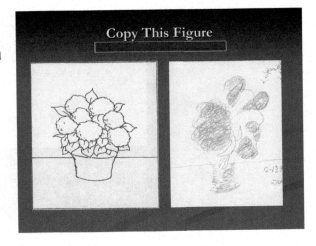

Fig. 14.11 A patient with a right frontal stroke attempts the Draw a Clock Test and to copy Luria's Ramparts with multiple perseverative errors and difficulty terminating each figural component

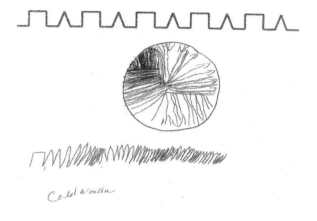

or figural component in the ramparts task to the next figural component. We might also reference some of Vincent van Gogh's paintings with perseverative or "compulsive" hash marks (see Fig. 12.6) and his poorly regulated anger displays, which may be more relevant to his frontal lobe epileptic disorder. In his case, we can appreciate the potential benefits of frontal lobe epilepsy in his artwork, energy level, and emotional passion (see Naifeh and Smith 2011).

Perseverative disorders appear to be more prominent with dorsolateral frontal and with orbitofrontal dysfunction (see Fuster 2008; see also Morrison and Baxter 2012). Inertia from left frontal dysfunction may include perseverative phonemes, words, or phrases and may include repetitive self-statements. One woman, following her anterior left middle cerebral artery stroke, appeared frustrated and increasingly embarrassed as she repeatedly asked the diagnostician, "May I have this dance?" These perseverative aspects of a nonfluent dysphasia may well correspond with perseverative graphics where letters, words, and/or phrases may be repetitively presented in the patient's writings. Figure 14.12 provides a sample of one patient's

Fig. 14.12 Motor dysgraphia with perseverative errors following a left frontal lobe stroke

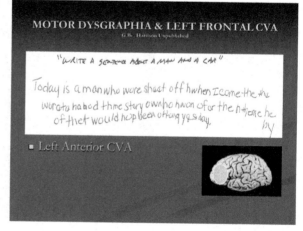

graphics following a left frontal lobe stroke. Some of the graphical mistakes in this sample appear to be perseverative errors.

Inertia may result from left frontal pathology, with the positive affective behaviors of laughter or gelastic lability, and from right frontal pathology with perseverative negative affectivity, including sadness with basal ganglia involvement and anger with orbitofrontal involvement (e.g., Fulwiler et al. 2012). The assessment of these systems might consist simply of saying the word "sad" and waiting several seconds as the patient with diminished subcortical right frontal function decompensates into a crying episode (see Oettinger 1913). The same individual might easily be redirected to a positive valence through the statement "It is a beautiful day today!" Many clinicians struggle to be effective with angry, reactively violence-prone individuals possibly referred by the courts or by other significant authority figures. Patient decompensation for anger begins with the induction posed by the first inquiry of "so what makes you angry?" At this point, the patient may begin to perseverate on this theme to the point of terminating therapy and having very negative attributions toward the therapist. Indeed, psychologists do not have a terrific track record serving patients with anger disorders and many interventions, for the violence-prone patients are limited to one or two sessions! Also, relevant here is the relative likelihood that an individual, sensitive to external control issues and prone to anger, would seek counseling or psychotherapy. Cynical life views and social anxiety among other issues may reduce the probability of seeking out professional care of this sort and the likelihood of compliance with intervention efforts.

Poorly regulated laughter or gelastic lability has seldom been the subject of scientific inquiry; although it was demonstrated to result from a brain disorder early on by Constantini in 1910 (see Oettinger 1913). This may be the case due to societal or cultural associations with laughter or humor as a positive trait, whereas rumination or reactive anger or sadness may be more readily perceived as an emotional disorder. Nonetheless, affective deregulation of laughter is a common occurrence from brain damage and from heightened social emotional or frontal stress. Each year, university staff members and students develop heightened laughter during high stress periods, including final exams. This laughter may indeed be incongruent with the underlying emotions of fatigue and mild depression. But clinically, gelastic lability must be recognized as it contributes to the syndrome analysis and more readily identifies the stressors that the patient is susceptible to decompensation. Contextual therapy may be implemented to minimize the stressors, which specifically challenge this functional neural system. In some cases, this may be implemented early on with combined vocational counseling and with work-hardening therapies to maximize success on return to work.

One young woman had just completed her doctoral degree in veterinary medicine as the first person in her family to not only finish high school but also to far exceed these goals as she was accepting her appointment as a resident in an academic medical setting. With all of the stressors associated with degree completion, packing, and relocation to her new job, she developed a small stroke in her right frontal lobe. By the neuropsychologist's impressions, this stroke would have been manageable given a little time and therapy, perhaps. But, unfortunately, with the

Fig. 14.13 Total time spent laughing during baseline, the discussion of neutral affect content, and the discussion of negative valence content or sad themes across three replications

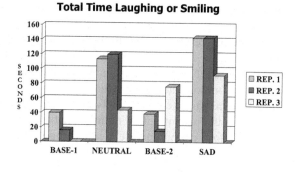

Fig. 14.14 Galvanic skin responses at the left and the right hand as a function of emotional display with happy affect corresponding with increased right hemibody activation

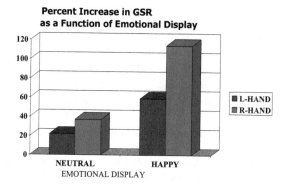

acute onset of right frontal dysfunction, she lost strength at the left hemibody, fell to the left, and severely contused her left frontal lobe. She was now severely depressed with multiple adjustment issues, the discussion of which resulted in uncontrolled laughter and, at times, mirror activation of laughter systems within the doctoral students that were working on this case.

Using an ABAB reversal design with three replications (Demakis et al. 1994), George Demakis explored the laughter behavior as a function of the emotional content of the clinical interview. The patient had previously identified and rated the content as "neutral" or as "sad" in affective valence, with the latter including statements about her loss of job, the disappointment involved, and the multiple adjustment issues that she must deal with across settings. Note that expressed speech was, by itself, a stressor for the left frontal region, with even neutral content eliciting gelastic responses. However, the discussion of negative valence and emotionally provocative sad themes yielded heightened laughter and gelastic behavior. Concurrent with the behavioral display of laughter, skin conductance responses (Galvanic skin responses) were recorded from the left and the right hemibody. As depicted in Figs. 14.13 and 14.14, laughter was recorded concurrent with lability or deregulation of the Galvanic skin response at the right hemibody. The results provide for some initial evidence that laughter may be incongruent with experienced emotion, with increased laughter resulting from the processing of negative affective stress. Laughter deregulation appears to be an indication of diminished capacity of left

Fig. 14.15 Relationship of fluency type (verbal and figural) to functional neuroanatomy. (Adapted with permission from http://www. brains.rad.msu.edu and http:// brainmuseum.org, supported by the US National Science Foundation)

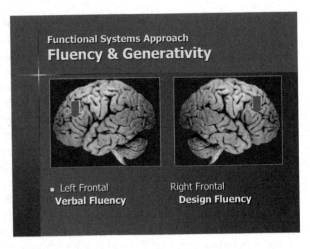

frontal or bifrontal lobe systems to manage the dual-task challenge of regulating expressed speech concurrent with maintaining the stability of these positive affective expressions (laughter).

Executive Functions

The assessment of frontal lobe integrity by the neuropsychologist includes various measures of executive functions. While there exists some controversy over what aspects of the examination should be included under the general rubric of the executive functions, these will be defined herein to include behavioral speed and sequencing, fluency or generativity, go/no-go behavior, working memory (e.g., Hasher and Zacks 1988), and cognitive flexibility (see also Stuss and Knight 2013). This includes aspects involved in shifting intention or the deployment of strategies that must vary with the changing contingencies involved in the task.

Many studies have shown the importance of the dorsolateral prefrontal cortex in the performance of cognitive tasks that are reliant on working memory (see Barbas et al. 2013) and related tasks such as executive function and the implementation of goal-directed behavior (see Reuter-Lorenz 2013). This is one of the last brain regions to develop (Passler et al. 1985; Bystron et al. 2008; see also Barbas et al. 2013). This developmental delay in maturation, apparent in comparisons with the posterior cerebral regions, is now a well-established fact (see Teffer and Semendeferi 2012). The critical role of this region (Brodmann's area 46) in cognition has been demonstrated in measures of working memory not dependent on current sensory perceptions of the outside world. This area is also somewhat distinct in that it does not directly mediate motor responses. Instead, it must integrate these neural representations and guide goal-directed behavior (see Morrison and Baxter 2012). Interestingly, Schmeichel and Demaree (2010) note in their review of the literature that people with heightened working memory capacity follow instructions better and better regulate their emotions. Moreover, these researchers provide evidence that individuals with higher working memory capacity show more self-enhancement and less negative emotion following negative feedback.

Working memory has been described as "the ability to keep events in mind" (Goldman-Rakic 1995) and as such it is constantly updated rather than representing long-term memory of events or places. Miller (2000) defines the cognitive domain of the dorsolateral prefrontal cortex as the capacity to "acquire and implement the 'rules of the game' needed to achieve a given goal in a given situation," which is further complicated by the ongoing requirement to modify the rules. Also relevant to the nature of the cognitive contribution of this brain region is the element of timing, which is mediated by area 46, as the process must be organized and executed in an appropriate sequence if the goal is to be achieved (see Fuster 2008). Morrison and Baxter (2012) note that "these are perhaps the most complex cognitive tasks that are performed by the cerebral cortex, and it is no surprise that the dorsolateral prefrontal cortex is greatly expanded in humans and in nonhuman primates compared with other mammals."

Moreover, the complexity of these neuronal circuits has been shown to be highly vulnerable to aging (e.g., Rapp and Amaral 1989; Raz et al. 1997) and may be more so than the hippocampus. This interpretation receives bidirectional support in research since the cognitive functions attributed to this region are among the first to be affected by the aging process. Morrison and Baxter (2012) argue that "such functions, and particularly the establishment of ever-changing rules to guide goal-directed behavior, may require an extraordinary level of synaptic plasticity, perhaps more than in any other brain region." Such short-term memory frameworks require extensive synaptic plasticity, which might explain why the cognitive functions that are mediated by this region are so vulnerable to the aging process.

Literatures exist on impairments in executive functions in the elderly, which seem to promote stereotypical attributions toward aging, including the term *cognitive rigidity*. Hasher and Zacks (1988; see Park and McDonough 2013) proposed a different perspective in which inefficient inhibitory mechanisms limited working memory capacity, resulting in impaired abilities such as maintaining or switching attention to relevant information. Park and McDonough (2013) note that research findings from neuropsychological testing and structural imaging show that resource-demanding tasks rely heavily on the frontal cortex. Furthermore, these brain regions show some of the greatest age-related decline in structural volume (Raz et al. 2005), consistent with the notion of a shrinking pool of resources. Reduced functional capacity in these neural structures is found with aging on combined measures of memory capacity (see Reuter-Lorenz 2013). Cappell et al. (2010) report parallel evidence indicating that older adults reach their maximal level of performance and neural activation of these regions at lower objective levels of demand than do younger adults. But generally, the functions being assessed involve the intention to, and the appreciation of, rule-regulated behaviors or strategies, with the ability to shift the approach to the task as the rules are changed.

Foland-Ross et al. (2013) note that depression symptomotology and vulnerability to onset and recurrence of depressive episodes appear to be a function of working memory capacity. Recurrent negative thoughts or rumination occur with depressive episodes and maintain negative affect (Nolen-Hoeksema et al. 2008). These frontal systems are capacity-limited in maintaining intention or awareness to the task, strategy, or rule requirements for successful completion of the task. They are relevant to dual or multiple concurrent processing demands (Kinsbourne and Cook 1971; Kinsbourne and Hicks 1978; Harrison 1991) with a narrowing of the intentional field perhaps to a single task or dimension with damage to these systems. Deficits in these systems may provide a functional neuroanatomical foundation for ruminative thoughts (Kelley et al. 2013).

One of the most consistent findings in major depressive disorder, for example, is the preeminent role of negative affective themes, where the individual has difficulty in disengaging from negatively valenced content (see Foland-Ross et al. 2013). Godlewska and colleagues (Godlewska et al. 2012) note that patients with acute depression manifest a range of negative biases in the processing of emotional information, which are believed to contribute to the etiology and maintenance of the depressed state. Depressed patients selectively recall more negative, self-related emotional information on memory tasks (Bradley and Matthews 1983; Bradley et al. 1995) and demonstrate a negatively biased perception of key social signals such as emotional facial expressions (Gur et al. 1992; Bouhuys 1999). Moreover, emotional biases such as these have been associated with pathological responses across a network of neural areas involved in emotional processing, including an increased response of the amygdala to negative facial expressions in depressed patients compared to matched controls (Sheline et al. 2001; Victor et al. 2010).

Research has consistently shown a negative emotional bias, with the depressed patient having difficulty disengaging from the processing of negative emotional material (e.g., Fritzsche et al. 2010; Harrison and Gorelczenko 1990; Gotlib, Krasnoperova et al. 2004; see also Crews and Harrison 1995; see also Shenal et al. 2003; see also Cox and Harrison 2008b; however, see Crews et al. 1999). Inhibitory deficits in the initial processing of irrelevant negative information have also been described (Goeleven et al. 2006) with a relative inability of depressed individuals to remove task-irrelevant negative thoughts and memories from working memory (Berman et al. 2011; Levens and Gotlib 2010; see also Foland-Ross et al. 2013). Joorman and Gotlib (2010) argue that this inflexibility in working memory with a failure to expel negative material functions to sustain perseverative thoughts and emotional preoccupation that characterizes major depressive disorder. In this fashion, working memory influences or regulates the experience of emotions (see Isen 1984).

Rumination appears to be a consistent and perpetual display of affective inertia or emotional perseveration along a specific valence, where the valence is relevant to the dysfunctional neural system involved (Kelley et al. 2013). For example, right orbitofrontal lesions promote working memory deficits in anger or hostile themes (e.g., Fulwiler et al. 2012; Holland et al. 2012; Mitchell and Harrison 2010; Demaree and Harrison 1996, 1997a, b; Mollet and Harrison 2007a; Everhart et al. 1995; Everhart and Harrison 1995, 1996) and right lentiform lesions promote sad emotional inertia (e.g., Ross and Rush 1981; Constantini 1910; cited in Oettinger 1913; see also Parvizi et al. 2009). Moreover, transcranial direct current stimulation over the right frontal region has been used to produce anger rumination (Kelley et al. 2013).

It is clear in this respect that working memory capacity is limited and especially with damage or pathology within these neural systems. Dual concurrent processing demands may no longer be possible and the individual may lose flexibility in thought or emotion with damage to these systems. Pathology or incapacity in working memory systems promotes perseverative processing and inflexibility in thought or emotion. Difficulties disengaging from irrelevant or redundant negative content also prevent the individual from processing new, relevant information that might facilitate the flexible reappraisal or reinterpretation of events (Siemer 2005). Consistent with this conclusion, Demeyer et al. (2012) found that rumination or difficulties in the disengagement from processing irrelevant, negative emotional content, mediate the worsening of depressive symptoms at follow-up evaluation.

Foland-Ross et al. (2013) provide a parsimonious conclusion for the review of this line of research stating that "…getting 'stuck' processing negative information is likely to have deleterious effects on mood." Clinical interventions often try to weaken perseverative thoughts and emotions and the evidenced-based interventions are many (e.g., Acceptance and Commitment Therapy-ACT; Hayes et al. 1999; see Crews and Harrison 1995; see also Shenal et al. 2003; see also Cox and Harrison 2008b). Efforts to promote the functional integrity of these systems will benefit from practical efforts to diminish processing load. This might include efforts to minimize dual or multiple concurrent processing demands (e.g., multiple negative emotional events) and efforts to prompt or redirect to a neutral or mildly positive emotional valence.

The Category Test originally introduced by Halstead and revised by Reitan as a subtest within the Halstead–Reitan neuropsychological battery is purported to be one of the more sensitive indicators of brain dysfunction in that battery (see Lezak et al. 2012). The successful completion of the Category Test requires specific mental abilities: abstract concept formation, learning capacity, adaptive skill, and cognitive flexibility. Relatedly, the assessment of executive functions is made to be more sensitive through the presentation of dual or multiple concurrent task demands and through increments in the duration of time on the task where vigilance or concentration demands may be increasingly relevant. Thus, freedom from distractibility may be assessed within the demands of an executive function's test battery.

Many "frontal lobe tests" exist in multiple forms to improve their sensitivity (e.g., through increased vigilance demands or time on the test) and to alter their relative sensitivity to detecting dysfunction within the left or the right frontal lobe (e.g., Delis et al. 2001; see Strauss et al. 2006; Lezak et al. 2004; see also Lezak et al. 2012; see also Milberg et al. 2009). Examples of evidenced-based tests sensitive to frontal lobe executive deficits include measures of speed and sequencing such as the Trail Making Test Form A and Form B (Reitan 1958). This test was developed by Partington in 1938 (see Partington and Leiter 1949) as a measure of divided attention. The form A requires simple sequencing within a set of numbers, connecting them in numerical order, whereas the later version requires alternate sequencing among two sets of stimuli (numbers and letters). Thus, the form B is generally considered more sensitive to frontal dysfunction and the reduced ability to perform the dual task demands of sequencing and alternating the sequence across the two sets of stimuli. Investigation of electroencephalographic activity implicated frontal lobe activation during performance on the Trail Making Test (Segalowitz et al. 1992). Moll and colleagues (Moll et al. 2002) used a verbal adaptation of the Trail Making Test and found that the set-shifting component of alternating letters of the alphabet and consecutive numbers activated the left dorsolateral prefrontal cortex and supplementary motor area/cingulate sulcus, which are areas that are consistently shown to be sensitive to executive functioning and particularly cognitive flexibility (e.g., Zakzanis et al. 2005).

Many versions of the test exist, including the Comprehensive Trail Making Test (Moses 2004; Reynolds 2002; Kahn et al. 2012), the Color Trail Making Test (D'Elia et al. 1996), and Paul Foster's Figure Trail Making Test (Foster et al. 2013). The Color Trail Making Test was developed specifically so that there would be less potential for language and cultural bias. For example, the Color Trails Test substitutes colors in place of letters of the alphabet. The Figure Trail Making Test was developed to reduce the linguistic demands and to shift the requirements to a figural design-drawing activity. This version involves the alternate sequencing of design drawings with the expectation that it may be more sensitive to right frontal function. The spatial span and letter–number sequencing subtests from the Wechsler Memory Scale III (Wechsler 1997) and the spatial addition subtest from the Wechsler Memory Scale IV (Wechsler 2009) might provide additional examples.

The evaluation of flexibility in thought and the ability to implement a change in strategy secondary to the changing contingencies in the setting might be assessed by instruments such as the Wisconsin Card Sorting Test (Berg 1948; Psychological Assessment Resources 2003). This test requires the subject to pick a response based on one or another dimensional strategy, such as color, shape, number, and location. Once the effective strategy has been acquired, the strategy is shifted by the examiner such that choices, which would have previously been correct, are now incorrect. Frontal lobe executive deficits present with inflexibility in implementing a new strategy and perhaps perseveration on a now ineffective approach to the problem. It has oft been said that change is hard to implement. But, this may reach a pathological level of rigidity in the individual with frontal lobe incapacity.

Tests of fluency and generativity have long been known, within our literatures, to be sensitive measures of executive functions. Consistent with Broca's initial claim that "we speak with the left brain," verbal fluency measures such as the Thurston Word Fluency Test (Thurstone 1938; Thurstone and Thurstone 1949) and the Controlled Oral Word Association Test (e.g., Phelps et al. 1997; Baldo et al. 2006; Baldo, Shimamura et al. 2001) have been used to assess left frontal brain systems. For example, a functional magnetic resonance study of normal participants during their performance of the Controlled Oral Word Association Test revealed that the left frontal lobe (middle and inferior frontal gyri) was more activated than the right (Wood et al. 2001).

Fig. 14.16 Relationship between right frontal activation levels on the electroencephalogram and performance on the Ruff Figural Fluency Test (RFFT)

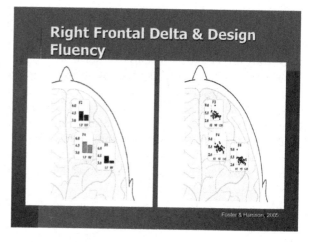

In contrast, figural fluency or design fluency measures have been shown to be more sensitive to right frontal brain systems (Ruff et al. 1994; Ruff et al. 1986). The Ruff Figural Fluency Test (RFFT; Ruff 1996) requires the patient to generate as many unique designs as possible, connecting two or more dots from within a five-dot matrix. There are multiple forms where the arrangement of the dots varies and the data are collected over multiple trials. Figure 14.16 shows the relationship between figural fluency and electrical activation over the right frontal lobe using quantitative electroencephalography (Foster et al. 2005). It should be noted that delta magnitude is commonly used as an indicator of pathology on the electroencephalogram and is associated with deactivation or diminished arousal at those brain regions.

The results from Paul Foster's project indicate that decreased right frontal lobe activation or cortical synchrony (increased delta magnitude) corresponds with decreased figural fluency using the RFFT. Moreover, in related work (e.g., Williamson and Harrison 2003; Holland et al. 2011; Mitchell and Harrison 2010), hostile, violence-prone individuals have shown diminished activation over right frontal regions and they have been reliably deficient in their performance on the RFFT (see Fig. 14.17). Thus, the deficits in figural fluency may, to some extent, predict

Fig. 14.17 Perseverative errors indicating right frontal lobe deficits in hostile, violence-prone individuals

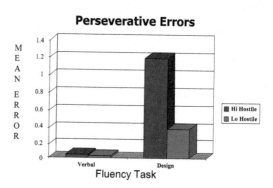

diminished regulation over anger and other functions attributed to these frontal regions. This certainly may include reactive increments to right frontal stress in sympathetic measures of arousal, including cardiovascular measures of blood pressure and heart rate, circulating blood glucose levels, and skin conductance. Moreover, verbal fluency challenge or left frontal stress has been used to reliably lower systolic blood pressure and heart rate in hostile individuals, whereas figural fluency challenge or right frontal stress reliably increases these measures (Williamson and Harrison 2003; Mitchell and Harrison 2010). Oxygen saturation levels should vary in a consistent fashion with dyspnea, lowered saturation levels, and elevated sympathetic tone on the RFFT (see Hu and Harrison 2013a). But, support for these predictions likely will be contingent upon the use of subjects with minimal frontal capacity.

Executive functions are more commonly expressed within the everyday context of managing the self safely within social, emotional, and behavioral settings. One man provides an example of safety management concerns as he started his car, drove through the garage, across his backyard and through his fence, across the alley and through his neighbor's fence, across his neighbor's yard, and into his neighbor's bedroom injuring two people as they slept in their bed. The stress and excitement likely increased his antigravity posturing with extensor pressure on the throttle and a diminished ability to shift to the alternate behavior of applying pressure to the brake pedal. Clearly, cognitive and behavioral flexibility are safety-related constructs predictive of relocation to a nursing care or assisted living facility (Cahn-Weiner et al. 2000).

Go/no go procedures are routinely used to investigate regulatory control or executive function deficits resulting from frontal lobe pathology or incapacity (Drewe 1975). One procedural example (Luria 1980; Christensen 1979) involves a tapping task with the instructions "When I tap once, you tap twice and when I tap twice, you tap once." Simple shifts in this sequence may reveal executive deficits with inertia, indicated by perseverative tapping patterns. The go/no-go task provides a simple paradigm with which to investigate brain activation during operations involving response inhibition and response competition. The frontal lobe regions involved in the error processing system appear to overlap with brain areas implicated in the formulation and execution of articulatory plans (Menon et al. 2001).

Another man was brought into the office by his spouse. When asked how the neuropsychologist might be of help, he was adamant that he did not have any problems. He had no idea why his spouse had brought him in, as he expressed some anger-related control issues over his wife's actions. These were clues for the evaluation, of course. But he finally acknowledged that he did have some problems mowing his lawn. He noticed that it now took him about an hour and a half to gas the mower and to get it ready for the yard work. But, when he started mowing, he continued well onto his neighbor's yard until his wife shouted "Norman, get back in your yard!" At this point, he would turn around and mow across his yard and well into the other neighbor's yard until his wife shouted again, and so forth. Clearly, this is a good neighbor to have, at least in Florida! These executive deficits were apparent within the context of extrapersonal space or geographical boundaries and occurred with minimal appreciation of his deficits (anosognosia). The findings are also potentially relevant to rule-regulated behavior within and across these social/geographical boundaries. The evaluation was sufficient to confirm a right frontal lobe problem.

In contrast, those executive deficits, which arise from left frontal dysfunction, may be more readily appreciated as grammatical, logical, and/or linguistic in origin or in the expression of an ideational dyspraxia (De Renzi et al. 1968; Rothi et al. 1997). Speech, for example, may be tangential or perseverative at the phonemic, phrase, sentence, or paragraph level. The latter involves a deficit in the organizational regulation of sequential movements. Moreover, the left frontal patients are characteristically aware of their deficits to the point of frustration and lowered self-esteem. These frontal systems include cortical regions essential in the sequential regulation of ongoing movement. In addition, these cortical regions intimately interact with subcortical regions (e.g., the basal ganglia) critical for the intentional initiation or motivational desire to respond. Energy level and speed of response are relevant to ideational dyspraxia and to other features presenting with left frontal lobe pathology.

An engineering professor, who was also an avid woodworker, provides an illustration of the construct within the constructional domain where combinatory activities must be well organized and sequenced to complete a project. The gentleman spent one summer after his right frontal lobe stroke attempting to build a simple doghouse. Each week during visitations, the lumber would have been assembled into a different and nonfunctional array. He would lose his temper and question that others might be "messing around" with his project. Eventually, in disgust, he threw the lumber across the yard and donated his tools to an assistive living facility. The doghouse took on many shapes over these weeks, but the architecture was never organized into a meaningful design. But, Paul Foster (Foster et al. 2011, 2012; see also Roth et al. 2013) has demonstrated increased probability of infrequent associations with diminished frontal capacity. Therefore, it is remarkable to expect that incapacity here may at least promote creative expressions (less frequent associations) and the generation of somewhat novel designs or combinations of the workable materials. In this sense, one person's disability might be expressed as an opportunity in another. History provides many examples and Vincent van Gogh might be one of these (see Naifeh and Smith 2011).

Effortful Control

Efforts to study self-regulation have often focused on behavioral and cognitive processes, which may be more generally considered as executive functions. Executive function refers to attention shifting, working memory, and inhibitory control processes that are critical in organizational planning, behavioral sequencing, problem solving, and goal-directed activity. Executive function is similar to effortful control in that it also refers in part to the ability to inhibit a prepotent or dominant response in favor of a less salient response. That is, effortful control considers the appetitive or aversive nature of the conditions under which control is required, while the primary focus of work on executive function concerns the deployment of cognitive processes under conditions that are essentially affectively neutral. At a neural systems level, this integration is represented in an approach that examines the role of interconnected prefrontal brain structures in emotional reactivity, emotion regulation, attention, and cognitive control (Bush et al. 2000; Groenewegen and Uylings 2000).

To some extent, the work on executive functions extends into research on effortful control. Though the distinction may appear to lack some clarity, the assessment of effortful control may include self-soothing behavior and regulatory function involving emotional displays. Effortful control more specifically refers to the ability to inhibit a dominant or prepotent response in favor of a subdominant or less salient response. This might be assessed within one or another theoretical framework of emotional functions such as the approach/withdrawal model (Kinsbourne 1978; Davidson 1995; Heller et al. 1998). In this sense, effortful control allows for the regulation of approach and withdrawal tendencies in the face of immediate cues for reinforcement or punishment and emotion-related regulatory control (e.g., Raver 2004) and the development of social competence. We might regulate our thoughts, emotions, and actions or choose to forego immediate gratification or reward for a larger reward following a delay or longer-term investment approach. Planning ahead, resisting distractions, and goal-oriented behavior are all fundamental to the self-regulation process. Moreover, impairments in effortful control would support instability of relationships, impulsivity, and difficulties in controlling emotion. Each of these has substantial bases within the literature on disorders of the frontal lobe (see Oades 1998; see also Emond et al. 2009; see also Granacher 2008; Helfinstein and Poldrack 2012).

Adaptive goal-directed behavior involves monitoring of ongoing actions and performance outcomes and subsequent adjustments of behavior and learning. Ridderinkhof et al. (2004) provide a review and meta-analysis of primate and human studies of the cortical interactions that subserve the recruitment and implementation of such cognitive control and suggest a critical role for the posterior medial frontal cortex in performance monitoring and the implementation of associated adjustments in cognitive control. The authors conclude that much of the posterior medial frontal cortex, including Brodmann's areas 6, 8, 24, and 32, is reliably engaged following the detection of response conflict, errors, and unfavorable outcomes. The authors point to the direct link between activity in this area and subsequent adjustments in performance. Moreover, they conclude that critical functional interactions exist between the posterior medial prefrontal cortex and the lateral prefrontal cortex, such that monitoring-related posterior prefrontal cortex activity serves as a signal that engages regulatory processes in the lateral prefrontal cortex to implement performance adjustments. This relationship is reminiscent of the perseverative disorders, which have resulted from lesions of the lateral prefrontal cortex. Further, the relationship is supported by findings of cognitive inflexibility and the failure to implement new strategies subsequent to shifting cognitive demands with damage to the lateral prefrontal cortex.

The authors draw three conclusions from the meta-analysis. First, performance monitoring is associated with posterior medial prefrontal cortical activation. Second, Brodmann's area 32 represents the area undergoing pronounced activation for all types of monitored events, suggesting the importance of this area for a unified performance-monitoring function. Third, activations related to pre-response conflict and uncertainty occur more often in Brodmann's area 8 and less often in area 24 than do activations associated with errors and negative feedback. Activation foci, which the authors associate with reduced probabilities of obtaining reward, cluster slightly more dorsally than foci associated with errors and failures to obtain anticipated reward. This more general monitoring function provides for the capacity of this region to signal the need for adjustment to alter the probable outcomes of subsequent responses. The authors conclude that the evidence indicates a tight link between activity in this area and subsequent adjustments in performance, suggesting that the posterior medial prefrontal cortex signals other brain regions that changes in cognitive control are needed. The brain region most directly implicated in effecting these control adjustments is the lateral prefrontal cortex. Thus, monitoring-related medial prefrontal cortical activity appears to serve as a signal that engages control processes in the lateral prefrontal cortex essential for adaptive behavior involving self-regulation or regulatory control.

Self-regulation relates to the control of emotions, feelings, instincts, needs, motivations, and desire. It is often beneficial to have the ability to delay gratification, control response inclinations or desires, and to pursue goals. Moreover, the failure to self-regulate is associated with maladaptive behaviors including intoxicant abuse, drunkenness, gluttony, aspects of domestic violence, arguments, interpersonal conflicts, and so forth. Rostowski and Rostowska (2012) conclude in their review that these functions are very important in all aspects of physical and mental health and that failure of these systems underlies health and mental disorders of all kinds. Ultimately, a better understanding of the factors promoting both successful implementation and failures of self-regulation can bring valid insight into these health problems and their treatments (see Engelberg and Sjoberg 2005; see also Cacioppo et al. 2007).

A large body of scientific evidence on the prefrontal cortex has helped to illuminate the regional specificity for various cognitive functions, including cognitive control and decision making. Recently, Gläscher et al. (2012) elaborated on this specificity, providing detailed causal evidence for functional–anatomical specificity in the human prefrontal cortex. These investigators used a neuropsychological approach with a unique data set accrued over several decades. They applied voxel-based lesion–symptom mapping in 344 individuals with focal brain lesions (165 involving the prefrontal cortex) who had been tested on a comprehensive battery of neuropsychological tasks. Two distinct functional–anatomical networks were revealed within the prefrontal cortex. One of these associated with cognitive control (response inhibition, conflict monitoring, and switching), which included the dorsolateral prefrontal cortex and the anterior cingulate cortex. The second network was associated with value-based decision making and included the orbitofrontal, ventromedial, and frontopolar cortex. The authors also note that cognitive control tasks share a common performance factor related to set shifting that was linked to the rostral anterior cingulate cortex. Instead, regions in the ventral prefrontal cortex were required for decision making.

Damage to the prefrontal cortex is related to social errors, including tactless actions and the commission of faux pas in social or interpersonal settings secondary to diminished emotional self-control, reduced emotional downregulation, and errors in the recognition of other people's emotions. This characteristically presents with a lack of empathy or intuitive appraisal of the social setting, emotional reactivity or instability, and poorly regulated adjustments of autonomic functions (see Lieberman 2007; see also Rostowski and Rostowska 2012; Williamson and Harrison 2003; Holland et al. 2012). Instead, the functions of the orbitofrontal cortex alter amygdala activation and kindling phenomenon. Within the right cerebral hemisphere, this has a moderating effect on anger and reactive aggression, fostering instead a calming process and quieting of the emotional activation.

Rostowski and Rostowska (2012) point out that this allows for the making of alternative decisions, which as a consequence of effortful control or self-regulation may be compliant or subordinate to value systems and strategies to promote individual or interpersonal relations. These authors conclude that "the orbitofrontal cortex can support manifestations of appropriate and correct forms of social behavior through numerous and different forms of constant monitoring of this behavior, especially in its relation to and its compatibility with social norms, as well as limiting the degree to which emotions influence cognitive processes." Damage or deactivation of the orbitofrontal cortex can weaken or reduce these processes of monitoring one's behavioral appropriateness and insight into oneself. Moreover, reduced capacity at the orbitofrontal cortex constrains the possibilities of inducing the social emotions of shame, uneasiness, or the feeling of having committed a tactless action. Beyond these emotions are others which motivate the effort to make corrective actions for those who were improper, unsuitable, and socially incorrect.

The discussion of effortful control necessarily includes the potential for carryover effects of the experience or learning derived from these events (see Beer et al. 2006; Beer 2007; see also Rostowski and Rostowska, 2012). These individuals display a tendency not only to act improperly but also to induce incorrect emotions, where the emotion functions to strengthen rather than correct the inaccurate social behaviors. In the case of anger attributions, these may become principled or rationale, supporting the rightful bases for the aggressive actions toward another and strengthening the misattribution to the precipitating source. It might also be appreciated that, in some cases, the source may fall victim to the high hostile or reactively aggressive individual with orbitorfrontal pathology or incapacity. This may be fundamental to some forms of bullying behavior, teasing, humiliating others, or extremes involving physically abusive behavior or assault. These systems are relevant to some forms of psychopathy and even may be present in marital conflicts (Beer 2006, 2007; Beer et al. 2006).

Effortful control mechanisms are easily altered in traumatic brain injury and the emotional sequelae are well established (Hart et al. 2012). Irritability and anger management problems are estimated to affect at least two thirds of those moderate to severe brain injury patients (van Zomeren and van den Burg 1985; Kim et al. 1999). These effects persist or even worsen over the first year of recovery (van Zomeren and van den Burg 1985; Hanks et al. 1999). Poorly managed or regulated anger after brain injury has a broad impact on family, social relationships, and outcomes within the community (Kim et al. 1999). They alter vocational outcomes, including employment (Delmonico et al. 1998). Impaired effortful control over temper adds significantly to the caregiver's burden (Marsh et al. 1998a, 1998b) and potentially results in the exclusion of the individual from access to treatment programs. Hart and colleagues (2012) refer to disruption of these brain mechanisms leading to a "vicious cycle" of increasing isolation and loss of sources of social support. They further note that anger dysregulation after traumatic brain injury, and especially that involving the orbitofrontal and temporal regions, is among the most common reasons for neuroleptic and major tranquilizer use, which also have serious iatragenic adverse effects limiting recovery (see also Silver and Arciniegas 2007).

Additional evidence exists for the importance of right frontal regions in social cooperation in contrast to oppositional social competition. For example, researchers (Cui et al. 2012) used near-infrared spectroscopy recording techniques to simultaneously measure brain activity in two individuals playing a computer-based cooperation game side by side. Calculation of interbrain activity coherence patterns between the two participants revealed that the coherence between signals generated by participants' right superior frontal cortices increased during cooperation, but not during competition. Increased coherence of brain activity at the right frontal region was also associated with better cooperative performance. Importantly, the authors argue that this technique, and others like it, heralds the emergence of a new field of social neuroscience.

The development of neuroimaging technologies, especially, has deepened our understanding of neurocognitive processes underlying social behavior. This is evident in existing lines of research on theory of mind (Gallagher and Frith 2003), moral judgment (Greene et al. 2001), trust (King-Casas et al. 2005), and agency (Tomlin et al. 2006). Also relevant here are contributions in understanding structural and functional differences that exist in the brains of individuals with deficits in social cognition, including, for example, childhood autism (Shultz et al. 2000; Shultz 2005; Dapretto 2006; Hadjikhani et al. 2004; Pierce et al. 2001; see also Just, Kellera, Malavea, Kanab, Varmac, in press) and conduct disorder (see Cubillo et al. 2012). These technological advances continue and hold much promise for revealing the functional neural systems responsible for individual differences and the methodologies and interventions which may improve or maximize the integrity of the system and of the individual attempting to overcome one or another disadvantage.

Motor and Premotor Syndromes

The motor strip or motor cortex consists of large pyramidal cells with the cell bodies in layer five of this cortex. The dendritic fields extend up through layer one at the surface where there is a substantial degree of overlap across the dendritic fields of

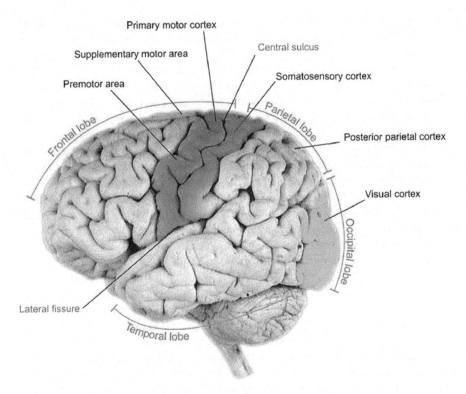

Fig. 14.18 Lateral view of the left cerebral hemisphere showing the motor, premotor, and supplementary motor cortices. Originally published in *Brain–Computer Interfaces: A Gentle Introduction,* Graimann et al. 2010, p. 12

multiple cells. Stimulation of this outer most layer yields complex motor activation, with increasing specificity found upon lowering the electrode into deeper cortical layers (see Rothwell et al. 1991). Collectively, these cells are referred to as the upper motor neurons, which descend via the corticospinal or corticofugal projections down through the corona radiata and through the external and internal capsules. They eventually control the motor units via synaptic transmission onto second-order alpha motor neurons, which exit the spinal cord projecting on toward and into the motor units within the skeletal muscle fibers. Damage at the upper motor neuron level has been related to paresis or plegia and, subsequently, to hyperreflexia or altered muscle tone (e.g., Futagi et al. 2012). This region of the frontal lobe makes up Luria's primary projection cortex for this modality and is under the direct influence of the surrounding secondary association cortices, referred to collectively as the premotor cortex (see Fig. 14.18).

The premotor cortex has been found to activate shortly before the motor cortex (see Rizzolatti et al. 1996; see also Fogassi and Semone 2013) and to play a significant role in the planning or preparation for movement. Processing here is critical to properly sequenced or fractionated movements of our body parts in the conduct of volitional activities. Less evidence is available for the role of these regions in regulating peristalsis and cardiovascular and cardiopulmonary coordination. However, the shared role is likely (e.g., Foster et al. 2010). The premotor cortex is intimately interactive with the basal ganglia (e.g., Oguri et al. 2013) allowing for the initiation, intent or desire for movement. These processes are conducted in transitional movements concurrent with those which allow for kinesthetic feedback on the melody of movement, through interactions with the somatosensory cortices. As the desire or intent for the initiation of movement develops, the planning and coordination for implementing the activity follows and ultimately with feedback from the somatosensory regions to insure that the planned activity was carried out in a coordinated fashion, allowing for moderation of the ongoing movements.

The intimate interconnectivity with the extrapyramidal dopaminergic systems of the basal ganglia and with the nucleus accumbens within the brainstem tegmentum conveys importance not only in motor intent but also for the intrinsic reward systems underlying intentional movement. This system would be subject to the history of the stimulus contingencies or behavioral consequences of action. For example, disruption of these pathways (see Öngür and Price 2000) within the left frontal lobe may render the individual with reduced social initiation concurrent with a diminished appreciation of social rewards. Thus, the individual may have reduced social initiation (see Scott and Schoenberg 2011; see also Grossman 2002), reduced fluency or sparcity in expressed speech, and lowered self-esteem.

The motor and the premotor cortices maintain a topographical arrangement roughly corresponding to the parts of the body with the face and oral motor apparatus located along the ventral paths and with the arm, hand, and fingers; and eventually the leg, feet, and toes located more dorsally in this homunculus. Again, the historical developments, which were influential in the appreciation of these arrangements, include the stimulation techniques and activities of Fritsch and Hitzig (Fritsch and Hitzig 1870) and also of Charles Scott Sherrington's student Wilder Penfield (e.g., Penfiedl and Boldrey 1937; Penfield and Rasmussen 1950; Penfield 1967), along with his friend Herbert Jasper. Woolsey (1952; see also Woolsey, Erickson and Gibson 1979) provided some of our earliest cortical maps (e.g., Woolsey 1933, 1952; Woolsey et al. 1979) using evoked potentials with substantial technological savvy. He also contributed influential comparative data across species. Also, multiple lesion studies identified these homunculi, including Broca's case studies (Broca 1861) and others. Lesion of the premotor cortices may be demonstrated in the functional loss of the fractionation of movement or discoordination of the muscle groups involved in the ongoing activity. This might include aspects of executive functions underlying movement proper as in a rapid alternating movement or even a rhythm-tapping task (see Luria 1980).

The evaluation of rapid oscillating movements is specific to the representation of the body across the homunculi within each cerebral hemisphere. Thus, rapid alternating movement tasks include finger tapping, fist to palm alternation, finger to thumb alternation, speech alternations in repetition of sequential sounds (e.g., "lah-pah-kah"), and foot tapping. Alternating flexion and extension of the mandible and guttural contractions underlie the production of phonemic and verbal sounds at the base of the premotor homunculus within the left inferior posterior frontal region. But, they may be evaluated also through the movements expressed in the oscillations of the violinist's fingers back and forth upon the instrument's strings. Impairments in these coordinated or sequentially processed movements are referred to within the context of frontocerebellar dysfunction as *dysdiadochokinesia* and more broadly in relation to the *ideational dyspraxias* (see Heilman and Gonzalez Rothi 2012a) as opposed to the *gestural or ideomotor dyspraxias* more common with parietal lobe damage.

Damage within the premotor regions may result in both a diminished desire and an intent to implement action patterns for that body region located on the homunculus at the location of the lesion. This effect may worsen with time as the accumulated learning history for that body region differs from that of the contralateral body region under control of the intact homologous premotor cortex. More specifically, the contralateral body region may be used less when compared to features of the motor neglect. But, the movements of this body region are discoordinated, poorly organized, and disturbed in kinesthetic melody. Thus, the feedback reflects inadequacies in the performance of the body part concurrent with diminished desire or intent to perform the underlying movements. With the active use of that body part, the learning history differentially reinforces the use of the unaffected side of the body potentiating the neglect rather than ameliorating the effects with recovery of functions (see *Constraint-induced movement therapy*: see Miltner et al 1999; Bowden et al. 2013; Liepert et al. 1998; see also Wittenburg and Schaechter 2009). Pavlovian features might contribute to this developing response bias, through release (from inhibition) of the homologous brain regions contralateral to the lesion location.

The subcortical dopaminergic systems of the left frontal lobe may be differentially involved in processing social approach and positive affect in opposition to the more socially avoidant and affectively negative neural structures within the right frontal lobe. Poorly effected social engagements might also be sensitive to a differential learning history and such might be the case with social anxiety syndrome. It is not hard to imagine a differential learning process as a function of two oppositional and interactive frontal lobes, where social embarrassment and or lowered self-esteem might release negative and cynical affective systems more prone to social avoidance or withdrawal along with elevations in sympathetic tone. This conceptual framework allows for double dissociation (Teuber 1955; Kinsbourne 1971; Luria 1973, 1980) of the intervention approaches to activate or to promote processing within the desired functional neural systems and to deactivate or diminish the activation of the nonpreferred functional neural systems. For example, social skills training, positive socialization, and expanded social support might promote left frontal activation just as stress management, deep breathing, and progressive muscle relaxation might deactivate or diminish the activation of negative affect analysis and sympathetic drive.

To counteract these potentially cumulative effects with experience, consent may be acquired and the unaffected extremity may be carefully restrained to foster the development of improved and coordinated movement with the affected limb (e.g., see Taub et al. 2002; however, see Bowden et al. 2013). The patient may mislocate or misuse the limb, initially, with efforts to overcome the restraint. But, with continued therapy and supportive positive rationale, the neglect may resolve. This may result subsequent to the required activation of the affected premotor cortex and through the effects of experience or learning; as with the integration of the aesthetic or somatic basis for coordinated movement. The intent and initiation of movement at the unaffected extremity may further extinguish the intent and initiation of the affected extremity via inhibition processes across the corpus callosum. Activation of the left premotor cortex may be expected to diminish activation of the homologous region of the premotor cortex in the opposing brain. Again by analogy, the lateralized affective processing systems might be found in the clinical variants of exposure therapies for social anxiety disorder. Although motor extinction is typical or common with relative motor impairments at one extremity or the other, motor generalization may occur with more severe pathology. Motor generalization across the corpus callosum to the unaffected limb is common with exertion efforts to move the more severely compromised extremity.

Broca's area, at the left inferior posterior frontal lobe, is part of this system. Activation of Broca's area is associated with the organized and well-regulated fluent expression of speech, wherein phonemic articulation sequences are fluently presented without articulation or perseverative errors (see Duffau 2012). Activation of the region corresponds with the desire or the intention to talk and the temporal sequence of activation spreads ultimately to the motor cortex or final motor pathway to the oral motor apparatus, including the tongue, lips, and surrounding anatomical regions. Partially, due to the rich interface with the dopaminergic systems of the basal ganglia and to the reward circuits of the nucleus accumbens, these premotor regions are involved in the desire or the intent to perform or to engage in activities. Moreover, the preferential relationship for engaging in activities within the contralateral hemispace is largely maintained and expressed by activation of these brain regions. A notable example might be drawn from the dorsolateral premotor and prefrontal regions described as the *frontal eye fields*. Lesion, functional incapacity, or excessive stressor demands over the right frontal eye field characteristically results in a right gaze bias, ipsilateral to the lesioned frontal lobe (Guitton et al. 1985; Ladavas et al. 1997; Pierrot-Deseilligny et al. 1991) and heightened distractor responsivity at the right side (e.g., Suzuki and Gottlieb 2013). Also, transcranial direct current stimulation over the right frontal lobe promotes anger rumination (Kelley et al. 2013), revealing the combined role of the premotor cortex in the regulatory control over behavior, cognition, and emotion.

Distractibility: Frontal Eye Fields

Presentation of discrete visual events can generate activity in the superior colliculus within as little as 40 ms (Guitton 1992) and electrical stimulation of the superior colliculus establishes an ipsilateral saccade response with minimal delay on the order of just 20 ms (Sparks 1986). Saccadic eye movements are rapid reflexive visuomotor responses reaching stereotypical velocities of 900° per second (Wurtz and Goldberg 1989) often triggered by the abrupt appearance of a visual stimulus within the peripheral visual fields. These saccadic eye movements provide for the relocation of the peripheral challenge onto the foveal area of the retina for improved acuity and perceptual analysis (Carpenter and Reddi 2000). Frontal lobe regulatory control and decision-making processes which modulate this response are largely derived from the frontal eye fields (Holmes 1938; Guitton et al. 1985; Walker et al. 1998; Suzuki and Gottlieb 2013) with downregulation of a rightward saccade by the ipsilateral right frontal eye field and with downregulation or inhibition of a leftward saccade by the ipsilateral left frontal eye field. In contrast, intentional direction of left gaze is instrumented by the contralateral right frontal eye field, whereas intentional rightward gaze is instrumented by the contralateral left frontal eye field.

Clinically, there appears to be substantial necessity for the integrity of this tissue in order to complete smooth and persistent directional eye movements to the contralateral side. Indeed, it appears that directed head movements require minimal integrity of these regions in comparisons with that necessary for directed eye movements. Thus, recovery of function, as with an intentional or motor neglect is initially approached in therapy through increments in the saliency or novelty of the sensory information originating from that hemispace to take advantage of the second functional unit and the integrity of the sensory systems within the posterior brain regions. Secondly, directional head movements toward the neglected hemispace may be functionally intact or recover prior to the ability to move the eyes or to persist in directional gaze in that direction. Thus, the head movements may be prompted and rehearsed to compensate for the motor neglect. Eventually, the therapeutic goal becomes one of persisting in lateral gaze and in smooth visual pursuit within that hemispace. By extension, the goal might be maintenance of eye contact in a gaze-aversive individual (e.g., with autism spectrum disorder). The measurement of gaze or persistence might be further analyzed within left and right hemispace or concurrent with the affective valence (positive or negative).

Frontal eye field dysfunction may be evident in the form of rapid eye movements (REM) to the right, secondary to damage in the right frontal eye field and to the left, secondary to left frontal eye field damage. Bilateral frontal eye field dysfunction may present with bilateral motor neglect and poorly regulated REM. These behavioral deficits correspond with deactivation or diminished metabolic activation within these regions. In contrast, heightened activation or metabolic increments may yield the oppositional eye movements contralateral to the side of the elevated cortical activity. This may occur clinically with seizure activity proximal to the premotor or prefrontal regions and with the intent for lateralized eye movements away from the seizure focus. But, the location along the homunculus extending across these areas is relevant, where activation of the ventral components of the premotor or prefrontal regions may be evidenced in directional jaw movements as with a pulling of the mandible in a direction opposite to the basal frontal or orbitofrontal seizure.

In contrast to the cortical regions responsible for directional gaze and visual pursuit, a lesion of the dopaminergic system (e.g., see Suzuki and Gottlieb 2013) within the substantia nigra and/or basal ganglia may produce depletion in the initiation of eye movements. This might be expressed more broadly in the Parkinsonian patient, for example, with masked faces, locked gaze, sparse blink rate, and bradykinesia (Karson 1983; see also Bologna et al. 2012). But, ultimately, the presenting features may vary as a function of the Parkinsonian syndrome with right hemibody onset featuring left frontal dysfunction as previously described and with left hemibody onset yielding alternative features consistent with a right frontal syndrome (Foster et al. 2010; Foster et al. 2010, 2011). Right hemibody onset of Parkinson's disease appears to be associated with more severe depressive and anxiety symptoms, but only when disease duration is considered (Foster et al. 2013). Thus, the extent of pathology may be relevant in the appreciation of the functional features in the patient's presentation consistent with cerebral lateralization.

Cerebral balance theory may be relevant to the interpretation of the positional location of the head and eyes in the evaluation of a patient with a frontal lobe disorder. Relative activation of one or the other frontal lobe may be the initial clue in the assessment with directional gaze bias to the left, secondary to relative right frontal activation and with directional gaze bias to the right with relative activation of the left frontal region. Furthermore, head, neck, or truncal rotation about the neuraxis may present toward an ipsilesional position. More directly stated, the left frontal lobe appears *intent* on interacting toward and within the right hemispace, whereas the right frontal lobe appears *intent* to interface with the left hemispace. Intentional bias may be manipulated through demands reflecting the associated specializations of either brain. For example, time urgency or rapid pacing demands on the game clock yields a rightward hemispatial bias in soccer goalies' approach to the ball just as does approach for a romantic kiss (Roskes et al. 2011). Appetitive behavior yields a rightward bias on approach (Harmon-Jones and Gable 2009) just as predatory behavior might yield a leftward directional bias (see Ghirlanda and Vallortigara 2004). This may generalize to other species with some 77% of walruses having a right flipper preference when feeding (Levermann et al. 2003), whereas toads are reported to have a left forelimb feeding bias (Sovrano 2007; see also Corbalis 2010). Indeed, empirical research shows that lateralization can affect both when predators are detected (Lippolis et al. 2002) and the direction in which prey tends to escape (Cantalupo et al. 1995).

Lateralized emotional processing bias has been demonstrated across species and across multiple categories or dimensional aspects of the stimuli being processed. It has been detected in the intentional responses to the event(s) evident in directional response bias and it has been found within the perceptual analyzers of the second functional unit. Even the common QWERTY keyboard has been used to demonstrate directional emotional processing bias as a function of positional location of the keys within the left or the right side of the keyboard (Jasmin and Casanto 2012). These researchers found that words spelled with more letters on the right side of the keyboard, and others with more letters on the left side, influence our evaluations of the emotional valence of the words. This relationship was significant across English, Spanish, and Dutch languages, with words containing more right-sided letters rated as more positive in valence than words with more left-sided letters.

However, multiple brain regions may participate in these positional biases and ultimately in the postural placements of our body parts. The parietal eye fields appear to be organized along gaze-centered coordinates such that goal-related activity (e.g., reaching for an object) is remapped when the eyes move (Medendorp et al. 2003). This might be expressed by the description of an individual with a right parietal lesion where the patient has diminished awareness of body part location within extrapersonal space and presents with a *proprioceptive* torticollis (Harrison et al. 1985). Alternatively, a *spasmodic* torticollis may be reflected in rotation of the head about the neuraxis and often with the spasm in the sternocleidomastoid region secondary to an upper motor neuron or combined extrapyramidal motor lesion. The therapeutic goal may initially be that the patient will maintain the head, neck, and truck upright and at midline. But, the intervention modality and the progression of therapeutic techniques might better be derived from the integrity of surrounding brain regions. Moreover, the spasmodic features may be approached in therapy somewhat like the dystonia seen elsewhere in the body, through ranging techniques and exertional therapy, to strengthen the oppositional antagonistic muscle group(s).

The broader role of the frontal regions in the regulatory control or inhibition of reflexes may be demonstrated within this system. Brain-stem reflexes, through pathways involving the pons at the brain-stem tegmentum (floor) and the superior colliculi at the brain-stem tectum (roof), provide for an ipsilateral saccade to a provocative visual stimulus at the left or the right visual field. By way of example, the birdbrain consists largely of the superior colliculus, which may directly facilitate the orienting response to find the worm. The right frontal eye field activates not only for an intentional leftward gaze but also for the inhibition of an ipsilateral reflexive saccade (however, see Suzuki and Gottlieb 2013). This system is intimate to any discussion within the context of distractibility or selective attention. The role of the frontal lobes in the suppression of reflexive ocular behavior was appreciated early on (Holmes 1938; Guitton et al. 1985; Walker, Husain et al. 1998). Decreased activation or a functional lesion within the right frontal eye field may release an ipsilateral right gaze bias and concurrently disinhibit the homologous left frontal eye field promoting rightward-directed gaze. Moreover, it may release the ipsilateral saccadic reflex originating within brain-stem regions, including the right tectum at the superior colliculus (e.g., see Suzuki and Gottlieb 2013).

This lesion, though, implies relative distractibility toward the right or the side ipsilateral to the lesion. Thus, right frontal lobe deactivation in a child may present with heightened distractibility to the right hemispace with an inability to override the saccadic reflexes to right-sided visual events. With the appreciation of the asymmetry of brain function, a more refined assessment may be conducted, including the recognition that the treatment for distractibility may be restricted to the lateralization of the brain disorder. This child might benefit from the provision of minimal distraction within the right hemispace, perhaps by placing the child in the classroom setting adjacent to the wall at her right. Moreover, therapeutic intervention would involve exertional eye movements toward and within left hemispace, again facilitated by placement against the wall at the child's right side. The therapy might be more confrontational perhaps, through visual motor exercises with lateral pursuit and persistence components built into this technique. The focus, in this example, is on intentional and motor systems. The understanding of the integrity of this functional unit will partially depend on the use of double dissociation techniques in the neuropsychological evaluation (Teuber 1955; Kinsbourne 1971), which might establish functional integrity of the sensory modalities represented within the second functional unit of each cerebral hemisphere.

A potential consideration for any task demanding the allocation of metabolic resources or *activation* of the regulatory brain systems of the frontal region is that the task itself is a stressor, by definition, for that brain region (Kinsbourne and Cook 1971; Kinsbourne and Hicks 1978; Harrison 1991). If the cellular network is contributing to ongoing respiratory function or, perhaps, to the regulatory control over sympathetic activation and anger, then the dual-tasking demands of the therapy task may result in deregulation or decompensation of these secondary functions. If this is expected and monitored for, with caution, then all may go well during the intervention. However, if the frontal lobe stressor decompensates the regulation of other functions and there is naïve appreciation of the dual concurrent task demands, the outcome may be negative or costly in some unexpected way.

Delayed-Response Deficits

Cerebral asymmetry exists in the opposing functions of rapidly initiating movement or rapidly stopping movement, with these functions being implemented by a prefrontal basal ganglia network including the presupplementary motor area, the inferior frontal gyrus, and the basal ganglia (Malmo 1942; see Aron et al. 2007b; Chambers et al. 2009; Hampshire et al. 2010; Chikazoe 2010; see also Cai et al. 2012; see also Oguri et al. 2013). Impairment in right frontal lobe function is implicated in clinical presentations of delayed response deficits involving initiation without adequate delay or in "go–no-go" situations where implementation of a "no-go" response is required, fundamentally derived from social rules and/or pragmatics. Functional magnetic imaging studies (e.g., Aron and Poldrack 2006; Aron et al. 2007a; Boecker et al. 2010) and neurophysiological studies (e.g., Swann et al. 2012) show activation in the right presupplementary motor area which increases on trials involving the successful cancellation of a prepotent response prior to and during response inhibition or stopping. Also, studies of epilepsy patients and Parkinson's patients using macroelectrode stimulation techniques have demonstrated an arrest of ongoing vocal or manual movements with stimulation of the presupplementary motor area (e.g., Luders et al. 1988; Swann et al. 2012).

Bechara (2004) suggests that there may be several mechanisms of impulse control or response inhibition that can be measured by different tasks and attributed to different neural regions. These different mechanisms of impulse control include motor impulsiveness or initiation without adequate delay. This has been proposed in humans to have several forms (Evenden 1999) including impulsive preparation, which involves making a response before all the necessary information has been obtained, and impulsive execution, which involves quick action without thinking (Evenden 1999). Moreover, Bechara (2004) draws a distinction between two subtypes of motor impulsiveness based on functional neuroimaging and clinical lesion studies. These consist of motor impulsiveness of a nonaffective nature, reflecting an inability to inhibit a prepotent response that is outside of the emotional domains. Stuss and colleagues have presented performance findings on the Stroop Test from a large number of patients with lesion of the prefrontal cortex (Stuss et al. 2001), which indicate that this general area (cingulate and mesial aspect of the superior frontal gyrus) may be critical for the type of response inhibition required by the Stroop.

Secondly, emotional impulsivity reflects an inability to inhibit a prepotent response that is affective in nature, perhaps based upon previously acquired reward associations. Bechara (2004) suggests that go/no-go tasks, delayed alternation tasks, and reversal learning tasks are "prime examples of paradigms that measure this type of behavioral control." Beyond the assessment of motor impulsiveness, the author suggests that there may be a similar mechanism at the "thought" or "short-term memory" level, reflecting an inability to inhibit a recurrent thought held in working memory. Perseveration on the Wisconsin Card Sorting Task (WCST) and an inability to shift attentional sets provide reasonable measures of this type of deficit in impulse control (Dias et al. 1996, Milner 1963). Bechara (2004) argues for the importance of the lateral frontal and anterior insular cortices as critical for the regulatory control over this type of impulsiveness, based on functional neuroimaging (Konishi et al. 1999; Lombardi et al. 1999) as well as clinical lesion studies (Anderson et al. 1991; Milner 1963) using the WCST.

Delayed-response deficits also reflect a temporal component that has gone awry. The left hemisphere appears specialized for the rapid sequential analysis, comprehension, and expression of information as might occur with linguistic or propositional speech processing or with the implementation and use of familiar tools. Relative left cerebral activation and a right hemispatial bias are expected in approach-motivated activities and time-urgent demands requiring rapid pacing. The right cerebral hemisphere appears specialized for appreciating the need for caution and perhaps more holistic analysis of the perceptual array. These differences may be expressed more directly in the nature of delayed-response deficits resulting from the relative activation of one or the other frontal lobe. These may present in the brain-damaged patient, where right frontal deactivation may leave the patient with a preponderance of rapid or ballistic initiation of behavior afforded by the left frontal region (left frontal release).

Many people refer to such behavioral initiation as "impulsive," though this language is not clearly definable, and lends itself to nonscientific origins. Instead, the delayed-response deficit more common after right frontal or orbitofrontal injury is one of initiation without adequate delay or the failed imposition of a time delay necessary to properly and discretely organize the behavioral, cognitive, or affective display (see Oades 1998; see also Thompson-Schill et al. 2005; see also Emond et al. 2009; see also Granacher 2008; Helfinstein and Poldrack 2012). Thus, the behavior may appear reactive, disorganized, or abrupt and may have unexpected consequences, including safety management issues and the like. Alternatively, through the initiation of a rapidly paced ballistic response, the individual may poorly manage the social or emotional context and appear out of control or poorly regulated in some respect, including the poor implementation of rule-regulated behaviors lacking social propriety or pragmatics (see Granacher 2008; see also Anderson et al. 1999; Eslinger, Flaherty-Craig and Benton 2004; see also Yeates et al. 2012; see also Helfinstein & Poldrack 2012; see also Damasio and Anderson 2003; Damasio et al. 2012; see also Hu and Harrison 2013b; see also Keenan et al. 2003; see also Tompkins 2012). Others note the associated insensitivity to future consequences of behavior subsequent to prefrontal damage (e.g., Bechara et al. 1994).

A common complaint of the individual with left frontal deactivation or damage is "my get up and go got up and went." This may be most clearly the case with decremental loss of the dopaminergic areas of the left frontal lobe, where behavioral slowing or inertia is often dense and with presenting features of chronic fatigue. More severe dysfunction in this region may leave the patient on a substantially different time basis for interacting with others and with the contextual events within the world, beyond the social array. One might appreciate the temporal disconnection on observation of a patient within a hospital setting, for example, where staff might ask the patient if he would like some food or a treat, perhaps. The staff member, with a busy schedule and many to care for, then leaves the room followed finally by the patient's response to the question, "yes." But, alas no one is there to hear the response. This disconnect may be debilitating and deeply frustrating for the patient with behavioral slowing or bradykinesia. It seems as if the entire world is working in a different temporal dimension, lending the patient to increasing social isolation, and aggravating a dysphoric disposition with sparsity for positive affect or pleasure in life. Even milder forms of initiation deficits in speech may promote increased anxiety and lowered self-esteem. This point resonates with evidence showing relative deactivation of the left frontal region and right frontal activation with anxiety.

Substantial scientific evidence relates the affective dimension of happiness, high levels of self-esteem, and energetic initiation fundamentally derived from dopaminergic or left basal ganglia brain systems and the left frontal region more broadly defined. Yet, the most overlooked construct, altered by pathology in this region, is simply the time base derived from these neural networks (e.g., Holland et al. 2007, 2009; see Allman and Meck 2012). This time base may run fast with relative right frontal deactivation and with right frontal/orbitofrontal pathology. Release of the left brain, via right frontal damage, reveals somewhat of a "happy-go-lucky" brain, devoid of caution or appreciation of social proprieties or concerns, as it seems to fail to appreciate risk within the external threat setting. This might even be of vital import in urgent settings, where we must "throw caution to the wind." With its demise, we see the loss of energy, a slow temporal basis for initiation, social apathy (see Scott and Schoenberg 2011; see also Grossman 2002), increasing caution, and lowered self-esteem. Relatedly, Heath Demaree (Demaree et al. 2005) presented the theoretical position incorporating *dominance,* where chronic relative left and right frontal activation would be associated with feelings of dominance and submission, respectively.

There exists a relationship between these altered characteristics of the individual suffering a brain disorder and the attributions of those family members and health-care professionals working with the patient, which is often emotionally negative. The slow, low-energy level, left frontal lobe patient is described by others as lacking motivation or active and fully engaged participation in the recovery or rehabilitation efforts. The attributions reflect negative intonations that the patient "wants me to do everything for him." It may be useful for the caregiver or staff member to appreciate the mirroring that may occur in response to the social encounter with someone, expressing diminished or compromised integrity for these brain regions. It may also help to physically provide an altered time base to cue the caregiver for the distinct differences between the patient's and their own time base. This might result from sample time estimation procedures (e.g., Holland et al. 2010) to empirically determine the extent of slowing in the patient. Providing additional time may require the patience of a saint. However, it may at least prompt and remind others of this fundamental processing difference, with the hope of reducing frustration on everyone's part.

The homologous frontal regions of the right brain normally provide regulatory control to implement slow and cautious rule-regulated behaviors, including social proprieties, control over negative affective displays, and pragmatics. Others note the associated insensitivity to future consequences of behavior subsequent to prefrontal damage (e.g., Bechara et al. 1994). When lesioned, the patient may show left frontal release symptoms corresponding with the now dominant position of the intact left frontal systems, charged with heightened energy levels and rapid processing speed free of the caution and slowing mechanisms of the lesioned right frontal lobe. This time-base disorder, then, is one of rapid, ballistic initiation. But, the patient's attributions are fundamentally derived from their perceptual appreciation of elapsed time. One of the early discoveries associated with hostility and rage-related behavior, but also with cardiovascular reactivity and cardiovascular disease, was the temporal construct of *time urgency*. With diminished right frontal capacity, the individual is set apart from others in the domain of time management and estimation of elapsed time, where the clocking mechanisms are accelerated. Thus, the perception is that more time has elapsed than might be realistic or accurate when held to the standards of a calibrated timepiece. Many of us have appreciated this relationship between time urgency and anger management deficits at the traffic intersection, where the angry persons behind us are honking their horn having perceived that our initiation was substantially behind their schedule.

Many brain-injured patients have been evaluated for urinary incontinence on the hospital ward. When the nursing staff or others charged with patient's care are interviewed, attributions are commonly expressed that this individual may be doing this *"intentionally,"* a clue to the examiner that frontal systems may be involved. The assessment involves double dissociation techniques to determine whether or not the patient has urgency or sensation signaling the need to void. Somatosensory sensitivity, reception, and comprehension of bodily signals indicating a filling bladder are very helpful and support an improved prognosis as might be expected. Once the integrity of these systems is established, the examiner may evaluate the third functional unit for regulatory control over bladder reflexes and voiding. But, most relevant to the discussion at hand are the temporal processing constraints and attributions of the patient and, especially, with evidence for dysfunctional right frontal mechanisms. This patient might describe their attributions on interview, stating that "I pushed the call bell and I waited and waited and waited." "No one came." "So I waited some more." "Finally, I said to hell with them and (voided) on the bed!"

Assessing the individual for time estimation may reveal an accelerated clock and the failure to fully appreciate the personal relevance of social proprieties or social rules. Others note the associated insensitivity to future consequences of behavior subsequent to prefrontal damage (e.g., Bechara et al. 1994). This may establish the foundational basis for subsequent interventions relevant to the temporal contingencies involved and the interindividual discrepancies (nurse vs. patient; or caregiver vs. patient) which promote discord or poor communication. The patient with an accelerated clock may derive attributions that she/he has waited for longer than she/he should have been asked to wait. Control issues may be promoted as the dominant theme, with negative attributions toward the staff for their response to expressed needs for care. Education of the caregiver or family member of these temporal management problems, through the provision of supporting time estimation data, can sometimes be helpful and diffuse the situation a bit. The integrity of the patient's verbal acknowledgment of the rules, though, against a backdrop of rule violations, supports another literature here on social anarchy, resulting from right prefrontal damage (see Beer 2006, 2007; Beer et al. 2006). Similar findings exist in the literature on the ventrolateral prefrontal region or orbitofrontal cortex using nonhuman primates (e.g., Agustín-Pavón et al. 2012).

It might be useful for the family member or caregiver to appreciate that stressors specific to these right frontal brain regions may further decompensate the patient and amplify these management problems. Stressors, which may be expected to substantially decompensate someone with diminished right frontal capacity, include anger or fear provocation; the perception of control issues or being forced to do something; elevated prosody, volume, or "motherese" (Bunce and Harrison 1991); and implied or directed threats in the broadest sense of the word. These may include negative facial expressions or tone of voice as conveyed through others' prosodic expressions or the conveyance of dislike through abrupt mannerisms or expressions. Active right frontal stressors also include those yielding elevations in sympathetic tone as with incremental systolic blood pressure, heart rate, sweating, and/or facial engorgement and this may include those changes derived from exertional activities. Moreover, the dual-tasking demands of a noisy environment conveying multiple distractions and negative affectivity of others may decompensate these systems and accelerate the clock.

The clock may be altered through perseverative inertia for some. Many individuals try to cope with disorganization syndromes and rapid pacing or hyperkinesis from right frontal brain disorders. Many have decompensated in the performance of behavioral chains and where, initially, the task performance was functional but where the activities accelerated on repetition to the point of being temporally out of control. One housewife comes to mind who would initiate house cleaning with everything seemingly under control. But, as she continued on these exertional activities, she would be going faster and faster. This would continue until she was rapidly pacing with elevated heart rate, sweating, and the appearance of being compulsively out of control. She improved with the implementation of frequent relaxing breaks or time-out periods in order to return to baseline levels. Indeed, she seemed capable of managing up to 5 min of house cleaning at a time through this approach. She would also benefit from interventions designed to promote a quiescent state of parasympathetic dominance, including deep breathing and progressive muscle relaxation techniques. She preferred environmental design and modified her architecture with garden spaces inside and outside of her house, which she had fitted with a Koi pond and pleasant waterfalls.

Clocking mechanisms have been successfully manipulated even as a function of the concurrent performance on standardized neuropsychological testing involving designs or figural fluency in contrast to verbal fluency tasking. Initially, the predictions involved a test of functional capacity theory (see Klineburger and Harrison 2013) where the dual task demands over the left or the right frontal region would reduce the regulatory control over the time-based estimate. Thus, Kate Holland was able to accelerate the time base with the concurrent performance of the figural fluency test and, to decelerate the time base with the concurrent performance of the verbal fluency test (Holland et al. 2007; see also Holland et al. 2010). These diametrically opposite effects on time estimation or temporal processing are reminiscent of the diametrically opposite effects recorded in heart rate, systolic blood pressure, and skin conductance as a function of left or of right frontal task demands. Specifically, verbal fluency demands, positive affective learning, and happy facial expressions have lowered these indices of sympathetic drive, whereas figural fluency, negative affective learning, and angry facial expressions have increased these measures (Herridge et al. 2004; Mollet and Harrison 2007a, b; Williamson and Harrison 2003; Holland et al. 2011).

Interestingly, the evidence for cerebral asymmetry differences, where right frontal pathology relates to initiating abruptly or "impulsively" and where left frontal pathology relates to behavioral slowing or bradykinesia and amotivational states, is also evident in research on neurotransmitter systems (see Welberg 2013). Serotonergic activation in the raphe nucleus has been related to the ability to wait for delayed rewards in rats. Miyazaki et al. (2012) used a laboratory task requiring rats to remain in a fixed posture at a reward site to receive food or water reward. Serotonergic inhibition at the dorsal raphe nucleus resulted in delayed response deficits where the rats failed to wait for a reward if there was a longer delay interval. In contrast, no effect was appreciated under conditions in which the delay was short. Also, the authors concluded that the inhibition of the raphe site did not affect cognitive or motor functions. The authors concluded that activation of serotonin neurons is essential for delay in initiating reward behaviors.

Also, Haleem (2012) concludes from review of the literature that decreases in serotonin neurotransmission at postsynaptic sites (subsequent to dietary restriction) lead to hyperactivity, depression, and behavioral impulsivity. A meta-analytic review of the literature (Baglioni et al. 2011) revealed a link between depression and elevated arousal with insomnia. This evidence is consistent with potentially over-lapping areas of research showing lateralization of serotonin and/or heightened se-rotonergic responsiveness within right cerebral systems (e.g., see Fitzgerald 2012). Furthermore, sex differences appear to exist in temporal processing speed. Sex dif-ferences in neurocognitive performance have repeatedly been reported in adults and in adolescence, with a female superiority in processing speed (Camarata and Wood-cock 2006; Waber, De Moor et al. 2007) and a male superiority in perceptual analy-sis and working memory (Waber et al. 2007; Reynolds et al. 2008). Van Deurzen et al. (2012) investigated the relationship between response time and depression, providing evidence that, in girls, enhanced response time variability is associated with affective problems. These authors further note the potential relevance of these findings, since twice as many women than men are found to suffer from depression and a gender difference starts to emerge in adolescence (Hankin and Abramson 1999; Angold and Costello 2006).

Hypokinesis and Mutism: Supplementary Motor

Earlier on, the basic speech systems were discussed, including the neural archi-tecture interfacing Broca's and Wernicke's areas via the arcuate fasciculus. This functional anatomy has been found to relate to expressed speech fluency and to receptive speech comprehension and semantic relevance. Perhaps one of the most interesting areas within the frontal lobes is the supplementary motor cortex (see Fig. 14.18) overlying the cingulate gyrus and the behavioral effects resulting from lesion, incapacity, or destruction of this tissue. The supplementary motor region is found on the midline above the anterior cingulate gyrus and within the distribution of the anterior cerebral artery, substantially removed from the traditional speech regions both in cortical and in vascular anatomy. However, lesion of the left or bilat-eral supplementary motor cortex often presents with mutism (however, see Fontaine et al. 2002), followed perhaps by recovery to soft, hoarse, whispered speech. This occurs with akinesis at its worst, often recovering to hypokinesis over time.

Patients suffering from akinetic mutism are characterized by unresponsiveness contrasting with the superficial appearance of alertness. The eyes are open, appar-ently following the examiner, but without speech or movement and without com-munication with the examiner or others. Poppen (1939) and Dandy (1946) provided reports on early cases of akinetic mutism following bilateral anterior cerebral artery ablation. Another patient was described by Skultety (1968) subsequent to a right anterior cerebral artery clipping to remove a falx meningioma but with probable thrombosis at the left anterior cerebral artery distribution (see Freemon 1971). The left anterior cerebral artery distribution appears to be a requisite area for pathology with lesion overlay affecting the supplementary motor cortex and often including the left cingulate gyrus and/or the basal ganglia (e.g., Fontaine et al. 2002; see also Duffau 2012).

One adolescent boy was seen after a "suicide attempt" with a bullet entering the left midsagittal region of the skull from above. The neuropsychologist was a bit suspicious of the entry wound from above and at the left side of his head as he was a right-handed male and he remained apprehensive on the unit. The lesion extended down through the supplementary motor region and through the cingulate gyrus rendering him akinetic/hypokinetic and essentially mute. He was effectively unable to share his circumstances with the team. But, over time, his fear of the mother's boyfriend was revealing. He was eventually able to confirm that he had been shot, while on his knees, as his speech recovered to a soft, hoarse whisper.

The evaluation of the supplementary motor region will overlap with brain areas critical in the assessment of the arousal systems, including the underlying anterior cingulate gyrus associated with the classic startle response (e.g., Pissiota et al. 2003; Levenson et al. 2012) and the dorsomedial thalamocortical projections. Neuroanatomical studies of the cortical modulation of the startle reflex demonstrate that exaggeration of the startle reflex arises with loss or diminished capacity of the "top-down" inhibitory or regulatory functions of the anterior cingulate cortex on either the reflexive brain-stem startle circuit or on the amygdale or both (Yeomans and Frankland1995; Sanchez-Navarro et al. 2005; Thayer and Brosschot 2005; Klineburger and Harrison 2012). The cingulate and supplementary motor cortices are probable sites processing extreme arousal-related events, including those providing a release of reflexive freezing responses and the Pavlovian-conditioned emotional response or CER. The supplementary motor cortex has been linked to the performance or initiation of motor sequences from memory rather than the performance of movements guided by a visual cue. This system appears to prefer a specific sequence of movements that have been learned. In this sense, it plays a prominent role in the initiation of actions under internal volitional control, where the memory sequence is established, rather than being responsive, to external stimulus-driven cues (Shima and Tanji 1998).

A young woman was evaluated initially in a stuporous state with akinesis and mutism. She had answered the doorbell at her home with her baby in her arms when her boy friend placed a small caliber revolver between her eyes and extirpated the supplementary motor cortex. The family expressed concerns about her inability to speak. But, the prognosis at that point for speech was rather good, since she had no evidence of direct trauma to the classic speech systems. The neuropsychologist expressed optimism for her recovery to soft, whispered speech. But, the prognosis for her lower extremities was less optimistic with ablation of the dorsomedial motor and premotor cortices. She was a country music enthusiast. She confirmed her comprehension of speech and her preference for entertainment as she initiated a bit more with the music. She gradually recovered meaningful, but with low volume, hoarse-whispered speech and remained hypokinetic over the 6-month period that she was followed.

To the diagnostician and therapist alike, the appreciation of the proximal functional systems with the bladder, genitalia, and contralateral lower extremity may be useful. Also, the evaluation is relevant to the startle response, the freezing response, and tonic immobility with proximity or overlay on the cingulate gyrus.

The prognosis will vary as a function of the lesion overlay on the medial left basal ganglia where social rewards, initiation, and energy level may be affected (see Scott and Schoenberg 2011; see also Grossman 2002). Therapeutic intervention might manipulate the arousal systems with substantial controversy over the potential for hyperarousal with a hypokinetic state. But my first intervention efforts with akinetic mutism as a postdoctoral fellow were more effective with arousal reduction therapies, perhaps.

Another young man was a recent university graduate found hanging, with pants down, from the ceiling beam. He was initially diagnosed with an anoxic encephalopathy but, following the completion of the neuropsychological evaluation, his secondary diagnosis was akinetic mutism. He recovered slowly at a regional head injury facility where, at follow up, he was found to be rotational in his walking pattern possibly indicating lateralization in his cerebral activation patterns. Relaxation techniques and, earlier on, hypnosis had sometimes been used successfully as an intervention for mutism with theoretical underpinnings involving arousal mechanisms (hyperarousal vs. hypoarousal states). Some evidence has been in support of a stuporous arousal state with this syndrome, whereas diametrically opposite theoretical interpretations have received some support. Following relaxation techniques and a gentle massage, this man who appeared stuporous, was now conversing fluently. He would also wake up reliably alert, verbal, and seemingly normal asking if "...NC State really won the national basketball championship." This happened in 1984; the year of his anoxic event.

Mutism might also be present within the developmental context of infancy and childhood. A beautiful 8-year-old girl was brought in by her parents with concerns that "she doesn't really seem to talk." She appeared normal in every respect with loving parents and a pleasant demeanor. The interview revealed problems with bladder control and bedwetting. She was a bit hypoactive and was unsteady at her lower extremities. She was impersistent on motor tasking and vocalization. Her speech was expressed as a soft, low-volume whisper. But, she was essentially mute. Palpation of her head, along with the underlying symptoms, led to the discovery of a neural tube defect with an incomplete closure over the midsagittal frontal region. The family was referred to a neurosurgeon and they opted to close the previously undiscovered skull defect.

Another possible developmental mutism was evident in the 16 April 2007 killings at Virginia Tech, perhaps. Seung-Hui Cho was diagnosed with mutism in middle school, which was associated with severe anxiety and major depressive disorder. He received treatment and continued to receive therapy and special education support until his junior year of high school. At Virginia Tech, it might be reasonable that some may have noted his mutism as a lack of participation. In retrospect, and with the benefit of hindsight, the social isolation and performance deficits from this disorder may have fueled the flame of passionate anger within this young man, subsequent to public embarrassment and having "lost face" in front of his peers.

Less is known of the functions of the right supplementary motor region (e.g., Heilman et al. 2004). However, patients generally show an alteration of pitch in speech toward the higher frequencies and staccato speech with lesion extension into the subcortical regions. Moreover, deactivation of these regions more realistically relates to hyperkinesis and a rapid or ballistic initiation style. Although hyperkinesis looks much different in an elderly person in comparison with that seen in early childhood, the basic syndrome appears at least similar. The therapists and caregiver may appreciate this, as the patient may be a management issue in the office, in the therapy setting, and in the hospital room where safety is a concern.

One hyperkinetic elderly woman, following recovery from her right anterior cerebral artery cerebrovascular accident, was delighted with her discharge and return to home, having never fully appreciated her stroke or functional deficits in the first place. Upon return to home, she prepared the first meal for her family as she made 40 bologna sandwiches for her husband and adult son. By her husband's accounting of this episode, her son laughed! The father said that he made him eat all of the remaining sandwiches over the next several days!

Part V
Diagnostic Nomenclature

Chapter 15
Neuropsychopathology

Some psychologists and psychiatrists have been slow, or even at times reticent, to embrace the neuroscience findings providing for the anatomical bases of emotion, cognition, and behavior. Others have been equally oppositional to the dynamic responses and structural changes within and between neural systems as a function of experience and environmental influences. Much of the language, including the diagnostic terminology within the field, was developed prior to the appreciation of these discoveries and reflects substantial heterogeneity of variance of brain functions. For example, the term depression may encompass multiple brain disorders, involving disparate areas of the brain (Crews and Harrison 1995; Shenal et al. 2003; Crews et al. 1999; Bruder et al. 2012). Even the more robust literatures on brain pathology have been ignored until recently, for example, including the dementias, dysphasias, and dyslexias.

Allen Frances, the chairperson for the Diagnostic and Statistical Manual of Mental Disorders, Fourth Edition Task Force (DSM-IV: American Psychiatric Association 2000), presided over an enormous expansion of the "mental disorder" vocabulary. Yet, the inherent flaws of this ever-expanding language base were summarized by Frances in plain language. "There is no definition of a mental disorder. It's bullshit. I mean, you just can't define it" (Greenberg 2011; see also Kirk et al. 2013). With specific relevance to the new Diagnostic and Statistical Manual of Mental Disorders, Fifth Edition (DSM-V; American Psychiatric Association 2013), he charges that it lacks sufficient scientific support, defies clinical common sense, and was prepared without adequate consideration of risk–benefit ratios and the economic cost of expanding the reach of psychiatry (Science Codex, May 20, 2013). Dan Blazer (2013) of Duke University, who served on the DSM-V task force, states that "we're basically drawing artificial lines, and the body and the mind do not work like that." Revolution is at hand when archaic language and diagnostic categories defy more specific understandings of the nature and location of components of the disorders as understood within functional neural systems theory.

The fundamental problem is captured to some extent by Damasio (1994, pp. 40). He states: "The distinction between disorders of 'brain' and 'mind,' between

© Springer International Publishing Switzerland 2015
D. W. Harrison, *Brain Asymmetry and Neural Systems,*
DOI 10.1007/978-3-319-13069-9_15

'neurological' problems and 'psychological' or 'psychiatric' ones, is an unfortunate cultural inheritance that permeates society and medicine. It reflects a basic ignorance of the relation between brain and mind. Diseases of the brain are seen as tragedies visited on people who cannot be blamed for their condition, while diseases of the mind, especially those that affect conduct and emotion, are seen as social inconveniences for which sufferers have much to answer. Individuals are to be blamed for their character flaws, defective emotional modulation, and so on; lack of willpower is supposed to be the primary problem."

These problems may be accentuated with the layperson's take on "psychological" problems as something separate from the disorders of the body and where the onus is on the individual as a faulty person or, worse still, a lunatic or nut of psychiatric proportions. Many individuals experience hallucinations or formesthesias as discussed earlier in these writings. Knowing the functional neuroanatomy of the sensory system affected and the laterality of the projections and the emotional valence of the event may be critical in providing a framework or rationale for the patient to feel comfortable in revealing these "secrets."The historical attributions within the field were disturbing and relevant to competency as an individual, rather than to the diagnostic utility afforded and its relevance to therapeutic intervention—other than with antipsychotic medication. These individuals may achieve solace in the knowledge that the brain cells are alive and overactive as opposed to the psychological attributions. They may also be comforted in the statistical basis where about one in five patients experience these, at least temporarily, following a stroke (Walters et al. 2006).

Thomas Insel, director of the National Institute of Mental Health, largely dismissed the current language used in the DSM- V, for lacking "validity" and for its foundational bases drawn from opinions rather than from science. The diagnostic manual was described as "a dictionary, creating a set of labels and defining each.... Patients with mental disorders deserve better." he writes in a blog (see Earley 2013, USA Today, May 24, 2013). Under Insel's leadership, the National Institute of Mental Health will no longer attend to the Diagnostic and Statistical Manual of Mental Disorders as it allocates monies for scientific research. Insel clarified the position of the National Institute of Mental Health further, stating that "mental disorders are biological disorders involving brain circuits." This position statement initially seems to exclude environmental activation of biological circuits. However, from this foundational view, the institute will focus its support of scientific research, and language, on identifying the neural systems responsible for thoughts, emotions, and behaviors and causal mechanisms, fundamental to the environmental array and experiences, which alter their functional integrity.

The purpose of the present section is not to provide a functional analysis of all of the varied disorders described by language systems not directly based on brain function. Instead, it is encouraged that problems ascribed to "psychopathology" be held to higher standards, where conceptual arguments and pathological constructs are based on neural systems theory. This perspective does not require brain damage for the expression of psychopathology. Rather, the stress of life and learning history are conceived within the functional neural system processing the demands or requirements of that specific stressor and where the capacity for processing may be inadequate with altered metabolic processes within these systems. From this perspective, alternative attributions might better be rendered to understand anxiety, depression, and fearful states.

This is evident for the disorders of diminished emotional sensitivity found with deactivation within the right parietal systems (Heller 1990; Heller et al. 1998; Heilman et al. 1978; Heilman and Bowers 1990; Bruder et al. 2012). It is also be evident in the stress-related disorders, like that with social anxiety and depression (Everhart et al. 2002; Crews et al. 1995); with affective disorders like hostility; with the posttraumatic disorders (Rhodes et al. 2013; Demaree et al. 2002); and even with childhood learning disability (Huntzinger and Harrison 1992) and childhood depression (Emerson et al. 2005).Also, neuroanatomical change underlies posttraumatic stress disorders, where exposure to traumatic events exceeding the functional capacity of specific circuits results in alterations and perhaps damage of the circuit(s) involved (e.g., Knutson et al. 2013; Sartory et al. 2013).

The weight of valid constructs derived from other theoretical perspectives would survive as integral to a living brain system and its environmental array, whereas the ever more complex and potentially invalid constructs might be gradually discarded. For therapeutic intervention, a look under the hood would be revealing. If, by analogy, we took our car to the mechanic and she cared not about the system involved (e.g., fuel or electrical), then a more heterogeneous diagnosis might involve "treating the whole car." Certainly, this less tailored intervention would ultimately be costly and maybe less precise (e.g., replacing the starter). Upon accepting the language of neuroscience, psychologists may have language which allows for communication with our colleagues in the other sciences, where all may benefit from a more universal language system; and one ultimately built upon anatomy. This substrate meets the demands of science, broadly defined, as it is visible, quantifiable, and subject to manipulation.

Instead, the present framework will be applied to depression by the way of example. Subsequent to this discussion, a framework of language based on functional cerebral systems will be provided. The key to this system is the appreciation of dynamic functional brain systems, wherein stress is relevant to the brain system or region where altered metabolic activation may take place and where capacity limitations may be relevant. Thus, the conceptual framework does not require a physical lesion of brain cells for a diagnosis of a pathological state. Instead, altered metabolic function or imbalance may be sufficient for a psychological problem. But, the nature of the disordered system will in many cases be identifiable within the diagnostic system derived from pathological brain function. It is reasonable to expect that this perspective may provide for specificity in the disorder and associated features derived from proximal brain regions. The labeling of a patient with the diagnosis of schizophrenia might be improved by identifying the neologisms, word salad speech, semantic paraphasic errors, poor linguistic carryover, and positive affective auditory hallucinations as a receptive dysphasia originating from the left superior posterior temporal gyrus and underlying brain stem structures.A psychotic disorder characterized by social impropriety, negative affective deregulation, perseveration on negative affective themes, social anarchy, and highly tangential emotional themes might be understood within the neuroanatomical domain of the right prefrontal region.

Earlier on, Brian Shenal and W. David Crews III provided a framework for the analysis of depression from within quadrant theory (Crews and Harrison 1995; Shenal, Harrison and Demaree 2003). However, the nature and specifics of the depression differ with the relative activation of the left and right brain and with the relative activation and integrity of the three functional units within each brain. From this perspective, left frontal incapacity or diminished metabolic rate may feature a loss of energy, diminished positive emotion, reduced self-esteem, attributions of self-blame, somatic concerns, diminished social approach behaviors, and diminished verbal fluency. Research supports that this may result from multiple origins, ranging from a series of failure experiences to altered catecholamine levels (dopamine and norepinephrine) and to loss of cellular integrity of this brain region.

Right frontal incapacity or diminished metabolic rate may feature negative emotional reactivity or lability with crying (right basal ganglia; e.g., Ross and Rush 1981; Constantini 1910; cited in Oettinger 1913; see also Parvizi et al. 2009) or anger deregulation and lability (right orbitofrontal; Fulwiler et al. 2012; Everhart, Demaree et al. 1995; Everhart and Harrison 1995, 1996; Demaree and Harrison 1996, 1997b). The later, along with anxiety or fear disorders (e.g., Agustín-Pavón et al.2012), might feature perceived external control issues or threats originating from extrapersonal space and with anosognosia or diminished appreciation of that individual's role as causal to the emotional disturbance. Research again supports that these may result from multiple origins, ranging from negative emotional stress to control issues originating from extrapersonal space, diminished serotonergic resources, and loss of the cellular integrity of this brain region.

Right posterior incapacity or diminished metabolic rate may feature emotional emptiness along with flattened or bland affect (Heilman et al. 1978; see also Heilman et al. 2012). Elevated intensity in the perception of negative emotional events might result from increments in the metabolic rate of these same brain regions, whereas the topographical specificity would relate to the sensory system affected. The disturbance would relate to arousal level and to extrapersonal space to include delusions of space and confusion for places and faces. Altered activation of the fusiform gyrus, by way of example, might correspond with gaze aversion in someone otherwise diagnosed with a more heterogeneous label of autism spectrum disorder (Hadjikhani et al. 2004).

The left posterior regions provide for psychological disturbances in body awareness, postural integrity underlying ideomotor or gestural praxis (e.g., Vingerhoets et al. 2012), and delusions of personal space. Reception or comprehension of language might be disturbed with dyslexia and dysgraphia, verbal hyperfluency, hypergraphia, and auditory verbal or somatic hallucinations. To facilitate some thought in this regard, a sample of empirically supported diagnostic categories relevant to clinical neuropsychopathology has been included in Table 15.1. It is of

Table 15.1 A sample of empirically supported diagnostic categories for clinical neuropsychopathology

Aphasia	Aprosodia	
Expressive	Expressive	
Receptive	Receptive	
Mixed	Mixed	
Global	Global	
Transcortical expressive	Transcortical expressive	
Transcortical receptive	Transcortical receptive	
Transcortical mixed	Transcortical mixed	
Alexia	Agraphia	
Expressive	Expressive	
Receptive	Receptive	
Mixed	Mixed	
Global	Global	
Transcortical expressive	Transcortical expressive	
Transcortical receptive	Transcortical receptive	
Transcortical mixed	Transcortical mixed	
	Alexia with agraphia	
	Alexia without agraphia	
Agnosia	Affective agnosia	Formesthesia dysesthesia
Visual	Visual affective	Visual
Letter	Auditory affective	Auditory
Line	Somatic affective	Tactile
Facial		Temperature
Prosopagnosia		Pain
Pattern		Vestibular
Symbol		
Auditory		
Word		
Affect		
Tactile		
Astereognosis		
Agraphesthesia		
Temperature		
Pain		
Vestibular		
Apraxia	Affective apraxia	Constructional apraxia
Gestural	Gestural	Motor
Ideational	Ideational	Sensory
Ideomotor	Ideomotor	Mixed

Table 15.1 (continued)

Mixed	Mixed	Global
Global	Global	
		Geographical confusion
		Left–right confusion

some relevance here, that earlier Diagnostic and Statistical Manuals excluded disorders of speech and language and even the dementia disorders. Without integration of findings drawn from the neurosciences, surely more delays are inevitable, along with the growing burden of new language (e.g., receptive aphasia vs. receptive speech disorder).

Part VI
Functional Brain Asymmetry

Chapter 16
Two Brains, Two Kidneys, Two Lungs: Functional Brain Asymmetry

Roger Sperry, in his 1981 Nobel lecture, summarized the prevailing view of the right hemisphere when he began studying it:

> The right hemisphere was not only mute and agraphic, but also dyslexic, word deaf and apraxic, and lacking generally in higher cognitive function.

But Sperry's research with split-brain patients was, at least for him, convincing evidence that the right hemisphere was not a mere accessory. The view that the right hemisphere is a minor, less important player (or "relatively retarded," as Sperry colorfully put it in his 1981 Nobel Laureate address published in 1982; Sperry 1982) was no longer tenable:

> Everything we have seen indicates that the surgery has left these people with two separate minds, that is, two separate spheres of consciousness. What is experienced in the right hemisphere seems to lie entirely outside the realm of awareness of the left hemisphere. This mental division has been demonstrated in regard to perception, cognition, volition, learning and memory. Sperry (1966, p. 299)

The weight and the essence of the world rest squarely on the shoulders of the right brain, as it watches over all else, including its literal and loquacious brother. There exists the functional and anatomical equivalent of two brains (Gazzaniga et al. 1962; Sperry 1966, 1982; Gazzaniga 1998, 2000; see also Springer and Deutsch 1998), much like the dual representatives of most other structures in the human body, including two kidneys, two lungs, and so forth. The two cerebral hemispheres are set on top of their respective brain stems, with each brain having its own complement of brain-stem structures. This arrangement is somewhat reminiscent of the appearance of a cauliflower with a large head on top of the stem. Several explanations for the emergence of hemispheric specializations have been proposed, including an enhancement of an individual's ability to perform two different tasks at the same time (Rogers et al. 2004), an increase in neural capacity due to an avoidance of unnecessary duplication of neural networks (Vallortigara 2006), and the greater speed of uni-hemispheric processing since no interhemispheric transfer via the corpus callosum is required (see Ocklenburg and Güntürkün 2012; Ringo et al. 1994).

© Springer International Publishing Switzerland 2015
D. W. Harrison, *Brain Asymmetry and Neural Systems*,
DOI 10.1007/978-3-319-13069-9_16

The modern view of brain lateralization draws upon the findings of clinical lesion studies derived from early work in patients with unilateral brain damage. Provocative findings of cerebral asymmetry also accrued from more recent work in split-brain patients (Gazzaniga 1998, 2000). It is no longer questionable in science that each cerebral hemisphere has become specialized for different but complementary processes that together account for aspects of behavior, cognition, and emotion. This specialization of function provides a neuroanatomical mechanism to reduce processing time and energetic expenditure associated with transmitting information over long distances. This seems more relevant from an evolutionary standpoint with increasing brain size required to accommodate newer and more adaptive functions. Moreover, this interpretation is supported by evidence that an increase in brain volume in primates is accompanied by a decrease in the relative size of the corpus callosum and anterior commissure, and an increase in local intrahemispheric circuitry (Rilling and Insel 1999; Herculano-Houzel et al. 2010; cited in Mutha et al. 2013).

Mutha et al. (2013) argue along with Gazzaniga (2000) that "the development of specialized local circuits lateralized to a single hemisphere could allow the emergence of greater behavioral complexity without incurring the cost of always coupling the two hemispheres for every aspect of neural processing." It is clear from studies of lateralization of cognitive and perceptual processes that each hemisphere contributes unique mechanisms to the control of any given function. For example, language comprehension recruits the left hemisphere for lexical, semantic, and syntactic processing, and the right hemisphere for processing its emotional and nonverbal features such as prosody (e.g., Grimshaw 1998; see also Heilman et al. 2012). Mutha et al. (2013) further note the dependence of visual perception on the synthesis of global features of a stimulus, which occurs largely in the right hemisphere, and characterization of the details of the same stimulus, which occurs primarily in the left hemisphere (e.g., Delis et al.1986; Weissman and Woldorff 2005). The distinctiveness of processing styles conveys well beyond the distinction of handedness. For example, Mutha et al. (2013) found specialized mechanisms in the functional use of each side of the body consistent with the functional specializations of the brain largely in control of the extremity. Specifically, these authors found that the left hemisphere provides predictive control mechanisms necessary for the shape and the direction of movement, while the right hemisphere stabilizes the extremity (arm) at a desired or intended position, through specifying the impedance around that position and relative to the body axis.

At this point in the history of neuroscience, several rather clear anatomical distinctions can be identified in comparisons of the left and the right brain (Amunts 2010; see also Ocklenburg and Güntürkün 2012). But these discrepancies are not easily discerned. Much of the effort at gross inspection of the nervous system early on resulted, instead, in the impression of a seeming lack of morphological distinctiveness. In these gross inspections, there were clear problems in distinguishing the structures of the left brain from those of the right brain. They look alike. More careful inspection reveals a characteristic counterclockwise torque or "petalia," which in humans is expressed with extension of the left parieto-occipital region beyond that of the right hemisphere and with extension of the right frontal region beyond that of the left hemisphere (Chiu and Damasio 1980), for example.

Beyond this, at once superficial inspection, major functional differences have been uncovered in how each brain processes sensory events from our world and how perception and expression differ between them. They do not see the world the same way. They do not feel the world the same way. They are specialized for different aspects of learning and memory. They differ remarkably in basic functions including emotions, cognition, and behavior. In the following sections, some of these differences will be discussed. To understand these differences is to understand much of the nature of various brain disorders. To better understand these two brains and the rather precarious balance between them, we may provide an improved foundation for our attributions towards others' behaviors and especially those with a disturbed equilibrium or imbalance between the two brains. Also, the activation or metabolic rate of each brain is dynamically responsive to provocation, as for example, in response to threat or with more extreme activation states such as those underlying confusion and agitation.

An elementary distinction between the two brains can be made initially in the rational, logical, and linguistic features of the left brain, which is deeply distinguished from the intuitive, spatial, and emotional features of the right brain. The weight of the world may fall differentially on the right brain and it is suited for meeting these needs and responsibilities. It is adept at distinguishing a lie from the truth in its reliance on subtle nonverbal nuance rather than what has been verbally presented in the presentation of the case (Etcoff et al. 2000). In some nonlinguistic ways, the left brain appears naïve in comparison with the intellect and capacity of the right brain, which sits more comfortably in the evolution of the mammalian and primate line. Pencil and paper tests, including questionnaires and measures of intelligence, fall short of measuring this brain in its capacity as it appreciates the gist of the situation within a heartbeat and broadly expands our depth of knowledge beyond the linguistic and rational discourse. Our intuitive capacity and nonverbal perception of the truth or the essence are the source of our most authentic "gut feelings" and our most creative capacities.

Efficient recognition of salient events signaling threat or beneficial returns to the individual is crucial to survival and these cues may be discordant with verbal dialogue or they may arise from other than human species or events, where language does not provide the relevant cues (see Larson et al. 2012). Indeed, preferential processing of potential threat cues and reward has been well demonstrated (e.g., O" hman and Mineka 2001; Gable and Harmon-Jones 2008). Beyond those contextual events conveying critical import to survival, Lorenz (1943; see Larson et al. 2012) argued that the even the configural aspects of a human infant's head and facial configuration evoke adult care. Research such as this and that of Darwin (1872/1998) and Ekman (1973; Ekman et al. 1990), which has identified universal emotional facial expressions, indicates that very elementary features convey connotations of strong survival value. Larson et al. (2012) further note that "…pioneering work by ethologists such as Tinbergen demonstrated that even in animals well down the phylogenetic hierarchy, there exist mechanisms for prioritized responding to very primitive survival-relevant cues (EiblEibesfeldt 1989)." Based on this work, Larson et al. (1990) used an implicit association test to examine emotional associations between three simple shapes (downward- and upward-pointing triangles, circles) and pleasant, unpleasant, and neutral scenes. Participants were significantly faster to categorize downward-pointing triangles as unpleasant compared to neutral or pleasant categorization. These findings were specific to downward-pointing shapes containing an acute angle, supporting the hypothesis that simple geometric forms convey emotion and that this perception does not require explicit judgment or language processing.

To a great extent, psychology and modern society have placed the higher premium on verbal intellect and rationale or logical linguistic thought with a devaluation of our more original, instinctive, nonrational, and holistic ways of knowing. To lose touch with these mechanisms is, by definition, an emotional disorder and may preclude our search for a verbal analysis of what is processed nonlinguistically by the "emotional brain." This disconnect is obvious as many right brain issues defy efforts to place them into words. Albert Einstein, though heralded as the most highly regarded scientist in recent times, rejected this approach as ineffective and amiss with the capacity of the human for creative thought. In this context he said, "The intuitive mind is a sacred gift and the rationale mind is a faithful servant. We have created a society that honors the servant and has forgotten the gift."

Einstein was born on March 14, 1879 and shares the first three digits of his birthday (3.14) with the mathematical constant "pi." He lived in Princeton during the later years of his life with an office on campus, though he was not a university faculty member. He was on the faculty at the Institute for Advanced Study and remained a contributing member of the larger intellectual community of Princeton. He was born in Ulm, Württemberg, Germany in 1879, and developed the special and general theories of relativity. In 1921, he won the Nobel Prize for physics for his explanation of the photoelectric effect. He died on April 18, 1955, in Princeton, New Jersey.

Disconnection Syndromes

Gross inspection of a debrided brain will reveal a system of interconnections with "cables" traversing across the corpus callosum (e.g., the splenium), interfacing one region of the left or the right brain with the homologous region in the contralateral cerebral hemisphere. Also revealed by debridement are longitudinal connections within each brain interfacing posterior brain regions with the frontal lobe. Incremental access to these large highways within and between the cerebral hemispheres is available as processing moves from the primary projection area(s) to the secondary and tertiary association cortices (e.g., Hofer and Frahm 2006). Within each functional unit are still other interconnections providing for the interface of one sensory modality with another sensory modality. Still other "U-connections" provide for tight influences among interactive brain regions. This is most clearly exemplified in the U-connection between the premotor cortex and the somatosensory cortex (e.g., see Cappe et al. 2012), where planned movement is altered by the dynamic feedback on positional changes or locations of body parts involved in these coordinated expressions or ideational praxis.

Appreciation of the functional role of these connections provides a basis for understanding one or another "disconnection syndrome." Historically, the discovery of these disconnection syndromes has contributed much to our understanding of how the brain works on even a broader scale and through the combined interactions of functional neural networks or systems. Severing of the corpus callosum, connecting the association areas of one cerebral hemisphere with the homologous regions within the other, produces a "split-brain syndrome." This was first appreciated as the "alien-limb syndrome" (e.g., Bejot et al. 2008), though the disconnection extends substantially beyond our limbs, to include a disconnection between our emotions and the logical verbal discourse of the other brain. Although massive in effect, this disconnect is seldom appreciated by the layperson and by the health-care professional. To elucidate on these disorders, one must first know what questions to ask and what to look for in this individual. Having this knowledge may open the door to a new enlightenment and to new opportunities for discovery and intervention.

Alien-Limb Syndrome

The neuropsychologist Roger Wolcott Sperry received the Nobel Prize in 1981, along with the visual researchers David Hunter Hubel and Torsten Nils Wiesel (see Berlucchi 2006), for his research on patients who had undergone complete corpus callosotomies (i.e., split-brain operations) for the purposes of controlling intractable epilepsy. Sperry stated (1966, p. 299) that "Everything we have seen indicates that the surgery has left these people with two separate minds, that is, two separate spheres of consciousness. What is experienced in the right hemisphere seems to lie entirely outside the realm of awareness of the left hemisphere. This mental division has been demonstrated in regard to perception, cognition, volition, learning and memory." Redemption for the right hemisphere would come slowly as the talking brain has been the priority. Roger Sperry, in his 1981 Nobel lecture, summarized the prevailing view of the right hemisphere when he began studying it: "The right hemisphere was not only mute and agraphic, but also dyslexic, word deaf and apraxic, and lacking generally in higher cognitive function." But Sperry's research with split-brain patients was, at least for him, convincing evidence that the right hemisphere was not a mere accessory. The view that the right hemisphere is a minor, less important player (or "relatively retarded," as Sperry colorfully put it in his 1981 Nobel Laureate address published in 1982; Sperry 1982) was no longer tenable.

Although the layperson and many health-care professionals would never appreciate that they are in the presence of a patient with a split-brain syndrome (Gazzaniga 1998, 2000, 2013), the disorder may indeed be among the most provocative in its impact on the patient and in its potential impact on progress within neuroscience.

Though at once sharing the responsibilities of all that a person is or is able to do as a person, these separated cerebral hemispheres are clearly independent in their thoughts, feelings, and actions. Just as two roommates may coexist within an apartment in harmony, these two hemispheres may coexist in harmony. Or, they may not!

That the two brains are capable of quite independent processing is clearly evident in oppositional syndromes where the two brains are pitted, one against the other, following disconnection of the communication corridor known as the corpus callosum (Gazzaniga et al. 1962; Sperry 1966, 1982; Gazzaniga 1998, 2000; see also Springer and Deutsch 1998). For the linguistically communicative left hemisphere, this may be identified in verbal complaints, confirmed by our observations, of an alien limb. The other hemisphere controlling the limb may defy the wishes of the conversant left hemisphere and even attempt to inflict pain or harm on its cerebral brother. Patients with alien-hand syndrome may spontaneously grasp objects and even other people (see Scepkowski and Cronin-Golumb 2003; see also Husain 2013). These individuals are able to talk about their hand making these movements, but the cerebral hemisphere discussing the alien limb reports that *it* (left hemisphere inferred) is not controlling them, and instead that it feels that the movements are being controlled by an external agent (right hemisphere inferred).

One man was caught trying to strangle his wife with his left upper extremity, while his right upper extremity fought to save her life, with one hand upon the other in active opposition. This act was carried out by two oppositional brains, like two separate individuals embroiled in a desperate struggle, even with vital implications. In his interviews, he was adamant that he had no desire to harm even a hair on her head. Nor would he even entertain such an idea. In his logic and in his linguistic expressions of his logic, he loved his wife and had positive thoughts about her. But, these were the musings and expressions of the brain that was doing the talking. This was the brain that had intervened to protect his wife, through the interface at his right arm and hand. This was not the "guilty" brain! The discussions of the "attempted murder" or of the physical assault might have been as relevant if the inquiry was of another person, removed from the animosity, the anger, and the sense of deep hatred and anguish that drove the attack and that controlled the left side of the body. The initial clues came instead from facial expressions, directional gaze, body movements at each side, and from sympathetic tone, perhaps.

Substantial controversy arose in the case originally from the disbelief that a man could distort the clear and evident truth that he had been caught in the act of harming his wife by attempting to strangle her. The skepticism was paramount based on the implications drawn from the assessment of this man. You might imagine the raised eyebrows when the brain scans revealed a "butterfly tumor" with agenesis of the corpus callosum effectively rendering the two brains disconnected, isolated from one another, and even oppositional in their thought without communication access along this interhemispheric cable. The talking brain, instead, now was reasonably found to be innocent of the assault and motives and even protective in its efforts to save his wife from the sinister (left sided) limb.

Oligodendroglia cells wrap around these long axons and function in very complex ways, much like "mother cells." They are, themselves, influenced by the associative analysis of many astrocytes which control neurogenesis and which influence all decision-making activities prior to accessing the neuronal highways (see Koob 2009). At an elementary level, they function to speed conduction of the action potential along the neuron, where they are more common on longer axons that are traveling some distance for communication with other areas of the nervous system. These longer axons originating from pyramidal cell bodies are most common in the longitudinal tract connecting the front and back parts of each brain, in the descending pyramidal motor system connecting the motor cortex with lower motor neurons (alpha motor neurons) influencing muscle contractions, and in the corpus callosum where the cells interface with the homologous brain areas within the left and the right cerebral hemisphere. The myelin secreted by these cells allows for "saltatory conduction," which is roughly 20 times faster than conduction of the action potential down an unmyelinated axon (see Bartzokis 2012). Glial cells function in many ways, including assistance with the nutritional requirements of neurons. They have a high metabolic rate and are the source of most primary brain tumors or gliomas. In this case, the tumor has an affinity for the glial cells covering the corpus callosum and, on exposure, appears like a butterfly resting between the left and the right cerebral hemispheres.

The corpus callosum is subject to other pathological processes beyond tumor genesis, including cerebrovascular accidents involving one or another division of the anterior cerebral artery and specifically the pericallosal and callosomarginal arteries. Alternatively, one or another division of the posterior cerebral artery might be affected, disconnecting communication at the posterior brain regions. The alien-arm syndrome is most shockingly apparent with a complete dissection of the corpus callosum. But, permutations exist for disconnection of any homologous region of the left and of the right hemisphere. By example, the latter might occur with a lesion to the splenium of the corpus callosum. This might disconnect the parietal–occipital–temporal regions of one hemisphere from the occipital input of the other brain. This is evident in the syndrome of alexia without agraphia (Wernicke 1874; Kleist 1934; see also Ardila 2012) discussed elsewhere in these writings, where the patient may write but be unable to read what she/he has just written. A stroke in the distribution of the posterior cerebral artery of either brain may encroach upon the splenium of the corpus callosum, effectively disconnecting visual analyzers and their distinct perceptual specializations.

The neuropsychologist might, upon occasion, be prompted to assess for an alien-arm syndrome as she/he arrives at the hospital early in the morning with a newer therapist or nurse stating that the patient is "inappropriate" or that "his behavior is unacceptable." On inquiry, the behavior that is unacceptable is commonly at the left hemibody, such as heterosexual advances or groping with the left hand or arm. The clinical expression of this brain disorder is known as "the alien-arm syndrome." In actuality, though, it may present at the upper and/or the lower extremity on the left side. The probability of an alien limb complaint differs between the upper and the lower extremity (e.g., Ito et al. 2013), since dorsal lesions within the axial or mesial brain regions more readily yield paresis or plegia in the lower extremity. But, the probability of alien features in the lower extremity may increase during the recovery phase and with improved lower-extremity mobility. These behaviors are frequently socially inappropriate with gestures or groping characteristics of the right brain (e.g., see social pragmatics and proprieties). They may convey negative affect in their content, reflecting not only improprieties but also coercion or control, emotional content, and postural gesturing. For example, one man was described after an infarction within the bilateral anterior cerebral artery distribution (Bejot et al. 2008). Though quite out of character by family accounts of his prior disposition, this man's left hand would expose his genitals and engage in public masturbation with that extremity. All the while, he verbally expressed his condemnation of these behaviors and his anxiety of having been a part of the behavioral display.

One fearful woman suffering a remote right-sided posterior cerebral artery cerebrovascular accident was experiencing left visual field hallucinations of "mean and scary" people as she tried to kick at them with her left leg. Everyone knew to watch out for that left leg and to approach her from her less apprehensive right hemispatial processing system (her left hemisphere). In the hospital or skilled-care setting, the odds are good that the nursing and therapy staff, along with concerned friends and family, are accessing the patient at their left hemispace and left hemibody. Simply placing the bed with others accessing the patient's more positive, socially engaging, and less fearful left brain through the right visual, auditory, and somatosensory pathways and right hemispace may be clinically significant in reducing negative outcomes. Moreover, approach from within the left hemispace and possibly manipulating the patient's left side for care and cleaning may activate sympathetic drive with incremental heart rate (even tachycardia), blood pressure, glucose mobilization, and so forth, which may further destabilize or threaten the patient's medical status and care.

Now, on discussion of the limb with this person, the therapist/practitioner may hear a different account from that reported by others. One successful businessman and community leader stated "Doctor you must do something about that arm. It is getting me in trouble! My wife is angry and leaving me because of what it is doing. I would never do anything like that." Regardless, the left hemisphere usually has its accounting of what is going on at the ipsilateral left side of the body and at the left hemispace. Interestingly, sometimes an inappropriate smirk at the left hemiface might be appreciated while the left brain tells *its* story much like an uncooperative observer where the guilty brain is relishing in the duress of its loquacious roommate (the left hemisphere).

One woman had been referred with a diagnosis of "kleptomania" with recurrent theft, which she adamantly denied in her verbal accounts during the interview. Following her around the hospital allowed for observation of the thievery or stealing behavior by the left hand and arm, within the left hemispace, and at the left visual field. In her logical accounting of this on confrontation, she expressed her attributions that "someone or something was doing this" to her, still adamant that she would not steal. The left brain that was responding to the verbal inquiry had not performed these activities and was generally not aware, though it appeared to be somewhat perplexed. To say that she was "in denial" would be incorrect as there was no recognition by this brain of the event or the motive. The brain that was talking was innocent, naïve, and honest! It had little to no experiential or cognitive basis for appreciating the behaviors and motives of the other brain and with isolation of one cerebral hemisphere from the other.

The alien arm may also engage in harmful self-destructive activities through injurious behavior directed to the ipsilateral hemibody (right hemibody). This might torture the left hemisphere, whereas the right hemisphere would be relatively exempt from pain. One woman complained in a flat affective intonation as the right brain pulled her hair out on the right side of the head, the side with relative projections to the left brain expressing the complaint. The left arm and hand would extend over to the right side of the head, grab a handful of hair belonging to the other brain, and pull. To say that these two brains were oppositional might be an understatement. The verbal expressions from the left brain provided complaints and confirmed discomfort induced by these activities. But, the verbal arguments made to others within the social array (doctors and therapists) were arguably less relevant to the socially avoidant and less loquacious right brain effecting the pain. An exception might be readily appreciated if there are tactile formesthesias at the left hemibody. For example, one woman with combined damage to the corpus callosum and right thalamus repetitively hit herself at the left hemiface and arm where she felt "snakes biting" her.

The disconnection may or may not be emotional in nature and some researchers appreciate the motor components of grasping objects with the inappropriate hand (see Husain 2013). An example of a more circumspect motor integrative disorder might be provided by another patient following a stroke within the distribution of the anterior cerebral artery. The resulting lesion effectively disconnected the two frontal lobes and altered bimanual integration activities. When asked to transfer the pencil in his right hand over to his left hand, he could not complete the task and appeared confused. Similarly, he could not take his right hand and place it over into the space belonging to the other cerebral hemisphere (left hemispace). Also, these failures to transfer material or objects across to the other hemibody resulted with left-hand placements. Finally, he mastered the task at his level, as each hand was able to find his mouth. Thus, he would relay materials or objects, such as the pencil in the palm of his right hand, by first moving it to his mouth, where it would be held in place for the other hand to find it. Throughout his recovery, and with therapy interventions for this, he gradually expanded the overlapping spaces of each cerebral hemisphere, providing for bimanual activities at improved levels of competency.

This "alien-limb syndrome" should not be confused with "limb alienation syndrome." The latter does not result exclusively from a disconnection between the two brains, but rather from substantial damage to the right cerebral hemisphere. More often with a left upper extremity plegia, the intact left hemisphere must account for this cumbersome and orphaned left arm, now attached to the body but devoid of ownership. The resulting account may be expressed in the form of alienation or disgust with the left arm and the attribution that "Someone stuck it on me. It is not mine. I want to throw it away or get rid of it." One man asked in earnest "why did the doctor cut my arm off and attach this thing to me?" Instead, the therapists will want the patient to learn to care for it, to protect it from the spokes of the wheelchair, and to begin to integrate the extremity into a unified concept of the self. This is best accomplished by placing the orphaned limb within the right hemispace and through

Fig. 16.1 Complaints and a drawing of a left-sided hallucination in a woman with chronic left-leg alienation, recurrent falls, and multiple efforts at repair and replacement of her left hip. *TBI* traumatic brain injury

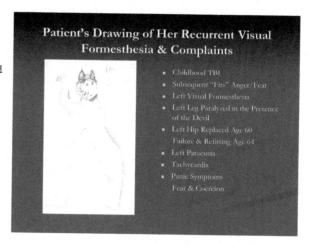

Patient's Drawing of Her Recurrent Visual Formesthesia & Complaints

- Childhood TBI
- Subsequent "Fits" Anger/Fear
- Left Visual Formesthesia
- Left Leg Paralyzed in the Presence of the Devil
- Left Hip Replaced Age 60 Failure & Refitting Age 64
- Left Paracusia
- Tachycardia
- Panic Symptoms Fear & Coercion

multimodal interactions including visualization, active touching, and manipulation using the right hand and arm, and of more relevance here, the left brain.

An alien leg is less common, but failure to appreciate this problem may confound efforts to manage the limb in recovery. Figure 16.1 provides some descriptors of a woman with chronic left leg problems and feelings of alienation, following a traumatic injury to the back of her right brain in childhood. The confusion for the leg and alienation with it were not appreciated until after her second left hip replacement, secondary to continued falls and misplacement or separation of the implant. She had complex symptoms largely involving her left hemibody and left hemispace, including a recurrent left visual formesthesia of a devil that would spring up and come at her. Her father's voice heralded the impending assault and the intent of the devil "he is going to kill us!" The left leg would become paralyzed with the onset of the hallucinations.

Group studies on alien-arm syndrome are virtually nonexistent. However, some more detailed investigation is available from Riddoch and her colleagues (Riddoch et al. 1998; Humphreys and Riddoch 2000; see McBride et al. 2012). They provided evidence that patients with an alien right hand can correctly pick up a cup with the left hand as long as the cup's handle is also on the left. However, when the handle is on the right, the patients are more likely to grasp the cup with the right hand, despite instructions to the contrary. These "interference" errors were reduced when the task was to point to the object, rather than grasp, and also when the objects were inverted. Therefore, it might be argued that these patients respond according to well-learned action associations rather than according to the verbal directions that they were given. The action afforded by the object was disrupted when the object was inverted, or when the action required was not the one usually made to the object (pointing instead of grasping), so fewer interference errors were reported under these conditions.

In a review of automatic motor activation in the executive control of action, McBride and colleagues (McBride et al. 2012) report that alien-limb syndromes are most often associated with focal lesions to the medial frontal lobes (particularly the supplementary motor area; e.g., Lhermitte 1983; Boccardi et al. 2002), damage to the corpus callosum (e.g., Biran and Chatterjee 2004), and following medial parietal lesions subsequent to posterior cerebral artery stroke (e.g., Coulthard et al. 2007; Bartolo et al. 2011). They note that the syndrome is increasingly recognized in patients with corticobasal degeneration, a slowly progressive neurodegenerative condition which affects cortical regions as well as the basal ganglia (e.g., Murray et al. 2007; Tiwari and Amar 2008).

When the neural correlates of unwanted movements were investigated using functional imaging techniques in a patient with alien-limb syndrome with corticobasal degeneration, the researchers (Schaefer et al. 2010; see also Husain 2013) reported that voluntary and alien movements activated similar brain regions, including motor and parietal cortices. However, the right inferior frontal gyrus, which has been associated with inhibitory control of motor responses (e.g., Swann et al. 2012; Hampshire et al. 2010), was activated only during alien movements. McBride and colleagues conclude that right frontal activation may reflect unsuccessful attempts to inhibit alien movements, highlighting the impact of automatically afforded actions, where alien-limb patients might find them particularly difficult to inhibit.

The present discussion of a disconnection between the cerebral hemispheres has, for the sake of simplicity and demonstration, focused on the actions of the left side of the body that occur without the awareness of the left brain and without access to the language systems and verbal linguistic mediation. Equally important, though, are the cognitive and emotional disadvantages of diminished communication and analysis by a functional and/or structural disconnection between the cerebral hemispheres.

Many other variants likely exist for diminished communication between the cerebral hemispheres, which may present as psychopathology of one sort or of another. Social and linguistic deficits have been documented in individuals with agenesis of the corpus callosum (e.g., Symington et al. 2010; see also Paul 2011). Using the child behavior checklist, Badaruddin and colleagues (Badaruddin et al. 2007) accessed a data set ($n = 733$) of individuals with a community diagnosis of agenesis of the corpus callosum. Evaluating a subset of high-functioning children, aged 6–11 years, they found that 39% exceeded clinical cutoffs for social problems and that 48% exceeded clinical cutoffs for attentional difficulties.

Autism spectrum disorders are clinically defined by a constellation of deficits in communication, social skills, and repetitive interests and behaviors (American Psychiatric Association 2000). Autism spectrum disorders are known to have numerous etiologies, including structural brain malformations. One area of research has investigated the brain malformation consisting of agenesis of the corpus callosum. For example, Lau and colleagues (Lau et al. 2013) measured the occurrence of autism traits in a cohort ($n = 106$) of individuals with agenesis of the corpus callosum. Forty-five percent of children, 35% of adolescents, and 18% of adults with agenesis of the corpus callosum exceeded the predetermined autism screening cutoff. The authors also note that magnetoencephalography measures of resting-state functional connectivity of the right superior temporal gyrus were inversely correlated with performance on the autism spectrum quotient's imagination domain.

Chapter 17
Self-Awareness

For many years, self-awareness was beyond scientific inquiry and lost in language that defied investigation using the scientific method. This included very diffuse language like conscious or unconscious thought, which was largely the language of the Freudian and Neo-Freudian era, historically. These diffuse attributions might be replaced with evidence of inaccessibility of one or of another brain region to the language-processing systems in the left cerebral hemisphere, for example. But, with the development of neuropsychological methods, scientific language evolved with relevance to the role of differing cerebral regions in the awareness of self. Heath Demaree (Demaree and Harrison 1997) initially discussed this in a model of self-awareness of deficits underlying the emotional disorders. Each brain contributes a somewhat different understanding of our self and evidence for this is fundamental to several of our theories of spatial analysis (Heilman et al. 1995; Heilman and Van Den Abell 1980; Heilman, Watson and Valenstein 2012). One example might be the differential specialization of each brain for the processing of space, with superiority of left cerebral systems in personal spatial analysis (e.g., body parts and gestures) and with superiority of right cerebral systems in extrapersonal space (e.g., the role of others in my attributions of causality). Paul Foster extended this model of spatial analysis to emotional valence (Foster et al. 2008).

Anosognosia and Anosodiaphoria

Clinical neuroscientists have had access to a methodological technique, which allows for a temporary or reversible lesion of one or the other cerebral hemisphere, through the administration of a barbiturate to one or the other carotid artery ascending toward the brain along the front of the neck. This artery, on each side, bifurcates with the internal carotid artery branches merging alongside the Circle of Willis at the top of the brain stem. From here, much of the cerebral blood flow, and especially that within the middle cerebral artery distribution, is supplied to the left brain

© Springer International Publishing Switzerland 2015
D. W. Harrison, *Brain Asymmetry and Neural Systems,*
DOI 10.1007/978-3-319-13069-9_17

(left carotid artery) and the right brain (right carotid artery). This Wada technique allows for one or the other cerebral hemisphere to be put to sleep for the purpose of uncovering or releasing the functions served by the contralateral, and still awake, cerebrum.

Kirt Goldstein (1952) may have been the first to report what he termed a "catastrophic response" associated with injury to the left frontal lobe. With deactivation of the left brain, the subject may experience a "catastrophic reaction" wherein he or she is quite aware of the resulting functional deficits, including paralysis of the contralateral limb(s). Indeed, elevated somatic concerns and actively seeking help for one's problems is a hallmark for relative left cerebral dysfunction or deactivation. Sodium Amytal (amobarbital sodium) injection studies found that injection at the left carotid artery produced dysphoric behaviors, pessimistic statements, guilt, somatic complaints and worries about the future (Rossi and Rosadini 1967; Silberman and Weingartner 1986). Quite the opposite effect corresponds to deactivation of the right cerebral hemisphere, where the subject displays "indifference" to the deficits or a lack of appreciation or concern for functional loss resulting from the reversible

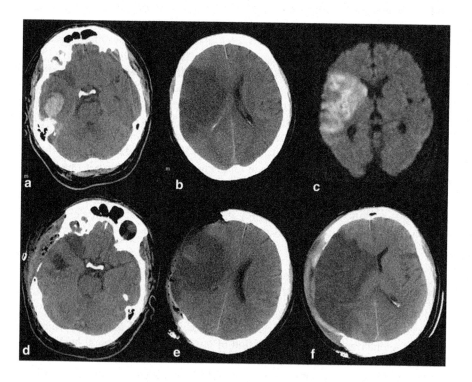

Fig. 17.1 MRI depicting right middle cerebral artery distribution cerebrovascular accident in a patient with anosognosia and anosodiaphoria. *MRI* magnetic resonance imaging (Originally published by Schirmer, Clemens M., *Decompressive Craniectomy,* Neurocritical Care, Volume 8 Issue 3, p. 459)

lesion. Injection at the right carotid artery produced euphoric reactions consisting of smiling, laughing, mimicry, and a sense of well-being. Note that injection within one or the other hemisphere was thought to result in a release of the opposite brain from contralateral inhibitory influences (Silberman and Weingartner 1986).

These laboratory findings generalize to patients suffering acute cerebrovascular accidents of the left or right cerebral hemisphere, where left-sided cerebrovascular accidents typically correspond with awareness of one's functional deficits and an urgent admission to the emergency room with the patient actively seeking help. These features of left cerebral dysfunction contrast dramatically with the features of an acute right-sided cerebrovascular accident. Indeed, many of these patients fail to appreciate their deficits and do not seek help or medical care. Often, the patient with a right-sided cerebrovascular accident is admitted some time after the event with others (e.g., family members or employers) bringing the patient to the hospital stating that "something is wrong with him," in opposition to the patient's report. Individuals with a right middle cerebral artery distribution cerebrovascular accident (see Fig. 17.1) somewhat characteristically present with anosognosia and anosodiaphoria as do those with injury to these brain regions from other mechanisms (Pia et al. 2004; Orfei et al. 2007; see also Prigatano 2010; see also Gerafi 2011).

A construction worker was remodeling the exterior of a physician's office with the use of nail guns and 16d nails using a multilevel scaffold. The worker above him sat his gun down on the scaffold, which was at the level of this young man's head. The gun discharged with the nail completely imbedded into the man's right parietal lobe extending down toward the brain stem. This young man went in to see the physician at the encouragement of his coworkers, appreciating the blood at the right side of the head. The physician did not appreciate that there was a nail in the worker's brain, and the patient did not appreciate that he had a problem. Indeed, he stated to the physician that he was fine. Following the placement of sutures to the scalp laceration, he promptly returned to work; until his deteriorating condition was appreciated by others now demanded that he go to the emergency room. Family members and the workers' compensation case manager were informed of his safety risk, predicting an injury should he return to work and to the upper-level scaffolding. The case manager, though, was under pressure to return her cases to work, and she did. It was shortly thereafter, unfortunately, when he again presented to the office, but now with multiple, and seemingly preventable, physical injuries and disabilities.

Eventually, on the rehabilitation unit, the patient with this disorder may continue to fail to acknowledge functional deficits corresponding with a right-sided brain injury. Many of those with right-sided brain disorders will continue to explain away even the most severe functional deficits stating that "I shouldn't be doing these silly therapy activities." "I wouldn't have any problems, if I were not in the damned hospital!" Some may remember the Presidential Press Secretary James Brady after the attempted assassination of President Ronald Reagan on March 30, 1981 (Shaffer and Henry 1981; see also Abrams 1994). The first of six bullets fired from John Hinckley, Jr.'s gun hit Mr. Brady in the head. During his recovery in the hospital and on national news, he appeared to be somewhat irritated that the president might not ask him to return to his former job as the presidential press secretary of the USA. He maintained his language abilities but seemed instead to have poor insight into his deficits and the insurmountable odds against his return to his former occupation. Consistent with his former demeanor and good sense of humor, the president's words to his surgeons were "I hope you're all Republicans," and to his wife he said, "Honey, I forgot to duck" (Britton 2009).

Anosognosia is the failure to recognize one's deficits with a lack of insight or underestimation of sensory, perceptual, motor, cognitive, and/or affective deficits as the result of brain pathology (Orfei et al. 2007). This disorder was first named by Joseph Babinski in 1918 (Babinski 1918; see Prigatano 2010; see also Starkstein and Tranel 2012). Although anosognosia has been studied primarily in patients suffering right-sided cerebrovascular accidents with left hemiplegia (e.g., Pia et al. 2004; Berti et al. 2005), the diagnostic term is relevant to a much broader array of problems for which the patient seems to lack insight or the recognition of their deficits. However, confounding variables have been identified which may ultimately affect diagnosis and the interpretation of laterality effects (Starkstein et al. 2010). Among these are variables associated with the reliance upon verbal reports from the patient which might exclude some of those with aphasic disorders resulting from left hemisphere damage or those suffering from dementia processes, perhaps.

Geschwind (1965a, b) provided an anatomical explanation for anosognosia in relation to left hemiplegia based upon a classic disconnection syndrome. In this case, the critical lesion might be to the corpus callosum yielding a disconnection from the right brain and the presumed awareness of the hemiplegic limb to the language analyzers of the left cerebral hemisphere, which would be required for the verbal processing of the event. The left brain would have awareness for its right hemibody and a general lack of awareness for the left hemibody, which is instead a function of right cerebral systems. Also relevant here would be the emotional appreciation of the deficit generally derived from right cerebral systems analysis and comprehension. Thus, recovery might begin with teaching the left brain to say that there is a problem, based upon a logical and rational verbal analysis. This might well be accomplished long before the patient actually *believes* that there is a defect, reflecting depth of emotional processing and understanding perhaps somewhat beyond the basic processing specializations of left cerebral systems.

Pia et al. (2004) provide an extensive meta-analytic review of some 85 studies published from 1938 to 2001 on the underlying anatomy of anosognosia. There is general agreement that anosognosia arises most frequently from lesions within the right hemisphere. The authors suggest that damage to frontal, parietal, or temporal cortex is sufficient to produce anosognosia, but also these symptoms may arise from damage to specific subcortical structures within these regions. Anosognosia was found to be more common following lesions to frontal and parietal cortex without significant subcortical involvement, when compared to temporal lobe cortical lesions without significant subcortical damage. Anosognosia was most common following frontoparietal lesions as part of a cortico-subcortical network involved in the awareness of and production of motor acts.

This conclusion, however, is relevant to a potential historical bias in the assessment of anosognosia for left hemiplegia or the lack of awareness of motor defect. Also, it is reasonable to expect that although somatosensory awareness of motor acts may be primary in the detection of a disorder in limb movement other sensory modalities play an integral role in this system. For example, failure of the limb to move under intentional conditions might also be assessed within visual analyzers if they remain intact. Moreover, the assessment of the relative integrity of each sensory and motor system is fundamental to neuropsychological interventions, where compensatory mechanisms are brought into play within the rehabilitation context.

In another project (Berti et al. 2005), using anatomical neuroimaging techniques and people with right-hemisphere lesions and left hemiplegia, results indicated that anosognosia was characterized by lesions to premotor cortex (area 6 and 44), primary motor cortex, and somatosensory regions. Also differentially involved were Brodmann's area 46 and the insular region. Berti et al. (2005) reasoned that the intentional aspects of movement originates from more than a single cortical region and likely involves different premotor areas. Damage to these areas may distort the intended representation of the motor act altering the process of initiation, planning, and monitoring of activity from the affected body parts. Thus, separate processes may contribute to altered awareness or appreciation of deficits in the anosognostic patient.

Karnath et al. (2005) also used neuroimaging techniques, including magnetic resonance imaging and computed tomography for a right hemisphere lesion location analysis of patients with and without anosognosia for their left hemiplegia. Lesions in both groups involved parietal and temporal cortical areas, the insular region, and the deep white matter regions. A lesion subtraction analysis was performed between the groups to identify the regions critical for anosognosia of left hemiplegia. The results indicated that the right posterior insula was usually damaged in the anosognostic group compared with those without anosognosia. This finding was substantiated by others (Vocat et al. 2010) during the initial stages of recovery from stroke, where damage to the insula is a common finding with anosognosia. Also, these results attest to the underlying neural architecture, where the posterior insular region shares connections to the primary and secondary somatosensory cortex, the premotor cortex, the prefrontal cortex, and the superior and inferior temporal cortex (see Appelros et al. 2007). Regardless, the lesion location of the participants in the Karnath et al. (2005) project was not circumscribed and included other structures extending beyond the posterior insula, such as the temporal and/or parietal cortex, the basal ganglia, and deep white matter regions (see Gerafi 2011).

Therapy for this disorder may be initiated through the resources of the intact left cerebral hemisphere. More specifically, the patient is provided the verbal and logical rationale for the conclusion that there must be a substantial brain disorder, and this rationale is overrehearsed. The initial phase of treatment is successful at the point that the patient can state his deficits *verbally* as have been described by his therapists or counselors. Yet, the patient with a right-brain disorder, when asked if he actually *believes* what his therapists have told him, if honest, may state, "no, I don't think that it is true." Thus, the belief or emotional impact of a loss of function is assigned to the damaged right brain with an inability to fully appreciate loss or the impact of the damage to this hemisphere. The emotional belief or depth of understanding that there is a problem with me appears to require the negative affective analysis of the right brain. For many, the weight and the essence of the world rest squarely on the shoulders of this right brain as it watches over all else, including its literal and loquacious brother!

This diagnostic distinction is predictive of outcome or prognosis as the patient with right-sided brain dysfunction may more readily "get back on the horse" or, more realistically, return to work perhaps. This may actually be indicative of a successful outcome, overall. However, this patient more readily returns to the hospital secondary to a safety failure, including fall injuries, driving accidents, mechanical safety violations, and such. Also, if the spouse or caregiver is overly heroic and fragile, as might be the case with an elderly woman caring for a physically large man, the risk may compromise the patient with loss of the caregiver and with both the patient and the caregiver requiring medical care (e.g., with a fractured hip, general anesthesia, and repair). For many people with right-brain disorders, the eventual outcome will hinge on supervision or safety interventions designed to minimize risk for the patient and the caregiver.

Family members and the less experienced health-care professional may not appreciate the safety risk if they rely on the patient's verbal report. Once the anosognosia has resolved, the patient may verbally state "I am not safe driving the car or cutting down trees with my chainsaw." But, devoid of the emotional appreciation of the safety risk and the implementation of caution by the right brain, this patient will return to high-risk behaviors without appropriate caution. Thus, the onus is more specifically on others, including the caregiver and health professionals charged with the patient's care. In many cases, the caregiver of a patient with a right-brain disorder faces the challenge of contributing to the safety management issue at the direct expense of dealing with active control issues and emotional discomfort, for example, in managing the activities of a fiercely independent parent with no real appreciation of his deficits. In contrast, the burden on the caregiver and the health-care providers, working instead with a left-brain disorder, often is derived from processing numerous somatic complaints, self-critical concern, apathy, and/or amotivational features (see Scott and Schoenberg 2011; see also Grossman 2002) and perhaps with aggravation of a metabolic syndrome (i.e., obesity; Anthony et al. 2006) with weight gain and inactivity or hypokinesis.

Another way to look at this issue of the differential appreciation of risk by the two brains and, likewise, the appropriate implementation of caution, is to appreciate the supervision needs of the left and the right cerebral hemispheres. Within the context of preparation of the patient and caregiver for discharge from a rehabilitation unit, the multidisciplinary team will meet and derive a recommendation for the level of care required in the supervision and assistance of the patient for basic safety and the completion of functional activities of daily living. The supervision and assistance needs of the patient with a left-hemisphere disorder are more typically based on confusion with body parts or praxis, speech-related deficits perhaps, and behavioral slowing or initiation deficits. So risk is certainly an issue. However, by and large, a patient with integrity of right cerebral systems and orbitofrontal regions will implement caution in their activities and appreciate their safety risk. In contrast, the patient with a right-hemisphere disorder, and especially with a dysexecutive disorder of the right frontal lobe, may be unsafe and initiate behaviors "impulsively," with diminished organizational preparation and a general failure to appreciate that they have a safety problem. More simply said, individuals with functional and structural integrity of the right cerebral hemisphere can generally take care of themselves, whereas the unsupervised *happy go lucky and loquacious* left brain is a formula for trouble and sometimes for disaster.

This phenomenon raises an interesting issue for the psychologist and for science in general as much of our current efforts in research are investments in self-report questionnaires or other verbal inquiry. This is more clearly the case in clinical psychology where it may be acceptable to "simply ask the patient." This approach has reached the pinnacle of clinical practice mandates, where the professional progeny of the earlier behavioral psychologists, focused on objective behavior, now consider subjective self-report indices to be "objective measures." The assumption in science has long been that the left brain was the dominant brain because of its capacity for speech and logical and literal analysis. However, this talking and logical brain may not be able to take care of itself any more than a young child.

It is the richness and the depth of processing by the right brain that contributes more to our appreciation of risk and implementation of appropriate caution. It may be that this largely nonverbal brain is indeed the dominant brain by comparison and one which benefits from a long evolutionary history. It will activate in situations of threat or risk and it may, indeed, take care of me and insure my safety or my family member's safety. To this end, it is somewhat reminiscent of the male gorilla at play with other family members. Although seemingly involved with playful activities, the alpha male stops to check the horizon periodically. It is always alert for potential threats to you or to your group and, if ever there is conflict, your right brain is the one on which you may rely. It will know what to do if your child is threatened or if there is an intruder in the house, and it will know when you should not be behind the steering wheel of the automobile or the bar of a chainsaw.

One man sought a rapid release from the rehabilitation setting and prior to completion of his therapy protocol, after his right-sided cerebrovascular accident. In short order, he used his chainsaw to fell a large oak tree. His adolescent son was assisting him as the tree landed on the family's house. Another young woman with a large right-sided stroke returned to the cab of her tractor-trailer rig despite her verbal acknowledgment that these vocational activities were no longer an option. Another man returned to his political campaign with minimal appreciation of his left-sided paralysis, geographical confusion, facial agnosia, dysprosodia, and emotional disorder. At the point of his abrupt departure, the therapists were working on the goal of dressing his left hemibody with maximum cuing and redirection to this task.

The differential aspirations of the two brains may play out in ways not previously appreciated by many. One of the most difficult areas of clinical science has been in the treatment of alcoholism with recurrent consumption despite substantial costs to the individual, to the family, and to society. One interesting aspect of this form of substance abuse consists of the corresponding associations with negative affect or anger management issues. Scientists and nonscientists alike have long recognized a two-way association between alcohol consumption and violent or aggressive behavior (Reiss and Roth 1994; Moss and Tarter 1993; Roizen 1997; Cherpitel, Martin et al. 2013). Many spouse abusers and child abusers have expressed their worst traits following a heavy drinking episode. This negative emotional relationship with alcohol, featuring anger and/or violence, may co-occur with poor insight or a failure to appreciate one's deficits. For example, Witt et al. (2013) conducted a systematic review and meta-analysis of studies that reported risk factors associated with violence using six electronic databases (CINAHL, EBSCO, EMBASE, Global Health, PsycINFO, PUBMED) and Google Scholar. The studies were selective, in that they identified violence in adults diagnosed, using Diagnostic and Statistical Manual of Mental Disorders (DSM) or International Classification of Disease (ICD) criteria, with schizophrenia and other psychoses. There were 110 eligible studies reporting on 45,533 individuals, 8439 (18.5 %) of whom were violent. The authors found that the risk factors included hostile behavior, recent drug misuse, nonadherence with psychological therapies, poor impulse control scores, recent substance misuse, recent alcohol misuse, and nonadherence with medication.

For generations, other's attributions have been expressed that the alcoholic is *in denial*. But, this attribution may well be incorrect, based on what we have learned about right-brain disorders, negative emotions (anger), and self-awareness of deficits. From this perspective, it may be reasonable for future researchers to ask questions based on functional cerebral laterality. It is possible, for example, that the poor appreciation of one's deficits along with the positive emotional bias ("happy, go lucky") of the left cerebrum may play out in drinking disorders. The left brain may be characterized by optimistic anticipation of future events with positive emotional attributions, whereas the right brain may be prepared for the negative reflection on past failures (see Manuck et al. 2000). If so, then this theoretical position may be applied for an improved understanding of alcoholism to provide a tentative explanation for this seemingly illogical behavior. From this perspective, the individual returns once again to the bottle with the optimism and lack-of-caution characteristic of the left hemisphere for the positive or beneficial effects associated with drinking, rather than a cautious appreciation of the cost involved in these behaviors.

The impact of anosognosia may be a function of the impact on the person with this syndrome, and the potential for the individual's influence or legacy to alter history. An interesting account of this is provided by Edwin Weinstein (Weinstein 1981), a neuropsychiatrist charged with reviewing the medical history of President Woodrow Wilson. He writes: "The symptoms indicated that Wilson suffered an occlusion of the right middle cerebral artery, which resulted in a complete paralysis of the left side of the body, and a left homonymous hemianopsia—a loss of vision in the left half fields of both eyes. Because he had already lost vision in his left eye from his stroke in 1906, he had clear vision only in the temporal (outer) half field of his right eye." He continued in the description of this brain disorder stating, "Following his stroke, the outstanding feature of the President's behavior was his denial of his incapacity." Wilson appreciated no loss of capacity other than acknowledging his cane as his "third leg." But, his associates noticed changes in his personality as he became increasingly suspicious or paranoid. His brother-in-law from his first marriage wrote that Wilson "would be seized with what, to a normal person, would seem to be inexplicable outbursts of emotion." He was described as easily angered and by anyone questioning his competency with the later days of his presidency described as a "graveyard of fired associates" (Morris 2010).

One professional musician (pianist) who had developed her own television show featuring celebrity interviews, including Elizabeth Taylor, was recovering from a large right-sided cerebrovascular accident. She had a left upper extremity plegia, dramatic right gaze preference, and a dysmusia from her stroke. But, her left brain appreciated no real deficits. She held her right arm out straight in front of the neuropsychologist as she stated with robust and assertive clarity, that she would have no difficulty playing her piano at all! She stated that, "I have two strong arms and hands and they are well coordinated." The talking left brain seemed to have no real appreciation of the paralyzed left arm and hand. Nor did it have any appreciation for the loss of musical-processing capability with the loss of its sister and roommate... the right cerebral hemisphere.

Body Part Awareness and Gestures

Parietal lesions are associated with disorders of body schema, personal space, and left–right orientation (DeRenzi 1982). The left brain contributes to self-awareness primarily for proximal space or specifically body part awareness (see Goldenberg 2000). A brain disorder referred to as autotopagnosia is defined by the inability to recognize one's body parts. However, the term itself is controversial with its use in reference to both one's own body parts and the body parts of others. In the present discussion, the reference is restricted to the identification of one's own body parts. This awareness involves the appreciation of individual body parts and their relative or relational location with other body regions. This includes left–right awareness and aspects of the relative size of different body regions. This awareness appears intimate to the movement disorders referred to as the ideomotor or gestural dyspraxias. Persons with damage, especially to the left parietal region, often evidence impaired body part awareness and ideomotor or gestural dyspraxia. Ultimately, these postural placement errors may be poorly appreciated. Yet, the patient with the brain disorder and the family members may be critical to safety. Careful assessment and appreciation of these problems may prove useful in the prediction of a fall and the probability of readmission to the medical center. This may especially be evident with placement errors for the lower extremities.

In their research review, Ferri and colleagues (Ferri et al. 2012) conclude that the existent literature on body representation ascribes a crucial role to the parietal cortices in the assignment of body identity (Hodzic et al. 2009; Sugiura et al. 2006; Uddin, Kaplan et al. 2005), in the representation of the position of body parts in space and, collectively, the body schema (Creem-Regehr et al. 2007; Bonda et al. 1995).

They contrast the functional aspects of cerebral laterality stating that right parietal lesions often give rise to a deficit of body identity such as personal neglect (Committeri et al. 2007; Guariglia and Antonucci 1992; Bisiach et al. 1986), whereas left parietal lesions lead to a deficit in localizing body parts in relation to one's own or others' body as with autotopagnosia and heterotopagnosia, respectively (Cleret de Langavant et al. 2009; Sirigu et al. 1991; Semenza 1988; Ogden 1985).

Vingerhoets et al. (2012) investigated the effect of which hand was used and handedness on the cerebral lateralization of gesturing learned movements for tool use. Fourteen right-handed and fourteen left-handed volunteers performed unimanual and bimanual tool-use gestures with their dominant or nondominant hand during functional magnetic resonance imagery. The authors report a left hemispheric lateralization of function in the right- and left-handed group, regardless of which hand(s) performed the task. Asymmetry was most marked in the dorsolateral prefrontal cortex, the premotor cortex (Exner's hand area), and the superior and inferior parietal lobules. No significant difference was found in the asymmetric cerebral activation patterns between left- and right-handers during unimanual tool-use gesturing. Bimanual tool use showed increased left premotor and posterior parietal activation in left- and right-handers. Lateralization indices of the 10 % most active voxels within these brain regions were calculated for each individual in a contrast that compared all tool versus all control conditions. Left-handers showed a significantly reduced overall lateralization index when compared with right-handers, largely due to diminished asymmetry at the parietal lobe (superior and inferior parietal lobules). The authors conclude that the recollection and expression of learned gestures recruits a similar left-lateralized activation pattern in right and left-handed individuals. Handedness was found to only influence the strength, and not the side, of the lateralization, with left-handers showing a reduced degree of asymmetry that was most readily observed over the posterior parietal region.

One patient comes to mind as an example of left parietal pathology, specifically. This was a 13–year-old boy who was large for his age having reached a towering 6 ft 3 in. in height. He was late on returning home with his mother expressing her irritation over the phone and demanding his rapid return to their house. During his return on his bike, he developed right hemiparesis and was subsequently shown on computerized tomography (CT) scan to have a large left parietal cerebrovascular accident within the middle cerebral artery distribution. On the neuropsychological evaluation, he identified his elbow as his "nose" and his hand, extending around his index finger toward his thumb in a curvilinear manner, as his "ear" with substantial body part confusion.

In rehabilitation of a body confusion disorder, therapy typically begins with locating the correct body parts prior to implementing therapeutic demands for the proper postural placement of these body parts underlying more complex praxis. Subsequent to this, therapeutic interventions attempt to assist the patient in learning to sequence body postures for more complex movement or for the successful completion of functional activities of daily living. With the specialization of the right brain in peripersonal or extrapersonal space (see below), therapy was initiated by placing the young man in a supine position on top of white butcher paper and working with him to trace his body with a magic marker onto the paper. Once completed, the paper was placed in front of him on the opposing wall in *extrapersonal space*. Using right brain visual–spatial techniques, he located various body parts with substantial accuracy. The challenge for the boy and his therapists then became the integration of recognized body parts from extrapersonal space into the self-concept of his body parts within *personal space*. Progress was gradual and cumulative in this regard with the young man eventually improving postural or gestural praxis and sequencing these for more complex movements.

Limb Alienation Syndrome

Damage to the right brain may provide a red flag for the clinician with a need to assess for various alterations in emotional processing and the understanding of aspects of our identity that make up the self-concept. Limb alienation syndrome is a sometimes robust misattribution of disgust, anger, or fear towards our own body parts, following a right-brain injury. The patient may confirm on inquiry "that thing is not part of my body!" in reference to their left arm. The explanations may reflect somatosensory delusions using more antiquated language where the limb is viewed as having been attached and against the patient's will or preference. An engineering professor, after his right-sided cerebrovascular accident, explained that he had been forcefully and violently disconnected from his wife (previously attached to the left arm) by a silkworm missile. He expressed his spatial delusions with the belief that he had been relocated "from California" to the Virginia facility and against his will and directives. He expressed his anger toward his left hemibody, as an appendage placed on him against his will. He continued to have delusions of being moved around the Virginia Medical Center coercively and beyond his own control. But, of course, he had experienced some relocation within the hospital setting. Paranoia does frequently have some factual basis in reality, but with distortion (see Bentall and Udachina 2013).

Chapter 18
Attributions and Appraisal

Perceptual appraisal and attributions toward one's self or toward others for causality appear to differ as a function of the relative activation of the two brains. This was expressed early on with the balance or valence theories (Ehrlichman 1987; Silberman and Weingartner 1986; Davidson and Fox 1982) where relative activation of the right cerebral hemisphere was found to correspond with negative affective appraisal of sadness, anger, fear, or disgust. In contrast, relative activation of the left cerebral hemisphere was found to correspond with either a somewhat bland, neutral appraisal of events, ranging to happy or interesting attributions toward others and toward the self. The preponderance of evidence suggests that these cognitive, emotional, and social derivations are lateralized to the brain making them and to the specializations of that cerebral hemisphere.

Research indicates a negative attribution bias for those sitting within your left hemispace or left visual field in varied settings (e.g., Jaeger et al. 1987; Harrison and Gorelczenko 1990). Interestingly, this might include a classroom where you are within the instructor's left hemispace. It might also include a business meeting with half of the attendees at the left and the other half at the right side. While the research shows these effects to be statistically reliable, they are generally considered to be small and unlikely to determine your grade in the class, your promotion, or your job assignments. However, with dynamic activation of the right brain through an emotional induction, perhaps, or through one or another background sympathetic agonist, the effect may become more meaningful (see Mitchell and Harrison 2010). Also, this might be especially the case, if your instructor or employer is highly hostile and where affective stress or pain may differentially activate his/her right brain with the intensity of the negative affective attributions toward you now becoming potentially meaningful (Demaree and Harrison 1997b).

These affective attributions and perceptual appraisals differ with more extreme and inappropriate activation of either the left or right brain's sensory processing regions (the second functional unit). Rob Walters looked at the evaluations of about 150 patients recovering from a recent cerebrovascular accident (Walters et al. 2006). Roughly one in five was found to be experiencing sensory events (hallucinations

© Springer International Publishing Switzerland 2015
D. W. Harrison, *Brain Asymmetry and Neural Systems*,
DOI 10.1007/978-3-319-13069-9_18

or formesthesias) that would result from inappropriate activation of a brain region, rather than from a complete death of these cells. In the majority of cases, these events appear to resolve rather rapidly during the recovery process. The clinical inquiry for these events originating within the right brain must be persistent and the examiner must be observant as the patient's access to left brain verbal processing centers may be minimal and, especially, if the pathology disconnects the right cerebral hemisphere's access to the left hemisphere via the corpus callosum. This would be more probable with a stroke within the axial circulation as with occlusion of the posterior cerebral artery, which supplies the visual projection cortex but also the splenium of the corpus callosum. But, diminished access to the verbal analyzers within the left hemisphere is a prominent, if not fundamental, feature inherent to the functional analysis performed by the right hemisphere. It is often very difficult to express these and other negative affective events through logical linguistic speech. The overwhelming body of evidence indicates that the neural processing systems within the right brain simply do not process information in this manner! Right cerebral systems, instead, appear more capable of single word utterances, expletives, emotional sounds, emotional facial expressions, and directed gaze toward the left side (with activation).

The clinical feature indicating the occurrence of hallucinations originating within the right posterior cerebral region may be agitation or extreme fear concurrent with elevated sympathetic tone. This might correspond with cardiovascular features of elevated blood pressure and heart rate often with a cascade into tachycardia. Sweating is one of our better indicators of elevated sympathetic tone (e.g., skin conductance responses; e.g., Elie and Guiheneuc 1990). A slow reflex depolarization of the skin, the so-called galvanic response, is known to occur to an arousing stimulus. This response is thought to originate from synchronized activation of sweat glands as a response to a volley discharge in efferent sympathetic nerve fibers (e.g., Elie and Guiheneuc 1990) under cerebral control (e.g., Critchley et al. 2000). But, overt behaviors may provide the clue with evidence for a preoccupation for the left hemibody or left hemispace with facial expressions conveying fear or apprehension or perhaps anger or disgust. Abnormal activation of these posterior right cerebral regions corresponds with attributions that someone or something (often a devil-like creature) is trying to do them harm. Also common are negative attributions involving the perception of significant control issues conveying threat or the intent to harm. There is substantial evidence that the right hemisphere is specialized for processing faces and places. Thus, the control issues may involve delusional beliefs that someone or something is trying to force them into another space/place against their will. These spatial delusions include the perception of being put in a place or forced through space all while experiencing a loss of control over the self during these events.

With practice, the patient may become increasingly capable in his or her ability to verbally identify these events, their emotional valence, directional movement, and specific location within the left hemispace. More than 90% of the negative affective visual formesthesias may be located within the left hemispace or specifically within the left visual field (see Fig. 18.1; Walters et al. 2006; Mollet et al. 2007). Negative affective attributions or appraisal of these events include perceptions of

Fig. 18.1 Positive and negative affective attributions corresponding with hallucinations within the left or right visual fields

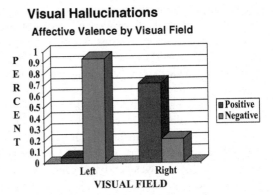

the devil, drug dealers, people trying to do them harm, snakes, and similar events. Quite the opposite attributions present with hallucinations occurring within the right visual field or right hemispace (left brain). Specific descriptions by the patients include the visualization of "happy children," "funny looking donkeys," "a little man walking up the wall," "pretty balloons," "funny little frogs," and "happy people that like me." For many, these events are experienced with great interest and enjoyment with a preference to observe the events and without negative affective symptoms or cardiovascular features as identified originating from the other hemisphere.

One woman demonstrated both the affective attributions and appraisal relevant to this discussion and the features of anosognosia described somewhat earlier on, following a right thalamic stroke (see Fig. 18.2). She developed an abrupt onset of

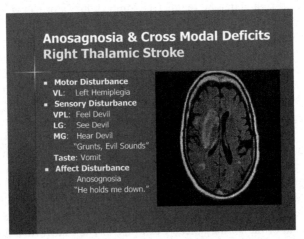

Fig. 18.2 Brain MRI of a patient with right thalamic stroke developing acute left hemiplegia with attributions that "the devil was holding me down" corresponding with anosognosia and left-sided multimodal hallucinations. The anatomical relationships include the motor nucleus (*VL*), the somatosensory nucleus (*VPL*), the visual nucleus (*LG*), and auditory nucleus (*MG*) of the thalamus. LG lateral geniculate, MG medial geniculate, MRI magnetic resonance imaging, VL ventral lateral, VPL ventral posterolateral

left-sided paralysis. She was unable to appreciate that the deficits resulting in her inability to get off the floor were due to problems with her body. The attribution or appraisal of the event was, instead, that the "devil was holding me down against my will." She failed to know her functional deficits consisting of left hemiplegia. Instead, she made external attributions to the manipulations of the devil on her person. In addition to these left-sided somatosensory events, she saw this demon at her left, was able to hear him grunting and making evil noises, and could taste him. She describes the odors of hell consisting of dead bodies, sulfur, and the taste of vomit. She further stated her attributions that he had urinated on her while she was on the floor! During her recovery, she experienced episodic left hemiplegia. She remarked that her "...left arm was paralyzed each time that he came around," conveying her perceptual attribution that the origin of the functional loss was from an external source within extrapersonal space.

Olfactory hallucinations were suffered by a brilliant 38-year-old composer along with severe headaches several months before he died of a right temporal lobe glioblastoma multiforme (see Waxman 2010, pp. 232). A colleague confirmed that George Gershwin had experienced an "...indefinable burning smell..." "...at the end of the second concert...." Some describe these olfactory hallucinations originating within right cerebral and brainstem regions as "the odors of hell." Many have experienced the odor of burning or rotting corpses, burning sulfur, and/or putrid vomit. With improved knowledge of functional neural systems, a syndrome analysis might have easily revealed a left upper quadrantanopsia with the tumor impacting the optic radiations projecting through the right temporal lobe carrying visual information from the left visual field. More relevant with this composer might be the expectation or clinical ruling out for an associated dysmusia and altered affective comprehension for emotional sound (auditory affective agnosia) with functional alterations of the right temporal region (see Limb 2006; Zatorre et al. 1992; Zatorre et al. 1994; Nan and Friederici 2012). With the proximity of the lesion to auditory affective processors in the right temporal lobe, nonverbal hallucinations of fearful sounds (screams, groans, drums) might also be probable. Instead, his physicians attributed these symptoms to "a neurotic disorder" and referred him for psychotherapy several months prior to the onset of coma and subsequently his death.

Another woman was preoccupied at the left side of the office, where she confirmed her fear of a "five foot, four inch tall black devil with a red tail." This evil entity was threatening her. She stated that "he looks like a bear." These events consisted of auditory and visual formesthesias with the devil "telling me to go with him." This was perceived as a threat or control issue as she did not want to go with him. She expected that he "would tell me to die." She was clear that he was trying to make her do things that she did not want to do. However, she did not hear him talking or speaking words, per se. She stated that "I think that he may be wetting my bed (instead of me)." She noted "I can't use my left arm when he is around." She also described spatial delusions where "he moves me around the hospital...I don't like it." The size was interesting as she was clear that he was large. But, there was space enough on top of the sink (where he was standing) for only a 2-ft. devil, at most. She was able to confirm the logic of this argument, but remained adamant that he was 5 ft 4 in. tall! Her visual representations of this multimodal creature are presented in Fig. 18.3.

Fig. 18.3 Patient's drawing of the devil at her left side standing on the sink in the neuropsychologist's office. She expresses her attributions that the devil is the one wetting her bed! She notes her inability to use her left arm, when he is around. The figure is viewed as threatening and coercive

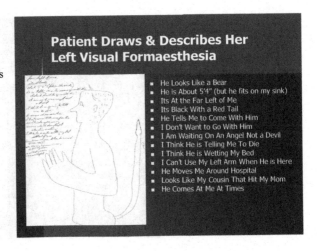

A young boy had a substantial effect on the reputation of one clinician in town. His well-educated parents (engineering professors) brought the boy to the neuropsychological assessment team with a previous diagnosis of "Asperger's syndrome." The neuropsychological evaluation confirmed right occipital and temporal lobe functional deficits with social apprehension or fear with acute and replicable onset of his symptoms following left-sided visual oscillation. The parents and the previous diagnostician were unaware that he was having threatening and fearful visual hallucinations. Again, the right brain does not talk. You generally have to look at the affective and directional behavior for the initial clues! Well, the therapy recommendations included "no video games for awhile" with the rapid visual (and auditory) oscillation substantially sufficient and fully adequate to induce the events just described. Years later, this unfortunate clinician is still known, unfairly, by the children in the town as the doctor who will take your video games away. What an awful reputation to have!

Formesthesia may be less common following left brain syndromes, but these certainly do occur with some frequency. The left hemisphere favors a positive appraisal bias. These events yield "interesting" attributions or enjoyable pleasant and positive attributions, which may impact reporting frequency, perhaps. Figure 18.4 depicts focal areas of encephalomalacia specifically at the dorsal somatosensory cortex (tactile), the superior temporal gyrus (auditory), and the occipital lobe (visual) in a patient after surgical repair of a skull fracture. This gentleman experienced positive affective components with the vision of an angel at the right side "doing good things" and with tactile formesthesias of "interesting things crawling up my right leg." Further, he experienced "clicking sounds" originating at the right hemispace. By inference, an encephalomalacia a bit posterior to the auditory projection cortex might have resulted in hearing people talking and enjoyment in listening to them.

The distinction between the attributions derived from left and right brain hallucinations are made a bit clearer with the features presented by a man with left thalamic lacunar infarctions. This gentleman appeared to see food, chew the food, enjoy its flavor and texture, and manipulate it with his hands through movements

Fig. 18.4 Positive attributions of right-sided sensory hallucinations originating from focal encephalomalacia at the left occipital, left parietal, and left temporal regions

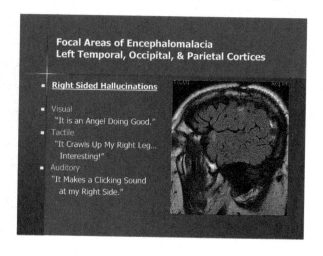

of the extremities, which correspond with eating behaviors. The affective display was one of quiet enjoyment of these appetitive behaviors, while others looked on in astonishment. These multimodal and actionable hallucinations may occur in generally toxic states, following anesthesia or anoxia, with kidney failure, dehydration, overmedication, and a multitude of other brain disorders. For another man, a left thalamic infarction was fully sufficient for the occurrence of these seemingly enjoyable and multimodal events.

Following a traumatic brain injury with bilateral temporal occipital injury, a woman described her right-sided formesthesias of dead relatives and friends who talked to her with reassurance, calming her fear, and helping her to feel secure. These occurred with perceptions of bright light or sunshine and a warm, quiescent state of peace, by her descriptions. However, the right temporal lobe encephalomalacia appeared to promote pervasive apprehension and anger as she was forced away from this pleasant, peaceful place and back to "this world," a place in which she acknowledged feelings of great fear and insecurity. She described her mate as an anchor and her primary source for security since the accident. She was calmed by maintaining nearly constant physical contact with him through somatosensory afferents. In his absence, she would place his shirt close to her face potentially to augment olfactory afferents and pheromones from her man.

Clearly, some of these examples of the differential attributions and appraisals of the left and the right brain reflect acute to subacute pathology as might be evident after a stroke, perhaps. In other scenarios, the attributions and appraisals appear to be more chronic manifestations of these differences between the perceptual biases of the two cerebral hemispheres. The influences of negative emotional biases on health, behavior, and cognition have been investigated in a number of experiments (e.g., Frasure-Smith et al. 1993; Shenal and Harrison 2003; Shimojima et al. 2003; Sirois and Burg 2003). In the context of this research, hostility has arisen as one of the most frequently examined emotional constructs in health psychology due to its correlation with the development of cardiovascular disease and more recently with the metabolic syndrome (see also Anthony et al. 2006). Anxiety, fear, or apprehension may follow closely behind in terms of the research areas where efforts are expended on these emotional disorders via psychological investigations.

Behaviorally, hostility is described as an attitude that motivates aggressive behavior toward objects and people (Spielberger et al. 1985). It is a negative emotional trait that encompasses cynicism, suspiciousness toward others, and proneness to anger (Prkachin and Silverman 2002). Physiologically, high hostility may result in overactivation of the sympathetic nervous system (Keefe et al. 1986) and hyperreactivity to environmental (Frederickson et al. 2000) and laboratory stress (Demaree and Harrison 1997b; Klineburger and Harrison 2012). Therefore, even milder attributional biases and appraisal differences may have a large and costly impact on the individual and their family and, in the broader context, on society at large. In this respect, highly hostile individuals have demonstrated negative appraisals and cynical, hostile, and angry perceptual attributional bias toward auditory, visual, somatosensory, and vestibular events and defensive or reflexive motor responses in facial expressions and antigravity posturing (e.g., Futagi et al. 2012). The attributional biases in these experiments have been largely within the left hemispace and at the left hemibody with origin within the right cerebral systems (see Cox et al. 2012; see also Carmona and Harrison 2009; Mitchell and Harrison 2010; see also Holland et al. 2012).

Godlewska and colleagues (Godlewska et al. 2012) note that patients with acute depression manifest a range of negative attributional biases in the processing of emotional information, which are believed to contribute to the etiology and maintenance of the depressed state. When compared to healthy controls, depressed patients selectively recall more negative, self-related emotional information on memory tasks (Bradley and Matthews 1983; Bradley et al. 1995) and demonstrate a negatively biased perceptual appraisal of key social signals such as emotional facial expressions (Gur et al. 1992; Bouhuys et al. 1999). Moreover, emotional biases such as these have been associated with pathological responses across a network of neural areas involved in emotional processing, including an increased response of the amygdala to negative facial expressions in depressed patients compared to matched controls (Sheline et al. 2001; Victor et al. 2010).

Harmer and colleagues proposed that the therapeutic effect derived from antidepressant medications might be mediated by early reversal of these negative emotional biases (Harmer et al. 2009). Using healthy volunteers, the authors report that 7-day treatment with the selective serotonin reuptake inhibitor (SSRI) citalopram, and the selective noradrenalin reuptake inhibitor reboxetine diminished the recognition of negative emotional faces, increased recall of positive self-referential words, and attenuated the amygdala response to fearful faces as measured by the functional magnetic resonance imaging. Similarly, studies in depressed patients have shown attenuation of the amygdala response to sad and fearful facial expressions during SSRI treatment (e.g., Victor et al. 2010; see also Haleem 2012).

Chapter 19
Confabulation of Speech, Faces, and Places

With mesial temporal lobe lesion, incapacity and/or metabolic extremes affecting the hippocampal and perihippocampal regions, anterograde memory deficits may be present with confabulation (e.g., Gilboa and Verfaellie 2010). However, substantial evidence supports a loss of the temporal tag with confabulation following orbitofrontal or basal frontal lesions and possibly with disconnection from the hippocampus (see Hirstein 2009). Confabulation may appear to the caregiver and therapists to be something along the lines of pathological lying and dissimilation. But, there are many ways to cross-validate these features. Ultimately, this might be accomplished through the demonstration of a functional metabolic or structural lesion to the brain systems responsible for the assimilation, consolidation, and storage of new information into memory. The processes are evident in learning with anterograde memory deficits and poor consolidation or carryover of new information resulting from damage or functional denigration of these neural systems (e.g., Fernaeus et al. 2013).

The hippocampus and fornix are arranged in a curvilinear fashion, swinging like a ram's horn from the anterior temporal region and immediately proximal to the amygdala, around through the subcortical occipital and parietal regions. The "ram's horn" culminates proximal to the origins of the secondary memory system at the dorsomedial thalamus (see Heilman and Valenstein 2003, 2012; e.g., Unsworth et al. 2012). The hippocampus has long been considered the primary memory system essential for learning new information, through the processes of long-term potentiation (Bliss and Lomo 1973; Alger 1984), consolidation (e.g., Guzowski et al. 2000), and reverberation (see Wang 2001) of the neural circuit. The consolidation process may occur as the hippocampus maintains the persistence or long-term potentiation of sensory information for a substantial period of time as with short-term memory. This appears to be augmented by kindling phenomenon at the anterior end of the hippocampus through emotional activation of the amygdala (e.g., Goodard and Douglas 1975; Kellett and Kokkinidis 2004; see also McGaugh 2004).

Bliss and Lomo (1973) discovered that high-frequency stimulation of neurons in the hippocampus results in a lasting increase in synapse strength, known as

© Springer International Publishing Switzerland 2015
D. W. Harrison, *Brain Asymmetry and Neural Systems,*
DOI 10.1007/978-3-319-13069-9_19

long-term potentiation. The long-term potentiation relies upon glutamate receptors (*N*-methyl-D-aspartate, NMDA) and blocking the long-term potentiation may eliminate learning or the consolidation of new information. Moreover, stimulus persistence at the hippocampal region may be augmented by kindling phenomena at the amygdala, if emotional learning is underway (see McGaugh 2004). Dysfunction in this system may convey with impersistence of the memory trace to the extent that the information is insufficient for acquisition or consolidation into more permanent memory stores. Alternatively, reduced regulatory control with functional incapacity of the frontal region may heighten persistence and promote rumination or perseverative themes, perhaps (see Sotres-Bayon et al. 2004; Lee and Choi 2012).

Much effort has been expended to improve or to facilitate hippocampal neurogenesis, another specialized function of this brain region. Pereira et al. (2007) used a magnetic resonance imaging approach to generate cerebral blood volume maps over time in the hippocampal formation of exercising mice. Among all hippocampal subregions, exercise was found to have a primary effect on dentate gyrus cerebral blood volume, the only subregion that supports adult neurogenesis. Moreover, exercise-induced increases in dentate gyrus cerebral blood volume were found to correlate with postmortem measurements of neurogenesis. Secondly, using similar magnetic resonance imaging technologies, the authors generated cerebral blood volume maps over time in the hippocampal formation of exercising humans. Similar to the findings with mice, exercise had a primary effect on dentate gyrus cerebral blood volume and these changes in cerebral blood volume were found to selectively correlate with cardiopulmonary and cognitive function. Taken together, these findings show that dentate gyrus cerebral blood volume provides an imaging correlate of exercise-induced neurogenesis and that exercise differentially targets the dentate gyrus—a hippocampal subregion important for memory and one which is implicated in cognitive aging and dementia (see Morrison and Baxter 2012).

The bulk of the hippocampus lies within the temporal lobe, where it has direct access below the auditory cortex for the acquisition of sounds, including linguistic speech sounds in the left temporal lobe and prosodic or melodic sounds within the right temporal lobe. But, the hippocampus travels toward the occipital lobe for the acquisition of visual information as with the lexical memory systems in the left occipital and temporal region (e.g., Vogel et al. 2012; Jobard et al. 2003; Turkeltaub et al. 2002; Bolger et al. 2005; Vigneau et al. 2006; see also McCandliss et al. 2003) and with the visual pattern or spatial memory systems of the right occipital and temporal region (see Kanwisher and Dilks, in press). Continuation into the parietal lobe allows for acquisition of somatic memories with postures underlying functional praxis within the left parietal region and with specialization for emotional postures and body sensations involved in heightened arousal at the right parietal region (Chewning et al. 1998; Heilman et al. 1995; Jeerakathil and Kirk 1994; Jeong et al. 2006; Heilman and Valenstein 1979; Ota et al. 2001; Hillis et al. 2005; Watson et al. 1994). This right parietal structure may be necessary and sufficient for the learning or acquisition of extrapersonal spatial locations (e.g., Capotosto et al. 2012) and those locations distant from the body proper. Also, this map maintains its dimensional integrity with repositioning of our body or eyes (see parietal eye fields).

This involves the consolidation of geographical coordinates, including those located at specific radial directions of the compass and more simply known as east, west, north, and south. These provide for internal three-dimensional representations of space distal to our bodies. The shifting of dynamic perspective may provide perceptual depth as with a hologram image (see Pribram 1991). The three-dimensional cytoarchitectural locations within our brain allow for this extrapersonal representation of space with replication of the dimensional coordinates in the pattern of neuronal networks activated from interacting with and traversing the terrain. A potentially important functional homologue presents here with the right parietal lobe specialized for east and west spatial comprehension (Paterson and Zangwill 1944; Benton et al. 1974, Walsh 1978), whereas the left parietal region appears specialized for left and right spatial comprehension (Gerstmann 1940, 1957; see also Cabeza et al. 2012). The former involves knowledge and comprehension of extrapersonal space and the latter involves knowledge more intimate to our body and awareness of personal space.

Damage within the right parietal region may evidence in geographical east–west confusion or spatial agnosia (Paterson and Zangwill 1944; Benton et al. 1974, Walsh 1978), whereas damage within the homologous left parietal region may evidence in left–right confusion (Gerstmann 1940, 1957; see also Cabeza et al. 2012) and confusion of the body parts and their fundamental postures or gestures (Heilman 1973). Language may ultimately bring both extrapersonal space and personal space under discussion. But, the depth of processing and analysis performed by the right hemisphere likely extends well beyond the level of the language processing systems providing access to bodily systems well beyond the linguistic or propositional speech systems. These regions stand ready to engage even under vital challenge or threat, where words no longer prevail and where language may prove ineffective.

Early research on the hippocampus involved the isolation of a slab of this tissue, extending from the hippocampus out to the cortex (Alger 1984). These cells are remarkable anatomically, in that they are, in essence, designed as reverberatory circuits to promote the persistence or the long-term potentiation of the iconic or echoic or haptic trace arising from the transduction of the sensory events within our environment. Indeed, the activation of this slab of tissue, maintained in a petri dish in a bath of physiological saline, was sufficient to result in repetitive action potentials of indefinite duration. The cell body within the hippocampus would be activated with a minute electrical current with the action potential moving toward the cortex. But, the axonal fibers of these cells bifurcate and the collateral axons return to stimulate that cell's soma once again—and again! These *reverberatory neurons* (Iijima et al. 1996) are self-stimulating in a fashion to perpetuate the sensory memory trace.

The repetitive and reverberating activation then may promote persistence of the memory trace available for consolidation or learning (Kinsbourne and Wood 1975; O'Keefe and Nadel 1978; Squire 1992; Eichenbaum and Cohen 2001). These hippocampal cells are cholinergic and cholinergic agonists (e.g., acetylcholinesterase inhibitors) may promote improved consolidation of the trace or learning (Lewis and Shute 1967; Woolf et al. 1984). Choline and the vitamin B complex are critical here in the manufacture of acetylcholine, the primary neurotransmitter of the hippocampus essential for learning and memory. Choline, often grouped with vitamin B, is the limiting factor for the production of acetylcholine, whereas acetate is relatively cheap and readily available for acetylcholine synthesis. Relevant to the college student or anyone wanting to promote learning and memory, is the functional relationship between ethyl alcohol consumption and the depletion of vitamin B (Morgan 1982; Martin et al. 2004). Vitamin B complex promotes learning and memory and its depletion provides for one of the nutritional dementias, secondary to chronic or repetitive alcohol consumption. Often the younger college students invest substantial resources in their tuition in order to learn, only to counteract these processes through their social activities with excessive alcohol consumption. Nutritional supplements might be helpful in this regard, and even in advance of the social or "the bar crawl."

The persistence and reverberatory action of hippocampal cells may be augmented by kindling phenomena (Goddard and Douglas 1975), which are observed with the activation of the amygdaloid bodies. The amygdala, located at the anterior medial temporal lobe, occupies a strategic anatomical location proximal to the anterior end of the hippocampus with projections from the cingulate gyrus. These connections provide access with startle stimuli or salient environmental events to systems facilitating habituation to the novel and/or salient stimulus (e.g., Levenson et al. 2012). But, the amygdala appears to function in kindling phenomena, where affective primes may accumulate and persist resulting in heightened intensity of the affective state with these events. Negative affective kindling may be critical to survival, insuring the consolidation and duration of emotional memories. Kindling phenomena from the amygdala may augment the persistence of information within these primary memory circuits partially as a function of the affective intensity of the event. Moreover, laterality research has focused on the asymmetry of emotional learning as a function of the affective valence. Strong negative emotional augmentation of learning or consolidation has commonly been ascribed to the neural architecture of the right brain (e.g., LeDoux 1992; Abercrombie et al. 1998).

The secondary memory system consists substantially of the bidirectional dorsomedial thalamic projections up to the frontal lobe within each brain (see Heilman and Valenstein 2003, 2012; e.g., Unsworth et al. 2012). This memory system provides for the maintenance of intention to the task and freedom from distractibility, allowing for concentration or focused intention to the task. Dysfunction in the secondary memory system may result in increased distractibility for what that brain is distracted by, as with tangential verbiage for the left brain and as with tangential spatial designs or emotional themes for the right brain. These systems are late to develop, though. Indeed, the more common complaint among freshman students for learning in their classes is that they cannot concentrate. They frequently report reading the same paragraph over and over again, failing to focus their intent on the material and without carryover of the material being learned. This improves as the student continues beyond their freshman year and with their advancing development of the frontal lobes. These dorsomedial frontal projections are potentiated through noradrenergic agonists and interactions with the locus coeruleus in the brain stem (Radley et al. 2008).

Learning is very much the function of the entire brain. Indeed, one might argue that learning is what the brain is all about, regardless of the brain area involved or the functional part of a cerebral system. The hippocampus or primary learning system allows for acquisition or learning (facts and events), whereas the frontal mechanisms allow for vigilance and concentration or intent to the task free from distraction or tangential variations in cognition (e.g., see Fuster 1980, 2004; see Barbas et al. 2013; Suzuki and Gottlieb 2013). With more severe hippocampal dysfunction or disconnection with the frontal region or with older and, perhaps, compensated lesions, patients may confabulate or seem to make up memories to fill the void (see Herstein 2009). The right temporal lobe patient or the individual with a deep posterior right cerebral lesion may confabulate for faces and places. The discussion might be initiated by saying to the patient "Do you remember me?" Before the conversation is finished, the patient might be discussing the many experiences together over the years, even though it was a first encounter. The focus is for the familiarity of a face and for the familiarity or experiences with a place. These confabulations for space or places may occur with fear and coercive complaints of being relocated or placed there against their desire with delusionary content.

Milner et al. note specific impairments in epileptic patients undergoing unilateral brain operations for the relief of epilepsy. Impairments are revealed on a variety of spatial learning and spatial memory tasks after right anterior temporal lobectomy, but only if the ablation encroached extensively upon the hippocampus and/or the parahippocampal gyrus. The tasks sampled ranged from simple delayed recall of the position of a point on a line (Corsi 1972; Rains and Milner 1994) to more complex ones, such as stylus maze learning. These demonstrations extend to visual (Milner 1965), tactile (Corkin 1965), and spatial conditioning or associative learning (Petrides 1985). Milner notes that the results for spatial memory are in marked contrast to those with visual patterns, including faces, abstract designs, or the figurative detail in representational drawings, where an impairment in recognition memory is demonstrable after right temporal lobe removal even when the hippocampal region is spared (Kimura 1963; Milner 1968; Burke and Nolan 1988; Pigott and Milner 1993). They also contrast with those for verbal material, where memory impairment is typically seen after left-sided but not right-sided temporal lobectomy.

Individuals prone to *panic attacks* with unbridled fear may be sensitive to increments in carbon dioxide levels (see Homnick 2012) and right frontal incapacity with dyspnea to stress or on exertion. These provide two fundamental mechanisms for the treatment of panic through manipulation of oxygen saturation levels and through minimal right frontal stress. Deep breathing may be helpful. But, these interpretations are contraindicated for the oft tendered "just breath into a paper bag" advice to increase carbon dioxide levels. Confabulation for space or place may also appear to be a disorder of orientation or mental status with *confusion for place* being the clue for a right brain disorder. Luria (1973) may have been one of the first to describe these disorders in neuropsychological terms in his book *The Working Brain*. He described lesion-induced spatial delusions in the temporoparietal cortical areas, whereby patients believed that they were simultaneously present in two separate towns. Benton and Tranel (1993), in their review of left and right hemisphere disease patients, further suggested that right-sided lesions may decrease topographical memory and the ability to recognize familiar surroundings. These findings are extended by the longitudinal investigations of taxicab drivers in London, where anatomical augmentation or structural enhancement occurs within these regions with topological experience over time (Woollett and Maquire 2011).

One construction contractor expressed extreme anger and homicidal thoughts after a head injury in a motor vehicle accident. He stated clearly that his anger was the result of being forced back into his "box." His attributions of coercion provide insights into the workings of the right cerebral hemisphere and its relative specialization for external threat. His spatial delusions involved movement to another place and a location which he related to something like nirvana with peace and calm. The delusions involved a forced or coerced relocation "back into this box" and expressed feelings of being hurtled through extrapersonal space. In his theoretical paper on vestibular spatial processing and emotion, Joseph Carmona (Carmona et al. 2009) provides relevant discussion. Joseph notes that the patient reported extreme hostility and homicidal cognitions associated with loss of spatial control during these delusions. The electroencephalogram was recorded during an anger induction phase, and the patient was encouraged to imagine a spatial episode. Results indicated marked right hemisphere beta activation at temporoparietal electrode sites during the session. He continues stating that the case described above illustrates the possibility that spatial processing interactions with anger or hostility may converge on a theme of control or coercion.

The spatial delusions occurred with generally impoverished learning and severe deficits in memory for human faces and places. The evaluation involved double dissociation techniques (Teuber 1955; Luria 1973; Luria 1980) and the use of syndrome analysis techniques (see Tonkonogy and Puente 2009; Luria 1980; see also Shenal et al. 2001), along with normative comparisons on standardized neuropsychological measures, and quantitative electroencephalography. This three-pronged assessment approach provides a substantial basis for syndrome analysis and for the localization of brain disorders. This approach benefits from converging evidence from each of the three assessment methodologies and stands solid on its legs, something like a three-legged stool, by analogy.

The findings from the quantitative electroencephalogram indicated decreased activation of the right hemisphere relative to the left. Specifically, there was a decreased beta activity over the right temporal and parietal region, which was consistent with the findings from the neurobehavioral status exam and from the standardized test results in the neuropsychological evaluation. The results were interpreted as supporting the role of the right hemisphere, and more specifically, the right parietal and temporal mediation of awareness of one's self in relation to one's location in extrapersonal space (Everhart et al. 2001). A similar patient may have been discussed in a case study by Nighoghossian et al. (1992). They described a patient with a right internal capsule infarct whose spatial delusion consisted of sensations of traveling through European cities on various days. He insisted on leaving the house, despite the protests of his family.

The patient suffering these problems may be at substantially increased risk for related problems due to the proximity of the brain regions involved and the now untrustworthy functions of those brain regions. One issue here with the spatial confusion, coercive feelings, and the like, is the risk for elopement. Some with right posterior lesions may walk into space with spatial confusion and travel indefinitely without appreciating that they are lost or that they are in the wrong place. A young businesswoman and chief executive officer (CEO) of her company was recovering from a right posterior brain disorder. She had been asked by her case manager to travel from Charlottesville, Virginia to an office in Salem, Virginia. After becoming concerned that she missed her appointment and after some follow-up investigation, she was found several states over, having driven or travelled through space as a person *lost in space*. All were grateful for her cell phone technologies, which provided the resources necessary to track her down. Pathology in this neural system might also be evident in the elderly dementia patients found on the freeway walking indefinitely and confabulating their directions and activities.

This is quite different from the rapid pacing and hyperkinesis that might be expressed in right frontal lobe dementia patients, where they may typically walk toward and from distinct landmarks in the setting. The right frontal lobe damaged patient may be characterized by descriptors like "impulsive" or "ballistic" or as having poor insight to safety. But, the secondary memory system in the right frontal region, if deactivated or dysfunctional, may lead to tangential variants of activity, initiated abruptly, without cautionary discretion. This may appear to the caregiver or therapist as indiscretion, with the attribution being that the patient is not following the rules or that the patient is a management issue. Although these are functionally accurate accountings of the patient's behavior, it may be helpful to understand that this is a brain disorder and one affecting the brakes or executive controls, which would normally prevent the initiation out into space without discretion or caution. This has survival value in many settings and may not only put the patient at risk but also the person trying to assist or manage the patient's care needs.

Confabulation with the left temporal lobe patient or the patient with deep posterior left cerebral lesions may consist of nonsense accountings for what has been said to them or the basic activities that they have been involved in during the day. This patient might be asked "What did you have for breakfast this morning?" The response might be "ham and eggs," initially. But, when asked again after 5 min, the patient might say "pancakes with syrup. And, yes, they were good." This can be most frustrating to the caregiver or therapist as the individual is communicative and speech appears to be functional, but the patient may be unable to learn or to recall the information provided them verbally.

In the absence of carryover for verbal information with a patient who appears to be verbally capable, the inference might be that this individual is not taking therapy seriously or that they are not trying. But, the caregiver will appreciate that the hippocampal lesion is demonstrative of a subcortical dysphasia, wherein the cortical speech analyzers of Broca's and Wernicke's areas may be intact, but where their connections with the primary memory system may be defunct. This might produce an impersistence of the memory trace for speech sounds and a failure to acquire or consolidate the verbal information (e.g., Morris, Friston, Bu¨chel, Frith, Young, Calder, and Dolan; also see review by Hamann 2001). Working memory may remain intact, for example, but result in a flattened acquisition curve, where no or little improvement or carryover is observed across multiple learning trials (e.g., Fernaeus et al. 2013). This may be more remarkable for the acquisition of events and facts, whereas procedural memory requiring the memory of rules, habits, or skills may be more affected with damage to the basal ganglia or striatum (Saint-Cyr et al. 1988; Packard and Poldrack 2003; see also Squire 2004). This might account for the bradykinesia with Parkinson's disease, for example.

The integrity of these systems, involved in the acquisition of new information, may be partially assessed using standard test instruments to measure verbal list learning. The Rey Auditory Verbal Learning Test (RAVLT: Rey 1964; Shapiro and Harrison 1990; Strauss et al. 2006; Fernaeus et al. 2013) and the California Verbal Learning Test (CVLT: Delis et al. 1987; Delis et al. 2000; Silveri et al. 2013) have been extensively used to assess for these purposes. Verbal learning or acquisition is assessed using one or another test of verbal paired associate learning and/or logical verbal information. The latter may be assessed through the recall of a short story (see Wechsler et al. 2009). Following a delay, the resiliency of these memory traces might be assessed a second time through delayed recall of a story, delayed paired associates, and so forth. Examples of this common technique are available through the Repeatable Battery for the Assessment of Neuropsychological Status (RBANS: Randolph 1998) and also through the Wechsler Memory Scale-IV (WMS-IV: Wechsler 2009; Wechsler et al. 2009). The consolidation of the memory trace differs from access to the memory. Problems with the latter have been related to secondary memory system deficits (e.g., Unsworth et al. 2012), where the integrity of the primary memory system may be demonstrated through intact recognition of the material among distracter items. Discrepancy analysis with comparisons among recall and recognition measures may be useful for this purpose.

Nonverbal learning and memory, including face encoding appear to be consistently implicated with right medial temporal function while word encoding is underpinned by the left hippocampus (e.g., Kelley et al. 1998; see Kalapouzos and Nyberg 2010). These functions may be evaluated through many alternative standardized tests administered with a focus upon one or another sensory modality. However, the majority are like those used within the public school system and consist of pencil and paper tests administered through auditory and/or visual modalities to the exclusion of the remaining sensory modalities. Commonly used tests designed to be less dependent upon auditory linguistic processing include the Rey Complex Figure Copy Test (Loring et al. 1990), Musical Tones & Melodies (e.g., Denman 1987), and the visual–spatial subtests of the WMS-IV (Wechsler et al. 2009) or with the earlier Wechsler Memory Scale-III (Wechsler 1997). The later instrument includes several of these subtests along with the deletion of facial recognition subtests (Faces I and Faces II) from the newer memory scale. However, the scientific basis for the WMS-IV was initially questioned as measurement theory rather than brain theory may drive the commercial production of this instrument (e.g., Loring and Bauer 2010).

Clearly, there is a need within neuropsychology for additional progress in the development of learning and memory tests to assess the acquisition, recognition, and recall within other sensory modalities and with multimodal processing demands. Often overlooked are impairments in the sensory processing and acquisition of information conveyed through the modalities of taste, smell, temperature, touch and pressure, pain, and vestibular function. Also, relevant here would be progress in the development of instrumentation to assess for altered thresholds or sensitivity across the left and the right hemibody and within left and right hemispace.

Hugdahl (1988) (see also Kimura 1967; Ley and Bryden 1982; Bryden and MacRae 1989; Pollmann 2010; Hugdahl et al. 2009; Hugdahl 2003, 2012) encourages the use of sensory extinction tests and specifically the dichotic listening procedures for the assessment of functional differences in cerebral laterality. However, tests assessing acquisition or carryover within each of the sensory modalities are needed. Moreover, one might envision the development of neuropsychological assessment instruments providing for comparison among the sensory modalities using normative data. Discrepancies between the patient's performance and the normative comparisons might provide for an objective basis in defining the relative integrity of lateralized neural systems processing each sensory modality, in addition to providing a basis for the evaluation of recovery of function within and among these neural systems. Some progress has been made in nomothetic-based comparisons, using one or another form of "discrepancy analysis" to evaluate and compare standard score performance indices which are discrepant with those of another subtest or scale. However, these too are scored on group data and may be less sensitive than within-subject performance comparisons.

Normative comparisons are typically based on group statistics, reflecting the mean and standard deviation of a representative sample of individuals from the group. The standard comparisons are largely devoid of within-subject comparisons specific to the individual with, for example, a localized loss of function. Paul Satz (1993) addressed this issue in his "threshold theory" and pointed out the regression to average in a high-functioning individual after a significant brain injury. Arguably, no tool is more powerful and specific in the assessment and localization of brain injury than the double dissociation techniques of Teuber (1955; see also Luria 1980; see also Tonkonogy and Puente 2009). In this scenario, the individual is used as his/her own control or comparison in the assessment of loss or stability of functions. Just as the strength of the left hand may be assessed using comparisons of the same individual's right hand, brain areas processing propositional speech expression may be compared with areas processing nonpropositional speech expression in homologous comparisons of functional neuroanatomy. Visual processing of linguistic images may be compared with visual pattern processing, and so forth.

Beyond the comparisons made within and among the sensory modalities, the evaluation of learning and memory may be further extended through the assessment of carryover or consolidation of emotional material processed within each sensory modality. Earlier, Kathy Snyder (Snyder and Harrison 1997; Snyder et al. 1998; Everhart et al. 2005) developed the Auditory Affective Verbal Learning Test (AAVLT). This test assesses the acquisition of a 15-item word list presented over

Fig. 19.1 Affectively valenced word examples for the AAVLT

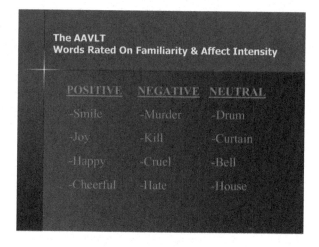

Fig. 19.2 Hostile violence-prone individuals may be disabled for learning positive or neutral information but with resilient capacity for recalling emotionally negative material

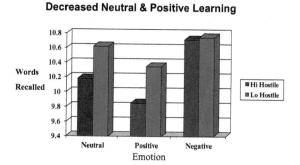

five consecutive learning trials. Thus, the format is familiar to many in its similarity to the RAVLT (Rey 1964; Shapiro and Harrison 1990; Strauss et al. 2006) and the CVLT (Delis et al. 1987). However, the lists consist of positive, negative, and neutral emotional content drawn from a collection of words normed on their affect valence, affect intensity, and their frequency of use within the English language. The lists were developed using the Toglia and Battig (1978) index of word norms. Examples are provided in Fig. 19.1.

This instrument has been used not only for the investigation of learning within positive, negative, and neutral affective categories but also for the investigation of the impact of the learning material on autonomic nervous system functions (Mollet and Harrison 2007a, b), where blood pressure and heart rate may be directionally controlled as a function of the affective valence of the items learned. Simply learning the positive list has been found to reliably lower mean arterial blood pressure and heart rate (Snyder et al. 1998), whereas learning the negative affective list reliably increases blood pressure and heart rate or sympathetic tone. Moreover, the evaluation of hostile violence-prone individuals reveals a learning disability on acquisition of positive and neutral information, whereas no disability is present in the acquisition of negative affective information (see Fig. 19.2; Everhart et al. 2008b). Additionally, Everhart et al. (2003) reported increased low-alpha power, using quantitative electroencephalography, in high hostiles during negative list learning. Low alpha power is inversely correlated with brain activation (however, see review by Klimesch 2012). The authors suggested that high hostiles may be more familiar with negative affective words, leading to decreased cerebral activation during learning. However, more recent research (Everhart et al. 2008b) indicates reduced arousal in high hostiles but with relative right cerebral activation regardless of list valence. Erik Everhart reviews some of the research using this instrument within his chapter on the AAVLT (Everhart et al. 2008a).

Chapter 20
Logical Linguistic and Affective Prosodic Speech

Propositional Speech

Most intimate to the localization controversy was Paul Broca's (1861) discovery of the role of the left inferior posterior frontal region or frontal operculum in expressed speech and the derivation of the fluency construct in the neuropsychological literature. The impact of this discovery was taken to the next level with the subsequent legendary conclusion, derived by Broca, when he stated "We speak with the left brain." Later, Wernicke (1874) extended our knowledge of the logical, linguistic analysis of speech with his presentation of a fluent patient with impaired comprehension or impaired meaning as a result of a superior left temporal lobe lesion. It is now fully common within the neurosciences to appreciate the left brain's specialization for logical linguistic or "propositional" speech; the role of "Broca's area" in fluent speech expression; and the role of "Wernicke's area" in speech comprehension or the provision of meaningful speech. Indeed, the layperson is instructed to appreciate changes in speech as a primary warning feature of an impending or actual stroke where a rapid presentation to the emergency room and perhaps treatment with a blood thinner such as tissue plasminogen activator (tPA) may alter the outcome or prognosis as with a nonhemorrhagic event or with regional cerebral infarction.

About 97% of right-handed people have language lateralized to the left hemisphere, with another 3% expressing a right-hemispheric or bilateral representation. Among left-handed people, language may be found within the left brain in about 70% of this group with bilateral representation appreciated in the remaining 30% (Coren 1992). The lateralization bias appears to be somewhat larger for speech production than for speech comprehension (Corballis 1998). Similarly, signed languages are also represented predominantly in these left cerebral systems (Corina

---

"It is not a logical world. If it were, men would ride side saddle." Little Jimmy Dickens, 92-year-old star of the Grand Ole Opry, 2013

© Springer International Publishing Switzerland 2015
D. W. Harrison, *Brain Asymmetry and Neural Systems*,
DOI 10.1007/978-3-319-13069-9_20

377

1998; Hickok and Bellugi 2000; see also Corballis 2010). Sign language, though, appears to involve more right cerebral processing, perhaps secondary to the coding of spatial relationships during these activities (Emmorey et al. 2002).

Although the history of expressive or nonfluent aphasia is most directly tied to the demonstration of a structural lesion at Broca's area, more commonly the neuropsychologist is involved in diagnosing a functional lesion, wherein the brain scan appears normal or insensitive to the behavioral speech deficit in question. This should not be surprising to anyone who has ever had their car repaired by a mechanic. If the mechanic took a picture of your automobile now running like an old clunker and, after looking at the picture, said that "The car looks fine...nothing showed up in the pictures," you would probably appreciate that you need further diagnostics performed on the vehicle, if not a new mechanic. A picture simply does not tell you how the car runs! Of course, the reverse argument holds: If the deficit is apparent in the image, the specific location, the type of pathology involved, and the arterial distribution of the lesion provide indications of probable cognitive, behavioral, and emotional features to be confirmed or disconfirmed on the neuropsychological evaluation.

The more sensitive measure of brain dysfunction may be derived from behavioral, affective, and cognitive measures. But, other methodologies are available to look at indices reflective of the functional state of the underlying tissue or brain systems (see Spenser et al. 1995). This might include positron emission tomography (PET), functional magnetic resonance imagery (fMRI), and the quantitative electroencephalogram (QEEG; Knight 1997; Moore et al. 1999; Hoffman, Lubar et al. 1999; Shenal et al. 2001; Thatcher et al. 1999) to mention but a few.

One younger man was referred to for neuropsychological services (Foster and Harrison 2001) with a long-standing expressive speech deficit, wherein he was nonfluent in the expression of logical linguistic or propositional speech. Figure 20.1 displays the delta magnitude (MV) recorded over his left frontal region at electrode sites F3 and F7. The recordings were taken during a quiet baseline state and subsequent to the imposition of the frontal stressor, which consisted of efforts to engage in expressed speech. Recall again that delta magnitude is the most commonly used bandwidth to identify a sluggish brain region with pathological electrical activation. It is common during deep sleep stages 3 and 4 but less so during awake episodes in normal brains. Delta waves are very slow, ranging from 1 to 4 Hz, and are of high

Fig. 20.1 Delta magnitude recorded over the left frontal region (F3 and F7) during quiet (no expressed speech) and during speech expression

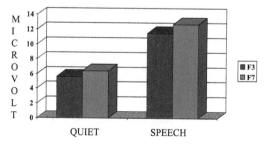

amplitude. In contrast, highly aroused awake states may be characterized by beta activation with low amplitude waveforms above 12 Hz. The patient demonstrates significant and pathological slowing of the quantitative encephalographic record on efforts to express speech as recorded over Broca's area.

For speech to be understood, signals from primary auditory projection pathways must first make bilateral contact with receptive areas in the superior temporal gyrus that are sensitive to frequency and amplitude modulated signals (e.g., see Sharma and Dorman 2012). Hickok and Poeppel (2007) specifically suggested that this information is directed in a ventral stream to the cortical region in the superior temporal sulcus and in the posterior inferior temporal lobe that serves as an interface between auditory-based representations and meaning derived from linguistic sounds. Simultaneously, signals are sent along the dorsal stream to frontal regions involved in motor planning via an area at the boundary between the parietal and temporal lobes. The left frontal lobe contributes the expression or fluency components to linguistic speech, whereas the left temporal lobe contributes differentially to the comprehension and meaning of linguistic speech. The specific feature of language wherein one comprehends speech, though, is not exclusive to the reception and analysis of linguistic input via the auditory pathways. Indeed, disruption of this localized cerebral region will alter the meaning of the speech produced or expressed by that individual through its forward contributions to the frontal lobe and eventually the motor output regions of that person's brain. So, the region is critical to the assignment of meaning and coherence of a sequentially processed series of phonemic sounds (see Duffau 2012). If this region is altered in its activation or metabolic level, through either a functional and presumably reversible change in the dynamic state of the region or through a lesion with cellular loss, then meaningful speech analysis, comprehension, and production may be altered (see Hugdahl 201; see Mesulam, 1990, 2000). This may be apparent in the production of semantic errors altering the meaning of expressed speech and in the semantic alteration of the patient's comprehension of speech, all the while preserving, and in transcortical lesions, perhaps increasing output (e.g., transcortical receptive dysphasia).

Moreover, Wernicke's area is not isolated from sensory processing within other modalities initiated far afield from the auditory analyzer. Information received at the primary projection area for vision and that received at the primary projection area for audition, for example, move toward higher order more complex analysis through the primary projection cortex on to the secondary association cortex for each modality and, then, eventually through the crossmodal interface (e.g., within the angular gyrus or tertiary association cortices) for the associated auditory schema or sounds corresponding with the visual stimulus event. Thus, a visual grapheme presenting to the occipital cortex and fusiform gyrus of the left hemisphere (e.g., Park et al. 2012) may elicit the corresponding auditory sound for that letter as information moves toward Wernicke's area. Yet, appreciate here that the auditory meaning or sound-based logical comprehension of that visual or of that tactile event is derived from the proper electrochemical and metabolic activation of Wernicke's area. Based on this fundamental component of the theory, one might appreciate that language or speech specifically, is interfaced with visual analyzers and with parts of the brain more directly dedicated to body part awareness, and postural placement or gestural configurations (i.e., the parietal lobe and somatosensory analyzers) found within the dorsal stream extending through the parietal region.

Once this component of speech corresponding to Luria's second functional unit has been processed, and the meaningful basis for expressions derived, then information flow continues through the intrahemispheric, longitudinal pathway toward the frontal lobe. The primary, and most directly efficient, pathway for this flow of information toward the expressive speech systems is via the arcuate fasciculus, a small cable essentially connecting the auditory analyzer in the temporal lobe to the premotor facial and lingual cortex in the frontal lobe. This small bundle of axons is most narrowly assessed in the analysis of speech repetition or disorders of repetitive speech. The examiner might ask the patient to "repeat what I say" followed by a commonly used passage within the industry (e.g., "No ifs ands or buts about it" or "The man drove the car."). Disconnection of the access to Broca's area from the second functional unit component consisting of Wernicke's area and Heschl's gyrus may result in a disturbance in repetition with relatively preserved comprehension. But this loop, along the arcuate fasciculus, is more significant diagnostically when the pathway is intact concurrent with dysfunction within and between the speech regions surrounding the pathway as may occur with the transcortical dysphasias. The hallmark or distinguishing feature of the transcortical dysphasias is intact repetition, oft referred to as echolalia. Echolalia is distinguished through the analysis of the transcortical motor dysphasia with impaired fluency and the analysis of transcortical receptive dysphasia with impaired comprehension, but with integrity for the pathway allowing for the echo or repetition of phrases or even sentences presented to the patient. This circuit, indeed, appears as a loop from auditory input to motor output, yielding an echo-like response.

Two patients may be relevant to discussion of this diagnostic category. One suffered a devastating catastrophic event perhaps most relevant to the men reading this book. The patient's "honey do list" included a specific and oft requested repair of the garage door, which seemed to hang up at the most awkward times. This young mother had her baby in her arms as she struggled with the door with the car running in the garage. When it finally broke loose, it struck her on the head rendering her unconscious and at the rear of the running vehicle's exhaust, as the door closed behind her. She presented unconscious at the medical center with carbon monoxide exposure and an anoxic encephalopathy. As the neuropsychological evaluation of this woman was performed, essentially with flat electrical activation on the electroencephalogram as with brain death, she maintained the ability to occasionally repeat speech passages with echolalia. Interestingly, this new mother maintained one additional component of speech a the examiner initiated nursery rhymes like "Mary had a little lamb," she could sometimes complete these over rehearsed passages by saying "who's fleece was white as snow."

The second patient, by way of example, was a distinguished attorney who had assisted the labor leader Jimmy Hoffa with his legal affairs. By popular legend or factual account depending on your perspective, this labor boss was actively affiliated or involved with the mafia. This became the focus of several major news organizations and of the Federal Bureau of Investigation when Hoffa disappeared and when decades later, at the writing of this book, Hoffa's body had still not been found. The attorney now was advancing in age and was seen with an active brain disorder involving his speech and/or language systems. The nurses seemed to enjoy his presence perhaps partially because what they presented him, via their speech, echoed back to them via his speech. For example, the nurses might say "good morning" and the patient would echo back "good morning." They would say "it is a beautiful day" and the patient would echo "it is a beautiful day." I confess my own motives were a bit more sinister as I worked up to what I thought was the critical question. This man echoed much of what I said to him with the most notable exception of my question about "…what happened to Jimmy Hoffa…," where he clearly stated "Oh, I don't know about that man!"

Nonpropositional Speech

For the most part, and especially in men, the right brain appears largely devoid of logical sequentially processed language. But as Kenneth Heilman and others have noted (e.g., Heilman et al. 1975; Ross and Mesulum 1979; Tucker et al. 1977; Heilman et al. 2012), the right brain most definitely contributes to vocal communication, with damage in the right brain resulting in one or another of the aprosodias. The right brain contributes much of the prosody or affective intonation in speech (Heilman et al. 1975; see also Bourguignon et al. 2012), otherwise referred to as "nonpropositional" speech components. In addition, the right brain may, in many cases, provide a rich assortment of expletives, grunts, groans, screams, and emotional sounds or pantomimed gestures (e.g., Harrison et al. 1990). Even Broca's famous patient Tan, after some frustration with the perseverative utterance "tan, tan, tan," would activate the right brain speech systems with the expletive "goddamit!" The right brain not only provides for the majority of prosodic intonation but also the right frontal lobe appears critical in the regulation of the volume of speech. Thus, an expressive dysprosodia, resulting from right frontal lobe dysfunction (see below), may present with socially inappropriate volume, which may convey anger or emotional intensity within the social milieu.

The expression of subtle affective nuance or emotional prosodic variation is attributed to the right frontal lobe, perhaps as a regulatory function over reflexes, for example. Indeed, right frontal stress or exceeding the capacity of this region to regulate sympathetic cardiovascular functions may result in hyperreflexia of the vocal apparatus, resulting in altered or elevated pitch or in altered or elevated volume or intensity of vocal expressions. Therapeutic interventions may seek a reduction in right frontal stressors (e.g., control issues, anger provocation, and negative affective induction). But, also therapies for the expressive dysprosodias may involve interventions originally developed for the hyperreflexic disorders expressed at other body parts, such as flexor tone at the upper extremities, where extensor exercises might be in order or where activation and exercise of oppositional muscle groups (antagonistic to the hyperreflexic muscle group) may prove beneficial. If by analogy, hyperreflexic responses of the vocal apparatus present with elevated volume or pitch in speech, then oppositional therapies and contextual therapies might be considered. For example, an expressive dysprosodia may be apparent in a contextual setting with heightened background noise levels, where the patient is now inappropriately loud in comparison with others against the same background. Noise reduction techniques, educational awareness, or reduced vocal motor tone through muscle therapy techniques may be helpful. One interesting variant of the dysprosodic disorders is that of the *foreign-accent syndrome* (e.g., Gurd et al. 2001) The impact of a stroke, for example, may be sufficient to alter the prosody, pitch, and tempo of speech to the extent that the patient may acquire a foreign accent.

By analogy from the aphasic disorders derived from left cerebral pathology, research on prosodic or affective intonation in speech eventually followed within a similar functional neuronal system with disturbance in auditory affect recognition or receptive dysprosodia resulting from alterations at the right temporal region (Heilman et al. 1975). Such a patient might develop an auditory affect agnosia or general failure to understand affectively intoned speech and melody (see Heilman et al. 2012). By extension, the expressive dysprosodias were found to more commonly result from cortical or subcortical involvement of the right frontal region (Ross and Mesulam 1979). For instance, patients with right hemisphere lesions may speak in a monotone voice. Transcortical dysprosodia (motor, receptive, and/or mixed transcortical dysprosodias) may leave repetition of emotional sounds intact with disturbance in the other component(s) within this functional neural system. This may be more loudly appreciated in vocalizations, where the moaning and/or wailing patient produces hyperprosodic or exaggerated emotional sounds with irrelevant emotional conveyance and decreased meaning (e.g., transcortical receptive dysprosodia; Harrison et al. 1991). In normal individuals, the processing or rehearsal of affective words may facilitate right hemisphere performance measures with activation and interfere with left hemispheric processing (Demakis and Harrison 1994; Bunce and Harrison 1991).

Chapter 21
Social Approach and Social Avoidance

Activation of the left brain appears to correspond with social approach behaviors, whereas activation of the right brain has primarily been associated with socially avoidant behavior or withdrawal. The social approach behaviors resulting with left frontal activation seem consistent with the desire to speak and the act of speech expression derived from this brain region. This would also appear consistent with the positive affect resulting with activation at this region, where it may promote some verbal discourse and continuity of the contact sufficient for socialization within the setting. Activation of the left frontal lobe occurs with an infant's approach toward the mother and continues to correspond with approach behavior through adulthood (see Cox and Harrison 2008a, b, c, 2012; Fox and Davidson 1987, 1988; Davidson 1995). Moreover, approach motivation is linked to relative left-hemispheric brain activation, which in turn leads to more right-oriented behavior by approach-motivated individuals (Roskes et al. 2011). Roskes and colleagues manipulated the subject's motivation with some motivated toward avoidance and some motivated toward approach behavior. Performance on a line-bisection task under low or high time pressure revealed right-oriented judgments only in those motivated for approach and under high time pressure or rapid pacing demands.

Activation of the right frontal lobe has been found with mildly negative affective expression and with social avoidance or withdrawal (e.g., Kelley et al. 2013). This may be functional in many ways to diminish the intensity and the duration of the negative affective interaction. The act of withdrawal from a mildly negative social encounter, or from a less socially desirable individual, might even have survival value and especially prior to additional provocation where the emotion may gain momentum and where aggressive encounters may avalanche into social approach behavior with anger. Avoidance motivation with activation of the right hemisphere appears to result in a similar left hemispatial bias and with heightened potential for negative affective analysis and caution.

The fundamental problem in the theoretical developments attempting to explain these functional differences originated in the discussion of overt anger or "anger-

© Springer International Publishing Switzerland 2015
D. W. Harrison, *Brain Asymmetry and Neural Systems,*
DOI 10.1007/978-3-319-13069-9_21

out" behavior where anger resulting from right brain activation does correspond with social approach behavior resulting in directed close proximity aggression Harmon-Jones and Allen 1998; Harmon-Jones et al. 2003a; Cox and Harrison 2008a, c 2012; Kelley et al. 2013). In one laboratory, David Cox put this to the test, with the findings supporting the prominent view of "anger in" resulting with right frontal activation, but with "anger-out" behavior resulting from concurrent activation of *both* the left and the right frontal lobes (Cox and Harrison 2008a, c). David's findings leave open the question of energy level, dominance, and submission, the valence of the attack, and self-attributions, which may vary with combined activation scenarios.

Anger may present in milder forms, and even with diminished self-esteem, through facetious or sarcastic humor. Although many would endorse these forms of humor as sometimes deeply funny, the slant is toward the negative and even to the hostile affective valence. Sarcasm conveys a harsh or bitter derision or irony. The sarcastic statement or innuendo may clearly implicate its victim and even torture that individual through negative affective humor. Sarcastic humor has been found with relative decreased activation of the left frontal lobe, with diminished self-esteem, and with depression. Right hemisphere damage may eliminate the ability to appreciate sarcasm and with a prominent role for the right temporal region (e.g., Shamay-Tsoory et al. 2005). These findings may have direct relevance to the lateralization of emotion and specifically the dominance–submission model discussed below (Demaree et al. 2005).

Behavioral asymmetries are well documented for approach-motivated humans and nonhumans alike with a right-oriented bias reflecting activation of the left hemisphere along with the cognitive, affective, temporal, and spatial response biases of that brain. Just as the chicken might turn its right eye toward the kernel of corn, while elevating the left eye in a vigilant position against the hawk or fox, toads attempting to catch prey are more likely to flick their tongue to their right side than to their left side (Vallortigara et al. 1998); and dogs wag their tail toward the right at the sight of their owner (Quaranta et al. 2007). The approach behavior of romantic partners also varies reliably as they turn their head to the right to kiss twice as often as they turn it to the left (Güntürkün 2003). Relatedly, approach-motivated humans show a rightward bias on the line-bisectiontest (Friedman and Förster 2005; Nash et al. 2010), which has been rather consistently associated with left-hemispheric brain activation in the literature (Harmon-Jones 2003a, b).

Motivational models of emotion focus on motor or behavioral activation responses altered by an emotion. Approach behaviors or states lead to greater left hemispheric activation, while withdrawal behaviors or states lead to greater right hemispheric activation at the frontal cortex (Davidson 1993, 2000). Gray (2001) stated that the distinction between the emotional states of approach and withdrawal is conceptually one of the better-validated constructs in emotion research. In contrast to the valence model, left frontal activation is not associated with positive valence, but rather a behavioral approach state, whereas right frontal activation is associated with a behavioral withdrawal state and not negative valence, per se (Harmon-Jones 2004a). This distinction gains importance due to the fact that certain emotions, such as anger, have a negative valence but may produce behavioral approach rather than withdrawal.

Davidson's (1993) motivational model describes activation of the left frontal lobe as resulting in approach-related behavior, while activation of the right frontal lobe is associated with withdrawal-related behavior. Davidson (2003b) states that left-sided prefrontal cortex activation is required for the initiation of behavior related to appetitive goals and that hypoactivation of the left prefrontal cortex may result in depression (see also Demaree et al. 1995). Alternatively, right-sided prefrontal cortex activation occurs with behavioral inhibition and vigilance that is related with negative or aversive emotional states and traits. Gray (1990) behavioral activation system and a behavioral inhibition system for emotion where the behavioral activation system relates to "hope" and "happiness" emotions, for example, while the behavioral inhibition system is related to emotions such as "anxiety" and "fear." Personality measures relating to the behavioral activation and behavioral inhibition systems correlate significantly with anterior brain asymmetry which indicates approach or withdrawal states (Harmon-Jones and Allen 1997; Sutton and Davidson 1997).

Anxiety disorders, in addition to more general elevations in anxiety, have been found with diminished frontal lobe activity, particularly at the left frontal region (Akiyoshi et al. 2003; Everhart et al. 2002). Numerous investigations using electroencephalography (EEG) to assess brain function have reported relative left frontal deactivation and right frontal activation in anxious individuals (Baving et al. 2002; Blackhart et al. 2006; Foster and Harrison 2002a; Nitschke et al. 1999; Papousek and Schulter 2002; Wiedemann et al. 1999). The meta-analysis by Thibodeau et al. (2006) provides the conclusion that anxiety is associated with a relative increase in right frontal electroencephalographic activity. Research has also indicated relative right frontal activation and left frontal deactivation in patients suffering generalized anxiety disorder (Mathews et al. 2004). Patients with generalized anxiety disorder in this project were also found to possess a higher N-acetylaspartate/creatine ratio at the right dorsolateral prefrontal cortex. The finding of relative right frontal activation in comparisons with that at the left frontal region extends to patients suffering panic disorder (Akiyoshi et al. 2003; Wiedemann et al. 1999).

Instead, Heilman and Gilmore (1998) describe an approach/withdrawal system dependent on intrahemispheric interactions between the anterior and posterior brain regions. These authors state that the right posterior cerebral hemisphere has a special role in motor activation or for the preparation to respond to the stimulus array. The frontal lobes are viewed as mediators of avoidance behaviors, whereas the parietal lobes are seen as mediators of approach behaviors. This theoretical position receives empirical support with evidence from lesion studies. Lesions of the right frontal lobe lead to the inability to inhibit the initiation of responses, manual grasp responses, and inappropriate approach behaviors. Consequently, frontal lobe lesions may produce approach behavior as a result of disinhibition of the parietal lobe. From this theoretical account, the parietal lobe (which mediates approach) is normally inhibited by the frontal lobe. With lesioning or deactivation of the frontal lobe the parietal lobe becomes disinhibited, producing excessive approach behavior. Lesions of the parietal lobe leads to neglect, deviations of eye, head, and arm movements, inability to respond, and withdrawal behaviors (Heilman and Gilmore,

1998). Heightened withdrawal behavior may be a result of increased activation of the frontal lobe secondary to a decrease in parietal activation, leading to inhibition or suppression of approach behavior. Schutter et al. (2001) found support for parietal mediation of approach behavior through measurement of asymmetrical activation of electrical activity in the EEG.

Integration of the valence and motivational models of emotion has been difficult due to emotions like anger or hostility, which are negative in valence, but which can produce approach behaviors. High-trait measures of anger, hostility, and aggression along with anger induction correlate with increased baseline levels of left relative to right frontal activation (Harmon-Jones and Allen, 1998; Harmon-Jones, 2004b; Harmon-Jones and Sigelman, 2001). Harmon-Jones (2004a) suggests that anger generates approach behaviors that are aimed at resolving the anger, including acts of aggression. However, work done by Harmon-Jones is potentially at odds with prior research indicating right hemisphere functions in negative emotion (i.e., Demaree et al. 2002; Foster and Harrison, 2004; Burton and Labar 1999; Blair et al. 1999).

Gina Mitchell (Mollet and Harrison 2006; Mitchell and Harrison 2010) provides the following considerations. Gina states that to overcome the discrepancies it may be necessary to look at cerebral activation in brain areas other than the frontal lobe s. In two case studies of patients with hostility and anger problems, it was found that hostility resulted from deactivation of the right frontal lobe and increased activation of the right temporal parietal region (Everhart et al. 1995; Demaree and Harrison 1996). These cases might best be interpreted with caution as they are single subjects conveying extreme anger disorders. Demaree and Harrison (1997) found activation of the right temporal region in high hostile participants in response to a pain stressor as evidenced by changes in dichotic listening. These results indicate that the right temporal region is important for anger. Waldstein et al. (2000) found negative emotion induction to correspond with bilateral activation of the frontal lobes and predominately in the endorsement of anger. Waldstein et al. suggest that anger may be related to either right or left frontal activation depending on how an individual handles emotion. Gina further states that anger may be more likely to activate the left frontal lobe with outwardly expressed anger through approach behaviors (see also Cox and Harrison 2008a, c; also Cox and Harrison 2012). Individuals suppressing anger may be more likely to activate the right frontal lobe as a result of anger suppression and withdrawal (Waldstein et al. 2000). Regardless, individual differences in emotional style may be more relevant in determining cerebral activation differences and their presence may be a contributing factor to some of the controversy in the literature.

Gina continues to note that an integration of Davidson's (1993) model and Heilman and Gilmore's (1998) model may provide the most parsimonious explanation for cerebral activation patterns concurrent with anger. Anger produces changes in both the anterior and posterior brain that are associated with behavioral approach or withdrawal. Additionally, it would lend support to Borod's (1992) addition to the valence model, indicating the importance of both the right and left frontal lobe s, and the right posterior cerebral regions in emotion. Other emotions may be better served through this approach to emotion as well. In accordance with Davidson's model, depression is most often associated with relative right frontal activation or left frontal hypoactivation (Baehr et al. 1998; Davidson et al. 1995) and produces social isolation, low-self esteem, hypoactivity and withdrawal behaviors.

Cerebrovascular accidents restricted to the left frontal lobe may be associated with sadness and depression (Gainotti 1972; Hama et al. 2006; Morris et al. 1996; Shimoda and Robinson, 1999). Depression severity has been related to the proximity of the lesion to the frontal pole (Robinson et al. 1984). Even without lesion pathology, depression is associated with volumetric hemispheric frontal lobe asymmetry (Kumar et al. 2000). Electroencephalographic investigations have demonstrated relative left frontal deactivation and right frontal activation in depressed individuals (Debener et al. 2000; Henriques and Davidson 1991; Schaffer et al. 1983), including individuals with a history of depression (Henriques and Davidson 1991; Vuga et al. 2006). Also, a meta-analysis of these findings resulted in the conclusion that depression was related to a relative increase in right frontal activation using electroencephalographic measures (Thibodeau et al. 2006). Additional evidence suggests that depression may also be concurrent with suppression of the right temporal-parietal cortex (see Heller 1990). Incorporation of Heilman and Gilmore's (1998) model is necessary here to account for noted changes in right posterior cortex during anger or hostility and depression.

Further, this helps to account for other behavioral correlates of depression such as decreased arousal and decreased performance on spatial tasks that require the implementation or recruitment of the right parietal lobe (Henriques and Davidson 1997; Bruder et al. 2012). The three models that are presented here offer interesting and differential views of cerebral activation in emotion and pain. Evidence supporting each model should be carefully considered in order to advance theory in this area. For example, the right hemisphere model and the valence model have received support primarily by data derived from patients with brain damage and/ or stroke (e.g., Heilman et al. 1975; Ross et al., 1981; Adolphs et al. 1996; Adolphs 2001; Borod et al. 2002; Heilman et al. 2004). However, this perspective is supported even on multigenerational studies using electroencephalography in combination with magnetic resonance imagery (Bruder et al. 2012). On the other hand, the motivational model has primarily been investigated through the examination of anger (Harmon-Jones and Allen, 1998; Harmon-Jones 2004b; Harmon-Jones and Sigelman, 2001), an emotion that is typically viewed as having high negative emotionality and a high arousal component. Furthermore, different methodological approaches (behavioral vs. EEG) have served each model differently. Demaree and Harrison (1997) found right hemisphere temporal lobe activation after stress in high hostiles using a behavioral measure, while Harmon-Jones and Allen (1998) found left frontal activation with anger using the electroencephalogram (see Holland et al. and Harrison, 2012; see also Mitchell and Harrison 2010).

Chapter 22
Positive and Negative Emotion

The more prominent theories on the hemispheric specialization or lateralization of emotion include the *right hemisphere* (Borod et al. 1983; Heilman and Bowers 1990), the *valence or balance* (Tucker 1981; Tucker and Williamson 1984; Davidson and Fox 1982; Ehrlichman 1987; Silberman and Weingartner 1986; Davidson 1993; Ekman et al. 1990), and the *behavioral inhibition system/behavioral activation system* (BIS/BAS: Gray 1982; McNaughton and Gray 2001; Davidson 1995; Harmon-Jones et al. 2003, 2004), or *approach–withdrawal* models. *Manuck's model* of optimistic anticipation of future events versus negative reflection on the past events is relevant here in the discussion of cerebral laterality (Manuck et al. 2000). Heath Demaree and others contributed to this list with the addition of the *dominance–submission model* (Demaree et al. 2005) and the *quadrant model* of emotional processing (Shenal et al. 2003; Foster et al. 2008; Carmona et al. 2009). An attempt is made here to extend these emotion theories here to include the temporal duration of emotional memory. Specifically, this model briefly addresses the concept of transient duration for neutral memories and the persistence, over time, of emotional memories.

Right Hemisphere Model

With the exception of Heilman's right hemisphere model, much of the focus in emotion research has been on the frontal lobes. A major distinction might be drawn between the right hemisphere model and the others, in that this model attributes a primary role of the right brain to emotional processing, comprehension, and expression independent of valence. This link between emotion and the right brain was made some time ago, including Mills' (1912a, b) observation of decreased emotional expression with unilateral right hemispheric lesions. Affective indifference secondary to right hemisphere lesion was appreciated by Babinski (1914) and by others (Denny-Brown et al. 1952). From this perspective, emotions are substantially lateralized to the right hemisphere. Although some level of processing might be

© Springer International Publishing Switzerland 2015
D. W. Harrison, *Brain Asymmetry and Neural Systems,*
DOI 10.1007/978-3-319-13069-9_22

accomplished by the left brain, any deeply intense emotion appears to require right cerebral and brain-stem systems as these neural systems provide a fundamental basis for underlying arousal or activation in emotion.

One multigenerational longitudinal study identified a significant role for the right hemisphere in depression. Using electroencephalography, the authors (Weissman et al. 2005) initially reported an alpha asymmetry with less right than left parietal activity in family members who were at a high risk for developing depressive disorder (Bruder et al. 2007). Klimesch (2012) reviews the evidence for alpha activity and concludes that it largely reflects cortical inhibition. For example, alpha power, rather than cortical desynchrony, is larger over visual cortices when attention is focused on the auditory part of a compound auditory–visual stimulus (Foxe et al. 1998). Similarly, alpha power is larger over parietal regions (dorsal stream) when a task engages the ventral stream (Jokisch and Jensen 2007). Moreover, in cueing and hemifield tasks, alpha power is heightened over the ipsilateral than the contralateral hemisphere (e.g., Sauseng et al. 2009; Kelly et al. 2006).

Neuroanatomical support for the earlier finding of an association between familial risk for depression and cerebral asymmetries followed in subsequent research (Peterson et al. 2009), where cortical thickness was compared between the high- and low-risk groups. Cortical thinning was found across large expanses of the lateral aspects of the parietal, posterior–temporal, and frontal cortices of the right hemisphere in the high-risk group. The authors offer that the convergent findings of less cortical activity over the right parietal sites using electroencephalography in individuals at high risk for depression may derive from cortical thinning.

Additional support for this was found in a continuation of this multigenerational study of individuals at risk for depression (Bruder et al. 2012). Seventy-five subjects from the electroencephalagraphic study underwent magnetic resonance scanning about 5 years later. The high-risk participants were the biological descendants of probands having major depression, whereas the low-risk participants were descendants of individuals without history of depressive disorder. The authors used voxel-wise correlations of cortical thickness and alpha power with age and gender covariates. High risk was associated with reduced activity over the right parietal region, which was associated with cortical thinning. Moreover, just as diminished left frontal cortical activity has been thought to reflect reduced activity in approach-related systems, right parietal hypoactivity has been long associated with emotional underarousal in major depressive disorder (e.g., Jaworska et al. 2012).

In the right hemisphere model, Heilman and colleagues provide for a distinct role of the frontal lobe in the regulatory control and expression of emotion, whereas the posterior brain regions are critical for the sensory analysis, reception, and comprehension of emotion. The right brain appears specialized for the processing of the negative emotional valences, including anger, sadness, and fear. Moreover, the right brain is viewed by many neuroscientists as the "emotional brain" (Heilman and Bowers 1990). With this perspective, damage to the right brain may, by definition, involve an emotional disorder of one form or another. Deep right frontal lesions of the basal ganglia often yield sad lability with crying (e.g., Ross and Rush 1981; Constantini 1910; cited in Oettinger 1913; see also Parvizi et al. 2009). Right orbitofrontal lesions often yield incremental reactivity to anger or rage behavior (e.g., Agustín-Pavón et al. 2012; see also Fulwiler et al. 2012). Fear or profound apprehension often corresponds with posterior temporal and inferior temporal lobe pathology. This may result from damage to the fusiform area or regions critical to the recognition of familiar and/or trustworthy faces. Also, a deep white matter lesion may disconnect these areas, including the fusiform gyrus, from the regulatory control or inhibition of activation by the frontal lobe via the longitudinal tract. Facial recognition may be distorted with visual form esthesias within the left hemispace and often of feared facial images connoting the devil or those trying to do them harm (Walters et al. 2006; Mollet et al. 2007; see also Lutterveld and Ford 2012). In contrast, deactivation of the second functional unit or the back of the right brain (parietal lobe) often results in diminished emotionality or flat, bland affect (e.g., Heilman et al. 1978).

Pathological laughing or crying may result from the release of cortical inhibition of upper brain-stem centers that integrate the motor activation patterns involved in laughing and crying (Woodworth and Sherrington 1904; Wilson 1923). Thus, emotional lability may be considered an essential part of the pseudobulbar palsy syndrome resulting as a consequence of lesions in the corticobulbar pathways (Woodworth and Sherrington 1904). Ross and Rush (1981) propose that pathological affect may result in patients with lesions of the right inferior frontal lobe in association with a major depressive disorder. McCullagh and colleagues (McCullagh et al. 1999) also provide early evidence for the role of the prefrontal cortex in the pathophysiology of pathological affect with emotional lability. Others (e.g., Parvizi et al. 2001; see also Parvizi et al. 2009) suggest that the critical lesions eliciting pathological crying are located along fronto-pontocerebellar pathways.

Right-sided brain damage frequently presents with flattened or bland affect expression (see Heilman et al. 2012). But, the key issue here is in the distinction between affective deregulation or *reactive* emotionality as opposed to the diminished arousal and diminished sympathetic tone that may exist in the patient with a right parietal lobe lesion, for example (Heilman 1997; Heller 1993; Heller et al. 1998). Some individuals are able to appreciate emotional emptiness following these lesions. But, far more often than this, the patient is largely unaware of their emotional disability and bland affective presentation as the lesion typically results along with impoverished awareness of deficits or anosognosia and anosodiaphoria (see Prigatano 2010). Woe is the student or committee member where the instructor or group leader presents with a bland monotone facial and prosodic speech expression, as this truly can challenge that person's ability to remain aroused and attentive in the captive setting. *Neuronal mirroring* provides one's best clue of the emotional features of the other person in the social setting. So, emotionally flat, low arousal levels may be echoed in the expression of similar features in the observer with mirror activation of these brain regions.

A local physician, after his fall and right hemisphere brain injury, reported that he could no longer feel any intensity with his emotions. Nor could he appreciate his wife's emotional displays. The spouse reported feeling emotionally bankrupt in the relationship, whereas he had always been a warm and loving partner. This man conveyed a flat nonresponsive affect similar perhaps to patients in the initial study on facial expression deficits with right-sided lesions by Buck and Duffy (1980). Borod's group has demonstrated this deficit in facial expression in right-sided stroke patients across several studies (e.g., Borod et al. 1985, 1986). The lack of depth in the emotional response of these individuals was demonstrated by Heilman et al. (1978) using skin conductance responses (SCR), a relatively pure measure of sympathetic tone. These right-sided stroke patients were nonresponsive to emotionally provocative stimuli. Interestingly, the lack of emotionality has been associated with classic psychopathy (Hare and Quinn 1971), where the public is looking for remorse as a basic indication of being human in the context of the acts that are perpetrated and especially where another has been harmed. These patients, though, maintain the facial motor controls necessary for normal linguistic expressions or verbal communication against the background of flattened affective facial expression and diminished sympathetic drive.

Therapy with this couple involved establishing contractual agreements based on what each partner valued from this relationship. This did involve a regular investment in time spent talking directly with his wife, instead of listening to the television. It also involved efforts to learn and to produce basic facial configurations, which might convey emotional concern, empathy, and interest to his partner during these meetings. Although some progress was made in this respect that was clearly quantifiable, the spouse remained emotionally distraught, needing to broaden her social support network to fill the residual emotional void. Nothing is more clear in the rehabilitation literature than the issue of caregiver stress with these individuals often at substantial risk for stress-related ailments and injuries as they support the patient's needs in the hospital and home setting (Pearlin et al. 1990). Indeed, the variable more important to the patient's prognosis, overall, is the continuity of care and support by the caregiver (Brodaty et al. 1993).

Although many families are large and many family members are concerned and caring, the burden of support largely falls on a single caregiver and this individual may eventually appreciate oppositional interactions, rather than supportive interactions, with other family members, perhaps with differences in opinion as to the best approach to care. This is one of the more difficult and stressful activities in our society. Many of these individuals have multiple responsibilities for the care of their own children and the care of their parent, for example. The patient's prognosis is improved not just through patient care, though this is where the majority of effort is expended, but through care and support of the caregiver with efforts to promote the quality of life for these individuals. Through these efforts, we may promote continuity of care and support of our patients and for the long haul.

All too often, the caregiver enters the hospital shortly after the patient is admitted with deterioration of health or physical injury, which can be related to the trauma of care. Imagine the demoralizing impact to the patient and to the broader family unit when an overly heroic caregiver falls and breaks her hip trying to prevent the patient from falling. With the caregiver out of the picture and needing community resources, the patient might be removed from the home setting for care in a nursing facility. Although this has many benefits, many are clear that this is the least desired outcome. Instead the patient may relate this with terminal life events, going to the nursing care facility prior to death, rather than with expectations for recovery.

Another issue warrants research as there are potential benefits of the right parietal stroke and diminished emotionality. In some cases, the patient with flat affect and minimal affective regulatory demands may actually have reduced cardiovascular risk or even more stable glucose levels along with reduced anger or depression. Elevated and/or reactive sympathetic tone as expressed through elevated blood pressure and heart rate may benefit from this lesion, which reduces sympathetic tone and arousal. A middle-aged woman with chronic depression culminating in a suicide attempt with a gun to the right side of her head was evaluated in a rehabilitation unit. The bullet went up through the right parietal lobe exiting the vertex of the skull. The social work department was immediately responsive to the suicidal risk issues and sought advice based on the neuropsychological evaluation. Imagine their surprise when they began to appreciate that the deep and long-standing negative emotions were reduced or eliminated, at least in the short run, through extirpation of the right parietal lobe. The patient was flat and essentially nonresponsive to negative emotional provocation. Of course, this example is not meant to imply that this is a method of inoculation. But, sometimes the negative effects of brain damage come along with a few positive effects, if we look for them.

More dramatic, though, is the individual with deregulated emotional states. This may involve negative emotional regulation through right frontal executive controls. It may involve positive emotional regulation via left frontal systems if it is mild in intensity, such as a smile or a cursory sense of humor. However, if the positive affect is deeply robust or intense, then the right cerebral hemisphere is likely involved. Thus, gelastic lability or pathological laughter is associated with left frontal or more often with bifrontal disinhibition (Demakis et al. 1994; Parvizi et al. 2001, 2009). Regulatory control for anger is more associated with right orbitofrontal activity with increased inhibitory regulation via the uncinate fasciculus over the right amygdale (e.g., Fulwiler et al. 2012). This system has been a primary focus in the laboratory with hostile and violence-prone individuals (Harrison and Gorelczenko 1990; Demaree and Harrison 1997b; Cox and Harrison 2008a,b,c; Carmona et al. 2009; Herridge et al. 1997; Holland et al. 2007; Mitchell and Harrison 2010; Herridge et al. 2004; Holland et al. 2001).

This is partly due to the function of these cells in regulating anger. But, also this region appears to be involved in regulating sympathetic activation of the heart through increased heart rate, systolic blood pressure, and sympathetic tone, more broadly defined. More specifically, beyond the capacity of these cells to maintain anger control, the disinhibition may present with an expressive dysprosodia with an avalanche in the volume of speech and reactive increments in blood pressure and heart rate. Moreover, this region appears to be directly involved in the regulatory control over cardiovascular reflexes, glucose mobilization, oxygen saturation, time/temporal management, and possibly cholesterol mobilization as these may help insure survival for the fight! But, chronic destabilization of these response systems may result in cardiovascular disease (Siegel 1985; Smith and Pope 1990; Smith et al. 2004; Suls and Bunde 2005; Mitchell and Harrison 2010; Walters and

Harrison 2013a, b; see also Cox and Harrison 2008b; see also Anthony et al. 2006). It is also evident in research findings that the metabolic syndrome may result with alterations in the reward systems of the left hemisphere and especially with increased abdominal fat (Luo et al. 2013). One example here might be the labile response to negative affective stressors (anger) with poorly controlled glucose mobilization or clinically with diabetes mellitus (Walters and Harrison 2013a, b).

Lesion or deactivation of the lentiform bodies deep within the right frontal lobe may result in deregulation over crying (see Parvizi et al. 2009). Following a functional (metabolic) or structural lesion here, simply saying the word "sad" and waiting several seconds may be fully sufficient to decompensate the patient, resulting in crying behavior and perseveration or rumination on a sad affective theme. Now, with cerebral mirroring in others that are concerned and caring people, the problem may well be aggravated or accentuated. More specifically, when a person has difficulty controlling their emotion with crying behavior others might elevate their sad prosody and empathetic expressions, substantially increasing the stress on the patient's emotional processing systems to manage even more sad emotional provocations. This includes the prosodic input to the patient, through tone of voice and affective facial displays congruent with the patient's sadness. This is not, by itself, depression. Instead, these brain regions appear to be critical for putting the brakes on the expression of sad emotion and for redirecting the emotional content to another valence. So, this is emotional inertia or impaired regulatory control as defined elsewhere on the discussion of the frontal lobes. Such a patient can very often be rendered free of the perseverative theme and crying behavior simply by switching the affect to a neutral or slightly positive valence. Simply saying "the weather is nice today" may be adequate to redirect the emotion, providing a prompt to a more satisfactory emotional display for the social setting.

An example here would be a minister who suffered a deep right frontal stroke, only to find himself deregulated in the expression of crying. Just talking about sad things or even asking "are you sad" would decompensate his ability to control his expressions. On return to the pulpit, he was most debilitated by sad or emotional gospel hymns, which when played would yield uncontrolled crying. Now all of this became much more difficult to manage as his supportive, but misinformed, parishioners tried to let him know how much they cared. This made it virtually impossible for him to do his job and he was looking at disability. With education and explanation of the syndrome, though, enough progress was made so that he was able to continue work and to gradually increase his tolerance for sad emotional themes. Again, spastic or hyperreflexic laughter has long been associated with descending frontal fibers and the lentiform bodies (see Parvizi et al. 2009).

Emotional decompensation is typically more transient or brief in duration with a positive emotional valence than with a negative emotional valence. The left brain is specialized for the rapid sequential processing and elimination of information, whereas the right brain is a bit more persistent and less forgiving or forgetting. The Behavioral Neuroscience Laboratory at Virginia Tech estimates that the subjects participating in their research projects have more than 50 positive events occurring in their interactions with other people each day. People smile at the participants. They say nice things to them. They are friendly to them. They hold the door for the participants. They ask them how they are doing (expressing concern). For the majority of these people, there are relatively few negative events present each day and, in most cases, the number of these events is small relative to the number of positive affective events experienced. But, they report taking the negative emotional material or experiences to bed with them, dreaming about them, and processing these events throughout the night. Neither is the right brain dismissive of these events as they may later be of relevance as the individual becomes attuned to the evolving threat, should the material be meaningful to survival or safety in the broader sense.

Among the negative emotional valences, the effect with anger deregulation appears more persistent with priming or kindling of the emotion over time. This may be due to the mechanisms of the right amygdale in promoting emotional memory where the valence is highly relevant to survival and with the decompensation resulting from the perception of threat or loss of control in the setting (e.g., Morris, Friston, Bu"chel, Frith, Young, Calder, & Dolan; also see review by Hamann 2001). Although controversial, the persistence of anger and fear may be recurrent throughout the night promoting high activation REM sleep within the neural systems and especially those of the right brain. The recurrent processing promotes consolidation of the event in memory but also provides for lighter sleep and the increased probability of awakening throughout the night. Though these features are disturbing with depression or with a brain disorder, the evolutionary significance of these brain systems and their role in survival is clearly important.

The left brain is capable of one primary emotional valence that we know of and that is happy emotion (however, see approach-related emotions: e.g., Kelley et al. 2013). The neural substrates of the left brain provide for a smile and it may display a sense of humor or interest. Schiff and MacDonald (1990), for example, reported that experiencing positive emotions generated greater changes at the right side of the face, while negative emotions generated greater changes at the left side of the face. Consistent with this evidence, Lee et al. (2004) report significant activation at the left dorsolateral frontal and temporal lobes in response to positive emotions and at the right dorsolateral frontal and temporal lobes in response to negative emotions.

Moreover, emotions expressed by the left frontal region appear less intense and less persistent. Emotional intensity for any valence, including happiness, anger, sadness, and fear appears to demand participation by the right hemisphere (e.g., Sackeim and Gur 1978), just as approach-related emotions appear to demand participation by the left frontal region. Relative deactivation of the left frontal lobe may occur with increased negative valence underlying the humor. The right brain may also prefer an emotionally negative or hostile slant on the happiness or humor, for example, with specialization for cynical or sarcastic humor or caustically facetious humor. It may be "funny." However, the humor may be costly to someone. It may serve the purpose to degrade the social prowess of the target or convey some degree of rule violation within the broad or specific social context. The perceptual bias of the right cerebral hemisphere leans toward negative affect. It also appears to relish in socially inappropriate humor violating basic social proprieties or pragmatics. It might also find the observance or participation in harm to others or to animals humorous, when logically this is inconsistent with many social or cultural rules and expectations.

The Valence or Balance Model

The *valence model,* also referred to as the *balance model,* has largely been based on research investigating lateralized frontal lobe activation during positive or negative emotional valences. Evidence for hemispheric differences in emotional processing is also written on our faces, in the differential emotional expressivity of the two sides of the face (Darwin 1872). Relative activation of the left frontal lobe has been associated with positive affect, whereas relative activation of the right frontal lobe has been associated with negative affect in normal subjects. But, in a comprehensive review of emotional valence in facial expressions, Borod and colleagues (1997) found that the balance in emotional expression was not symmetrical. Overall, the left hemiface appears to be more involved in emotional expression in general (e.g., Blackburn and Schirillo 2012) even though the left hemiface appears more involved specifically in negative emotional expressions (Indersmitten and Gur 2003; Rhodes et al. 2013). This finding appears to hold on investigation of facial movement during emotion. The muscles of the left hemiface move more than their counterparts on the right during emotional expression (e.g., Dimberg and Petterson 2000; Nicholls et al. 2004). These asymmetries in facial expression are thought to arise because the muscles of the lower face are largely controlled by the contralateral hemisphere (Brodal 1981).

Interestingly, the left cheek is more often prominently displayed than the right cheek in portraits and photographs (McManus and Humphrey 1973; Powell and Schirillo 2009). This leftward bias is strongest when the model wants to display emotion, but is eliminated when concealing emotion (Nicholls et al. 1999). Many studies have demonstrated that emotions are rated as more expressive when they are displayed on the left side of the face (Powell and Schirillo 2009) and individuals who are more emotionally expressive are more likely to present the left cheek when posing for a portrait (Nicholls et al. 2002). Portraits featuring the left face are judged as more emotional (Nicholls et al. 2002) and the leftward posing bias also appears to be stronger among females (Nicholls et al. 2002). The preference for turning the left cheek in portraits appears to be the result of right hemisphere activation. Thomas and colleagues (Thomas et al. 2012) found that, overall, politicians were more likely to display the left cheek in their official photographs, consistent with prior reports of a leftward posing bias in portraiture. However, conservative politicians were significantly more likely to display the left-cheek bias than were liberal politicians.

Bob Rhodes compared left and right hemifacial electromyography (facial muscle action potentials) in hostile violence-prone individuals and those scoring low on hostility measures with much positive social affective cognition. Both low and high hostiles had heightened electromyographic or facial muscle activation over the left hemiface. However, the hostile group displayed significantly greater facial dystonia evident in elevated facial tension. This was remarkable in the left-sided facial dystonia of the hostile individuals. Matt Herridge (Herridge et al. 1997) investigated the lateralization of sympathetic drive to the right cerebral hemisphere as a function of posed facial expressions with elevated sympathetic tone in hostiles and with the production of negative facial configurations, through contraction of the corrugator muscle. Hostiles poorly habituated to sympathetic tone derived from negative facial configurations, whereas low hostiles poorly habituated to positive facial configurations.

George Demakis (Demakis et al. 1994) also looked at this as a function of lateralized frontal lobe damage. Pathological laughter or gelastic lability, in this case, occurred with significant elevation in skin conductance at the right hemibody (left frontal lobe damage). In general, the expression of positive emotion in normal subjects has resulted with left hemisphere activation based on these behavioral indices. Similarly, on the perceptual side, negative faces are often identified more rapidly or more accurately when presented within the left visual field with projections to the right hemisphere (Ley and Bryden 1979; Everhart and Harrison 2000; Harrison and Gorelczenko 1990; Harrison et al. 1990). Najt et al. (2013) recently reviewed the visual half-field studies of hemispheric specialization in facial emotion perception and reevaluated the empirical evidence with respect to each of the partly conflicting hypothesis derived from the dominant theoretical proposals. These authors conclude that their findings indicate a left visual field/right hemisphere advantage for the perception of angry, fearful, and sad facial expressions, whereas a right visual field/left hemisphere perceptual advantage exists for happy facial expressions. Furthermore, they conclude that the findings for the perception of specific facial emotions do not fully support the right hemisphere hypothesis, the valence-specific hypothesis, or the approach/withdrawal model. Instead, they advance a position based on their systematic literature review and the results from their study that a consistent left visual field/right hemisphere advantage exists only for a subset of emotional faces, expressing negative emotional configurations. This conclusion advances a "negative (only) valence model." These include facial expressions of anger, fear, and sadness. The processing of happy faces may be superior within the right visual field/left hemisphere. Alternatively, the right hemisphere model may still hold, if the emotional valence is intense.

Much of the research supporting the valence model has come from the electroencephalographic evaluation of activation over the frontal lobes. Here again, left frontal activation has occurred with positive emotional states, whereas right frontal activation has occurred with negative emotional states (e.g., Davidson and Fox 1982; Davidson and Henriques 2000; Ekman and Davidson 1993). These findings largely parallel those of neuroimaging studies (Wager et al. 2003; Bench et al. 1992). These affective differences in the lateralization of positive and negative emotion are present early in the developmental course. Fox and colleagues (Fox and Davidson 1987; Fox and Davidson 1988; Fox et al. 1995) have provided evidence of left frontal activation to positive affective events in infants and right frontal activation to negative affective events. Missing from these theoretical arguments is the potential relevance of movement out into extrapersonal space as opposed to withdrawal or narrowing of the behavioral fields to the proximal or personal spatial domain.

However, much of this research has involved only modest levels of stress or activation of the frontal lobe(s), whereas much of our clinical interest in these systems relates to those situations involving extreme levels of stress and where the metabolic range of the regulatory cells has been exceeded as may occur with a rage, panic, cry, or other deregulated and reactive states. This limitation on the valence model may be great, since this research is characteristic of an institutional setting where only modest stress is likely to be approved by an institutional review board or human subjects committee. Therefore, research that involves exceeding the capacity of these systems would not be approved and might be most difficult to handle in an ethical and concerned fashion. This would be even more difficult to justify with vulnerable subjects (e.g., children and individuals with psychopathology). However, life provides this level of challenge and even more. This is especially the case on vital challenge or confrontation. But, it might be evident in those among us with diminished capacity under stress that would be considered mild to moderate for others.

Life does not require the approval of an institutional review board or human subjects committee. Naturally occurring events, including the loss of a loved one or catastrophic events may easily exceed the metabolic capacity of these brain regions and possibly result in profound and unbridled anger, rage, fear, or sadness. If an analogy can be made between these brain systems and muscle systems, then we might easily see that resistance or regulatory efforts in the form of activation, as in the valence theory, are useful up to a point or level of resistance. After that level is exceeded, the metabolic or electrical activation should be acutely and robustly diminished as in the point at which the muscle is about to tear from our bones from limb exertion. This acute loss of regulatory control may have survival value as the suppression of these emotions is thrown to the wind and where we our desperate in our struggle against these events. These are some of the derivations of capacity theory (e.g., Carmona et al. 2009; Mitchell and Harrison 2010; Williamson and Harrison 2003; Holland et al. 2011; Walters and Harrison 2013a; see also Collins and Koechlin 2012; Klineburger and Harrison 2013).

One example may be proffered here from a younger man with developmental speech deficits, consisting of a combined expressive dysphasia and expressive dysprosodia (Williamson et al. 2000). By his history and his parent's confirmation, he had been nonfluent for both components of the speech examination for his entire life despite no recollection of trauma or significant medical events. As part of the assessment, a quantitative electroencephalogram was obtained prior to and subsequent to the stressful demands of producing speech output. Delta magnitude was recorded as the bandwidth commonly used to assess for pathological processes within the cerebral hemispheres, where this bandwidth, otherwise, might be heightened in the deepest dimensions of non-REM sleep in a normal individual. The results provide evidence of the frontal regions shutting down under stress (see Fig. 22.1). Capacity theory holds that this function would release or "unbridle" the posterior brain regions from frontal lobe regulatory or inhibitory controls via the longitudinal tracts. Also relevant, though, are the horizontal tracts with inferred far-field effects across the corpus callosum and specifically at the homologous cortical region(s).

Philip Klineburger (Klineburger and Harrison 2013) aptly notes in this regard, that the system must be efficient in its metabolic reallocation or energy utilization. Philip argues that increased metabolic activity in one region of the brain should

Fig. 22.1 Evidence for frontal lobe areas deactivating as a function of mild stress from initiating speech in conversation

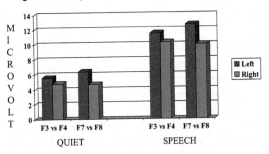

precipitate decrements in metabolic activity in the oppositional brain regions. Through this argument, the capacity theory is now integrated with the assumptions of quadrant theory with activation dynamics yielding far-field effects on multiple brain regions. From this perspective, any efforts at strict localization of function, again, must be qualified by the demands of the equipotentialist and where the imprint of experience or expression is broad based in its effect on seemingly disparate brain systems. Moreover, Luria (1980) and Hughlings Jackson (1874) would add the vertical dimension within these functional neural systems with the three basic functional units and with hierarchical analysis and processing, respectively. It would appear reasonable that other permutations exist where activation of one region of an oppositional neural system might elicit reciprocal activation of the region opposed to it, perhaps.

Behavioral Inhibition System/Behavioral Activation System

The BIS/BAS model was originally proposed by Jeffrey Gray based on nonhuman animal research (Gray 1981, 1982, 1987, 1990, 1994; Gray et al. 1997) with many efforts to adapt the model to human research, including the development of the BIS/BAS scales (Carver and White 1994). These efforts have focused on self-report assessment instruments or questionnaires rather than on more direct behavioral measures. Thus, any effort to generalize here is a bit worrisome. Regardless, lateralized EEG findings have been reported with higher BAS scores corresponding with left frontal activation (Harmon-Jones and Allen 1997), whereas higher BIS scores correspond with right frontal activation (Sutton and Davidson 1997). An alternative approach might be to evaluate the predictions drawn from this theory using behavioral measures as demonstrated by Gray. There is a fundamentally close relationship between the BIS/BAS model and the behaviorally more specific approach/withdrawal model discussed below.

Approach/Withdrawal Model

One of the fundamental roles of the lateralized brain is to establish the basic reaction guidelines for approach and withdrawal ("common sense") which, although balanced, provide for a dynamic concerted shift in favor of the appropriate bias (to approach and engage or to avoid) for any given situation. This system was obviously somewhat skewed for one man who was completing his neuropsychological evaluation after 8 days of hospitalization for snakebites. The man described his actions and those of the snake as follows. He said "I saw a copperhead, so I went over and I kicked it. It bit me. So, I kicked it again. It bit me again." The snake had obviously bested the man in his efforts to engage and in this instance as an act of aggression. But, either combatant might have done well to avoid or to escape the conflagration, rather than to initiate and to redress the near fatal engagement. The man had taken some heart, though, from the explanation provided by his physician. He reflected on the explanation quoting the doctor as follows. "Snakes typically avoid or escape the presence of humans, except during the dog days of summer, when a cloud forms over the snakes eyes and they are unable to see."

Kinsbourne (1978) suggested a fundamental relationship reflecting cerebral lateralization, wherein there was specialization of the left hemisphere for approach behavior and specialization of the right hemisphere for withdrawal behavior. This was evaluated earlier on with rodents in a T-maze, for example, with a string tied to the animal's tail. The degree of approach or avoidance conditioning within the variants of the maze was literally recorded by measuring the length of the string, where longer strings allowed for inference that the animal ventured further out toward the reward and, presumably secondary to elevated positive emotion. Approach and withdrawal appear to reflect components of different emotions corresponding with the specializations for those emotions by distinct neuroanatomical architecture within the left and right cerebral hemispheres.

The approach–withdrawal model draws on the same theoretical foundation as the valence model with frontal activation asymmetries being the overriding element of interest. However, a distinction may be drawn where the determination of emotionality is derived based on whether the emotion tends to provoke approach behaviors or withdrawal behaviors, rather than examining behaviors in terms of positive or negative affective valence. This model proposes that emotions that tend to provoke withdrawal behaviors will result with relative activation of the right frontal region, while approach behaviors will result with relative activation of the left frontal region (Davidson 1995). Electroencephalographic studies of clinical populations provide support for the model, where relative right activation or left deactivation of depressed patients is related to the degree of depressive symptom exhibition and particularly withdrawal type behaviors (Heller et al. 1998; Henriques and Davidson 1991; Robinson and Downhill 1995). For example, Jaworska et al. (2012) found that unmedicated depressed adults were characterized by lower activity in the left frontal cortical region implicated in approach/positive affective tendencies as well as diffuse cortical hypoarousal associated with right parietal inactivity, though sex differences emerged in their comparisons.

Additional empirical support is derived from electroencephalography studies demonstrating frontal lobe cortical patterns of activation which correlate with measures of BIS/BAS, consistent with the predictions of the model (Harmon-Jones 2001; Hewig et al. 2004; Wacker, Heldmann et al. 2003). Substantial research exists in support of the BIS/BAS predictions that anger is lateralized primarily to relative activation of the left frontal region with numerous studies indicating relative activation of the left frontal region in response to anger provocation (Lane et al. 1997; Kelley et al. 2013). For example, Wacker and colleagues (2003) demonstrated that relative activation patterns of the left anterior region in response to imagining a past event that elicited anger were strongly correlated with scores on measures of BAS. The Harmon-Jones group has provided evidence indicating that individuals who believe that they will have the opportunity to engage in an angry response to an elicitor demonstrate greater relative activation of the left anterior region, compared to individuals who believe that they will not have the opportunity to respond to the same stressor (Harmon-Jones et al. 2003). Consistent with the postulations of the BIS/BAS model, it would appear logical that the expression of anger would be considered as an approach behavior and lead to relative activation of the left anterior region, while inhibition of such behaviors would be considered withdrawal behaviors and would lead to relative activation of the right anterior region. For example, Kelley et al. (2013) manipulated cortical activation using direct current transcranial stimulation. This project provided additional evidence that anger associates with greater relative left frontal cortical activity predicting approach-oriented aggressive action, whereas anger associates with greater relative right frontal cortical activity predicting inhibited rumination.

Building from the evidence produced by Harmon-Jones and Allen (1998) regarding the utility of resting anterior cortical activation to predict trait anger, researchers have shown that resting relative activation of the left anterior region corresponds with measures of BAS and trait anger (Hewig et al. 2004). However, it has been proposed that there is a relationship between BAS and positive affect, causing some researchers to question whether the observed patterns of anterior cortical activation are due to the BAS system or positive affect (Carver and White 1994). Moreover, in the demonstrable case of relatively increased left frontal activity one may appreciate the inference for reduced relative right frontal activity, where diminished regulatory control over negative emotion (e.g., Agustín-Pavón et al. 2012), anger (e.g., Fulwiler et al. 2012), and sympathetic control (Demakis et al. 1994; Herridge et al. 1997; Carmona et al. 2008) are fundamental features. Thus, part of the final analysis may relate to the loss of regulatory control over anger with deactivation of the right frontal/orbitofrontal region or relative activation of the left frontal region or both. Interestingly, quadrant theory extends beyond these two areas with dynamic activation across the cerebral quadrants and raises the specter for indirect diagonal relationships between the left frontal and right posterior quadrants and the right frontal and left posterior quadrants (Foster et al. 2008; Shenal et al. 2003).

Also, the subcortical structures of the basal ganglia and brain-stem dopaminergic pathways appear to be most influential. Wacker and colleagues (Wacker et al. 2013) note that several models describe an underlying motivational system for the modulation of incentive salience, reward sensitivity, and behavioral approach (e.g., Depue and Collins 1999). Moreover, they argue that dopaminergic transmission has been firmly established in the central role underlying these functional features of this behavioral approach system (for a review, see Depue and Collins 1999; cited in Wacker et al. 2013). The authors further note that parallel findings, largely based on electroencephalogram studies in humans (e.g., Davidson 1998) support the similar role of the approach system "that is activated by the perception of goals, initiates goal-directed behavior toward those goals, and is associated with approach-related motivational and emotional states (i.e., desire, wanting, enthusiasm, pregoal-attain-ment positive affect, but also anger; see Harmon-Jones 2004)." They conclude from their review that the neuroanatomical foundation for this approach system is found to encompass not only regions of the left prefrontal cortex but also the dopaminer-gic circuitry central to the BAS.

Optimistic Anticipation of Future Events and Negative Reflection on Past Events

The left brain appears to be a rapid sequential processor with high energy level and positive attributions and expressions best related to the initiation of activity with a positive anticipatory disposition, free of negative affective concerns or cautions. This brain seems to be a "happy go lucky" brain free from the weight of negative reflections on the past, including those events weighted toward the negative va-lences through failure experiences, embarrassment, injury, or a negative impact that we may have on others through our expressions or behaviors. It will return to home alone (e.g., after a right-sided cerebrovascular accident) without the resources of the right brain and without concern or caution for its safety. It will try to drive a car or run a table saw, and all without appreciation of the deficits and safety or supervision needs for these activities.

   In contrast, the weight and the essence of the world may rest on the right brain, ever-accumulating cautions from the rules that we acquire from society, our par-ents, and from our previous failures. This includes historical accounts where reflection may diminish the probability of our attempting these activities again and where a lesson, once learned, is a lesson for life. Many have mused over the ever-apparent negative consequences of alcoholism for the individual and for their family, whereas the same behavior occurs again and again and with indifference by the alcoholic. These behaviors have been described as originating from a patient "in denial." For the clinical neuropsychologist, though, this conveys anosognosia and anosodiaphoria, where only the latter reflects denial or disbelief and where the former reflects a failure to recognize one's problems with diminished insight into their deficits. Otherwise said, one must first recognize one's deficits, if one is to deny that they exist.

   An example might be found with a patient suffering from obsessive–compulsive disorder (by her therapist's diagnostic system). Her obsession with germs and dis-ease-related contamination seems to be supported after the loss of a family member to infectious disease. Her therapist relates that she, and many others with these symptoms, seems unable to learn that her concerns are excessive and that they negatively impact her family. Carryover of therapeutic statements and discussions appear to be minimal. Within this context, the individual here may be presenting with anosodiaphoria and earlier on with anosognosia (see Prigatano 2010). The anosognosia is treated when the patient can repeat the therapist's verbal arguments and rationale. But, this left-brain-related function often has minimal impact on the belief. The failure to change the belief is anosodiaphoria, a hallmark of right cerebral dysfunction. Moreover, in this example, the learning disability is expressly related to carryover or acquisition of the emotion and learning that one does not need to be as cautious as they feel like they must be for safety. This possible relationship of obsessive–compulsive disorder symptoms with right cerebral pathology receives some additional support from a case study of a man suffering from obsessive–compulsive disorder by his therapist's diagnosis. Quantitative electroencephalography demonstrated significant slowing over the right frontal region (Harrison, unpublished).

These emotions and behaviors may reflect other aspects of a right brain disorder beyond the anger and beyond the poor insight to deficits. Relative activation of the left brain may convey, through release mechanisms, the expectations of the individual with overriding positive anticipation of future events related to alcohol consumption. Release of this system would promote optimism rather than reflection on the harm and negative outcomes produced by drinking in the past. These may include the loss of your job, your family, and financial stability, as examples. In contrast, individuals with a larger right frontal region in the anterior cingulate gyrus (capacity inferred) describe themselves as experiencing heightened worry about potential problems, fearfulness in the face of uncertainty, shyness with strangers, and fatigability. These differences in temperamental disposition to fear and anticipatory worry in men and women, then at least correspond to anatomical asymmetries underlying capacity within the medial frontal regions and specifically at the anterior cingulate (Pujol et al. 2002; see also Amunts 2010).

By analogy, this positive optimism may be expressed during overt aggression, where one perceives that they will win and that they will inflict harm on the other person, rather than that they will be harmed. Some degree of self-righteousness and feelings of invulnerability may present in situations where the anger occurs with social approach behavior and self-attributions that this is a principled act. One patient with an anger disorder was adamant that if she did not address the error made by the cook at a local quick food restaurant when her order was incorrectly filled (a hamburger without pickles), others might receive similar treatment. Whether or not this is a valid attribution, the anger that she conveyed in her facial expression, prosody, and elevated volume of speech more probably activated the sales person's right brain via neuronal mirroring. This would potentially lower the ability to communicate logically and linguistically through her propositional speech systems via contralateral inhibition across the corpus callosum (by example). Instead, the sales person *would remember* the anger and control issues and that she did not like this woman. The circularity might result in even further decrements in the ability to comprehend future orders from the customer or, at the extreme, a refusal to provide service to this customer.

Dominance and Submission Model

Heath Demaree (Demaree et al. 2005; see also Cox and Harrison 2008b) in a review of prominent neuropsychological theories of emotion, wherein brain laterality is foundational to the functional constructs, acknowledges the processing specialization of each brain and offers a compelling revision or extension to these theories. He argues for using "dominance" as an important construct underlying frontal asymmetry. Dominance is defined as "feelings of control and influence over everyday situations, events, and relationships versus feelings of being controlled and influenced by circumstances and others" (Mehrabian 1994, p. 2). Moreover, the energetic, socially interactive, and positive behavioral and affective dimensions resulting and derived from left frontal activation would provide the individual strength by association with a social or politically based group within that setting. Socially avoidant, negative affective interpersonal behaviors with inefficient logical linguistic verbal interactions with others would render the individual relatively isolated, alone and with dysphoric affect, wherein negative affective bias and self-reflection might promote a submissive posture. This would include the political social array in opposition to this individual where the group would compromise that individual's efforts even for fairness or justice, if that be the goal. Chronic right frontal activation would be derived from situations, including chronic social stress or oppositional states with negative attributions.

Heath notes that depression involves downward shifts on the dominance dimension (Mehrabian 1995; Plutchik 1993). Depressive symptoms relate to a diminished sense of control and self-assurance and increments in feelings of being beaten, overwhelmed, helpless, and hopeless. Social dominance has been decreased in nonhuman animals with serotonergic antagonists (e.g., Raleigh et al. 1991). Moreover, clinical data suggest that depressed patients taking serotonin-specific reuptake inhibitors (SSRIs) show increments in personality attributes related to dominance, including boldness and assertiveness (Brody et al. 2000). Dominance plays a further role in the distinction between internalizing and externalizing behavioral problems, where internalizing problems relate to low dominance and where externalizing problems relate to high dominance. Further discussion of an integrated theoretical position with consideration of the spatial processing biases of each brain (extrapersonal space/right brain and personal space/left brain) may be in order now with converging evidence at hand. The reader is further directed to the discussion of proximal or personal space and distal or extrapersonal space to better understand the role of the left brain in the former and the role of the right brain in the latter.

This framework provides for a bias in the assignment of attribution blame either to the self or to others. Moreover, research indicates that approach emotions, broadly defined, are rated as relatively high dominance, whereas withdrawal emotions are rated as low dominance. In this sense, approach-related anger, along with approach-related happiness, would associate with feelings of dominance. Avoidant-related fear and disgust would associate with feelings of submissiveness. In the derivation of this position, activation of one or of the other frontal region may provide for a distinction of relevance with implications for psychotherapeutic and pharmacologic interventions, respectively. The latter interventional category receives support through converging lines of discovery with lateralization of neurotransmitter systems in these asymmetrical and sometimes oppositional left and right cerebral hemispheres. Dopaminergic activation increases informational redundancy, whereas noradrenergic arousal facilitates orienting to novelty. Moreover, evidence that these neurotransmitter pathways are lateralized in the human brain is consistent with the left hemisphere's specialization for complex motor operations involved in hand and tool use (e.g., Bracci et al. 2012) and the right hemisphere's specialization for the integration of bilateral perceptual input (Tucker and Williamson 1984).

Transient Neutral and Persistence Emotional Memory Model

Processing differences between the two brains indicate that negative affective memories may persist indefinitely and beyond that duration derived from neutral and even modestly positive memories. Paul Foster and colleagues (Foster et al. 2012) developed the Emotional Controlled Oral Word Association Test (e-COWAT) to investigate these differences concurrently with the sympathetic responses derived from recalled memories differing in affective valence. Completion of the test requires the recall of positive, negative, and neutral memory events and, subsequently, the generation of words (fluency) related to the memory. Interestingly, the average age of the memory varied in a diametrically opposite direction to the number of words recalled across the three affect conditions. Neutral memories reflected the most recent age for the memory of events, where the *age of the memory* increased with happy words and where the oldest or most persistent associations with memory events were found for the negative emotion or sad condition. Cerebral laterality theory has long held that the left brain's linguistic processing systems are specialized for rapid temporal sequencing and diminished persistence of the residual trace. This rapid sequential processing style has been related to the linguistic processing demands of the left cerebral hemisphere and where persistent memories of negative affective events may indeed have survival value, an attribute ascribed to the right cerebral hemisphere. Moreover, the results from this investigation are reminiscent of earlier discussions of the relative role of the right hemisphere in persistence, slow temporal processing, negative affect, and sympathetic tone. Indeed, it was proposed early on (Hebb 1949, p. 70) "…that the persistence or repetition of a reverberatory activity (or 'trace') tends to induce lasting cellular changes that add to its stability…."

In the same project, sympathetic nervous system arousal was assessed using skin conductance measures and heart rate. Heightened sympathetic tone was found using skin conductance, where emotional word recall and production (both for happy and for sad words) yielded reliably increased skin conductance over the neutral word generation condition. This relationship between the emotional associations and sympathetic tone yielded some interpretive consistency with the aforementioned relationships for persistence or age of the memory trace. More specifically, age of the memory was significantly related to measurement of sympathetic tone with older memories not only predictive of affective valence (negative emotional words) but also predictive of increased sympathetic tone on measures of skin conductance.

In contrast to these comparisons providing partial support for the theoretical predictions, divergent findings resulted from the specific analyses of the number of happy words recalled as a function of heart rate as opposed to skin conductance measures. Earlier on, Kathy Snyder provided evidence for the impact of happy word processing significantly lowering blood pressure and heart rate using the Auditory Affective Verbal Learning Test (AAVLT: Snyder and Harrison 1997; Snyder et al. 1998; see also Mollet and Harrison 2007a, b). This instrument has been used to investigate emotional learning within positive, negative, and neutral valences and the effect of rehearsal of positively or negatively valenced material on autonomic nervous system functions, where blood pressure and heart rate may be directionally controlled as a function of the affective valence of the items learned. Simply learning the positive list has been found to reliably lower blood pressure and heart rate, whereas learning the negative affective list reliably increases blood pressure and heart rate or sympathetic tone. Moreover, the evaluation of hostile violence-prone individuals reveals deficient carryover on the acquisition of positive and neutral information, whereas normal carryover or learning is present in the acquisition of negative affective information (Everhart et al. 2005, 2008a, b; Mollet and Harrison 2007a, b).

The diametrically opposite impact of positive and negative affective learning on the AAVLT on the branches of the autonomic nervous system was reproduced using facial muscle configuration memory with posed facial muscle contractions (Herridge et al. 1997). Contraction of the corrugator muscle alone increased skin conductance whereas contraction of the zygomatic muscle alone (smile) lowered skin conductance. Consistent with these earlier investigations, the recollection of happy memories and the generation of event-associated words lowered skin conductance. Diametrically opposite results were found for measures of heart rate, where the production of more words was positively related to faster heart rate. Methodological differences between the present project and the earlier work may be partially responsible. Specifically, in the earlier manipulations cardiovascular measures (heart rate and blood pressure) were taken prior to and subsequent to the processing of the affective words. In the present study, the recording of heart rate was continuous and concurrent with the production of the happy words. Thus, the cardiovascular and cardiopulmonary demands which were accrued in the support of ongoing speech/word production may have impacted the heart rate measures in a divergent fashion with that of the skin conductance recordings.

Dynamic Functional Capacity Model

The emotion literature is mature in its ability to implicate several distinct brain regions which contribute to emotional processing and to altered or variable emotional states. Nonetheless, it is currently lacking a model that specifically explains how these functional neural systems temporally and dynamically interact to produce intensely pleasurable or intensely negative emotions. A conceptual model, the dynamic functional capacity theory, has been proposed that provides a foundation for the further understanding of how brain regions interact to produce intense and unbridled evoked emotional experiences. The dynamic functional capacity theory (Williamson and Harrison 2003; see Carmona et al. 2009; Mitchell and Harrison 2010; Holland et al. 2011; Walters and Harrison 2013a; see also Collins and Koechlin 2012; see also Klineburger and Harrison 2013) details that brain regions mediating emotion and arousal regulation have a limited functional capacity that can be exceeded by intense stimuli. The prefrontal cortex is hypothesized to abruptly deactivate when this happens, resulting in the inhibitory release of sensory and association cortices, the limbic system, the reward circuit, and the brain-stem reticular activating system, causing an "unbridled" activation of these areas. This process underlies the production or release of extremely intense emotional experiences and/ or expressions.

The dynamic functional capacity theory is consistent with Luria's integration of the fundamental views of the localizationist and equipotentialist into a conceptual understanding of functional neural systems theory where geographically distant and functionally distinct brain regions work in a concerted and orchestrated fashion. Beyond this, though, the dynamic functional capacity theory is a neuropsychological model which specifically attempts to provide a basis for the experience of intense and briefly unregulated emotional states of one or of another affective valence. Earlier, Joseph Carmona (Carmona et al. 2009) proposed a "functional capacity" model in which the frontal lobes provide for a dimensionally limited regulatory role over arousal, emotion, and autonomic responses to provocative events and contextual settings. The neural substrates of the prefrontal cortex which allow for and effect instrumental control over posterior and subcortical brain regions have capacity limitations as they rely on cellular resources that are finite; a basic property of any biological or cellular constellation.

The dynamic functional capacity theory proposes that the prefrontal regions activate for regulatory control over arousal and emotional perception fundamentally engaged in response to the sensory perceptual and contextual arrays and the stressful demands of the setting. This basic tenet is fundamental to existing theoretical explanations for emotion and the literature in support of this thesis has been discussed in previous sections (e.g., balance theory). The functional integrity of the prefrontal region to manage these stressors is not unlimited though and is indeed susceptible to real-life excesses, which may extend well beyond the biological limits of this tissue. The limitations of these neural circuits are acknowledged and the analogy is derived from other bodily tissues (e.g., striate muscles) which may, under extreme or excessive load, rather abruptly diminish their efforts at resistance or control. The biological and therefore functional limitations of the prefrontal regions under extreme stress ultimately reflect finite cellular resources although the possibility for more active disengagement derived from intentional decision is not outside of the scope of this theoretical perspective. But, at the simplistic level the functional capacity of the frontal regions can be exceeded when these resources are depleted.

The reliance of cognitive functions upon finite resources is established in related literature. For example, Baumeister's strength model of self-control (Baumeister et al. 2007; see Denson et al. 2012) posits that self-control depends on a limited energy resource that is vulnerable to deterioration over time from repeated or continuous exertion. Baumeister provides evidence that self-regulation decreases glucose levels and that decreased glucose levels mediate self-regulation failure. The model also proposes that self-regulation draws on a common domain-general resource (Wagner and Heatherton 2010), implicating the anterior prefrontal cortex and its broad based and flexible processing capacity. Self-control relies on the interactions of brain regions similar to that described in the dynamic functional capacity theory, including the inhibitory control mechanisms of the prefrontal cortex over subcortical structures associated with reward and emotion. Indeed, self-regulation involves a balance between subcortical brain regions representing the reward, salience, and emotional value of a stimulus, and prefrontal regions associated with self-control (Heatherton and Wagner 2011; see also Wagner et al. 2012). In his review, Klineburger (Klineburger and Harrison 2013) highlights the relevance of finite resources underlying functional capacity. He proposes that the prefrontal cortex has exceptionally high energy demands beyond that of other brain regions. Supporting this, in the human brain, there are systematic differences in blood flow comparisons among different areas with a maximum over the frontal lobes (20–30% above average) and a minimum over the occipital lobes (Ingvar and Lassen 1977).

Additional evidence for resource limitations can be found in the neural efficiency model (Haier et al. 1988). Haier and colleagues used positron emission tomography and found negative correlations between Raven's Advanced Progressive Matrices scores and absolute regional metabolic rates. Metabolic rate varied as a function of brain region across individuals as a function of intelligence, indicating that the brains of more intelligent individuals consumed less energy. In this research, intelligence varied with the efficiency of brain function in energy utilization, rather than to how strongly the brain worked. These authors suggested that this capacity could derive from the selective idling of brain areas that are irrelevant for successful task performance, in addition to the more focused use of specific task-relevant areas. Thus, neural efficiency in the prefrontal cortex and the brain regions it must regulate may be a contributing factor in functional capacity. Neural efficiency would be derived from restricted or precise energy expenditure and the functional effect produced with less resource consumption.

Excessive capacity demands, beyond that reasonable for the maintenance of self-regulation or effortful control, may disinhibit interconnected posterior and subcortical brain regions contributing to an elevation in the experience of the emotion. This results in the more exquisite experience of the activated valence through the augmentation of the sensory–perceptual systems, the autonomic nervous system, and the arousal systems via inhibitory release type mechanisms among others (e.g., kindling phenomenon). The theory generally predicts an abrupt loss of regulatory control, releasing more primitive systems, and intense positive or negative emotional experiences and/or expressive displays. Depending on the neural systems involved, these might include rage-anger, panic, or euphoria-jubilation. It is also meaningful that these experiences and the displays that may accompany them achieve a level of intensity and a corresponding level of release or disinhibition that may reflect even a response to a vital threat, though a euphoric release of pleasure is also part and parcel of this proposed mechanism.

The dynamic functional capacity theory proposes that the loss of effortful regulatory control or intentional inhibition involves excessive functional demands placed upon the prefrontal cortex. Klineburger (Klineburger and Harrison 2013) specifically proposes that the process involved in processing demands beyond capacity limitations involves four separate but intimately related phases. The first phase involves the activation of subcortical and sensory cortices in response to a stimulus. In the second phase, the prefrontal cortex activates in order to process the demands of effortful or regulatory control, including the inhibition of the resulting emotional and arousal responses. This phase results in a somewhat global increase in brain activation and a state marked by large energy expenditure. Klineburger elaborates that the functional capacity of the prefrontal cortex is exceeded during the third phase when resources are exhausted and/or when neuroanatomical integrity is insufficient and tissue damage or destruction is eminent or probable. This phase is characterized by an abrupt, temporary deactivation of the prefrontal cortex and loss of effortful, intentional, regulatory control over other brain regions. This phase is further characterized by the concurrent release and subsequent activation of the now deregulated brain regions. Moreover, this loss of governmental control may be sufficient to reflect an otherwise "unbridled" state, specific to the neural systems which are released and to their specific roles in the emotional valence experienced/expressed. Finally, in the fourth phase, after resources have been replenished or at the point that the stimulus is withdrawn, the prefrontal cortex approaches a return to baseline levels.

Klineburger (Klineburger and Harrison 2013) acknowledges the well-established regulatory role of the prefrontal cortex but extends the theoretical account to that where capacity limitation serves an adaptive, self-protective function unique to this brain region. He states that, specifically, the prefrontal cortex attempts to prevent excitotoxicity associated with extremely intense activation demands by deactivating or disconnecting itself through recurrent inhibitory systems and through remaining temporarily dormant. Two mechanisms can be appreciated with the first relevant to the time necessary for rest, replenishment, and repair following times of stress (see Caccioppo and Bernston 2007). Secondly, the adaptive loss of regulatory control under pivotal life event stressor(s) may allow for a heightened activation or diminished suppression of more primitive brain regions (see Arnsten et al. 2010) adept at survival and potentially at all cost. This potentially adaptive response might lower sensory thresholds (Isaac and DeVito 1958) and may allow more creative responses via spreading activation mechanisms (see Dietrich 2004; Foster et al. 2011, 2012a, 2012b).

A functional biobehavioral analogy might be found in skeletal muscle mechanisms involving reflexive control and release mechanisms (see Willis and Coggeshall 2004). Just as a muscle will respond to slight stretch (e.g., activation of the flower spray endings in the myotube) and to moderate stretch (e.g., activation of the 1A annulospiral endings in the nuclear bag) through reflexive contraction of that skeletal muscle to counteract or resist the stress, it will respond to extreme and potentially damaging stress (e.g., activation of the Golgi tendon organ), instead, with reflex relaxation of the muscle giving way to the stressor and saving the muscle and tendon from damage. As the muscle tissue in this analogy will fail under a load that it is unable to support, so too will the prefrontal cortex give way or fail in self-regulation or effortful self-control. This process, then, will provide a release of subcortical and brain-stem arousal mechanisms (see Luria's first functional unit) and of posterior sensory projection and association cortices (see Luria's second functional unit).

Extending on this analogy, Klineburger (Klineburger and Harrison 2013) states

Just as individuals of varying strength can lift varying amounts of weight, individuals of varying prefrontal functional capacity can handle varying amounts of emotion and stress. Thus, an individual's functional capacity level reflects a relatively stable trait. Individuals with high functional capacity are able to handle high levels of stress; handle strong emotional stimuli; and regulate strong emotional impulses. Like bodybuilders lifting heavy weights, they may seek out and thrive in contexts of high demand. They also possess a wide range in which they can functionally operate because they have a high "ceiling" or "threshold"in their capacity. Those with low functional capacity may be able to operate under circumstances of low stress and emotion but are unable to operate under moderate or high levels of stress and emotional provocation. Compared to strong bodybuilders, low capacity individuals are like novice bodybuilders, tiring after fewer repetitions (or less weight). It is hypothesized that their functional range may be narrower than high capacity individuals, reflecting a more limited range in which they can functionally operate. This means that changing environmental demands are more likely to exceed their capacity.

The capacity of the prefrontal cortex is held subject to those factors that contribute to individual differences in this region or the areas that fall subject to its regulatory control. This would certainly include traumatic brain injury, cerebrovascular accidents, and hypoxic events, for example. One of the more common outcomes of traumatic brain injury is damage to the basal or orbitofrontal regions of the prefrontal cortex (Eslinger et al. 1996) and the myelinated tracts connecting the prefrontal regions with subcortical structures of the limbic system and arousal-related structures of the brain stem (e.g., Bigler 2004; Wilde et al. 2006). It is well established that damage to this region, whatever the mechanism of injury, results in executive function deficits and impaired self-regulation over emotion (e.g., see Hart et al. 2012). Irritability and anger management problems are estimated to affect at least two thirds of those with moderate to severe traumatic brain injury patients (van Zomeren and van den Burg 1985; Kim et al. 1999). These effects persist or even worsen over the first year of recovery (van Zomeren and van den Burg 1985; Hanks et al. 1999). This includes evidence of generalized anxiety disorder along with diminished myelination of the uncinate fasciculus which connects the orbitofrontal region to the amygdala (Phan et al. 2009) or lesser activation of the orbitofrontal cortex to anxiety-inducing stimuli (Kent et al. 2005; see Klineburger and Harrison 2013).

These findings are at least consistent with the long-established relationship where the orbitofrontal prefrontal region is critically involved in the downregulation of the amygdala (Milad et al. 2007). Right orbitofrontal lesions often yield incremental reactivity to anger or rage behavior (e.g., Agustín-Pavón et al. 2012; see also Fulwiler et al. 2012). Patients with panic disorder show lesser activation in the orbital frontal cortex in response to anxiety-inducing stimuli (Kent et al. 2005). Abnormalities in the orbitofrontal prefrontal cortex in the right hemisphere, in particular, are shared by several different anxiety disorders (Rauch et al. 1997).

On the other hand, the capacity of the frontal regions may vary with the activity of the remaining functional neural systems, including those within Luria's second functional unit for reception, comprehension, and analysis of sensory information (parieto-occipitotemporal regions) and those within Luria's first functional unit modulating arousal levels. For example, a substantial body of evidence acquired across decades of research has emphasized the importance of the amygdala in fear (Davis et al. 2010; Adolphs and Tranel 2000). Davis et al. (2010) note that amygdala-restricted manipulations interfere with the acquisition and recall of conditioned fear and other forms of anxiety-related behaviors in nonhuman animals. Feinstein and colleagues (Feinstein et al. 2013) discuss the importance of findings in humans with focal bilateral amygdala lesions. Although rare, these cases do exist and they have proved crucial for understanding the role of the human amygdala in fear.

The authors refer to the most intensively studied case of patient SM, whose amygdala damage relates to the Urbach–Wiethe disease. Earlier research established that patient SM does not condition to aversive stimuli (Bechara et al. 1995), fails to recognize fearful faces (Adolphs and Tranel 2000) and demonstrates a marked absence of fear during exposure to a variety of fear-provoking stimuli, including life-threatening traumatic events (Feinstein et al. 2011). Feinstein and colleagues (Feinstein et al. 2013) extend the findings beyond SM, noting that patients with similar lesions have largely yielded similar results (e.g., Adolphs et al. 1999; Becker et al. 2012). One interesting exception has been found with SM and other patients with bilateral amygdale damage, where inhalation of 35 % carbon monoxide evoked not only fear but also panic attacks. Relevant here are theoretical propositions for functional neuroanatomical systems devoted to fear evoked from external threats and those which arise internally secondary to diminished oxygen saturation levels, perhaps, or from elevations in carbon monoxide levels.

Also evident here is the role of more dorsal prefrontal regions in the regulation of other subcortical structures, including the lentiform bodies involved in crying and laughter (e.g., Ross and Rush 1981; Constantini 1910; cited in Oettinger 1913; see also Parvizi et al. 2009). Dolan and colleagues (Dolan et al. 1994) found that the neuropsychological symptoms in depression, including cognitive deficits, were associated with profound hypometabolism in the medial prefrontal cortex. Consistent with these findings, bipolar and unipolar depressives are characterized by decreases in cerebral blood flow and the rate of glucose metabolism in the prefrontal cortex (Drevets et al. 1997).

Thus, deficits in neuroanatomical integrity and metabolism are among the mechanisms which contribute to diminished functional capacity by restricting resources and decreasing the efficiency of communication between the prefrontal cortex and other brain regions (see Klineburger and Harrison 2013). Galliot and Baumeister (2007) showed that functional capacity levels may vary over time within an individual where self-regulation relies on adequate levels of circulating blood glucose that are temporarily reduced by tasks that require effortful self-regulation. Resource depletion of brain glucose stores below that adequate for self-regulation is necessarily sufficient for the dynamic transition into an altered functional state of operation with ipsilateral inhibitory release over subcortical regions and posterior sensory cortices via longitudinal tracts (e.g., the arcuate fasciculus). In his review of these mechanisms, Klineburger (Klineburger and Harrison 2013) states that this "unbridled" release from regulatory control allows for uninhibited activation of these regions, further intensifying the emotional experience and arousal evoked from the event.

Exceeding capacity may allow more primitive brain regions to control behavior in the face of extreme stress or danger and may also lead to the inhibitory release of reflexes that may be beneficial in a survival scenario. Arnsten and colleagues (2009; Arnsten et al. 2010) propose that, in response to danger, the prefrontal cortex can be rapidly taken "off-line" to switch control of behavior to more primitive brain regions that mediate instinctive reactions. For example, high levels of catecholamine release during stress exposure contribute to the disconnection of prefrontal cortex networks, while augmenting activation of the amygdala and related structures. The authors propose that exceeding capacity may allow for increased arousal levels, precipitating the increased metabolism of stored fats into glucose to be released into the bloodstream and ultimately to the brain in order to replenish temporarily low glucose levels within the central nervous system.

The dynamic functional capacity theory proposes that very low levels of blood glucose may cause the prefrontal cortex to disengage or to be taken "off-line" and that this may serve a broader and more protective role in the reduction of excitotoxicity. Relatedly, Arnsten and colleagues (Arnsten 2009; Arnsten et al. 2010) have proposed a theory of "rapid neuroplasticity" that claims the prefrontal cortex is able to rapidly deactivate or disconnect from other brain regions by way of recursive inhibitory connections within this tissue. More specifically, she and her colleagues propose a mechanism inherent to the prefrontal cortex that weakens network connections for the prevention of overexcitability. Moreover, the proposal claims that there is a negative feedback mechanism that may function to prevent seizures in prefrontal cortex microcircuits. The dynamic functional capacity theory holds that a low glucose level is one precursor for exceeding capacity. Severe hypoglycemia causes neuronal death and cognitive impairment secondary to excitotoxicity and damage to the genetic structure of the cell. Additionally, forebrain oligodendrocytes are highly vulnerable to excitotoxicity (McDonald et al. 1998; see Klineburger and Harrison 2013). Also, glial cells clearly play a role in the regulation of neural activity at the synapse and in the establishment of axonal connections among neurons (see Koob 2009).

The dynamic functional capacity model appears relevant to aging research where cognitive decline has long been coupled with the evidence for reduced neural capacity or cellular aging and degeneration. Gerontological studies parametrically manipulating cognitive load have revealed that as task difficulty increases, older adults reach a maximum level of neural function sooner than do young adults. The effect is prominent especially within the prefrontal cortex (Cappell et al. 2010; Nagel et al. 2009; see Park and McDonough 2013). Based upon their review of this literature, Park and McDonough (2013) conclude that neural resources reach a ceiling for activation earlier and at lower levels of difficulty in older adults than in young adults. Evidence for this has been incorporated within existing cognitive theory on aging. Among this is the compensation-related utilization of neural circuits hypothesis or CRUNCH model (Reuter-Lorenz and Cappell 2008; see Park and McDonough, 2013). According to the authors, the CRUNCH model aligns closely with a limited resource theory of aging, which postulates an ever-shrinking passive pool of (cognitive) resources with age that is increasingly inadequate to maintain cognitive function. Beyond this, the CRUNCH model proposes that the brain is responding dynamically to cognitive challenge by selectively activating more neural resources than younger adults in order to maintain performance accuracy and efficiency. Missing here are the predictions derived from dynamic functional capacity model for situational variants which exceed capacity and the fundamental biological foundations for these predictions.

Chapter 23
Light and Dark

There is overwhelming evidence of cerebral asymmetry in humans and many other species (e.g., Gazzaniga 1998, 2000; see Hugdahl and Westerhausen 2010; see also Hudgahl 2012; see also Ocklenburg and Güntürkün 2012). For example, the asymmetry of brain function is evident early on in the avian brain (Denenberg 1981; see also Nottebohm 1977), including the lateralization of attack and copulation responses (e.g., Howard et al. 1980). It is commonly thought that lateralization of brain function in humans and other species is an invariant aspect of genetic expression (e.g., Denenberg 1981). However, exposure to illumination appears to be instrumental in this process with developmental prospects for reversal of dominant asymmetry patterns and subsequently for the moderation of laterality effects with light.

Early on, Rogers and Anson (1979) suggested that light experience may have an important role in establishing lateralization in the chicken forebrain, because after day 17 of incubation the embryo is oriented in the egg such that the left eye is occluded by the chicken's wing and body, while the right eye is next to the air sac and exposed to light input (see Freeman and Vince 1974). The chicken's optic nerve decussates completely at the optic chiasm such that light entering the right eye stimulates developmental processes in the left hemisphere in advance of the right and so forth. Moreover, Rogers (1982) showed that chickens hatched from eggs incubated in darkness fail to show functional asymmetry of the forebrain for the asymmetrical control of attack and copulation behaviors.

The amount of ambient light reaching embryos during development greatly influences the lateralization of adults. This effect has been demonstrated in multiple species including chicks, zebrafish (see Bibost et al. 2013), and also live-bearing fish. For example, using live-bearing fish, Dadda and Bisazza (2012) measured behavioral lateralization in 10-day-old goldbelly topminnows born from females

---

"The night is a haven and darkness the harbor both for solitude and for shame." (Author unknown)"Every one is a moon, and has a dark side which he never shows to anybody." Mark Twain (1835–1910)

© Springer International Publishing Switzerland 2015
D. W. Harrison, *Brain Asymmetry and Neural Systems*,
DOI 10.1007/978-3-319-13069-9_23

that had been maintained at high or low light intensitiesduring pregnancy. Fish from high light treatment were significantly lateralized in both visual and motor tests while fish exposed to low light intensities were not.

Ambient sensory conditions are often overlooked in neuropsychology and foremost among these influences are the effects of illumination levels. Light-responsive pathology is common across cultures and this is evident in migraine syndromes with photophobic responses or the exacerbation of pain through ambient lighting levels or glare. Pain and light reflect separate sensory modalities which may interact at multiple brain levels and with close proximity at the thalamic nuclei. Retinal projections have been found to the pulvinar and centromedian nuclei within the thalamus (Noseda et al.2010). Moreover, brain-mapping evidence exists for direct optic nerve to pulvinar connections in the thalamus using diffusion magnetic resonance tractography (Maleki et al. 2012).

The presence of this pathway has implications for photophobia, a somewhat common cross-modal sensory phenomena linking pain with ambient light level. The pulvinar receives trigeminal pain-sensitive neurons innervating vascular and dural structures, providing a link between sensory modalities critical for allodynia (Burstein et al. 2010). Allodynia refers specifically to this connection between the pain modality and a seemingly separate sensory modality which does not normally provoke or exacerbate pain (see Merskey and Bogduk 1994). In this example, potentially excruciating pain may result subsequent to incremental activation within the visual system with bright light exposure. This visual pathway is not one that is directly involved in perceptual acuity, in contrast with the perceptual projections from the retina to the lateral geniculate nucleus of the thalamus and on to the occipital cortex. Despite this clinical relationship commonly observed between light and pain, the vast majority of the research on ambient lighting attests to the beneficial effects of exposure, and to multiple maladies, which may result from inadequate intensity and/or duration of light during the day.

The efficacy of bright light therapy is particularly well documented for winter depression, a condition in which the administration of bright light is generally regarded as the first line of treatment (Golden et al. 2005). Also, it has been acknowledged for its low cost, its home-based intervention appeal, and a much reduced iatrogenic impact than that resulting from a pharmacotherapy approach (Terman and Terman 2005). The potential for the implementation of light therapy as a low-cost intervention within the broader architecture of an institutional or residential facility is provocative. The low cost of the intervention sits among the broader list of favorable features to be wrought from ambient or contextual therapy interventions of this sort (see Garre-Olmo et al. 2012).

The utility of ambient light therapy for winter depression has considerable intuitive appeal and evidence-based support, as seasonal changes in mood are experienced by much of the population and these variations in mood have been demonstrated as a function of light (Golder and Macy 2011). Moreover, accumulating evidence indicates that bright light might be equally efficacious for nonseasonal depression (Terman and Terman2005;Kripke 1998; Tuunainen et al. 2004), as well as other morbidities, including disturbed sleep (Kohyama 2011; Phipps-Nelson

et al. 2006; Phipps-Nelson et al. 2006a, 2006b), perinatal depression (see Crow-
ley and Youngstedt 2012), fatigue (Ancoli-Israel et al. 2011; Rastad et al. 2011),
Parkinson's disease (Paus et al.2007) and neuroendocrine abnormalities (Salgado-
Delgado et al. 2011).

There is a need for additional targeted research on the functional cerebral laterality differences which may be found to be altered by ambient lighting and/or noise
levels and the potential differences between the cerebral hemispheres and their underlying brainstem structures in their response to these conditions. The potential for
discovery is supported through existing literatures where arousal (Isaac and DeVito
1958; Münch et al. 2012), affective (Rosenthal et al. 1984; Golder and Macy 2011),
and cardiovascular differences (Scheer et al. 1999; Harrison and Kelly 1989; Smolders et al. 2012) rather consistently emerge as a function of lighting and/or noise levels. The evidence includes the differential impact of light on the serotonergic brain
systems, including the raphe (e.g., Cagampang et al. 1993), the suprachiasmatic
nucleus of the hypothalamus (Ginty et al.1993), and the pineal body (Wurtman et al.
1963). These are light-processing systems within the brain that are largely outside of
perception or without the acuity associated with the classic visual projections onto
the occipital lobe. The pineal body not only looks like a remnant of a third eye, it also
has similarities across species underlying photophilic and photophobic behavior patterns. Moreover, the pinealocytes look like cells found within the retinas. However,
this structure is located on the back of the brainstem and close to the superior colliculus, another system critical in the perception of ambient lighting and our arousal
responses to light and dark ambient conditions (Kallman and Isaac 1980).

Each of these light-responsive brain regions plays a role in serotonin production. The pineal body, for example, alters serotonin as a derivative of the hormone
melatonin. Among the neurotransmitter systems, none has been more clearly associated with anger, depression, and anxiety than the serotonergic system, although
helplessness was earlier found to relate to levels and depletion of the catecholamine
norepinephrine (Schildkraut 1965). Indeed, statistically significant differences in
serotonin have been found even among those attempting suicide and those who
were successful in these activities. Lowered levels were found in the successful
group (Stanley and Mann 1983).

Many families appreciate this in their battle with cancer where they will meet
wonderful folks fighting the same battles. Many oncologists acknowledge that no
entirely adequate cure has been discovered sufficient to eliminate cancer or perhaps
even the remnant probability of recurrence in a more general sense. Nonetheless, we
may appreciate that substantial progress has been made in the treatment and control
of nausea resulting from chemotherapy intervention. The chemotherapy setting is a
most remarkable one. To begin with, the patient is essentially physically restrained
via the intravenous pumps, concurrent with the infusion of toxic chemicals into their
blood stream. This might be a good formula for the development of posttraumatic
stress disorder as it shares many commonalities with the research methods used in
nonhuman animal research to study the phenomenon. Jason Parker (Parker 2010),
using a murine model, demonstrated that physical restraint used in the development
of the nonhuman animal posttraumatic stress disorder, results in slower wound heal-

ing times. Thus, by extension, he argued for the relevance of physical restraint in the development of human posttraumatic stress disorder, and especially with concomitant battle or surgical wounds. The impact of posttraumatic stress disorder on wound healing in humans has been assessed with evidence for delayed healing (e.g., Wilson et al. 2011). However, experimental manipulation of perceived or actual loss of control or the attribution of external control through physical restraint has seldom been investigated in research on humans.

In visiting with these fine folks undergoing group chemotherapy sessions, and frequently with family present along with them, it is apparent that many are depressed. The attribution of many oncologists and lay people, though, is that "they are depressed because they have cancer." But, with the appreciation of the progress that has occurred within the oncology field, which has specific relevance to the control of nausea, we can ascribe much of this success to the concurrent administration of *Aloxi* (palonosetron) in the chemotherapy setting. Aloxi is a serotonergic antagonist (De Leon 2006)! Thus, it remains a reasonable question for research that the depression is at least partly iatrogenic or arising as a side effect of the pharmacological intervention for nausea. An analogy may exist in research on pain control, where Vicodin (acetaminophen and hydrocodone) for pain management has been found to be oppositional to serotonergic therapies for depression. The iatrogenic effects here might easily be lost or discounted in a depressed patient with a pain disorder (see Mollet and Harrison 2006).

Less well recognized is that, in industrial societies, average levels of exposure to bright light for the general population (> 1000 lux) average only about 1 h per day even in good weather conditions (Espiritu et al. 1994), and these levels might be inadequate for mood regulation and for circadian entrainment in susceptible individuals. The possibility that mood dysfunction is related to light exposure is suggested by the higher prevalence of depression identified as a seasonal affect disorder at increased latitude and where there is a reduced sun exposure to ultraviolet light (UVB) radiation. Light has been found to provide for circadian entrainment and even to facilitate sleep onset and deeper sleep periods (Lewy et al. 1987). Bright light exposure sometimes on the order of 2000–10,000 lux has been used not only in circadian entrainment (e.g., see Kondratova and Kondratov 2012) but also in the treatment of seasonal affect disorder or SAD (Rosenthal et al. 1984; Alden and Harrison 1993). In their review of light therapy techniques for SAD, Lall et al. (2012) conclude that morning light exposure is most effective in producing beneficial effects on the depression.

Parker and Brotchie (2011) state that "The prevailing theory is that seasonal affect disorder is related to the shorter daily photoperiod of winter, with suboptimal light input to retinal photoreceptors disturbing the optimal functioning of the suprachiasmatic nucleus and its interactions with the pineal gland and melatonin secretion." Rod and cone photoreceptors within the retina not only provide for a visual map of the surroundings but also process distinctly separate, nonimage-forming functions. Perhaps the most important afferents to the suprachiasmatic nucleus arise directly from photoreceptors in the retina forming the retinohypothalamic tract (Hattar et al. 2002). Hattar and colleagues discovered that the retinal rod and cone cells are not required for light-based circadian entrainment, but that there exists a subset of retinal cells (2500 of a total of 100,000 cells) containing a light-sensing retinaldehyde-based pigment, melanopsin (Ecker et al. 2010), which is involved in this process. This light-evoked activation from the complimentary action of rods, cones, and intrinsically photoreceptive ganglion cells drives the behavioral and physiological mechanisms of circadian photoentrainment, the pupillary light reflex, pineal melatonin suppression, and sleep (Peirson et al. 2009; Lall 2012).

Hattar et al. (2002) acknowledge an important connection between the suprachiasmatic nucleus of the hypothalamus and the pineal gland. Melatonin, which is the only known hormonal output from the pineal gland, affects the suprachiasmatic nucleus via inhibition of its activation or neuronal firing (Borjigin et al. 1999; see also Bjorvatn and Pallesen 2009). Thus, the functional relationship between the suprachiasmatic nucleus and the pineal gland seem appears to be bidirectional. Ablation of the suprachiasmatic nucleus in mammals has been shown to eliminate circadian rhythms, and transplantation restores the rhythm to the period of the donor animal (Ralph et al. 1990). Although research investigating the effect of light therapy on SAD has produced inconsistent findings, it has been suggested that light and the circadian regulation of the suprachiasmatic nucleus are closely related to the pathophysiology of seasonal affect disorder and possibly other forms of psychopathology. This position ultimately implicates dysfunction within the serotoninergic systems via a process of depletion of this neurotransmitter (Wilkins et al.2006).

Psychologists are generally aware of the heightened risk for depression in the months of December through at least February. Moreover, many appreciate that these are high-risk periods with reduced symptoms during the months of increased duration and more intense lighting levels. Depression may be treated, in some, through the provision of alternative lighting sources (Alden and Harrison 1993) with successful intervention reported typically with broad spectrum light bulbs between 2000 and 10,000 lux. Ten thousand lux is perceived as a painful stimulus to many. So, lower intensities are desirable. Also, the issue of ultraviolet exposure is important with efforts to restrict or eliminate this bandwidth to reduce cataract formation and damage to the skin.

More relevant to this discussion of functional differences between the cerebral hemispheres, though, is the issue of the potential for asymmetrical influences of light on functional neuroanatomy. We have attempted to address this at the Virginia Tech Behavioral Neuroscience Laboratory and in medical centers within our region. In our sample of psychiatric and rehabilitation facilities along the eastern USA, the lighting levels varied from 1 to 40 ft candles. Yes, the 40-ft candles were measured in the nursing and administrative areas. It is reasonable that this would be a minimal requirement for reading, as 40-ft candles are equivalent to that measured in a typical office with the lights turned off! This office would have windows allowing for some indirect lighting in the room. Some of the participating psychiatric facilities where illumination levels were recorded had severely depressed patients and patients with altered arousal level and confusion. The rehabilitation units also had severely compromised patients, all attempting to recover and to rehabilitate within these restricted lighting environments.

Therapists might attest to the positive feedback that may be recruited from residential, institutionalized, or hospitalized individuals simply from taking the patient out of doors and into the sunshine and the fresh air. One man serves, for example, as he appeared violence prone and threatening to his nursing staff. This man was quadriparetic but his skills, which were exquisitely developed as a Navy Seal, terrified the staff to the point that his primary care nurse refused to return to work. Assigned to his care at this point, I received unwarranted praise for my intervention efforts to enjoy the out of doors, instead of having each day filled with indoor therapies and confrontations, and with the performance demands placed on him by his therapists. In actuality, this was a highly disciplined man even after suffering severe head and spinal injuries. It may ultimately be hard to exceed these natural settings and out of doors for promoting recovery and rehabilitation of our patients, and especially if they have been confined to a dimly lit medical or nursing care setting for an extended period of time.

A responsive administration at one major medical center altered their lighting levels in their rehabilitation unit when they were provided with feedback on ambient sensory conditions at their center. They remodeled the facility allowing for natural lighting to flood the therapy areas with the sun rising on one side of the unit and setting on the other. They chose to provide additional lighting in the remainder of the unit. But, this was provided through fluorescent bulbs. The level of morale at the rehabilitation center was monitored through self-reports of the therapists. But, there were many confounding variables, precluding journal publication even though the findings appeared to be substantially in support of increased illumination and improved morale. Controversy exists, though, and even as it did within some early psychiatric settings, where light was thought to increase agitation.

Hitler modeled his eugenics program after one of our Virginia psychiatric facilities (Video: The Lynchburg Story 1993). Visitors to this facility in recent times might be surprised that some of the black paint spread over the windows and lighting implements are still present decades later and in an active treatment facility. Interestingly, a pervasive deficiency in vitamin D was found among the residents, raising concerns for a possible contributory role of the institution in the psychopathology of the residents. The majority of bodily tissues have vitamin D receptors and the brain is no exception, including both neuronal and glial cells. Eyles et al.(2005) report the presence of vitamin D receptors in the prefrontal cortex, hippocampus, cingulategyrus, thalamus, hypothalamus, and substantianigra. This evidence is of potentially significant theoretical import, since many of these regions have been implicated in the pathophysiology of depression (see Drevets et al.2008).

As reviewed by Holick (2007; see also Parker and Brotchie 2011), our primary source of vitamin D is from exposure to sunlight. Most vitamin D is produced in the body by exposure and penetration of the skin to UVB radiation from sunlight and with a wavelength of 290–320 nm. Parker and Brotchie (2011) note that variations in age, skin color, latitude, time of day of exposure, and time of the year affect the actual amount of exposure. Following penetration of the skin, solar UVB radiation (wavelength 290–315 nm) converts 7-dehydrocholesterol to pre-vitamin D_3, which is rapidly converted to vitamin D_3. While their review provides evidence from multiple projects supporting the role for vitamin D in depression, there were studies not in support of this conclusion. Moreover, they point out that depressed individuals in the positive studies may be less likely to go outside or to engage in outdoor activities, thus lowering their levels of vitamin D as a result of depression, rather than as a putative causal agent.

Another relevant literature exists more prominently on the elderly and with early to mid-stage dementia (Prinz and Raskind 1978; Morin and Gramling 1989; Gilley et al. 1995). Many of these people suffer from an affective disorder with increasing agitation, anger,or fear during the transition into the evening hours. Collectively, this phenomenon has been referred to as "sundowning" with implications for lowered light levels, circadian rhythms, and other potentially relevant factors. Often these episodes occur with minimal staff support and after therapies have been terminated, with the patient alone in their room at the end of the day. There are many potential correlates with these affective disorders, including ambient noise levels within the institutional setting with higher noise levels related to increased agitation and lowered quality of life (e.g., Garre-Olmo et al. 2012).

Lighting levels are predictive not only of the quality of life but also of social interactions and agitation. Negative emotion with reduced lighting alters the down regulation of melatonin by light resulting in deregulated or disrupted circadian rhythms. Fear or profound agitation may be promoted in the patient with hallucinations or formesthesias, where these sensory events may be more difficult to suppress in the quiet and dim lighting of one's room. The chemical restraints that may be required to deal with a sundowning patient may aggravate the brain disorder with iatrogenic effects. This might, by way of example, include Haldol (haloperidol) with the potential for cumulative destruction of dopaminergic cells and perhaps even tardive dyskinesia with continued use. This has been a topic of investigation for some of our neurochemistry colleagues at Virginia Polytechnic Institute (e.g., Subramanyam et al. 1991).

Elderly patients with dementia frequently experience disturbances of mood, behavior, sleep, socialization, and activities of daily living along with cognitive decline. All of these function more generally to heighten caregiver burden and ultimately the need for institutionalization (see Banerjee et al. 2003). These limitations restrict treatment possibilities providing opportunities for the use of alternative techniques such as the manipulation of contextual variables as part of a broad-spectrum intervention plan (Harrison et al. 1989; Harrison et al. 1990; Alden et al. 1991; Shapiro et al. 1996). Bright light exposure may be part of this contextual therapy approach with substantial evidence for circadian alterations with dementia (Swaab et al. 1985). The suprachiasmatic nucleus of the hypothalamus is a critical circadian pacemaker highly responsive to ambient lighting and melatonin (Moore 1996) and circadian entrainment occurs with bright light exposure (Czeisler et al. 1989). Elderly dementia patients may experience attenuated circadian synchronization and especially problematic are care settings with reduced or suboptimal levels of light exposure and patients with suboptimal melatonin production.

When bright light has been used as a contextual therapy, it has been shown to ameliorate behavioral (Mishima et al. 1994) and sleep (Van Someren et al.2002) disturbances attributed to the dementia process rather than to the environmental or sensory limitations within the setting. Also, exogenous melatonin has been shown to increase dendritic complexity following 14 days of treatment (Ramirez-Rodriquez et al. 2011) and to facilitate connections among neuronal cell assemblies through increased dendritic length and an increased number of dendrite arborizations in hippocampal cultures incubated with melatonin for 6 h (Dominguez-Alonso et al. 2012). This increase in complexity would be expected to facilitate neuronal connections and associative strength among neuronal assemblies, and support for this hypothesis was recently found using human subjects (Fosteret al. in press). Foster and colleagues (in press) found some evidentiary support from their literature review, where melatonin appears to promote improved learning and memory. Moreover, exogenous melatonin administration 50 min prior to testing was found to significantly improve word generation associative strength (spreading activation among neuronal circuits) on a verbal fluency test requiring the production of words beginning with specific letters (Foster et al. in press).

Light therapy and melatonin treatments have been used for some time now to reset the circadian clock in humans. Several studies have reported promising effects of combined light and melatonin therapy on the progression of neuropsychiatric disease in the aged (see Kondratova and Kondratov 2012; see also Most, Scheltens and Van Someren 2010). Melatonin treatment in Alzheimer's disease patients has resulted in mixed findings with no significant improvements in sleep quality found in comparisons between the melatonin-treated groups and the control groups in double-blind, randomized, placebo-controlled studies (Singer et al. 2003; Gehrman et al. 2009; Serfaty et al. 2002). However, in their review of the literature, Kondratova and Kondratov (2012) point out that the effects of melatonin are dependent on the phase of the circadian cycle, and that combinations of melatonin and bright light therapy significantly improve the quality of sleep and daytime activity of patients according to other studies (Mishima et al. 2000; Dowlinget al. 2008). Mild cognitive impairment is often used as a neuropsychological marker for neurodegenerative disease and/or Alzheimer's dementia. Significant improvement was found when individuals diagnosed with mild cognitive impairment were treated either using melatonin or melatonin in combination with light therapy (Cardinali et al. 2010). The authors also point out that, in murine models of Alzheimer's disease, treatment with melatonin inhibited oxidative and amyloid-β pathology, protected against cognitive deficits and neurodegeneration, and increased survival (Feng et al.2004; Matsubara et al. 2003; see Kondratova and Kondratov 2012).

In a double-blind, placebo-controlled randomized trial evaluating a combination of light and melatonin on a daily basis for an average of 15 months, researchers (Rixt et al. 2008) used indirect ceiling-mounted, whole-day bright light. The authors conclude that light reduced the cognitive deficits in the dementia patients by 5 % without reductions in the progression of the cognitive decline, an effect consistent with the dominant pharmacological treatment using the acetylcholinesterase inhibitors (Courtney et al. 2004). Light reduced depressive symptoms by a relative 19 % and attenuated the gradual increase in functional limitations by 53 %. A similar finding supporting treatment efficacy over time by 2 % was found for its effect on sleep duration. The authors conclude that "…the simple measure of increasing the illumination level in group care facilities ameliorated symptoms of disturbed cognition, mood, behavior, functional abilities, and sleep. Melatonin improved sleep, but its long-term use by elderly individuals can only be recommended in combination with light to suppress adverse effects on mood. The long-term application of whole-day bright light did not have adverse effects, on the contrary, and could be considered for use in care facilities for elderly individuals with dementia."

In one project (Harrison et al. 1990), a group of dementia patients was investigated with hyperfluent dysprosodia characterized by excessive and disruptive emotional vocalizations. Such patients spend much of their time, moaning or wailing , in their hospital wards resulting in diminished morale among the nursing staff and other patients. It is clearly important to achieve an accurate diagnosis of these brain disorders. It is somewhat common for the patient with a vocalization disorder to be moved close to the nursing station for intense monitoring and pain control by implementing medication requests through the physician. It is useful to the nursing staff, caregiver, therapists, and family members to understand that this is a speech disorder and not, necessarily, a pain disorder requiring intervention. The moaning and wailing patient might cease their disruptive behaviors on visitation or engagement in conversation about the weather or whatever, and arguably from the activation of the left hemisphere in processing linguistic content or propositional speech. Subsequent to the end of the social engagement or conversation, the moaning and wailing start up again.

In this project, combinations of the patient's routine illumination level versus bright light and low versus moderate intensity white noise (i.e., 45 vs. 75 dB) were used while recording the frequency and the intensity of the patient's vocalizations. The duration of visual-orienting responses within the left and the right hemispace was recorded as an additional indication of relative right or left hemisphere activation. Directional gaze toward the left or the right provides a prominent index of relative right or left frontal eye field activation (Guitton et al. 1985; Ladavas et al. 1997; Pierrot-Deseilligny et al. 1991; see also Suzuki and Gottlieb 2013). This recording was done using an overhead camera with the patient's nose as the pointer on a grid overlay to quantify rotation of the head about the neuraxis to the left or right side. The results indicated that bright light reliably increased logical linguistic or propositional speech sounds concurrent with the duration of the visual-orienting responses toward the right hemispace. In contrast, ambient noise reliably increased the intensity of emotional tone, decreased logical linguistic speech utterances, and increased the duration of time spent orienting toward the left hemispace.

The results, though tentative, were suggestive of lateralized activation systems and the potential for environmental modification of disruptive emotional behaviors and even reductions in monitoring and restraint requirements. Right-sided orienting responses were increased in dementia patients with bright light exposure in a separate investigation in this line of research (Alden et al. 1991). Related research used elevated white noise conditions to increase right brain activation at least through that indicated by sympathetic drive with elevations in blood pressure and heart rate (Harrison and Kelly 1989). Increased arousal level in normal elderly subjects reliably improved arithmetical performance to levels statistically equivalent with the younger college-age comparisons, but at the cost of increased blood pressure and heart rate in the elderly group. The elevated blood pressure may be useful to promote improved perfusion through the capillary walls in the aging cerebrovascular system, although this interpretation remains controversial. But, activation or heightened arousal may be differentially a function of the right parietal temporal region (Heilman et al. 1978; Heller et al. 1992; see Heilman et al. 2012) and with a more universal arousal response secondary to activation of the right hemisphere and sympathetic tone.

John Alden provided evidence of bright light producing a significant effect in cerebral lateralization of depressed women with an improved ability to shift their intentional focus to the left ear on a dichotic listening task (Alden and Harrison 1993). In a related pilot project (Harrison et al. 2003), ambient sensory conditions were manipulated with patients recovering from stroke to investigate the possibility of left cerebral activation to bright light using speech processing and dichotic presentations to the left and right ear. Dichotic listening techniques are commonly used in the neuropsychological evaluation as well-established indices for cerebral laterality (see Hugdahl 1988, 2003, 2012; Hugdahl et al. 2009). In this project, bright light (2500 lux) resulted in improved speech sound detection at the right ear in both left- and right-sided stroke patients in comparison to performance during a condition of dim light (350 lux). The left- and right-sided cerebrovascular accident patients remained statistically below the no cerebrovascular accident patients in these comparisons. However, with increased lighting the left-sided stroke group was statistically equivalent to the right-sided stroke group in processing speech sounds. The results were interpreted within the theoretical proposal of left cerebral activation to heightened levels of illumination. Release of right cerebral systems might be predicted for this theoretical proposal on darkening of the ambient setting or context. These contextual interventions and manipulations were proposed early on by Walt Isaac (Isaac and DeVito 1958; Isaac and Reed 1961; Kallman and Isaac 1980; Delay et al. 1978; Lowther and Isaac 1976). However, they remain largely unexplored through the present within the realm of neuropsychology and rehabilitation.

Even functional cerebral laterality, as measured by grip strength using a hand dynamometer, has been shown to be susceptible to ambient illumination levels. Handgrip strength in right-handed college-aged women was significantly stronger at the preferred right hand in bright light conditions, whereas symmetry in grip strength was found in these women during low light conditions (Shapiro et al. 1996). In another project investigating functional motor systems with elderly dementia patients suffering from chronic bruxism, bright lighting conditions provided for a significant reduction in the target behavior (Harrison et al. 1989). Although some initial success has resulted from these efforts, the effects seem more remarkable in the brain damaged or demented patients, perhaps. To date, we have been relatively unsuccessful in our development of methodology necessary to show these effects in normal, otherwise healthy, college samples.

Beyond these comparisons of left or right brain activation are those findings of character and morality under differing sensory conditions. Though it may be common sense for the layperson, science is beginning to show that we behave in a more negative affective manner to low light conditions. Sherman and Clore (2009) provide evidence using a simple Stroop task. Three studies examined automatic associations between words with moral and immoral meanings and the colors black and white. The speed of color naming in a Stroop task was faster when words in black concerned immorality (e.g., greed), rather than morality, and when words in white concerned morality (e.g., honesty), rather than immorality. In addition, priming immorality by having participant's hand-copy an unethical statement speeded identification of words in the black font. Relevant behaviors might include social improprieties, deceitfulness, or dishonorable behavior of one form or another, whereas "in the light of day" we may show more positive emotions and honorable social behavior. We have long been aware of negative affective events associated with dim lighting, possibly to include criminal activities, robbery, assault, and so forth. But, these associations may be stereotypical and alternative positive relationships also certainly exist. Moreover, negative emotional events occur on bright and hot summer days. However, the better predictor of negative emotion here, and specifically anger and violence, appears to be elevated temperature (heat; e.g., Anderson 2001; see also Agnew2012) and not illumination level.

A caution might be mentioned here in that violent behavior increases with increments in temperature as with heat in the summer months. This has long been appreciated and is even addressed in Shakespeare's Romeo and Juliet as Anderson (2001; see also Anderson 1989) recalls Act 3, Scene 1."I pray thee, good Mercutio, let's retire. The day is hot, the Capulets abroad, and, if we meet, we shall not 'scape a brawl, for now, these hot days, is the mad blood stirring." This temperature and aggression relationship has been largely linear, with higher temperatures related to heightened levels of aggression (Anderson 1987) and this is true up to a point. The point being that temperature and beyond, at which it is simply too hot to fight and, alas, we await another day to harm the scoundrel! Temperature-related events should not be confused with illumination levels as the two factors may be orthogonal as per our discussions (above). Conflicting findings on the neuropsychological and rehabilitation effects of bright light might easily be confounded by heat if the light source also produces heat, as well it may.

The geriatric literature has appreciated *sundowning* in dementia patients for decades now. However, the preferred interpretations have focused on limited staffing and such, rather than on the possibility of shifting cerebral asymmetries with dynamic and relative activation of negative emotional processing systems during low light conditions and the related shift in emotional valence to fear or apprehension with the transition to darkness. Also, predatory behavior, among the aggressive, may increase with these transitions and out of the light of day. The national criminal justice reference system (Painter and Farrington 1997) investigated the effect of lighting on criminal behavior, for example. The investigation found that for all crime in the experimental area, the crime prevalence (total number of victimizations) decreased by 23% after improved street lighting was installed compared to the 12 months prior to the installation of the lights. In the control area, the crime prevalence decreased by only 3%. The relationship between darkness and the sinister is deeply rooted in our language and cultures. The term "sinister" may simply mean left handed. But, the functional anatomical relationship between the right hemisphere and the processing of the left-sided sensory event is now well established as is the lateralization of fearful hallucinations and negative affective processing biases to the right brain (Walters et al. 2006; Mollet et al. 2007). These relationships are common even to our language where "sinister" is left handed and where negative emotion is "viewing the world in a dark light!" In his reference to negative emotion, Mark Twain (1835–1910) describes the human condition stating "Every one is a moon, and has a dark side which he never shows to anybody."

Chapter 24
Arousal Theory

Classic Arousal Theory

The single most ubiquitous construct in neuroscience is that of arousal or activation and no brain system contributes more to these functions than the mesencephalic reticular formation. This diffusely projecting neural network originating in the midbrain determines the activation or arousal level of higher-level brain regions ipsilateral to the side activated by electrical stimulation (Moruzzi and Magoun 1949; French 1958). This system receives sensory information, including ambient lighting and sound levels, altering the sensitivity or the intensity of information conveyed through the primary sensory projection pathways extending through the thalamic relays and onto the primary projection cortices for each modality (Isaac 1960; see also Heilman et al. 2003). Thus, arousal levels, as witnessed initially through behavioral activation and cage activity levels (Jacobsen 1931; Richter and Hines 1938) and later through cortical desynchrony of the electroencephalogram, may be altered directly as a function of ambient sensory conditions (Isaac and DeVito 1958; Isaac 1960). In turn, this activating system appears to be regulated through the descending influences of the frontal lobes (French et al. 1954; Isaac and DeVito 1958).

One of the critical breakthroughs in this classic research was Walter Isaac's initial insight and demonstration that this arousal system functions in response to our contextual sensory array. This was an important argument for anyone interested in the ethical treatment of laboratory animals, as invasive electrical stimulation or lesion placement was no longer required to manipulate this system within the neurosciences. Walt demonstrated that the system could be manipulated more naturally, and noninvasively, through natural lighting or sound manipulations or through other sensory modalities as exists within the contextual setting. Walt demonstrated an understanding of this system far ahead of his time and with direct implications for the treatment of arousal disorders such as the hyperkinetic or hypokinetic syndromes, stupor and coma, and attention deficit hyperkinetic disorder. More specifically, behavioral overarousal from this theoretical perspective may result from overactiva-

© Springer International Publishing Switzerland 2015
D. W. Harrison, *Brain Asymmetry and Neural Systems,*
DOI 10.1007/978-3-319-13069-9_24

tion either through the reticular formation's influence on higher cerebral systems or through the inadequate frontal regulatory control or inhibition of the reticular formation.

Prefrontal lobectomized diurnal monkeys, in light, were known to be hyperactive (Jacobsen 1931; Richter and Hines 1938; Ruch and Shenkin 1943). This was demonstrated again on the measurement of cage activity levels. However, as predicted by arousal theory, the activity level of these monkeys was found to be influenced by the ambient sensory levels of lighting and of noise. More specifically, reductions in sensory input into the reticular formations of these primates lowered arousal level corresponding to the oppositional influences that would normally result from frontal lobe integrity and regulatory control over these systems. Moreover, the sensory effects across the modalities of sound and lighting level were interactive with light being a necessary concomitant of auditory stimulation in producing elevated activity level in the prefrontal lobectomized monkey (Isaac and DeVito 1958).

Arousal or activation theory (e.g., Hebb 1955, 1959; Lindsley 1960) provides for the lowering of arousal, through reductions in sensory input or through habituation processes (Sokolov 1960; Scott 1966) as might occur with prolonged exposure to the environment or with the development of a stable state, for example, after relocation to a novel environment. Novel and irrelevant sensory events or heightened ambient sensory conditions may, in contrast, increase arousal level or even result in overarousal with respect to an optimal state (Harrison et al. 1990; Harrison et al. 2003; Harrison and Kelly 1989; Harrison and Pavlik 1983; Harrison and Isaac 1984). Such an event is common with the elderly patient relocated several times from home to hospital, to assisted living, to nursing care facility. Overarousal may result, early on, following relocation stress with environmental novelty and dishabituation processes active. The same individual may be substantially underaroused at a later date and following habituation to the environmental setting. The inference here is that the needs of the patient, at least within the arousal dimension, may vary substantially early, as opposed to later, in the relocation or dishabituation process.

There may be an inversion of this process within the institutional setting, where there is a buzz of excitement and determination of staff members to meet the medical and multidisciplinary team demands, completing multiple evaluations, immediately upon the patient's relocation to a new setting and away from the safety and security of the patient's home. These demands surely aggravate control issues and may decompensate a patient prone to paranoid ideation or one easily confused by overstimulation. But, within this dimension, and wherever feasible, the patient may benefit more, early after relocation stress, through efforts to implement routine; through the provision of quiet; and through efforts to improve the perception of internal control as opposed to control by others or by extraneous events. Promoting the patient's need to accommodate to the setting is seldom a concern within a medical model focused on diagnosis and testing. But, whenever possible, these efforts may be rewarded in the promotion of medical stability through neuropsychological interventions to develop and to maintain a stable state.

One might be tempted to think that these relocation effects would dissipate with advanced dementia, where the patient appears to only minimally appreciate their surroundings. But in unpublished research (Zicafoose et al. 1988), relocation of advanced dementia patients from one ward to another within the same large veteran's medical center demonstrated significant degradation in verbal learning on the Rey Auditory Verbal Learning Test (RAVLT) along with reliable increases in blood pressure and heart rate using automated assessment apparatus and procedures consistent with the guidelines and standards of the Association for the Advancement of Medical Instrumentation (AMI), the American National Standards Institue (ANSI), the American Heart Association (AHA), and the National High Blood Pressure Education Program (NHBPEP; Harrison et al. 1988; Harrison and Kelly 1987; Kelly and Harrison 1994). Dishabituation confounds even the assessment of rapidly repeated blood pressure and heart rate measures and with significantly elevated dishabituation confounds in the elderly when compared with younger groups (Harrison and Edwards 1988). Moreover, older subjects require extended exposure to habituate to novelty or contextual change (Harrison and Isaac 1984; Harrison and Pavlik 1983).

In contrast, the patient who is well familiar and stable within their routine environment, with underarousal in this setting, may benefit from staff and family efforts to enrich the environment a bit. This might be promoted through the appropriate engagement in preferred activities and potentially out of doors if the weather is good. The recreation therapist may be instrumental in this respect, through community reentry activities and the like. But, the family is most reactively present following acute decline or relocation to a care facility. These efforts may be tapered off following the initial 30-day crisis period. In this example, the engaged efforts of the family may be welcomed more following the implementation of routine and the acquisition of stability in this new setting.

Classic arousal theory developed over time from a simple two-stage model to one of increasing complexity, where the synergistic and antagonistic influences of distinct neural systems and neurotransmitters systems became evident. The initial two-stage model classically provides for alterations in arousal through the diffusely projecting nuclei of the brain-stem reticular formation and through the thalamic system. Alterations in arousal result partly as a function of sensory input projecting to the brain-stem reticular activating system. The thalamic system was discovered in 1942 by Dempsey and Morison (Dempsey and Morison 1942a, 1942b; Dempsey and Morison 1943) when low-frequency stimulation of the unit evoked widely distributed, high-voltage, cortical waves, characterized by long latencies and an initially progressive voltage increment.

These recruitment responses produced high-voltage synchronous waveforms, which resembled the spindle bursts produced by the administration of barbiturate ananesthesia. Moreover, the evidence indicated that the neural pathways were the same and that the brain stem mediated the arousal response evident in the electroencephalogram (Dempsey and Morison 1942a, 1942b; Dempsey and Morison 1943; Lindsley et al. 1949; Moruzzi and Magoun 1949; see also Starzl and Magoun 1951). The significance of this neural pathway (reticular formation and the thalamic system) was even more dramatic with Jasper's (Jasper and Droogleever-Fortuyn 1946; Jasper 1949; Hunter and Jasper 1949) discovery that the diffusely projecting

Fig. 24.1 The two-pathway model with the reticular formation promoting arousal and with the frontal lobe promoting regulatory control over this activation

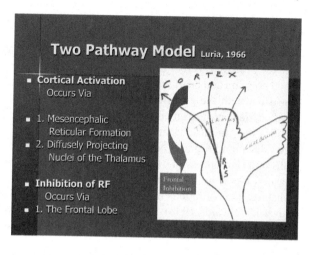

thalamic system served as the subcortical pacemaker driving high-voltage cortical events with epilepsy. In opposition to this ascending reticular activating system, the moderation of arousal level was performed through the descending and regulatory influences of the frontal lobes (Isaac and DeVito 1958; French et al. 1954). This relatively simple model is depicted in Fig. 24.1.

Functionally, for the family member and for the therapist working with an individual with an arousal disorder, understanding these oppositional systems may provide some clues on practical approaches to assist the patient with an arousal-related disorder and may also alter the attributions one has derived towards this person. First, it may be helpful to appreciate that relocation effects or disruption in routine may stress the capacity of the frontal regulatory control systems with decompensation of the patient to the point that they may look worse than they actually are from a functional standpoint. Heightened arousal, from disruption of the patient's routine of daily living, may result in the dementia patient appearing substantially more confused and the poorly regulated patient appearing substantially more deregulated. Such an individual may benefit from the implementation of routine in daily activities, the provision of a quiet, nondistracting environment, and possibly the provision of familiar items from home such as a shawl or pictures of family members. But, smells are important. So, presentation of familiar odors and pheromones is important and washing the materials might better be avoided, if appropriate. The sight, touch, and smell of the family dog might also be useful in the form of animal-assisted therapies (see Nimer and Lundahl 2007).

Secondly, the arousal disorder may be identified as one of overarousal or underarousal and this is often a function of the lesion location within the right frontal lobe (overarousal) or within the brain stem (underarousal). Thus, sensory stimulation or contextual therapies may be a consideration for the brain-stem patient with hypoarousal (e.g., Wood 1991), whereas lowered intensities of ambient sensory input such as with a quiet, nondistracting environment may be helpful to the deregulated hyperaroused patient. This distinction, between the frontal lobe and the brain-stem lesioned patient, provides for a basic clinical "rule of thumb" for the implementation of a therapeutic environment. The approach is vested within theory and evidenced based in outcome. However, there is complexity in the arousal systems and a careful diagnosis may provide clarity for the intervention.

One basic distinction is that which results from an amotivational apathetic frontal lobe syndrome (see Scott and Schoenberg 2011; see also Grossman 2002; Hu and Harrison 2013b; see also Damasio et al. 2012; see also Granacher 2008) usually with origin of the lesion in the medial frontal lobe and, especially, with diminished activation within the dopaminergic systems of the left frontal lobe (the basal ganglia and striate bodies). A second basic distinction exists at the brain-stem level, where an inferior pontine lesion in the raphe may diminish serotonergic activation and result in some degree of insomnia and with diminished access to deep and restful stage III and stage IV sleep. This would be oppositional at the brain-stem level to a lesion in the upper pons (reticular formation) or dorsomedial thalamic region (diffusely projecting thalamic nuclei of the reticular activating system). The former results in the classic syndrome of decreased arousal level ranging from stupor to even coma. The latter results in a sleep disorder with access only to the lightest level of sleep (stage I) and where the patient would have no appreciation of sleep, conveying an exhausted and hypoaroused appearance. It may also be oppositional to a lesion in the locus coeruleus and noradrenergic influences promoting rapid eye movement (REM) sleep.

Early on, Bremer (1937, 1938) posited an arousal theory whereby sensory input was instrumental to two general arousal states, and more specifically that associated with wakefulness as opposed to sleep state. In classic research, Bremer developed the *cerveau isole,* a surgical midcollicular transection (between the superior colliculi and the inferior colliculi) largely above the sensory afferents to the brain stem via the cranial nerves. As predicted, the elimination of sensory input to the cerebral hemispheres resulted in a chronically "sleeping" comatose animal. In contrast, the surgical transection known as the *encephale isole* left sensory path-

Fig. 24.2 The cerveau isole and encephale isole resulted in "permanent sleep" and "almost normal sleep," respectively

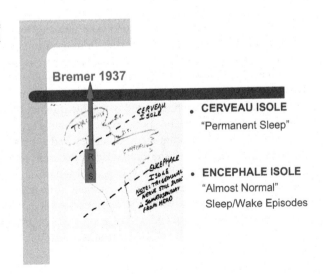

ways intact through the cranial nerves and with access to the cerebral hemispheres (see Fig. 24.2). As predicted, retaining sensory access to the cerebral hemispheres resulted in near-normal arousal patterns of alternating sleep and wakefulness. This supported Bremer's claim that wakefulness was a function of sensory input, whereas the elimination of sensory input was evidenced through sleep behavior.

But all is not simple in science, and systematic research led to the implementation of a midpontine transection (the Batani transaction) between the two previous planes of section (Winn 2001, p. 149). With this lesion, in stark contrast to the predictions of Bremer, the animals developed insomnia. It was partially through this manipulation that the pontine reticular formation, essential for high-activation states and wakefulness, was discovered. Another structure in the lower pons, the raphe, was found to be oppositional to the reticular formation and to promote deep sleep. The reticular formation largely uses acetylcholine for heightened arousal (e.g., Lewis and Shute 1967) with the serotonergic influences of the raphe (e.g., Cagampang et al. 1993; Miyazaki et al. 2012) promoting deep sleep. Dorsal to these structures, the locus coeruleus provides for noradrenergic arousal and sleep influences, largely through the inhibition of the raphe, and indirectly through the agonistic influence overarousal functions driven by the reticular formation. Additional serotonergic agonists are found through the area prostrema with facilitation of the raphe and with sensitivity to the circulating indolamines in the blood.

This system might be promoted in the evening, before a stressful workday, by serving turkey dinner and with elevation of circulating tryptophan levels. Incremental levels of circulating tryptophan in the bloodstream appear to facilitate the activation of the area prostrema and, indirectly, the raphe promoting a night of deep and restful sleep! One dedicated teacher discovered this, inadvertently, through her efforts to feed and nurture every student that she has known over the years. After serving a large group of graduate and undergraduate students Thanksgiving dinner consisting of lots of turkey, folks settled down in front of the fireplace and the rest is history. It was only awhile before the guests (some near strangers) were sleeping peacefully in her family room! The serotonergic structures of the raphe are activated indirectly through indolamine agonists in the bloodstream interacting with the area prostrema. They have also been found to be activated by slow rhythmic stimulation, as from the warmth of the radiant heat from the fire upon your face or even from petting a cat (see Siegel 1979). The implication here is that even bright light stimulation must be controlled and specific, if it is to heighten arousal level. Activation of the reticular formation with bright light therapy to promote arousal may be doomed to failure if the light is confounded with slow rhythmic oscillation from warmth or possibly the rhythmic beat of ocean waves on the beach, for example.

These oppositional brain-stem structures and others, including the substantia nigra (dopaminergic) and the dorsomedial thalamic regions, contribute to a balanced and oscillating sleep system (e.g., Hobson et al. 1975; Riemann et al. 2012). The balance among these systems is intimate to literatures not only on sleep and wakefulness but also on depression, anxiety, hostility, and other affective concerns. The classic arousal theory under discussion, though, provides both for a direct perceptual pathway for each sensory modality projecting onto its respective cortical region and for the indirect, and in some ways parallel, influences of the reticular formation on these sensory projection and association pathways.

The basic arousal model to be considered next is presented in Fig. 24.3. Elsewhere in these readings, the primary projection pathways for several major sensory modalities were discussed. Sensory input from each of these modalities (e.g., vision, audition, somaesthesis) travels from the sensory receptor, projecting onto the specific thalamic nucleus for that modality. Thus, the ganglion cells originating at the retina, in the visual pathway, synapse at the lateral geniculate nucleus (LGN) of the thalamus; audition will synapse at the medial geniculate nucleus (MGN) of the thalamus; and the body senses of touch, pressure, warmth, cold, and pain synapse at the VPL nucleus of the thalamus. Although not personally a fan of abbreviations like VPL, this particular abbreviation is most useful. The reader will appreciate this abbreviation instead of the actual structural name, which is the "nucleus ventralis posterolateralis of the thalamus!" We can all save our breath, in this case, using the abbreviation and we may avoid having our tongue tied into knots with these lingual gymnastics!

These thalamic relays continue on within each brain up to the primary projection cortex for each of the respective sensory modalities. This functional neuroanatomy is depicted in the right side of Fig. 24.3. The sensory projections travel on to the primary projection areas (sensory cortex). From this point, an increasingly complex analysis and comprehension of the sensory data results from processing in the surrounding association cortices and all under the direction of the glial cells with spreading activation not only along dendritic fields but also with calcium flux between and among astrocytes (see Koob 2009). Ultimately, the information flow gains access to the longitudinal tracts and conveys toward the third functional unit, the frontal lobes. This is depicted in the upper part of Fig. 24.3.

The sensory and motor projection pathways are tightly grouped together at the thalamic nuclei, and a small lesion here may directly (via the projections) and indirectly (via the arousal systems) influence the function of disparate cortical areas providing one or another of the thalamic brain syndromes, which are discussed else-

Fig. 24.3 The oppositional role of the reticular formation and frontal lobe over the processing of sensory information through their projections and onto association cortices

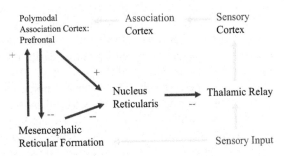

Arousal Theory
see Heilman, et al., 1994; 2003 p#310

Polymodal Association Cortex: Prefrontal

Association Cortex

Sensory Cortex

Nucleus Reticularis

Thalamic Relay

Mesencephalic Reticular Formation

Sensory Input

where in these writings. Parsimony will provide a clue to the diagnostician where a small thalamic lesion might be more tenable than one extending far afield and across a broad cortical region. Fluctuating arousal levels with pseudocortical symptoms waxing and waning within and/or between sessions might provide another.

As depicted at the bottom of Fig. 24.3, there are redundant sensory projections onto the mesencephalic reticular formation. Sensory activation of the reticular formation promotes inhibition of an inhibitory structure, the nucleus reticularis, yielding a release or activation of the primary sensory projections traveling through the thalamus on their way to the cortical projection areas (Jasper 1949; Moruzzi and Magoun 1949; Starzl and Magoun 1951). More specifically, increased ambient lighting may have an incremental activating effect, via the mesencephalic reticular formation, resulting in heightened perceptual intensity of sound at the auditory projections or perhaps touch or pain at the somatosensory projections. Diminished frontal lobe regulatory control over these arousal systems can have a similar effect, resulting in incremental activation and poorly regulated sensory projections and ultimately dishabituation to the environmental setting. Thus, pain might be made to be more intense through increments in ambient noise (bottom up) or through diminished frontal capacity (top down). The latter might be effected through dual or complex tasking demands or functional activities weighing down the regulatory capacity of the right frontal region, for example.

This theory coupled with capacity theory predicts that diminished frontal lobe sufficiency supports overarousal and poor habituation to the setting. This might be seen in one or more of the executive or regulatory deficits discussed in the section on the frontal lobes. Moreover, as demonstrated by Isaac and colleagues (Isaac and DeVito1958; Isaac and Reed 1961; Kallman and Isaac 1980), elevated activity levels may be diminished by working at the other end of the model, through decreased sensory input into the reticular formation. Sensory thresholds or sensitivity might be affected or manipulated by either diminished frontal lobe capacity or frontal lobe stress (Woods et al. 2013) and through the manipulation of ambient sensory conditions within other sensory modalities (e.g., Delay et al. 1978). These systems have also been shown to be susceptible to pharmacological manipulation altering sensory thresholds (e.g., Goetsch and Isaac 1983; Delay et al. 1979; Lowther and Isaac 1976).

The individual with a frontal lobe syndrome may lose either or both of the regulatory pathways which dampen or diminish the intensity of sensory information within the brain. This might result in lowered somatosensory thresholds, where the clothing on their body is irritating. Lowered auditory thresholds might be expressed by a patient bothered by noises easily tolerated by others. Sensitivity to lights might also be increased (Burstein et al. 2010; see also Merskey and Bogduk 1994). These perceptually more intense ambient events may decompensate some and may promote related symptoms for others (e.g., ocular migraine with photophobia; skin sensitivity disorders with allergic-like responses). One young woman with a right frontal lobe syndrome expressed social improprieties, delayed response deficits (impulsive), and heightened somatic sensitivity. The latter was expressed with her great discomfort in her clothing. She had scarring on her skin from repetitive scratching and irritation and she painted her body with calamine lotion, while wearing only enough clothing to suffice.

An example from the other end of the arousal system might be that of the family of a stuporous or even comatose patient and their efforts to assist their loved one out of the coma or stuporous state. Often, family members and therapists attempt this in essence through sensory stimulation procedures. But, many have been frustrated, at least initially, in these efforts. If the lesion disconnects the arousal systems from the descending pathways, the arousal system may be disinhibited in some way. This includes a loss of inhibition over the thalamic radiations resulting in the recruitment of high-voltage cortical synchrony (e.g., Dempsey and Morison 1942a, 1942b; Dempsey and Morison 1943; see also Peter-Derex et al. 2012), which in the damaged brain may promote seizure activity (Jasper et al. 1046; Jasper 1949; Hunter and Jasper 1949). Moreover, the patient recovering from a coma or stupor is often agitated and even aggressive with poor safety management in their compromised state. Thus, the prudent physician might maintain the coma through barbiturates or other pharmacological techniques and sometimes without the family or therapists appreciating these interventions, which are at least oppositional to their efforts to achieve an alert and aroused state.

Communication among the team of therapists, physicians, and family or friends may be most helpful at this point and to establish a coordinated effort eventually to decrease the medication management, promote appropriate recovery of arousal, and maintain optimal safety. But, the agitated patient with an arousal disorder may benefit more from the provision of a care unit or facility designed specifically for the management of coma and stupor and where agitation or loud vocalization may have a minimal negative impact on other patients in the area. There are also many administrative concerns in the management of these patients that far exceed the scope of these writings.

Cerebral asymmetry in these neural systems is evident within vast literatures on arousal and emotional disorders. One man, for example, was in a stuporous to comatose state, where sensory stimulation therapy involved passive range of his extremities. The range of his left-sided extremities resulted in agitation, sympathetic drive, and fitful foul-mouthed expressions at the loudest volume. Imagine the initial dismay expressed by the patient's family and the medical staff, when the passive range of his right-sided extremities resulted in positive affective facial displays, tone of voice, and cooing sounds like a baby.

Chapter 25
Right Hemisphere and Arousal

The basic arousal theory, discussed above, is largely axial in nature with brain-stem reticular formation activation of the arousal response and with the regulatory inhibition or modulation of arousal, through the descending fibers originating in the prefrontal region. Similarities, though, were apparent to some that the bilateral prefrontal lobectomized hyperkinesis demonstrated in nonhuman primates was more comparable, perhaps, to a right prefrontal lesion in a human (personal communication Walt Isaac). This relationship in lateralized energy level and regulatory control is discussed elsewhere in these writings. By the 1980s and through the contributions of Kenneth Heilman, Wendy Heller, and many others (e.g., see Bruder et al. 2012), it became clear that these brain systems were lateralized and especially in humans. Ken's group was able to demonstrate this through several discussions within the neuroscience literature. But, most notable were the applications of the classic arousal theories discussed above to more modern constructs involved in unilateral neglect disorders (see Heilman et al. 1978, 2003; see Heilman and Gonzalez Rothi 2012b) and in the lateralization of emotion preferentially to the right cerebral hemisphere (Heilman et al. 2012; see Heilman and Valenstein 2012).

With lateralization of the negative emotional valences and the high-intensity affective states of anger and fear to the right brain, it became increasingly clear that the underlying arousal system must be lateralized. French (1958) and Moruzzi and Magoun (1949) had appreciated differences in the persistence of activation with electrical stimulation of the right mesencephalic reticular formation resulting in increased persistence in activation. Even warning stimuli delivered to the right brain, as opposed to the left brain presentation, reduced reaction time to imperative stimuli (Heilman and Van Den Abell 1979) supporting the role of the right hemisphere in high-arousal and defensive states. Heilman et al. (2003) refer to the regulatory role of the frontal lobes in moderating emotional valence. Others identified a more specific role for the left frontal region regulating positive emotional valences and for the right frontal region over negative emotional valences (e.g., Davidson and Fox 1982). However, the impact on arousal theory (see also Heller 1993) more directly, was through Heilman's appreciation that "The right hemisphere, and especially the

© Springer International Publishing Switzerland 2015
D. W. Harrison, *Brain Asymmetry and Neural Systems,*
DOI 10.1007/978-3-319-13069-9_25

parietal lobe, appears to have a strong excitatory role on arousal and the left hemisphere an inhibitory role." Moreover, Kenneth Heilman and his coauthors state that "Arousal is mediated by the mesencephalic and diencephalic portions of the reticular activating system. However, the cortex, and especially the right inferior parietal cortex, appears to modulate these arousal systems."

For the family member, caregiver, or therapist, arousal disorders are the more fundamental characteristic of brain-stem lesions when the patient presents with a hypoaroused state. The more severe arousal disorder, and that which may result in a more guarded prognosis, is that demonstrated by the patient with a combined brain-stem lesion affecting the reticular formation and a lesion within the right inferior parietal region. Hyperaroused states are more common with deactivation or lesion of the right prefrontal and mesial prefrontal regions as a result of diminished regulatory control over the reticular formation via the descending or corticofugal projections and the concurrent decrease in regulatory control over the right posterior brain regions via the longitudinal tract (see quadrant theory: Foster et al. 2008; Shenal et al. 2003; Carmona et al. 2009). Of course, extension of this theoretical perspective with deactivation of the right frontal region would release a relative activation in the homologous left frontal region(s). Even patients suffering from Parkinson's disease may differ with onset of symptoms at the left or at the right hemibody and as a function of disease duration. Foster et al. (2013), for example, found increased energy levels as a function of disease duration in those Parkinson's patients with left hemibody onset of symptoms.

The regulatory control over both anger and fear may be a function of the dynamic capacity of the right frontal region to concurrently process negative affective stress challenge, while maintaining control over sympathetic tone to convey a stable state in our facial expressions, social interactions, cardiovascular dynamics, and in our glucose and cholesterol levels. But, anger and aggressive interactions require not only the "fuel for the fire" but also substantial oxygen saturation for cellular performance and metabolic processes. With inadequate oxygen saturation and with incremental carbon dioxide levels, the regulatory demands are incremental and eventually pose an immediate threat to survival activating panic symptoms and efforts to withdraw or escape the setting, and at all cost. Elevated carbon dioxide levels have been reported with panic (Biber and Alkln 1999; see also Homnick 2012) and incremental carbon dioxide levels may precipitate the panic attack (Maddock 2010; see also Esquivel et al. 2010). If this argument is maintained through research endeavors (see Hu and Harrison 2013a), then the treatment of panic might include increased oxygen saturation, whereas the treatment of anger or rage may include elevation of glucose or insulin support for perfusion of the cells as with hypoglycemic or hyperglycemic states (Walters and Harrison 2013a, 2013b). These relationships are interrelated where high levels of oxygen saturation deplete carbon dioxide saturation levels and potentially alter cognition and emotional status (e.g., Meuret et al. 2010, 2012).

The interpretation of such effects attributable to one or another component of a functional neural system may vary with the activity of the remaining functional neural systems, including those within Luria's second functional unit for reception, comprehension, and analysis of sensory information (parieto–occipito–temporal regions) and those within Luria's first functional unit modulating arousal levels. Amygdala-restricted manipulations interfere with the acquisition and recall of conditioned fear and other forms of anxiety-related behaviors in nonhuman animals. Feinstein and colleagues (Feinstein et al. 2013; see also Adolphs et al. 1999; Becker et al. 2012) provide evidence where amygdala damage effectively eliminates conditioning to aversive stimuli (Bechara et al. 1995), the ability to recognize fearful faces (Adolphs and Tranel 2000) and a marked absence of fear during exposure to a variety of fear-provoking stimuli, including life-threatening traumatic events (Feinstein et al. 2011). One interesting exception has been found, where patients with bilateral amygdala damage evidenced not only fear but also panic attacks following inhalation of 35 % carbon monoxide (Feinstein et al. 2013). Relevant here are the theoretical propositions for functional neuroanatomical systems devoted to fear evoked from external threats and those which arise internally secondary to diminished oxygen saturation levels, perhaps, or from elevations in carbon monoxide levels.

Also relevant to this discussion are brain-stem arousal mechanisms, largely affected by sensory modulation (e.g., see Damasio and Carvalho 2013); and right posterior brain mechanisms, largely involved in the activation and comprehension of threatening or negatively valenced emotional stimuli. An emotional disorder is implied for the patient with right inferior parietal brain pathology (e.g., see Damasio et al. 2013). Such a person may experience a diminished arousal reaction to provocative affective events. This may include the emotional domain of maintaining integrity in one's relationship with their spouse, caregiver, or perhaps with their therapist. Laypeople describe such a patient as "obstinate" or "stubborn" with a lack of appreciation for others' emotions as expressed in their facial expressions or through their tone of voice. Indeed, comprehension of emotionally provocative events is compromised in these individuals (Heilman et al. 1978; Heilman et al. 1975).

Chapter 26
Parasympathetic and Sympathetic Tone

Sweet Memories and Rapid Heart Rate

In several nutritional neuroscience projects, W. David Crews III investigated the impact of antioxidants and dietary interventions on neurocognitive performance. In these investigations, memory and executive functions were assessed along with blood assays and measures of cardiovascular functions, including heart rate. Neurocognitive and psychophysiological effects were found subsequent to the use of *Ginkgo biloba* (Crews et al. 2005), cranberry juice versus placebo (Crews et al. 2005), and other putative antioxidant sources. A favorite investigation in this line of research was undoubtedly "the Hershey's dark chocolate project." One thousand five hundred pounds of Hershey's antioxidant-rich chocolate, a placebo product, and 800 pounds of Hershey's Cocoa arrived on the loading docks at Virginia Tech's Roanoke Higher Education Center, formerly the home of Norfolk and Southern Railways executives. The rich treasure was carefully divided into packets for the participants to add to their dietary regime.

Recruitment of subjects for this demanding project was a matter of holding back all of the potential volunteers from across the Commonwealth of Virginia. Television crews and radio disc jockeys were all on board to promote the research, I think partially in hopes of receiving some free chocolate, which was generously donated by The Hershey Company. Indeed, an unrequested call was received from an "on-the-air" disc jockey that went something like this "Is it true…that you…are giving away…free Hershey's chocolate…for participating…in your research project (emphasis added)?" But, one of the primary findings of this research was simply the promotion of sympathetic tone with elevated heart rate from the consumption of dark chocolate. With the antioxidant effects put aside, chocolate has caffeine in it, which appeared to promote increased heart rate (Crews et al. 2008). David reviewed the effects of cocoa- and chocolate-related products on neurocognitive functioning in his chapter in the book entitled *Chocolate in Health and Nutrition* edited by Watson, Preedy, and Zibadi (Crews et al. 2013).

© Springer International Publishing Switzerland 2015
D. W. Harrison, *Brain Asymmetry and Neural Systems*,
DOI 10.1007/978-3-319-13069-9_26

Cerebral Laterality and the Autonomic Nervous System

The left and right cerebral hemispheres appear to differ in their parasympathetic and sympathetic nervous system roles, perhaps having direct implications for cardiovascular and cardiopulmonary functioning. The evidence supports a right hemispheric role in activation of the sympathetic nervous system, whereas the left hemisphere is primarily associated with activation of the parasympathetic nervous system. The relative lateralization of the parasympathetic and sympathetic nervous systems to the left and right hemispheres, respectively, is supported by the results of numerous investigations, including those examining the cardiovascular effects of unilateral insular stimulation (Oppenheimer and Cachetto 1990; Oppenheimer et al. 1992; Hoffman and Rasmussen 1953), left- and right-sided cerebrovascular accident (Heilman et al. 1978), and unilateral intracarotid sodium amobarbital injections or Wada technique (Zamrini et al. 1990; Hilz et al. 2001). These findings appear consistent with the subsequent demonstration of stronger connections between the right hemisphere and the lateral (sympathetic) than medial (parasympathetic) hypothalamus by Lemaire's group (Lemaire et al. 2011; see Burtis et al. 2013).

For instance, Oppenheimer et al. (1992) found bradycardia and depressor responses following stimulation of the left insula, but tachycardia and pressor responses following stimulation of the right insula (see also James et al. 2013). Furthermore, reduced heart rate results from right hemisphere strokes (Andersson and Finset 1998), whereas left insular strokes are associated with increased cardiac sympathetic tone (Oppenheimer et al. 1996). Unilateral intracarotid sodium amobarbital injections of the left hemisphere have been used to generate increases in heart rate and blood pressure, whereas intracarotid sodium amobarbital inactivation of the right hemisphere results in decreases in heart rate and blood pressure (Hilz et al. 2001; Zamrini et al. 1990). Yoon et al. (1997) found that following left hemisphere inactivation the low-frequency/high-frequency ratio power spectral analysis of heart rate increased significantly, indicating a shift in sympathovagal balance towards sympathetic dominance.

Beyond the regulatory functions of the frontal lobes, pathological activation of the second functional unit yields similar effects. For example, right temporal lobe seizures may be evident in cardiovascular presentations, including tachycardia and arrhythmia (Marshall et al. 1983). Neuroanatomic connections between the left and the right cerebral hemisphere and the heart provide links that allow cardiac arrhythmias, tachycardia, bradycardia, or syncope to occur in response to regional brain activation. Davis and Natelson (1993) provide a review of the pathogenesis of malignant cardiac arrhythmias, where brain mechanisms linked to cardiovascular functions have been shown to produce arrhythmia both experimentally and clinically with specific examples, including stroke, epilepsy, and environmental stress. Moreover, these authors focus on sympathetic activation where they hypothesize that the individual with a diseased heart has a greater likelihood of experiencing cardiac arrhythmia and sudden cardiac death when the neurocardiac axis is activated. They conclude based on their review of possible brain-related mechanisms promoting arrhythmias, that the nervous system directs the events leading to cardiac damage by raising catecholamine levels.

One woman in her 30s, with a doctoral degree in law, presented with memory complaints a few years subsequent to a head injury with loss of consciousness. She had been diagnosed with bipolar disorder and she appreciated episodic anger, which she could not easily control. She had recurrent or episodic bouts of tachycardia and these episodes would often awaken her from sleep. Her episodes consisted of leftward vection and she would drift to the right in ambulation and sometimes with falls to the right. She described migraine headaches centered over her right frontal region. More careful interview revealed left upper extremity tactile formesthesia with fear or apprehension accompanying these events. She reported olfactory hallucinations with prior episodes. She confirmed visual formesthesia with fear as these images moved abruptly within her peripheral vision.

Following the completion of the standardized neuropsychological test battery, including assessment of memory functions, she was assessed using a neurobehavioral status exam. These procedures revealed enlargement of the left bicep with a distinct muscular cord or flexor synergy. Though she had been stable in her autonomic responses and denied vectional complaints throughout the testing, cautious and gentle passive range of the left arm was enough to decompensate her with the abrupt change in her pleasant emotional demeanor and now with pathological crying and gelastic lability. She confirmed the onset of the leftward vection with nausea, increased somatic warmth with sympathetic drive, and profound negative emotional feelings. Moreover, the allocation of additional right frontal resources through leftward-directed visual gaze and pursuit aggravated these symptoms, whereas slight rightward gaze was useful in the abatement of symptoms. She had also learned that close physical contact with her spouse at her right hemispace and hemibody was useful in promoting emotional stability, reducing sympathetic drive, and in resolving the nausea and dizziness. This basal right frontal syndrome provides an example of the release or loss of regulatory control over right posterior cerebral systems charged with negative affective perceptions, leftward vection and abdominal synergy, and cardiac tempo with episodic tachycardia, bodily warmth, and such. Her memory scores confirmed only the episodic nature of the memory disorder as right temporal release appeared to suppress the processing and consolidation of verbal, linguistic memory.

Cerebral asymmetries in autonomic nervous system function might be appreciated by way of example using a patient with elevated parasympathetic drive associated with left temporal and parietal involvement. Comparison might be drawn in this case from a well-educated man who had worked for the federal government and served actively in the military, where he had been a recipient of the Purple Heart. His admission to the medical center ultimately would identify an embolic cerebrovascular accident. However, the autonomic nervous system events included acute low blood pressure, bradycardia, and atrial fibrillation, and these events were recurrent on the rehabilitation unit. Interestingly, these cardiovascular events were precipitated by auditory and by visual–verbal or linguistic processing demands in his therapies. With oscillating speech sounds and propositional speech processing demands, the man would become preoccupied with events at the right hemispace. He would show increased semantic paraphasic errors in speech and speech content would vary eventually to "word salad" content. He readily confirmed that he was experiencing very positive emotions and enjoyment from the visual formesthesias and auditory paracusia at the right hemispace. His facial configuration expressed positive emotion and enjoyment consistent with his verbal accounting of his emotions during these events.

Hemispheric lateralization of the parasympathetic and sympathetic nervous systems finds support from the results of investigations examining cerebral asymmetry of the associated neurotransmitters. Acetylcholine is the primary parasympathetic postganglionic neurotransmitter, whereas norepinephrine is the primary sympathetic postganglionic neurotransmitter (Iversen et al. 2000; Loewy 1990; Nestler et al. 2001). Earlier, Wittling and Genzel (1995) proposed greater involvement of the right brain in noradrenergic neurotransmitter activity. Research supports this proposition, as norepinephrine is found to be strongly lateralized to the right hemisphere (Oke et al. 1978). Acetylcholine distribution is also asymmetrical with greater left hemisphere representation for this neurotransmitter system (Glick et al. 1982; Kononenko 1980). Moreover, several reviews of the literature have supported the asymmetrical lateralization of norepinephrine to the right brain and acetylcholine to the left brain (Liotti and Tucker 1995; Tucker and Williamson 1984; Wittling 1995).

Given these lateral asymmetries in norepinephrine and acetylcholine, medications that augment or interfere with the actions of these central nervous system neurotransmitters should have an effect on cardiovascular functions. Accordingly, the administration of atomoxetine, a central nervous system norepinephrine reuptake inhibitor, is found with increased heart rate (Kelly et al. 2005). Further, administration of the cholinergic antagonist scopolamine is associated with increased heart rate (Ebert et al. 2001) and the cholinergic agonist pyridostigmine results in increased heart rate variability (Nobrega et al. 2001; Soares et al. 2004), an index of parasympathetic influences on cardiac functioning.

We have proposed a quadrant model of the cerebral control of cardiovascular functioning, with the cerebral regulation of cardiovascular functioning varying along both longitudinal (anterior to posterior) and lateral (left-to-right) neuroanatomical axes (Foster and Harrison 2006). Specifically, we have proposed that, given the inhibitory influence of the frontal lobes over the posterior regions of the brain, the left frontal lobe is charged with the inhibition or regulatory control over parasympathetic influences on cardiovascular functioning and the left posterior regions with the excitation of, or increases in, parasympathetic tone. Conversely, the right frontal lobe is charged with the inhibition or regulatory control over sympathetic influences on cardiovascular functioning and the right posterior regions with the excitation of, or increases in, sympathetic tone. Paul Foster (Foster et al. 2010) provided support for the model in a study in which relative left frontal lobe activation was related to an increased baseline heart rate. Moreover, resting heart rate in this project became continually lower as the degree of asymmetry shifted towards heightened relative right frontal lobe activation. The opposite pattern was found for asymmetry in activation over the posterior cerebral regions. Relative activation of the left posterior region was related to decreased blood pressure. Moreover, resting blood pressure increased as the degree of asymmetry shifted towards heightened relative activation of the right posterior region.

Partial support for the quadrant model was found in a related project using quantitative electroencephalography (Foster et al. 2008), where the relationship between indices of lateral and longitudinal asymmetry of cerebral activity and resting heart rate and blood pressure was investigated. Specifically, negative correlations were found between resting heart rate and low beta magnitude asymmetry at the frontal poles and dorsolateral frontal regions along with high beta magnitude at the dorsolateral frontal region. Regarding asymmetry along the longitudinal axis, a significant negative correlation was found between resting systolic blood pressure and low beta magnitude asymmetry for the left frontal and left parietal regions. Additionally, significant positive correlations were found between resting heart rate and low beta magnitude asymmetry for the right frontal and right parietal regions. Hence, the significant correlations in this study were in accordance with the a priori predictions derived from the quadrant model (Foster and Harrison 2006).

The findings indicate that patterns of asymmetry across the left and right frontal and then across the right frontal and right parietal regions are involved in resting heart rate. Specifically, asymmetry in activity favoring the left frontal lobe is predictive of higher resting heart rate. As this asymmetry shifts to relatively higher right frontal lobe activation, resting heart rate decreases. The relationship between resting heart rate and right frontal lobe activity then continues along the longitudinal neuroanatomical axis such that asymmetry favoring right frontal lobe activity is predictive of lower resting heart rate. As activity decreased at the right frontal lobe and increased at the right parietal region, resting heart rate increased. Thus, the findings provide support for a division of responsibility between the left and right frontal and posterior regions in regulating cardiovascular functioning. The findings are also consistent with the view that the left hemisphere is involved in parasympathetic influences on cardiovascular functioning and that the right hemisphere is involved in sympathetic influences. Relatedly, Hilz et al. (2006) provide support for the inhibitory role of the right frontal region over sympathetic activation as right ventromedial prefrontal cortical activity corresponds with relative activation of the parasympathetic nervous system, whereas right ventromedial prefrontal lesions result in cardiovascular activation with emotional stimuli (see also Meerwijk et al. 2012).

Parkinson's disease may represent a unique opportunity to study the cerebral lateralization of the sympathetic and parasympathetic nervous systems in the regulation of cardiovascular functioning. Parkinsonian motor symptoms typically begin in an asymmetrical fashion (Hoehn and Yahr 1967; Hughes et al. 1992; Leentjens et al. 2002; Rajput et al. 1993; Shulman et al. 2001), which is mirrored by asymmetry in the underlying neuropathological processes (Huang et al. 2001; Kempster et al. 1989; Kumar et al. 2003; Leenders et al. 1990; Tatsch et al. 1997). Later in the disease process, when there is bilateral motor involvement, asymmetry continues to exist both in motor dysfunction (Fross et al. 1987; Lee et al. 1995; Martin and Calne 1987) and neuropathologically (Tatsch et al. 1997). Further, Parkinson's disease is known to affect not only the dopaminergic system but also the cholinergic and noradrenergic systems (Emre 2003; Zgaljardic et al. 2004).

Given the asymmetrical neuropathological involvement throughout the course of Parkinson's disease and the hemispheric asymmetries in the control of the parasympathetic and sympathetic nervous systems, differences in heart rate and blood pressure may be present in Parkinson's patients with left versus right hemibody onset, and these cardiovascular changes might have potentially important clinical implications. Partial support for this was found (Foster et al. 2010) where left-hemibody-onset Parkinson's patients presented with higher resting heart rate than did the right hemibody onset Parkinson's patients. Systolic blood pressure readings though were oppositional to the heart rate findings, potentially indicating a need for more sophisticated or precise measures of cardiovascular influences by the sympathetic and parasympathetic branches of the autonomic nervous system.

The right brain has been extensively related to the generation and the regulation of sympathetic tone, whereas the left brain has been related to parasympathetic tone (Wittling 1990, 1995; Wittling and Genzel 1995). If one thinks about the two most prominent negative emotional valences, anger and fear, it is easy to appreciate the associated involvement of these right brain regions in the generation of sympathetic tone (the second functional unit) and the regulation of sympathetic tone (the third functional unit—frontal lobe region). Anger and fear result in corresponding increments in systolic blood pressure, heart rate, skin conductance responses, and the mobilization of glucose and cholesterol. All of which may increase our likelihood of surviving a significant threat, to include fleeing from the threat or the engagement in fighting or defensive behaviors. Cholesterol helps to slow or stop the bleeding if we are injured, whereas the glucose and cardiovascular mobilization provide the energy and increased perfusion of oxygen and nutrients necessary for combat or for escape. Under appropriate conditions, this system will keep us alive (normal function). However, we are increasingly, and painfully, aware that with chronic activation of these systems, resulting from an anger disorder or hostility, we are at a heightened risk for cardiovascular disease, diabetes, and the metabolic syndrome (Siegel 1985; Smith and Pope 1990; Smith et al. 2004; Suls and Bunde 2005; Mitchell and Harrison 2010; Walters and Harrison 2013a, 2013b; see also Cox and Harrison 2008b; see also Anthony et al. 2006). It is also evident in research findings that the metabolic syndrome may result with alterations in the reward systems of the left hemisphere and especially with increased abdominal fat (Luo et al. 2013).

The literature also supports a right hemispheric model of hostility for healthy individuals as evidenced using positron emission tomography (PET) to assess brain activation to anger inductions (Kimbral et al. 1999; Weisz et al. 2001). Kimbrell et al. (1999) used PET to measure regional cerebral blood flow changes as a function of the emotional response. Subsequent to the anger induction, the participants displayed significant increases in activation at the right thalamic and the right temporal regions, whereas there was significant deactivation at the right frontal regions. Kimbrell et al. concluded that transient levels of anger provide unique regional brain activation, which includes a relative deactivation of the right frontal lobe. Using depth electrodes in patients with epilepsy with aggressive behavior, Saint-Hilaire et al.

(1981) recorded epileptic activity localized to the right hippocampal region during spontaneous aggression. These researchers also used electroencephalographic recordings in aggressive patients with temporal seizures to show that the epileptic activity during spontaneous epileptic aggressive behavior localizes first in the right amygdala, then in the right temporal cortex, right hippocampus, and parahippocampal gyrus, and then reaches the anterior and median cingulate gyrus and right supplementary motor area.

By extension of this discussion, damage or incapacity within the right orbitofrontal region may result in reduced regulatory control over the expression of negative emotion (e.g., Agustín-Pavón et al. 2012; see also Fulwiler et al. 2012), anger, or aggressive behaviors. Reduced regulatory control from this region may be evident in an expressive dysprosodia with an avalanche of speech volume, uncharacteristic shouting, or loud abrasive features. Moreover, the loss of regulatory control over anger and fear, for example, may reflect a concurrent loss of control over the cardiovascular system with corresponding reactivity in systolic blood pressure, heart rate (Harrison and Emerson 1990; Demaree et al. 2000; Herridge et al. 2004; Williamson and Harrison 2003; Everhart et al. 2008; Shenal and Harrison 2004; Rhodes et al. 2013; see also Foster et al. 2008), glucose (Walters and Harrison 2013a, 2013b), skin conductance (Herridge et al. 1997), facial expression (Rhodes et al. 2013), and temporal lobe activation on the quantitative electroencephalogram (Mitchell and Harrison 2010).

Interestingly, the orbitofrontal region is seen in a lateral view of the brain proximal to the motor cortex. The motor cortex provides for directed efferent motor control with a topographical representation of the body wherein facial motor control is located inferior to the other body regions and at the basal region of the brain. This topographical anatomy provides for proximal brain areas in control of facial expressions and in control of anger (right orbitofrontal). Our language provides the clue to this where *losing face* occurs with the loss of emotional control or loss of a stable self-regulated state. The functional neural anatomy for the face and for anger control are likely consistent with Kinsbourne's (Kinsbourne and Hicks 1978) concept of *functional cerebral space* and where concurrent tasking demands may result in interference or a loss of control over the secondary task performed by that brain region. The dual-task challenge in this case might be the maintenance of regulatory control for anger and sympathetic control of the heart, while concurrently maintaining a stable, well-regulated facial expression.

Herath et al. (2001) used functional magnetic imagery during the performance of dual reaction time tasks. Performance of the dual task was found to activate cortical regions in excess of those activated by the performance of component single tasks. Moreover, these investigators reported that dual-task interference was specifically associated with increased activity in a cortical field located within the right inferior frontal gyrus. This area has been previously implicated in emotion regulation and hostility, lending to the notion that areas of the brain that regulate emotion in particular, may be subject to the burden of dual-task interference effects.

The tasking demands might simply involve contracting facial muscles involved in the emotion. Matthew Herridge (Herridge et al. 1997) did investigate this as did others (Ekman et al. 1983). Matthew found that contracting the corrugator muscle elevated sympathetic tone using a skin conductance measure. Hostile, violence-prone individuals maintained elevated skin conductance at the left hemibody with reduced habituation at the left hemibody to repeated contractions of this muscle. Low hostile individuals evidenced reduced sympathetic tone and a reduced rate of habituation at the right hemibody. Earlier on, George Demakis (Demakis et al. 1994) had demonstrated elevated skin conductance responses at the right hemibody with positive affective displays of spastic laughter. Hostiles maintain elevated cardiovascular measures of sympathetic tone as indicated by heart rate, blood pressure, and skin conductance measurements. The results support the role of the right orbitofrontal region in the regulatory control over negative emotion (e.g., Agustín-Pavón et al. 2012), anger (Fulwiler et al. 2012), and the dual concurrent management demands for regulating facial expressions and sympathetic drive. Moreover, the results support the conclusion of diminished resources at this region in hostile, violence-prone men. Subsequently, we provide evidence of diminished frontal capacity in hostile men, whereas facial dystonia was evident in the electromyogram recordings in this group and especially at the left hemiface (relatively diminished right frontal capacity; Rhodes et al. 2013).

The possibility for facial postures depicting one or another affective valence to influence cardiovascular function and stress responses was also investigated in a project where the participants' positive facial expressions were covertly manipulated. Kraft and Pressman (2012) asked the participants to complete two different stressful tasks while holding chopsticks in their mouths producing a Duchenne smile, a standard smile, or a neutral expression. One group was made aware of the manipulation by asking them to smile with covert manipulation of the comparison group using the chopsticks. Interestingly, both the covert and the posed facial configuration depicting a smile reliably lowered heart rates during stress recovery in comparisons with the neutral facial configuration. These findings were taken as evidence for the role of facial expressions to directly manipulate or alter cardiovascular function and, in this example, to promote diminished sympathetic drive or augmentation of parasympathetic drive.

Further support for the right hemispheric involvement in the regulation of cardiovascular processes has been found by Weisz et al. (2001). Baroreceptor stimulation (through a neck suction device) led to a significant regional cerebral blood flow (rCBF) increase in the anterior, inferior part of the lateral prefrontal cortex and only in the right hemisphere, thereby implicating the right frontal lobe in sympathetic activity in normal men. Weisz et al. (2001) concluded that the right hemisphere plays a larger role than the left hemisphere in baroreceptor regulation. Although this research is not specific to hostility, it demonstrates the role of the right frontal lobe in sympathetic activation.

Paul Foster (Foster et al. 2012) put this theoretical position to the test in a recent project where directional predictions were made for cardiovascular functions resulting from left- or right-hand somatosensory stimulation. Such stimulation of the fingers (Francis et al. 2000; McGlone et al. 2002) or forearm (Coghill et al. 1994) has been shown to activate at least the somatosensory components of the insular region (e.g., Mazzola et al. 2012; see also Foster et al. 2012). Vibrotactile stimulation at the left hand induced positive change scores in heart rate, whereas stimulation at the right hand induced negative change scores on this measure and the effect was larger with stimulation at the left hand. The comparison was statistically reliable. Only partial support for the hypothesis was obtained, though, as systolic blood pressure was not affected using these methods.

Paul Foster (Foster et al. 2012) provides additional evidence in his research on affective memory where fluency and generativity of words related to positive or negative memories were evaluated using the Emotional Controlled Oral Word Association Test (e-COWAT). Sympathetic tone as measured using skin conductance values was significantly increased for emotional memories over that of words produced from neutral memories. Skin conductance was negatively correlated with the number of words produced associated with happy memories. Moreover, the age of the memory was assessed with negatively valenced memories being of significantly older origin. These relationships were potentially consistent with the aforementioned theoretical perspectives. The persistence of negative memories over extended time periods may have survival value. Yet, the persistence of negative memories may be integral to some forms of psychopathology, as, for example, with posttraumatic stress disorders (PTSD).

The left brain has been implicated in the promotion of quiescent states, the lowering of blood pressure and, secondary to pathology, bradycardia or lowered blood pressure. Though the evidence is less provocative at this point in our neuroscience than is the role of the right brain in sympathetic activation, the left brain may be more specialized for parasympathetic activation as, for example, through the crossed vagal control of the right atrium in the heart and the vagal mediation of stomach contractions and the excretion of stomach acids for digestive purposes. Rob Walters was able to provide evidence for glucose mobilization in response to right frontal lobe stress (Walters and Harrison 2013a, 2013b). Regulatory control here generalizes with deregulated anger expression and elevated speech volume secondary to exceeding the stress management capacity of this system. Initial evidence supports the relative role of the left frontal lobe in the regulation of glucose absorption (Holland et al. 2011a, b; Holland et al. 2012; see also Holland et al. 2011c), but again the science is somewhat more speculative for the role of the left brain in parasympathetic functions.

The need for glucose mobilization during anger or fear-related states where high energy levels would have survival value on entering the fight or on efforts to escape the threat stimulus appears obvious. But, if the role of the left brain is oppositional to that of the right brain and if the left brain is sufficient in the promotion of quiescent states, then digestive processes following the consumption of a meal should be principally a left frontal stressor (e.g, Holland et al. 2011a, b). Kate Holland and David Cox (Cox et al. 2008; Holland et al. 2010) evaluated electrical activation of the left and right frontal lobes using quantitative electroencephalography before and after the consumption of a submarine sandwich. Also, assessed was the impact of digestion on verbal fluency measures, blood pressure, and heart rate (Holland et al. 2011a; see also Holland et al. 2011b). As you might imagine, it was not difficult

Fig. 26.1 Beta magnitude in microvolts supporting the hypothesis that food digestion is a relative left frontal stressor

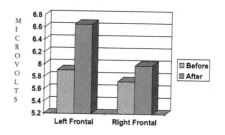

Electrical Activation (Beta Magnitude) Before & After a Meal

to find volunteer participants on a college campus for this free meal! The a priori findings were partially supported in these initial projects. Some of the findings are presented in Fig. 26.1

The role of the left brain in promoting quiescent states was supported by this finding of left frontal electrical activation with heightened arousal to the demands of digesting a meal (see also Holland et al. 2011a). Also, reliable effects were found for verbal fluency and for blood pressure. These processes may well be relevant in the broader considerations of the functions of the left frontal lobe within the context of social approach behavior, positive emotion, high energy level, and optimistic anticipation of future events. The question remains unanswered, at this point, what the effects of left frontal deactivation may be on food consumption where the energy level is low secondary to fatigue; where positive affect is diminished; and where social approach behaviors and optimistic anticipation are replaced by social avoidance and reflection on negative past events. These functions clearly have potential relevance to eating disorders and possibly to the metabolic syndrome and obesity. Laterality questions might be answered here with the primary approach to depression including selective serotonin reuptake inhibitors (SSRIs) and selective serotonin–norepinephrine reuptake inhibitors (SSNRIs) and where these medications have side effects, including weight gain (e.g., Harvey and Bouwer 2000). But, at this intersection in our science, it appears that the right frontal, amygdala, and insular region regulate the mobilization of glucose into the blood stream, whereas the left frontal, amygdala, and insular region may regulate the absorption of glucose from the blood stream (e.g., Luo et al. 2013). It also seems reasonable that brain utilization of glucose and insulin may relate to peripheral utilization of these substances.

For example, Anthony et al. (2006) measured brain glucose metabolism using PET in seven insulin-resistant and seven insulin-sensitive men during suppression of endogenous insulin by somatostatin, with and without an insulin infusion. Insulin was related to heightened metabolism in ventral striatum and prefrontal cortex, whereas decreased metabolism was found in right amygdala/hippocampus relative to global brain measures. Frontal mechanisms were also indicated in comparisons of the insulin-resistant subjects, where insulin's effect was reduced in the ventral striatum and prefrontal cortex. The authors conclude that brain and peripheral insulin resistance is especially related in cerebral regions subserving appetite and reward. Further, they postulate that a diminished link between food intake control mechanisms and those mechanisms which control energy balance may contribute to the development of obesity and insulin resistance.

Continued theoretical development here might be promoted from an integration of the earlier groundbreaking accomplishments in the understanding of the differential involvement of the cerebral hemispheres in spatial analysis both within the extrapersonal and within the personal spatial domains (Chewning et al. 1998; Heilman et al. 1995; Jeerakathil and Kirk 1994; Jeong et al. 2006; Heilman and Valenstein 1979; Ota et al. 2001; Hillis et al. 2005; Watson et al. 1994). Briefly, the right hemisphere has been demonstrated to play a preeminent role in the processing of extrapersonal space and this might be involved in threat detection within the left and the right hemispace. With threat activation from within extrapersonal spatial perception, sympathetic tone elevates in response to the threat and to promote survival as with incremental blood pressure, heart rate, sugar mobilization, oxygen saturation, and such. Importantly though, existing theory of spatial analysis provides for the ability of the left hemisphere to process a limited area and that specifically at the right hemispace. Thus, damage to the right hemisphere might predictably yield a left hemineglect syndrome with preservation of spatial analysis at the right hemispace.

Building on spatial processing theory and by efforts to extend it to the allocation of autonomic resources, we might expect that damage to the right brain would correspond with a general or broad-spectrum reduction in sympathetic arousal, where the impact would be appreciated at the left hemibody and hemispace. It may follow that the left brain is capable of processing some aspects of sympathetic tone and specifically at the right hemibody. This proposal receives some initial support with functional magnetic resonance imaging evidence for a primary role of right cerebral systems in skin conductance responses, whereas bilateral involvement was evident on this reward generation task (Critchley et al. 2000). Regions that covaried with increased skin conductance responses included right orbitofrontal cortex, right anterior insula, right inferior parietal lobule (see also Tranel and Damasio 1994), right fusiform gyrus, and left lingual gyrus. Earlier, George Demakis (Demakis et al. 1994) provided evidence for the restricted role of the left frontal region with sympathetic tone recorded in galvanic skin responses which were elevated specifically at the right hemibody with pathological laughter. Moreover, Matthew Herridge (Herridge et al. 1997) demonstrated laterality in the rate of habituation of galvanic responses recorded at the left and right hemibody as a function of posed facial expressions, where positive expressions and low hostility yielded prolonged sympathetic activation at the right hemibody. Negative facial expressions and high hostility yielded prolonged sympathetic activation at the left hemibody.

If these theoretical propositions are valid, then we have similar bases for understanding the lateralization of parasympathetic systems to the left cerebral hemisphere. But more importantly, this theoretical proposition does not support symmetry in function(s) but it does provide for functional balance between the two cerebral hemispheres. In other words, the proposition is for contributions to sympathetic tone from both the right and the left cerebral hemispheres and for contributions to parasympathetic tone from both the left and the right cerebral hemispheres. But, instead of symmetry providing for balance of functional control within the autonomic nervous system, it is proffered that the right brain would be preeminently capable in all sympathetic functions at both the left and the right hemibody augmented by sympathetic capabilities of the left brain more restricted to the right hemibody and its systems. Conversely, it is proffered that the left brain would be preeminently capable in all parasympathetic functions at both the right and the left hemibody augmented by parasympathetic capabilities of the right brain more restricted to the left hemibody and its systems. If so, then damage to the left brain may diminish parasympathetic functions more broadly, whereas the right cerebral hemisphere may be able to effect parasympathetic functions more specifically at the left hemibody. Regulatory control systems originating in frontal regions would work to maintain sympathetic drive (right frontal) or parasympathetic drive (left frontal), through these balanced but not completely symmetrical systems.

Converging evidence for laterality of the autonomic nervous system has also been found using pupillometric measures in response to ambient lighting within either the left or the right eyecup. Perhaps the most notable redundant pathway for vision is found in the ganglionic projections of the retina to the superior colliculus. Perry and Cowey (1984) concluded from converging lines of evidence that not more than 10%, and perhaps as few as 7% of the retinal ganglion cells project to the superior colliculus. These authors also found that the division of the crossed and uncrossed projections is asymmetric. They estimate, as did Pollack and Hickey (1979), that about 70% of the retinal projections to the superior colliculus come from the contralateral eye. Using healthy participants with either their left, right, or neither eye patched, Burtis et al. (in press) recorded pupillary sizes with infrared pupillography as an index of sympathetic drive. Pupillary dilation was significantly greater with left than with right eye viewing. The authors conclude that the larger pupil with left monocular viewing may have been caused by relative deactivation of the left hemispheric-mediated parasympathetic activity versus activation of the right hemispheric-mediated sympathetic activity.

It is also relevant that immune functions are influenced by cerebral asymmetry. The evidence in support of this connection between brain lateralization and the immune system is partially provided by animal research in which the effects of unilateral cortical lesions on immune system responses were evaluated (Renoux et al. 1983). Renoux and colleagues demonstrated that relative right-sided brain activation, following partial ablation of the left frontoparietal cortex of mice, decreased immune responses, whereas homologous lesions either had no effect or increased immune responses. Kang et al. (1991) and later Davidson et al. (1999) found activation differences reflected in prefrontal asymmetry in healthy individuals were related to basal natural killer cell function, where higher levels of natural killer cell function were found with left frontal activation subjects than with right frontal activation subjects.

Changes in lymphocyte proliferation and natural killer cell activity have been related to negative life events specifically among individuals with heightened left frontal activation (Liang et al. 1997). Also, Koh et al. (2012), using blastogenic responses to phytohemagglutinin as a measure of immunity and single photon emission computed tomography as a measure of cerebral perfusion, concluded that decreased cerebral blood flow in the left inferior parietal lobule and the left supramarginal gyrus of patients with undifferentiated somatoform disorder was predictive of reduced blastogenic responses to phytohemagglutinin regardless of sex and age. From a capacity theory perspective, this system might provide for mobilization of immune defenses, specifically relevant to the nature of the stressor and the specific brain regions processing the stress. Moreover, these findings directly implicate changes in emotional processing with immune system mobilization. These mechanisms and the dual processing specializations of the frontal lobe and regions under their regulatory control are relevant to both divergent and convergent relationships between emotional valence and health.

Chapter 27
Fast Energetic "Happy-Go-Lucky" and Slow "Cautious" Response Styles

The left brain is specialized for the rapid processing of information in a logical serial fashion. This is exemplified in the processing advantage of the left brain for speech sounds, where a rapid series of phonemic sounds is analyzed and comprehended as well as organized for expression or repetition through a series of sequential motor movements (see Duffau 2012). Moreover, the rapid temporal transitions required for speech favor the development of stronger left hemispheric responses over time due to the left hemisphere's processing bias for rapid sequential analysis (e.g., Boemio et al. 2005; Zatorre and Belin 2001; see also Glasel et al. 2011). Although we ascribe great intellectual advantage to the left brain, this may be a misinterpretation or misattribution as our measures of intelligence are often designed for success in school using verbal or linguistic measures and rapid sequential processing styles. Instead, the right brain is slow-paced and cautious allowing us to proceed when the determination has been made that the behaviors to engage in are safe and also appropriate for the setting and the rules of engagement (e.g., Aron et al. 2007b; Chambers et al. 2009; Hampshire et al. 2010; Chikazoe 2010; Cai et al. 2012). This analysis involves the social proprieties, social and political rules and implications, and just plain pragmatics of the behavior, should it be implemented or initiated in this contextual setting or in another. The right cerebral hemisphere provides for evaluation, more globally than sequentially, at a higher level of analysis beyond that which may be assessed by any extant intelligence test using pencil-and-paper measures.

We may rely on our healthy right brain to make even life or death decisions within one or another context as it will be able to perceive and to appreciate the more subtle nuance in a facial expression or another's tone of voice or postural mannerisms. Serial linguistic processing capabilities of the left brain certainly add much to the equation. But, by itself the left brain may be unsafe and shallow in its analyses and operative behaviors. It is better known for its high energy level and positive affective style, rapid processing style, serial or sequential analysis capabilities, and preference for social approach related interactions. This is a "happy-go-lucky brain." If left alone, it may initiate in a rapid or ballistic fashion without caution and without propriety with little respect for pragmatics, like turn taking. In a rehabilitation

© Springer International Publishing Switzerland 2015
D. W. Harrison, *Brain Asymmetry and Neural Systems*,
DOI 10.1007/978-3-319-13069-9_27

setting, it may also initiate up out of the wheelchair for toileting or personal care, resulting in a fall and, perhaps, additional injuries.

If the effect of lesion or damage of one frontal lobe is to disinhibit or "release" the other frontal lobe, then a release of the left frontal lobe may energize the person with freedom from the depth of appreciation of their deficits, whereas release of the right frontal lobe may yield a slow and cautious individual with somatic complaints, self-depreciating concerns or the deep appreciation of what ails them. The patient with damage to the deep structures of the left frontal lobe might state this in their own language as follows: "My get up and go, got up and went!" Well-intentioned family members and therapists might find themselves describing this person with attributions that she/he "won't even try." But, this inertia is specifically character-ized by behavioral slowing and difficulty in response initiation. This syndrome cor-responds with the amotivational and apathetic features (see Scott and Schoenberg 2011; see also Grossman 2002) described elsewhere in these writings. Quadrant theory also predicts that right frontal deactivation would release right parietotem-poral regions with predictable increases in arousal level (e.g., Heilman et al 1978; Heilman and Valenstein 1979; see also Heilman et al. 2012), increased activation into extrapersonal space (e.g., hyperkinesis), increased sympathetic drive, and ex-ternal attributions rather than self-awareness, per se.

The patient with right frontal lobe damage may benefit from elevated energy level or "get up and go." But, with medial lesions this disinhibition may range to hyperki-nesis or "impulsivity" with delayed response deficits featuring initiation without ad-equate delay for the appropriate organization of the response and the appreciation of the risks involved (see Oades 1998; see also Emond et al. 2009; see also Granacher 2008; Helfinstein and Poldrack 2012; Cai et al. 2012). The family member and the therapist may be responsible for understanding that the initiation of behaviors with-out caution, in a rapid and energetic fashion, may well put the patient or others at risk of harm. This more energetic patient may initiate activities into extrapersonal space, including out of their wheelchair for toileting without assistance or the engagement in other rapidly initiated approach responses. In such an individual, the onus may fall differentially upon others in the setting for providing adequate supervision and assistance. The challenge with such an individual may be to promote safe indepen-dence with the provision of opportunities to make mistakes that are not costly to the health of the patient or to those providing the care. Much needs to be done on the engineering side in this respect. A clear example presents with the wheelchair, which may be better designed for tripping the patient with bulky metal foot supports and such. Hospital and nursing care facilities frequently elevate the bed for ease of as-sisting the patient with cleaning and transfer activities. But, when left in this elevated position, the "impulsive" patient may be set up for a nasty fall.

The engineering needs are enormous as many of these patients will reenter the hospital after a safety failure at home or in other settings (see Snow et al. 2009). Many may suffer injurious falls or lose hard-fought recovery ground while in the hospital. Creative approaches are needed that do not increase liability issues for the care facility but, instead, promote a safe recovery. The accelerated time clock of the patient with release of the left frontal lobe and/or release of the right temporal parietal region, resulting from a right frontal lobe lesion, may pose other problems for family and therapists as the time base for action frequently differs. The patient may perceive others as slow and lazy perhaps, whereas others may attribute the patient's time-urgent demands to callous insensitivity or worse. Avoiding injury may ultimately favor successful outcomes, self-sufficiency, and independence.

Time pressure with rapid pacing demands has been shown to alter behavioral orientation, resulting in a seldom-appreciated spatial processing bias relevant to safety. In one experiment (Roskes et al. 2011), the motivation of participants was manipulated such that some were motivated toward avoidance and some were motivated toward approach. Participants then took part in a line-bisection task that was performed under low or high time pressure demands. Results indicated that subjects motivated for approach under high time pressure demands made more right-oriented judgments. Of interest to sports fans, these results were replicated in an archival analysis of goalie actions during FIFA Soccer World Cup penalty shootouts that showed goalies dove to the right more often when their teams were behind, than when they were tied or ahead.

This rapidly paced clock may be a residual effect of diminished capacity within the right frontal lobe and, as such, may correspond with deregulated negative affect such as anger. Many have experienced the traffic light scenario at an intersection, where the person behind them thinks that they are extraordinarily slow in initiating forward motion. The driver's attribution toward the agent provocateur is easily a mirror response of agitation, if not anger. This is not inconsistent with research where a fast or deregulated time clock occurs with a deregulated negative emotional state of anger and where anger is a social approach behavior with high energy (Harmon-Jones and Allen 1998; Harmon-Jones 2003a, 2003b; Kelley et al. 2013; Cox and Harrison 2008a, b, c, 2012), feelings of dominance (Demaree et al. 2005), and potentially improprieties in behavior. Kate Holland has shown that this clock may be purposefully or inadvertently altered through imposed left or right frontal stressors (Holland et al. 2007, 2009).

The clock might also be altered during dishabituation conditions involving intentional resources with recovery of stable-state interval estimation and temporally based behaviors subsequent to habituation to the stressor. Also, older age groups may require extended exposure for recovery of habituation or stable states secondary to disruption in their routine (Harrison and Isaac 1984; Harrison and Pavlik 1983; Harrison and Edwards 1988). Prolonged recovery time for temporal responding on an interval-based operant schedule has also been demonstrated as a function of aging using a conditioned emotional response paradigm and a nonhuman animal model (Harrison et al. 1984). Right frontal stressors demanding the concurrent regulatory control over sympathetic tone and negative emotionality with threat or perceived loss of control are sufficient for this purpose. Also, hostile violence-prone individuals may have a predisposition for this along with being prone to develop cardiovascular disease. The heart with its regular beat and tempo and with destabilization and mobilization to threat or anger may not only function as the effector meeting the demands of the brain but also in its afferent role providing vagal and other forms of feedback to the brain.

Bradykinesia or behavioral slowing is one hallmark of left frontal deactivation (e.g., Foster et al. 2010). This is a common presentation with Parkinson's disease and especially following the onset of symptoms, including tremor, at the right hemibody. The observation of timing deficits in patients with focal basal ganglia lesions provides strong evidence for the involvement of the caudate and putamen in timing and time perception (Schwartze et al. 2011; however, see Coslett et al. 2010). The loss of dopaminergic systems of the left frontal lobe with this disease process may result in a "locked-in syndrome," where the initiation of basic movements such as that needed to shave, eat breakfast, and kiss your wife can no longer be taken for granted. These structures might be potentiated with Sinemet (Carbidopa–Levodopa), the precursor of dopamine. But, this is no cure and the beneficial effects appear to be transient. Moreover, if the dopaminergic system is potentiated beyond the titration point, then iatrogenic effects, including hallucinations, are common. But even with the patient suffering from Parkinson's disease, where we typically associate the disease process with bradykinesia and immobility, the clinical course may differ with the onset of symptoms at the left or right hemibody and as a function of disease duration. Foster et al. (2013), for example, found increased energy levels as a function of disease duration in those Parkinson's patients with left hemibody onset of symptoms.

In their review of models of time perception, Allman and Meck (2012) note that the experience of time is fundamental to "making sense of the world," where clinical neurobehavioral or psychiatric disorders of behavior and/or cognition relate to difficulties with these processes (e.g., schizophrenia and autism). They note that although no disorder has been solely characterized as a disorder of timing, pathophysiological differences have been reported and particularly in individuals with Parkinson's disease, schizophrenia, autism, and attention-deficit hyperactivity disorder. Moreover, different clinical populations display characteristic timing functions specific to their diagnostic classification as evident with the temporal distinctions between depression and mania (Se´vigny et al. 2003; Bschor et al. 2004). Even the classic neuropsychological syndromes such as neglect disorder may be temporally bound as a case study of one patient, after parietal occipital damage, showed variant performance between impaired delayed and preserved immediate grasp responses (Rossit et al. 2011).

Attention-deficit disorder seems relevant to the temporal dimension with the essential features of the disorder being described as impulsivity, hyperactivity, and inattention. Toplak et al. (2006) (see also Allman and Meck, 2012) provide a comprehensive review of temporal processing research with this population with variable findings typically supporting reduced sensitivity on time interval threshold tasks and underestimation of time duration. Parent ratings of these children also appear to reflect a consensus that "sense of time" is poor and poorly managed, although the exact nature of the deficit remains to be determined. Perhaps also relevant by example, children with autism were compared with controls on a temporal bisection task in one project (Allman et al. 2011). The time-based estimates of the children with autism were significantly shorter and the degree of deviation or inaccuracy was found to correlate with scores on diagnostic tests for language and communication, and for working memory. Those participants with the "worst" communication and working memory functions produced reliably shorter estimates. The authors conclude that individuals with autism tend to truncate or shorten time and produce more variable estimates of time passage.

Oscillations within neural circuits provide the brain its own fundamental temporal structure (Buzsaki and Draguhn 2004; see also Kung et al. 2013) and there is substantial evidence that these oscillations are subject to entrainment or manipulation through environmental events (see Klimesch 2012). Within the rehabilitation or medical center setting, time management skills are a pervasive presenting problem for patients recovering from one or another brain disorder. Temporal management and the perception of the passage of time appear to be intimate to discussions of executive function deficits or disorders of inhibitory control. There is a long history of intervention procedures to promote improved temporal processing, including the use of timers, time lines, schedules, prompts, and routine in daily activities. In their discussion of corticostriatal regions subserving interval timing, Gu and Meck (2011) report beneficial effects of various forms of rhythmic entrainment. Computer-based rhythm and timing training have proved beneficial with attention-deficit hyperactivity disorder (Leisman et al. 2010; Leisman and Melillo 2011; see Allman and Meck 2012). Relatedly, interhemispheric training-induced coordination and coherence of oscillation frequencies within thalamocortical pathways may facilitate improvement in autism spectrum disorders (Melillo and Leisman 2009). Moreover, rhythmic auditory stimulation via music therapy has yielded improvements in gait speed, stride length, and depression indices associated with Parkinson's disease (Paccehetti et al. 2000; Hayashi et al. 2006: cited in Allman and Meck 2012).

Allman and Meck (2012) focus their review of temporal processing on two primary models of interval timing: the striatal beat frequency model (Matell and Meck 2004) and a more traditional information processing conceptual frame known as the scalar expectancy theory (Gibbon and Church 1984). The neuroanatomical focus of this time-based neuronal system was on the interactions between the cortex and basal ganglia, including the dorsal striatum receiving primary inputs from the premotor and prefrontal cortices (e.g., Draganski et al. 2008; Ford et al. 2013; Oguri et al. 2013) and the ventral striatum receiving afferents from the orbitofrontal cortex (e.g., Coull et al. 2011). Secondly, the frontocerebellar system was discussed as a critical system for precise temporal processing and the role of this circuit in coordinated sequencing is well established. Practically speaking, damage within one or the other cerebellar hemisphere may at first appear to represent a frontal lobe syndrome prior to the localization of the syndrome with confirmation of brain-stem pathology on examination. This is evident with delayed response deficits with a failure to stop secondary to right frontocerebellar pathology and with behavioral slowing, perhaps, with left frontocerebellar pathology (see Aron et al. 2007b; Chambers et al. 2009; Hampshire et al. 2010; Chikazoe 2010; see also Allman and Meck 2012; see also Cai et al. 2012).

Chapter 28
Personal, Peripersonal, and Extrapersonal Space

In the process of acquiring information about the world, the sense organs are not passive receptacles of energy information. Instead, the sensory systems have been found to actively probe and search the extrapersonal space to update the internal representations of the world and the contents of emotional and cognitive schema (Droogleever-Fortuyn 1979). Gitelman et al. (2002) note that perturbations in this process, usually resulting from right hemisphere lesions, lead to the syndrome of left hemispatial neglect. Left hemispatial neglect has been most frequently associated with lesion sites involving the parietal and premotor cortices and, less frequently, the cingulate region, the thalamus, or the basal ganglia of the right hemisphere (Heilman and Valenstein 1972, 1979; Heilman and Van Den Abell 1980; Mesulam 1981; Heilman et al. 2012). These regions are interconnected with each other and collectively give rise to a large-scale neural network subserving numerous facets of spatial attention (see Gitelman et al. 2002).

The right brain appears to be specialized for spatial analysis and comprehension for the regions outside of our body and its parts. It is particularly adept at processing faces and places in contrast to the linguistic capabilities of the left brain. Kenneth Heilman has contributed much to our understanding of spatial processing differences between the two brains, through the identification of personal and peripersonal or extrapersonal space and through the appreciation and articulation of the contributions of each brain to the processing and awareness of these locations and spatial arrays. Whereas the left brain appears specialized for the awareness of body parts and positions or "personal space," the right brain appears specialized for peripersonal space or the location of our body and of other objects or events within the external world (Heilman et al. 1995; Foster et al. 2008). Thus, geographical awareness corresponds with right brain integrity and geographical confusion may occur with damage here (e.g., Paterson and Zangwill 1944; Benton et al. 1974, De Renzi 1982; Fisher 1982). In contrast, left–right confusion (Gerstmann 1940, 1957; see also Cabeza et al. 2012), body-part confusion, or gestural confusion (gestural dyspraxia; Heilman 1973) correspond with damage or deactivation in the left brain.

© Springer International Publishing Switzerland 2015
D. W. Harrison, *Brain Asymmetry and Neural Systems,*
DOI 10.1007/978-3-319-13069-9_28

The left brain appears capable of processing proximal space at the left and the right hemibody, whereas the right brain may be more limited in this respect to its proximal space at the left hemibody. Partially due to this arrangement of specialized attentional and intentional allocations of space, a lesion of the left brain appears to more directly impact praxis, including ideomotor or gestural (Luria's second functional unit) and ideational or sequential components (Luria's 3rd functional unit). This appears to follow with the proximal allocation of resources to the body and the sum of its volitional parts or skeletal movements. The right brain appears more able to process the attentional and intentional demands of the entirety of the extrapersonal space at both the left and right hemispaces, whereas the left brain may be more limited to processing within the right hemispace. Thus, a right cerebral hemisphere lesion appears to lend itself to a left sensory (Luria's second functional unit) or motor (Luria's third functional unit) hemineglect syndrome and specifically the neglect of extrapersonal space at the left side of our world. The limitations on the left brain in its capabilities for processing extrapersonal space may restrict the impact of left brain damage to only a minimal effect on neglect within the right hemispace. Moreover, the diagnosis and measurement of a right hemineglect of extrapersonal space may be more difficult, with the impact being more subtle in its expression. These asymmetries in the specialization for space within the intrapersonal or proximal spatial dimensions and the extrapersonal or distal spatial dimensions are displayed in Fig 28.1.

The tertiary association cortex at the parietal–occipital–temporal region appears specialized for the analysis and comprehension of space through integration among the primary sensory modalities. Lesion of this area in the right brain may result in extrapersonal deficits (however, see Roth et al. 2013) as with sensory constructional dyspraxia or dressing dyspraxia. East–west confusion may result from a lesion of the right parietal region (Paterson and Zangwill 1944; Walsh 1978), whereas lesions of the homologous region at the left hemisphere appear to be more demonstrative of left–right confusion and body-part confusion (Gerstmann 1940, 1957; see

Fig. 28.1 A model of the hemispatial processing for each brain within intrapersonal and extrapersonal space. (see Heilman et al. 1995; Heilman et al. 2003)

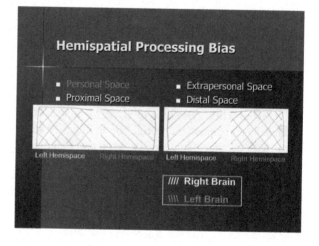

also Cabeza et al. 2012). Similarities might be appreciated in the function of these homologous brain regions. But, the distinction lies in the appreciation and comprehension of near or proximal space as exists at the body or the appreciation of space external and distant from the body as within extrapersonal or peripersonal space. Functional magnetic resonance studies have provided evidence consistent with the clinical findings where the dorsal parieto-occipital region activates especially to stimuli within extrapersonal or peripersonal space and associated with the dorsal pathway to premotor regions involved in the planning and execution of reaching movements (Gallivan et al. 2009; see also Aimola et al. 2012).

Pribram (1991) argued that the spatial array within our environment may exist through a replication of the array within our neural architecture. The three-dimensional environmental array, processed from one or another perceptual point or perspective, exists as a concurrent representation of these elements and as a Gestalt or holistic integration within the three-dimensional neural network. The functional anatomy is part of a neural net, where the vantage point of the perceiver varies across multiple dimensions and where these dimensional changes in perspective provide clues to the multidimensional shape of the object or scene. This perspective or vantage point approach, when wedded to a dimensionally complex neural network perspective, provides for dynamics and provides an exponential increase in the comprehension of shapes and places within extrapersonal space. Karl provides an analogy here where the variant of perspective contributes to perception much like a hologram. The hologram achieves apparent dimensionality beyond its image, through manipulations of perspective or vantage point. No area of the brain contributes more to this than the spatial analyzers within the right parietal region and the right parietal–occipital–temporal region. This might be evident within the parietal eye fields (see Suzuki and Gottlieb 2013), where three-dimensional perceptual constants may be maintained concurrent with positional change or relocation of the eyes and directional gaze (Medendorp et al. 2003).

Karl retired from Stanford University to build his "Brain Institute" just down the road from Behavioral Neuroscience Laboratory at Virginia Polytechnic Institute. He was instrumental in bringing scientists and philosophers together from many areas of specialization to discuss and imagine the brain and its functions. He was to appreciate space more specifically, when his good friend rearranged his library based on the color of the book's cover rather than Karl's arrangement by subject area. This required appreciation of the logical basis for the spatial arrangement as distinct from the color of the book's cover. Karl appreciated spatial arrangements through his photography and, in the auditory modality, through his love of the violin.

The immediate frontier in research on personal and extrapersonal space is to extend this inquiry into the role of specific brain regions and especially the frontal lobes in the control of organ systems, including the adrenal gland, the thyroid gland (e.g., hyper and hypothyroid disorders), pancreatic insulin functions (e.g., hyper and hypoglycemic disorders), the stomach (e.g., acid secretion and contraction and absorption disorders), and cardiovascular functions (e.g., bradycardia and tachycardia; low and high blood pressure), and vasodynamics. Sympathetic drive prepares us for movement into, through, or out of some extrapersonal space and parasympathetic drive for processing and the conduct of personal space activities. The right frontal lobe seems disposed to regulate the incremental arousal states associated with sympathetic drive, whereas the left frontal lobe seems disposed to regulate quiescent states of low energy, absorption of digestive products (e.g, Holland et al. 2011), and parasympathetic drive (e.g., Holland et al. 2011).

Though the research is still lacking here, a patient with a diagnosis of hyperthyroidism with elevated energy and possibly hyperkinesis may benefit from the assessment of a right frontal lobe disorder, whereas a patient with chronic fatigue, diminished energy or "get up and go" may be diagnosed with hypothyroidism and indicates the need for a rule out of a left frontal lobe brain disorder. Right temporal lobe seizures may be evident in cardiovascular presentations, including tachycardia and arrhythmia (Marshall et al. 1983) and syncope may develop with seizure focus or elevated activation within the left temporal lobe. With an improved understanding of the role of functional brain systems in all organ functions, the focus of intervention might be redirected as with the management of right frontal lobe stressors prior to radiation implants to target the thyroid gland for destruction. This less invasive approach may prove effective for some patients and serve to minimize the iatrogenic effects of radiation seeds or surgery.

The appreciation of asymmetries in spatial processing, where the left brain is more apt to comprehend and to develop organized responses to proximal spatial demands and where the right brain is more apt to comprehend and to develop organized responses to extrapersonal or distal spatial demands, came with advances in the appreciation of asymmetries in emotional processing. This includes a role for the right brain in processing fear and anger along with its vigilance over distal space where threats may appear "on the horizon" and where a defensive or aggressive response might be mounted with activation of sympathetic drive and mobilization of the cardiovascular, cardiopulmonary, and musculoskeletal systems. Integration of these multidimensional constructs, within the functional cerebral systems of the right brain, allows a conceptual basis for understanding their intimate relationship one unto the other. Coercive attributions of external control and of others trying to do me harm would be close in their anatomy to the bodily responses necessary to mount the response. In the active state, this would require increased blood pressure, increased heart rate, circulating blood glucose and cholesterol, and oxygen saturation. The elevation in cholesterol levels might even reduce the spillage of blood, should the person be injured secondary to rapid clotting time.

Foster et al. (2008) provided an example of the hemispatial specialization for emotion, wedding emotion to the three-dimensional spatial array through the quadrant model. Here both left and right hemispace and vertical spatial dimensions (close and far) were found to reliably differ with emotional valence. Using pegs labeled with one or another emotional valence (positive and negative), he asked normal subjects to place them on a board using Velcro attachments. Subjects reliably segregated positive and negative affective pegs across left and right hemispace and within the distal and proximal axis. This unique procedural design was the first, to our knowledge, to provide evidence that emotions can bias attentional allocation, whereas numerous research investigations have examined how attention is biased for emotional stimuli.

Maguire et al. (2000), using brain imaging techniques, showed that part of the hippocampus is enlarged in taxi drivers compared with the general public, and that drivers with more experience have larger hippocampi than those with less experience. Indeed, there was a significant correlation between the length of time an individual had spent as a taxi driver and the volume of the right hippocampus. The London taxi drivers are required to learn a large number of places and the most direct routes between them. More recently, Maguire (Woollett and Maguire 2011) extended these findings, and by some accounts stunned the scientific community, by demonstrating neurogenerative properties of experience. Specifically, geographical experiences (e.g., finding shortcuts and such) make an individual's hippocampus grow. It was found that the total volume of the hippocampus remained constant in the control group in comparisons with the taxi drivers. But, the growth in one region, the posterior portion of the hippocampus, was found to be at the expense of the anterior portion. Functional differences were also found with a reduced ability to perform on the Rey-Osterrieth Complex Figure Test (see Lezak et al. 2012). The results relate to spatial memory in our discussion. However, they are more exciting in the demonstration of the effects of experience to alter brain anatomy and capacity. The dentate gyrus is remarkable in this respect with neurogenesis evident especially in this brain region (Kuhn et al. 1996). It is also evident that traumatic experiences, exceeding the capacity of the neural circuitry involved (see Carmona et al. 2009; Williamson and Harrison 2003; Foster et al. 2008; Mitchell and Harrison 2010; see also Klineburger and Harrison 2013), may produce diametrically opposite effects and potentially neurodegenerative changes (e.g., Knutson et al. 2013; Sartory et al. 2013: see also Parker 2010).

Chapter 29
Relationships: Proximal and Distal

Substantial progress has been made in the study of relationships (see Hazan and Shaver 1994) and even in shared cognitive processes among groups. There are many potential variations of this theme worthy of discussion that remain largely outside of scientific inquiry, currently. One logical extension of laboratory efforts to understand emotion is the aspect of being held close to someone in a relationship or being identified as more distant in the relationship. Whether or not one or the other cerebral hemisphere is specialized for the depth of a relationship remains unclear. The left brain may be specialized for those we *hold near and dear* emotionally, with the presentation of a somewhat universal positive attribution bias to these individuals. Being emotionally close to another person may promote expectations of forgiveness for incidental or actual transgressions, at least up until the moment that the cerebral processing advantage shifts to negative valences and the processing bias of the right hemisphere. At the simplest level, the processing bias of the right brain does not appear to hold others close. The attribution bias here appears to be negative or distanced from these individuals outside the group, unless it is a more deeply positive relationship that may need to be protected or defended under assault. But, the interdependence of these neural systems is evident as one influences the other. A simple example might hold as evidence that the visual spatial analysis and perception of physical size and strength of a threatening individual are diminished in the presence of a friend (Holbrook and Fessler 2013).

Each individual's perception is influenced not only by the presence of close or supportive group members but also by the perceptual analysis and comprehension of others who may not hold us in high esteem. In some situations, you might find yourself essentially *walking into another's right brain*. Once there, the negativity afforded to your behavior provides for a persistent and possibly endless cycle of opposition and even anger. This proposal indicates that we easily walk into another person's left brain and easily walk out again, through the dissolution of trust or violation of support and positive regard for this individual. More poignant to this discussion, though, is that once a person has appeared on the radar screen of the

© Springer International Publishing Switzerland 2015
D. W. Harrison, *Brain Asymmetry and Neural Systems,*
DOI 10.1007/978-3-319-13069-9_29

negative emotional appraisal systems at the right brain and once assimilated into its architecture, it may be difficult to get out!

The neural architecture and physiological specializations of the right brain may remain ever vigilant and ever distrustful, unless one is able to work through the offensive issues or secondary to forgiveness, perhaps. If that person is held to be dear and if there is a violation of this relationship and trust, then the negativity afforded to that individual may be great with substantial depth and persistence. Some may attest to this after the dissolution of a close marital relationship, followed by a heated divorce. Clinton Comer (Comer and Harrison 2013) has provided a potential theoretical basis for this with integration of opponent process theory into modern neuropsychological theory on emotional processing. Also relevant, by way of example, would be the evidence for the persistence of negative memories (Foster and Harrison 2004; see also Foster et al. 2011, 2012) and the severity of depression as a function of the left or right hemibody onset in Parkinson's disease (Foster et al. 2010, 2011)

Another possible variation in this relationship theme can be found in our approach to professional guidelines in authorship. Any of the logical linguistic contributions of scientists or others relevant to written accounts and available through previous publication may be acknowledged through a written citation or reference. This provides for substantial freedom in writing and promotes the generation of additional written words through subsequent publication. These verbal linguistic citations are largely the products of the left brain as the right brain is somewhat devoid of language and logical linguistic processing mechanisms. In contrast, the products, constructed works, and drawings of the right brain are not readily accessible for use by others through the provision of verbal acknowledgement or citation simply placed in textual verbiage. Instead, we must ask for permission and acquire a copyright release to use the artwork or the pictures that are deemed desirable for secondhand use. This relationship holds if we would like to use the *whole* of the written project as in a *copy or picture* of the text or publication, which would amount to plagiarism or a violation of another's *ownership* of the project.

That we hold our friends "*close*" and our enemies "*at a distance*" seems to fit the basic framework of theory, where the right cerebral hemisphere is specialized for both negative emotion and extrapersonal space at a distance from me. The vigilance over this space provides a constant level of threat assessment, where those viewed as in opposition to one's self may be monitored by a cautious and protective brain, whereas those close to us may enjoy a more cavalier side of us, wherein our friendship is more tolerant and "happy go lucky." Even a recent president distinguished these sides saying "You are either for us or against us", highlighting the aspect of our political alliances, where we stand either together in opposition to our enemies or separately with distance conveyed along with the negative emotion.

Chapter 30
Right Hemi-aging Theory

The one universal finding across aging research is an increased heterogeneity of variance with increasing age. Thus, any conclusions based upon the aging construct must be circumspect and with all caution related to sweeping generalities or conclusions on what aging is or what changes accompany the aging process. Cross-sectional age comparisons may result in findings more relevant to differing experiential backgrounds or the "cohort effect" (see Sigelman and Rider 2011). Moreover, age comparisons based upon chronological age differences seem to be somewhat independent of the aging construct. The factual basis of this confound might be entirely evident in the day to day encounters of a clinical neuropsychologist working with patients of differing ages. By way of example, one neuropsychologist was completing the evaluation of a 105-year-old gentleman who was actively dating. The ladies seemed to swoon in his presence with his military uniform; still a sharp fit on his angular frame. He was solid on his feet, coordinated in his gait, and well mannered in his social demeanor and pragmatics. In so many ways he was arguably more youthful than the 13-year-old boy that the neuropsychologist was following for a cerebrovascular accident. The older man was, however, extremely hard of hearing and refused his appliance to amplify sounds. Upon inquiry as to why he would not wear his hearing aid, he responded that he had "already heard everything that he wanted to hear."

To a large extent, the early conception of the aging brain that emerged from behavioral research and from structural imaging studies portrayed the aging process as one where the individual was a victim of passive deterioration and decline with age, with preservation of a few functional domains. Contrary to this view, subsequent research has demonstrated that the aging brain is an adaptive and plastic structure, responding in a dynamic fashion to exertional stress (e.g., Pereira et al. 2007; see also Morrison and Baxter 2012), cognitive challenge (e.g., Maguire et al. 2000), and to developmental structural demands (Glasel et al. 2011; see also Aeby et al. 2012) or the demands of adjusting to deterioration itself. Evidence of this nature has fundamentally changed traditional views of cognitive aging. In addition, modern neural

© Springer International Publishing Switzerland 2015
D. W. Harrison, *Brain Asymmetry and Neural Systems*,
DOI 10.1007/978-3-319-13069-9_30

theories of aging (e.g., see Park and McDonough 2013) provide a stronger emphasis on dynamic changes and compensatory mechanisms related to brain plasticity and neural reorganization. To some extent this has been evidenced by the resurgence of interest and research designed to improve cognition through enhancement of neural structures or reorganization of functional neuronal circuitry.

Some evidence indicates that the right cerebral hemisphere may be relatively precocial at birth, referring to a relative ability to "hit the ground running" at birth or before. This would lend itself, by necessity, to the demands of urgent survival states as in the mobilization of the sympathetic nervous system with a gasp for air, with hunger demands, and with emotional irritability as with an angry or urgent cry to signal the imperative needs of the newborn. The child might then have a relative advantage in processing extrapersonal spatial cues as with sensorimotor development and emotional bonding with the parent's or the caregiver's face and with the processing of emotional spatial arrays, including the prosodic variations of motherese or "baby talk" (Bunce and Harrison 1991). Propositional or logical linguistic speech develops subsequent to this from the perspective of right hemi-aging theory. One father recalled his children as babies, where there would be the occasional bout of irritability. The boy might seem inconsolable and his wife would talk to the boy without effect. But, as he attempted emotional facial postures and affective prosody he would, sometimes, gain immediate attention and, if he was lucky, a quieting of the emotional expressions!

The right superior temporal region is better known for the processing of emotional or prosodic sounds (Heilman et al. 1975; Ross and Monnot 2008; see also Hickok and Poeppel 2007; see also Bourguignon et al. 2012) and for its contributions to facial comprehension and recognition with close associations with the occipitotemporal region and the fusiform gyrus (Pallis 1955; McCarthy et al. 1997; Turk-Browne et al. 2010; see also Kanwisher and Dilks, in press). Others have demonstrated a functional role for this brain region in high-level "theory of mind" or social perception and interindividual communicative behavior, where inferences are drawn relevant to communicative intentions, the generation of communicative action, and the prediction of forthcoming communicative intent (e.g., Carrington and Bailey 2009; Haxby et al. 2002; see Shallice and Cooper 2011; see also Cabeza et al. 2012).

In their review of the literature on early development, Aeby and colleagues (Aeby et al. 2012) note that newborns are known from behavioral studies to be able to recognize the voice of their mother from the voice of a stranger (Mehler et al. 1988). Moreover, research on newborns using magnetoencephalography techniques has shown that they are already sensitive to prosodic cues in language at birth (Sambeth et al. 2008). Aeby et al. (2012) also provide evidence of early developmental resources for facial processing, noting that newborns are known from behavioral studies to preferentially process human over nonhuman faces (Johnson et al. 1991) and that they are able to imitate facial gestures (Meltzoff and Moore 1977).

Using positron emission tomography with 2-month-old infants, Tzourio-Mazover and colleagues (Tzourio-Mazoyer et al. 2002) demonstrated activation patterns at the temporal lobe and ventral occipitotemporal facial processing region similar to that recorded in adults (see also Kanwisher & Dilks, in press). The under-lying arousal component essential for empathy and neuronal mirroring is present in newborns as they have been shown to express discrete emotions at birth (Izard 1982). Also, by 10 weeks of age, they are capable of imitating expressions of fear, sadness, and surprise (Haviland 1987). Interestingly, based on the well-established contribution of the right temporal parietal region to arousal (Heilman et al. 1978; see also Heilman and Valenstein 2012), Aeby et al. (2012) conclude that the lat-eralization of the arousal systems to the right brain may influence the differential maturity of each cerebral hemisphere favoring the right brain. More specifically, they state that the evidence suggests that the microstructural modifications in brain structures around the right temporal and occipital regions "observed between 35 and 43 weeks of gestation in preterm neonates could contribute to the functional maturation of these brain regions with increasing age, in a period of life where voices, prosody and faces represent extremely salient stimuli."

They review evidence of asymmetric maturation of these right cerebral systems noting that between 12 and 14 weeks of gestation, several genes are asymmetrically expressed toward the right side in the human temporal cortex (Sun et al. 2005). The right superior temporal sulcus appears earlier in development (Chi et al. 1977; Dubois et al. 2008). Dubois et al. (2008) have shown that this region is deeper with-in the right hemisphere in premature infants. Others find significant lateralization favoring the right brain in normal term newborns where cortical surface expansion of the right hemisphere is larger than that of the left hemisphere from 0 to 1 year of age (Hill et al. 2010). Evidence for lateralization favoring this region is present in 3-month-old infants and the asymmetry increases faster than the general growth of the brain in this time frame with profound microstructural changes in the right su-perior temporal sulcus (Glasel et al. 2011; see also Aeby et al. 2012). These findings appear to be more relevant to regional differences within the brain and complexity has been found also favoring the left hemisphere.

Language processing is subserved by a neural network involving the left cerebral hemisphere, which has been shown to be substantially lateralized in normal adults. Blumstein and Amso (2013) argue that a fixed neural architecture for language is one that would be present and robust at birth and constant or invariant over the life span. Instead, they report functional magnetic resonance imagery findings from newborn infants which appear to challenge this notion in support of dynamic devel-opmental changes. For example, Perani et al. (2011) paired whole brain functional magnetic resonance imagery and diffusion tensor imaging during auditory language input in 1- and 2-day-old newborns. Language processing was less lateralized in the newborns, with activation found at the bilateral temporal regions as well as the left inferior frontal gyrus in newborns in response to speech. The authors elaborate here in that the activation was indeed increased in the primary and secondary auditory cortex on the right side, a finding which is at once inconsistent with the adult model and which supports a modular bias to process within the right cerebral hemisphere after birth. Blumstein and Amso (2013) note that heightened interhemispheric con-nectivity rather than intrahemispheric connectivity is found in infants relative to adults in the neural language network (Obleser et al. 2007). These developmental differences in the hemispheric laterality of language extend well into childhood as evident in other work examining sentence processing in 6-year-old children (Friederici et al. 2011).

If the right brain is precocial and relatively mature at birth in comparison to the left brain, then there might be biological bases upon which to predict the earlier demise or shortened life span of the more rapidly developing regions within the right cerebral hemisphere. This concept has some support within developmental and biological psychology (see Gibson and Petersen 2011). The distinction between altricial and precocial species is sometimes focused on longevity or life span differences. The right hemi-aging theory did indeed receive substantial attention in geropsychology within the 1970s and 1980s, but was abandoned as the distinction drawn between the left and right brains, at that time, was thought to be identified by differences on the intelligence test, where verbal intelligence was considered a left brain measure and where nonverbal intelligence was considered a right brain measure. Though these distinctions are not without empirical support, the nonverbal measures of intelligence relied heavily on performance measures. Therefore, a decline in nonverbal intelligence relative to verbal intelligence with aging was largely based on performance deficits where the measures were heavily confounded by response speed, for example.

There is evidence that this theory of right hemi-aging should not be so readily discarded. Evidence for right hemi-aging has emerged partially based on the findings in early infancy and childhood and also with some aging research, where measures of cerebral laterality were used instead of a reliance on the standard measures of intelligence. Numerous studies have supported the hypothesis that functions of the right hemisphere decline differentially with age (Lapidot 1987; Rastatter and McGuire 1990; Riege et al. 1980). Lapidot (1987), for example, found visual smooth pursuit to be restricted toward and within the left hemispace in older adults, a finding suggesting deficits in right frontal eye field capacity as a function of age. Visual smooth pursuit toward and within right hemispace was unremarkable in age-related comparisons.

Much of the research on cognitive decline with aging has focused on the construct of working memory. Reuter-Lorenz (2013) notes from her review of the literature that an important discovery pertaining to the aging of working memory and other cognitive domains is that "even when younger and older adults perform equally, imaging measures indicate that their brains function differently." This author cites the relationship between working memory load and activation, where reduced capacity or capacity limitation with aging may advanced heightened activation states for the processing of minimal cognitive loads. For example, minimal verbal working memory demands, consisting of about two to four items, generally produce matched accuracy performance abilities across age groups. However, older adults are reported to show heightened activity over younger adults particularly in right frontal regions typically critical to executive control operations (Cappell et al. 2010; Mattay et al. 2006; Schneider-Garces et al. 2010; see Reuter-Lorenz 2013). With incremental load, younger adults show increasing activation in these frontal regions, whereas reduced activation and lower performance results are found with older adults. These findings support increasing capacity limitations with aging and specifically in right frontal regions where cognitive decline relates to working memory load.

Moreover, decrements in hemispheric asymmetry have been consistently reported during healthy aging and with decline related to disease processes such as Alzheimer's (e.g., Long et al. 2012). Rather consistent findings of reduced cerebral asymmetry with aging have been found using electroencephalography (Bellis et al. 2000; Deslandes et al. 2008), near-infrared spectroscopy (e.g., Herrmann et al. 2006), behavioral (Reuter-Lorenz et al. 1999) and structural and functional magnetic resonance imagery (e.g., Cabeza 2002; Cabeza et al. 2004; Li et al. 2009; Long et al. 2012).

Functional neuroanatomical evidence supports right hemi-aging. For example, in a meta-analysis of functional magnetic resonance imaging research investigating hippocampal atrophy in individuals with mild cognitive impairment, Alzheimer's disease patients, and aging controls, a significant left-less-than-right asymmetry pattern was found (Shi et al. 2009). Significant atrophy was found in both the left (effect size, 0.92; 95% confidence interval, 0.72–1.11) and right (effect size, 0.78; 95% confidence interval, 0.57–0.98) hippocampus, which was lower than that in Alzheimer's disease patients (effect size,1.60, 95% confidence interval, 1.37–1.84, in left; effect size, 1.52, 95% confidence interval, 1.31–1.72, in right). When the authors compared aging controls, the average volume reduction weighted by sample size was 12.9 and 11.1% in left and right hippocampus in mild cognitive impairment, and 24.2 and 23.1% in left and right hippocampus in Alzheimer's disease patients, respectively. The findings revealed a bilateral hippocampal volume loss in mild cognitive impairment with the extent of atrophy less than that in Alzheimer's disease patients. Moreover, comparisons of left and right hippocampal volume revealed a consistent left-less-than-right asymmetry pattern, but with different extents of atrophy in the control (effect size, 0.39), mild cognitive impairment (effect size, 0.56), and Alzheimer's disease (effect size, 0.30) groups.

Some have argued that reduced asymmetry impedes communication and the exchange of information across the hemispheres, resulting in impaired working efficiency (Oertel et al. 2010). However, this interpretation is potentially flawed as reduced cerebral lateralization is a frequent finding in women (McGlone 1978, 1980; Harrison et al. 1990; see Good et al. 2001; see also Sommer et al. 2004; see also Wallentin 2009; Shin et al. 2005; see also Bibost et al. 2013). Moreover, many projects fail to control for sex differences in laterality, which clearly may confound these interpretations (e.g., Long et al. 2012). This potential confound may be more relevant towards the later parts of the life span sampled, where there is a shift in survival demographics favoring women.

Attenuation of emotional processing skills in the elderly may support the right hemi-aging theory. Christine McDowell (McDowell and Harrison 1993) evaluated the accuracy of emotional face recognition across age groups using a photo album with standardized emotional faces. The younger and older groups were found to have equivalent accuracy scores in recognizing happy faces, whereas the group of healthy elder participants showed significant impairments in the recognition of negative affective valences (sadness, anger, and fear). In a related project, Billings et al. (1993) found that younger and elderly women differed in the identification of neutral facial stimuli, revealing a heightened positive emotional bias and specifically within the left visual field. These findings were replicated and extended in a subsequent project (McDowell et al. 1994), where elderly men and women were reliably less accurate in the identification of negative affective faces with equivalent performance evident in comparison with younger participants in the identification of positive affective faces. A significant positive affective bias was also found on processing neutral faces within the right visual field (left cerebral hemisphere). Park and McDonough (2013) point out that regions of ventral visual cortex (e.g., fusiform face area) that are specialized for face and scene processing become less specialized with advanced age (Grady et al. 1992; Park et al. 2004). These authors also note that subsequent studies have revealed a relationship between decrements in processing speed and processing specialization. Specifically, decreases in processing specialization are associated with decreases in processing speed and other higher-order cognitive processes in older adults (Park et al. 2010).

John Alden (Alden et al. 1997) extended the cerebral laterality research to the auditory modality using a dichotic listening task. These procedures provide for the concurrent presentation of different sounds to each ear. The test is sensitive to cerebral laterality with the "dominant" brain region, within one or the other cerebral hemisphere, extinguishing the activation and receptive processing of the homologous brain region (Hugdahl 2003, 2012; Hugdahl et al. 2009). Thus, relative activation or integrity of the left auditory cortex might result in a right ear processing advantage under concurrent bilateral auditory processing demands. Younger and elderly women were asked to shift their intentional focus to the left or the right ear in comparisons with an initial unfocused condition. The right hemi-aging theory was supported on the intentional task, in which older women were significantly less able to allocate resources to left hemispatial processing but not to the processing of information from the right hemispace and right ear (however, see Hugdahl 2012).

Chapter 31
Right Hemisphere and Pain

If the basic property of protoplasm is irritability (Barnes 1910), then the basic sensory modality critical to our survival may well be pain. Surely, it is fundamental to the human condition. The arrival and interpretation of these signals are intimate to defensive and offensive gestures, emotionality, and deeply pervasive memories to insure that we minimize the creation or replication of this consequence which results from many ill-conceived activities and pursuits. Failures are evident, though, in the public media in the stiff human competition for the "Darwin Awards" and for "America's Funniest Home Videos." We simply do not do many of the activities that gave us pain when we last tried them. Nature simply has an effective parenting style to promote rapid acquisition of emotional responses to pain and their associated contingencies (e.g., Harrison et al. 1984) along with long duration and persistent memories. If you touch a hot stone and get burned, you may indeed be less likely to touch it the next time when the opportunity knocks.

As a graduate of the doctoral program at the University of Georgia, I became aware of some of the early student exploits that influenced Crawford Long in his discovery of anesthetics (Long 1849). During my time at Georgia, students and faculty would head down to the Oconee River to O'Malley's for socials or just to relax on the deck overlooking the river. But, some hold that, in Long's day, folks would meet down by the river for ether and laughing gas frolics. These activities might be morally questionable for some. But the consequence of these "frolics" would, in many ways, compare with the better accomplishments of many hours spent in the laboratory engaged in the most rigorous of scientific inquiries. The discovery would be more influential on the battle field or on the surgical table. Indeed, ether and laughing gas were identified by Long for their anesthetic properties from simply witnessing an injury at the party without the concomitant appreciation of pain.

This was of great import as few options existed at the time for analgesia short of "biting the bullet" or a block of wood. This seemed to work well for John Wayne in his many movies, where he stood his ground as a "real man" for men everywhere to measure themselves against. John might be shot in the leg, bite the bullet as the doc cut it out with a knife, and be back in the saddle on his horse at sundown to ride into

© Springer International Publishing Switzerland 2015
D. W. Harrison, *Brain Asymmetry and Neural Systems*,
DOI 10.1007/978-3-319-13069-9_31

the glorious western sunset! Interestingly, the European Caucasian at one time held great hopes for Curare to be a magic elixir for the elimination of pain. Curare derivatives had been used by Native Americans and others with Curare-tipped darts or arrows capable of bringing down prey and foe with muscle paralysis resulting from this cholinergic antagonist (see Lee 2005). Curare effectively blocks acetylcholine, the neurotransmitter used by the body for all skeletal or striate muscle contractions to move the limbs. As the story goes, Curare was tried on the battlefield, where no one screamed or evidenced pain behavior, during amputations and such. Finally, someone survived the surgery, conveying the reality of exquisite pain with an inability to scream or to move the vocal motor apparatus to express the pain more overtly in actionable behaviors.

The thesis for this section of text is that the right cerebral hemisphere appears specialized for pain and activation of pain-related concomitants, including incremental blood pressure, sympathetic tone, negative emotion, escape or avoidance behavior, caution, and depletion of serotonergic stores. In approaching this argument, I am reminded of Walter Isaac's submission of his research to a journal editor in which he stated that "cats are nocturnal." The editor returned the manuscript to Walt asking for a reference as to this factual statement. After the laboratory team searched the literature finding no such reference, Walt resubmitted the manuscript stating simply that cats "forage at night, feed at night, and have sex at night." The editor responded to this submission saying "…but, so do I!" Through this debate, Walt published his manuscript and his conclusion that cats are, indeed, nocturnal animals (Isaac and Reed 1961).

By analogy, a right cerebral hemisphere specialization for pain might be argued through the common pain-related dimensions occurring within the autonomic nervous system, behavior, affect, social, and speech systems. Pain is most closely featured within the autonomic nervous system, through incremental activation of the sympathetic branch to include increased blood pressure, heart rate, and sweating or skin conductance. There is evidence indicating that increased blood pressure may serve to reduce pain (see Mollet and Harrison 2006; al'Absi and Petersen 2003; see also Bruehl and Chung 2004) and evidence for this has been seen in the laboratory with individuals at risk for cardiovascular disease, through baseline hostility, with highly reactive cardiovascular responses to pain (e.g., Herridge et al. 2004). Earlier in these writings, classic arousal theory was discussed and the revolutionary developments in this theory through the contributions of Wendy Heller (Heller 1990, 1993; Heller et al. 1997), Kenneth Heilman (Heilman et al. 1978; Heilman et al. 2003, 2012), and others demonstrating cerebral asymmetry in arousal with a prominent role ascribed to the right brain. Incremental arousal was found to be a contribution of the right parietal region with the somatosensory system found within this region (e.g., Mazzola et al. 2012). Relevant here is the flattened or bland affective state that results from deactivation or lesion of the right parietal area and especially the inferior region proximal to the somatic facial area. Pain most typically yields incremental arousal suggesting the underlying architecture to be within the right hemisphere.

Gina Mitchell (Mollet and Harrison 2006) appreciated the intimate relationship between emotional processing systems and pain processing systems in her application of quadrant theory to these processes. Ultimately, Gina would integrate these views within the capacity theory also appreciating the sensory perceptual analyzers within the posterior brain and the regulatory control over these areas imposed by the frontal lobe (Mitchell and Harrison 2010; see also Meerwijk et al. 2012). The layperson and neuroscientist alike clearly appreciate the redundancy of chronic pain and depression (see Meerwijk et al. 2012). The intimate relationship between acute pain and anger or fear is appreciated by all those sharing the human experience. Pain may release reflexive responses with sympathetic nervous system activation, pupil dilatation, incremental blood pressure and heart rate, and possibly aggressive or defensive behaviors. The degree of intimacy among these functions is equivalent to, or beyond that of, the now commonsense level acknowledgement of the relationship between logical, linguistic speech and the left brain. Yet, the zeitgeist is not unlike that experienced at the time that Broca made his, now famous, accolade stating that "we speak with the left brain."

A compassionate understanding of these functional neural systems promises to provide new perspectives in the understanding and potentially the treatment of pain disorders and, in turn, disorders of affective valence and regulatory control. Relative activation of each of the cerebral quadrants and the processing demands imposed over the frontal regions may provide new insights and new therapies from a fresh perspective. These observations and arguments may be worthy of extension to other literatures on social or physical trauma, perhaps. Relatively little attention has been directed to what some have called "psychological pain" or grief despite a well-known association between these experiences and depression and as a precursor for suicidal behavior. Meerwijk et al. (2012) performed an overview of studies on brain function related to the actual experience or recall of an autobiographical event involved in "psychological pain." Based on their review, they proposed a tentative neural network for grief or "psychological pain" that includes the thalamus, anterior and posterior cingulate cortex, prefrontal cortex, cerebellum, and parahippocampal gyrus. They conclude that overlap exists with the circuits involved in "physical pain," but with a markedly reduced role for the insula, caudate, and putamen during psychological pain.

By way of discussion, chronic pain is a strong predictor of depression and anger, as is the depletion of serotonin. Both serotonin (see Fitzgerald 2012) and pain (see Mollet and Harrison 2006) show evidence for lateralization to the right cerebral hemisphere, indicating a possible role of the asymmetrical neurochemical properties of the left and right hemispheres accounting for differences in emotional processing. One variant of this lateralization hypothesis indicates that serotonin is biased toward activating a majority of right hemispheric brain regions or functions more so than in the left hemisphere, with norepinephrine showing an opposite left hemispheric bias. Another variant of the hypothesis is that serotonin not only tends to activate the right hemisphere but that it also functions through deactivation of the left hemisphere (e.g., Smith et al. 2009; see Fitzgerald 2012). This take on the fundamental theory of neurotransmitter asymmetries is consistent with cerebral balance theory where significant deactivation of one cerebral hemisphere provides for a release of the opposing hemisphere and where heightened activity provides for suppression or inhibition of the homologous brain regions within the opposite cerebral hemisphere. In this example, significant activation of the right hemisphere also deactivates the left hemisphere.

Norepenephrine appears to have the opposite functional pattern with activation of the left hemisphere and potentially altering cerebral balance mechanisms. Dopamine, a neurochemical associated with positive affect (see Ashby et al. 1999) is found at higher levels in the left hemisphere (see Tucker and Williamson 1984; see also Wittling 1995). Alternatively, the right hemisphere may use more serotonin (sees Tucker and Williamson 1984; see also Wittling 1995). Altered levels of serotonin are associated with negative affect such as, hostility, aggression (Cleare and Bond 1997), and depression (Flory et al. 2004). This is more clearly the case in men who have demonstrated heightened cerebral laterality in comparisons with women (e.g., Kimura 1999; McGlone 1978, 1980). Moreover, some evidence indicates that men may have twice the level of serotonin found in women (Nishizawa et al. 1997; Stein et al. 2012), where women suffer more depressive complaints and are twice as likely to be diagnosed with depression (Burt and Stein 2002). Women receive 78 % of the selective serotonin reuptake inhibitors (SSRIs) prescribed for mood disorders (Kessler 2003; see Haleem 2012). Also, women are more likely to seek help for chronic pain. Negative affective stress and pain may deplete these serotonergic reserves, altering emotional state. Thus, effective stress management and pain management may elevate serotonin levels, depleting the intensity of the depressive response.

Part VII
Brain Pathology

Chapter 32
Circulatory Systems

For any of the myriad types of brain pathology, there may be signal features that help in the diagnosis of the disease or the identification of the underlying process which ultimately may be displayed in the presenting symptoms. However, there are no absolute symptoms for any specific origin of pathology as the clinical features will be largely derived from the area or part(s) of the brain involved and the functional neural systems which are ultimately affected. The time course and prognosis may vary too with the type of pathology. The purpose of the writings which follow is not to provide, in any way, a comprehensive accounting of the types of injuries or insults which might afflict the brain, but rather to provide an initial discussion of the more common types of pathology and their basic features. Ultimately, the impact of the pathology will relate to many complex and interacting aspects of pathological presentations, including the acute versus chronic, the abrupt versus insidious, the adult versus developmental, and underlying issues of the extent of medical complexity. We begin with a discussion of vascular pathology with a focus on cerebrovascular accidents associated with specific arterial distributions. A presentation of the basic elements involved in traumatic brain injury and post-concussive syndrome follows. This is followed by a discussion of neurodegenerative disease or dementia processes. Subsequent to this is an introduction to the convulsive disorders or seizures.

For any specific or more complex presentation of pathology, the neuropsychologist may contribute, as part of a broader and multidisciplinary team, to better understand and to elucidate upon the following diagnostic and care-related issues. A primary focus of the evaluation is to localize the brain regions affected in order to specify the neuropsychological syndromes resulting from the insult. From this syndrome analysis and lesion localization effort, the individual's rehabilitation needs may be more clearly understood and ultimately shared with a cohesive multidisciplinary therapy team. This basic clinical derivation of the diagnosis and identification of spared and affected brain regions can improve the therapeutic milieu and ideally with coordinated efforts across the team of contributors. The therapists closely involved with the patient during the rehabilitation efforts often include the

© Springer International Publishing Switzerland 2015
D. W. Harrison, *Brain Asymmetry and Neural Systems*,
DOI 10.1007/978-3-319-13069-9_32

speech pathologist, the occupational therapist, the physical therapist, the recreation or community reentry specialist, social worker, clinical psychologist, nurse, physical medicine, neurologist, physiatrist, and other disciplines.

The neuropsychologist may be able to identify the extent of damage and determine if the brain cells are dead, malnourished, deactivated, or compressed, for example. The evaluative process often involves efforts to establish estimates of premorbid cognitive status. This might be estimated from any and all available sources, including the patient's family, educational history, occupational accomplishments, and other historical evidence. Related to these determinations, the neuropsychologist may address the prognosis and time course for recovery in distinguishing a degenerative process or active and potentially progressive pathological process from those warranting a more optimistic prediction for recovery of function(s). Through these efforts, the neuropsychologist ultimately is a teacher and assists in the education of all of the rehabilitation team members and in the education and therapeutic interventions with the family and the caregiver. This process extends in the provision of baseline neurocognitive, affective, and behavioral markers and often using standardized test batteries from which future comparisons may be made in the assessment of progress. The evidence may well support the recovery of functions as opposed to the progression of symptoms over time. Regardless, the evidence for recovery or progressive decline across temporally spaced evaluations will be a primary resource for the team of providers and certainly for the family faced with imposed needs for planning and preparation to provide for care and maybe ongoing financial planning for supervision and/or assistance needs.

The type of pathological process involved in any specific brain disorder is complex and may include focal or more general malnutrition of brain cells, ischemic events such as that following a thromboembolic occlusion of cerebral arteries, hemorrhagic events, and neoplastic events with tumor genesis, and electroconvulsive episodes. Malnutrition may result in the brain cells becoming edematous or watery, and this might be evident in a Wernicke–Korsakoff syndrome as a nutritional dementia from alcoholism. Neoplastic growth often provides for aberrant and dysfunctional cellular processes well beyond the tumor focus identified on the computerized tomography (CT) or the magnetic resonance imaging (MRI) scan of the head. Cerebral hemorrhage may result in contact-related brain cell death with iron killing these cells on contact. In addition, there may be space-occupying effects from the bolus with significant pressure effects on surrounding brain regions perhaps in some 15–20 % of strokes. Far-field effects of compression may be evident often at the brain stem region with altered mentation and arousal or lethargy or perhaps with compression at the basal skull regions.

Traumatic brain injury may tender diffuse injury effects through the complex physical forces conveyed through the gelatinous brain material. This may be expressed in the somewhat focal coup and contrecoup injuries with head trauma and through diffuse axonal injury from torsion, shearing force, and stretching of neuron and glial pathways and interconnections. Glial scarring may add to the complexity of a brain injury through the development of an irritative focus and potentially one sufficient to evoke an epileptic or electroconvulsive event. Seizures also provide for high-voltage spikes and electrical activation at the site of origin, which may inhibit oppositional brain systems. Seizure-related events may not only reflect active neuropathology but also progressive pathology through multiple mechanisms. This was recognized early on (see Engel 2013) with the formation of a "mirror focus" at the homologous region within the other brain conveyed across the corpus callosum. Pathological processes often undermine the integrity of other processes, potentially leading to a cascade of pathology and instability in the system at least initially following a brain injury. From this dynamic state, the brain injured may be eventually stabilized and able to more fully participate in rehabilitative therapies and the recovery process.

Vascular Pathology

According to the US Center for Disease Control and Prevention (Go et al. 2013), stroke is considered to be the leading cause of long-term disability and the third leading cause of death. Each year, ≈795,000 people continue to experience a new or recurrent stroke (ischemic or hemorrhagic) accounting for an estimated 143,000 deaths in 2008. Approximately, 610,000 of these are first attacks and 185,000 are recurrent attacks. In 2009, stroke caused≈1 of every 19 deaths in the USA. On average, every 40 s, someone in the USA has a stroke and someone dies from a stroke approximately every 4 min. The number of deaths from a stroke as a percentage of the population appears to have declined substantially over the past 30 years. Racial disparities still exist with the death rate from stroke nearly 50% greater for African Americans than for other racial groups. Native Americans and Hispanics have lower death rates (see Fig. 32.1). It appears noteworthy that stroke kills more than twice as many American women every year as breast cancer. More women than men die from stroke, and the risk for women may be higher due to longer life expectancies than that enjoyed by men. Women appear to suffer greater general disability after stroke than men, whereas men may suffer greater focal disability. For example, a woman may have a better prognosis than a man for a focal stroke in Broca's area in terms of recovery of expressive speech functions. However, a woman may be more likely to experience general disability from a fall and fractured hip subsequent to the stroke. Women aged 45–54 also appear to be part of a surge in strokes due to increased risk factors such as the metabolic syndrome and/or obesity. Cultural differences also emerge in the literature. For example, one project appreciated the relative risk factors with Hispanic women (Luo et al. 2013).

Imaging Issues

A variety of imaging resources are increasingly available to the neuropsychologist, ranging from the more traditional CT scan, to the MRI scan, to the positron emis-

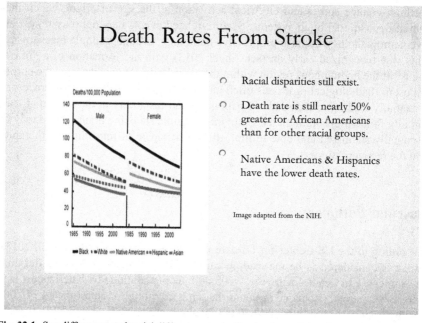

Death Rates From Stroke

Deaths/100,000 Population

○   Racial disparities still exist.

○   Death rate is still nearly 50% greater for African Americans than for other racial groups.

○   Native Americans & Hispanics have the lower death rates.

Image adapted from the NIH.

Fig. 32.1 Sex differences and racial differences in death rates from stroke. (Image is adapted from the National Institutes of Health)

sion tomography (PET) scan, to the functional MRI (fMRI) scan, to an increasing array of even more exotic and potentially helpful technologies. As of these writings though, CT scans and MRI scans were fundamental to the evaluation and commonly available in addition to one or another type of angiogram, including the magnetic resonance angiogram (MRA; see Fig. 32.2). These technologies are most useful in situations where the brain pathology is evident in the image and provides for a focal or more generalized accounting of the brain regions affected by the precipitating or more insidious type of pathological event.

In the hands of a skilled radiologist, much information may be derived from even the common CT scan, which can inform the neuropsychological evaluation and intervention efforts. Temporal factors are relevant in many cases where the behavior, cognition, and affect may be substantially clear and evident in the pathological presentation but where the scan may be unremarkable. Over appreciation of an unremarkable scan in the face of behaviorally evident pathology may be somewhat like taking your automobile to a mechanic with obvious electrical or mechanical problems. After the image is developed and the mechanic returns to tell you that "nothing showed up in the pictures…that the scans are normal," you might better seek out a new mechanic than base your judgments and actions upon the pictures. Often the scans must be repeated before pathology is finally evident, and in some cases, the pictures remain unremarkable in the face of clinically definable pathology. The images are simply more useful when they allow us to visualize a structural anomaly or area of damage.

Fig. 32.2 An MRA image depicting the ascending arterial tree within the distribution of the left and right brain. MRA magnetic resonance angiography (Adapted from image originally published by Tsai, Chung-Fen, *Cerebral infarction in acute anemia,* Journal of *Neurology,* Volume 257 Issue 12, p. 2049)

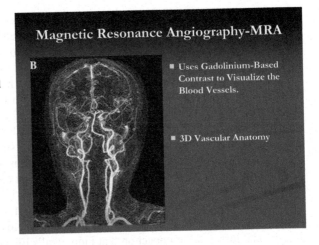

Following an ischemic stroke, an early CT scan may be grossly normal as edema and the area of infarction have not yet well developed sufficient to be identified on the image. More subtle indications may include the loss of the gray–white matter differentiation. With progression, the ischemic event may be evident in additional changes, possibly including ventricular effacement, midline shift between the

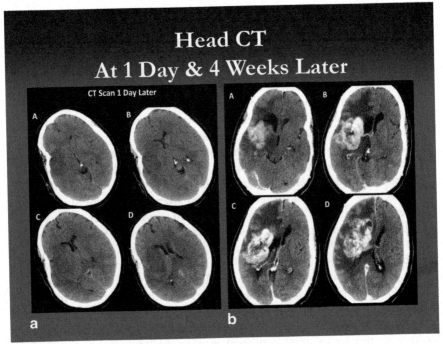

Fig. 32.3 Repeat head CT on day 1 and day 3 after a left middle cerebral artery distribution cerebrovascular accident. *CT* computerized tomography

cerebral hemispheres, and/or hypodensity within areas depicted on the scan (see Fig. 32.3). Thus, both the pathological process involved with the ischemic event and the developmental course for the stroke may be slowly evident in subsequent images or maybe not evident at all, in some cases.

The imaging techniques themselves are not without hazard or iatrogenic implications. Some evidence supports the stronger radiation hazard of the CT scan as a cancer risk and especially for younger age groups, for example. Moreover, the angiogram may involve the insertion of a radioactive bolus into the bloodstream with a risk for iatrogenic stroke. Although the risk for stroke from angiogram appears minimal nationwide (2–5 %), the risk does appear to vary from one facility to the next. Also, this discussion points to the potential discrepancies between the scientific assessment of risk and the perspective on risk for the individual suffering an iatrogenic or nosocomial event. For science, the risk is assessed as a potential impact to the broader population of individuals exposed to medical procedures. But, the personal loss and the impact on that individual's life and upon the family may weigh substantially larger in the scheme of things. The statistical risk assessments hardly hold weight if you or your family member is the one who experiences the induced stroke from an elective diagnostic study and if the assessment did not clearly weigh in the provision of a different approach to care or intervention.

Vascular Diseases

One Friday morning, I met a young married couple at my office with somewhat atypical complaints surrounding the episodic loss of consciousness by the man and specifically "on the dance floor." Interestingly, the couple complained that the man would pass out while "slow dancing" rather than "fast dancing." As I worked through the myriad possibilities for the potential cause of these events, I decided to ask for a demonstration. The events were recurrent for the couple, and we were at least within a medical center where support was easily available, if needed. The man was rather short and the woman was rather tall and, as I came to appreciate, she was also rather well endowed. Upon placing his neck between her breasts as they embraced for the "slow dance," indeed he fainted and declined to the floor. The angiogram confirmed arterial disease, and he soon elected to undergo a bilateral endarterectomy in hopes of opening the carotid arteries for improved blood flow. Another man would pass out "in front of the mirror shaving." Indeed, as soon as he pushed the razor tight against the left side of his neck, he would lose consciousness, which again provided a clue for his underlying vascular pathology and supported his subsequent decision with his physician to undergo an endarterectomy procedure.

The progression of vascular pathology ultimately expressed in a cerebrovascular event or stroke may develop over many years as with chronic atrial fibrillation, arteriovascular disease, cardiovascular disease, or congestive heart failure. It may also develop more abruptly as with some embolic strokes, for example, following a period of uncontrolled atrial fibrillation or with valvular cardiac origin. The typical presentation and recovery from a stroke begins with an acute onset of symptoms derived from the brain region affected and perhaps from responsive vascular dynamics to the event. This is followed by a period of stabilization of symptoms and subsequently by slow improvement or resolution of symptoms.

There are notable exceptions to this model, however. These include chronic hypertensive pathology with the development of multiple, primarily, white matter lesion sites and progressive decline unless the blood pressure can be lowered and possibly with the physician's efforts to thin the blood. Another exception is found with cerebral vasculitis, which may underlie progressive neural pathology. Also relevant, by way of example, is the individual involved in contact sports with repeated shear force and concussive trauma, which may result in multiple cerebrovascular lesions and eventually underlie a "punch-drunk syndrome," perhaps. In one assessment, an elderly woman was seen for a dementia process that paled by comparison with that of her son who had planned to be the woman's caregiver. It was soon apparent that the National Football League (NFL) star was suffering more significant pathology than was the demented mother and that the discharge of the patient from the hospital would need to incorporate planning for the care of both the patient and her son. A further exception to the recovery model described above may be found with vasculitis and especially with a chronic or progressive vasculitis.

Types of Vascular Pathology

A university professor, close friend, and colleague of mine shared her experiences with me involving a more acute and treatable vasculitis resulting from a spider bite over her vertebral artery at the posterior right side of her neck. Subsequent to the bite and the spread of the inflammatory process, she began to experience left-sided visual hallucinations accompanied by apprehension and mildly paranoid content. She was knowledgeable of brain function, allowing her to make informed attributions for her state of affairs and to seek appropriate medical interventions and, in this case, with antibiotics. Had she been less well informed and knowledgeable of her own cerebral functions, the outcome may have been substantially more dramatic with psychiatric admission, diagnosis of an acute schizoaffective disorder, embarrassment, and possibly the loss of her job. But she remained vigilant with vasculitis often signaling a more chronic disease process and potentially with inflammation spreading throughout the vascular distribution across multiple body regions and organ systems.

A somewhat less insidious form of vascular pathology may arise from a spontaneous intracranial hemorrhage following rupture of a dilated artery or aneurysm. The brain is wrapped tightly by the meninges, which resemble a sausage skin, perhaps. This "skin" consists of three primary layers with the tough and protective dura mater ("tough mother") located close to the skull and with the middle layer or arachnoid layer serving as a boundary division between substances which are away from direct contact with brain tissue and substances which are in close proximity to the pia mater ("little mother") and more directly with the brain matter. Thus, a subarachnoid hemorrhage may expose brain tissue to blood and its iron contents, resulting in damage to that area of the brain, whereas an extradural hemorrhage may exert its effects on brain tissue somewhat indirectly through pressure effects. Moreover, the neural surgeon may be able to withdraw this bolus of blood via aspiration using a cannula or other technique to relieve the pressure or it may resolve through more natural processes over time. The subarachnoid hemorrhage provides for multiple methods of damaging brain tissue where perhaps 50% of the damage may result from pressure effects and where perhaps 50% of the damage is direct from the blood to the neural tissue. Hemorrhage and ischemic events are not independent. The bleed may result in reflex stenosis or vasodynamics which may only gradually resolve during recovery.

Blood Supply to the Brain and Brain Stem

The arterial vascular tree providing oxygen- and nutrient-rich blood to the brain originates from four primary arteries ascending up the neck and eventually to the cerebral hemispheres. The two carotid arteries ascend up the ventral surface at each side of the neck just as the vertebral arteries ascend up the posterior surface at each side of the neck. Figure 32.4 provides a view of the arterial blood supply and the principal arteries supplying the brain. On viewing the image, one can appreciate that some of the arteries bifurcate forming collateral arteries, whereas others merge together to form a unitary blood vessel. The principal example of the latter situation would be the basilar artery, situated at midline over the ventral pons or brain stem. The basilar artery is formed from the emergence of the two vertebral arteries ascending the posterior neck region. These arteries move toward the anterior surface and merge forming the basilar artery, which will again bifurcate to form the inferior and the superior cerebellar arteries and eventually with more dramatic flair the bifurcations will form the circle of Willis near the thalamus and optic chiasm. The circle of Willis will provide additional bifurcations with the posterior communicating arteries extending toward the mesial occipital lobe and with the anterior communicating arteries extending toward the mesial frontal lobe structures (e.g., the cingulate gyrus and supplementary motor cortex).

Much of the blood flow and arteriovascular dynamics for the axial brain regions, including the brain stem and cerebellum and the midline cerebral structures and corpus callosum, arise from the vertebral arteries at the back of the neck and eventually from their emergence into the basilar artery at the front of the pons. The vertebrobasilar system supplies blood perfusion to the inner ear, brain stem, and cerebellum and rupture or occlusion of this supply is frequently recognized from the accompanying symptoms of truncal ataxia, dysmetria, nystagmus, vertigo, nausea, and possibly vomiting (e.g., see Chakor and Eklare 2012). Indeed, this constellation of symptoms is oft referred to as a *vertebrobasilar syndrome*. The diagnosis of this syndrome will characteristically involve the examination of lateral

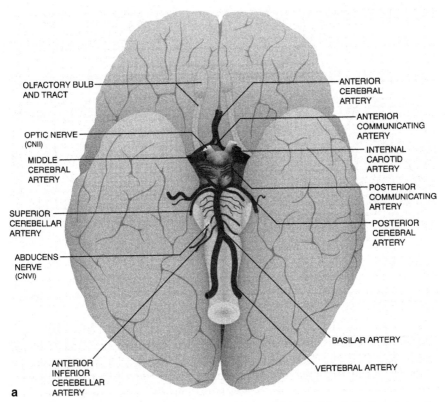

OLFACTORY BULB AND TRACT

OPTIC NERVE (CNII)

MIDDLE CEREBRAL ARTERY

SUPERIOR CEREBELLAR ARTERY

ABDUCENS NERVE (CNVI)

a

ANTERIOR INFERIOR CEREBELLAR ARTERY

ANTERIOR CEREBRAL ARTERY

ANTERIOR COMMUNICATING ARTERY

INTERNAL CAROTID ARTERY

POSTERIOR COMMUNICATING ARTERY

POSTERIOR CEREBRAL ARTERY

BASILAR ARTERY

VERTEBRAL ARTERY

Fig. 32.4 Arterial blood supply to the brain. (Originally published in *Metabolic Encephalopathy,* McCandless and David 2009, p. 9)

and vertical components of visual smooth pursuit. The examination will include measures of visual convergence on approach from the left and from the right side. Also assessed will be indices of truncal coordination, ambulatory gait, and balance. A lesion within these pathways may alter frontocerebellar functions affecting coordinated and intentional movements. Dysmetria might be initially appreciated on the finger-to-nose task with poorly sequenced or coordinated movements of the arm, hand, and finger in reaching out to touch the examiner's finger and upon return to touching the patient's own nose. Limb coordination is also likely to be affected with dysmetria or dysdiadochokinesia. This might be assessed using a variety of tasks. Some of these were described by Luria (Christensen 1979), including the fist–palm alteration task, the finger-to-thumb task (sequentially tapping each finger with the thumb: unimanual and bimanual versions), the finger tapping test (Wertham 1929; a rate measure), and many other measures of coordinated movement.

The Romberg task is part of this assessment. The patient is asked to stand with feet together followed by tasking the patient with the eyes closed. Of course, the patient is potentially at risk during these procedures and efforts must be made to ensure safety. Initially, when the eyes are open, truncal stability might be maintained

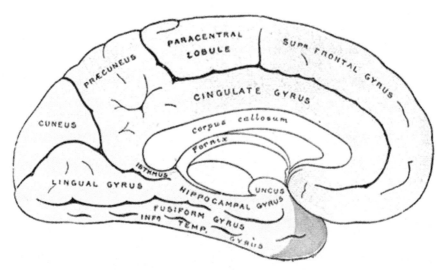

Fig. 32.5 A midsagittal view of the left cerebral hemisphere showing the distribution of the anterior cerebral artery (blue) and the posterior cerebral artery (yellow). (Originally published in *Gray's Anatomy of the Human Body*)

through three primary modalities, including vision, proprioception, and vestibular senses. With mild damage in this system, the patient may be able to compensate for vestibular or proprioceptive disturbance using visual cues. With the eyes closed, heightened demand is placed on the functional integrity of the vestibular and proprioceptive modalities, and truncal sway or instability may arise upon presenting them with this challenge. With more severe disability secondary to proprioceptive or vestibular damage or with a midline cerebellar lesion, truncal instability and/or ataxia may be appreciated along with the multitude of safety risks in performing even routine functional activities of daily living. Many of these patients will be at an increased risk for falls related to disturbances in and among these functional systems. Damage to the lower brain stem structures approximating the medulla might present within vital systems, possibly including respiratory hyperreflexia with hiccups and/or belching behavior, orthostatic hypotension, or other symptom arrays.

Ultimately, the midline or axial regions of each cerebral hemisphere are supplied by the anterior and posterior cerebral arteries (see Fig. 32.5). Much of the blood flow for the lateral structures within the cerebral hemispheres will be supplied from the carotid arteries traveling initially up the front of the neck with the internal carotid arteries emerging alongside of the circle of Willis. From this emergent location, the carotids appear ideally situated to fill the branches of the ipsilateral left or right middle cerebral artery. The middle cerebral artery distribution within each brain provides the majority of the blood supply to the lateral structures, including the Broca's and Wernicke's classic areas for components of propositional speech, the face, and the hand (see Fig. 32.6).

Regardless, the focus of the current discussion on intracranial hemorrhage will benefit from the appreciation of this anatomy, which is much like that a plumber

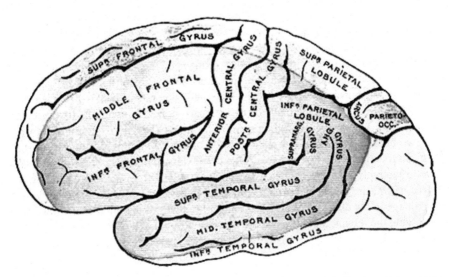

Fig. 32.6 A lateral view of the left cerebral hemisphere showing the distribution of the middle cerebral artery (red). (Originally published in *Gray's Anatomy of the Human Body*)

might have to deal with when looking at multiple pipes being reduced to a single pipe, which in turn divides into multiple pipes once again. This might well be a plumber's nightmare as the management of physical hydraulic forces of blood traveling through these pipes would appear to be most difficult. But, this is more the case in the mammalian body with a dynamically responsive cardiovascular system to maintain blood pressures sufficient for the diversity of activities with which we may engage. Something surely may break and the common or logical places for this to occur are often at the points of emergence or mergence of two arteries. This provides for a heightened probability of hemorrhagic stroke around the circle of Willis and often at the anterior communicating arteries or at the basilar artery. The blood vessel wall may be weakened, leading to rupture with incremental pressures and dynamic pressures, perhaps.

With rupture of an anterior communicating artery aneurysm, the neuropsychological presentation may initially involve features related to damage in the bilateral mesial frontal regions and mesial subcortical structures, including the corpus callosum via the callosomarginal and pericallosal arteries and their bifurcations. Bilateral infarction of the anterior cerebral arteries feeding off of the circle of Willis may produce one or another variant of the alien limb syndrome (e.g., Bejot et al. 2008) from damage to the corpus callosum. The cingulate gyrus might easily be affected along with arousal-related projections from the thalamic and amygdala regions. The assessment of the integrity of the startle response and the orienting responses might be appropriate. This might initially require a rule out in the assessment for "split-brain syndromes" restricted to a disconnection of the frontal lobes one from the other. Akinetic mutism (e.g., Fontaine et al. 2002; see also Duffau 2012) might be present and especially with occlusion of the left anterior cerebral artery (Poppen

1939). Lower body regions may be affected and specifically the lower extremities, the bladder, and the genitalia.

Cerebrovascular hemorrhagic events affecting the posterior cerebral arterial distribution might require more focused assessment of the visual projections, visual perceptive analysis, hippocampal and perihippocampal integrity, and perhaps loss of sensory awareness relatively restricted to the lower, rather than to the upper, contralateral extremity, and the bladder region or genitalia. The splenium of the corpus callosum may be at risk with the potential for a disconnection syndrome between the two brains. A classic example of the functional pathology resulting from such a lesion location would be the syndrome of alexia without dysgraphia. This lesion may effectively disconnect the two occipital cortices from access to the angular gyrus within the left cerebral hemisphere. In contrast, the cortical pathology akin to damage in the distribution of the posterior branches of the middle cerebral artery supplying the left angular gyrus would be the syndrome of dyslexia with dysgraphia (Gerstmann 1940, 1957; see also Cabeza et al. 2012; see also Ardila 2012).

Rupture of the arterial divisions emanating from the basilar artery commonly involve cerebellar damage with functional concerns for balance, coordination, truncal ataxia, dysmetria (see Chakor and Eklare 2012), and ultimately an increased probability of falling and fall-related injuries. Ischemic changes may be assessed on the neuropsychological evaluation further up the arterial tree in measures of functional integrity of axial brain structures, including the occipital lobes, and through the assessment of visual agnosias. Lower body structures may again be at relative risk, including the lower extremities, bladder, genitalia, and the midline structures of the corpus callosum. But, the effects of intracranial hemorrhage may ultimately signal vital concerns for the neural surgeon as pressure effects may compress lower brain structures involved in vital functions and reflexive responses for blood pressure, respiration, and such.

Less common are spontaneous intracranial hemorrhages which directly involve the internal carotid arteries. Rupture of an external carotid artery was witnessed by a former graduate student who always sat on the front row of his church. A man arrived earlier and sat on the front row, replacing the student who now took his seat behind him and in the second row. The hemorrhagic event apparently occurred following an abrupt stretching of the neck during the service. Fortunately, the former student was also an emergency medical technician and was able to assist in the management of the hemorrhage. However, loss of blood flow or pressure effects from an intracerebral bleed affecting the distributions of the middle cerebral artery within either cerebral hemisphere may signal assessment needs for the contralateral upper extremity and facial regions with relative sparing of the lower extremity. Assessment of pathology within the distribution of the middle cerebral artery may include alterations in the integrity of the propositional speech systems and linguistic analyzers supplied by the middle cerebral artery ascending the left cerebral hemisphere. It may include alterations in the integrity of the nonpropositional speech systems and prosodic analyzers supplied by the middle cerebral artery ascending the right cerebral hemisphere. This artery, within each brain, provides branches supplying both the second and third functional units, requiring differential diagnosis of intentional and executive functions of the frontal lobes (see Stuss and Knight 2013) and the sensory receptive and comprehension capabilities of the lateral convexities of the parietal, temporal, and occipital regions.

Cerebral ischemia more often results from mechanical obstruction as with a thrombotic event, for example, which may ultimately cause a cerebral infarction or result in necrotic tissue. Thrombosis may result from local plague or clot formation along an artery due to scarring of the vessel wall, perhaps. Embolic events occur with any foreign matter traveling through the vessel and eventually with occlusion of the vessel, preventing blood flow and ultimately oxygen saturationand nutritional support for the tissue downstream from the clot. This sometimes occurs as an episodic event as a transient ischemic attack. However, the bolus may be sufficient to produce more permanent effects on the tissue feed by the vessel in the form of cerebral infarction. The onset may be acute or it may be more insidious in nature. The latter presentation may follow the ontogenesis of arterial stenosis or narrowing of the arterial wall with progressively insufficient blood flow. Cerebral damage results with any situation, acute or chronic, which results in inadequate blood flow, a lack of adequate oxygen saturation, and/or nutritional deprivation. Eventually, this pathology may be identified as an encephalomalaciaor softening of the brain.

The diagnosis and appreciation of the likely functional pathology resulting from cerebrovascular insult is again topological secondary to the specific distributions of the anterior cerebral artery, the posterior cerebral artery, and the middle cerebral artery within each brain and the neuropsychological contributions of each of these regions. Knowledge of the functional differences and processing specializations attributable to each brain is critical to assessment, therapeutic intervention, and education of therapists, the family, and the patient. More specifically, knowledge of the functional features derived from cerebral regions supplied by each cerebral artery provides a prior expectations and hypotheses for the examination, which may be investigated through double dissociation techniques (Teuber 1955; Kinsbourne 1971; Luria 1973, 1980).

Anterior Cerebral Artery

Figure 32.7 provides images of the locations and divisions of the anterior cerebral artery along with imaging data depicting left-sided pathology on MRI and MRA scans. The anterior cerebral artery differentially supplies the motor and premotor cortices with projections to the lower extremities, genitalia, and bladder. It supplies the supplementary motor cortices where damage has been implicated in akinetic mutism or bradykinesia syndromes for the left frontal region and hyperkinesis syndromes and impulsivity for right frontal pathology (see Oades 1998; see also Emond et al. 2009; see also Granacher 2008; Helfinstein and Poldrack 2012). It ultimately supplies the corpus callosum via the callosomarginal and pericallosal divisions, providing a necessary rule out for disconnection syndromes between the frontal lobes of each brain. This may produce one or more variants of the alien limb syndrome or affect the integration of bilateral motor movements where the left and right hemibody must function together in well-timed, coordinated tasking in an orchestrated fashion. The orbitofrontal branch, upon occlusion, may be relevant to

Anterior Cerebral Artery

Anterior Cerebral Artery to Medial Surface

Stroke May Disconnect Frontal Lobes

May Damage Corpus Callosum

Divisions of ACA

Callosomarginal

Pericallosal

Orbitofrontal

Frontopolar

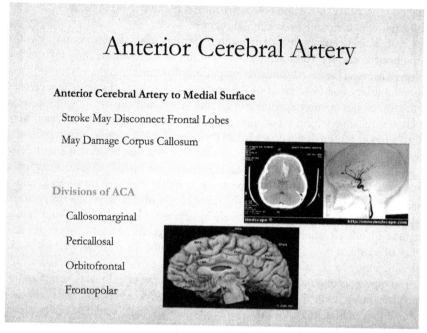

Fig. 32.7 Locations and divisions of the anterior cerebral artery along with imaging data depicting left-sided pathology on MRI and MRA scans. *MRI* magnetic resonance imaging, *MRA* magnetic resonance angiography (Adapted from images copyrighted by Springer Science + Business Media New York)

diminished regulatory control over the amygdala via the uncinate and arcuate fasciculi connecting these brain regions.

The literature is more robust on reduced negative emotional regulation with right orbitofrontal damage (e.g., Agustín-Pavón et al. 2012), where incapacity within the right orbitofrontal region may result in hyperreflexive or reactive expression of anger or aggressive behaviors (e.g., Fulwiler et al. 2012). Reduced regulatory control from this region may be evident in nonpropositional speech impairment. Damage may be reflected in an expressive dysprosodia possibly somewhat flat or sparse in prosody with an avalanche of speech volume or loud abrasive features subsequent to emotional stress or provocation. Moreover, the loss of regulatory control over anger and fear, for example, may reflect a concurrent loss of control over the cardiovascular system with corresponding reactivity in systolic blood pressure, heart rate (Harrison and Emerson 1990; Demaree et al. 2000; Herridge et al. 2004; Williamson and Harrison 2003; Everhart et al. 2008; Shenal and Harrison 2004; Rhodes et al. 2013; see also Foster et al. 2008), glucose (Walters and Harrison 2013a, b), skin conductance (Herridge et al. 1997), facial expression (Rhodes et al. 2013), and temporal lobe activation on the quantitative electroencephalogram (Mitchell and Harrison 2010). (The thalamoperforating branch may, upon occlusion, yield one or another thalamic syndrome.)

Damage within the distribution of the frontopolar artery may raise assessment questions for behavioral, social, and affective features often altered by pathology in these regions. Much of the research on depression has implicated the left frontal poles and basal ganglia, which may be affected by a stroke in this arterial distribution. Social and emotional sequelae from right orbitofrontal involvement and damage at the right frontal pole include social anarchy and social improprieties, respectively. Such a patient may be able to state and discuss the social rules guiding public behaviors, whereas they may find that the rules do not apply to them (the social anarchist). The individual with impaired social pragmatics and/or improprieties may display a most curious lack of care in public behaviors for turn taking, appreciation of social cues, and even more provocative improprieties such as picking their nose in public, inadequacies in basic hygiene, and related malfeasance (e.g., Eslinger et al. 2004; Tompkins 2012; see also Yeates et al. 2012).

Middle Cerebral Artery

The divisions of the middle cerebral artery most notably include the arterial feeds to the frontoparietal region. Two prominent branches of this artery are seen in the prerolandic and the rolandic arteries. Damage within the distribution of these branches of the middle cerebral artery characteristically results in relative functional deficits at the upper body regions, including the face and hand contralateral to the brain pathology. This corresponds with a relative sparing and improved prognosis for recovery of the contralateral lower extremity. This may be of considerable importance to the person suffering a stroke here, for example, as many of these folks will place the imperative upon recovery of the lower extremity to enable ambulation and freedom of movement without a wheelchair or major assistive device. The lower extremity may instead be more susceptible to a stroke or pathological processes within the distribution of the anterior cerebral artery with limb projections arising from the contralateral and midline frontal region.

Involvement of the motor cortices may provide for muscle weakness or paresis ranging to paralysis or plegia of the contralateral body regions. This is often followed by spasticity, dystonia, or hyperreflexia of the body region with antigravity posturing subsequent to upper motor neuron damage (e.g., Futagi et al. 2012). Pathology within the frontal lobe distribution of the middle cerebral artery supplying the motor cortex may produce a relative loss of function at the contralateral face and upper extremity (hand, arm, and shoulder), resulting in a right hemiplegia (severe) or right hemiparesis with subsequent dystonia apparent in heightened flexor tone or resistance to passive range of motion. This may be seen in a grasp reflex at the hand and flexor tendencies for the limb. Watershed events may convey antigravity posturing at the contralateral lower extremity with extensor posturing and possibly arching of the contralateral back and postural distortion. However, the prognosis and severity of dysfunction are again related to relative location of the face and upper extremity within the distribution of the branches of the middle cerebral artery and with the functional integrity of the lower extremity determined more by the integrity of the mesial cerebral regions supplied by the anterior cerebral artery.

This relationship persists as we move into the premotor regions of the frontal lobe and the prerolandic arterial branch. Premotor changes may be assessed in measures of motor coordination and integrated motor sequencings. Examples may include discoordinated performance in rapid alternating movement tasks, including fist-to-palm sequencing, finger-to-thumb sequencing, and disrupted intentional or volitional movements. The tasks used by the neuropsychologist to assess for premotor involvement may be discussed in the literatures on executive functions (see Stuss and Knight 2013) or even effortful control, whereas the assessment of the motor cortex relies more on measures of strength (e.g., using a dynamometer) or tonicity (e.g., passive range of motion). The premotor deficits resulting from pathology within the distribution of the middle cerebral artery are largely conveyed at the contralateral face and upper extremity, whereas these deficits at the lower extremity may be prominent with involvement within the distribution of the anterior cerebral artery.

A posterior cerebral feed from the middle cerebral artery travels along the Sylvian fissure to the posterior temporal and inferior parietal region. The artery is named after the cerebral region that it ultimately supplies, which is familiar to us as the angular gyrus. Thus, the angular artery is relevant to neuropsychological deficits within the left posterior temporal gyrus, including Wernicke's dysphasia or deficits in comprehending speech where a "word salad" of fluent discourse may be expressed by the patient. The homologous branches of the right middle cerebral artery, upon occlusion, raise the specter for auditory affect agnosia, receptive dysprosodia, and dysmusia.

The more distal branches feed the angular gyrus. This area within the left cerebral hemisphere is rich in neuropsychological functions which, when lost, are more familiar as "learning disabilities." This angular gyrus within the left cerebral hemisphere is critical for reading, writing, and the performance of arithmetical operations or calculations. Pathology here may underlie dyslexia with dysgraphia and/or dyscalculia, whereas a cerebrovascular accident within the left posterior cerebral artery raises concerns for dyslexia without dysgraphia. The latter involves severing or disconnection of the occipital cortices from the angular gyrus just as the former involves the demise of the cortical region itself. Pathology within this branch of the left middle cerebral artery will prompt the neuropsychological evaluation for components of Gerstmann's syndrome (Gerstmann 1940, 1957; see also Cabeza et al. 2012), and this includes the assessment for ideomotor dyspraxias.

Similar pathology within this branch of the right middle cerebral artery will prompt assessment for facial agnosia, geographical confusion, and constructional dyspraxia. Also relevant here and especially from discrete lesions is the possibility of a disconnection syndrome as the tertiary association area within the angular gyrus conveys access of one sensory system with the emergence of analysis from the associated sensory processing completed within each modality. Visual analysis may be disconnected from somatosensory analysis and from auditory analysis with pathology cutting the links within the dorsal and ventral pathways, respectively. These "disconnection syndromes" are, to some, the "holy grail" for the neuropsychologist where knowledge of functional cerebral systems allows for discovery of provocative clinical syndromes which otherwise may go unnoticed or undiscovered.

Infarcts within the distribution of the middle cerebral arteries are mostly associated with ipsilateral internal carotid artery disease (Wodarz 1980; see also Joinlambert et al. 2012). This includes vascular compromise secondary to internal carotid artery or middle cerebral artery stenosis. However, Joinlambert et al. (2012) identified several other factors leading to hemodynamic compromise. Among these were bradycardia, atrial fibrillation, hypotension, dehydration, anemia, and heart failure. The authors conclude that cardiovascular monitors, such as blood pressure, are important in the disease process to insure the maintainance of sufficient brain perfusion.

Posterior Cerebral Artery

The posterior cerebral artery projects along the midline at the back of each cerebral hemisphere extending into the deep-seated cerebral structures along the watershed area within each cerebral hemisphere (see Fig. 32.8). Branches of the posterior cerebral artery are most remarkable for their provision of oxygen saturation and nutrition to the primary visual projection cortices within either cerebral hemisphere known specifically as the striate or calcarine cortex. The visual projections are topographically arranged across the left visual field within the right cerebral hemisphere and across the right visual field within the left cerebral hemisphere. This cortex is found at midline within each occipital lobe with foveal or center field projections found near the outer convexities and with more and more peripheral visual field projections lying deeper (rostral) within these projections. Thus, the specific location of an infarct within the distribution of the posterior cerebral artery signifies the probable location of the corresponding visual field defect with deeper lesions affecting the peripheral visual field.

Beyond the expectations for visual field defects arising from pathology within the posterior cerebral artery distribution supplying the visual projections and primary occipital cortex, visual agnosia of one or another form is part of the differential diagnostic concerns (e.g., Martinaud et al. 2012). Visual agnosias arise from pathology within the posterior cerebral artery distribution supplying the secondary and tertiary association cortices with increasing complexity of the analytical deficit as the lesion extends into more complex association cortex at the tertiary areas. However, the tertiary association cortices extend into the arterial watershed areas where they may benefit from dual arterial blood supplies (e.g., posterior cerebral artery and middle cerebral artery territories). Specific cortical regions have been identified which respond to faces at the fusiform face area (Puce et al. 1995; Kanwisher et al. 1997) and at the occipital face area (Gauthier et al. 2000). Visual word images are processed within the visual word form area (Cohen et al. 2000). Visual processing of body parts occurs within the extrastriate body area (Downing et al. 2001), buildings and scenes within the parahippocampal place area (Epstein and Kanwisher 1998), and common objects in the lateral occipital complex (Malach et al. 1995).

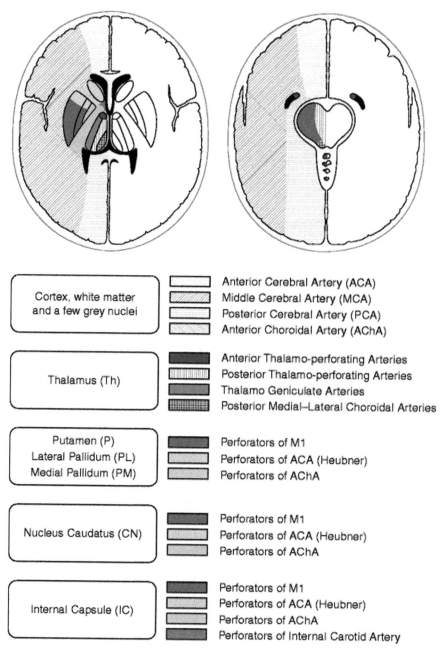

Fig. 32.8 Vascular territories and watershed areas of the cerebral arteries. (Originally published in *Vascular Territories,* Bradac and Gianni Boris 2011, p. 88)

Moreover, cerebrovascular accidents in these regions commonly provoke elevated activation of the surrounding cellular structures with perhaps one in four or five patients developing visual hallucinations. The examiner will do well to specifically inquire about the location of the image and the directional movement of the image as functional correlates for the pathologic lesion. For the left posterior cerebral artery distribution, the visual paresthesias or visual formesthesias will characteristically appear within the right visual field and convey an emotional overlay of interest of pleasure, whereas the events generated from pathology within the supply distribution of the right posterior cerebral artery are typically within the left or "sinister" visual field and accompanied by negative emotional attributions (fear, threats, control issues; Walters et al. 2006) and elevations in sympathetic drive (tachycardia, sweating, dynamic elevation in blood glucose). Episodic elevations in sympathetic tone with tachycardia, heightened blood pressure, or a surge in glucose levels may provide the initial diagnostic clues for a right occipital temporal event, and this may occur with behavioral features of fear or apprehension. Oscillating visual events within the pathological visual field may lower the threshold and increase the probability for these events and caution may be warranted in the type and location of rapidly fluctuating luminance or illumination levels.

The rostral or anterior feeds from the posterior cerebral artery within each brain supply the splenium of the corpus callosum. If the cerebrovascular accident encroaches upon the splenium, then a cerebral disconnect may result where each brain is absent of the visual information and processing specialization of the homologous brain region. This was expressed earlier on in the positions taken by Hughlings Jackson (1879) and A. R. Luria (1966, 1973, 1980) on the location of the lesion in dyslexia. If the reading problem results essentially from a visual deficit, then the diagnosis and treatment would be for one or another of the *visual agnosias* or anopsias. This might include visual letter recognition deficits subsequent to a left occipital or fusiform lesion (e.g., Park et al. 2012). In contrast, lesions within the language processing areas supplied by the left middle cerebral artery result in one or another of the *aphasic alexias* (see Luria 1980; see also Ardila 2012). The second diagnostic system involves the basic distinction between dyslexia of subcortical origin and dyslexia of cortical origin. This second diagnostic system (e.g., Imtiaz et al. 2001) distinguishes among *dyslexia with dysgraphia*(a cortical lesion within the distribution of the middle cerebral artery) and *dyslexia without dysgraphia*(a subcortical lesion within the distribution of the posterior cerebral artery).

The posterior cerebral artery supplies the midline areas of the temporal lobe through its anterior temporal artery branch extending to supply the amygdala. This structure, of course, is richly associated with emotional learning and emotional kindling phenomena. Cerebrovascular accidents encroaching on this region may provide a rich set of diagnostic concerns for the neuropsychologist and the area is historically associated with what were once more broadly considered to be psychiatric disorders. The posterior temporal artery branch of the posterior cerebral artery supplies the hippocampus where cerebrovascular accident may impact the consolidation or learning of new information within that cerebral hemisphere. This might affect verbal learning with infarction of the left posterior cerebral artery and the learning of faces or places with infarction of the right posterior cerebral artery.

Other branches of the posterior cerebral artery include the parieto-occipital artery, where obstruction may yield a visual processing defect with respect to the contralateral lower visual quadrant or specifically a lower quadrantanopsia for that half-field.

Hemispheric Laterality and Stroke

Hedna et al. (2013) raised the concern that probability of stroke may differ for the left and right cerebral hemisphere. The authors point out structural and mechanical differences in the vascular supply to each brain, where velocity differences exist in the carotid circulation and direct branching of the left common carotid artery from the aorta. Based on these anatomical and mechanical differences, they assessed the probability that large-vessel ischemia (including cardioembolism) is more common in the territorial distribution of the left middle cerebral artery. The analyses included 317 ischemic stroke patients, excluding individuals with hemorrhagic stroke, stroke of undetermined etiology, cryptogenic stroke, and bilateral ischemic strokes. Laterality and vascular distribution were correlated with outcomes using a logistic regression model. The etiologies of the large-vessel strokes were atherosclerosis and cardioembolism.

Statistically, significant differences were found between the hemispheres for the overall event frequency, mortality, National Institutes of Health Stroke Scale (NIHSS) score, Glasgow Coma Scale score, and rate of mechanical thrombectomy interventions. The authors report that "left hemispheric strokes (54%) were more common than right hemispheric strokes (46%; $p=0.0073$), and had higher admission NIHSS scores ($p=0.011$), increased mortality ($p=0.0339$), and higher endovascular intervention rates ($p \leq 0.0001$)." Not only are ischemic strokes more frequent in the distribution of the left middle cerebral artery (122 vs. 97; $p=0.0003$), they are also often related to a worse outcome than their right hemispheric counterparts. Among these strokes, those impacting the left middle cerebral artery distribution were most common. Although the anatomical and mechanical contributions to hemispheric laterality differences in stroke, alternative explanations for the findings may exist. Among these are neuropsychological features of right hemispheric stroke, where insight to deficits is poor with clinical expectations for anosognosia and anosodiaphoria. Indeed, an individual poor insight to deficits may be unlikely or less likely to seek help or medical intervention. The extent to which right cerebral stroke goes unnoticed or undiagnosed remains to be established through rigorous statistical analysis.

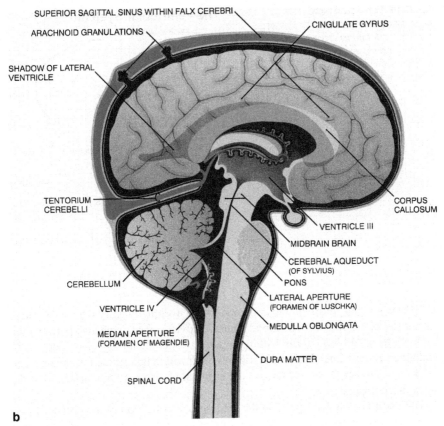

SUPERIOR SAGITTAL SINUS WITHIN FALX CEREBRI

CINGULATE GYRUS

ARACHNOID GRANULATIONS

SHADOW OF LATERAL
VENTRICLE

TENTORIUM
CEREBELLI

CORPUS
CALLOSUM

VENTRICLE III

MIDBRAIN BRAIN

CEREBRAL AQUEDUCT
(OF SYLVIUS)

CEREBELLUM

PONS

VENTRICLE IV

LATERAL APERTURE
(FORAMEN OF LUSCHKA)

MEDULLA OBLONGATA

MEDIAN APERTURE
(FORAMEN OF MAGENDIE)

DURA MATTER

SPINAL CORD

b

Fig. 32.9 The ventricular system of the brain. (Originally published in *Metabolic Encephalopathy*, McCandless and David 2009, p. 4)

Ventricular Pathology

Nervous tissues are constantly bathed and supported by "waters" in the form of protein-free cerebrospinal fluid, a substance similar to blood plasma. Cerebrospinal fluid is produced by the choroid plexus, a system of membranes lining the ventricles, which includes the pia mater, blood capillaries, and ependymal cells. Collectively, this apparatus produces about 20 ml of cerebrospinal fluid per hour. This fluid collects in the lateral ventricles, within the cerebral hemispheres, and flows downward through openings (foramen) and into the third ventricle above the roof of the mouth (see Fig. 32.9). Subsequently, the "waters" flow into the fourth ventricle within the brain stem. Additional cerebrospinal fluid bathes the central canal within the spinal cord. However, much of the "waters" from the brain now exit as used or "dirty" cerebrospinal fluid through the foramen of Luschka and Magendie (see Fig. 32.10). This exodus from the innermost regions of the brain merges the fluid into the cister-

Fig. 32.10 Left lateral view
of the ventricular system
showing the foraminal access
to each ventricle and out to
the cisternae via the foramen
of Luschka and Magendie

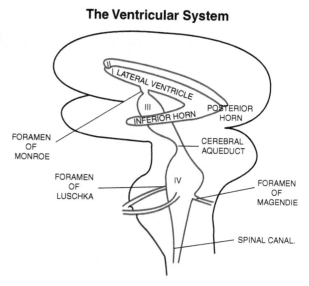

The Ventricular System

LATERAL VENTRICLE

III

INFERIOR HORN

POSTERIOR HORN

FORAMEN OF MONROE

CEREBRAL AQUEDUCT

FORAMEN OF LUSCHKA

IV

FORAMEN OF MAGENDIE

SPINAL CANAL.

nae (sewer), which surround the brain and which eventually carry the waste waters toward the blood sinuses. The blood sinuses provide for an exchange between the fresh oxygenated blood borne within the arterial system and the spent, deoxygenated blood and used cerebrospinal fluid traveling within the venous system of blood vessles on the way to the kidney, for the elimination of waste products, and to the lungs for reoxygenation.

The brain is wrapped tightly in the meninges, a collection of membranes reminiscent of a sausage skin. The meninges provide structural support and protection for the Jell-O-like consistency of brain matter, which is also supported by the fluid containing network of ventricles, cisterns, and blood sinuses. The outer most layer of the meninges is the dura mater or "tough mother," which consists of the periostial layer (innervated and vascularized) and the meningeal layer (flat cells; not vascularized). The separation between the dural layers forms the chamber which functions as a blood sinus containing dirty used and deoxygenated blood. This blood will flow down from these chambers into the jugular veins descending at the front of the neck and onto the kidney and lungs for cleaning and reoxygenation.

The middle layer of the meninges is the arachnoid layer, which is nonvascular and noninnervated. The chamber formed above the arachnoid layer and the below the dura then is the blood sinus. Below the arachnoid layer are the cisterns filled with spent cerebrospinal fluid. The innermost layer of the meninges, adhering closely to brain tissue, is the pia mater or "little mother," a highly vascularized and nutritive membrane forming part of the choroid plexus. Cerebrospinal fluid, under pressure, is forced through the arachnoid villa and into the blood sinus, where it mixes with the used blood supply and eventually descends down the jugular veins.

This ventricular system very much resembles a series of lakes (ventricles) where the cerebrospinal fluid moves, under pressure, through the system, eventually pass-

Fig. 32.11 The flow of cerebrospinal fluid through the ventricular system into the cisterns and eventually joining the venous blood supply in the sinus

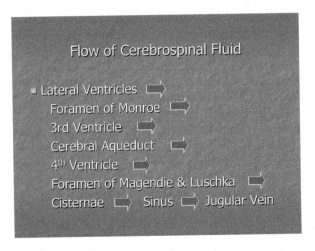

ing to the cisterns on the outside of the brain, and eventually being forced out of the cisterns and into the blood sinuses for drainage down the jugular veins. The "wash" accumulates toxic substances and by-products of the multitude of metabolic functions performed by the nervous system. The waste products include those that result from the breakdown of neurotransmitters or chemical degradation through enzymatic action (e.g., cortisol and the catecholamines) and which accumulate in the cerebrospinal fluid as a function of turnover process. Some of these by-products may be acquired from the urine or from the venous blood to assess the relationships between cerebral dynamics and psychological functions (e.g., reactive anger, depression, schizophrenia). These by-products also mark the ebb and flow of metabolic functions across the day in the form of circadian rhythms or dynamic processing demands. The spent cerebrospinal fluid acquired along its pathway leading to the elimination of these by-products and toxins from the body (e.g., urine) provides a ready resource to the neuroscientist and psychophysiologist or psychopharmacologist investigating state- and trait-dependent relationships in behavioral, cognitive, and emotional processing.

The flow of cerebrospinal fluid through the ventricular system is depicted in Fig. 32.11. Fluid produced by the choroid plexus lining the ventricles maintains a steady hydraulic force or pressure relieved by inherently leaky nature of this system. Some of the fluid leaks out at the level of the spinal cord around nerve roots and peripheral nerves exiting at regular levels through the vertebral column. But, much of the pressure is relieved at each ventricle with dynamic flow into the lower-level ventricle. Ultimately, the fluid must exit the ventricular system through the foramen of Magendie and Luschka and enter the cisterns. Brain pathology and neuropsychological deficits may arise from any process which impedes the flow of the fluid and thereby affects the pressure levels at the wall of the ventricle and emanating via physical forces against the soft and vulnerable neural tissues. Elevated cerebrospinal fluid pressures are reflected in the clinical disorder of hydrocephaly, and this may occur with relative restriction to one or another ventricle, one or the other cerebral hemisphere, or, more commonly, through generalized pressure effects. Hydrocephalus may be visualized using imagery techniques, with enlarged ventricles in the absence of enlarged cortical sulci, for example. Pressure effects may also be evident in midline shift of the midsagittal sulcus separating the cerebral hemispheres and, commonly, with compression of the brain stem. Clinical correlates frequently present with lethargy, irritability, and/or confusion.

Cerebrospinal fluid functions collectively to minimize damage to the central nervous system. It functions as part of the blood–brain barrier and eliminates toxic substances and spent by-products from neuronal and glial metabolism. Ultimately, it moves these substances toward the venous drainage. However, simply blocking a foramen may essentially inflate the ventricle above the blockage and potentially crush or injure surrounding brain systems via physical forces and compression.

Chapter 33
Neurodegenerative Disorders

The integrity of neuropsychological systems is subject to progressive decline as a function of multiple underlying dementia processes, defined by neurocognitive decline in memory, judgment, language, complex motor skills, and other neuropsychological processes. The dementias involve progressive neurodegenerative changes over time subsequent to neuronal or glial cell loss or malfunction. Although multiple maladies may underlie neurodegenerative decline, Alzheimer's disease is considered the most common cause of dementia, representing about 60% of all dementias identified at the clinical assessment. Estimates of the number of people affected by Alzheimer's disease in the USA vary according to the National Institute on Aging, with numbers ranging from 2.4 to 4.5 million, with the discrepancies drawn from operational differences in diagnostic criterion and measurement (Rodgers 2008). Other neurodegenerative disorders include vascular dementia, arising from recurrent stroke or vascular pathologies, dementia with Lewy bodies, alcoholic or substance use dementia, post-concussive dementia subsequent to head trauma, frontotemporal dementia, and many other progressive or recurrent brain disorders.

The following summary statement on demographics and incidence is provided from the Alzheimer's Foundation of America (Powers et al. 2008): "The incidence of dementia doubles approximately every five years in individuals between the ages of 65 and 95 and by some estimates may reach nearly 50 percent by age 85 (Evans et al. 1989). Alzheimer's disease is the most common cause of dementia among people aged 65 and older. Alzheimer's disease is not a normal part of aging; however, age is the greatest known risk factor. And with the older population on the threshold of a boom, dementia is an especially significant issue (Rodgers 2008). A 2005 United States Census Bureau report on aging in the United States notes that the population age 65 and older in 2030 is expected to be twice as large as in 2000, growing from 35 million to 72 million and representing nearly 20 percent of the United States population at the latter date (Wan et al. 2005). According to the latest government statistics, Alzheimer's disease is now the sixth leading cause of death in

© Springer International Publishing Switzerland 2015
D. W. Harrison, *Brain Asymmetry and Neural Systems*,
DOI 10.1007/978-3-319-13069-9_33

Fig. 33.1 Anatomical correlates of Alzheimer's disease

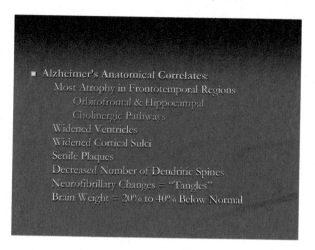

- Alzheimer's Anatomical Correlates:
    Most Atrophy in Frontotemporal Regions
        Orbitofrontal & Hippocampal
        Cholinergic Pathways
    Widened Ventricles
    Widened Cortical Sulci
    Senile Plaques
    Decreased Number of Dendritic Spines
    Neurofibrillary Changes = "Tangles"
    Brain Weight = 20% to 40% Below Normal

the United States, rising one notch from seventh place in 2005, while the number of deaths from other chronic conditions, including diabetes, declined in 2006."

Although the demographics of aging are initially encouraging with more people living longer, there remains concern that the quality of life may decline both for the elderly individual with mild cognitive impairment or more advanced dementia and for the caregiver with increased demands for assistance. Incremental gains in life span or longevity may ultimately translate to incremental changes in the associated disabilities and financial burden for the family and for society at large. Otherwise said, the population may be older but less healthy, where the burden of care distributes across those willing to provide assistance and through the mandates of government.

Disruption of cholinergic projections has been related to several dementia subtypes and especially to Alzheimer's disease. Indeed, Alzheimer's disease has been identified as a "cholinergic dementia" (see Coyle et al. 1983) with loss of cholinergic pathways correlating with cognitive decline. Cholinergic systems are also affected in subcortical vascular dementia, whereas the processes appear to differ. For example, Kim et al. (2013) found atrophy within the cholinergic projections of the nucleus basilis of Meynert as the predominant contributor to cognitive impairments in Alzheimer's disease, whereas the cognitive dysfunction of subcortical ischemic vascular dementia was related to compromise of subcortical cholinergic fibers rather than to the nucleus itself. The regions affected initially appear to be at the hippocampal systems involved in memory consolidation and at the basal forebrain or orbitofrontal region. Some of the predominant anatomical changes with Alzheimer's disease are identified in Fig. 33.1.

Fig. 33.2 General etiological
categories of dementia

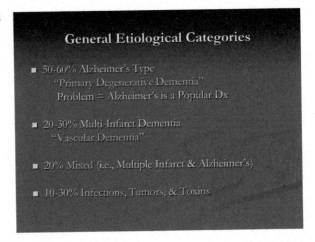

Figure 33.2 provides for comparison of the general etiological categories of dementia. Alzheimer's disease or "primary degenerative dementia" accounts for about 50–60% of the dementia cases (see Francis et al. 1999), whereas about 20–30% of the cases may be identified as multi-infarct dementia. Perhaps, some 20% of the cases reflect mixed dementia disorders, leaving somewhere between 10 and 30% of the dementias arising from other causative origins, including infection, neoplasm, and toxins.

Chapter 34
Traumatic Brain Injury

It was in a June 1934 exhibition game when the pitcher Ray White of the Norfolk Tars threw a fastball striking the Yankees' Lou Gehrig above the right eye, knocking him unconscious (see Fig. 34.1). Although the head injury was not a singular event of this kind for Gehrig, it was instrumental in raising some controversy decades later that "Lou Gehrig's disease," otherwise known as amyotrophic lateral sclerosis or "ALS," was at least confounded by the history of multiple head injuries of traumatic origin (McKee et al. 2010; see also Bernstein 2013). A connection was found in a project investigating the effects of brain trauma in National Football League players (McKee et al. 2010). Fourteen former players over the past 50 years were found who had been diagnosed with ALS. No causal relationship was found between head injury and the eventually fatal motor neuron disease. Also, the researchers did conclude that the players did not in fact have the disease (ALS). However, they did find evidence indicating that brain trauma can cause brain degeneration resulting in an "ALS-like disorder" (Haley 2003). Beyond this conclusion, the authors went further to suggest that Lou Gehrig himself might have suffered from something other than ALS as a result of the multiple blows to the head which he had sustained across his career (McKee et al. 2009).

Sports-related traumatic brain injuries are all too common. Yet, the victims of sports injuries largely remain anonymous. One university football star was referred for a neuropsychological evaluation with concerns for an abrupt onset of difficulties in his college coursework with failing grades. The speech deficiencies that had been attributed to ethnicity or cultural differences in language were instead found to be the content of a global aphasic disorder, subsequent to a vicious blow to the head in spring football camp. The young man was being discharged from the football program and expelled from his degree-granting program for playing his best on the field and doing his best in his academic studies. Though previously he was heralded on the football field for the sport, the awareness and appreciation of traumatic brain injury was inadequate, as he returned to his home without the abilities to comprehend or to express speech.

© Springer International Publishing Switzerland 2015
D. W. Harrison, *Brain Asymmetry and Neural Systems,*
DOI 10.1007/978-3-319-13069-9_34

The Yankees' Lou Gehrig was helped off the field after being struck in the head by a fastball from pitcher Ray White of the Norfolk Tars in a June 1934 exhibition game. The ball caught Gehrig above the right eye, knocking him unconscious.

Fig. 34.1 On June 29, 1934, Lou Gehrig is helped off the field after being struck in the head by an inside fastball thrown by Norfolk Tars pitcher Ray White at a Yankees–Tars exhibition game at Bain Field. The ball caught Gehrig above the right eye, knocking him unconscious. Lenny Goodman of Norfolk is shown carrying the bucket in the front of this shot. Doc. Earle V. Painter, Yankees trainer is the gentleman on the left. Copyright release courtesy The Virginian-Pilot

In another episode in college sports, the remarkably talented quarterback had a recognizable head injury (based on observations of his game play), which had been the content for didactic discussion on the neuropsychology practicum team. This apparent injury occurred 2 weeks prior to the quarterback's dismissal from the team for unsportsmanlike conduct, where he had kicked a defensive lineman in the head following the end of the play. Shortly thereafter, he appeared again in the local news for brandishing a firearm at a McDonald's Restaurant near his hometown. All of this appeared remarkably out of character for this tried, tested, and formerly disciplined team leader.

Indeed, it is all too common for college teams to lose some of their better athletes following spring training and subsequent to the seasonal effects and their efforts to earn a starting position on the team. Although these examples, and others like them, represent case studies as presented here, they serve to provide examples of some of the defining features common to traumatic brain injuries:irritability, poor emotional regulation, and depression. With transducers located in the football players' helmets sometimes measuring over 100 Gs (Duma 2012) and with the potential for as many as 2000 such concussive blows to the helmet (and presumable the head and brain), traumatic brain injury may represent a likely rather than an improbable event. Indeed, these negative outcomes of our most popular sports seem all the more probable and expected. Many clinicians and many of those within neuroscience have held onto mythical views towards head injury, including the notion that it requires a loss of consciousness or at least an altered level of consciousness; that the location of the injury is restricted to the area of impact to the head; and that these injuries are restricted in time to the event and a subsequent recovery period extending across a 6-month window.

Instead, head injury may render the individual subject to active and potentially progressive decline which, in some cases, may only be appreciated later in life. For example, many individuals diagnosed with "Alzheimer's" or "dementia" are notable for their distinct history of prior head trauma (e.g., Lee et al. 2013). The physical forces, rather than focal impact at the point of contact, may be substantially more displaced affecting the integrity of diffuse and generalized brain systems. These changes may be evident in diffuse axonal injury and dendritic damage arising from torsion, stretching, and shearing forces. The injury may be aggravated by subsequent edema or swelling of the brain tissues, compression of the fluid containing spaces, and alteration of the circulation of cerebrospinal fluid and blood flow or ischemic events.

The secondary consequences of head injury surely include, for some, the loss of established relationships (e.g., with their spouse or significant other), and negative impact upon employment or career advancement. These aspects of traumatic brain injury are substantial with a long history of generally poor appreciation by society at large. This is perhaps more evident in the development of awareness for these features within the National Football League (e.g., Hart et al. 2013; McKee et al. 2010; see also Bernstein 2013). For many though, and perhaps for the majority of those with traumatic brain injury, a good recovery and return to a previous lifestyle may be anticipated through a gradual restoration of normal brain functions and processes.

Military conflicts across the globe have deepened the crisis in traumatic brain injury management and treatment and neuropsychological services are often in high demand within the Veterans Administration Medical Centers across the country. Current and recent military conflicts have seen the effective use of improvised explosive devices or "IEDs" as weapons frequently used by combatants and terrorists (Duckworth et al. 2012; Gupta and Przekwas 2013). Indeed, blast-induced traumatic brain injury has become the "signature wound of the war on terror" (Bhattacharjee 2008). Gupta and Przekwas (2013) note, in their review of the literature, that 20% of service members or 320,000 of the deployed force potentially suffer from traumatic brain injury (Tanielian and Jaycox 2008: RAND report). Blast injuries were found to account for about 70% of wounded service member's injuries. About 80% of these were classified as mild traumatic brain injuries, with penetrating or severe head injuries accounting for some 2.8% of the wounds. Although the majority of mild traumatic brain injury cases are expected to recover, persistent symptoms are to be expected after injury, including chronic dizziness, fatigue, headaches, and impaired memory or cognition (e.g., Heltemes et al. 2012).

Beyond the apparent risks from contact sports and soccer, half of all traumatic brain injuries arise from accidents involving automobiles, motorcycles, bicycles, and pedestrians (e.g., see Asemota et al. 2013). Traumatic brain injury remains the leading cause of death and disability in children and young adults in the USA. In those individuals 75 years of age and older, the majority of head injuries are caused by falls. Some 20 % of traumatic brain injuries are due to violence, including child abuse and gunshots. About 3 % arise secondary to sports-related injuries.

The defining criteria for the diagnosis of traumatic brain injury are derived from evidence of neuropsychological deficits following the assessment of learning and memory, executive functions, and the identification or exclusion of specific neuro-psychological syndromes discussed elsewhere in this book. Even mild concussive injury may result in permanent brain damage. Thus, the diagnosis of head injury is one which identifies an active and potentially progressive disorder, often requiring follow-up evaluation on a yearly basis not to exceed a 5-year interval. Although these guidelines are crude, they have served many individuals over the years and provide an ongoing interface between the patient and their health-care providers. These relationships may be useful in their implied role for monitoring and detection. They may also prove useful in supportive care and referral, including the treatment of depressive disorders and the provision of counseling, if needed. Clearly, the resources extend beyond these professionals and more importantly to the caregiver and to community resources for those with head injury, including the Brain Injury Society active in many communities and on the World Wide Web.

References

Abdelrahman, R. S. (2012). Prevention of shivering during regional anaesthesia: Comparison of Midazolam, Midazolam plus ketamine, Tramadol, and Tramadol plus Ketamine. *Life Science Journal, 9*(2), 132–139.

Abercrombie, H. C., Schaefer, S. M., Larson, C. L., Oakes, T. R., Lindgren, K. A., Holden, J. E., Perlman, S. B., Turski, P. A., Krahn, D. D., Benca, R. M., & Davidson, R. J. (1998). Metabolic rate in the right amygdala predicts negative effect in depressed patients. *NeuroReport, 9*, 3301–3307.

Abrams, H. (1994). *The President has been shot: Confusion, disability, and the 25th Amendment.* Redwood City: Stanford University Press.

Adair, J. C., Na, D. L., Schwartz, R. L., & Heilman, K. M. (2003). Caloric stimulation in neglect: Evaluation of response as a function of neglect type. *Journal of the International Neuropsychological Society, 9*, 983–988.

Adolphs, R., & Tranel, D. (2000). *The amygdala* (ed. J. P. Aggleton, pp. 587–630). New York: Oxford University Press.

Adolphs, R., Damasio, H., Tranel, D., & Damasio, A. R. (1996). Cortical systems for the recognition of emotion in facial expressions. *Journal of Neuroscience, 16*(23), 7678–7687.

Adolphs, R., Tranel, D., Hamann, S., Young, A. W., Calder, A. J., Phelps, E. A., & Damasio, A. R. (1999). Recognition of facial emotion in nine individuals with bilateral amygdale damage. *Neuropsychologia, 37*(10), 1111–1117.

Adolphs, R., Jansari, A., & Tranel, D. (2001). Hemispheric perception of emotional valence from facial expressions. *Neuropsychology, 15*(4), 516–524.

Adrian, E. D., & Matthews, B. H. C. (1934). The Berger rhythm: Potential changes from the occipital lobes in man. *Brain, 4*, 355–385.

Aeby, A., Van Bogaert, P., David, P., Balériaux, D., Vermeylen, D., Metens, T., & De Tiège, X. (2012). Nonlinear microstructural changes in the right superior temporal sulcus and lateral occipitotemporal gyrus between 35 and 43 weeks in the preterm brain. *NeuroImage, 63*(1), 104–110. (ISSN 1053-8119, 10.1016/j.neuroimage.2012.06.013).

Agnew, R. (2012). Dire forecast: A theoretical model of the impact of climate change on crime. *Theoretical Criminology, 16*(1), 21–42.

Agrawal, Y., Carey, J., Della Santina, C., Schubert, M., & Minor, L. (2009). Disorders of balance and vestibular function in US adults. *Archive of Internal Medicine, 169*, 938–944.

Aguirre, G. K., & D'Esposito, M. (1999). Topographical disorientation: A synthesis and taxonomy. *Brain, 122*, 1613–1628.

Aguirre, G. K., Dettre, J. A., & Alsop, D. C., et al. (1996). The parahippocampus subserves topographical learning in man. *Cerebral Cortex, 6*, 823–829.

Agustín-Pavón, C., Braesicke, K., Shiba, Y., Santangelo, A. M., Mikheenko, Y., Cockroft, G., Asma, F., Clarke, H., Man, M., & Roberts, A. C. (2012). Lesions of ventrolateral prefrontal or anterior orbitofrontal cortex in primates heighten negative emotion. *Biological Psychiatry, 72*(4), 266–272. doi:10.1016/j.biopsych.2012.03.007.

© Springer International Publishing Switzerland 2015
D. W. Harrison, *Brain Asymmetry and Neural Systems,*
DOI 10.1007/978-3-319-13069-9

Aiello, M., Jacquin-Courtois, S., Merola, S., Ottaviani, T., Tomaiuolo, F., Bueti, D., Rossetti, Y., & Doricchi, F. (2012). No inherent left and right side in human 'mental number line': Evidence from right brain damage. *Brain, 135*(8), 2492–2505. doi:10.1093/brain/aws114.

Aimola, L., Schindler, I., Simone, A. M., & Venneri, A. (2012). Near and far space neglect: Task sensitivity and anatomical substrates. *Neuropsychologia, 50*(6), 1115–1123.

Akiyoshi, J., Hieda, K., Aoki, Y., & Nagayama, H. (2003). Frontal brain hypoactivity as a biological substrate of anxiety in patients with panic disorders. *Neuropsychobiology, 47*, 165–170.

al'Absi, M., & Petersen, K. L. (2003). Blood pressure but not cortisol mediates stress effects on subsequent pain perception in healthy men and women. *Pain, 106*(3), 285–295.

Albrecht, J., Demmel, M., Schöpf, V., Kleemann, A. M., Kopietz, R., May, J., & Wiesmann, M. (2011). Smelling chemosensory signals of males in anxious versus nonanxious condition increases state anxiety of female subjects. *Chemical Senses, 36*, 19–27.

Alden, J. D., & Harrison, D. W. (1993). An initial investigation of bright light and depression: A neuropsychological perspective. *Bulletin of the Psychonomic Society, 31*(6), 621–623.

Alden, J. D., Crews, W. D., & Harrison, D. W. (1991). Cerebral asymmetry in dementia: Effect of context on hemi-attention. *Perceptual and Motor Skills, 72*, 802.

Alden, J. D., Harrison, D. W., Snyder, K. A., & Everhart, D. E. (1997). Age differences in intention to left and right hemispace using a dichotic listening paradigm. *Neuropsychiatry, Neuropsychology, and Behavioral Neurology, 10*, 239–242.

Alger, B. E. (1984). Hippocampus. Electrophysiological studies of epileptiform activity in vitro. In R. Dingledine (Ed.), *Brain slices* (pp. 155–199). New York: Plenum Press.

Allman, M. J., & Meck, W. H. (2012). Pathophysiological distortions in time perception and timed performance. *Brain, 135*(3), 656–677 (First published online September 15, 2011.). doi:10.1093/brain/awr210).

Allman, M. J., DeLeon, I. G., & Wearden, J. H. (2011). A psychophysical assessment of timing in individuals with autism. *American Journal on Intellectual and Developmental Disabilities, 116*, 165–178.

Altenmuller, E., Schurmann, K., Lim, V. K., & Parlitz, D. (2002). Hits to the left, flops to the right: Different emotions during listening to music are reflected in cortical lateralization patterns. *Neuropsychologia, 40*(13), 2242–2256.

Amedi, A., Raz, N., Pianka, P., Malach, R., & Zohary, E. (2003). Early 'visual' cortex activation correlates with superior verbal memory performance in the blind. *Nature Neuroscience, 6*(7), 758–766. doi:10.1038/nn1072.

American Forces Press Service. (2006). Army activates first interrogation Battalion. An April press release.

American Psychiatric Association. (2000). *Diagnostic criteria from DSM-IV-TR*. Washington, DC: American Psychiatric Association.

American Psychiatric Association. (2013). *Diagnostic and statistical manual of mental disorders* (5th ed.). Washington, DC: American Psychiatric Association.

Amunts, K. (2010). Structural indices of asymmetry. In K. Hugdahl & R. Westerhausen (Eds.), *The two halves of the brain: Information processing in the cerebral hemispheres* (pp. 145–175). Cambridge: MIT Press.

Anand, B. K., & Brobeck, J. R. (1951). Hypothalamic control of food intake in rats and cats. *Yale Journal of Biology and Medicine, 24*, 123–140.

Anand, B. K., Dua, S., & Schoenberg, K. (1955). Hypothalamic control of food intake in cats and monkeys. *The Journal of Physiology, 127*, 143–152.

Ancoli-Israel, S., Rissling, M., Trofimenko, V., Natarajan, L., Parker, B., Lawton, S., Desan, P., & Liu, L. (2011). Light treatment prevents fatigue in women undergoing chemotherapy for breast cancer. *Support Care Cancer, 20*(6), 1211–1219.

Anderson, C. A. (1987). Temperature and aggression: Effects on quarterly, yearly, and city rates of violent and nonviolent crime. *Journal of Personality and Social Psychology, 52*(6), 1161–1173.

Anderson, C. A. (1989). Temperature and aggression: Ubiquitous effects of heat on occurrence of human violence. *Psychological Bulletin, 106*(1), 74–96. doi:10.1037/0033-2909.106.1.74.

Anderson, C. A. (2001). Heat and violence. *Current Directions in Psychological Science, 10*(1), 33–38.

Anderson, S. W., Bechara, A., Damasio, H., Tranel, D., & Damasio, A. R. (1999a). Impairment of social and moral behaviour related to early damage in human prefrontal cortex. *Nature Neuroscience, 2*, 1032–1037.

Anderson, J. M., Gilmore, R., Roper, S., Crosson, B., Bauer, R. M., Nadeau, S., Beversdorf, D. Q., Cibula, J., Rogish, M. III, Kortencamp, S., Hughes, J. D., Gonzalez Rothi, L. J., & Heilman, K. M. (1999b). Conduction aphasia and the arcuate fasciculus: A reexamination of the Wernicke–Geschwind model. *Brain and Language, 70*(1), 1–12.

Anderson, A. K., Christoff, K., Stappen, I., Panitz, D., Ghahremani, D. G., et al. (2003). Dissociated neural representations of intensity and valence in human olfaction. *Nature Neuroscience, 6*, 196–202.

Andersson, S., & Finset, A. (1998). Heart rate and skin conductance reactivity to brief psychological stress in brain damaged patients. *Journal of Psychosomatic Research, 44*, 645–656.

Andrews, G. D., & Lavin, A. (2006). Methylphenidate increases cortical excitability via activation of alpha-2 noradrenergic receptors. *Neuropsychopharmacology, 31*(3), 594–601.

Anthony, K., Reed, L. J., Dunn, J. T., Bingham, E., Hopkins, D., Marsden, P. K., & Amiel, S. A. (2006). Attenuation of insulin-evoked responses in brain networks controlling appetite and reward in insulin resistance the cerebral basis for impaired control of food intake in metabolic syndrome? *Diabetes, 55*(11), 2986–2992.

Anton, G. (1899). Über die Selbstwahrnehmung der Herderkrankungen des Gehirns durch den Kranken bei Rindenblindheit und Rindentaubheit. *Arch Psychiatrie Nervenkrankh, 32*, 86–127.

Appelros, P., Karlsson, G. M., & Hennerdal, S. (2007). Anosognosia versus unilateral neglect. Coexistence and their relations to age, stroke severity, lesion site and cognition. *European Journal of Neurology, 14*, 54–59. doi:10.1111/j.1468-1331.2006.01544.x.

Applebaum, D., Fowler, S., Fiedler, N., Osinubi, O., & Robson, M. (2010). The impact of environmental factors on nursing stress, job satisfaction, and turnover intention. *Journal of Nursing Administration, 40*(7/8), 323–328.

Archer, T., Ogren, S. O., Johansson, G., & Ross S. B. (1984). The effect of acute zimeldine and alaproclate administration on the acquisition of two-way active avoidance: Comparison with other antidepressant agents, test of selectivity and sub-chronic studies. *Psychopharmacology, 84*(2), 188–195.

Ardila, A. (2012). Neuropsychology of writing. In E. L. Grigorenko, E. Mambrino, & D. D. Preiss (Eds.), *Writing: A mosaic of new perspectives* (pp. 311–483). New York: Taylor & Francis.

Arnsten, A. F. (2009). Stress signaling pathways that impair prefrontal cortex and function. *Nature Reviews: Neuroscience, 10*, 410–422.

Arnsten, A. F., Paspalas, P., Gamo, N., Yang, Y., & Wang, M. (2010). Dynamic network connectivity: A new form of neuroplasticity. *Trends in Cognitive Sciences, 14*, 365–375.

Aron, A. R., & Poldrack, R. A. (2006). Cortical and subcortical contributions to stop signal response inhibition: Role of the subthalamic nucleus. *The Journal of Neuroscience, 26*, 2424–2433.

Aron, A. R., Behrens, T. E., Smith, S., Frank, M. J., & Poldrack, R. A. (2007a). Triangulating a cognitive control network using diffusion-weighted magnetic resonance imaging (MRI) and functional MRI. *The Journal of Neuroscience, 27*, 3743–3752.

Aron, A. R., Durston, S., Eagle, D. M., Logan, G. D., Stinear, C. M., & Stuphorn, V. (2007b). Converging evidence for a fronto-basal-ganglia network for inhibitory control of action and cognition. *The Journal of Neuroscience, 27*, 11860–11864.

Asemota, A. O., George, B. P., Bowman, S. M., Haider, A. H., & Schneider, E. B. (2013). Causes and trends in traumatic brain injury for United States adolescents. *Journal of Neurotrauma, 30*(2), 67–75.

Ashburn, A., Stack, E., Pickering, R. M., & Ward, C. D. (2001). A community-dwelling sample of people with Parkinson's disease: characteristics of fallers and non-fallers. *Age Ageing, 30*, 47–52.

Ashby, G. F., Isen, A. M., & Turken, A. U. (1999). A neuropsychological model of positive affect and its influence on cognition. *Psychological Review, 106*(3), 529–550.

Astur, R. S., Taylor, L. B., Mamelak, A. N., Philpott, L., & Sutherland, R. J. (2002). Humans with hippocampus damage display severe spatial memory impairments in a virtual Morris water maze. *Behavioral Brain Research, 132,* 77–84.

Atkinson, R. C., & Shiffrin, R. M. (1968). Human memory: A proposed system and its control processes. In K. W. Spence & J. T. Spence (Eds.), *The psychology of learning and motivation* (Vol. 2, pp. 89–195). New York: Academic Press.

Babinski, J. (1914). Contributions of cerebral hemispheric organization in the study of mental troubles. *Review Neurologique, 27,* 845–848.

Babinski, J. (1918). Anosognosie. *Rev Neurol (Paris), 31,* 365–367.

Badaruddin, D. H., Andrews, G. L., Bolte, S., Schilmoeller, K. J., Schilmoeller, G., Paul, L. K., et al. (2007). Social and behavioral problems of children with agenesis of the corpus callosum. *Child Psychiatry and Human Development, 38*(4), 287–302. doi:10.1007/s10578-007-0065-6.

Baehr, E. I., Rosenfield, J. P., Baehr, R., & Earnest, C. (1998). Comparison of two EEG asymmetry indices in depressed patients vs. normal controls. *International Journal of Psychophysiology, 31,* 89–92.

Baglioni, C., Battagliese, G., Feige, B., et al. (2011). Insomnia as a predictor of depression: A meta-analytic evaluation of longitudinal epidemiological studies. *Journal of Affective Disorders, 135,* 10–19.

Baker, C. I., Hutchison, T. L., & Kanwisher, N. (2007). Does the fusiform face area contain subregions highly selective for nonfaces? *Nature Neuroscience, 10*(1), 3–4.

Baldo, J. V., Shimamura, A. P., Delis, D. C., Kramer, J., & Kaplan, E. (2001). Verbal and design fluency in patients with frontal lobe lesions. *Journal of the International Neuropsychological Society, 7,* 586–596.

Baldo, J. V., Schwartz, S., Wilkins, D., & Dronkers, N. F. (2006). Role of frontal versus temporal cortex in verbal fluency as revealed by voxel-based lesion symptom mapping. *Journal of the International Neuropsychological Society, 12*(6), 896–900.

Baloh, R. W. (2001). Prosper Meniere and his disease. *Archives of Neurology, 58,* 1151–1156.

Banerjee, S., Murray, J., Foley, B., Atkins, L., Schneider, J., & Mann, A. (2003). Predictors of institutionalisation in people with dementia. *Journal of Neurology, Neurosurgery, and Psychiatry, 74*(9), 1315–1316.

Barbas, H., Bunce, J. G., & Medalla, M. (2013). Prefrontal pathways that control attention. In D. T. Stuss & R. T. Knight (Eds.), *Principles of frontal lobe function* (2nd ed.). New York: Oxford University Press.

Barnes, C. R. (1910). *The American Naturalist, 44*(522), 321–332. Published by: The University of Chicago Press for The American Society of Naturalists, Chicago, IL.

Baron-Cohen, S. (2003). *The truth about the male & female brain: The essential difference.* New York: Basic Books.

Baron-Cohen, S., & Wheelwright, S. (2004). The empathy quotient: an investigation of adults with Asperger syndrome or high functioning autism, and normal sex differences. *Journal of Autism and Developmental Disorders, 34,* 163–175.

Barrett, A. M., Crossono, J. B., Crucian, G. P., & Heilman, K. M. (2000a). Horizontal line bisections in upper and lower body space. *Journal of the International Neuropsychological Society, 6,* 455–459.

Barrett, L. F., Lane, R. D., Sechrest, L., & Schwartz, G. E. (2000b). Sex differences in emotional awareness. *Personality and Social Psychology Bulletin, 26,* 1027–1035.

Bartholow, R. (1874). Experimental investigations into the function of the human brain. *American Journal of the Medical Sciences, 134,* 305–313.

Bartolo, M., Zucchella, C., Pichiecchio, A., Pucci, E., Sandrini, G., & Sinforiani, E. (2011). Alien hand syndrome in left posterior stroke. *Neurological Sciences, 32,* 483–486. doi:10.1007/s10072-011-0490-y.

Bartoshuk, L. M. (1968). Water taste in man. *Attention, Perception, & Psychophysics, 3*(1), 69–72.

Bartoshuk, L. M., Rennert, K., Rodin, J., & Stevens, J. C. (1982). Effects of temperature on the perceived sweetness of sucrose. *Physiology and Behavior, 28,* 905–910.

Bartzokis, G. (2012). Neuroglialpharmacology: Myelination as a shared mechanism of action of psychotropic treatments. *Neuropharmacology, 62*(7), 2137–2153. ISSN 0028-3908, doi:10.1016/j.neuropharm.2012.01.015.

Baumeister, R. F., Vohs, K. D., & Tice, D. M. (2007). The strength model of self control. *Current Directions in Psychological Science, 16,* 351–355.

Bavelier, D., & Neville, H. J. (2002). Cross-modal plasticity: Where and how? *Nature Reviews Neuroscience, 3*(6), 443–452.

Baving, L., Laucht, M., & Schmidt, M. H. (2002). Frontal brain activation in anxious school children. *Journal of Child Psychology and Psychiatry, 43,* 265–274.

Bayne, T., & Levy, N. (2005). Amputees by choice: Body integrity identity disorder and the ethics of amputation. *Journal of Applied Philosophy, 22*(1), 75–86.

Bear, M. F., Connors, B. W., & Paradiso, M. A. (2007). *Neuroscience: Exploring the brain* (3rd ed.). Philadelphia: Lippincott Williams & Wilkins.

Beccuti, G., & Pannain, S. (2011). Sleep and obesity. *Current Opinion in Clinical Nutrition and Metabolic Care, 14*(4), 402–412.

Bechara, A. (2004). The role of emotion in decision-making: Evidence from neurological patients with orbitofrontal damage. *Brain and Cognition, 55*(1), 30–40.

Bechara, A., Damasio, A. R., Damasio, H., & Anderson, S. W. (1994). Insensitivity to future consequences following damage to human prefrontal cortex. *Cognition, 50,* 7–15.

Bechara, A., Tranel, D., Damasio, H., Adolphs, R., Rockland, C., & Damasio, A. R. (1995). Double dissociation of conditioning and declarative knowledge relative to the amygdala and hippocampus in humans. *Science, 269,* 1115–1118.

Bechara, A., Damasio, H., & Damasio, A. R. (2000). Emotion, decision-making, and the orbitofrontal cortex. *Cerebral Cortex, 10,* 295–307.

Bechara, A., Tranel, D., & Damasio, A. R. (2002). The somatic marker hypothesis and decision-making. In F. Boller & J. Grafman (Eds.), *Handbook of neuropsychology: Frontal lobes* (Vol. 7, 2nd ed., pp. 117–143). Amsterdam: Elsevier.

Becker, D., & McDonald III, J. W. (2012). Approaches to repairing the damaged spinal cord: Overview. *Handbook of Clinical Neurology Series, 109,* 445–461 (Spinal Cord Injuries E-Book).

Becker, B., Mihov, Y., Scheele, D., Kendrick, K. M., Feinstein, J. S., Matusch, A., & Hurlemann, R. (2012). Fear processing and social networking in the absence of a functional amygdala. *Biological Psychiatry, 72,* 70–77.

Bedny, M., Pascual-Leone, A., Dodell-Feder, D., Fedorenko, E., & Saxe, R. (2011). Language processing in the occipital cortex of congenitally blind adults. *Proceedings of the National Academy of Sciences of the United States of America, 108,* 4429–4434.

Beebe, J. A., & Lang, C. E. (2008). Absence of a proximal to distal gradient of motor deficits in the upper extremity early after stroke. *Clinical Neurophysiology, 119*(9), 2074–2085. doi:10.1016/j.clinph.2008.04.293.

Beer, J. (2006). Orbitofrontal cortex and social regulation. In J. Cacioppo, P. Visser, & C. Pickett (Eds.), *Social neuroscience. People thinking about people* (pp. 153–166). Cambridge: MIT Press.

Beer, J. (2007). The importance of emotion-social cognition interactions for social functioning: Insights from orbitofrontal cortex. In E. Harmon-Jones & P. Winkielman (Eds.), *Social neuroscience* (pp. 15–30). New York: The Guilford Press.

Beer, J., John, O., Scabini, D., & Knight, R. (2006). Orbitofrontal cortex and social behavior: Integrating self-monitoring and emotion-cognition interactions. *Journal of Cognitive Neuroscience, 18,* 871–879.

Behrendt, R. P. (2003). Hallucinations: Synchronisation of thalamocortical γ oscillations underconstrained by sensory input. *Consciousness and Cognition, 12*(3), 413–451. ISSN 1053-8100, doi:10.1016/S1053-8100(03)00017-5.

Behrendt, R. P. (2012). Consciousness, memory, and hallucinations. In J. D. Blom & I. E. C. Sommer (Eds.), *Hallucinations* (pp. 17–31). New York: Springer. doi:10.1007/978-1-4614-0959-5_3.

Behrman, M., Avidan, G., Gao, F., & Black, S. (2007). Structural imaging reveals anatomical alterations in inferotemporal cortex in congenital prosopagnosia. *Cortex, 17,* 2354–2363.

Bejot, Y., Caillier, M., Osseby, G., Didi, R., Salem, D. B., Moreau, T., & Giroud, M. (2008). Involuntary masturbation and hemiballismus after bilateral anterior cerebral artery infarction. *Clinical Neurology and Neurosurgery, 100,* 190–193.

Bell, C. (1811). An idea of a new anatomy of the brain; submitted for the observations of his friends. A privately printed pamphlet. London: Strahan & Preston.

Bellis, T., Nocol, T., & Kraus, N. (2000). Aging affects hemispheric asymmetry in the neural representation of speech sounds. *Journal of Neuroscience, 20,* 791–797.

Bench, C. J., Friston, K. J., Brown, R. G., Scott, L. C., Frackowiak, R. S., & Dolan, R. J. (1992). The anatomy of melancholia-focal abnormalities of cerebral blood flow in major depression. *Psychological Medicine, 22,* 607–615.

Bender, M. B. (1952). *Disorders in perception.* Springfield: Charles C. Thomas.

Bensafi, M., Iannilli, E., Poncelet, J., Seo, H.-S., Gerber, J., et al. (2012). Dissociated representations of pleasant and unpleasant olfacto-trigeminal mixtures: An fMRI study. *PLoS One, 7*(6), e38358. doi:10.1371/journal.pone.0038358.

Bentall, R. P., & Udachina, A. (2013). Social cognition and the dynamics of paranoid ideation. In *Social cognition in schizophrenia: From evidence to treatment* (pp. 215–244).

Benton, A. L., & Tranel, D. (1993). Visuoperceptual, visuospatial, and visuoconstructive disorders. In K. M. Heilman & E. Valenstein (Eds.), *Clinical neuropsychology* (3rd ed., pp. 165–213). New York: Oxford University Press.

Benton, A. L., Levin, H. S., & Van Allen, M. W. (1974). Geographic orientation in patients with unilateral cerebral disease. *Neuropsychologia, 12*(2), 183–191.

Berg, E. A. (1948). A simple objective technique for measuring flexibility in thinking. *Journal of General Psychology, 39,* 15–22.

Berlucchi, G. (2006). Revisiting the 1981 Nobel Prize to Roger Sperry, David Hubel, and Torsten Wiesel on the occasion of the centennial of the Prize to Golgi and Cajal. *Journal of the History of the Neurosciences, 15*(4), 369–375.

Berman, M. G., Nee, D. E., Casement, M., Kim, H. S., Deldin, P., Kross, E., & Jonides, J. (2011). Neural and behavioral effects of interference resolution in depression and rumination. *Cognitive, Affective, & Behavioral Neuroscience, 11,* 85–96.

Bernstein, A. L. (2013). Into the red zone: How the National Football League's Quest to curb concussions and concussion-related injuries could affect players' legal recovery. *Seton Hall Journal of Sports and Entertainment Law, 22*(2), 9.

Berti, A., Bottini, G., Gandola, M., Pia, L., Smania, N., Stracciari, A., & Paulesu, E. (2005). Shared cortical anatomy for motor awareness and motor control. *Science, 309,* 488–491. doi:10.1126/science.1110625.

Bhattacharje, Y. (2008). Neuro- science. Shell shock revisited: Solving the puzzle of blast trauma. *Science, 319,* 406–408. doi:10.1126/science.319.5862.406.

Biber, B., & Alkln, T. (1999). Panic disorder subtypes: Differential responses to CO_2 challenge. *American Journal of Psychiatry, 156,* 739–744.

Bibost, A. L., Kydd, E., & Brown, C. (2013). The effect of sex and early environment on the lateralization of the rainbowfish. In D. Csermely & L. Regolin (Eds.), *Behavioral lateralization in vertebrates* (pp. 9–24). Berlin: Springer. doi:10.1007/978-3-642-30203-9_2.

Bigler, E. D. (2004). Neuropsychological results and neuropathological findings at autopsy in a case of mild traumatic brain injury. *Journal of the International Neuropsychological Society, 5,* 794–806.

Billings, L. S., Harrison, D. W., & Alden, J. D. (1993). Age differences among women in the functional asymmetry for bias in facial affect perception. *Bulletin of the Psychonomic Society, 31*(4), 317–320.

Binder, J. R., Medler, D. A., Desai, R., Conant, L. L., & Liebenthal, E. (2005a). Some neurophysiological constraints on models of word naming. *Neuroimage, 27,* 677–693.

Biran, I., & Chatterjee, A. (2004). Alien hand syndrome. *Archives of Neurology, 61,* 292–294. doi:10.1001/archneur.61.2.292.

Bisiach, E., Vallar, G., Perani, D., Papagno, C., & Berti, A. (1986). Unawareness of disease following lesions of the right hemisphere: Anosognosia for hemiplegia and anosognosia for hemianopia. *Neuropsychologia, 24,* 471–482.

Bjorvatn, B., & Pallesen, S. (2009). A practical approach to circadian rhythm sleep disorders. *Sleep Medicine Reviews, 13,* 47–60.

Blackburn, K., & Schirillo, J. (2012). Emotive hemispheric differences measured in real-life portraits using pupil diameter and subjective aesthetic preferences. *Experimental Brain Research, 219,* 447–455. doi:10.1007/s00221-012-3091-y.

Blackford, J. U., Allen, A. H., Cowan, R. L., & Avery, S. N. (2012). Amygdala and hippocampus fail to habituate to faces in individuals with an inhibited temperament. *Social Cognitive Affective Neuroscience.* (First published online January 19, 2012) doi:10.1093/scan/nsr078.

Blackhart, G. C., Minnix, J. A., & Kline, J. P. (2006). Can EEG asymmetry patterns predict future development of anxiety and depression? A preliminary study. *Biological Psychology, 72,* 46–50.

Blair, R. J. R., Morris, J. S., Frith, C. D., Perrett, D. I., & Dolan, R. J. (1999). Dissociable neural responses to facial expressions of sadness and anger. *Brain, 122,* 883–893.

Blamey, P., Artieres, F., Başkent, D., Bergeron, F., Beynon, A., Burke, E., & Lazard, D. S. (2013). Factors affecting auditory performance of postlinguistically deaf adults using cochlear implants: An update with 2251 patients. *Audiology and Neurotology, 18*(1), 36–47.

Blazer, D. (2013). Neurocognitive disorders in DSM-5. *American Journal of Psychiatry, 170*(6), 585–587.

Bliss, T. V., & Lømo, T. (1973). Long-lasting potentiation of synaptic transmission in the dentate area of the anaesthetized rabbit following stimulation of the perforant path. *The Journal of Physiology, 232*(2), 331–356.

Blonder, L. X., Bowers, D., & Heilman, K. M. (1991). The role of the right hemisphere in emotional communication. *Brain, 114,* 1115–1127.

Blumstein, S. E., & Amso, D. (2013). Dynamic functional organization of language: Insights from functional neuroimaging. *Perspectives on Psychological Science, 8,* 44–48.

Boccardi, E., Della Sala, S., Motto, C., & Spinnler, H. (2002). Utilisation behaviour consequent to bilateral SMA softening. *Cortex, 38,* 239–308.

Boecker, H., Sprenger, T., Spilker, M. E., Henriksen, G., Koppenhoefer, M., Wagner, K. J., Valet, M., Berthele, A., & Tolle, T. R. (2008). The runner's high: Opioidergic mechanisms in the human brain. *Cerebral Cortex, 18,* 2523 (New York, N.Y.).

Boecker, M., Drueke, B., Vorhold, V., Knops, A., Philippen, B., & Gauggel, S. (2010). When response inhibition is followed by response reengagement: An event-related fMRI study. *Human Brain Mapping, 32,* 94–106.

Boemio, A., Fromm, S., Braun, A., & Poeppel, D. (2005). Hierarchical and asymmetric temporal sensitivity in human auditory cortices. *Nature Neuroscience, 8,* 389–395.

Bogner, J. A., & Corrigan, J. D. (1995). Epidemiology of agitation following brain injury. *Neurorehabilitation, 5,* 293–297.

Bogner, J. A., Corrigan, J. D., Fugate, L., Mysiw, W. J., & Clinchot, D. (2001). Role of agitation in prediction of outcomes after traumatic brain injury. *American Journal of Physical Medicine and Rehabilitation, 80*(9), 636–644.

Bohbot, V. D., Kalina, M., Stepankova, K., Spackova, N., Petrides, M., & Nadel, L. (1998). Spatial memory deficits in patients with lesions to the right hippocampus and to the right parahippocampal cortex. *Neuropsychologia, 36,* 1217–1238.

Bolger, D. J., Perfetti, C. A., & Schneider, W. (2005). Cross-cultural effect on the brain revisited: Universal structures plus writing system variation. *Human Brain Mapping, 25,* 92–104.

Bologna, M., Fasano, A., Modugno, N., Fabbrini, G., & Berardelli, A. (2012). Effects of subthalamic nucleus deep brain stimulation and l-dopa on blinking in Parkinson's disease. *Experimental Neurology, 235*(1), 265–272.

Bonda, E., Petrides, M., Frey, S., & Evans, A. (1995). Neural correlates of mental transformations of the body-in-space. *Proceedings of the National Academy of Sciences of the United States of America, 92,* 11180–11184.

Booth, J. R., Cho, S., Burman, D. D., & Bitan, T. (2007). Neural correlates of mapping from phonology to orthography in children performing an auditory spelling task. *Developmental Science, 10,* 441–451.

Borison, H. L., & Wang, S. C. (1953). Physiology and pharmacology of vomiting. *Pharmacological Reviews, 5,* 193–230.

Borjigin, J., Li, X., & Snyder, S. H. (1999). The pineal gland and melatonin: Molecular and pharmacologic regulation. *Annual Review of Pharmacology and Toxicology, 39,* 53–65.

Borod, J. C. (1992). Interhemispheric and intrahemispheric control of emotion: A focus on unilateral brain damage. *Journal of Consulting and Clinical Psychology, 60*(3), 339–348.

Borod, J. C., Koff, E., & Caron, H. (1983). Right hemispheric specialization for the expression and appreciation of emotion: A focus on the face. In E. Perecam (Ed.), *Cognitive functions in the right hemisphere* (pp. 83–110). New York: Academic Press.

Borod, J. C., Koff, E., Lorch, M. P., & Nicholas, M. (1985). Channels of emotional communication in patients with unilateral brain damage. *Archives of Neurology, 42,* 345–348.

Borod, J. C., Koff, E., Perlman-Lorch, J., & Nicholas, M. (1986). The expression and perception of facial emotions in brain-damaged patients. *Neuropsychologia, 24,* 169–180.

Borod, J. C., Haywood, C. S., & Koff, E. (1997). Neuropsychological aspects of facial asymmetry during emotional expression: A review of the normal adult literature. *Neuropsychology Review, 7*(1), 41–60.

Borod, J. C., Bloom, R. L., Brickman, A. M., Nakhutina, L., & Curko, E. A. (2002). Emotional processing deficits in individuals with unilateral brain damage. *Applied Neuropsychology, 9*(1), 23–36.

Bouhuys, A. L., Geerts, E., & Gordijn, M. C. (1999). Depressed patients perceptions of facial emotions in depressed and remitted states are associated with relapse: A longitudinal study. *Journal of Nervous Mental Disorders, 187,* 595–602.

Bouillaud, J. (1865). Sur la faculte du langage articule. *Bulletin de l Academie Nationale de Medecine, 30,* 752–768.

Bourguignon, M., De Tiège, X., de Beeck, M. O., Ligot, N., Paquier, P., Van Bogaert, P., Goldman, S., Hari, R., & Jousmäki, V. (2012). The pace of prosodic phrasing couples the listener's cortex to the reader's voice. *Human Brain Mapping.* http://dx.doi.org/10.1002/hbm.21442.

Bowden, M. G., Woodbury, M. L., & Duncan, P. W. (2013). Promoting neuroplasticity and recovery after stroke: Future directions for rehabilitation clinical trials. *Current Opinion in Neurology, 26*(1), 37–42.

Bowers, D., & Heilman, K. M. (1980). Pseudoneglect: Effects of hemispace on a tactile line bisection task. *Neuropsychologia, 18,* 491–498.

Bowers, D., Bauer, R. M., & Heilman, K. M. (1993). The nonverbal affect lexicon. *Neuropsychology, 7,* 433–444.

Bracci, S., Cavina-Pratesi, C., Ietswaart, M., Caramazza, A., & Peelen, M. V. (2012). Closely overlapping responses to tools and hands in left lateral occipitotemporal cortex. *Journal of Neurophysiology, 107*(5), 1443–1456; (published ahead of print November 30, 2011). doi:10.1152/jn.00619.2011.

Bradley, B., & Matthews, A. (1983). Negative self-schemata in clinical depression. *Clinical Psychology, 22,* 173–181.

Bradley, B. P., Mogg, K., & Williams, R. (1995). Implicit and explicit memory for emotion-congruent information in clinical depression and anxiety. *Behavioural Research and Therapeutics, 33,* 755–770.

Brain, R. (1941). Visual disorientation with special reference to lesions of the right cerebral hemisphere. *Brain, 64,* 244–272.

Brandt, T., & Dieterich, M. (1994). Vestibular syndromes in the roll plane: Topographic diagnosis from brainstem to cortex. *Annals of Neurology, 36,* 337–347. doi:10.1002/ana.410360304

Bremer, C. R. (1935). Cerveau isole et physiologie du sommeil. *Comptes Rendus de la Societe de Biologief Paris, 118,* 1235–1241.

Bremer, F., & Parmeggiani. (1985a). F. Bremer. Bull Acad. Roy. Med. Bio. 4 (1937), pp. 68–86 (cited by Parmeggiani. Brain mechanisms of sleep. In D. J. McGinty et al. (Eds.), *Brain mechanisms of sleep* (pp. 1–33). New York: Raven Press.).

Bremer, F., & Parmeggiani. (1985b). F. Bremer. Bull. Soc. Ital. Biol. Sper., 13 (1938), pp. 271–290 (cited by Parmeggiani. Brain mechanisms of sleep. In D. J. McGinty et al. (Eds.), *Brain mechanisms of sleep* (pp. 1–33). New York: Raven Press.).

Bridoux, A., Laloux, C., Derambure, P., Bordet, R., & Monaca Charley, C. (2012). The acute inhibition of rapid eye movement sleep by citalopram may impair spatial learning and passive avoidance in mice. *Journal of Neural Transmission*, 1–7. Published online. doi:10.1007/s00702-012-0901-0.

Britton, T. L. (2009). *Ronald Reagan*. Edina: Checkerboard Library.

Broca, P. (1861). Remarques sur le siege de la faculte du langage articule, suivies d'une observation d'aphemie (perte de la parole). *Bulletins et mémoires de la Société Anatomique de Paris, 36*, 330–357.

Broca, P. (1864). Cited in (M. Critchley. (1970). *Aphasiology and other aspects of language*. London: Edward Arnold.).

Brodal, A. (1981). *Neurological anatomy in relation to clinical medicine* (3rd ed.). New York: Oxford University Press.

Brodaty, H., McGilchrist, C., Harris, L., & Peters, K. E. (1993). Time until institutionalization and death in patients with dementia: Role of caregiver training and risk factors. *Archives of Neurology, 50*(6), 643.

Brodmann K. (1909). *Vergleichende Lokalisationslehre der Grosshirnrinde*. Leipzig: Johann Ambrosius Bart.

Brody, L. R., & Hall, J. A. (2000). Gender, emotion, and expression. In M. Lewis & J. M. Haviland-Jones (Eds.), *Handbook of emotions* (pp. 265–280). New York: Guilford Press.

Brody, A. L., Saxena, S., Fairbanks, L. A., Alborzian, S., Demaree, H. A., Maidment, K. M., et al. (2000). Personality changes in adult subjects with major depressive disorder or obsessive-compulsive disorder treated with paroxetine. *Journal of Clinical Psychiatry, 61*(5), 349–355.

Brookover, C. (1913). The nervus terminalis in adult man: A preliminary account from the Anatomical Department of the Medical Department of the University of Arkansas (pp. 131–135.)

Brooks, J. C., Nurmikko, T. J., Bimson, W. E., Singh, K. D., & Roberts, N. (2002). fMRI of thermal pain: Effects of stimulus laterality and attention. *Neuroimage, 15*, 293–301.

Brown J. (1974). *Aphasia, apraxia and agnosia*. Springfield: Charles C. Thomas.

Bruder, G. E., Tenke, C. E., Warner, V., & Weissman, M. M. (2007). Grandchildren at high and low risk for depression differ in EEG measures of regional brain asymmetry. *Biological Psychiatry, 62*, 1317–1323.

Bruder, G. E., Bansal, R., Tenke, C. E., Liu, J., Hao, X., Warner, V., Peterson, B. S., & Weissman, M. M. (2012). Relationship of resting EEG with anatomical MRI measures in individuals at high and low risk for depression. *Human Brain Mapping, 33*(6), 1325–1333. doi:10.1002/hbm.21284.

Bruehl, S., & Chung, Y. (2004). Interactions between the cardiovascular and pain regulatory systems: An updated review of mechanisms and possible alterations in chronic pain. *Neuroscience and Biobehavioral Reviews, 28*, 395–414.

Brunner, R. J., Kornhuber, H. H., Seemuller, E., Suger, G., & Wallesch, C. W. (1982). Basal ganglia participation in language pathology. *Brain and Language, 16*, 281–299.

Bryden, M. P., & MacCrae, L. (1989). Dichotic laterality effects obtained with emotional words. *Neuropsychiatry, Neuropsychology, & Behavioral Neurology, 1*(3), 171–176.

Bschor, T., Ising, M., Bauer, M., Lewitzka, U., Skerstupeit, M., Muller-Orlinghausen, B., et al. (2004). Time experience and time judgment in major depression, mania and healthy subjects. A controlled study of 93 subjects. *Acta Psychiatrica Scandinavica, 109*, 222–229.

Buck, L., & Axel, R. (1991). A novel multigene family may encode odorant receptors: A molecular basis for odor recognition. *Cell, 65*(1), 175–187.

Buck, R., & Duffy, R. J. (1980). Nonverbal communication of affect in brain-damaged patients. *Cortex, 16*(3), 351.

Bujas, Z. (1971). Electrical taste. In L. M. Beidler (Ed.), *Handbook of sensory physiology IV, chemical senses 2* (pp. 180–199). New York: Springer-Verlag.

Bunce, V. L., & Harrison, D. W. (1991). Child- or adult-directed speech and esteem: Effects on performance and arousal in elderly adults. *International Journal of Aging and Human Development, 32*(2), 125–134.

Bunney, J. N., & Potkin, S. G. (2008). Circadian abnormalities, molecular clock genes and chronobiological treatments in depression. *British Medical Bulletin, 86*, 23–32. doi:10.1093/bmb/ldn019.

Burke, T., & Nolan, J. R. M. (1988). Material specific memory deficits after unilateral temporal neocorticectomy. *Society for Neuroscience (Abstracts), 14*, 1289.

Burstein, R., Jakubowski, M., Garcia-Nicas, E., Kainz, V., Bajwa, Z., Hargreaves, R., Becerra, L., & Borsook, D. (2010). Thalamic sensitization transforms localized pain into widespread allodynia. *Annals of Neurology, 68*, 81–91.

Burt, V. K., & Stein, K. (2002). Epidemiology of depression throughout the female life cycle. *Journal Clinical Psychiatry, 63*(7), 9–15.

Burtis, D. B., Heilman, K. M., Mo, J., Wang, C., Lewis, G. F., Davila, M. I., Ding, M., Porges, S. W., & Williamson, J. B. (2014). The effects of constrained left versus right monocular viewing on the autonomic nervous system. *Biological Psychology, 100*, 79–85.

Burton, L. A., & Labar, D. (1999). Emotional status after right vs. left temporal lobectomy. *Seizure, 8*(2), 116–119.

Burton, H., Snyder, A. Z., Diamond, J. B., & Raichle, M. E. (2002). Adaptive changes in early and late blind: An fMRI study of verb generation to hear nouns. *Journal of Neurophysiology, 88*, 3359–3371.

Bush, G. W. (2011). George W. Bush stutters over and over—YouTube. www.youtube.com/watch?v=U6mnWImtRTo.

Bush, G., Luu, P., & Posner, M. I. (2000). Cognitive and emotional influences in the anterior cingulate cortex. *Trends in Cognitive Sciences, 4*, 215–222.

Butter, C. M., Snyder, D. R., & McDonald, J. A. (1970). Effects of orbital frontal lesions on aversive and aggressive behaviors in rhesus monkeys. *Journal of Comparative Physiology and Psychology, 72*, 132–144.

Buxbaum, L. J., et al. (2004). Hemispatial neglect: Subtypes, neuroanatomy, and disability. *Neurology, 62*, 749–756.

Buzsaki, G., & Draguhn, A. (2004). Neuronal oscillations in cortical networks. *Science, 304*, 1926–1929.

Bystron, I., Blakemore, C., & Rakic, P. (2008). Development of the human cerebral cortex: Boulder Committee revisited. *Nature Reviews. Neuroscience, 9*, 110–122.

Cabeza, R. (2002). Hemispheric asymmetry reduction in old adults: The HAROLD model. *Psychology and Aging, 17*, 85–100.

Cabeza, R., Daselaar, S., Dolcos, F., Prince, S., Budde, M., & Nyberg, L. (2004). Task-independent and task-specific age effects on brain activity during working memory, visual attention and episodic retrieval. *Cerebral Cortex, 14*, 364–375.

Cabeza, R., Ciaramelli, E., & Moscovitch, M. (2012). Cognitive contributions of the ventral parietal cortex: An integrative theoretical account. *Trends in Cognitive Sciences, 16*(6), 338–352. ISSN 1364-6613, doi:10.1016/j.tics.2012.04.008. http://www.sciencedirect.com/science/article/pii/S1364661312001015.

Cacioppo, J. T., & Bernston, G. G. (2007). The brain, homeostasis, and health: Balancing demands of the internal and external milieu. In H. S. Friedman & R. S. Cohen (Eds.), *Foundation of health psychology* (pp. 73–91). New York: Oxford University Press.

Cacioppo, J. T., Amaral, D., Blanchard, J., et al. (2007). Social neuroscience. Progress and implications for mental health. *Perspectives on Psychological Science, 2*, 99–123.

Cagampang, F. R. A., Yamazaki, S., Otori, Y., & Inouye, S. I. T. (1993). Serotonin in the raphe nuclei: Regulation by light and an endogenous pacemaker. *Neuroreport, 5*(1), 49–52.

Cahn-Weiner, D. A., Mallory, P. F., Boyle, P. A., Marran, M., & Salloway, S. (2000). Prediction of functional status from neuropsychological tests in community-dwelling elderly individuals. *The Clinical Neuropsychologist, 14*(2), 187–195.

Cai, W., George, J. S., Verbruggen, F., Chambers, C. D., & Aron, A. R. (2012). The role of the right presupplementary motor area in stopping action: Two studies with event-related transcranial magnetic stimulation. *Journal of Neurophysiology, 108*, 380–389 (First published 18 April 2012). doi:10.1152/jn.00132.2012.

Calabrò, R. S., Baglieri, A., Ferlazzo, E., Passari, S., Marino, S., & Bramanti, P. (2012). Neurofunctional assessment in a stroke patient with musical hallucinations. *Neurocase (The neural basis of cognition), 18*, 514–520. doi:10.1080/13554794.2011.633530.

Cammalleri, R., Gangitano, M., D'Amelio, M., et al. (1996). Transient topographical amnesia and cingulate cortex damage: A case report. *Neuropsychologia, 34*, 321–326.

Cannon, A., & Benton, A. L. (1969). Tactile perception of direction and number in patients with unilateral cerebral disease. *Neurology, 9*, 525–532.

Cannon, W. B., & Britton, S. W. (1925). Studies on the conditions of activity in endocrine glands: Pseudaffective medulliadrenal secretion. *American Journal of Physiology, 72*(2), 283–294.

Cantalupo, C., Bisazza, A., & Vallortigara, G. (1995). Lateralization of predator-evasion response in a teleost fish (*Girardinus falcatus*). Neuropsychologia, *33*, 1637–1646.

Capotosto, P., Corbetta, M., Romani, G. L., & Babiloni, C. (2012). Electrophysiological correlates of stimulus-driven reorienting deficits after interference with right parietal cortex during a spatial attention task: A TMS-EEG study. *Journal of Cognitive Neuroscience, 24*, 2363–2371.

Cappe, C., Rouiller, E. M., & Barone, P. (2012). Cortical and thalamic pathways for multisensory and sensorimotor interplay. In M. M. Murray & M. T. Wallace (Eds.), *The neural bases of multisensory processes*. Boca Raton: CRC Press (Chapter 2). http://www.ncbi.nlm.nih.gov/books/NBK92866.

Cappell, K. A., Gmeindl, L., & Reuter-Lorenz, P. A. (2010). Age differences in prefrontal recruitment during verbal working memory maintenance depend on memory load. *Cortex, 46*, 462–473. doi:10.1016/j.cortex.2009.11.009.

Cardinali, D. P., Furio, A. M., & Brusco, L. I. (2010). Clinical aspects of melatonin intervention in Alzheimer's disease progression. *Current Neuropharmacology, 8*, 218–227.

Carmona, J. E., Holland, A. K., Stratton, H. J., & Harrison, D. W. (2008). Sympathetic arousal to a vestibular stressor in high and low hostile men. *Brain and Cognition, 66*, 150–155.

Carmona, J. E., Holland, A. K., & Harrison, D. W. (2009). Extending the functional cerebral systems theory of emotion to the vestibular modality: A systematic and integrative approach. *Psychological Bulletin, 135*(2), 286–302.

Carney, D. R., & Mason, M. F. (2010). Decision making and testosterone: When the ends justify the means. *Journal of Experimental Social Psychology, 46*, 668–671.

Carpenter, R. H. S., & Reddi, B. A. J. (2000). The influence of urgency on decision time. *Nature Neuroscience, 3*(8), 827–834.

Carrington, S. J., & Bailey, A. J. (2009). Are there theory of mind regions in the brain? A review of the neuroimaging literature. *Human Brain Mapping, 30*, 2313–2335.

Carver, C. S., & White, T. L. (1994). Behavioral inhibition, behavioral activation, and affective responses to impending reward and punishment: The BIS/BAS Scales. *Journal of Personality and Social Psychology, 67*(2), 319–333.

Casey, D. E. (1991). Neuroleptic drug-induced extrapyramidal syndromes and tardive dyskinesia. *Schizophrenia Research, 4*(2), 109–120.

Chakor, R. T., & Eklare, N. (2012). Vertigo in cerebrovascular diseases. *Cerebrovascular Diseases, 4*(1), 46–53.

Chakrabarti, B., & Baron-Cohen, S. (2006). Empathizing: Neurocognitive developmental mechanisms and individual differences. *Progress in Brain Research, 156*, 403–417.

Chambers, C. D., Garavan, H., & Bellgrove, M. A. (2009). Insights into the neural basis of response inhibition from cognitive and clinical neuroscience. *Neuroscience and Biobehavioral Reviews, 33*, 631–646.

Chang, H. C., & Gaddum, J. H. (1933). Choline esters in tissue extracts. *The Journal of Physiology, 79*(3), 255–285.

Chen, R., Cohen, L. G., & Hallett, M. (2002). Nervous system reorganization following injury. *Neuroscience, 111*(4), 761–773.

Chen, D., Katdare, A., & Lucas, N. (2006). Chemosignals of fear enhance cognitive performance in humans. *Chemical Senses, 31,* 415–423.

Cherpitel, C. J., Martin, G., Macdonald, S., Brubacher, J. R., & Stenstrom, R. (2013). Alcohol and drug use as predictors of intentional injuries in two emergency departments in British Columbia. *The American Journal on Addictions, 22*(2), 87–92.

Chewning, J., Adair, J. C., Heilman, E. B., & Heilman, K. M. (1998). Attentional bias in normal subjects performing visual and tactile radial line bisections. *Neuropsychologia, 36,* 1097–1101.

Chi, J. G., Dooling, E. C., & Gilles, F. H. (1977). Left-right asymmetries of the temporal speech areas of the human fetus. *Archives of Neurology, 34,* 346–348.

Chiao, J. Y., & Immordino-Yang, M. H. (2013). Modularity and the cultural mind: Contributions of cultural neuroscience to cognitive theory. *Perspectives on Psychological Science, 8,* 56–61.

Chikazoe, J. (2010). Localizing performance of go/no-go tasks to prefrontal cortical subregions. *Current Opinion in Psychiatry, 23,* 267–272.

Chiu, H. C., & Damasio, A. R. (1980). Human cerebral asymmetries evaluated by computed tomography. *Journal of Neurology, Neurosurgery, and Psychiatry, 43,* 873–878.

Chouinard, S., Poulin, J., Stip, E., & Godbout, R. (2004). Sleep in untreated patients with schizophrenia: A meta-analysis. *Schizophrenia Bulletin, 30,* 957–967.

Christensen, A. L. (1979). *Luria's neuropsychological investigation* (2nd ed.). Copenhagen: Munksgaard.

Christoff, K., & Gabrielli, J. D. (2000). The frontopolar cortex and human cognition: Evidence for a rostrocaudal hierarchical organization within the human prefrontal cortex. *Psychobiology, 28,* 168–186.

Chu, S., & Downes, J. J. (2000). Odour-evoked autobiographical memories: Psychological investigations of proustian phenomena. *Chemical Senses, 25,* 111–116.

Church, J. A., Balota, D. A., Petersen, S. E., & Schlaggar, B. L. (2010). Manipulation of length and lexicality localizes the functional neuroanatomy of phonological processing in adult readers. *Journal of Cognitive Neuroscience, 23,* 1475–1493.

Cleare, A. J., & Bond, A. J. (1997). Does central serotonergic function correlate with aggression? A study using D-fenfluramine in healthy subjects. *Psychiatry Research, 69*(2–3), 89–95.

Cleret de Langavant, L., Trinkler, I., Cesaro, P., & Bachoud-Levi, A. C. (2009). Heterotopagnosia: When I point at parts of your body. *Neuropsychologia, 47,* 1745–1755.

Clint Eastwood. (n.d.). BrainyQuote.com. from BrainyQuote.com Web site: http://www.brainyquote.com/quotes/quotes/c/clinteastw446681.html. Accessed 9 April 2013.

Coghill, R. C., Collins, D. L., Neelin, P., Peters, P., & Evans, A. C. (1994a). Automatic 3-D inter-subject registration of MR volumetric data in standardized Talairach space. *Journal of Computer Assisted Tomography, 18,* 192–205.

Coghill, R. C., Talbot, J. D., Evans, A. C., Meyer, E., Gjedde, A., Bushnell, C., & Duncan, G. H. (1994b). Distributed processing of pain and vibration by the human brain. *Journal of Neuroscience, 14*(7), 4095–4106.

Cohen, L., Dehaene, S., Naccache, L., Lehericy, S., Dehaene-Lambertz, G., et al. (2000). The visual word form area: Spatial and temporal characterization of an initial stage of reading in normal subjects and posterior split-brain patients. *Brain, 123*(2), 291–307.

Cohen, L., Martinaud, O., Lemer, C., Lehericy, S., Samson, Y., Obadia, M., Slachevsky, A., & Dehaene, S. (2003). Visual word recognition in the left and right hemispheres: Anatomical and functional correlates of peripheral alexias. *Cerebral Cortex, 13,* 1313–1333.

Colebatch, J. G., & Gandevia, S. C. (1989). The distribution of muscular weakness in upper motor neuron lesions affecting the arm. *Brain, 112*(3), 749–763.

Collias, N. E. (1960). An ecological and functional classification of animal sounds. In W. E. Lanyon & W. N. Tavolga (Eds.), *Animal sounds and communication* (443 p.). Washington, DC: American Institute of Biological Sciences, (Pub. No. 7).

Collins, A., & Koechlin, E. (2012). Reasoning, learning, and creativity: Frontal lobe function and human decision-making. *PLoS Biology, 10*(3), e1001293. doi:10.1371/journal.pbio.1001293.

Comer, C., & Harrison, D. W. (2013). Resurrecting the opponent process theory using neuropsychological models: An integrative approach. Manuscript submitted for publication.

Committeri, G., Pitzalis, S., Galati, G., Patria, F., Pelle, G., Sabatini, U., et al. (2007). Neural bases of personal and extrapersonal neglect in humans. *Brain, 130,* 431–441.

Cooper, R. M., Bilash, M. A., & Zubek, J. P. (1959). The effect of age on taste sensitivity. *Journal of Gerontology, 14*(1), 56–58.

Corballis, M. C. (1998). Cerebral asymmetry: Motoring on. *Trends in Cognitive Sciences, 2,* 152–157.

Corballis, M. C. (2010). Handedness and cerebral asymmetry. In K. Hugdahl & R. Westerhausen (Eds.), *The two halves of the brain* (pp. 65–88). Cambridge: MIT Press.

Corbetta, M., & Shulman, G. L. (2002). Control of goal-directed and stimulus-driven attention in the brain. *Nature Reviews. Neuroscience, 3,* 201–215.

Corbetta, M., et al. (2008). The reorienting system of the human brain: from environment to theory of mind. *Neuron, 58,* 306–324.

Coren, S. (1992). *The left-hander syndrome: The causes and consequences of left-handedness.* New York: Free Press.

Corina, D. P. (1998). Aphasia in users of signed languages. In P. Coppens, Y. Lebrun, & A. Basso (Eds.), *Aphasia in atypical populations* (pp. 261–310). Mahwah: Erlbaum.

Corkin, S. (1965). Tactually-guided maze-learning in man: Effects of unilateral cortical excisions and bilateral hippocampal lesions. *Neuropsychologia, 3,* 339–351.

Corsi, P. M. (1972). *Human memory and the medial temporal region of the brain.* Ph.D. thesis, McGill University, Montreal, Quebec, Canada.

Cosetti, M. K., & Waltzman, S. B. (2011). Cochlear implants: Current status and future potential. *Expert Review of Medical Devices, 8*(3), 389–401.

Coslett, H. B., Wiener, M., & Chatterjee, A. (2010). Dissociable neural systems for timing: Evidence from subjects with basal ganglia lesions. *PLoS One, 5,* e10324.

Coull, J. T., Cheng, R. K., & Meck, W. H. (2011). Neuroanatomical and neurochemical substrates of timing. *Neuropsychopharmacology, 36,* 3–25.

Coulthard, E., Rudd, A., Playford, E. D., & Husain, M. (2007). Alien limb following posterior cerebral artery stroke: Failure to recognize internally generated movements? *Movement Disorders, 22,* 1498–1502. doi:10.1002/mds.21546.

Coupar, F., Pollock, A., Rowe, P., Weir, C., & Langhorne, P. (2012). Predictors of upper limb recovery after stroke: A systematic review and meta-analysis. *Clinical Rehabilitation, 26,* 4291–4313.

Courtney, S. M. (2004). Attention and cognitive control as emergent properties of information representations in working memory. *Cognitive Affective Behavioral Neuroscience, 4,* 501–516.

Courtney, C., Farrell, D., & Gray, R., et al. (2004). Long-term donepezil treatment in 565 patients with Alzheimer's disease (AD2000): Randomised double-blind trial. *Lancet, 363*(9427), 2105–2115.

Cox, D. E., & Harrison, D. W. (2008a). Fluency and anger expression styles: Evidence for a functional cerebral systems model of anger. *Archives of Clinical Neuropsychology, 23,* 663.

Cox, D. E, & Harrison, D. W. (2008b). Models of anger: contributions from psychophysiology, neuropsychology, and the cognitive behavioral perspective. *Brain Structure & Function, 212,* 371–385.

Cox, D. E., & Harrison, D. W. (2008c). Neuropsychological and psychophysiological correlates of anger expression styles. *Journal of the International Neuropsychological Society, 14,* 272.

Cox, D. E., & Harrison, D. W. (2012). The effect of anger expression style on cardiovascular responses to lateralized cognitive stressors. Manuscript submitted for publication.

Cox, D. E., & Heilman, K. (2011). Thalamic intentional aphasia: A failure of lexical semantic self-activation. *Neurocase, 17*(4), 313–317.

Cox, D., Meyers, E., & Sinha, P. (2004). Contextually evoked object-specific responses in human visual cortex. *Science, 304,* 115–117.

Cox, T. M., Ragen, T. J., Read, A. J., Vos, E., Baird, R. W., Balcomb, K., & Benner, L. (2006). Understanding the impacts of anthropogenic sound on beaked whales. *Journal of Cetacean Research and Management, 7*(3), 177–187.

Cox, D. E., Holland, A. K., Carmona, J. E., Golkow, K., Valentino, S. E., & Harrison, D. W. (2008). Lateralized frontal activation as a function of food consumption. *Journal of the International Neuropsychological Society, 14,* 272.

Cox, D. E., Foster, P. S., Rhodes, R. D., Demaree, H. E., Everhart, E., & Harrison, D. W. (2012). Cerebral asymmetry in emotion and cardiovascular regulation: A functional systems analysis. Manuscript submitted for publication.

Coyle, J. T., Price, D. L., & DeLong, M. R. (1983). Alzheimer's disease: A disorder of cortical cholinergic innervation. *Science, 219*(4589), 1184–1190.

Craig, A. D. (2002). How do you feel? Interoception: The sense of the physiological condition of the body. *Nature Reviews Neuroscience, 3,* 655–666.

Craig, A. D. (2009). How do you feel-now? The anterior insula and human awareness. *Nature Reviews. Neuroscience, 10,* 59–70.

Craig, A. D. (2011). Significance of the insula for the evolution of human awareness of feelings from the body. *Annals of the New York Academy of Sciences, 1225,* 72–82.

Crair, M. C., Gillespie, D. C., & Stryker, M. P. (1998). The role of visual experience in the development of columns in cat visual cortex. *Science, 279,* 566–570.

Creem-Regehr, S. H., Neil, J. A., & Yeh, H. J. (2007). Neural correlates of two imagined egocentric transformations. *Neuroimage, 35,* 916–927.

Crews, W. D., Jr. (2012). CogniCheck.com: CogniCheck Memory Screenings, CogniCheck, Inc., Lynchburg.

Crews, W. D., Jr., & Bonaventura, S. (1992). The endogenous opiate system, Naltrexone administration, and self-injurious behavior: A brief report. *Southeastern Review, 1*(2), 3–4.

Crews, W. D., Jr., & Harrison, D. W. (1995). The neuropsychology of depression and its implications for cognitive therapy. *Neuropsychology Review, 5*(2), 81–123.

Crews, W. D., Jr., Harrison, D. W., Rhodes, R. D., & Demaree, H. A. (1995). Hand fatigue asymmetry in the motor performances of women with depressed mood. *Neuropsychiatry, Neuropsychology, and Behavioral Neurology, 8,* 277–281.

Crews, W. D., Harrison, D. W., & Rhodes, R. D. (1999a). Neuropsychological test performances of young depressed outpatient women: An examination of executive functions. *Archives of Clinical Neuropsychology, 14,* 517–529.

Crews, W. D., Jr., Rhodes, R. D., Bonaventura, S. H., Rowe, F. B., & Goering, A. M. (1999b). Cessation of long-term Naltrexone administration: Longitudinal follow-ups. *Research in Developmental Disabilities, 20*(1), 23–30.

Crews, W. D., Jr., Harrison, D. W., Griffin, M. L., Addison, K., Yount, A. M., Giovenco, M. A., & Hazell, J. (2005a). A double-blinded, placebo-controlled, randomized trial of the neuropsychological efficacy of cranberry juice in a sample of cognitively intact older adults: Pilot study findings. *The Journal of Alternative and Complementary Medicine, 11*(2), 305–309.

Crews, W. D., Jr., Harrison, D. W., Griffin, M. L., Falwell, K. D., Crist, T., Longest, L., Hehemann, L., & Rey, S. T. (2005b). The neuropsychological efficacy of Ginkgo preparations in healthy and cognitively intact adults: A comprehensive review. *Herbal Gram, 67,* 43–62.

Crews, W. D., Harrison, D. W., & Wright, J. D. (2008). A double-blind, placebo-controlled, randomized trial of the efficacy of dark chocolate and cocoa on variables associated with neuropsychological functioning and cardiovascular health: Clinical findings from a sample of healthy, cognitively intact older adults. *American Journal of Clinical Nutrition, 87,* 872–880.

Crews, W. D., Harrison, D. W., Keiser, A. M., & Kunze, C. M. (2009). The memory screening outreach Program: Findings from a large community-based sample of middle-aged and older adults. *Journal of the American Geriatrics Society, 57*(9), 1697–1703.

Crews, W. D., Jr., Harrison, D. W., Gregory, K. P., Kim, B., & Darling, A. B. (2013). The effects of cocoa- and chocolate-related products on neurocognitive functioning. In R. R. Watson, V. R. Preedy, & S. Zibadi (Eds.), *Chocolate in health and nutrition* (pp. 369–379). New York: Humana Press.

Crisinel, A., & Spence, C. (2012). A fruity note: Crossmodal associations between odors and musical notes. *Chemical Senses, 37*(2), 151–158. doi:10.1093/chemse/bjr085.

Critchley, H. D., Elliott, R., Mathias, C. J., & Dolan, R. J. (2000). Neural activity relating to generation and representation of galvanic skin conductance responses: A functional magnetic resonance imaging study. *The Journal of Neuroscience, 20*(8), 3033–3040.

Crowley, S. K., & Youngstedt, S. D. (2012). Efficacy of light therapy for perinatal depression: A review. *Journal of Physiological Anthropology, 31,* 15.

Cruz, A., & Green, B. G. (2000). Thermal stimulation of taste. *Nature, 403,* 889–892.

Cubillo, A., Halari, R., Smith, A., Taylor, E., & Rubia, K. (2012). A review of fronto-striatal and fronto-cortical brain abnormalities in children and adults with Attention Deficit Hyperactivity Disorder (ADHD) and new evidence for dysfunction in adults with ADHD during motivation and attention. *Cortex, 48*(2), 194–215.

Cui, X., Bryant, D. M., & Reiss, A. L. (2012). NIRS-based hyperscanning reveals increased interpersonal coherence in superior frontal cortex during cooperation. *Neuroimage, 59*(3), 2430–2437.

Culbertson, J. C., & Hyndman, J. G. (1879). The Cincinnati Lancet and Clinic: A Weekly Journal of Medicine and Surgery, Volume III, p. 305.

Czeisler, C. A., Weitzman, E. D., Moore-Ede, M. C., Zimmerman, J. C., & Knauer, R. S. (1980). Human sleep: Its duration and organization depend on its circadian phase. *Science, 210*(4475), 1264–1267. http://dx.doi.org/10.1126/science.7434029, PMid:7434029.

Czeisler, C. A., Kronauer, R. E., Allan, J. S., Duffy, J. F., Jewett, M. E., Brown, E. N., & Ronda, J. M. (1989). Bright light induction of strong (type 0) resetting of the human circadian pacemaker. *Science, 244*(4910), 1328–1333. doi:10.1126/science.2734611.

Dabbs, J. M., Jr., & Hargrove, M. F. (1997). Age, testosterone, and behavior among female prison inmates. *Psychosomatic Medicine, 59,* 477–480.

Dabbs, J. M., Jr., Carr, T. S., Frady, R. L., & Riad, J. K. (1995). Testosterone, crime, and misbehavior among 692 male prison inmates. *Personality and Individual Differences, 9,* 269–275.

Dadda, M., & Bisazza, A. (2012). Prenatal light exposure affects development of behavioural lateralization in a livebearing fish. *Behavioural Processes, 91*(1), 115–118.

Dalton, P., Doolittle, N., & Breslin, P. A. (2002). Gender-specific induction of enhanced sensitivity to odors. *Nature Neuroscience, 5*(3), 199–200.

Damasio, A. R. (1994). *Descartes' error: Emotion, reason, and the human brain.* New York: Penguin Books.

Damasio, A. R. (2005). A modern Phineas Gage. Descartes' Error: Emotion, reason, and the human brain. ISBN 014303622X. (1st ed.: 1994).

Damasio, A. R., & Anderson, S. W. (2003). The frontal lobes. In K. M. Heilman & E. Valenstein (Eds.), *Clinical neuropsychology* (pp. 404–446). New York: Oxford University Press.

Damasio, A. R., & Carvalho, G. B. (2013). The nature of feelings: Evolutionary and neurobiological origins. *Nature Reviews Neuroscience, 14*(2), 143–152.

Damasio, A. R., Bellugi, U., Damasio, H., Poizner, H., & Van Gilder, J. (1986). Sign language aphasia during left-hemisphere amytal injection. *Nature, 322,* 363–365. doi:10.1038/322363a0.

Damasio, H., Grabowski, T., Frank, R., Galaburda, A. M., & Damasio, A. R. (1994). The return of Phineas Gage: Clues about the brain from the skull of a famous patient. *Science, 264*(5162), 1102–1105. doi:10.1126/science.8178168, PMID 8178168.

Damasio, A. R., Grabowski, T. J., Bechara, A., Damasio, H., Ponto, L. L. B., Parvizi, J., & Hichwa, R. D. (2000). Subcortical and cortical brain activity during the feeling of self-generated emotions. *Nature Neuroscience, 3,* 1049–1056.

Damasio, A. R., Anderson, S. W., & Tranel, D. (2012). The frontal lobes. In K. M. Heilman & E. Valenstein (Eds.), *Clinical neuropsychology* (pp. 417–465). New York: Oxford University Press.

Damasio, A., Damasio, H., & Tranel, D. (2013). Persistence of feelings and sentience after bilateral damage of the insula. *Cerebral Cortex, 23*(4), 833–846.

Dandy, W. E. (1946). The location of the conscious center in the brain; the corpus striatum. *Bulletin of the Johns Hopkins Hospital, 79,* 34–58.

Dapretto, M. (2006). Understanding emotions in others: Mirror neuron dysfunction in children with autism spectrum disorders. *Nature Neuroscience, 9*(1), 28–30.

Darwin, C. (1872). *The expression of the emotions in man and animals.* London: John Murray.

Darwin, C. (1872/1998). *The expression of the emotions in man and animals* (3rd ed.). New York: Oxford University Press.

Das, L. S. (2007). *Buddha is as Buddha does.* New York: Harper Collins Publisher.

Davidson, R. J. (1993). Cerebral asymmetry and emotion: Conceptual and methodological conundrums. *Cognition and Emotion, 7,* 115–138.

Davidson, R. J. (1995). Cerebral asymmetry, emotion, and affective style. In R. J. Davidson & K. Hugdahl (Eds.), *Brain asymmetry* (pp. 361–388). Cambridge: MIT Press.

Davidson, R. J. (1998). Anterior electrophysiological asymmetries, emotion, and depression: Conceptual and methodological conundrums. *Psychophysiology, 35,* 607–614.

Davidson, R. J. (2000). What does the prefrontal cortex "do" in affect: Perspectives on frontal EEG asymmetry research. *Biological Psychology, 67*(1–2), 219–234.

Davidson, R. J. (2003a). Seven sins in the study of emotion: Correctives from affective neuroscience. *Brain and Cognition, 52,* 129–132.

Davidson, R. J. (2003b). Affective neuroscience and psychophysiology: Toward a synthesis. *Psychophysiology, 40,* 655–665.

Davidson, R. J., & Fox, N. A. (1982). Asymmetrical brain activity discriminates between positive and negative affective stimuli in human infants. *Science, 218,* 1235–1237.

Davidson, R. J., Ekman, P., Saron, C. D., Senulis, J. A., & Friesen, W. V. (1990). Approach-withdrawal and cerebral asymmetry: Emotional expression and brain physiology I. *Journal of Personality and Social Psychology, 58*(2), 330–341.

Davidson, R. J., Coe, C. C., Dolski, I., & Donzella, B. (1999). Individual differences in prefrontal activation asymmetry predict natural killer cell activity at rest and in response to challenge. *Brain, Behavior, and Immunity, 13,* 93–108.

Davis, M. H. (1996). *Empathy-A social psychological approach.* Boulder: Westview.

Davis, A. M., & Natelson, B. H. (1993). Brain-heart interactions. The neurocardiology of arrhythmia and sudden cardiac death. *Texas Heart Institute Journal, 20*(3), 158–169.

Davis, M., Walker, D. L., Miles, L. & Grillon, C. (2010). *Neuropsychopharmacology, 35,* 105–135.

Davis, F. C., Knodt, A. R., Sporns, O., Lahey, B. B., Zald, D. H., Brigidi, B. D., & Hariri, A. R. (2012). Impulsivity and the modular organization of resting-state neural networks. *Cerebral Cortex* (first published online May 29, 2012). doi:10.1093/cercor/bhs126.

Debener, S., Beauducel, A., Nessler, D., Brocke, B., Heilemann, H., & Kayser, J. (2000). Is resting anterior EEG alpha asymmetry a trait marker for depression? *Neuropsychobiology, 41,* 31–37.

Deems, D. A., Doty, R. L., Settle, R. G., et al. (1991). Smell and taste disorders. A study of 750 patients from the University of Pennsylvania Smell and Taste Center. *Archives of Otolaryngology-Head & Neck Surgery, 117,* 519–528.

de Groot, J. H. B., Smeets, M. A. M., Kaldewaij, A., Duijndam, M. J. A., & Semin, G. R. (2012). Chemosignals communicate human emotions. *Psychological Science* (Published online before print September 27, 2012). doi:10.1177/0956797612445317.

Dehaene, S., & Cohen, L. (1997). Cerebral pathways for calculation: Double dissociation between rote verbal and quantitative knowledge of arithmetic. *Cortex, 33,* 219–250. doi:10.1016/j.neuropsychologia.2006.11.012.

Déjerine, J. (1891). Sur un cas de cécité verbale avec agraphie, suivi d'autopsie. *Mémoires de la Société de Biologie, 3,* 197–201.

Delay, E. R., Smith, E. S., & Isaac, W. (1978). Effects of illumination on auditory threshold. *Physiology & Behavior, 20,* 201–202.

Delay, E. R., Steiner, N. O., & Isaac, W. (1979). The effects of d-amphetamine and methylphenidate upon auditory threshold in the squirrel monkey. *Pharmacology, Biochemistry, & Behavior, 10,* 861–864.

De Leon, A. (2006). Palonosetron (Aloxi): A second-generation 5-HT(3) receptor antagonist for chemotherapy-induced nausea and vomiting. *Proceedings (Baylor University Medical Center), 19*(4), 413–416.

Delgado, J. M. R. (1969). *Physical control of the mind: Toward a psychocivilized society*. New York: Harper and Row.

Delgado, J. M. R. (1977–1978). Instrumentation, working hypotheses, and clinical aspects of neurostimulation. *Applied Neurophysiology, 40,* 88–110.

Delgutte, B., Joris, P. X., Litovsky, R. Y., & Yin, T. C. T. (1999). Receptive fields and binaural interactions for virtual-space stimuli in the cat inferior colliculus. *Journal of Neurophysiology, 81,* 2833–2851.

D'Elia, L. F., Satz, P., Uchiyama, C. L., & White, T. (1996). *Color trails test*. Odessa: Psychological Assessment Resources, Inc.

Delis, D. C., Robertson, L. C., & Efron, R. (1986). Hemispheric specialization of memory for visual hierarchical stimuli. *Neuropsychologia, 24,* 205–214.

Delis, D. C., Kramer, J., Kaplan, E., & Ober, B. A. (1987). *California Verbal Learning Test (CVLT) manual*. San Antonio: The Psychological Corporation.

Delis, D., Kaplan, E., Kramer, J., & Ober, B. (Eds.). (2000). *California Verbal Learning Test-II*. San Antonio: Psychological Corporation.

Delis, D. C., Kramer, J. H., & Kaplan, E. (2001). *The Delis-Kaplan executive function system*. San Antonio: The Psychological Corporation.

Della-Morte, D., & Rundek, T. (2012). Dizziness and vertigo. In M. Paciaroni, G. Agnelli, V. Caso, & J. Bogousslavsky (Eds.), *Manifestations of stroke. Frontiers of neurology and neuroscience* (Vol. 30, pp. 22–25). Basel: Karger.

Delmonico, R. L., Hanley-Peterson, P., & Englander, J. (1998). Group psychotherapy for persons with traumatic brain injury: Management of frustration and substance abuse. *The Journal of Head Trauma Rehabilitation, 13*(6), 10–22.

Demakis, G. J., & Harrison, D. W. (1994). Subvocal rehearsal of neutral and affective words interferes with left-hemisphere performance and facilitates right-hemisphere performance. *Psychobiology, 22,* 238–243.

Demakis, G. J., & Harrison, D. W. (1997). Relationships between verbal and nonverbal fluency measures: Implications for assessment of executive functioning. *Psychological Reports, 81,* 443–448.

Demakis, G. J., Herridge, M. L., & Harrison, D. W. (1994). Pathological display of positive affect: A bilateral electrodermal case study. *Neuropsychiatry, Neuropsychology, & Behavioral Neurology, 7*(3), 154–159.

Demaree, H. A., & Harrison, D. W. (1996). Case study: Topographical brain mapping in hostility following mild closed-head injury. *The International Journal of Neuroscience, 87,* 97–101.

Demaree, H. A., & Harrison, D. W. (1997a). A neuropsychological model relating self-awareness to hostility. *Neuropsychology Review, 7*(4), 171–185.

Demaree, H. A., & Harrison, D. W. (1997b). Physiological and neuropsychological correlates of hostility. *Neuropsychologia, 35*(10), 1405–1411.

Demaree, H. A., Crews, W. D., Jr., & Harrison, D. W. (1995). Topographical brain mapping in depression following mild closed head injury: A case study. *Journal of Neurotherapy, 1*(1), 38–43.

Demaree, H. A., Harrison, D. W., & Rhodes, R. D. (2000). Quantitative electroencephalographic analyses of cardiovascular regulation in low- and high-hostile men. *Psychobiology, 28,* 420–431.

Demaree, H. A., Higgins, D. A., Williamson, J. B., & Harrison, D. W. (2002). Asymmetry in hand grip strength and fatigue in low- and high-hostile men. *International Journal of Neuroscience, 112,* 415–428.

Demaree, H. A., Everhart, D. E., Youngstrom, E. A., & Harrison, D. W. (2005). Brain lateralization of emotional processing: Historical roots and a future incorporating "dominance." *Behavioral and Cognitive Neuroscience Reviews, 4*(1), 3–20.

Demeyer, I., De Lissnyder, E., Koster, E. H., & De Raedt, R. (2012). Rumination mediates the relationship between impaired cognitive control for emotional information and depressive symptoms: A prospective study in remitted depressed adults. *Behavior Research and Therapy, 50,* 292–297.

Dempsey, E. W., & Morison, R. S. (1942a). The production of rhythmically recurrent cortical potentials after localized thalamic stimulation. *American Journal of Physiology, 135*, 293–300.

Dempsey, E. W., & Morison, R. S. (1942b). The interaction of certain spontaneous and induced cortical potentials. *American Journal of Physiology, 135*, 301–308.

Dempsey, E. W., & Morison, R. S. (1943). The electrical activity of a thalamocortical relay system. *American Journal of Physiology, 138*, 283–296.

Denenberg, V. H. (1981). Hemispheric laterality in animals and the effects of early experience. *Behavioral and Brain Sciences, 4*, 1–21.

Denman, S. B. (1987). *Denman neuropsychology memory scale.* Charleston, S. C., Private Publishing.

Denny-Brown, D. (1956). Positive and negative aspects of cerebral cortical functions. *North Carolina Medical Journal, 17*, 295–303.

Denny-Brown, D., Meyer, J. S., & Horenstein, S. (1952). The significance of perceptual rivalry resulting from parietal lesions. *Brain, 75*, 434–471.

Denson, T. F., DeWall, C. N., & Finkel, E. J. (2012). Self control and aggression. *Current Directions in Psychological Science, 21*(1), 20–25. doi:10.1177/0963721411429451.

Depue, R. A., & Collins, P. F. (1999). Neurobiology of the structure of personality: Dopamine, facilitation of incentive motivation, and extraversion. *Behavioral and Brain Sciences, 22*, 491–569.

De Renzi, E. (1982). *Disorders of space exploration and cognition.* Chichester: Wiley.

De Renzi, E., Pieczuro, A., & Vignolo, L. A. (1968). Ideational apraxia: A quantitative study. *Neuropsychologia, 6*, 41–52.

Derntl, B., Schöpf, V., Kollndorfer, K., & Lanzenberger, R. (2013). Menstrual cycle phase and duration of oral contraception intake affect olfactory perception. *Chemical Senses, 38*(1), 67–75.

Descartes, R. (1637). *Discourse on method and related writings.* Penguin, 1999.

Descartes, R. (1649). Les passions de l'ame. In Descartes: Oeuvres et Lettres (Bibliotheque de la Pleiade), A. Bridoux (Ed.), Gallimard. 1969, p. 710.

Descartes, R. (1966–1983). In E. Lojacono (Ed.), *Opere scientifiche.* Torino: UTET.

Deslandes, A., de Moraes, H., Pompeu, F., Ribeiro, P., Cagy, M., Capitao, C., Alves, H., Piedade, R., & Laks, J. (2008). Electroencephalographic frontal asymmetry and depressive symptoms in the elderly. *Biological Psychology, 79*, 317–322.

Dias, R., Robbins, T. W., & Roberts, A. C. (1996). Dissociation in prefrontal cortex of affective and attentional shifts. *Nature, 380*, 69–72.

Dias, R., Robbins, T. W., & Roberts, A. C. (1997). Dissociable forms of inhibitory control within prefrontal cortex with an analog of the Wisconsin card sort test: Restrictions to novel situations and independence from "on-line" processing. *Journal of Neuroscience, 17*, 9285–9297.

Dickens, James Cecil "Little Jimmy Dickens". (2013). Star of the Grand Ole Opry. Nashville, TN.

Diekelmann, S., & Born, J. (2010). The memory function of sleep. *Nature Review Neuroscience, 11*(2), 114–126.

Dietrich, A. (2004). The cognitive neuroscience of creativity. *Psychonomic Bulletin & Review, 11*(6), 1011–1026. doi:10.3758/BF03196731.

Dijk, D. J., & Czeisler, C. A. (1995). Contribution of the circadian pacemaker and the sleep homeostat to sleep propensity, sleep structure, electroencephalographic slow waves, and sleep spindle activity in humans. *Journal of Neuroscience, 15*(5 Pt 1), 3526–3538. PMid: 7751928.

Dimberg, U., & Petterson, M. (2000). Facial reactions to happy and angry facial expressions: Evidence for right hemisphere dominance. *Psychophysiology, 37*, 693–696.

Ditchburn, R. W., & Ginsborg, B. L. (1952). Vision with a stabilized retinal image. *Nature, 170*, 36–37.

Djouhri, L., Fang, X., Koutsikou, S., & Lawson, S. N. (2012). Partial nerve injury induces electrophysiological changes in conducting (uninjured) nociceptive and nonnociceptive DRG neurons: Possible relationships to aspects of peripheral neuropathic pain and paresthesias. *Pain, 153*(9), 1824–1836.

Dodrill, C. B. (1978). The hand dynamometer as a neuropsychological measure. *Journal of Consulting and Clinical Psychology, 46*(6), 1432–1435.

Dog, Wikipedia. (July 2005). The olfactory bulb in dogs is roughly forty times bigger than the olfactory bulb in humans, relative to total brain size, with 125–220 million smell-sensitive receptors. The bloodhound exceeds this standard with nearly 300 million receptors. Dogs can discriminate odors at concentrations nearly 100 million times lower than humans can. (Footnotes omitted); The Dog's Sense of Smell, UNP-66, at 1 (In fact, a dog has more than 220 million olfactory receptors in its nose, while humans have only 5 million.

Dolan, R. J., Bench, C. J., Brown, R. G., Scott, L. C., & Frackowiak, R. S. J. (1994). Neuropsychological dysfunction in depression: The relationship to regional cerebral blood flow. *Psychological Medicine, 24*, 849–857. doi:10.1017/S0033291700028944.

Dolder, C. R., & Nelson, M. H. (2008). Hypnosedative-induced complex behaviours: Incidence, mechanisms and management. *CNS Drugs, 22*(12), 1021–1036. doi:10.2165/0023210-200822120-00005. PMID 18998740.

Dominguez-Alonso, A., Ramirez-Rodriguez, G., & Benitez-King, G. (2012). Melatonin increases dendritogenesis in the hilus of hippocampal organotypic cultures. *Journal of Pineal Research, 52*, 427–436.

Doty, R. L. (1976). Reproductive endocrine influences on human nasal chemoreception: A review. In Doty, R. L. (Ed.), *Mammalian olfaction, reproductive processes, and behavior* (pp. 295–321). New York: Academic Press.

Doty, R. L. (1986). Gender and endocrine-related influences on human olfactory perception. In H. L. Meiselman & R. S. Rivlin (Eds.), *Clinical measurement of taste and smell* (pp. 377–413). New York: Macmillan.

Doty, R. L. (1995). *The smell identification test administration manual*. Haddon Heights: Sensonics Inc.

Doty, R. L., Shaman, P., & Dann, M. (1984). Development of the University of Pennsylvania Smell Identification Test a standardized microencapsulated test of olfactory function. *Physiology & Behavior, 32*, 489–502.

Doty, R. L., Frye, R. E., & Agrawal, U. (1989). Internal consistency and reliability of the fractionated and whole University of Pennsylvania Smell Identification Test. *Perception & Psychophysics, 45*, 381–384.

Dowling, G. A., et al. (2008). Melatonin and bright-light treatment for rest-activity disruption in institutionalized patients with Alzheimer's disease. *Journal of the American Geriatrics Society, 56*, 239–246.

Downing, P. E., Jiang, Y., Shuman, M., & Kanwisher, N. (2001a). A cortical area selective for visual processing of the human body. *Science, 293*, 2405–2407.

Downing, P. E., Jiang, Y., Shuman, M., & Kanwisher, N. A. (2001b). Cortical area selective for visual processing of the human body. *Science, 293*, 2470–2473.

Dozier, J., & Brown, W. D. (2012). Psychiatric disorders that affect sleep. *Fundamentals of Sleep Technology* (p. 228).

Draganski, B., Kherif, F., Kloppel, S., Cook, P. A., Alexander, D. C., Parker, G. J., et al. (2008). Evidence for segregated and integrative connectivity patterns in the human basal ganglia. *Journal of Neuroscience, 28*, 7143–7152. doi:10.1523/JNEUROSCI.1486-08.2008.

Drake, R. A. (1984). Lateral asymmetry of personal optimism. *Journal of Research in Personality, 18*, 497–507.

Drake, R. A. (1987). Effects of gaze manipulation on aesthetic judgments: Hemisphere priming of affect. *Acta Psychologica, 65*, 91–99.

Drevets, W. C., Price, J. L., Simpson, J. R., Todd, R. D., Reich, T., Vannier, M., & Raichle, M. E. (1997). Subgenual prefrontal cortex abnormalities in mood disorders. *Nature, 386*, 824–827.

Drevets, W. C., Price, J. L., & Furey, M. L. (2008). Brain structural and functional abnormalities in mood disorders: Implications for neurocircuitry models of depression. *Brain Structure and Function, 213*, 93–118.

Drewe, E. A. (1975). Go—No go learning after frontal lobe lesions in humans. *Cortex, 11*(1), 8–16.

Dubois, J., Benders, M., Cachia, A., Lazeyras, F., Ha-Vinh Leuchter, R., Sizonenko, S. V., Borradori-Tolsa, C., Mangin, J. F., & Huppi, P. S. (2008). Mapping the early cortical folding process in the preterm newborn brain. *Cerebral Cortex, 18*, 1444–1454.

Du Bois-Reymond, E. (1848). *Untersuchungen über thierische Elektricität, Erster Band.* Berlin: Georg Reimer. (The book can be downloaded from Max Planck Institute for the History of Science (806 pages, high resolution, ~180.24 MB)). http://vlp.mpiwg-berlin.mpg.de/library/pdf/lit92__Hi.pdf.

Duckworth, J. L., Grimes, J., & Ling, G. S. F. (2012). Pathophysiology of battlefield associated traumatic brain injury. *Pathophysiology* (Available online 14 June 2012). doi:10.1016/j.pathophys.2012.03.001, [Epub ahead of print].

Dudchenko, P. (2010). *Why people get lost: The psychology and neuroscience of spatial cognition.* Oxford: Oxford University Press. ISBN 0-19-921086-1. OCLC 791205815.

Duffau, H. (2012). The "frontal syndrome" revisited: Lessons from electrostimulation mapping studies. *Cortex, 48*(1), 120–131.

Duma, S. M. (2012). *Head acceleration measurements in Helmet-Helmet impacts and the youth population.* Ray Winston Daniel II (Doctoral dissertation, Virginia Polytechnic Institute and State University).

Duncan, W. C., Jr., Pettigrew, K. D., & Gillin, J. C. (1979). REM architecture changes in bipolar and unipolar depression. *American Journal of Psychiatry, 136*(11), 1424–1427.

Dunn, T. D. (1895). Double hemiplegia with double hemianopsia and loss of geographic centre. *Transactions, College of Physicians of Philadelphia, 17,* 45–56.

Earley, P. (24 May 2013). Mental illness manual no "Bible": Doctors, insurers rely on it for diagnoses. But it's neuroscience that offers the path to a cure. *USA Today,* 12A.

Eastwood, J. D., Smilek, D., & Merikle, P. M. (2001). Differential attentional guidance by unattended faces expressing positive and negative emotion. *Perception and Psychophysics, 63,* 1004–1013.

Ebert, U., Grossmann, M., Oertel, R., Gramatte, T., & Kirch, W. (2001). Pharmacokinetic pharmacodynamic modeling of the electroencephalogram effects of scopolamine in healthy volunteers. *Journal of Clinical Pharmacology, 41,* 51–60.

Ecker, J. L., Dumitrescu, O. N., Wong, K. Y., Alam, N. M., Chen, S. K., LeGates, T., Renna, J. M., Prusky, G. T., Berson, D. M., & Hattar, S. (2010). Melanopsin-expressing retinal ganglion-cell photoreceptors: Cellular diversity and role in pattern vision. *Neuron, 67*(1):49–60. doi:10.1016/j.neuron.2010.05.023.

Egan, M. F., Goldberg, T. E., Kolachana, B. S., Callicott, J. H., Mazzanti, C. M., Straub, R. E., Goldman, D., & Weinberger, D. R. (2001). Effect of COMT Val[108/158] Met genotype on frontal lobe function and risk for schizophrenia. *Proceedings of the National Academy of Sciences of the United States of America, 98*(12), 6917–6922 (Published online 2001 May 29). doi:10.1073/pnas.111134598.

Egeth, H. E., & Yantis, S. (1997). Visual attention: Control, representation, and time course. *Annual Review of Psychology, 48,* 269–297. doi:10.1146/annurev.psych.48.1.269.

Ehrlichman, H. (1987). Hemispheric asymmetry and positive-negative affect. In D. Ottoson (Ed.), *Duality and unity of the brain.* Hampshire: Macmillan.

Eibl-Eibesfeldt, I. (1989). *Human ethology.* New York: Aldine de Gruyter.

Eichenbaum, H., & Cohen, N. J. (2001). *From conditioning to conscious recollection: Memory systems of the brain.* Oxford: Oxford University Press.

Eimer, M., & Driver, J. (2001). Crossmodal links in endogenous ans exogenous spatial attention: Evidence from event-related brain potential studies. *Neuroscience and Biobehavioral Reviews, 25,* 497–511.

Ekman, P. (1973). Universal facial expressions in emotion. *Studia Psychologica, 15,* 140–147.

Ekman, P., Levenson, R. W., & Friesen, W. V. (1983). Autonomic nervous system activity distinguishes among emotions. *Science, 221*(4616), 1208–1210.

Ekman, P., Davidson, R. J., & Friesen, W. V. (1990). The Duchenne smile: Emotional expression and brain physiology. II. *Journal of Personality and Social Psychology, 58*(2), 342–353.

Ekstrom, A. D., Kahana, M. J., Caplan, J. B., Fields, T. A., Isham, E. A., Newman, E. L., et al. (2003). Cellular networks underlying human spatial navigation. *Nature, 25,* 184–187.

Elder, R. (2001). The science of attraction. *Journal of Hybrid Vigor,* Emory University, Issue 1.

Elie, B., & Guiheneuc, P. (1990). Sympathetic skin response: Normal results in different experimental conditions. *Electroencephalography and Clinical Neurophysiology, 76*(3), 258–267.

Elsberg, C. A., Levy, I., & Brewer, E. D. (1936). A new method for testing the sense of smell and for the establishment of olfactory values of odorous substances. *Science, 83*, 211–212. doi:10.1126/science.83.2148.211.

Emerson, C. S., & Harrison, D. W. (1990). Anger and denial as predictors of cardiovascular reactivity in women. *Journal of Psychopathology and Behavioral Assessment, 12*(4), 271–283.

Emerson, C. S., Harrison, D. W., & Everhart, D. E. (1999). Investigation of receptive affective prosodic ability in school-aged boys with and without depression. *Neuropsychiatry, Neuropsychology, and Behavioral Neurology, 12*, 102–109.

Emerson, C. S., Mollet, G. A., & Harrison, D. W. (2005). Grip strength asymmetry in depressed boys. *Neuropsychiatry, Neuropsychology, and Behavioral Neurology, 142*(2), 130–134.

Emmorey, K., Damasio, H., McCullough, S., Grabowski, T., Ponto, L. L. B., Hichwa, R. D., et al. (2002). Neural systems underlying spatial language in American sign language. *NeuroImage, 17*, 812–824.

Emond, V., Joyal, C., & Poissant, H. (2009). Structural and functional neuroanatomy of attention-deficit hyperactivity disorder (ADHD). *L'Encéphale, 35*(2), 107–114.

Emre, M. (2003). Dementia associated with Parkinson's disease. *Lancet Neurology, 2*, 229–237.

Engel, J., Jr. (2013). *Seizures and epilepsy* (Vol. 83). USA: Oxford University Press.

Engel, S. A., Rumelhart, D. E., Wandell, B. A., Lee, A. T., Glover, G. H., Chichilnisky, E. J., & Shadlen, M. N. (1994). fMRI of human visual cortex. *Nature, 369*, 525.

Engelberg, E., & Sjoberg, L. (2005). Emotional intelligence and inter-personal skills. In R. Schulze & R. Roberts (Eds.), *Emotional intelligence. An international handbook* (pp. 289–307). Massachusetts: Hogrefe & Huber Publishers.

Epple, G., & Herz, R. S. (1999). Ambient odors associated to failure influence cognitive performance in children. *Developmental Psychobiology, 35*, 103–107.

Epstein, R., & Kanwisher, N. (1998). A cortical representation of the local visual environment. *Nature, 392*(6676), 598–601.

Epstein, R., DeYoe, E. A., & Press, D. Z., et al. (2001). Neuropsychological evidence for a topographical learning mechanism in parahippocampal cortex. *Cognitive Neuropsychology, 18*, 481–508.

Eslinger, P. J., Grattan, L. M., & Geder, L. (1996). Neurologic and neuropsychiatric aspects of frontal lobe impairments in postconcussive syndrome. In M. Rizzo & D. Tranel (Eds.), *Head injury and postconcussive syndrome* (pp. 415–440). New York: Churchill Livingstone.

Eslinger, P. J., Flaherty-Craig, C. V., & Benton, A. L. (2004). Developmental outcomes after early prefrontal cortex damage. *Brain and Cognition, 55*, 84–103.

Espiritu, R. C., Kripke, D. F., Ancoli-Israel, S., Mowen, M. A., Mason, W. J., Fell, R. L., Klauber, M. R., & Kaplan, O. J. (1994). Low illumination experienced by San Diego adults: Association with atypical depressive symptoms. *Biological Psychiatry, 35*, 403–407.

Esquivel, G., Schruers, K., Maddock, R., Colasanti, A., & Griez, E. (2010). Review: Acids in the brain: A factor in panic? *Journal of Psychopharmacology, 24*(5), 639–647.

Etcoff, N. L., Ekman, P., Magee, J. J., & Frank, M. G. (2000). Lie detection and language comprehension. *Nature, 405*, 139.

Evans, D. A., Funkenstein, H. H., Albert, M. S., et al. (1989). Prevalence of Alzheimer's disease in a community population of older persons. Higher than previously reported. *JAMA, 262*(18), 2551–2556.

Evenden, J. (1999). Impulsivity: A discussion of clinical and experimental findings. *Journal of Psychopharmacology, 13*, 180–192.

Everhart, D. E., & Harrison, D. W. (1995). Hostility following right orbital frontal deactivation and right temporal activation. *Journal of Neurotherapy, 1*(2), 55–59.

Everhart, E., & Harrison, D. W. (1996). Hostility following right CVA: Support for right orbito-frontal deactivation and right temporal activation. *Journal of the International Neuropsychological Society, 2*, 65.

Everhart, D. E., & Harrison, D. W. (2000). Facial affect perception among anxious and non-anxious men. *Psychobiology, 28*(1), 90–98.

Everhart, E., Demaree, H. A., & Harrison, D. W. (1995). Topographical brain mapping: Hostility following closed head injury. *The Clinical Neuropsychologist, 9,* 280.

Everhart, D. E., Demaree, H. A., Harrison, D. W., & Williamson, J. B. (2001). "Delusions" of space: A case study utilizing topographical brain mapping and QEEG. *Journal of Neurotherapy, 4*(4), 19–29.

Everhart, D. E., Harrison, D. W., Shenal, B. V., Williamson, J., & Wuensch, K. L. (2002). Grip-strength, fatigue, and motor perseveration in anxious men without depression. *Neuropsychiatry, Neuropsychology, and Behavioral Neurology, 15*(2), 133–142.

Everhart, D. E., Demaree, H. A., & Wuensch, K. L. (2003). Healthy high-hostiles evidence low-alpha power (7.5-9.5 Hz) changes during negative affective learning. *Brain and Cognition, 52*(3), 334–342.

Everhart, D. E., Demaree, H. A., & Harrison, D. W. (2005). The merging of cognitive and affective neuroscience: Studies of the affective auditory verbal learning test. In A. V. Clark (Ed.), *Causes, role and influence of mood states.* Hauppauge: Nova Biomedical Books.

Everhart, D. E., Demaree, H. A., & Harrison, D. W. (2008a). Hostility and brain function: The impact of hostility on brain activity during affective verbal learning. In L. Sher (Ed.), *Psychological factors and cardiovascular disorders: The role of psychiatric pathology and maladaptive personality features.* Haupaugge: Nova Science Publishers.

Everhart, D. E., Demaree, H. A., & Harrison, D. W. (2008b). The influence of hostility on electroencephalographic activity and memory functioning during an affective memory task. *Clinical Neurophysiology, 119,* 134–143.

Exner, S. (1881). Untersuchungen über die lokalisation der Functionen in der Grosshirnrinde des Menschen. In W. Braunmüller (Ed.). Vienna.

Eyles, D. W., Smith, S., Kinobe, R., Hewison, M., & McGrath, J. J. (2005). Distribution of the vitamin D receptor and 1 alpha-hydroxylase in human brain. *Journal of Chemical Neuroanatomy, 29,* 21–30.

Farah, M. J., & Epstein, R. A. (2012). Disorders of visual-spatial perception and cognition. In K. Heilman & E. Valenstein (Eds.), *Clinical neuropsychology* (5th ed.). New York: Oxford University Press.

Feinstein, J. S., Adolphs, R., Damasio, A., & Tranel, D. (2011). The human amygdala and the induction and experience of fear. *Current Biology, 21,* 34–38.

Feinstein, J. S., Buzza, C., Hurlemann, R., Follmer, R. L., Dahdaleh, N. S., Coryell, W. H., & Wemmie, J. A. (2013). Fear and panic in humans with bilateral amygdala damage. *Nature Neuroscience.* doi:10.1038/nn.3323.

Feldberg, W., & Gaddum, J. H. (1934). The chemical transmitter at synapses in a sympathetic ganglion. *The Journal of Physiology, 81*(3), 305–319.

Feng, Z., et al. (2004). Melatonin alleviates behavioral deficits associated with apoptosis and cholinergic system dysfunction in the APP 695 transgenic mouse model of Alzheimer's disease. *Journal of Pineal Research, 37,* 129–136.

Fenske, M. J., & Eastwood, J. D. (2003). Modulation of focused attention by faces expressing emotion: Evidence from flanker tasks. *Emotion, 3,* 327–343.

Fenton, G. W. (1999). Neurosurgery for mental disorder: Past and present. *Advances in Psychiatric Treatment, 5,* 261–270.

Fergusson, D., Doucette, S., Cranley Glass, K., Shapiro, S., Healy, D., Hebert, P., & Hutton, B. (2005). Association between suicide attempts and selective serotonin reuptake inhibitors: Systematic review of randomized controlled trials. *British Medical Journal, 330*(396), 7488.

Fernaeus, S. E., Julin, P., Almqvist, O., & Wahlund, L. O. (2013). Medial temporal lobe volume predicts rate of learning in Rey-AVLT. *Advances in Alzheimer's Disease, 2*(1), 7–12. doi:10.4236/aad.2013.21002.

Ferrall, S. C., & Dallenbach, K. M. (1930). The analysis and synthesis of burning heat. *The American Journal of Psychology, 42*(1), 72–82.

Ferri, F., Frassinetti, F., Ardizzi, M., Costantini, M., & Gallese, V. (2012). A sensorimotor network for the bodily self. *Journal of Cognitive Neuroscience, 24*(7), 1584–1595.

Ferrier, D. (1876). *The functions of the brain* (2nd ed., London: Smith, Elder, 1886).

Feshchenko, V. A., Reinsel, R. A., & Veselis, R. A. (2001). Multiplicity of the alpha rhythm in normal humans. *Journal of Clinical Neurophysiology, 18*(4), 331–344.

Fessler, D. M., & Holbrook, C. (2013). Friends shrink foes, the presence of comrades decreases the envisioned physical formidability of an opponent. *Psychological Science, 24*(5), 797–802.

Fias, W., Lammertyn, J., Reynvoet, B., Dupont, P., & Orban, G. A. (2003). Parietal representation of symbolic and nonsymbolic magnitude. *Journal of Cognitive Neuroscience, 15,* 47–56. doi:10.1162/089892903321107819.

Field, T. (2010). Touch for socioemotional and physical well-being: A review. *Developmental Review, 30,* 367–383.

Fields, R. D., & Sutherland, S. (2013). Cells that makes you. *Scientific American (Mind), 24*(2), 20–20.

Fiez, J., Balota, D., Raichle, M., & Petersen, S. (1999). Effects of lexicality, frequency, and spelling-to-sound consistency on the functional anatomy of reading. *Neuron, 24,* 205–218.

Finucane, M. M., Stevens, G. A., Cowan, M. J., Danaei, G., Lin, J. K., Paciorek, C. J., et al. (2011). National, regional, and global trends in body-mass index since 1980: Systematic analysis of health examination surveys and epidemiological studies with 960 country-years and 9.1 million participants. *Lancet, 377*(9765), 557–567.

Fisher, C. M. (1982). Disorientation for place. *Archives of Neurology, 39*(1), 33.

Fitzgerald, P. J. (2012). Whose side are you on: Does serotonin preferentially activate the right hemisphere and norepinephrine the left? *Medical Hypotheses, 79*(2), 250–254. ISSN 0306-9877, 10.1016/j. mehy.2012.05.001. http://www.sciencedirect.com/science/article/pii/S0306987712002150.

Fletcher, H., & Munson, W. M. (1933). Loudness, its definition, measurement and calculation. *Journal of the Acoustical Society of America, 5,* 82–108.

Flores-Gutierrez, E. O., Díaz, J. L., Barrios, E. F., et al. (2007). Metabolic and electric brain patterns during pleasant and unpleasant emotions induced by music masterpieces. *International Journal of Psychophysiology, 65*(1), 69–84.

Flory, J. D., Manuck, S. B., Matthews, K. A., & Muldoon, M. F. (2004). Serotonergic function in the central nervous system is associated with daily ratings of positive mood. *Psychiatry Research, 129,* 11–19.

Foerster, O. (1936). The motor cortex in man in the light of Hughlings Jackson's doctrines. *Brain, 59,* 135–159.

Fogassi, L., & Simone, L. (2013). The mirror system in monkeys and humans and its possible motor-based functions. In M. J. Richardson, et al. (Eds.), *Progress in motor control* (pp. 87–110). New York: Springer.

Foland-Ross, L. C., Hamilton, J. P., Joormann, J., Berman, M. G., Jonides, J., & Gotlib, I. H. (2013). The neural basis of difficulties disengaging from negative irrelevant material in major depression. *Psychological Science.* doi:10.1177/0956797612457380.

Fontaine, D., Capelle, L., & Duffau, H. (2002). Somatotopy of the supplementary motor area: Evidence from correlation of the extent of surgical resection with the clinical patterns of deficit. *Neurosurgery, 50*(2), 297–303.

Ford, A. A., Triplett, W., Sudhyadhom, A., Gullett, J., McGregor, K., FitzGerald, D. B., & Crosson, B. (2013). Broca's area and its striatal and thalamic connections: A diffusion-MRI tractography study. *Frontiers in Neuroanatomy, 7,* 8.

Foster, P. S., & Harrison, D. W. (2001). Quantitative electroencephalographic outcome assessment of expressive dysphasia and dysprosodia. *Archives of Clinical Neuropsychology, 16,* 762.

Foster, P. S., & Harrison, D. W. (2002a). Quantitative electroencephalographic assessment of an individual with comorbid depression and panic attack. *Archives of Clinical Neuropsychology, 17,* 808–809.

Foster, P. S., & Harrison, D. W. (2002b). The relationship between magnitude of cerebral activation and intensity of emotional arousal. *International Journal of Neuroscience, 112,* 1463–1477.

Foster, P. S., & Harrison, D. W. (2004a). Cerebral correlates of varying ages of emotional memories. *Cognitive and Behavioral Neurology, 17,* 85–92.

Foster, P. S., & Harrison, D. W. (2004b). The covariation of cortical electrical activity and cardiovascular responding. *International Journal of Psychophysiology, 52,* 239–255.

Foster, P. S., & Harrison, D. W. (2006). Magnitude of cerebral asymmetry at rest: Covariation with baseline cardiovascular activity. *Brain and Cognition, 61,* 286–297.

Foster, P. S., Branch, K. K., Witt, J. C., Giovannetti, T., Libon, D., Heilman, K. M., & Drago, V. (in press). Acetylcholinesterase inhibitors reduce spreading activation in dementia. *Neuropsychologia.*

Foster, P. S., Campbell, R., Williams, M., Branch, K., Roosa, K., Orman, C., & Drago, V. (in press). Administration of exogenous melatonin; increases spreading activation in lexical memory networks. *Journal of Psychopharmacology.*

Foster, P. S., Williamson, J. B., & Harrison, D. W. (2005). The ruff figural fluency test: Heightened right frontal lobe delta activity as a function of performance. *Archives of Clinical Neuropsychology, 20,* 427–434.

Foster, P. S., Drago, V., Ferguson, B. J., & Harrison, D. W. (2008a). Cerebral moderation of cardiovascular functioning: A functional cerebral systems perspective. *Clinical Neurophysiology, 119,* 2846–2854.

Foster, P. S., Drago, V., Webster, D. G., Harrison, D. W., Crucian, G. P., & Heilman, K. M. (2008b). Emotional influences on spatial attention. *Neuropsychology, 22*(1), 127–135.

Foster, P. S., Drago, V., & Harrison, D. W. (2009). Assessment of nonverbal learning and memory using the design learning test. *The Journal of Psychology, 143*(1), 1–20.

Foster, P. S., Drago, V., Harrison, D. W., Skidmore, F., Crucian, G. P., & Heilman, K. M. (2010a). Influence of left versus right hemibody onset Parkinson's disease on cardiovascular control. *Laterality: Asymmetries of Body, Brain and Cognition, 16*(2), 164–173.

Foster, P. S., Drago, V., Yung, R. C., Skidmore, F. M., Skoblar, B., Shenal, B. V., Rhodes, R. D., & Heilman, K. M. (2010b). Anxiety affects working memory only in left hemibody onset Parkinson disease patients. *Cognitive and Behavioral Neurology, 23*(1), 14–18.

Foster, P. S., Drago, V., Crucian, G. P., Sullivan, W. K., Rhodes, R. D., Shenal, B. V., Skoblar, B., Skidmore, F. M., & Heilman, K. M. (2011a). Anxiety and depression severity are related to right but not left onset Parkinson's disease duration. *Journal of the Neurological Sciences, 305*(1–2), 131–135 (Epub 2011 Mar 21).

Foster, P. S., Yung, R. C., Branch, K. K., Stringer, K., Ferguson, B. J., Sullivan, W., & Drago, V. (2011b). Increased spreading activation in depression. *Brain and Cognition, 77*(2), 265–270.

Foster, P. S., Hubbard, T., Yung, R. C., Ferguson, B. J., Drago, V., & Harrison, D. W. (2012). Cerebral asymmetry in the control of cardiovascular functioning: Evidence from lateral vibrotactile stimulation. *Laterality: Asymmetries of Body, Brain and Cognition.* doi:10.1080/13576 50X.2011.631545.

Foster, P. S., Drago, V., & Harrison, D. W. (2013a). The figure trail making test: An alternative measure of executive functioning. Manuscript submitted for publication.

Foster, P. S., Drago, V., Mendez, K., Witt, J. C., Crucian, G. P., & Heilman, K. M. (2013b). Mood disturbances and cognitive functioning in Parkinson's disease: The effects of disease duration and side of onset of motor symptoms. *Journal of Clinical Exp Neuropsychol, 35,* 71.

Foster, P. S., Hubbard, T., Poole, J., Pridmore, M., Bell, C., & Harrison, D. W. (2013c). Spreading activation in emotional memory networks and the cumulative effects of somatic markers. Manuscript in preparation.

Fox, P. T. (1995). Spatial normalization: Origins, objectives, applications and alternatives. *Human Brain Mapping, 3,* 161–164.

Fox, N. A., & Davidson, R. J. (1987). Electroencephalogram asymmetry in response to the approach of a stranger and maternal separation in 10-month-old infants. *Developmental Psychology, 23*(2), 233–240.

Fox, N. A., & Davidson, R. J. (1988). Patterns of brain electrical activation during facial signs of emotion in 10-month-old infants. *Developmental Psychology, 24*(2), 230–236.

Fox, N. A., Rubin, K. H., Calkins, S. D., Marshall, T. R., Coplan, R. J., & Porges, S. W. (1995). Frontal activation asymmetry and social competence at four years of age. *Child Development, 66*(6), 1770–1784.

Foxe, J., et al. (1998). Parieto-occipital approximately 10 Hz activity reflects anticipatory stat of visual attention mechanisms. *Neuroreport, 9,* 3929–3933.

Francis, P. T., Palmer, A. M., Snape, M., & Wilcock, G. K. (1999). The cholinergic hypothesis of Alzheimer's disease: A review of progress. *Journal of Neurology, Neurosurgery & Psychiatry, 66*(2), 137–147.

Francis, A., Chandragiri, S., Rizivi, S., et al. (2000). Is Lorazepam a treatment of neuroleptic malignant syndrome? *CNS Spectrum, 5,* 54–57.

Franck-Emmanuel, R., Dufor, O., Giussani, C., Draper, L., & François Démonet, J. (2009). The graphemic/motor frontal area (GMFA): Exner's area revisited. *Annals of Neurology,* 1–19 (Online). doi:10.1002/ana.21661.

Frasure-Smith, N., Lesperance, F., & Talajic, M. (1993). Depression following myocardial infarction. Impact on 6-month survival. *Journal of the American Medical Association, 270*(15), 1819–1825.

Frederickson, B. L., Maynard, K. E., Helms, M. J., Haney, T. L., Siegler, I. C., & Barefoot, J. C. (2000). Hostility predicts magnitude and duration of blood pressure response to anger. *Journal of Behavioral Medicine, 23,* 229–243.

Freemon, F. R. (1971). Akinetic mutism and bilateral anterior cerebral artery occlusion. *Journal of Neurology, Neurosurgery and Psychiatry, 34,* 693–698.

Freeman, R. (2006). Cardiovascular manifestations of autonomic epilepsy. *Clinical Autonomic Research, 16,* 12–17.

Freeman, B. M., & Vince, M. A. (1974). *Development of the avian embryo.* London: Chapman & Hall.

French, J. D. (1958). The reticular formation. *Journal of Neurosurgery, 15*(1), 97–115.

French, J. D., Hernandez-Peon, R., & Livingston, R. B. (1954). Projections from cortex to cephalic brain stem (reticular formation) in monkey. *Journal of Neurophysiology, 18*(1), 74–98.

Freud, S. (1962). On the grounds for detaching a particular syndrome from neurasthenia under the description "anxiety neurosis." *The standard edition of the complete psychological works of Sigmund Freud* (Vol. 3, pp. 87–120). London: Hogarth Press (Original work published 1895).

Friederici, A. D., Brauer, J., & Lohmann, G. (2011). Maturation of the language network: From inter- to intrahemispheric connectivities. *PLoS One, 6,* e20726.

Friedman, R. S., & Förster, J. (2005). Effects of motivational cues on perceptual asymmetry: Implications for creativity and analytical problem solving. *Journal of Personality and Social Psychology, 88,* 263–275.

Friedman, B. H., & Thayer, J. F. (1998). Anxiety and autonomic flexibility: A cardiovascular approach. *Biological Psychology, 49,* 303–323.

Fritsch, G., & Hitzig, E. (1870). Uber die elektrische Erregbarkeit des Frosshirns. In *The cerebral cortex* (trans: G. von Bonin; pp. 73–96). Springfield: Thomas.

Fritzsche, A., Dahme, B., Gotlib, I. H., Joormann, J., Magnussen, H., Watz, H., & von Leupoldt, A. (2010). Specificity of cognitive biases in patients with current depression and remitted depression and in patients with asthma. *Psychological Medicine, 40,* 815–826.

Fross, R. D., Tsin, J. K. C., McLennan, D. J., Schultzer, M., & Calne, D. B. (1987). Asymmetry in Parkinson's disease. *Neurology, 37*(Suppl), 320.

Frysinger, R. C., & Harper, R. M. (1989). Cardiac and respiratory correlations with unit discharge in human amygdala and hippocampus. *Electroencephalography and Clinical Neurophysiology, 72,* 463–470.

Fuller, G. N., & Burger, P. C. (1990). Nervus terminalis (cranial nerve zero) in the adult human. *Clinical Neuropathology, 9*(6), 279–283.

Fulwiler, C., King, J. A., & Zhang, N. (2012). Amygdala-orbitofrontal resting-state functional connectivity is associated with trait anger. *Neuroreport, 23*(10), 606–610. doi:10.1097/WNR.0b013e3283551cfc.

Furth, G., & Smith, R. (2002). *Amputee identity disorder: Information, questions, answers, and recommendations about self-demand amputation.* Bloomington: 1st Books.

Fuster, J. M. (1980). *The prefrontal cortex.* New York: Raven Press.

Fuster, J. M. (2004). Upper processing stages of the perception-action cycle. *Trends in Cognitive Science, 8,* 143–145.

Fuster, J. M. (2008). *The prefrontal cortex* (4th ed.). London: Academic Press.

Futagi, Y., Toribe, Y., & Suzuki, Y. (2012). The grasp reflex and moro reflex in infants: Hierarchy of primitive reflex responses. *International Journal of Pediatrics,* Article ID 191562, 10 pages. doi:10.1155/2012/191562.

Gable, P. A., & Harmon-Jones, E. (2008). Approach-motivated positive affect reduces breadth of attention. *Psychological Science, 19,* 476–482.

Gadea, M., Gomez, C., Gonzalez-Bono, E., Espert, R., & Salvador, A. (2005). Increased cortisol and decreased right ear advantage (REA) in dichotic listening following a negative mood induction. *Psychoneuroendocrinology, 30,* 129–138.

Gainotti, G. (1972). Emotional behavior and hemispheric side of the lesion. *Cortex, 8,* 41–55.

Galambos, R. (1956). Suppression of auditory activity by stimulation of efferent fibers to the cochlea. *Journal of Neurophysiology, 19,* 424–437.

Gallagher, H. L., & Frith, C. D. (2003). Functional imaging of theory of mind. *Trends in Cognitive Sciences, 7,* 77–83 (Regul. Ed.).

Galliot, M. T., Plant, E. A., Butz, D. A., & Baumeister, R. F. (2007). Increasing self-regulatory strength can reduce the depleting effect of suppressing stereotypes. *Personality and Social Psychology Bulletin, 33,* 281–294.

Gallivan, J. P., Cavina-Pratesi, C., & Culham, J. C. (2009). Is that within reach? fMRI reveals that the human superior parieto-occipital cortex encodes objects reachable by the hand. *Journal of Neuroscience, 29,* 4381–4391.

Galvani, L. (1791). In *Bon. Sci. Art. Inst. Acad. Comm., 7,* 363–418, English Transl. by M. Glover Foley, 1953, Burndy.

Gannon, P. J. (2010). Evolutionary depth of human brain language areas. In K. Hugdahl & R. Westerhausen (Eds.), *The two halves of the brain* (pp. 37–63). Cambridge: MIT Press.

Gardner, R. A., & Gardner, B. T. (1969). Teaching sign language to a chimpanzee. *Science, 165,* 664–672.

Garner, D. M. (2002). Body image and anorexia nervosa. In T. F. Cash & T. Pruzinsky (Eds.), *Body image: A handbook of theory, research, and clinical practice.* New York: The Guilford Press.

Garre-Olmo, J., López-Pousa, S., Turon-Estrada, A., Juvinyà, D., Ballester, D., & Vilalta-Franch, J. (2012). Environmental determinants of quality of life in nursing home residents with severe dementia. *Journal of the American Geriatrics Society, 60*(7), 1230–1236.

Garrett, J. C., Harrison, D. W., & Kelly, P. L. (1989). Pupillometric assessment of arousal to sexual stimuli: Novelty effects or sexual preference? *Archives of Sexual Behavior, 18,* 191–201.

Gauthier, I., Tarr, M. J., Moylan, J., Skudlarski, P., Gore, J. C., et al. (2000). The fusiform "face area" is part of a network that processes faces at the individual level. *Journal of Cognitive Neuroscience, 12,* 495–504.

Gauthier, L. V., Taub, E., Perkins, C., Ortmann, M., Mark, V. W., & Uswatte, G. (2008). Remodeling the brain plastic structural brain changes produced by different motor therapies after stroke. *Stroke, 39*(5), 1520–1525.

Gazzaniga, M. S. (1998). The split brain revisited. *Scientific American, 279,* 50–55.

Gazzaniga, M. S. (2000). Cerebral specialization and interhemispheric communication: Does the corpus callosum enable the human condition? *Brain, 123*(7), 1293–1326.

Gazzaniga, M. S. (2012). Shifting gears: Seeking new approaches for mind/brain mechanisms. *Annual Review of Psychology.* 2012 Sep 17. [Epub ahead of print].

Gazzaniga, M. S., Bogen, J. E., & Sperry, R. W. (1962). Some functional effects of sectioning the cerebral commissures in man. *Proceedings of the National Academy of Sciences of the United States of America, 48,* 1765–1769.

Gazzaniga, M. S., Bogen, J. E., & Sperry, R. W. (1967). Dyspraxia following division of the cerebral commissures. *Archives of Neurology, 16,* 606–612.

Geerdink, Y., Aarts, P., & Geurts, A. C. (2013). Motor learning curve and long-term effectiveness of modified constraint-induced movement therapy in children with unilateral cerebral palsy: A randomized controlled trial. *Research in Developmental Disabilities, 34*(3), 923–931.

Gehrman, P. R., et al. (2009). Melatonin fails to improve sleep or agitation in double-blind randomized placebo-controlled trial of institutionalized patients with Alzheimer disease. *The American Journal of Geriatric Psychiatry, 17,* 166–169.

Geminiani, G., & Bottini, G. (1992). Mental representation and temporary recovery from unilateral neglect after vestibular stimulation. *Journal of Neurology, Neurosurgery, & Psychiatry, 55,* 332–333.

Gerafi, J. (2011). *Anosognosia for hemiplegia: Theoretical, clinical, and neural aspects.* Diss. University of Skövde.

Gerstmann, J. (1940). Syndrome of finger agnosia, disorientation for right and left, agraphia and acalculia. *Archives of Neurology and Psychiatry, 44,* 398–407.

Gerstmann, J. (1957). Some notes on the Gerstmann syndrome. *Neurology, 7,* 866–869 (Cleveland, Ohio).

Geschwind, N. (1965a). Disconnexion syndromes in animals and man. I. *Brain, 88,* 237–294. doi:10.1093/brain/88.2.237.

Geschwind, N. (1965b). Disconnexion syndromes in animals and man. II. *Brain, 88,* 585–644. doi:10.1093/brain/88.3.585.

Geschwind, N. (1982). Language and the brain. *Scientific American, 226,* 76–83.

Ghirlanda, S., & Vallortigara, G. (2004). The evolution of brain lateralization: A game-theoretical analysis of population structure. *Proceedings of the Royal Society of London, 271,* 853–857.

Gibbon, J., & Church, R. M. (1984). Sources of variance in an information processing theory of timing. In H. L. Roitblat, T. G. Bever, & H. S. Terrace (Eds.), *Animal cognition* (pp. 465–488). Hillsdale: Erlbaum.

Gibson, K., & Petersen, A. (2011). *Brain maturation & cognitive development: Comparative and cross-cultural perspectives.* Piscataway: Transaction Publishers, Rutgers-State University of New Jersey.

Gilbert, A. N., & Wysocki, C. J. (1991). Quantitative assessment of olfactory experience during pregnancy. *Psychosomatic Medicine, 53*(6), 693–700.

Gilboa, A., & Verfaellie, M. (2010). Telling it like it isn't: The cognitive neuroscience of confabulation. *Journal of the International Neuropsychological Society, 16*(6), 961–966.

Gilley, D. W., Wilson, R. S., Fleischman, D. A., Harrison, D. W., Goetz, C. G., & Tanner, C. M. (1995). Impact of Alzheimer's-type dementia and information source on the assessment of depression. *Psychological Assessment, 7*(1), 42–48.

Gilley, P. M., Sharma, A., Dorman, M., & Martin, K. (2006). Abnormalities in central auditory maturation in children with language-based learning problems. *Clinical Neurophysiology, 117*(9), 1949–1956.

Ginty, D. D., Kornhauser, J. M., Thompson, M. A., Bading, H., Mayo, K. E., Takahashi, J. S., & Greenberg, M. E. (1993). Regulation of CREB phosphorylation in the suprachiasmatic nucleus by light and a circadian clock. *Science, 260*(5105), 238–241.

Gitelman, D. R., Parrish, T. B., Friston, K. J., & Mesulam, M. (2002). Functional anatomy of visual search: regional segregations within the frontal eye fields and effective connectivity of the superior colliculus. *Neuroimage, 15*(4), 970–982.

Gläscher, J., Adolphs, R., Damasio, H., Bechara, A., Rudrauf, D., Calamia, M., Paul, L. K., & Tranel, D. (2012). Lesion mapping of cognitive control and value-based decision making in the prefrontal cortex. *PNAS* (Published online before print August 20, 2012). doi:10.1073/pnas.1206608109PNAS.

Glasel, H., Leroy, F., Dubois, J., Hertz-Pannier, L., Mangin, J. F., & Dehaene-Lambertz, G. (2011). A robust cerebral asymmetry in the infant brain: The rightward superior temporal sulcus. *Neuroimage, 58,* 716–723.

Glick, S. D., Ross, D. A., & Hough, L. B. (1982). Lateral asymmetry of neurotransmitters in human brain. *Brain Research, 234*(1), 53–63.

Gluck, M. A., Mercado, E., & Myers, C. E. (2014). *Learning and memory: From brain to behavior* (2nd ed.). New York: Worth Publishers.

Go, A. S., Mozaffarian, D., Roger, V. L., Benjamin, E. J., Berry, J. D., Borden, W. B., & Turner, M. B. (2013). Executive summary: Heart disease and stroke statistics-2013 update a report from the American Heart Association. *Circulation, 127*(1), 143–152.

Goddard, G. V., & Douglas, K. M. (1975). Does the engram model of kindling model the engram of normal long term memory? *Canadian Journal of Neurological Sciences, 2*, 385–394.

Godlewska, B. R., Norbury, R., Selvaraj, S., Cowen, P. J., & Harmer, C. J. (2012). Short-term SSRI treatment normalises amygdala hyperactivity in depressed patients. *Psychological Medicine*, 1–9. Cambridge University Press. doi:10.1017/S0033291712000591.

Goeleven, E., Raedt, R. D., Baert, S., & Koster, E. H. W. (2006). Deficient inhibition of emotional information in depression. *Journal of Affective Disorders, 93*, 149–157.

Goetsch, V. L., & Isaac, W. (1983). The effects of d-amphetamine on visual sensitivity in the rat. *European Journal of Pharmacology, 87*(4), 465–468.

Goh, J. O., Chee, M. W., Tan, J. C., Venkatraman, V., Hebrank, A., Leshikar, E. D., & Park, D. C. (2007). Age and culture modulate object processing and object-scene binding in the ventral visual area. *Cognitive, Affective, & Behavioral Neuroscience, 7*, 44–52.

Goh, J. O., Leshikar, E. D., Sutton, B. P., Tan, J. C., Sim, S. K., Hebrank, A. C., & Park, D. C. (2010). Culture differences in neural processing of faces and houses in the ventral visual cortex. *Social, Cognitive, and Affective Neuroscience, 5*, 227–235.

Goldberg, E. (2001). *The executive brain: Frontal lobes and the civilized mind.* Oxford: Oxford University Press.

Golden, R. N., Gaynes, B. N., Ekstrom, R. D., Hamer, R. M., Jacobsen, F. M., Suppes, T., Wisner, K. L., & Nemeroff, C. B. (2005). The efficacy of light therapy in the treatment of mood disorders: A review and meta-analysis of the evidence. *American Journal of Psychiatry, 162*, 656–662.

Goldenberg, G. (2000). Chapter 9: Disorders of body perception. In Martha J. Farah & Todd E. Feinberg (Eds.), *Patient-based approaches to cognitive neuroscience* (pp. 110–111). Cambridge: MIT Press (Print).

Golder, S. A., & Macy, M. W. (2011). Diurnal and seasonal mood vary with work, sleep, and day length across diverse cultures. *Science, 333*, 1878–1881.

Goldman-Rakic, P. S. (1995). Cellular basis of working memory. *Neuron, 14*, 477–485.

Goldstein, K. (1952). The effect of brain damage on the personality. *Psychiatry, 15*, 245–260.

Goldstein, L. H. (2013). Acquired disorders of bodily movement. In L. H. Goldstein & J. E. McNeil (Eds.), *Clinical neuropsychology: A practical guide to assessment.* Malden: Wiley-Blackwell.

Gollisch, T., & Meister, M. (2010). Eye smarter than scientists believed: Neural computations in circuits of the retina. *Neuron, 65*(2), 150–164.

Good, C. D., Johnsrude, I. S., Ashburner, J., Henson, R. N., Friston, K. J., & Frackowiak, R. S. (2001). Cerebral asymmetry and the effects of sex and handedness on brain structures: A voxel-based morphometric analysis of 465 normal adult human brains. *NeuroImage, 14*, 685–700.

Gordon, H. W., & Bogen, J. E. (1974). Hemispheric lateralization of singing after intracarotid sodium amylobarbitone. *Journal of Neurology, Neurosurgery, and Psychiatry, 37*, 727–738.

Gotlib, I. H., Krasnoperova, E., Yue, D. L., & Joormann, J. (2004). Attentional biases for negative interpersonal stimuli in clinical depression. *Journal of Abnormal Psychology, 113*, 127–135.

Grabowska, A., Marchewka, A., Seniów, J., Polanowska, K., Jednoróg, K., Królicki, L., et al. (2011). Emotionally negative stimuli can overcome attentional deficits in patients with visuo-spatial hemineglect. *Neuropsychologia, 49*, 3327–3337.

Grady, C. L., Haxby, J. V., Horwitz, B., Schapiro, M. B., Rapoport, S. I., Ungerleider, L. G., & Herscovitch, P. (1992). Dissociation of object and spatial vision in human extrastriate cortex: Age-related changes in activation of regional cerebral blood flow measured with water and positron emission tomography. *Journal of Cognitive Neuroscience, 4*, 23–34. doi:10.1162/jocn.1992.4.1.23.

Granacher, R. P., Jr. (2008). *Traumatic brain injury: Methods for clinical and forensic neuropsychiatric assessment* (2nd ed.). Boca Raton: CRC Press/Taylor Francis Group.

Grant, R., Ferguson, M. M., Strang, R., Turner, J. W., & Bone, I. (1987). Evoked taste thresholds in a normal population and the application of electrogustometry to trigeminal nerve disease. *Journal of Neurology, Neurosurgery & Psychiatry, 50*(1), 12–21.

Graves, W. W., Desai, R., Humphries, C., Seidenberg, M. S., & Binder, J. R. (2010). Neural systems for reading aloud: A multiparametric approach. *Cerebral Cortex, 20,* 1799–1815.

Gray, J. A. (1981). A critique of Eysenck's theory of personality. In H. J. Eysenck (Ed.), *A model for personality* (pp. 246–277). Berlin: Springer.

Gray, J. A. (1982). Precis of the neuropsychology of anxiety: An inquiry into the functions of the septo-hippocampal system. *The Behavioral and Brain Sciences, 5,* 469–534.

Gray, J. A. (1987). *The psychology of fear and stress* (2nd ed.). Cambridge: Cambridge University Press.

Gray, J. A. (1990). Brain systems that mediate both emotion and cognition. Special issue: Development of relationships between emotion and cognition. *Cognition and Emotion, 4,* 269–288.

Gray, J. A. (1994). Framework on taxonomy of psychiatric disorder. In H. M. Van Goozen, J. A. Sergeant, & N. E. Van De Poll (Eds.), *Emotions: Essays on emotion theory* (pp. 29–59). Hillsdale: Lawrence Erlbaum.

Gray, J. A. (2001). Emotional modulation of cognitive control: Approach-withdrawal states double-dissociate spatial from verbal two-back task performance. *Journal of Experimental Psychology, 130*(3), 436–452.

Gray, J. A., Moran, P. M., Grigoryan, G., Peters, S. L., Young, A. M., & Joseph, M. H. (1997). Latent inhibition: The nucleus accumbens connection revisited. *Behavioral Brain Research, 88*(1), 27–34.

Gray, M., Nagai, Y., & Critchley, H. D. (2012). Brain imaging of stress and cardiovascular responses. In P. Hjemdahl, A. Steptoe, & A. Rosengren (Eds.), *Stress and cardiovascular disease* (pp. 129–148). London: Springer. doi:10.1007/978-1-84882-419-5_8.

Graziano, M. S. A. (2009). *The intelligent movement machine: An ethological perspective on the primate motor system.* New York: Oxford University Press.

Green, A. C., Bærentsen, K. B., Stødkilde-Jørgensen, H., Roepstorff, A., & Vuust, P. (2012). Listen, learn, like! Dorsolateral prefrontal cortex involved in the mere exposure effect in music. *Neurology Research International, 2012,* 11 p. (Article ID 846270). doi:10.1155/2012/846270.

Greenberg, G. (January 2011). Inside the bottle to define mental illness. *Wired.* http://www.wired.com/magazine/2010/12/ff_dsmV/.

Greene, J. D., Sommerville, R. B., Nystrom, L. E., Darley, J. M., & Cohen, J. D. (2001). An fMRI investigation of emotional engagement in moral judgment. *Science, 293,* 2105–2108.

Grimshaw, G. M. (1998). Integration and interference in the cerebral hemispheres: Relations with hemispheric specialization. *Brain & Cognition, 36,* 108–127.

Groenewegen, H. J., & Uylings, H. B. M. (2000). The prefrontal cortex and the integration of sensory, limbic, and autonomic information. In H. Uylings, C. Van Eden, J. De Bruin, M. Feenstra, & C. Pennartz (Eds.), *Progress in brain research, Vol. 126, cognition, emotion and autonomic responses: The integrative role of the prefrontal cortex and limbic structures* (pp. 3–28). Amsterdam: Elsevier.

Grön, G., Wunderlich, A. P., Spitzer, M., Tomczak, R., & Riepe, M. W. (2000). Brain activation during human navigation: Gender-different neural networks as substrate of performance. *Nature Neuroscience, 3,* 404–408.

Grossman, M. (2002). Frontotemporal dementia: A review. *Journal of the International Neuropsychological Society, 8,* 566–583. doi:org/10.1017/S1355617702814357.

Gu, B.-M., & Meck, W. H. (2011). New persectives on Vierordt's law: Memory-mixing in ordinal temporal comparison tasks. In A. Vatakis, A. Esposito, F. Cummins, G. Papadelis, & M. Giagkou (Eds.), *Time and time perception, LNAI 6789* (pp. 67–78). Berlin: Springer-Verlag.

Guariglia, C., & Antonucci, G. (1992). Personal and extrapersonal space: A case of neglect dissociation. *Neuropsychologia, 30,* 1001–1009.

Guitton, D. (1992). *Eye movements* (ed. R. H. S. Carpenter, pp. 244–276). London: MacMillan.

Guitton, D., Buchtel, H. A., & Douglas, R. M. (1985). Frontal lobe lesions in man cause difficulties in suppressing reflexive glances and in generation of goal directed saccades. *Experimental Brain Research, 58,* 455–472.

Güntürkün, O. (2003). Human behavior: Adult persistence of head turning asymmetry. *Nature, 421,* 711.

Gupta, R. K., & Przekwas, A. (2013). Mathematical models of blast-induced TBI: Current status, challenges, and prospects. *Frontiers in Neurology, 4,* 59.

Gur, R. C., Erwin, R. J., Gur, R. E., Zwil, A. S., Heimberg, C., & Kraemer, H. C. (1992). Facial emotion discrimination, II, behavioural findings in depression. *Psychiatry Research, 42,* 241–251.

Gurd, J. M., Coleman, J. S., Costello, A., & Marshall, J. C. (2001). Organic or functional? A new case of foreign accent syndrome. *Cortex, 37,* 715–718.

Guzowski, J. F., Lyford, G. L., Stevenson, G. D., Houston, F. P., McGaugh, J. L., Worley, P. F., & Barnes, C. A. (2000). Inhibition of activity-dependent arc protein expression in the rat hippocampus impairs the maintenance of long-term potentiation and the consolidation of long-term memory. *The Journal of Neuroscience, 20*(11), 3993–4001.

Ha, J. W., Pyun, S. B., Hwang, Y. M., & Sim, H. (2012). Lateralization of cognitive functions in aphasia after right brain damage. *Yonsei Medical Journal, 53*(3), 486–494. http://dx.doi.org/10.3349/ymj.2012.53.3.486.

Habib, M., & Sirigu, A. (1987). Pure topographical disorientation: A definition and anatomical basis. *Cortex, 23,* 73–85.

Hadjikhani, N., Joseph, R. N., Snyder, J., Chabris, C. F., Clark, J., Steele, S., McGrath, L., Vangel, M., Aharon, I., Feczko, E., Harris, G. J., & Tager-Flusberg, H. (2004). Activation of the fusiform gyrus when individuals with autism spectrum disorder view faces. *Neuroimage, 22,* 1141–1150.

Haegler, K., Zernecke, R., Kleemann, A. M., Albrecht, J., Pollatos, O., Brückmann, H., & Wiesmann, M. (2010). No fear no risk! Human risk behavior is affected by chemosensory anxiety signals. *Neuropsychologia, 48,* 3901–3908.

Hafting, T., Fyhn, M., Molden, S., Moser, M. -B., & Moser, E. I. (2005). Microstructure of a spatial map in the entorhinal cortex. *Nature, 436,* 801–806.

Hahn-Holbrook, J., Holt-Lunstad, J., Holbrook, C., Coyne, S., & Lawson, E. T. (2011). Maternal defense: Breastfeeding, diminished stress and heightened aggression. *Psychological Science, 22*(10), 1288–1295.

Haier, R. J., Siegel, B., Nuechterlein, K. H., Hazlett, E., Wu, J., Paek, J., Browning, H. L., & Buchsbaum, M. S. (1988). Cortical glucose metabolism rate correlates of abstract reasoning and attention studied with positron emission tomography. *Intelligence, 12,* 199–217.

Haleem, D. J. (2012). Serotonin neurotransmission in anorexia nervosa. *Behavioural Pharmacology, 23,* 478–495.

Haley, R. W. (2003). Excess incidence of ALS in young Gulf War veterans. *Neurology, 61,* 750–756.

Hall, J. A. (1978). Gender effects in decoding nonverbal cues. *Psychological Bulletin, 85,* 845–857.

Hall, J. A., Cartet, J. D., & Horgan, T. G. (2000). Gender differences in the nonverbal communication of emotion. In A. H. Fischer (Ed.), *Gender and emotion: Social psychological perspectives* (pp. 97–117). Paris: Cambridge University Press.

Hama, S., Yamashita, H., Shigenobu, M., Watanabe, A., Kurisu, K., Yamawaki, S., & Kitaoka, T. (2007). Post-stroke affective or apathetic depression and lesion location: Left frontal lobe and bilateral basal ganglia. *European Archives of Psychiatry and Clinical Neuroscience, 257*(3), 149–152.

Hamann, S. (2001). Review: Cognitive and neural mechanisms of emotional memory. *Trends in Cognitive Sciences, 5*(9), 394–399.

Hampshire, A., Chamberlin, S. R., Monti, M. M., Duncan, J., & Owen, A. M. (2010). The role of the right inferior frontal gyrus: inhibition and attentional control. *Neuroimage, 50,* 1313–1319. doi:10.1016/j.neuroimage.2009.12.109.

Han, L., Ma, C., Liu, Q., Weng, H. J., Cui, Y., Tang, Z., Dong, X., et al. (2012). A subpopulation of nociceptors specifically linked to itch. *Nature Neuroscience, 16*(2), 174–182.

Hanks, R. A., Temkin, N., Machamer, J., & Dikmen, S. S. (1999). Emotional and behavioral adjustment after traumatic brain injury. *Archives of Physical Medicine and Rehabilitation, 80*(9), 991–997.

Hare, R. D., & Quinn, M. J. (1971). Psychopathy and autonomic conditioning. *Journal of Abnormal Psychology, 77,* 223–235.

Harlow, H. F. (1962). Development of affection in primates. In E. L. Bliss (Ed.), *Roots of behavior* (pp. 157–166). New York: Harper.

Harmer, C. J., Goodwin, G. M., & Cowen, P. J. (2009). Why doantidepressants take so long to work? A cognitive neuropsychological model of antidepressant drug action. *British Journal of Psychiatry, 195,* 102–108.

Harmon-Jones, E. (2003a). Anger, coping, and frontal cortical activity: The effect of coping potential on anger-induced left frontal activity. *Cognition & Emotion, 17,* 1–24.

Harmon-Jones, E. (2003b). Clarifying the emotive functions of asymmetrical frontal cortical activity. *Psychophysiology, 40,* 838–848.

Harmon-Jones E. (2004a). Contributions from research on anger and cognitive dissonance to understanding the motivational functions of asymmetrical frontal brain activity. *Biological Psychology, 67*(1–2), 51–76.

Harmon-Jones, E. (2004b). On the relationship of anterior brain activity and anger: Examining the role of attitude toward anger. *Cognition and Emotion, 18,* 337–361.

Harmon-Jones, E., & Allen, J. H. B. (1997). Behavioral activation sensitivity and resting frontal EEG asymmetry: Covariation of putative indicators related to risk for mood disorders. *Journal of Abnormal Psychology, 105*(1), 159–163.

Harmon-Jones, E., & Allen, J. H. B. (1998). Anger and frontal brain activity: EEG asymmetry consistent with approach motivation despite negative affective valence. *Journal of Personality & Social Psychology, 74,* 1310–1316.

Harmon-Jones, E., & Gable, P. A. (2009). Neural activity underlying the effect of approach-motivated positive affect on narrowed attention. *Psychological Science, 20*(4), 406–409.

Harmon-Jones, E., Sigelman J. D., Bohlig, A., & Harmon-Jones, C. (2003). Anger, coping, and frontal cortical activity: The effect of coping potential on anger-induced left frontal activity. *Cognition & Emotion, 17,* 1–24.

Harrison, D. W. (1990). The vascular orienting response and the law of initial values. *Biological Psychology, 31,* 149–155.

Harrison, D. W. (1991). Concurrent verbal interference of right and left proximal and distal upper extremity tapping. *Acta Psychologica, 76,* 121–132.

Harrison, D. W., & Edwards, M. C. (1988). Blood pressure reactivity and bias vary with age: A comparison of traditional versus automated methods. *Medical Instrumentation, 22,* 230–233.

Harrison, D. W., & Gorelczenko, P. (1990). Functional asymmetry for facial affect perception in high and low hostile men and women. *International Journal of Neuroscience, 55,* 89–97.

Harrison, D. W., & Isaac, W. (1984). Disruption and habituation of stable fixed-interval behavior in younger and older monkeys. *Physiology & Behavior, 32,* 341–344.

Harrison, D. W., & Kelly, P. L. (1989). Age differences in cardiovascular and cognitive performance under noise conditions. *Perceptual and Motor Skills, 69,* 547–554.

Harrison, D. W., & Pauly, R. S. (1990). Manual dexterity, strength, fatigue, and perseveration. *Perceptual and Motor Skills, 70,* 745–750.

Harrison, D. W., & Pavlik, W. B. (1983). Effects of age, exposure, pre-exposure, and noise conditions on variable-interval performance. *Behavioral and Neural Biology, 39,* 268–276.

Harrison, D. W., Westbrook, R. D., & Pavlik, W. B. (1984). CER acquisition and extinction in younger and older rats. *Bulletin of the Psychonomic Society, 22,* 217–220.

Harrison, D. W., Garrett, J., Henderson, D., & Adams, H. (1985). Visual and auditory feedback for head tilt and torsion in a spasmodic torticollis patient. *Behavioral Research and Therapy, 23*, 87–88.

Harrison, D. W., Kelly, P. L., Gavin, M., & Isaac, W. (1989a). Extracranial and digital vascular events: An analysis of methods in BVP amplitude and biofeedback research. *Biological Psychology, 29*, 1–9.

Harrison, D. W., Zicafoose, B., Francis, J., & Lanter, J. (1989b). Contextual therapy for chronic bruxism in patients with advanced dementia. *Neuropsychiatry, Neuropsychology and Behavioral Neurology, 2*(3),183–188.

Harrison, D. W., Alden, J. D., Lanter, J. J., & Zicafoose, B. F. (1990a). Sensory modification of nonpropositional speech: Excessive emotional vocalization disorder with dementia. *Neuropsychology, 4*, 215–221.

Harrison, D. W., Gorelczenko, P. M., & Cook, J. (1990b). Sex differences in the functional cerebral asymmetry for facial affect perception. *International Journal of Neuroscience, 52*, 11–16.

Harrison, D. W., Demaree, H. A., Higgins, D. A., & Williamson, J. B. (2002). Asymmetry in hand grip strength and fatigue in low- and high-hostile men. *International Journal of Neuroscience, 112*(4), 415–428.

Harrison, D. W., Beck, A. L., Vendemia, J. M., & Walters, R. P. (2003). Ambient sensory conditions: Modification of receptive speech deficits in left-side stroke patients using bright light. *Perceptual and Motor Skills, 96*, 623–624.

Hart, T., Vaccaro, M., Hays, C., & Maiuro, R. (2012). Anger self-management training for people with traumatic brain injury: A preliminary investigation. *Journal of Head Trauma Rehabilitation, 27*(2), 113–122. doi:10.1097/HTR.0b013e31820e686c.

Hart, J., Kraut, M. A., Womack, K. B., Strain, J., Didehbani, N., Bartz, E., Cullum, C. M., et al. (2013). Neuroimaging of cognitive dysfunction and depression in aging retired National Football League players: A cross-sectional study. Neurobiology of aging NFL players. *JAMA Neurology, 70*(3), 326–335.

Hartikainen, K. M., Ogawa, K. H., Soltani, M., & Knight, R. T. (2007). Emotionally arousing stimuli compete for attention with left hemispace. *Neuroreport, 18*, 1929–1933.

Hartley, J., Maguire, E. A., Spiers, H. J., & Burgess, N. (2003). The well-worn route and the path less traveled: Distinct neural bases of route following and way finding in humans. *Neuron, 7*, 877–888.

Harvey, B. H., & Bouwer, C. D. (2000). Neuropharmacology of paradoxic weight gain with selective serotonin reuptake inhibitors. *Clinical Neuropharmacology, 23*(2), 90–97.

Hasher, L., & Zacks, R. T. (1988). Working memory, comprehension, and aging: A review and a new view. In G. G. Bower (Ed.), *The psychology of learning and motivation* (Vol. 22, pp. 193–225). San Diego: Academic Press.

Hasselmo, M. E. (2006). The role of acetylcholine in learning and memory. *Current Opinion in Neurobiology, 16*(6), 710–715.

Hattar, S., Liao, H. W., Takao, M., Berson, D. M., & Yau, K. W. (2002). Melanopsin-containing retinal ganglion cells: Architecture, projections, and intrinsic photosensitivity. *Science, 295*(5557), 1065–1070.

Haviland, J. L. M. (1987). The induced affect response: Ten-week-old infants' responses to three emotion expressions. *Developmental Psychology, 23*, 97–104.

Haxby, J. V., Hoffman, E. A., & Gobbini, M. I. (2002). Human neural systems for face recognition and social communication. *Biological Psychiatry, 51*, 59–67.

Haxlett, E. A., Buchsbaum, M. S., Haznedar, M. M., Singer, M. B., Germans, M. K., Schnur, D. B., Jimenez, D. A., Buchsbaum, B. R., & Troyer, B. T. (1998). Prefrontal cortex glucose metabolism and startle eyeblink modification abnormalities in unmedicated schizophrenia patients. *Psychophysiology, 35*, 186–198.

Hayashi, A., Nagaoka, M., & Mizuno, Y. (2006). Music therapy in Parkinson's disease: Improvement of parkinsonian gait and depression with rhythmic auditory stimulation. *Parkinsonism & Related Disorders, 12*, S76.

Hayes, S. C., Strosahl, K. D., & Wilson, K. G. (1999). *Acceptance and commitment therapy: An experiential approach to behavior change*. New York: Guilford Press.

Hazan, C., & Shaver, P. R. (1994). Attachment as an organizational framework for research on close relationships. *Psychological Inquiry, 5*(1), 1–22.

Head, H., & Holmes, G. (1911). *Brain, 34,* 102–254.

Heath, Robert. http://www.wireheading.com/robert-heath.html. Robert Heath at Wireheading.

Heath, R. G. (1963) Electrical self-stimulation of the brain in man. *American Journal of Psychiatry, 120,* 571–577.

Heatherton, T. F., & Wagner, D. D. (2011). Cognitive neuroscience of self-regulation failure. *Trends in Cognitive Sciences, 15,* 132–139.

Hebb, D. O. (1949). *The organization of behaviour*. New York: Wiley. ISBN 978-0-471-36727-7.

Hebb, D. O. (1955). Drives and the CNS. *Psychological Review, 62,* 243–254.

Hebb, D. O. (1959). A neuropsychological theory. In S. Koch (Ed.), *Psychology: A study of science: Vol. 1. Sensory, perceptual, and physiological formulations* (pp. 622–643). New York: McGraw-Hill Book Company.

Hécaen, H., & Albert, M. L. (1978). *Human neuropsychology*. New York: Wiley.

Hécaen, H., Tzortzis, C., & Rondot, P. (1980). Loss of topographic memory with learning deficits. *Cortex, 16,* 525–542.

Hedna, V. S., Bodhit, A. N., Ansari, S., Falchook, A. D., Stead, L., Heilman, K. M., & Waters, M. F. (2013). Hemispheric differences in ischemic stroke: Is left-hemisphere stroke more common? *Journal of Clinical Neurology, 9*(2), 97–102.

Heilman, K. M. (1973). Ideational apraxia-a re-definition. *Brain, 96,* 861–864.

Heilman, K. M. (1997). The neurobiology of emotional experience. *Journal of Neuropsychiatry, 9,* 439–448.

Heilman, K. M., & Bowers, D. (1990). Neuropsychological studies of emotional changes induced by right and left-hemisphere lesions. In N. Stein, B. Leventhal, & T. Trabasso (Eds.), *Psychological and biological approaches to emotion* (pp. 97–114). Hillsdale: Lawrence Erlbaum.

Heilman, K. M., & Gilmore, R. L. (1998). Cortical influences in emotion. *Journal of Clinical Neurophysiology, 15*(5), 409–423.

Heilman, K. M., & Gonzalez Rothi, L. J. (2012a). Apraxia. In K. M. Heilman & E. Valenstein (Eds.), *Clinical neuropsychology* (5th ed.). New York: Oxford University Press.

Heilman, K. M., & Gonzalez Rothi, L. J. (2012b). Neglect and related disorders. In K. M. Heilman & E. Valenstein (Eds.), *Clinical neuropsychology* (5th ed.). New York: Oxford University Press.

Heilman, K. M., & Valenstein, E. (1972). Auditory neglect in man. *Archives of Neurology, 26,* 32–35.

Heilman, K. M., & Valenstein, E. (1979). Mechanisms underlying hemispatial neglect. *Annals of Neurology, 5,* 166–170.

Heilman, K. M., & Valenstein, E. (2003). *Clinical neuropsychology* (4th ed.). New York: Oxford University Press.

Heilman, K. M., & Valenstein, E. (2012). *Clinical neuropsychology* (5th ed.). New York: Oxford University Press.

Heilman, K. M., & Van Den Abell, T. (1980). Right hemisphere dominance for attention: The mechanism underlying hemispheric asymmetries of inattention (neglect). *Neurology, 30,* 327–330.

Heilman, K. M., Scholes, R., & Watson, R. T. (1975). Auditory affective agnosia: Disturbed comprehension of affective speech. *Journal of Neurology, Neurosurgery, and Psychiatry, 38,* 69–72.

Heilman, K. M., Schwartz, H., & Watson, R. T. (1978). Hypoarousal in patients with the neglect syndrome and emotional indifference. *Neurology, 28,* 229–232.

Heilman, K. M., Chatterjee, A., & Doty, L. C. (1995). Hemispheric asymmetries of near-far spatial attention. *Neuropsychology, 9,* 58–61.

Heilman, K. M., Watson, R. T., & Valenstein, E. (2003). Neglect and related disorders. In K. M. Heilman & E. Valenstein (Eds.), *Clinical neuropsychology* (4th ed.). New York: Oxford University Press.

Heilman, K. M., Leon, S. A., & Rosenbek, J. C. (2004). Affective aprosodia from a medial frontal stroke. *Brain and Language, 89*(3), 411–416.

Heilman, K., Blonder, L. X., Bowers, D., & Valenstein, E. (2012a). Emotional disorders associated with neurologic disease. In K. Heilman & E. Valenstein (Eds.), *Clinical neuropsychology* (5th ed.). New York: Oxford University Press.

Heilman, K. M., Leon, S. A., Burtis, D. B., Ashizawa, T., & Subramony, S. H. (2012b). Affective communication deficits associated with cerebellar degeneration. *Neurocase*. doi:10.1080/135 54794.2012.713496.

Heilman, K. M., Watson, R. T., & Valenstein, E. (2012c). Neglect and related disorders. In K. M. Heilman & E. Valenstein (Eds.), *Clinical neuropsychology* (5th ed.). New York: Oxford University Press.

Hein, G., & Knight, R. T. (2008). Superior temporal sulcus-it's my area: Or is it? *Journal of Cognitive Neuroscience, 20,* 2125–2136.

Helfinstein, S. M., & Poldrack, R. A. (2012). The young and the reckless. *Nature Neuroscience, 15,* 803–805. doi:10.1038/nn.3116.

Heller, W. (1990). The neuropsychology of emotion: Developmental patterns and implications for psychopathology. In N. Stein, B. L. Leventhal, & T. Trabasso (Eds.), *Psychological and biological approaches to emotion* (pp. 167–211). Hillsdale: Erlbaum.

Heller, W. (1993). Neuropsychological mechanisms of individual differences in emotion, personality, and arousal. *Neuropsychology, 7*(4), 476–489.

Heller, E. J. (2012). *Why you hear what you hear: An experiential approach to sound, music, and psychoacoustics.* Princeton: Princeton University Press.

Heller, W., Nitschke, J. B., & Lindsay, D. L. (1997). Neuropsychological correlates of arousal in self-reported emotion. *Cognition and Emotion, 11,* 383–402.

Heller, W., Nitschke, J. B., & Miller, G. A. (1998). Lateralization in emotion and emotional disorders. *Current Directions in Psychological Science, 7,* 26–32.

Hellige, J. B., Laeng, B., & Michimata, C. (2010). Processing asymmetries in the visual system. In K. Hugdahl & R. Westerhausen (Eds.), *The two halves of the brain: Information processing in the cerebral hemispheres.* Cambridge: MIT Press.

Helmholtz, H. L. F. (1850a). *Arch Anat Physiol, 57,* 276.

Helmholtz, H. L. F. (1850b). Measurements on the time of twitching of animal muscles and the velocity of propagation of nerve impulses. *Müllers Archiv.*

Helmholtz, H. L. F. (1850c). On the velocity of propagation of nerve impulses. *Comptes Rendus, XXX, XXXIII.*

Helmholtz, H. L. F. (1885). *On the sensations of tone as a physiological basis for the theory of music* (576 p). London: Longmans.

Heltemes, K. J., Holbrook, T. L., MacGregor, A. J., & Galarneau, M. R. (2012). Blast-related mild traumatic brain injury is associated with a decline in self-rated health amongst US military personnel. *Injury, 43,* 1990–1995. doi:10.1016/j.injury.2011.07.021.

Henriques, J. B., & Davidson, R. J. (1990). Regional brain electrical asymmetries discriminate between previously depressed and healthy control subjects. *Journal of Abnormal Psychology, 99,* 22–31.

Henriques, J. B., & Davidson, R. J. (1991). Left frontal hypoactivation in depression. *Journal of Abnormal Psychology, 100,* 535–545.

Henriques, J., & Davidson, R. J. (1997). Brain electrical asymmetries during cognitive task performance in depressed and nondepressed subjects. *Biological Psychiatry, 42,* 1039–1050.

Henschen, S. E. (1925). Special article-Clinical and anatomical contributions on brain pathology. *Archives of Neurology and Psychiatry, W,* 226–249.

Herath, P., Klingberg, T., Young, J., Amunts, K., & Roland, P. (2001). Neural correlates of dual task interference can be dissociated from those of divided attention: An fMRI study. *Cerebral Cortex, 11,* 796–805.

Herculano-Houzel, S., Mota, B., Wong, P., & Kaas, J. H. (2010). Connectivity-driven white matter scaling and folding in primate cerebral cortex. *Proceedings of the National Academy of Sciences of the United States of America, 107,* 19008–19013.

Hermsdörfer, J., Li, Y., Randerath, J., Goldenberg, G., & Johannsen, L. (2012). Tool use without a tool: Kinematic characteristics of pantomiming as compared to actual use and the effect of brain damage. *Experimental Brain Research, 218,* 201–214.

Herridge, M. L., Harrison, D. W., & Demaree, H. A. (1997). Hostility, facial configuration, and bilateral asymmetry on galvanic skin response. *Psychobiology, 25,* 71–76.

Herridge, M. L., Harrison, D. W., Mollet, G., & Shenal, B. (2004). Hostility and facial affect recognition: Effects of a cold pressor stressor on accuracy and cardiovascular reactivity. *Brain and Cognition, 55,* 564–571.

Herrmann, M., Walter, A., Ehlis, A.-C., & Fallgatter, A. (2006). Cerebral oxygenation changes in the prefrontal cortex: Effects of age and gender. *Neurobiology of Aging, 27,* 888–894.

Herrmann, R., Heflin, S. J., Hammond, T., Lee, B., Wang, J., Gainetdinov, R. R., & Arshavsky, V. Y. (2011). Rod vision is controlled by dopamine-dependent sensitization of rod bipolar cells by GABA. *Neuron, 72*(1), 101–110.

Herve, P. Y., Crivello, F., Perchey, G., Mazoyer, B., & Tzourio-Mazoyer, N. (2006). Handedness and cerebral anatomical asymmetries in young adult males. *NeuroImage, 29,* 1066–1079.

Herz, R., & Engen, T. (1996). Odor memory: Review and analysis. *Psychonomic Bulletin & Review, 3,* 300–313.

Hewing, J., Hagemann, D., Seifert, J., Naumann, E., & Bartussek, D. (2004). On the selective relation of frontal cortical asymmetry and anger-out versus anger-control. *Journal of Personality and Social Psychology, 87*(6), 926–939.

Hickok, G., & Bellugi, U. (2000). The signs of aphasia. In F. Boller & J. Grafman (Eds.), *Handbook of neuropsychology* (Vol. 3, 2nd ed., pp. 31–50). Amsterdam: Elsevier.

Hickok, G., & Poeppel, D. (2007). The cortical organization of speech processing. *Nature Reviews Neuroscience, 8,* 393–402.

Hickok, G., & Rogalsky, C. (2011). What does Broca's area activation to sentences reflect? *Journal of Cognitive Neuroscience, 23*(10), 2629–2631. doi:10.1162/jocn_a_00044.

Hickok, G., & Saberi, K. (2012). Redefining the functional organization of the planum temporale region: Space, objects, and sensory-motor integration. In *The human auditory cortex* (pp. 333–350). New York: Springer.

Hill, J., Dierker, D., Neil, J., Inder, T., Knutsen, A., Harwell, J., Coalson, T., & Van Essen, D. (2010). A surface-based analysis of hemispheric asymmetries and folding of cerebral cortex in term-born human infants. *The Journal of Neuroscience, 30,* 2268–2276.

Hillis, A. E., Newhart, M., Heidler, J., Barker, P. B., Herskovits, E. H., & Degaonkar, M. (2005). Anatomy of spatial attention: Insights from perfusion imaging and hemispatial neglect in acute stroke. *Journal of Neuroscience, 25,* 3161–3167.

Hilz, M. J., Dutsch, M., Perrine, K., Nelson, P. K., Rauhut, U., & Devinsky, O. (2001). Hemispheric influence on autonomic modulation and baroreflex sensitivity. *Annals of Neurology, 49,* 575–584.

Hilz, M. J., Devinsky, O., Szczepanska, H., Borod, J. C., Marthol, H., & Tutaj, M. (2006). Right ventromedial prefrontal lesions result in paradoxical cardiovascular activation with emotional stimuli. *Brain, 129*(Pt 12), 3343–3355. doi:10.1093/brain/awl299.

Hirstein, W. (2009). *Confabulation: Views from neuroscience, psychiatry, psychology, and philosophy.* New York: Oxford University Press.

Hobson, J. A., McCarley, R. W., & Wyzinski, P. W. (1975). Sleep cycle oscillation: Reciprocal discharge by two brainstem neuronal groups. *Science, 189,* 55–58.

Hobson, J. A., Lydic, R., & Baghdoyan, H. A. (1986). Evolving concepts of sleep cycle generation: From brain centers to neuronal populations. *Behavioral and Brain Sciences, 9,* 371–448.

Hochberg, L. R., et al. (2006). *Nature, 442*(7099), 164–171.

Hochberg, L. R., et al. (2012). *Nature, 485*(7398), 372–375.

Hodge, G. K., & Butcher, L. L. (1980). Pars compacta of the substantia nigra modulates motor activity but is not involved importantly in regulating food and water intake. *Naunyn Schmiedebergs Arch Pharmacol, 313*(1), 51–67.

Hodzic, A., Muckli, L., Singer, W., & Stirn, A. (2009). Cortical responses to self and others. *Human Brain Mapping, 30,* 951–962.

Hoehn, M. M., & Yahr, M. D. (1967). Parkinsonism: Onset, progression, and mortality. *Neurology, 17,* 427–442.

Hofer, S., & Frahm, J. (2006). Topography of the human corpus callosum revisited? Comprehensive fiber tractography using diffusion tensor magnetic resonance imaging. *Neuroimage, 32*(3), 989–994.

Hoffman, B. L., & Rasmussen, T. (1953). Stimulation studies of insular cortex of *Macaca Mulatta. Journal of Neurophysiology, 16,* 343–351.

Hoffman, D. A., Lubar, J. F., Thatcher, R. W., Sterman, M. B., Rosenfeld, P. J., Striefel, S., Trudeau, D., & Stockdale, S. (1999). Limitations of the American Academy of Neurology and American Clinical Neurophysiology Society paper on QEEG. *Journal of Neuropsychiatry and Clinical Neuroscience, 11*(3), 401–407.

Holbrook, C., & Fessler, D. M. (2013). Sizing up the threat: The envisioned physical formidability of terrorists tracks their leaders' failures and successes. *Cognition, 127*(1), 46–56.

Holick, M. F. (2007). Vitamin D deficiency. *The New England Journal of Medicine, 357,* 266–281.

Holland, A. K., Carmona, J. E., Cox, D. E., Belcher, L. T., Wolfe, S., & Harrison, D. W. (2007). Time estimation differences in low and high hostile men: Behavioral and physiological correlates of right frontal regulation of an internal clock. *Psychophysiology, 44*(S1), 72.

Holland, A. K., Carmona, J. E., Harrison, D. W., Hunt, S., Riner, B., & Latham, K. (2009). Extending the functional cerebral systems approach to time estimation: Examining age-related changes in right hemisphere regulation of an internal clock. *Archives of Clinical Neuropsychology, 24,* 441.

Holland, A. K., Carmona, J., Scott, M., Hardin, J., & Harrison, D. W. (2010). Differences in cerebral lateralization of time estimation abilities and cardiovascular reactivity in diabetic and nondiabetic older adults (Annual meeting of the International Neuropsychological Society, Acapulco, Mexico.). *Journal of the International Neuropsychological Society, 16,* 212.

Holland, A. K., Carmona, J. E., Smith, A. S., Catoe, A., Hardin, J. F., & Harrison, D. W. (2011a). Lateralized differences in left hemisphere activation as a function of digestive stress: Changes in verbal fluency performance and regulation of diastolic blood pressure before and after food ingestion. *Archives of Clinical Neuropsychology, 25*(6), 576.

Holland, A. K., Newton, S. E., Smith, A. S., Hinson, D., Carmona, J. E., Cox, D., & Harrison, D. W. (2011b). Left lateralized cerebral activation as a function of food absorption and cognitive task demands: Examining changes in beta magnitude using a dual concurrent task paradigm (Annual meeting of the International Neuropsychological Society, Boston, MA.). *Journal of the International Neuropsychological Society, 17,* 131.

Holland, A. K., Smith, A., Moseley, L., Riner, B., Carmona, J. E., & Harrison, D. W. (2011c). Physiological and behavioral correlates of obesity: Changes in fluency performance and regulation of sympathetic tone in normal weight and overweight men and women. *Journal of the International Neuropsychological Society, 17.*

Holland, A. K., Smith, A., Newton, S., Hinson, D., Obi-Johnson, B., Carmona, J. E., & Harrison, D. W. (2011d). Examining changes in regulation of parasympathetic tone as a function of age and performance on a left-lateralized cognitive task before and after undergoing digestive stress (Annual meeting of the International Neuropsychological Society, Boston, MA.). *Journal of the International Neuropsychological Society, 17,* 259.

Holland, A. K., Carmona, J. E., & Harrison, D. W. (2012a). An extension of the functional cerebral systems approach to hostility: A capacity model utilizing a dual concurrent task paradigm. *Journal of Clinical and Experimental Neuropsychology, 34*(1), 92–106. doi:10.1080/138033 95.2011.623119.

Holland, A. K., Newton, S. E., Bunting, J., Coe, M., Carmona, J., & Harrison, D. W. (2012b). Examining age-related changes in left hemisphere functional cerebral systems before and after exposure to pre-digestive stress: Evidence in support of the capacity model. *Psychophysiology, 45*(S1).

Hollander, E., Novotny, S., Hanratty, M., Yaffe, R., DeCaria, C., Aronowitz, B., et al. (2003). Oxytocin infusion reduces repetitive behaviors in adults with autistic and Asperger's disorders. *Neuropsychopharmacology, 28*(1), 193–198.

Holmes, G. (1938). The cerebral integration of ocular movements. *British Medical Journal, 2*, 107–112.

Homnick, D. N. (2012). Dyspnea. In R. D. Anbar (Ed.), *Functional respiratory disorders* (pp. 67–87). Book Series Editor: S. I. Rounds. New York: Humana Press.

Hood, D. C., & Finkelstein, M. A. (1986). Sensitivity to light. *Terminology, 5*, 27.

Howard, K. J., Rogers, L. J., & Boura, A. L. A. (1980). Functional lateralization of the chicken forebrain revealed by use of intracranial glutamate. *Brain Research, 188*, 369–382.

Hu, S. R., & Harrison, D. W. (2013a). Frontal lobe regulatory control of oxygen saturation: Support for the Limited Capacity theory. Manuscript in preparation.

Hu, S. R., & Harrison, D. W. (2013b). Traumatic frontal lobotomy and amotivational apathetic syndrome: Revisiting the historical limitations of standardized neuropsychological assessments. Manuscript submitted for publication.

Huang, W. S., Lin, S. Z., Lin, J. C., Wey, S. P., Ting, G., & Liu, R. S. (2001). Evaluation of early stage Parkinson's disease with 99mTc-TRODAT-1 imaging. *Journal of Nuclear Medicine, 42*, 1303–1308.

Hubel, D. H., & Wiesel, T. N. (1962). Receptive fields, binocular interaction and functional architecture in the cat's visual cortex. *Journal of Physiology, 160*(1), 106–154.

Hubel, D. H., & Wiesel, T. N. (1965). Receptive fields and functional architecture in two nonstriate visual areas (18 and 19) of the cat. *Journal of Neurophysiology, 28*(2), 229–289.

Hugdahl, K. (1988). *Handbook of dichotic listening: Theory, methods and research* (p. xii, 650 p.). Oxford: Wiley.

Hugdahl, K. (2003). Dichotic listening in the study of auditory laterality. In K. Hugdahl & R. J. Davidson (Eds.), *The asymmetrical brain* (pp. 441–476). Cambridge: MIT Press.

Hugdahl, K. (2002). Chapter four: Brain asymmetry and cognition. *Psychology at the turn of the millennium, Volume 1: Cognitive, biological and health perspectives, 86.*

Hugdahl, K., & Anderson, L. (1987). The "forced attention paradigm" in dichotic listening to CV syllables: A comparison between adults and children. *Cortex, 22*, 417–432.

Hugdahl, K., & Westerhausen, R. (Eds.). (2010). *The two halves of the brain: Information processing in the cerebral hemispheres.* Cambridge: MIT Press.

Hugdahl, K., Westerhausen, R., Alho, K., Medvedev, S., Laine, M., & Hamalainen, H. (2009). Attention and cognitive control: Unfolding the dichotic listening story. *Scandinavian Journal of Psychology, 50*, 11–22.

Hughes, A. J., Ben-Shlomo, Y., Daniel, S. E., & Lees, A. J. (1992). What features improve the accuracy of clinical diagnosis in Parkinson's disease: A clinicopathological study. *Neurology, 42*, 1142–1146.

Hughlings-Jackson, J. (1879). On affection of speech from disease of the brain. *Brain, 2*, 323–356.

Hummel, T., Sekinger, B., Wolf, S. R., Pauli, E., & Kobal, G. (1997). 'Sniffin' sticks': Olfactory performance assessed by the combined testing of odor identification, odor discrimination and olfactory threshold. *Chemical Senses, 22*, 39–52.

Humphreys, G. W., & Riddoch, M. J. (2000). One more cup of coffee for the road: Object-action assemblies, response blocking and response capture after frontal lobe damage. *Experimental Brain Research, 133*, 81–93.

Hunter, J., & Jasper, H. H. (1949). Effects of thalamic stimulation in unanesthetized animals. *The Journal of Electroencephalography and Clinical Neurophysiology, 1*, 305–324.

Huntzinger, R., & Harrison, D. W. (1992). Fluent versus nonfluent subtypes of dyslexic readers. Paper presented at the meeting of the International Neuropsychological Society, San Diego, CA.

Husain, M. (2013). Exaggerated object affordance and absent automatic inhibition in alien hand syndrome. *Cortex*, 1–15. Journal homepage: www.elsevier.com/locate/cortex, 3, 4.

Husain, M., & Nachev, P. (2007). Space and the parietal cortex. *Trends in Cognitive Sciences, 11,* 30–36.

Iaria, G., Chen, J. K., Guariglia, C., Ptito, A., & Petrides, M. (2007). Retrosplenial and hippocampal brain regions in human navigation: Complimentary functional contributions to the formation and use of cognitive maps. *European Journal of Neuroscience, 25,* 890–899.

Iijima, T., Witter, M. P., Ichikawa, M., Tominaga, T., Kajiwara, R., & Matsumoto, G. (1996). Entorhinal-hippocampal interactions revealed by real-time imaging. *Science, 272*(5265), 1176–1179.

Imtiaz, K. E., Nirodi, G., & Khaleeli, A. A. (2001). Alexia without agraphia: A century later. *International Journal of Clinical Practice, 55*(3), 225–226. (PMID 11351780).

Indersmitten, T., & Gur, R. C. (2003). Emotion processing in chimeric faces: Hemispheric asymmetries in expression and recognition of emotions. *The Journal of Neuroscience, 23*(9), 3820–3825.

Ingvar, D. H., & Lassen, N. A. (1977). Cerebral function, metabolism and circulation. *Acta Neurologica Scandinavia, 57*(3), 262–269.

Insel, T. R., & Shapiro, L. (1992). Oxytocin receptor distribution reflects social organization in monogamous and polygamous voles. *Proceedings of the National Academy of Sciences of the United States of America, 89,* 5981–5985.

Isaac, W., & DeVito, J. L. (1958). Effect of sensory stimulation on the activity of normal and prefrontal-lobectomized monkeys. *Journal of Comparative and Physiological Psychology, 51,* 172–174.

Isaac, W., & Reed, W. G., (1961). The effect of sensory stimulation on the activity of cats. *Journal of Comparative & Physiological Psychology, 54,* 677–678.

Isaac, W., & Troelstrup, R. (1969). Opposite effect of illumination and d-amphetamine upon activity in the squirrel monkey (Saimiri) and owl monkey (Aotes). *Psychopharmacologia, 15*(4), 260–264.

Isen, A. M. (1984). Toward understanding the role of affect in cognition. In R. S. Wyer & T. S. Srull (Eds.), *Handbook of social cognition* (pp. 179–236). Hillsdale: Erlbaum.

Ito, I., Ito, K., & Shindo, N. (2013). Left leg apraxia after anterior cerebral artery territory infarction: Functional analysis using single-photon emission computed tomography. *European Neurology, 69*(4), 252–256.

Iversen, S., Iversen, L., & Saper, C. B. (2000). The autonomic nervous system and the hypothalamus. In E. R. Kandel, J. H. Schwartz, & T. M. Jessell (Eds.), *Principles of neural science* (4th ed., pp. 960–981). New York: McGraw-Hill.

Izard, C. (1982). *Measuring emotions in infants and young children.* New York: Cambridge Press.

Jack, R. E., Caldara, R., & Schyns, P. G. (2011). Internal representations reveal cultural diversity in expectations of facial expressions of emotion. *Journal of Experimental Psychology: General, 141,* 19–25.

Jackson, J. H. (1874). Illustrations of diseases of the nervous system. *London Hospital Reports, 1*(4), 70–471.

Jackson, J. H. (1915). Clinical remarks on emotional and intellectual language in some cases of disease of the nervous system: Development of the difference between emotional and intellectual expression. *Brain, 38*(1–2), 43–47. doi:10.1093/brain/38.1-2.43.

Jackson, J. H. (1958). Epilepsy and epileptiform convulsions. In J. Taylor (Ed.), *Selected writings of John Hughlings Jackson* (Vol 1, pp. 2–6). New York: Basic Books.

Jackson, A. (2012). Neuroscience: Brain-controlled robot grabs attention. *Nature, 485,* 317–318. doi:10.1038/485317a.

Jacob, S., & McClintock, M. K. (2000). Psychological state and mood effects of steroidal chemosignals in women and men. *Hormones and Behavior, 37*(1), 57–78.

Jacob, S. N., & Nieder, A. (2009). Tuning to non-symbolic proportions in the human frontoparietal cortex. *European Journal of Neuroscience, 30,* 1432–1442. doi:10.1111/j.1460-9568.2009.06932.x.

Jacobsen, C. F. (1931). A study of cerebral function in learning. The frontal lobes. *Journal of Comparative Neurology, 52,* 271–340.

Jacobson, S., & Marcus, E. M. (2011). *Neuroanatomy for the neuroscientist* (2nd ed.). New York: Springer.

Jaeger, J., Borod, J. C., & Peselow, E. (1987). Depressed patients have atypical hemispace biases in the perception of emotional chimeric faces. *Journal of Abnormal Psychology, 96*(4), 321–324.

Jaffe, B. F. (1969). The incidence of ear diseases in the Navajo Indians. *The Laryngoscope, 79*(12), 2126–2134.

James, W. (1950). *The principles of psychology.* New York: Dover Publications.

James, C., Henderson, L., & Macefield, V. G. (2013). Real-time imaging of brain areas involved in the generation of spontaneous skin sympathetic nerve activity at rest. *NeuroImage, 74,* 188–194.

Jancke, L., & Shah, N. J. (2002). Does dichotic listening probe temporal lobe functions? *Neurology, 58,* 736–743.

Jancke, L., Buchanan, T. W., Lutz, K., & Shah, N. J. (2001). Focused and nonfocused attention in verbal and emotional dichotic listening: An fMRI study. *Brain and Language, 78,* 349–363.

Jancke, L., Specht, K., Shah, J. N., & Hugdahl, K. (2003). Focused attention in a simple dichotic listening task: An fMRI experiment. *Cognitive Brain Research, 16,* 257–266.

Jasmin, K., & Casanto, D. (2012). The QWERTY effect: How typing shapes the meanings of words. *Psychonomic Bulletin & Review.* doi:10.3758/s13423-012-0229-7 (published online March 03).

Jasper, H. H. (1949). Diffuse projection systems: The integrative action of the thalamic reticular system. *Electroencephalography and Clinical Neurophysiology, 1, 405–*419.

Jasper, H. H., & Droogleever-Fortuyn, J. (1946). Experimental studies on the functional anatomy of petit mal epilepsy. *Research Publication-Association for Research in Nervous and Mental Disease, 26,* 272–298.

Javal, É (1878). Essai sur la physiologie de la lecture. *Annales d'Oculistique, 80,* 61–73.

Jaworska, N., Blier, P., Fusee, W., & Knott, V., (2012). Alpha power, alpha asymmetry and anterior cingulate cortex activity in depressed males and females. *Journal of Psychiatric Research, 46*(11), 1483–1491.

Jeerakathil, T. J., & Kirk, A. (1994). A representational vertical bias. *Neurology, 44,* 703–706.

Jenkins, W. M., & Masterton, R. B. (1982). Sound localization: effects of unilateral lesions in central auditory system. *Journal of Neurophysiology, 47,* 987–1016.

Jeong, Y., Drago, V., & Heilman, K. M. (2006). Radial character-line bisection. *Cognitive and Behavioral Neurology, 19,* 105–108.

Jewell, G., & McCourt, M. E. (2000). Pseudoneglect: A review and meta-analysis of performance factors in line bisection tasks. *Neuropsychologia, 38,* 93–110.

Jobard, G., Crivello, F., & Tzourio-Mazoyer, N. (2003). Evaluation of the dual route theory of reading: A metanalysis of 35 neuroimaging studies. *Neuroimage, 20,* 693–712.

Johnson, M. H., Dziurawiec, S., Ellis, H., & Morton, J. (1991). The tracking of face-like stimuli by newborn infants and its subsequent decline. *Cognition, 40,* 1–21.

Johnson, M. H., Stewart, J., Humphries, S. A., & Chamove, A. S. (2012). Marathon runners' reaction to potassium iontophoretic experimental pain: Pain tolerance, pain threshold, coping and self-efficacy. *European Journal of Pain, 16*(5), 767–774.

Joinlambert, C., Saliou, G., Flamand-Roze, C., Masnou, P., Sarov, M., Souillard, R., Saliou-Théaudin, M., Guedj, T., Assayag, P., Ducreux, D., Adams, D., & Denier, C. (2012). Cortical borderzone infarcts: clinical features, causes and outcome. *Journal of Neurology, Neurosurgery & Psychiatry, 83*(8), 771–775. doi:10.1136/jnnp-2012-302401.

Jokisch, D., & Jensen, O. (2007). Modulation of gamma and alpha activity during a working memory task engaging the dorsal or ventral stream. *Journal of Neuroscience, 27,* 3244–3251.

Jones, B. E. (1993). The organization of central cholinergic systems and their functional importance in sleep-waking states. *Progress in Brain Research, 98,* 61–71.

Joormann, J., & Gotlib, I. H. (2008). Updating the contents of working memory in depression: Interference from irrelevant negative material. *Journal of Abnormal Psychology, 117,* 182–192.

Joseph, R. (2000). The secondary motor areas: Secondary motor area 6. From: Neuropsychiatry, neuropsychology, clinical neuroscience by Rhawn Joseph, Ph. D. New York: Academic Press.

Jouvet, M. (1965). The paradoxical phase of sleep. *International Journal of Neurology, 5*, 131–150.

Jouvet, M. (1972). The role of monoamines and acetylcholine-containing neurons in the regulation of the sleep-waking cycle. *Ergebn Physiology, 64*, 166–307.

Just, M. A., Kellera, T. A., Malavea, V. L., Kanab, R. K., & Varmac, S. (in press). Autism as a neural systems disorder: A theory of frontal-posterior underconnectivity. *Neuroscience and Biobehavioral Reviews,* 1–28.

Kahn, H. J., & Whitaker, H. A. (1991). Acalculia: An historical review of localization. *Brain and Cognition, 17*(2), 102–115.

Kahn, D. A., Riccio, C. A., & Reynolds, C. R. (2012). Comprehensive trail-making test: Gender and ethnic differences for ages 8–18 years old. *Applied Neuropsychology: Child, 1*(1), 53–56.

Kallman, M. D., & Isaac, W. (1980). Disruption of illumination dependent activity by superior colliculus destruction. *Physiology & Behavior, 25*(1), 45–47.

Kalpouzos, G., & Nyberg, L. (2010). Hemispheric asymmetry of memory. In K. Hugdahl & R. Westerhausen (Eds.), *The two halves of the brain: Information processing in the cerebral hemispheres* (pp. 145–175). Cambridge: MIT Press.

Kang, D.-H., Davidson, R. J., Coe, C. L., Wheeler, R. E., Tomarken, A. J., & Ershler, W. B. (1991). Frontal brain asymmetry and immune function. *Behavioral Neuroscience, 105*, 860–869.

Kanwisher, N., & Dilks, D. (in press). The functional organization of the ventral visual pathway in humans. In L. Chalupa & J. Werner (Eds.), *The new visual neurosciences.*

Kanwisher, N., McDermott, J., & Chun, M. (1997a). The fusiform face area: a module in human extrastriate cortex specialized for face perception. *The Journal of Neuroscience, 17*, 4302–4311.

Kanwisher, N., Woods, R. P., Iacoboni, M., & Mazziotta, J. C. (1997b). A locus in human extrastriate cortex for visual shape analysis. *Journal of Cognitive Neurosciences, 9*(1), 133–142.

Kappas, A., Hess, U., Barr, C. L., & Kleck, R. E. (1994). Angle of regard: The effect of vertical viewing angle on the perception of facial expressions. *Journal of Nonverbal Behavior, 18*(4), 263–280.

Karnath, H. O., Baier, B., & Nägele, T. (2005). Awareness of the functioning of one's own limbs mediated by the insular cortex? *The Journal of Neuroscience, 25*, 7134–7138. doi:10.1523/JNEUROSCI.1590-05.2005.

Karson, C. N. (1983). Spontaneous eye-blink rates and dopaminergic systems. *Brain, 106*, 643–653.

Katz, W. F., Garst, D. M., Briggs, R. W., Cheshkov, S., Ringe, W., Gopinath, K. S., Goyal, A., & Allen, G. (2012). Neural bases of the foreign accent syndrome: A functional magnetic resonance imaging case study. *Neurocase, 18*(3), 199–211. Epub 2011 Oct 20. PMID: 22011212.

Kawamura, Y., & Kare, M. R. (1987). *Umami: A basic taste.* New York: Marcel Dekker.

Keefe, F. J., Castell, P. J., & Blumenthal, J. A. (1986). Angina pectoris in type A and type B cardiac patients. *Pain, 27*(2), 211–218.

Keenan, J. P., Gallup, G. G., & Falk, D. (2003). *The face in the mirror: How we know who we are.* New York: HarperCollins.

Kellett, J., & Kokkinidis, L. (2004). Extinction deficit and fear reinstatement after electrical stimulation of the amygdala: Implications for kindling-associated fear and anxiety. *Neuroscience, 127*(2), 277–287.

Kelley, W. M., Miezin, F. M., McDermott, K. B., Buckner, R. L., Raichle, M. E., Cohen, N. J., et al. (1998). Hemispheric specialization in human dorsal frontal cortex and medial temporal lobe for verbal and nonverbal memory encoding. *Neuron, 20*, 927–936.

Kelley, N. J., Hortensius, R., & Harmon-Jones, E. (2013). When anger leads to rumination induction of relative right frontal cortical activity with transcranial direct current stimulation increases anger-related rumination. *Psychological Science, 24*, 475–481. doi:10.1177/0956797612457384.

Kelly, P. L., & Harrison, D. W. (1994). Home blood pressure monitoring: A survey of potential users. *Biomedical Instrumentation and Technology, 28*, 32–36.

Kelly, R. P., Yeo, K. P., Teng, C. H., Smith, B. P., Lowe, S., Soon, D., et al. (2005). Hemodynamic effects of acute administration of atomoxetine and methylphenidate. *Journal of Clinical Pharmacology, 45*, 851–855.

Kelly, S. P., et al. (2006). Increases in alpha oscillatory power reflect an active retinotopic mechanism for distracter suppression during sustained visuospatial attention. *Journal of Neurophysiology, 95*, 3844–3851.

Kempster, P. A., Gibb, W. R. G., Stern, G. M., & Lees, A. J. (1989). Asymmetry of substantia nigra neuronal loss in Parkinson's disease and its relevance to the mechanism of levodopa related motor fluctuations. *Journal of Neurology, Neurosurgery, and Psychiatry, 52*, 72–76.

Kenshalo, D. R., & Nafe, J. P. (1960). Receptive capacities of the skin, in Symposium on Cutaneous Sensitivity, U.S. Army Medical Research Laboratory Report No. 424, Fort Knox, Kentucky.

Kent, J. M., Coplan, J. D., Mawlawi, O., Martinez, J. M., Browne, S. T., & Slifstein, M. (2005). Prediction of panic response to a respiratory stimulant reduced orbitofrontal cerebral blood flow in panic disorder. *American Journal of Psychiatry, 162(7)*, 1379–1381.

Kessler, R. C. (2003). Epidemiology of women and depression. *Journal of Affective Disorder, 74*, 5–13.

Keysers, C., Kaas, J. H., & Gazzola, V. (2010). Somatosensation in social perception. *Nature Review Neuroscience, 11*, 417–428.

Kim, S., Manes, F., Kosier, T., Baruah, S., & Robinson, R. (1999). Irritability following traumatic brain injury. *The Journal of Nervous and Mental Disease, 187(6)*, 327–335.

Kim, H. J., Moon, W. J., & Han, S. H. (2013). Differential cholinergic pathway involvement in Alzheimer's disease and subcortical ischemic vascular dementia. *Journal of Alzheimer's Disease, 35(1)*, 129–136.

Kimbrell, T. A., George, M. S., Parekh, P. I., Ketter, T. A., Podell, D. M., Danielson, A. L., Repella, J. D., Benson, B. E., Willis, M. W., Herscovitch, P., & Post, R. M. (1999). Regional brain activity during transient self-induced anxiety and anger in healthy adults. *Biological Psychiatry, 46(4)*, 454–465.

Kimura, D. (1963). Right temporal-lobe damage. *Archives of Neurology, 8*, 264–271.

Kimura D. (1967). Functional asymmetry of the brain in dichotic listening. *Cortex, 3*, 163–168.

Kimura, D. (1999). *Sex and cognition, a Bradford book*. Cambridge: MIT Press.

King-Casas, B., Tomlin, D., Anen, C., Camerer, C. F., Quartz, S. R., & Montague, P. R. (2005). Getting to know you: Reputation and trust in a two-person economic exchange. *Science, 308*, 78–83.

Kinsbourne, M. (1971). Cognitive deficit: Experimental analysis. In J. L. McGaugh (Ed.), *Psychobiology*. New York: Academic Press.

Kinsbourne, M. (1978). Evolution of language in relation to lateral action. In M. Kinsbourne (Ed.), *Asymmetrical function of the brain* (pp. 553–556). New York: Cambridge University Press.

Kinsbourne, M., & Cook, J. (1971). Generalized and lateralized effects of concurrent verbalization on a unimanual skill. *Quarterly Journal of Experimental Psychology, 23*, 341–345.

Kinsbourne, M., & Hicks, R. E. (1978). Functional cerebral space: A model for overflow, transfer and interference effects in human performance. In J. Requin (Ed.), *Attention and performance VII* (pp. 345–362). New York: Academic Press.

Kinsbourne, M., & Wood, F. (1975). Short-term memory and the amnesic syndrome. In D. Deutsch & J. A. Deutsch (Eds.), *Short term memory* (pp. 257–291). New York: Academic Press.

Kirk, S. A., Gomory, T., & Cohen, D. (2013). *Mad science: Psychiatric coercion, diagnosis, and drugs*. New Brunswick: Transaction Publishers.

Kleist, K. (1934). *Gehirnpathologie*. Leipzig: Barth (Access to this work is difficult outside specialist libraries, but there seems to be a recent Spanish compilation of Kleist's work in Biblioteca Gador's History of Psychiatry series, if interested.).

Klimesch, W. (2012). Alpha-band oscillations, attention, and controlled access to stored information. *Trends in Cognitive Sciences, 16(12)*, 606–617.

Klineburger, P., & Harrison, D. W. (2013a). The acoustic startle response in high and low hostiles as a function of stress: Evidence for diminished frontal lobe capacity. Manuscript submitted for publication.

Klineburger, P. C., & Harrison, D. W. (2013b). The dynamic functional capacity theory: A neuro-psychological model of intense emotions. Manuscript submitted for publication.

Klinke, R., Kral, A., Heid, S., Tillein, J., & Hartmann, R. (1999). Recruitment of the auditory cortex in congenitally deaf cats by long-term cochlear electrostimulation. *Science, 285*(5434), 1729–1733.

Knight, R. T. (1997). Electrophysiological methods in behavioral neurology and neuropsychology. In T. E. Feinberg & M. J. Farah (Eds.), *Behavioral neurology and neuropsychology* (pp. 101–119). New York: McGraw-Hill Book Company.

Knight, R. T., Scabini, D., & Woods, D. L. (1989). Prefrontal cortex gating of auditory transmission in humans. *Brain Research, 504,* 338–342.

Knutson, B., & Gibbs, S. E. B. (2007). Linking nucleus accumbens dopamine and blood oxygenation. *Psychopharmacology, 191,* 813–822. doi:10.1007/s00213-006-0686-7.

Knutson, K. M., Rakowsky, S. T., Solomon, J., Krueger, F., Raymont, V., Tierney, M. C., & Grafman, J. (2013). Injured brain regions associated with anxiety in Vietnamveterans. *Neuropsychologia, 51*(4), 686–694.

Koch, G., Oliveri, M., Cheeran, B., Ruge, D., Lo Gerfo, E., Salerno, S., Torriero, S., Marconi, B., Mori, F., Driver, J., Rothwell, J. C., & Caltagirone, C. (2008). Hyperexcitability of parietal-motor functional connections in the intact left-hemisphere of patients with neglect. *Brain, 131,* 3147–3155.

Koechlin, E., & Summerfield, C. (2007). An information theoretical approach to prefrontal executive function. *Trends in Cognitive Science, 11,* 229–235.

Koh, K. B., Sohn, S.-H., Kang, J. I., Lee, Y.-J., & Lee, J. D. (2012). Relationship between neural activity and immunity in patients with undifferentiated somatoform disorder. *Psychiatry Research: Neuroimaging, 202*(3), 30, 252–256. ISSN 0925-4927, 10.1016.

Kohyama, J. (2011). Sleep health and asynchronization. *Brain & Development, 33,* 252–259.

Kojima, T., Sumitomo, J., Nishida, A., & Uchida, S. (2013). Changes of the human core body temperature rhythm and sleep structure by 6-hour phase advance treatment under a natural light-dark cycle. www.sleepscience.com. br, 16.

Kolb, B., & Whishaw, I. Q. (2014). *An introduction to brain and behavior* (4th ed.). New York: Worth Publishers.

Kondratova, A. A., & Kondratov, R. V. (2012). The circadian clock and pathology of the ageing brain. *Nature Reviews: Neuroscience, 13,* 325–335.

Konishi, S., Kawazu, M., Uchida, I., Kikyo, H., Asakura, I., & Miyashita, Y. (1999). Contribution of working memory to transient activation in human inferior prefrontal cortex during performance of the Wisconsin card sorting test. *Cerebral Cortex, 9,* 745–753.

Kononenko, V. S. (1980). Cholinesterase activity of nerve tissue as an indicator of asymmetry of brain centers. *Human Physiology, 6,* 194–199.

Koob, A. (2009). *The root of thought: Unlocking glia.* FT Science Press, Prentice Hall-Pearson Publishing Company.

Kopčo, N., Huang, S., Belliveau, J., Kopčo, N., Huang, S., Raij, T., Tengshe, C., & Ahveninen, J. (2012). Neuronal representations of distance in human auditory cortex. *PNAS, 109*(27), 11019–11024. doi:10.1073/pnas.1119496109.

Kosfeld, M., Heinrichs, M., Zak, P., Fischbacher, U., & Fehr, E. (2005). Oxytocin increased trust in humans. *Nature, 435*(7042), 673–676.

Kosnar, P. (2012). Gender differences in same and opposite sex mediated social touch: Affective responses to physical contact in a virtual environment. Thesis completed for the degree of Master of Science in Human Technology Interaction.

Kraepelin, E. (1919). *Dementia praecox and paraphrenia.* Edinburgh: E. S. Livingston.

Kraft, T. L., & Pressman, S. D. (2012). Grin and bear it: The influence of manipulated facial expression on the stress response. *Psychological Science* (first published on September 24, 2012). doi:10.1177/0956797612445312.

Krasteva, G., & Kummer, W. (2012). "Tasting" the airway lining fluid. *Histochemistry and Cell Biology, 138,* 365–383. doi:10.1007/s00418-012-0993-5.

Krauzlis, R. J., Lovejoy, L. P., & Zénon, A. (2013). Superior colliculus and visual spatial attention. *Annual Review of Neuroscience, 17,* 43.

Kripke, D. F. (1998). Light treatment for nonseasonal depression: Speed, efficacy, and combined treatment. *Journal of Affective Disorders, 49,* 109–117.

Kripke, D. F., Simons, R. N., Garfinkel, L., & Hammond, E. C. (1979). Short and long sleep and sleeping pills. Is increased mortality associated? *Archives of General Psychiatry, 36*(1), 103–116.

Kuchenbuch, A., Paraskevopoulos, E., Herholz, S. C., & Pantev, C. (2012). Electromagnetic correlates of musical expertise in processing of tone patterns. *PLoS One, 7*(1), e30171. doi:10.1371/journal.pone.0030171.

Kucyi, A., Moayedi, M., Weissman-Fogel, I., Hodaie, M., & Davis, K. D. (2012). Hemispheric asymmetry in white matter connectivity of the temporoparietal junction with the insula and prefrontal cortex. *PLoS One, 7*(4), e35589.

Kuhn, H. G., Dickinson-Anson, H., & Gage, F. H. (1996). Neurogenesis in the dentate gyrus of the adult rat: Age-related decrease of neuronal progenitor proliferation. *The Journal of Neuroscience, 76*(6), 2027–2033.

Kujawa, S. G., & Liberman, M. C. (1997). Conditioning-related protection from acoustic injury: Effects of chronic de-efferentation and sham surgery. *Journal of Neurophysiology, 78,* 3095–3106.

Kumar, A., Bilker, W., Lavretsky, H., & Gottlieb, G. (2000). Volumetric asymmetries in late-onset mood disorders: An attenuation of frontal asymmetry with depression severity. *Psychiatry Research, 100,* 41–47.

Kumar, A., Mann, S., Sossi, V., Ruth, T. J., Stoessl, A. J., Schulzer, M., et al. (2003). PET correlates of levodopa responses in asymmetric Parkinson's disease. *Brain, 126,* 2648–2655.

Kung, S. J., Chen, J. L., Zatorre, R. J., & Penhune, V. B. (2013). Interacting cortical and basal ganglia networks underlying finding and tapping to the musical beat. *Journal of Cognitive Neuroscience, 25*(3), 401–420.

Kupers, R., & Ptito, M. (2011). Insights from darkness: What the study of blindness has taught us about brain structure and function. *Progress in Brain Research, 192,* 17–31. doi:10.1016/B978-0-444-53355-5.00002-6.

Kupers, R. C., Gybels, J. M., & Gjedde, A. (2000). Positron emission tomography study of a chronic pain patient successfully treated with somatosensory thalamic stimulation. *Pain, 87,* 295–302.

Kwong, K. K., Belliveau, J. W., Chesler, D. A., Goldberg, I. E., Weisskoff, R. M., Poncelet, B. P., & Turner, R. (1992). Dynamic magnetic resonance imaging of human brain activity during primary sensory stimulation. *Proceedings of the National Academy of Sciences of the United States of America, 89,* 5675–5679.

Lacey, J. I., & Lacey, B. C. (1958). Verification and extension of the principle of autonomic response-stereotypy. *The American Journal of Psychology, 71*(1), 50–73. Published by: University of Illinois Press.

Ladavas, E., Zeloni, G., Zaccara, G., & Gangemi, P. (1997). Eye movements and orienting of attention in patients with visual neglect. *Journal of Cognitive Neuroscience, 9,* 67–74.

Lall, G. S., Atkinson, L. A., Corlett, S. A., Broadbridge, P. J., & Bonsall, D. R. (2012). Circadian entrainment and its role in depression: A mechanistic review. *Journal of Neural Transmission, 119*(10), 1085–1096.

Lam, R. W. (2006). Sleep disturbances and depression: A challenge for antidepressants. *International Clinical Psychopharmacology, 21*(Suppl 1), S25–S29.

Lane, R. D., Fink, G. R., & Chau, P. M. (1997). Neural activation during selective attention to subjective emotional responses. *Neuroreport, 8,* 3969–3972.

Lapidot, M. (1987). Does the brain age uniformly? Evidence from effects of smooth pursuit eye movements on verbal and visual tasks. *Journal of Gerontology, 42,* 329–331.

Larson, C. L., Aronoff, J., & Steuer, E. L. (2012). Simple geometric shapes are implicitly associated with affective value. *Motivation and Emotion, 36,* 404–413.

Lashley, K. S. (1929). Search for the engram. *Brain Mechanisms & Intelligence.* Chicago: The University Press.

Lashley, K. S. (1950). In search of the engram. *Symposia of the Society for Experimental Biology, 4,* 553–561.

Lau, Y. C., Hinkley, L. B., Bukshpun, P., Strominger, Z. A., Wakahiro, M. L., Baron-Cohen, S., & Marco, E. J. (2013). Autism traits in individuals with agenesis of the corpus callosum. *Journal of Autism and Developmental Disorders, 43*(5), 1106–1118.

Lauter, J. L., Herscovitch, P., Formby, C., & Raichle, M. E. (1985). Tonotopic organization in human auditory cortex revealed by positron emission tomography. *Hearing Research, 20*(3), 199–205 (Elsevier B.V.).

LeDoux, J. E. (1992). Emotion and the amygdala. In J. P. Aggleton (Ed.), *The amygdala: Neurobiological aspects of emotion, memory, and mental dysfunction.* New York: Wiley-Liss, Inc.

Lee, M. R. (2005). Curare: The South American arrow poison. *The Journal of the Royal College of Physicians of Edinburgh, 35,* 83–92.

Lee, Y. K., & Choi, J. S. (2012). Inactivation of the medial prefrontal cortex interferes with the expression but not the acquisition of differential fear conditioning in rats. *Experimental Neurobiology, 21*(1), 23–29. http://dx.doi.org/10.5607/en.2012.21.1.23.

Lee, M. C., Klassen, A. C., Heaney, L. M., & Resch, J. A. (1976). Respiratory rate and pattern disturbances in acute brain stem infarction. *Stroke, 7,* 382–385.

Lee, C. S., Schulzer, M., Mak, E., Hamerstad, J. P., Calne, S., & Calne, D. B. (1995). Patterns of asymmetry do not change over the course of idiopathic parkinsonism: Implications for pathogenesis. *Neurology, 45,* 435–439.

Lee, G. P., Meador, K. J., Loring, D. W., Allison, J. D., Brown, W. S., Paul, L. K., Pillai, J. J., & Lavin, T. B. (2004). Neural substrates of emotion as revealed by functional magnetic resonance imaging. *Cognitive and Behavioral Neurology, 17,* 9–17.

Lee, A. C. H., Buckley, M. J., Pegman, S. J., Spiers, H., Scahill, V. L., Gaffan, D., et al. (2005). Specialization in the medial temporal lobe for processing of objects and scenes. *Hippocampus, 15,* 782–797.

Lee, Y. K., Hou, S. W., Lee, C. C., Hsu, C. Y., Huang, Y. S., & Su, Y. C. (2013). Increased risk of dementia in patients with mild traumatic brain injury: A nationwide cohort study. *PloS One, 8*(5), e62422.

Leenders, K. L., Salmon, E. P., Tyrrell, P., Perani, D., Brooks, D. J., Sager, H., et al. (1990). The nigrostriatal dopaminergic system assessed in vivo by positron emission tomography in healthy volunteer subjects and patients with Parkinson's disease. *Archives of Neurology, 47,* 1290–1298.

Leentjens, A. F. G., Lousberg, R., & Verhey, F. R. J. (2002). Markers for depression in Parkinson's disease. *Acta Psychiatrica Scandinavica, 106,* 196–201.

Leisman, G., & Melillo, R. (2011). Effects of motor sequence training on attentional performance in ADHD children. *International Journal on Disability and Human Development.* Advance Access published on August 20, 2011. doi:10.1515/ijdhd.2010.043.

Leisman, G., Melillo, R., Thum, S., Ransom, M. A., Orlando, M., Tice, C., et al. (2010). The effect of hemisphere specific remediation strategies on the academic performance outcome of children with ADD/ADHD. *International Journal of Adolescent Medicine and Health, 22,* 275–283.

Lele, P. P., & Weddell, G. (1959). Sensory nerves of the cornea and cutaneous sensibility. *Experimental Neurology, 1,* 334–359.

Lemaire, J. J., Frew, A. J., McArthur, D., Gorgulho, A. A., Alger, J. R., Salomon, N., & De Salles, A. A. (2011). White matter connectivity of human hypothalamus. *Brain Research, 1371,* 43–64.

Leon, S. A., & Rodriguez, A. D. (2008). Aprosodia and its treatment. *Perspectives on Neurophysiology and Neurogenic Speech and Language Disorders, 18,* 266–272. doi:10.1044/nnsld18.2.66.

Leonard, M. K., Ramirez, N. F., Torres, C., Travis, K. E., Hatrak, M., Mayberry, R. I., & Halgren, E. (2012). Signed words in the congenitally deaf evoke typical late lexicosemantic responses with no early visual responses in left superior temporal cortex. *The Journal of Neuroscience, 32*(28), 9700–9705.

Levens, S. M., & Gotlib, I. H. (2010). Updating positive and negative stimuli in working memory in depression. *Journal of Experimental Psychology: General, 139,* 654–664.

Levenson, R. W., Ekman, P., & Ricard, M. (2012). Meditation and the startle response: A case study. *Emotion, 12*(3), 650–658. doi:10.1037/a0027472.

Levermann, N., Galatius, A., Ehlme, G., Rysgaard, S., & Born, E. W. (2003). Feeding behavior of free-ranging walruses with notes on apparent dextrality of flipper use. *BMC Ecology, 3*(1), 9.

Levine, D. N., Warach, J., & Farah, M. J. (1985). The two visual systems in mental imagery: dissociation of "what" and "where" in imagery disorders due to bilateral posterior cerebral lesions. *Neurology, 35,* 1010–1018.

Lewandowsky, M., & Stadelmann, E. (1908). Uber einen bemerkenswerten Fall von Hirnblutung und tiber Rechenstomngen bei Herderkrankung des Gehims. *Journal fiir Psychologie und Neurologie, 11,* 249–265.

Lewis, P. R., & Shute, C. C. (1967). The cholinergic limbic system: Projections to hippocampal formation, medial cortex, nuclei of the ascending cholinergic reticular system, and the subfornical organ and supra-optic crest. *Brain, 90,* 521–54.

Lewy, A. J., Sack, R. L., Miller, S., & Hoban, T. M. (1987). Antidepressant and circadian phase-shifting effects of light. *Science, 235,* 352–354.

Ley, R. G., & Bryden, M. P. (1979). Hemispheric differences in processing emotions in faces. *Brain and Language, 7,* 127–138.

Ley, R. G., & Bryden, M. P. (1982). A dissociation of right and left hemispheric effects for recognizing emotional tone and verbal content. *Brain and Cognition, 1*(1), 3–9.

Lezak, M. D., Howieson, D. B., & Loring, D. W. (2004). *Neuropsychological assessment* (4th ed.). New York: Oxford University Press.

Lezak, M. D., Howieson, D. B., Bigler, E. D., & Tranel, D. (2012). *Neuropsychological assessment* (5th ed.). New York: Oxford University Press.

Lhermitte, F. (1983). "Utilization behaviour" and its relation to lesions of the frontal lobes. *Brain, 106,* 237–255. doi:10.1093/brain/106.2.237.

Li, Z., Moore, A., Tyner, C., & Hu, X. (2009). Asymmetric connectivity reduction and its relationship to HAROLD in aging brain. *Brain Research, 1295,* 149–158.

Liang, S.-W., Jemerin, J. M., Tschann, J. M., Wara, D. W., & Boyce, W. T. (1997). Life events, frontal electroencephalogram laterality, and functional immune status after acute psychological stressors in adolescents. *Psychosomatic Medicine, 59,* 178–186.

Lieberman, M. (2007). The X and C-Systems: The neural basis of automatic and controlled social cognition. In E. Harmon-Jones & P. Winkielman (Eds.), *Social neuroscience* (pp. 290–315). New York: The Guilford Press.

Liegeois-Chauvel, C., Peretz, I., Babai, M., Laguitton, V., & Chauvel, P. (1998). Contribution of different cortical areas in the temporal lobes to music processing. *Brain, 121,* 1853–1867.

Liepert, J., Miltner, W. H., Bauder, H., et al. (1998). Motor cortex plasticity during constraint-induced movement therapy in stroke patients. *Neuroscience Letters, 250,* 5–8.

Limb, C. J. (2006). Structural and functional neural correlates of music perception. *The Anatomical Record. Part A, Discoveries in Molecular, Cellular, and Evolutionary Biology, 288,* 435–446.

Lin, L., Chen, G., Kuang, H., Wang, D., & Tsien, J. Z. (2007). Neural encoding of the concept of nest in the mouse brain. *Proceedings of the National Academy of Sciences of the United States of America, 104*(14), 6066–6071 (Publisher: National Academy of Sciences).

Lindsley, D. B. (1952). Psychological phenomena and the electroencephalogram. *Electroencephalography and Clinical Neurophysiology, 4,* 443–456.

Lindsley, D. B. (1960). Attention, consciousness, sleep, and wakefulness. In J. Field, H. W. Magoun, & V. E. Hall (Eds.), *Handbook of physiology: Neurophysiology* (Vol. III). Washington, DC: American Physiological Society.

Lindsley, D. B., Bowden, J. W., & Magoun, H. W. (1949). Effect upon the EEG of acute injury to the brain stem activating system. *Electroencephalography and Clinical Neurophysiology, 1,* 475–486.

Liotti, M., & Tucker, D. M. (1995). Emotion in asymmetric corticolimbic networks. In R. J. Davidson & K. Hugdahl (Eds.), *Brain asymmetry* (pp. 389–423). Cambridge: MIT Press.

Lippolis, G., Bisazza, A., Rogers, L. J., & Vallortigara, G. (2002). Lateralization of predator avoidance responses in three species of toads. *Laterality, 7,* 163–183.

Loeb, J. (1885). Die elementaren Storungen einfacher Funktionen nach ober-flachlicher, umschriebener Verletzung des Großhirns. *Pfluger Archive, 37,* 51–56.

Loewy, A. D. (1990). Anatomy of the autonomic nervous system: An overview. In A. D. Loewy & K. M. Spyer (Eds.), *Central regulation of autonomic functions* (pp. 3–16). New York: Oxford University Press.

Lombardi, W. J., Andreason, P. J., Sirocco, K. Y., Rio, D. E., Gross, R. E., Umhau, J. C., et al. (1999). Wisconsin card sorting test performance following head injury: Dorsolateral frontostriatal circuit activity predicts perseveration. *Journal of Clinical and Experimental Neuropsychology, 21,* 2–16.

Long, C. W. (1949). An account of the first use of sulphuric ether by inhalation as an anaesthetic in surgical operations. *Southern Medical and Surgical Journal, 5,* 705–713.

Long, R. R. (1977). Sensitivity of cutaneous cold fibers to noxious heat: Paradoxical cold discharge. *Journal of Neurophysiology, 40,* 489–502.

Long, X., Zhang, L., Liao, W., Jiang, C., Qiu, B., & The Alzheimer's Disease Neuroimaging Initiative. (2012). Distinct laterality alterations distinguish mild cognitive impairment and Alzheimer's disease from healthy aging: Statistical parametric mapping with high resolution MRI. *Human Brain Mapping.* doi:10.1002/hbm.22157.

Longo, D. L., Kasper, D. L., Jameson, J. L., Fauci, A. S., Hauser, S. L., & Loscalzo, J. (2012). Oxytocin. In Harrison's™ *Principles of Internal Medicine* (18th ed., pp. 1784–1784). New York: McGraw-Hill.

Loomis, J. M., & Lederman, S. J. (1986). Tactual perception. In K. Boff, L. Kaufman, & J. Thomas (Eds.), *Handbook of perception and human performance, vol. 1, sensory processes and perception.* New York: Wiley.

Lorenz, K. (1943). Die angeborenen Formen mo¨glicher Erfahrung (The innate forms of potential experience). *Zeitschrift fur Tierpsychologie, 5,* 233–519.

Loring, D. W., & Bauer, R. M. (2010). Testing the limits: Cautions and concerns regarding the new Wechsler IQ and memory scales. *Neurology, 74,* 685–690.

Loring, D. W., Martin, R. C., Meador, K. J., & Lee, G. P. (1990). Psychometric construction of the Rey–Osterrieth complex figure: Methodological considerations and interrater reliability. *Archives of Clinical Neuropsychology, 5,* 1–14.

Lovell, J., & Kruger, J. (1994). *Apollo 13.* New York: Mariner Books.

Lowenstein, O., & Loewenfeld, I. E. (1950a). Role of sympathetic and parasympathetic systems in reflex dilation of the pupil; pupillographic studies. *Archives of Neurology & Psychiatry, 64,* 313–340.

Lowenstein, O., & Loewenfeld, I. E. (1950b). Mutual role of sympathetic and parasympathetic in shaping of the papillary reflex to light; pupillographic studies. *Archives of Neurology & Psychiatry, 64,* 341–377.

Lowther, W. R., & Isaac, W. (1976). The effects of d-amphetamine and illumination on behaviors of the squirrel monkey. *Psychopharmacology, 50,* 231–235.

Lu, C. L., Wu, Y. T., Yeh, T. C., Chen, L. F., Chang, F. Y., Lee, S. D., Ho, L. T., & Hsieh, J. C. (2004). Neuronal correlates of gastric pain induced by fundus distension: A 3T-fMRI study. *Neurogastroenterology and Motility, 16,* 575–587.

Lu, J., Sherman, D., Devor, M., et al. (2006). A putative flip-flop switch for control of REM sleep. *Nature, 441,* 589–594.

Luders, H., Lesser, R. P., Dinner, D. S., Morris, H. H., Wyllie, E., & Godoy, J. (1988). Localization of cortical function: New information from extraoperative monitoring of patients with epilepsy. *Epilepsia, 29*(Suppl. 2), S56–S65.

Lundström, J. N., Boyle, J. A., Zatorre, R. J., & Jones-Gotman, M. (2008). Functional neuronal processing of body odors differs from that of similar common odors. *Cerebral Cortex, 18,* 1466–1474.

Luo, S., Romero, A., Adam, T. C., Hu, H. H., Monterosso, J., & Page, K. A. (2013). Abdominal fat is associated with a greater brain reward response to high-calorie food cues in hispanic women. *Obesity, 21*(10), 2029–2036.

Luria, A. R. (1966). *Higher cortical functions in man.* New York: Basic Books.

Luria, A. R. (1973). *The working brain: An introduction to neuropsychology.* New York: Basic Books, Inc., Publishers.

Luria, A. R. (1980). *Higher cortical functions in man* (2nd ed.). New York: Basic Books, Inc., Publishers.

Luria, A. R. (1987). *The mind of a mnemonist: A little book about a vast memory.* Cambridge: Harvard University Press. ISBN 0-674-57622-5.

Luria, A. R., & Solotaroff, L. (1987). *The man with a shattered world: The history of a brain wound.* Harvard University Press. *ISBN 0-674-54625-3.*

Lutterveld, R., & Ford, J. M. (2012). Neurophysiological research: EEG and MEG. In J. D. Blom & I. E. C. Sommer (Eds.), *Hallucinations* (pp. 283–295). Springer US. doi:10.1007/978-1-4614-0959-5_21.

Lyon, L. (May 2009). 7 criminal cases that involved the sleepwalking defense. US News and World Report.

Maas, J. B., Wherry, M. L., Axelrod, D. J., Hogan, B. R., & Blumin, J. A. (1998). Power sleep: The revolutionary program that prepares your mind for peak performance. William Morrow Paperbacks.

MacDonald, K., & MacDonald, T. (2010). The peptide that binds: A systematic review of oxytocin and its prosocial effects in humans. *Harvard Review of Psychiatry, 18*(1), 1–21.

Machado, M. L., Lelong-Boulouard, L., Smith, P. F., Freret, T., Philoxene, B., Denise, P., & Besnard, S. (2012). Influence of anxiety in spatial memory impairments related to the loss of vestibular function in rat. *Neuroscience, 218,* 161–169.

Maddock, R. (2010). Panic attacks as a problem of pH: Study casts new light on the brain mechanisms behind recurrent bouts of intense anxiety. Scientific American Mind Matters, May 18.

Magendie, F. (1822a). *Expériences sur les fonctions des racines des nerfs rachidiens. Journal de physiologie expérimentale et de pathologie, 2,* 276–279.

Magendie, F. (1822b). Expériences sur les fonctions des racines des nerfs qui naissent de la moëlle épinière. *Journal de physiologie expérimentale et de pathologie, 2,* 366–371.

Maggi, S., Siviero, P., Wetle, T., Besdine, R. W., Saugo, M., & Crepaldi, G. (2010). A multicenter survey on profile of care for hip fracture: Predictors of mortality and disability. *Osteoporosis International, 21*(2), 223–231.

Maguire, E. A., Burgess, N., Donnett, J. G., et al. (1998a). Knowing where and getting there: a human navigation network. *Science, 280,* 921–924.

Maguire, E. A., Frith, C. D., Burgess, N., et al. (1998b). Knowing where things are: parahippocampal involvement in encoding object locations in virtual large-scale space. *Journal of Cognitive Neuroscience, 10,* 61–76.

Maguire, E. A., Gadian, D. G., Johnsrude, I. S., Good, C. D., Ashburner, J., Frackowiak, R. S., & Frith, C. D. (2000). Navigation-related structural change in the hippocampi of taxi drivers. *PNAS, 97*(8), 4398–4403.

Maison, S. F., & Liberman, M. C. (2000). Predicting vulnerability to acoustic injury with a noninvasive assay of olivocochlear reflex strength. *The Journal of Neuroscience, 20*(12), 4701–4707.

Maisto, S. A., Galizio, M., & Connors, G. J. (2008). *Drug use and abuse* (5th ed.). Belmont: Thomson Wadsworth Publisher.

Malach, R., Reppas, J. B., Benson, R. R., Kwong, K. K., Jiang, H., et al. (1995). Object-related activity revealed by functional magnetic resonance imaging in human occipital cortex. *Proceedings of the National Academy of Sciences of the United States of America, 92,* 8135–8139.

Maleki, N., Becerra, L., Upadhyay, J., Burstein, R., & Borsook, D. (2012). Direct optic nerve pulvinar connections defined by diffusion MR tractography in humans: Implications for photophobia. *Human Brain Mapping, 33*(1), 75–88.

Malmierca, M. S., & Hackett, T. A. (2010). Structural organization of the ascending auditory path-way. In A. Rees & A. R. Palmer (Eds.), *The auditory brain* (pp. 9–41). Oxford: Oxford University Press.

Malmo, R. B. (1942). Interference factors in delayed response in monkeys after removal of the frontal lobe. *Journal of Neurophysiology, 5,* 295–308.

Malmo, R. B. (1959). Activation: A neuropsychological dimension. *Psychological Review, 66,* 367–386.

Malnic, B., Hirono, J., Sato, T., & Buck, L. B. (1999). Combinatorial receptor codes for odors. *Cell, 96,* 713–723.

Manuck, S. B., Jennings, R., Rabin, B. S., & Baum, A. (2000). *Behavior, health, & aging.* New Jersey: Lawrence Erlbaum Associates, Inc., Publishers.

Maquet, P., Péter, J. M., & Aerts, J., et al. (1996). Functional neuroanatomy of human rapid-eye-movement sleep and dreaming. *Nature, 383,* 163–166.

Marcé, L. V. (1856). Mémoire sur quelques observations de physiologie pathologique tendant a démontrer l'existence d'un principe coordinateur de l'écriture et ses rapports avec le principe coordinateur de la parole. *Compte-rendu de la Société de Biologie, Paris, 3,* 93–115.

Marcuse, H. (1904). Apraktiscke symotome bein linem fall von seniler demenz. *Zentralbl. Mervheik. Psychiatrie, 27,* 737–751.

Marfurt, C. F., Cox, J., Deek, S., & Dvorscak, L. (2010). Anatomy of the human corneal innervation. *Experimental Eye Research, 90,* 478–492.

Marie, P. (1907). Presentation de malades atteints d'anarthrie par lesion de l'hemisphere gauche du cerveau. *Bulletins et Memoires Societe Medicale des Hopitaux de Paris, 1,* 158–160.

Marien, P., & Verhoeven, J. (2007). Cerebellar involvement in motor speech planning: Some further evidence from foreign accent syndrome. *Folia Phoniatrica et Logopaedica, 59*(4), 210–217.

Maron, E., Nutt, D., & Shlik, J. (2012). . *Current Pharmaceutical Design, 18,* 5699–5708.

Marsh, N. V., Kersel, D. A., Havill, J. H., & Sleigh, J. W. (1998a). Caregiver burden at 6 months following severe traumatic brain injury. *Brain Injury, 12*(3), 225–238.

Marsh, N. V., Kersel, D. A., Havill, J. H., & Sleigh, J. W. (1998b). Caregiver burden at 1 year following severe traumatic brain injury. *Brain Injury, 12*(12), 1045–1059.

Marshall, D. W., Westmoreland, B. F., & Sharbrough, F. W. (1983). Ictal tachycardia during temporal lobe seizures. *Mayo Clinic Proceedings, 58*(7), 443–446.

Martella, D., Plaza, V., Este'vez, A. F., Castillo, A., & Fuentes, L. J. (2012). Minimizing sleep deprivation effects in healthy adults by differential outcomes. *Acta Psychologica (Amst), 139*(3), 391–396.

Martin, W. R. W., & Calne, D. B. (1987). Imaging techniques and movement disorders. In C. D. Marsden & S. Fahn (Eds.), *Movement disorders 2* (pp. 4–16). London: Butterworth.

Martin, P. R., Singleton, C. K., & Hiller-Sturmhöfel, S. (2004). *The role of thiamine deficiency in alcoholic brain disease.* National Institute on Alcohol Abuse and Alcoholism (NIAAA) of the National Institute of Health.

Martinaud, O., Pouliquen, D., Gérardin, E., Loubeyre, M., Hirsbein, D., et al. (2012). Visual agnosia and posterior cerebral artery infarcts: An anatomical-clinical study. *PLoS One, 7*(1), e30433. doi:10.1371/journal.pone.0030433.

Masland, R. H. (2012). Another blue neuron in the retina. *Nature Neuroscience, 15*(7), 930–931.

Matell, M. S., & Meck, W. H. (2004). Cortico-striatal circuits and interval timing: Coincidence detection of oscillatory processes. *Brain Research. Cognitive Brain Research, 21,* 139–170.

Mather, M., Cacioppo, J. T., & Kanwisher, N. (2013). Introduction to the special section: 20 years of fMRI-What has it done for understanding cognition? *Perspectives on Psychological Science, 8,* 41–43. doi:10.1177/1745691612469036.

Mathews, S. J., Mao, X., Coplan, J. D., Smith, E. L. P., Sackeim, H. A., Gorman, J. M., & Shungu, D. C. (2004). Dorsolateral prefrontal cortical pathology in generalized anxiety disorder: A proton magnetic resonance spectroscopic imaging study. *American Journal of Psychiatry, 161,* 1119–1121.

Mathewson, K. E., Prudhomme, C., Fabiani, M., Beck, D. M., Lleras, A., & Gratton, G. (2012). Making waves in the stream of consciousness: Entraining oscillations in EEG alpha and fluctuations in visual awareness with rhythmic visual stimulation. *Journal of Cognitive Neuroscience, 24*(12), 2321–2333.

Matsubara, E., et al. (2003). Melatonin increases survival and inhibits oxidative and amyloid pathology in a transgenic model of Alzheimer's disease. *Journal of Neurochemistry, 85,* 1101–1108.

Mattay, V. S., Fera, F., Tessitore, A., Hariri, A. R., Berman, K. F., Das, S., & Weinberger, D. R. (2006). Neurophysiological correlates of age-related changes in working memory capacity. *Neuroscience Letters, 392,* 32–37.

Maximus, V. (2010). Encyclopædia Britannica. http://www.britannica.com/EBchecked/topic/622119/Valerius-Maximus. Accessed 17 June 2010.

Mazzola, L., Faillenot, I., Barral, F., Mauguière, F., & Peyron, R. (2012a). Spatial segregation of somato-sensory and pain activations in the human operculo-insular cortex. *NeuroImage, 60*(1), 409–418.

Mazzola, L., Isnard, J., Peyron, R., & Mauguière, F. (2012b). Stimulation of the human cortex and the experience of pain: Wilder Penfield's observations revisited. *Brain, 135*(2), 631–640.

McBride, J., Boy, F., Husain, M., & Sumner, P. (2012). *Front Hum Neurosci., 6,* 82. Published online 2012 April 24. doi:10.3389/fnhum.2012.00082.

McCandliss, B. D., Cohen, L., & Dehaene, S. (2003). The visual word form area: Expertise for reading in the fusiform gyrus. *Trends in Cognitive Sciences, 7,* 293–299.

McCann, J., & Peppé, S. (2003). Prosody in autism spectrum disorders: A critical review. *International Journal of Language & Communication Disorders, 38,* 325–350. doi:10.1080/13682 8203100015420.

McCarthy, G., Puce, A., Gore, J. C., & Allison, T. (1997). Face-specific processing in the human fusiform gyrus. *Journal of Cognitive Neuroscience, 9*(5), 605–610.

McClintock, M. K. (1971). Menstrual synchrony and suppression. *Nature, 229,* 244–245.

McClung, C. A. (2007). Circadian genes, rhythms and the biology of mood disorders. *Pharmacology & Therapeutics, 114,* 222–232.

McClure, E. B. (2000). A meta-analytic review of sex differences in facial expression processing and their development in infants, children, and adolescents. *Psychological Bulletin, 126,* 424–453.

McCullagh, S., Moore, M., Gawel, M., & Feinstein, A. (1999). Pathological laughing and crying in amyotrophic lateral sclerosis: An association with prefrontal cognitive dysfunction. *Journal of the Neurological Sciences, 169,* 43–48.

McDonald, J. W., Sandy, A. P., Krzysztof, L. H., Choi, D. W., & Goldber, M. P. (1998). Oligodendrocytes from forebrain are highly vulnerable to AMPA/kainate receptor-mediated excitotoxicity. *Nature Medicine, 4,* 291–297. doi:10.1038/nm0398-291.

McDowell, C. L., Harrison, D. W., & Demaree, H. A. (1994). Is right hemisphere decline in the perception of emotion a function of aging? *International Journal of Neuroscience, 79*(1–2), 1–11.

McFie, J., Piercy, M., & Zangwill, O. (1950). Visual spatial agnosia associated with lesions of the right cerebral hemisphere. *Brain, 73,* 167–190.

McGaugh, J. L. (2004). The amygdala modulates the consolidation of memories of emotionally arousing experiences. *Annual Review of Neuroscience, 27,* 1–28. doi:10.1146/annurev.neuro.27.070203.144157.

McGlone, J. (1978). Sex differences in functional brain asymmetry. *Cortex, 14,* 122–128.

McGlone, J. (1980). McGlone, sex differences in human brain asymmetry: A critical survey. *Behavioral and Brain Sciences, 3*(2), 215–263.

McGlone, F., Kelly, E. F., Trulsson, M., Francis, S. T., Westlin, G., & Bowtell, R. (2002). Functional neuroimaging studies of human somatosensory cortex. *Behavioral Brain Research, 135,* 147–158.

McKee, A., Cantu, R., Nowinski, C., et al. (2009). Chronic traumatic encephalopathy in athletes: Progressive tauopathy after repetitive head injury. *Journal of Neuropathology and Experimental Neurology, 68,* 709–735.

McKee, A. C., Gavett, B. E., Stern, R. A., Nowinski, C. J., Cantu, R. C., Kowall, N. W., & Budson, A. E. (2010). TDP-43 proteinopathy and motor neuron disease in chronic traumatic encephalopathy. *Journal of Neuropathology and Experimental Neurology, 69*(9), 918.

McKiernan, B. J., Marcario, J. K., Karrer, J. H., & Cheney, P. D. (1998). Corticomotoneuronal postspike effects in shoulder, elbow, wrist, digit, and intrinsic hand muscles during a reach and prehension task. *Journal of Neurophysiology, 80*, 1961–1980.

McManus, I. C., & Humphrey, N. K. (1973). Turning the left cheek. *Nature, 243*, 271–272.

Mechelli, A., Gorno-Tempini, M. L., & Price, C. J. (2003). Neuroimaging studies of word and pseudoword reading: Consistencies, inconsistencies, and limitations. *Journal of Cognitive Neuroscience, 15*, 260–271.

Medendorp, W. P., Goltz, H. C., Vilis, T., & Crawford, J. D. (2003). Gaze-centered updating of visual space in human parietal cortex. *The Journal of Neuroscience, 23*(15), 6209–6214.

Meerwijk, E., Ford, J., & Weiss, S. (2012). Brain regions associated with psychological pain: Implications for a neural network and its relationship to physical pain. *Brain Imaging and Behavior*, 1–14. doi:10.1007/s11682-012-9179-y.

Mehler, J., Jusczyk, P., Lambertz, G., Halsted, N., Bertoncini, J., & Amiel-Tison, C. (1988). A precursor of language acquisition in young infants. *Cognition, 29*, 143–178.

Mehrabian, A. (1994). Manual for the revised trait dominance-submissiveness scale (TDS). Available from Albert Mehrabian, 1130 Alta Mesa Road, Monterey, CA, USA 93940.

Mehrabian, A. (1995). Distinguishing depression and trait anxiety in terms of basic dimensions of temperament. *Imagination, Cognition, and Personality, 15*(2), 133–143.

Melillo, R., & Leisman, G. (2009). Autistic spectrum disorders as functional disconnection syndrome. *Reviews in the Neurosciences, 20*, 111–131.

Meltzoff, A. N., & Moore, M. K. (1977). Imitation of facial and manual gestures by human neonates. *Science, 198*(4312), 75–78. http://www.jstor.org/stable/1744187.

Mennella, J. A., & Ventura, A. K. (2011). Early feeding: Setting the stage for healthy eating habits. *Nestle Nutr Workshop Ser Pediatric Program, 68*, 153–163.

Menon, V., & Uddin, L. Q. (2010). Saliency, switching, attention and control: A network model of insula function. *Brain Structure & Function, 214*, 655–667.

Menon, V., Adleman, E., White, C. D., Glover, G. H., & Reiss, A. L. (2001). Error-related brain activation during a Go/NoGo response inhibition task. *Human Brain Mapping, 12*, 131–143.

Mercuriale, G. (1588). De morbis puerorum tractatus locupletissimi.

Meredith, M. (2001). Human vomeronasal organ function: A critical review of best and worst cases. *Chemical Senses, 26*, 433–445.

Merskey, H., & Bogduk, N. (Eds.). (1994). *Classification of chronic pain*. Seattle: IASP Task Force on Taxonomy.

Mesulam, M.-M. (1981). A cortical network for directed attention and unilateral neglect. *Annals of Neurology, 10*, 309–325.

Mesulam, M.-M. (1990). Large-scale neurocognitive networks and distributed processing for attention, language, and memory. *Annals of Neurology, 28*, 598–613.

Mesulam, M.-M. (2000). *Principles of behavioral and cognitive neurology* (2nd ed.). New York: Oxford University Press.

Mesulam, M. M., & Mufson, E. J. (1982). Insula of the old world monkey. I. Architectonics in the insulo-orbitotemporal component of the paralimbic brain. *The Journal of Comparative Neurology, 212*, 1–22.

Meuret, A. E., Rosenfield, D., Seidel, A., Bhaskara, L., & Hofmann, S. G. (2010). Respiratory and cognitive mediators of treatment change in panic disorder: Evidence for intervention specificity. *Journal of Consulting and Clinical Psychology, 78*, 691–704.

Meuret, A. E., Seidel, A., Rosenfield, B., Hofmann, S. G., & Rosenfield, D. (2012). Does fear reactivity during exposure predict panic symptom reduction? *Journal of Consulting and Clinical Psychology, 80*, 773–785.

Mikkelson, B., & David, P. Good luck, Mr Gorsky!; at Snopes.com: Urban Legends Reference Pages.

Milad, M. R., Rauch, S., Schoenbaum, G., Gottfried, J. A., Murray, E. A., & Ramus, S. J. (2007). The role of the orbitofrontal cortex in anxiety disorders. *Annals of the New York Academy of Sciences, 1211,* 546–561.

Milberg, W. P., Hebben, N. A., & Kaplan, E. (2009). The Boston process approach to neuropsychological assessment. In I. Grant & K. Adams (Eds.), *Neuropsychological assessment of neuropsychiatric disorders* (3rd ed.). New York: Oxford University Press.

Miller, E. K. (2000). The prefrontal cortex and cognitive control. *Nature Reviews Neuroscience, 1,* 59–65.

Miller, E. K., & Cohen, J. D. (2001). An integration theory of prefrontal cortex function. *Annual Review Neuroscience, 24,* 167–202.

Mills, C. K. (1912a). The cerebral mechanisms of emotional expression. *Transactions of the College of Physicians of Philadelphia, 34,* 381–390.

Mills, C. K. (1912b). The cortical representation of emotion, with a discussion of some points in the general nervous mechanism of expression in its relation to organic nervous mental disease. *Proceedings of the American Medico-Psychological Association, 19,* 297–300.

Milner, B. (1963). Effects of different brain lesions on card sorting. *Archives of Neurology, 9,* 90–100.

Milner, B. (1965). Visually-guided maze-learning in man: Effects of bilateral hippocampal, bilateral frontal and unilateral cerebral lesions. *Neuropsychologia, 3,* 317–338.

Milner, B. (1968). Visual recognition and recall after right temporal-lobe excision in man. *Neuropsychologia, 6,* 191–209.

Milner, P. (1991). Brain stimulation reward: A review. *Canadian Journal of Psychology Outstanding Contributions Series, 45*(1), 1–36.

Miltner, W. H., Bauder, H., Sommer M., Dettmers, C., & Taub, E. (1999). Effects of constraint-induced movement therapy on patients with chronic motor deficits after stroke: A replication. *Stroke, 30,* 586–592.

Mishima, K., Okawa, M., Hishikawa, Y., Hozumi, S., Hori, H., & Takahashi, K. (1994). Morning bright light therapy for sleep and behavior disorders in elderly patients with dementia. *Acta Psychiatrica Scandinavica, 89*(1), 1–7.

Mishima, K., Okawa, M., Hozumi, S., & Hishikawa, Y. (2000). Supplementary administration of artificial bright light and melatonin as potent treatment for disorganized circadian rest-activity and dysfunctional autonomic and neuroendocrine systems in institutionalized demented elderly persons. *Chronobiology International, 17,* 419–432.

Mitchell, G. A., & Harrison, D. W. (2010). Neuropsychological effects of hostility and pain on emotion perception. *Journal of Clinical & Experimental Neuropsychology, 32*(2), 174–189.

Miyazaki, K. W., Miyazaki, K., & Doya, K. (2012). Activation of dorsal raphe serotonin neurons is necessary for waiting for delayed rewards. *Journal of Neuroscience, 32,* 10451–10457.

Moallem, S. A., Balali-Mood, K., Balali-Mood, M., & Balali-Mood, M. (2012). Opioids and opiates. In *Handbook of drug interactions* (pp. 159–191). Springer.

Moan, C. E., & Heath, R. G. (1972). Septal stimulation for the initiation of heterosexual activity in a homosexual male. *Journal of Behavior Therapy and Experimental Psychiatry, 3,* 23–30.

Moayedi, M., Weissman-Fogel, I., Crawley, A. P., Goldberg, M. B., Freeman, B. V., et al. (2011). Contribution of chronic pain and neuroticism to abnormal forebrain gray matter in patients with temporomandibular disorder. *Neuroimage, 55,* 277–286.

Modahl, C., Green, L., Fein, D., Morris, M., Waterhouse, L., Feinstein, C., et al. (1998). Plasma oxytocin levels in autistic children. *Biological Psychiatry, 43*(4), 270–277.

Moll, J., Moll, F. T., Bramati, I. E., & Andreiuolo, P. A. (2002). The cerebral correlates of set-shifting: An fMRI study of the trail making test. *Arqui Neuropsiquiatria, 60,* 900–905.

Mollet, G. A., & Harrison, D. W. (2006). Emotion and pain: A functional cerebral systems integration. *Neuropsychology Review, 16,* 99–121.

Mollet, G. A., & Harrison, D. W. (2007a). Affective verbal learning in hostility: An increased primacy effect and bias for negative emotional material. *Archives of Clinical Neuropsychology, 22*(1), 53–62.

Mollet, G. A., & Harrison, D. W. (2007b). Effects of hostility and stress on affective verbal learning in women. *International Journal of Neuroscience, 117*(1), 63–83.

Mollet, G. A., Harrison, D. W., Walters, R. P., & Foster, P. S. (2007). Asymmetry in the emotional content of lateralized multimodal hallucinations following right thalamic stroke. *Cognitive Neuropsychiatry, 12*(5), 422–436.

Moniz, E. (1949). *Tentatives Operatoires dans le Traitement de Certaines Psychoses.* Paris: Masson et Cie.

Monrad-Krohn, G. H. (1947). Dysprosody or altered 'Melody of Language'. *Brain, 70,* 405–415.

Montagne, B., Kessels, R. P. C., Frigerio, E., de Haan, E. H. F., & Perrett, D. I. (2005). Sex differences in the perception of affective facial expressions: Do men really lack emotional sensitivity? *Cognitive Processes, 6,* 136–141.

Montagu, A. (1971). *Touching: The human significance of the skin.* New York: Columbia University Press.

Montagu, A. (1986). *Touching: The human significance of the skin* (3rd ed.). New York: Harper & Row.

Moore, D. R. (1988). Auditory brainstem of the ferret: Sources of projections to the inferior colliculus. *The Journal of. Comparative Neurology, 269,* 342–354.

Moore, R. Y. (1996). Entrainment pathways and the functional organization of the circadian system. *Progress in Brain Research, 111,* 103–119.

Moore, T. M., Shenal, B. V., Rhodes, R. D., & Harrison, D. W. (1999). Forensic neuropsychological evaluations and quantitative electroencephalography (QEEG). *The Forensic Examiner,* March/April: 12–15.

Moreno-Torres, I., Berthier, M. L., Mar Cid, M. D., Green, C., Gutiérrez, A., García-Casares, N., & Carnero, C. (2012). Foreign accent syndrome: A multimodal evaluation in the search of neuroscience-driven treatments. *Neuropsychologia, 51*(3), 520–537.

Morgan, M. Y. (1982). Alcohol and nutrition. *British Medical Bulletins, 38,* 21–29.

Morin, C. M., & Gramling, S. E. (1989). Sleep patterns and aging: Comparison of older adults with and without insomnia complaints. *Psychology and Aging, 4*(3), 290–294.

Morris, E. (2010). The anosognosic's dilemma: Something's wrong but you'll never know what it is. *New York Times Opinionator.*

Morris, P. L., Robinson, R. G., Raphael, B., & Hopwood, M. J. (1996). Lesion location and post-stroke depression. *Journal of Neuropsychiatry Clinical Neuroscience, 8,* 399–403.

Morris, J. S., Friston, K. J., Büchel, C., Frith, C. D., Young, A. W., Calder, A. J., & Dolan, R. J. (1998). A neuromodulatory role for the human amygdale in processing emotional facial expressions. *Brain, 121,* 47–57.

Morrison, J. H., & Baxter, M. G. (2012). The ageing cortical synapse: Hallmarks and implications for cognitive decline. *Nature Reviews Neuroscience, 13,* 240–250. doi:10.1038/nrn3200.

Morrot, G., Brochet, F., & Dubourdieu, D. (2001). The color of odors. *Brain and Language, 79,* 309–320. doi:10.1006/brln.2001.2493.

Morton, E. S. (1977). On the occurrence and significance of motivation-structural rules in some bird and mammal sounds. *American Naturalist, 111*(981), 855–869.

Moruzzi, G., & Magoun, H. W. (1949). Brainstem reticular formation and activation of the EEG. *EEG and Clinical Neurophysiology, 1,* 455–473.

Moser, E. I., Kropff, E., & Moser, M. B. (2008). Place cells, grid cells, and the brain's spatial representation system. *Annual Review of Neuroscience, 31,* 69–89.

Moses, J. A. (2004). Test review: Comprehensive trail making test. *Archives of Clinical Neuropsychology, 19,* 703–708.

Moss, H. B., & Tarter, R. E. (1993). Substance abuse, aggression, and violence. *American Journal of Addiction, 2*(2), 149–160.

Most, E. I., Scheltens, P., & Van Someren, E. J. (2010). Prevention of depression and sleep disturbances in elderly with memory-problems by activation of the biological clock with light—A randomized clinical trial. *Trials, 11,* 19.

Mountcastle, V. B. (1957). Modality and topographic properties of single neurons of cat's somatic sensory cortex. *Journal of Neurophysiology, 20,* 408–434. PMID 13439410.

Mountcastle, V. B. (1978). An organizing principle for cerebral function: The unit model and the distributed system. In G. M. Edelman & V. B. Mountcastle (Eds.), *The mindful brain*. Cambridge: MIT Press.

Müller, J. (1831). Bestätigung des Bell'schen Lehrsatzes, dass die doppelten Wurzeln der Rückenmarksnerven verschiedene Functionen, durch neue und entscheidende Experimente. [Froriep's]Notizen aus dem Gebiete der Natur- und Heilkunde. Weimar, 30, 113–117, 129–134.

Munakata, Y., Snyder, H. R., & Chatham, C. H. (2012). Developing cognitive control: Three key transitions. *Current Directions in Psychological Science, 21*(2), 71–77. doi:10.1177/0963721412436807.

Münch, M., Linhart, F., Borisuit, A., Jaeggi, S. M., & Scartezzini, J. (2012). Effects of prior light exposure on early evening performance, subjective sleepiness, and hormonal secretion. *Behavioral Neuroscience, 126*(1), 196–203. doi:10.1037/a0026702.

Murray, R., Neumann, M., Forman, M. S., Farmer, J., Massimo, L., Rice, A., Miller, B. L., Johnson, J. K., Clark, C. M., Hurtig, H. I., Gorno-Tempini, M. L., Lee, V. M., Trojanowski, J. Q., & Grossman, M. (2007). Cognitive and motor assessment in autopsy-proven corticobasal degeneration. *Neurology, 68,* 1274–1283. doi:10.1212/01.wnl.0000259519.78480.c3.

Mutha, P. K., Haaland, K. Y., & Sainburg, R. L. (2013). Rethinking motor lateralization: Specialized but complementary mechanisms for motor control of each arm. *PloS One, 8*(3), e58582.

Naber, F. B., Poslawsky, I. E., van Ijzendoorn, M. H., van Engeland, H., & Bakermans-Kranenburg, M. J. (2012). Brief report: Oxytocin enhances paternal sensitivity to a child with autism: A double-blind within-subject experiment with intranasally administered oxytocin. *Journal of Autism and Developmental Disorders 43*(1), 224–229.

Nafe, J. P., & Wagoner, S. W. (1936). The experiences of warmth, cold, and heat. *Journal of Psychology, 22*(2), 421–431.

Nagel, I. E., Preuschhof, C., Li, S., Nyberg, L., Bäckman, L., Lindenberger, U., & Heekeren, H. R. (2009). Performance level modulates adult age differences in brain activation during spatial working memory. *Proceedings of the National Academy of Sciences of the United States of America, 106,* 22552–22557. doi:10.1073/pnas.0908238106.

Naifeh, S., & Smith, G. W. (2011). *Van Gogh: The life.* New York: Random House.

Najt, P., Bayer, U., & Hausmann, M. (2013). Models of hemispheric specialization in facial emotion perception-a reevaluation. *Emotion, 13,* 159–167.

Nan, Y., & Friederici, A. D. (2012). Differential roles of right temporal cortex and Broca's area in pitch processing: Evidence from music and Mandarin. *Human Brain Mapping*. ISSN 1097-0193, doi.org/10.1002/hbm.22046.

Narcisse, P., Bichot, M. T., & Heard, R. D. (2011). Stimulation of the nucleus accumbens as behavioral reward in awake behaving monkeys. *Journal of Neuroscience Methods, 199*(2), 265–272. ISSN 0165-0270, doi:10.1016/j.jneumeth.2011.05.025.

Nash, K., McGregor, I., & Inzlicht, M. (2010). Line bisection as a neural marker of approach motivation. *Psychophysiology, 47,* 979–983.

National Institute of Neurological Disorders and Stroke (NINDS). (2007). *Brain basics: Understanding sleep.* National Institutes of Health (NIH), Bethesda, MD. NIH Publication No.06-3440-c.

National Sleep Foundation. "Sleep in America" poll. (2008). http://www.sleepfoundation.org/sites/default/files/2008%20POLL%20SOF.PDF.

Nazarali, A. J., Gutkind, J. S., & Saavedra, J. M. (1987). Regulation of angiotensin II binding sites in the subfornical organ and other rat brain nuclei after water deprivation. *Cellular and Molecular Neurobiology, 7,* 447–455.

Neikrug, A. B., & Ancoli-Israel, S. (2010). Sleep disorders in older adults—A mini review. *Gerontology, 56*(2),181–189 (Epub 2009 Sep 9).

Nestler, E. J., Hyman, S. E., & Malenka, R. C. (2001). *Molecular neuropharmacology: A foundation for clinical neuroscience.* New York: McGraw-Hill.

Neumann, D., Zupan, B., Babbage, D., Radnovich, A., Tomita, M., Hammond, F., & Willer, B. (2012). Affect recognition, empathy and dysosmia following traumatic brain injury. *Archives of Physical Medicine and Rehabilitation, 93*(8), 1414–1420.

Newman, A. J., Bavelier, D., Corina, D., Jezzard, P., & Neville, H. J. (2002). A critical period for right hemisphere recruitment in American sign language processing. *Nature Neuroscience, 5*(1), 76–80.

New York Times. (5 May 2006). Kennedy to enter drug rehab after car crash; Congressman wrecked car near Capitol.

Nicholls, M. E. R., Clode, D., Wood, S. J., & Wood, A. G. (1999). Laterality of expression in portraiture: Putting your best cheek forward. *Proceedings of the Royal Society of London. Series B, 266,* 1517–1522.

Nicholls, M. E. R., Clode, D., Lindell, A. K., & Wood, A. G. (2002a). Which cheek to turn? The effect of gender and emotional expressivity on posing behavior. *Brain Cognition, 48,* 480–484.

Nicholls, M. E. R., Wolfgang, B. J., Clode, D., & Lindell, A. K. (2002b). The effect of left and right poses on the expression of facial emotion. *Neuropsychologia, 40,* 1662–1665.

Nicholls, M. E. R., Ellis, B. E., Clement, J. G., & Yoshino, M. (2004). Detecting hemifacial asymmetries in emotional expression with three-dimensional computerized image analysis. *Proceedings of the Royal Society of London, 271,* 663–668. doi:10.1098/rspb.

Niedermeyer, E. (1997). Alpha rhythms as physiological and abnormal phenomena. *International Journal of Psychophysiology, 26*(1–3), 31–49.

Niedermeyer, E. (2011). In D. L. Schomer & F. H. Lopez da Silva (Eds.), *Niedermeyer's electroencephalography: Basic principles, clinical applications, and related fields* (6th ed.). Philadelphia: Lippincott, Williams, & Wilkins a Wolters Kluwer Business.

Nielsen, B., Hyldig, T., Bidstrup, F., González-Alonso, J., & Christoffersen, G. R. J. (2001). Brain activity and fatigue during prolonged exercise in the heat. *Pflügers Archiv European Journal of Physiology, 442*(1), 41–48. doi:10.1007/s004240100515.

Nighoghossian, N., Trouillas, P., Vighetto, A., & Philippon, B. (1992). Spatial delirium following a right subcortical infarct with frontal deactivation. *Journal of Neurology, Neurosurgery, & Psychiatry, 55,* 334–335.

Nimer, J., & Lundahl, B. (2007). Animal-assisted therapy: A meta-analysis. *Anthrozoos, 20*(3), 225–238.

Nisbett, R. E., & Miyamoto, Y. (2005). The influence of culture: Holistic versus analytic perception. *Trends in Cognitive Science, 9,* 467–473.

Nisbett, R. E., Peng, K., Choi, I., & Norenzayan, A. (2001). Culture and systems of thought: Holistic vs. analytic cognition. *Psychological Review, 108,* 291–310.

Nishizawa, S., Benkelfat, C., Young, S. N., Leyton, M., Mzengeza, S., de Montigny, C., et al. (1997). Differences between males and females in rates of serotonin synthesis in human brain. *Proceedings of the National Academy of Sciences of the United States of America, 94,* 5308–5313.

Nitschke, J. B., Heller, W., Palmieri, P., & Miller, G. A. (1999). Contrasting patterns of brain activity in anxious apprehension and anxious arousal. *Psychophysiology, 36,* 628–637.

Nobrega, A. C., dos Reis, A. F., Moraes, R. S., Bastos, B. G., Ferlin, E. L., & Ribeiro, J. P. (2001). Enhancement of heart rate variability by cholinergic stimulation with pyridostigmine in healthy subjects. *Clinical Autonomic Research, 11,* 11–17.

Nolen-Hoeksema, S., Wisco, B. E., & Lyubomirsky, S. (2008). Rethinking rumination. *Perspectives on Psychological Science, 3,* 400–424.

Noseda, R., Kainz, V., Jakubowski, M., Gooley, J. J., Saper, C. B., Digre, K., & Burstein, R. (2010). A neural mechanism for exacerbation of headache by light. *Nature Neuroscience, 13,* 239–245.

Nottebohm, F. (1977). Asymmetries in neural control of vocalization in the canary. In S. Harnad, R. W. Doty, L. Goldstein, J. Jaynes, & G. Kranthamer (Eds.), *Lateralization in the nervous system* (pp. 23–44). New York: Academic.

NOVA. (1993). Stranger in the mirror. Production by BBC-TV in association with WGBH/Boston Educational Foundation.

Nunn, J. A., Graydon, F. J. X., Polkey, C. E., & Morris, R. G. (1999). Differential spatial memory impairment after right temporal lobectomy demonstrated using temporal titration. *Brain, 122,* 47–59.

Nybo, L. (2012). Brain temperature and exercise performance. *Experimental Physiology, 97,* 333–339. doi:10.1113/expphysiol.2011.062273.

Oades, R. D. (1998). Frontal, temporal and lateralized brain function in children with attention-deficit hyperactivity disorder: A psychophysiological and neuropsychological viewpoint on development. *Behavioral Brain Research, 94*(1), 83–95.

Obleser, J., Wise, R. J., Dresner, M., & Scott, S. K. (2007). Functional integration across brain regions improves speech perception under adverse listening conditions. *Journal of Neuroscience, 27,* 2283–2289.

Ocklenburg, S., & Güntürkün, O. (2012). Hemispheric asymmetries: The comparative view. *Frontiers in Psychology, 3*(5), 1–9.

O'Craven, K. M., Rosen, B. R., Kwong, K. K., Treisman, A., & Savoy, R. L. (1997). Voluntary attention modulates fMRI activity in human MT-MST. *Neuron, 18,* 591–598.

Oertel, V., Knöchel, C., Rotarska-Jagiela, A., Schönmeyer, R., Lindner, M., van de Ven, V., Haenschel, C., Uhlhaas, P., Maurer, K., & Linden, D. (2010). Reduced laterality as a trait marker of schizophrenia-Evidence from structural and functional neuroimaging. *The Journal of Neuroscience, 30,* 2289–2299.

Oettinger, B. (1913). A case of pseudobulbar paralysis. In T. L. Stedman (Ed.), *A Weekly Journal of Medicine and Surgery, 84*(17).

Ogden, J. A. (1985). Autotopagnosia. Occurrence in a patient without nominal aphasia and with an intact ability to point to parts of animals and objects. *Brain, 108,* 1009–1022.

Ogle, J. W. (1867). Aphasia and agraphia. *St. George's Hospital Reports, 2,* 83–122.

Oguri, T., Sawamoto, N., Tabu, H., Urayama, S. I., Matsuhashi, M., Matsukawa, N., & Fukuyama, H. (2013). Overlapping connections within the motor cortico-basal ganglia circuit: fMRI-tractography analysis. *NeuroImage, 78,* 358–362.

Öhman, A. E., & Mineka, S. (2001). Fears, phobias, and preparedness: Toward an evolved module of fear and fear learning. *Psychological Science, 108,* 438–522.

Oke, A., Keller, R., Mefford, I., & Adams, R. N. (1978). Lateralisation of norepinephrine in human thalamus. *Science, 200,* 1411–1413.

O'Keefe, J., & Nadel, L. (1978). *The hippocampus as a cognitive map.* Oxford: Oxford University Press.

Olds, J. (1975). Reward and drive neurons. *Brain Stimulation Reward, 1,* 1–30.

Olds, J., & Milner, P. (1954). Positive reinforcement produced by electrical stimulation of septal area and other regions of rat brain. *Journal of Comparative and Physiological Psychology, 47*(6), 419–427.

Oliver, D. L., & Huerta, M. F. (1992). Inferior and superior colliculi. In D. B. Webster, A. N. Popper, & R. R. Fay (Eds.), *The mammalian auditory pathway: Neuroanatomy* (pp. 168–221). New York: Springer-Verlag.

Öngür, D., & Price, J. L. (2000). The organization of networks within the orbital and medial prefrontal cortex of rats, monkeys and humans. *Cerebral Cortex, 10*(3), 206–219.

Öngür, D., Ferry, A. T., & Price, J. L. (2003). Architectonic subdivision of the human orbital and medial prefrontal cortex. *The Journal of Comparative Neurology, 460*(3), 425–449.

Oppenheim, H. (1885). Uber eine durch eine klinisch bisher nicht verwetete Untersuchungsmethode ermittelte Sensibilitatsstorung bei einseitigen Erkrakungen des Großhirns, *Neurologiches Centralblatt, 37,* 51–56.

Oppenheimer, S. M., & Cechetto, D. F. (1990). Cardiac chronotropic organization of the rat insular cortex. *Brain Research, 533,* 66–72.

Oppenheimer, S. M., Gelb, A., Girvin, J. P., & Hachinski, V. C. (1992). Cardiovascular effects of human insular cortex stimulation. *Neurology, 42,* 1727–1732.

Oppenheimer, S. M., Kedem, G., & Martin, W. M. (1996). Left-insular cortex lesions perturb cardiac autonomic tone in humans. *Clinical Autonomic Research, 6,* 131–140.

Orfei, M. D., Robinson, R. G., Prigatano, G. P., Starkstein, S., Rüsch, N., Bria, P, & Spaletta, G. (2007). Anosognosia for hemiplegia after stroke is a multifaceted phenomenon: A systematic review of the literature. *Brain, 130,* 3075–3090. doi:10.1093/brain/awm106.

Orton, L. D., Poon, P. W., & Rees, A. (2012). Deactivation of the inferior colliculus by cooling demonstrates intercollicular modulation of neuronal activity. *Frontiers in Neural Circuits, 6,* 100.

Ota, H., Fujii, T., Suzuki, K., Fukatsu, R., & Yamadori, A. (2001). Dissociation of body-centered and stimulus-centered representations in unilateral neglect. *Neurology, 57,* 2064–2069.

Paccehetti, C., Francesca, M., Aglieri, R., Fundaro`, C, Martignoni, E., & Nappi, G. (2000). Active music therapy in Parkinson's disease: An integrative method for motor and emotional rehabilitation. *Psychosomatic Medicine, 62,* 386–393.

Packard, M. G., & Poldrack, R. A.(2003). Competition among multiple memory systems: Converging evidence from animal and human brain studies. *Neuropsychologia, 41,* 245–251.

Pagel, J. F. (2008). *The limits of dream: A scientific exploration of the mind/brain interface.* Burlington: Academic Press.

Pai, M. C. (1997). Topographical disorientation: Two cases. *Journal of the Formosan Medical Association, 96,* 660–663.

Painter, K., & Farrington, D. P. (1997). *Crime reducing effect of improved street lighting: The Dudley Project.* National Criminal Justice Reference Service, United States Department of Justice.

Pallis, C. A. (1955). Impaired identification of faces and places with agnosia for colours. *Journal of Neurology, Neurosurgery and Psychiatry, 18,* 218–224.

Palmer, E., & Ashby, P. (1992). Corticospinal projections to upper limb motoneurones in humans. *The Journal of Physiology, 448,* 397–412.

Pandi-Perumal, S. R., Ruoti, R. R., & Kramer, M. (2007). *Sleep and psychosomatic medicine.* London: Informa Healthcare.

Papousek, I., & Schulter, G. (2002). Covariations of EEG asymmetries and emotional states indicate that activity at frontopolar locations is particularly affected by state factors. *Psychophysiology, 39,* 350–360.

Park, D. C., & McDonough, I. M. (2013). The dynamic aging mind: Revelations from functional neuroimaging research. *Perspectives on Psychological Science, 8,* 62–67.

Park, D. C., Polk, T. A., Park, R., Minear, M., Savage, A., & Smith, M. R. (2004). Aging reduces neural specialization in ventral visual cortex. *Proceedings of the National Academy of Sciences of the United States of America, 101,* 13091–13095. doi:10.1073/pnas.0405148101.

Park, J., Carp, J., Hebrank, A., Park, D. C., & Polk, T. A. (2010). Neural specificity predicts fluid processing ability in older adults. *Journal of Neuroscience, 30,* 9253–9259. doi:10.1523/JNEUROSCI.0853-10.2010.

Park, J., Hebrank, A., Polk, T. A., & Park, D. C. (2012). Neural dissociation of number from letter recognition and its relationship to parietal numerical processing. *Journal of Cognitive Neuroscience, 24*(1), 39–50 (Posted Online November 21, 2011). doi:10.1162/jocn_a_00085.

Parker, J. L. (2010). *Natural stressors, posttraumatic stress disorder, and wound healing, in a murine model.* Dissertation submitted to the faculty of the Virginia Polytechnic Institute and State University in partial fulfillment of the requirements for the degree of Doctor of Philosophy in Developmental and Biological Psychology.

Parker, G., & Brotchie, H. (2011). 'D' for depression: Any role for vitamin D? *Acta Psychiatrica Scandinavica, 124,* 243–249. doi:10.1111/j.1600-0447.2011.01705.x.

Parslow, D. M., Morris, R. G., Fleminger, S., Rahman, Q., Abrahams, S., & Recce, M. (2005). Allocentric spatial memory in humans with hippocampal lesions. *Acta Psychologica, 118,* 123–147.

Partington, J. E., & Leiter, R. G. (1949). Partington's pathway test. *Psychological Service Center Bulletin, 1,* 9–20.

Partonen, T. (1998). Vitamin D and serotonin in winter. *Medical Hypotheses, 51*(3), 267–268.

Parvizi, J., Anderson, S. W., Martin, C. O., Damasio, H., & Damasio, A. R. (2001). Pathological laughter and crying: A link to the cerebellum. *Brain, 124,* 1708–1719.

Parvizi, J., Coburn, K. L., Shillcutt, S. D., Coffey, C. E., Lauterbach, E. C., & Mendez, M. F. (2009). Neuroanatomy of pathological laughing and crying: A report of the American Neuropsychiatric Association Committee on Research. *The Journal of Neuropsychiatry and Clinical Neurosciences, 21,* 75–87.

Pasqualotto, A., & Proulx, M. J. (2012). The role of visual experience for the neural basis of spatial cognition. *Neuroscience and Biobehavioral Reviews, 36*(4), 1179–1187. doi:10.1016/j. neubiorev.2012.01.008.

Passier, M. A., Isaac, W., & Hynd, G. W. (1985). Neuropsychological development of behavior attributed to frontal lobe functioning in children. *Developmental Neuropsychology, 1,* 349–370.

Patterson, A., & Zangwill, O. L. (1944). Disorders of visual space perception associated with lesions of the right cerebral hemisphere. *Brain, 67,* 331–358.

Paul, L. K. (2011). Developmental malformation of the corpus callosum: A review of typical callosal development and examples of developmental disorders with callosal involvement. *Journal of Neurodevelopmental Disorders, 3*(1), 3–27. doi:10.1007/s11689-010-9059-y.

Pauli, P., Wiedemann, G., & Nickola, M. (1999). Pain sensitivity, cerebral laterality, and negative affect. *Pain, 80*(1), 359–364.

Paus, T., Giedd, J., et al. (1999). Structural maturation of neural pathways in children and adolescents: In vivo study. *Science, 283,* 1908–1911.

Paus, S., Schmitz-Hubsch, T., Wullner, U., Vogel, A., Klockgether, T., & Abele, M. (2007). Bright light therapy in Parkinson's disease: A pilot study. *Movement Disorders, 22,* 1495–1498.

Pavlidis, P., Gouveris, H., Antonia, A., Dimitrios, K., Georgios, A., & Georgios, K. (2013). Age-related changes in electrogustometry thresholds, tongue tip vascularization, density, and form of the fungiform papillae in humans. *Chemical Senses, 38*(1), 35–43. doi:10.1093/chemse/bjs076.

Paxinos, G., Huang, X. F., Sengul, G., & Watson, C. (2012). Organization of brainstem nuclei. In *The human nervous system* (pp. 260–327). Amsterdam: Elsevier Academic Press.

Pearce, J. M. S. (2008). The development of spinal cord anatomy. *European Neurology, 59*(6), 286–291.

Pearce, J. M. S. (2009). Marie-Jean-Pierre Flourens (1794–1867) and cortical localization. *European Neurology, 61*(5), 311–314 (Switzerland).

Pearlin, L. I., Mullan, J. T., Semple, S. J., & Skaff, M. M. (1990). Caregiving and the stress process: An overview of concepts and their measures. *The Gerontologist, 30*(5), 583–594.

Peirson, S. N., Halford, S., & Foster, R. G. (2009). The evolution of irradiance detection: Melanopsin and the non-visual opsins. *Philosophical Transactions of the Royal Society of London. Series B, Biological Sciences, 364*(1531), 2849–2865. doi:10.1098/rstb. 2009.0050.

Penfield, W. (1967). *The excitable cortex in conscious man* (pp. 1–5). Liverpool: Liverpool University Press.

Penfield, W., & Boldrey, E. (1937). Somatic motor and sensory representation in the cerebral cortex of man as studied by electrical stimulation. *Brain, 60,* 389–443.

Penfield, W., & Faulk, M. E., Jr. (1955). The insula: Further observations on its function. *Brain, 78*(4), 445–470. doi:10.1093/brain/78.4.445.

Penfield, W., & Rasmussen, T. (1950). *The cerebral cortex of man: A clinical study of localization of function.* New York: Macmillan.

Peppercorn, M. A., & Herzog, A. G. (1989). The spectrum of abdominal epilepsy in adults. *American Journal of Gastroenterology, 84*(10), 1294–1296.

Perani, D., Saccuman, M. C., Scifo, P., Anwander, A., Spada, D., Baldoli, C., & Friederici, A. D. (2011). Neural language networks at birth. *Proceedings of the National Academy of Sciences of the United States of America, 108,* 16056–16061.

Pereira, A. C., Huddleston, D. E., Brickman, A. M., Sosunov, A. A., Hen, R., McKhann, G. M., & Small, S. A. (2007). An in vivo correlate of exercise-induced neurogenesis in the adult dentate gyrus. *Proceedings of the National Academy of Sciences of the United States of America, 104*(13), 5638–5643.

Peretz, I., Kolinsky, R., Tramo, M., Hublet, C., Demeurisse, G., et al. (1994). Functional dissociations following bilateral lesions of auditory cortex. *Brain, 117,* 1283–1301.

Peretz, I., Champod, A. S., & Hyde, K. (2003). Varieties of musical disorders: The Montreal battery of evaluation of amusia. *Annals of the New York Academy of Science, 999,* 58–75.

Perez, D. L., Catenaccio, E., & Epstein, J. (2011). Confusion, hyperactive delirium, and secondary mania in right hemispheric strokes: A focused review of neuroanatomical correlates. *Journal of Neurology and Neurophysiology,* S1. doi:10.4172/2155-9562. S1-003.

Peritz, G. (1918). Zur Pathopsychologie des Rechnens. *Deutsche Zeitschrift für Nervenheilkunde, 61,* 234–340.

Perkins, F. M., & Kehlet, H. (2000). Chronic pain as an outcome of surgery: A review of predictive factors. *Anesthesiology, 93*(4), 1123–1133.

Perkins, N. M., & Tracey, D. J. (2000). Hyperalgesia due to nerve injury: Role of neutrophils. *Neuroscience, 101*(3), 745–757.

Perry, V. H., & Cowey, A. (1984). Retinal ganglion cells that project to the superior colliculus and pretectum in the macaque monkey. *Neuroscience, 12*(4), 1125–1137.

Peter-Derex, L., Comte, J. C., Mauguière, F., & Salin, P. A. (2012). Density and frequency caudo-rostral gradients of sleep spindles recorded in the human cortex. *Sleep, 35*(1), 69.

Petersen, S. E., Fox, P. T., Posner, M. I., Minton, M., & Raichle, M. E. (1988). Positron emission tomographic studies of the cortical anatomy of single-word processing. *Nature, 331*(6157), 585–589 (18 February 1988 @ Macmillan Magazines Ltd.).

Peterson, B. S., Warner, V., Bansal, R., Zhu, H., Hao, X., Liu, J., Durkin, K., Adams, P. B., Wickramaratne, P., & Weissman, M. M. (2009). Cortical thinning in persons at increased familial risk for major depression. *Proceedings of the National Academy of Sciences of the United States of America, 106,* 6273–6278.

Petrides, M. (1985). Deficits on conditional associative-learning tasks after frontal- and temporal-lobe lesions in man. *Neuropsychologia, 23,* 601–614.

Pfaffmann, C. (1941). *J. Cell. Comp. Physiol., 17(2),* 243–258.

Phan, K. L., Orlichenko, A., Boyd, E., Angstadt, M., Coccaro, E. F., & Liberzon, I. (2009). Preliminary evidence of white matter abnormality in the uncinate fasciculus in generalized social anxiety disorder. *Biological Psychiatry, 66*(7), 691–694.

Phelps, E. A., Hyder, F., Blamire, A. M., & Shulman, R. G. (1997). FMRI of the prefrontal cortex during overt verbal fluency. *Neuroreport, 8*(2), 561–565.

Phillips, K. (1996). *The broken mirror: Understanding and treating body dysmorphic disorder.* New York: Oxford University Press.

Phipps-Nelson, J., Redman, J. R., Dijk, D. J., & Rajaratnam, S. M. (2006). Daytime exposure to bright light, as compared to dim light, decreases sleepiness and improves psychomotor vigilance performance. *Sleep, 26,* 695–700.

Pia, L., Neppi-Modona, M., Raffaella, R., & Berti, A. (2004). The anatomy of anosognosia for hemiplegia: A meta-analysis. *Cortex, 40,* 367–377. doi:10.1016/S0010-9452(08)70131-X.

Piazza, M., Izard, V., Pinel, P., Le Bihan, D., & Dehaene, S. (2004). Tuning curves for approximate numerosity in the human intraparietal sulcus. *Neuron, 44,* 547–555. doi:10.1016/j.neuron.2004.10.014.

Piazza, M., Pinel, P., Le Bihan, D., & Dehaene, S. (2007). A magnitude code common to numerosities and number symbols in human intraparietal cortex. *Neuron, 53,* 293–305. doi:10.1016/j.neuron.2006.11.022.

Pick, A. (1905). *Studien uber motorische apraxia und ihre mahestenhende erscheinungen.* Leipzig: Deuticke.

Pick, A. (1919). Uber Anderungen des Sprachcharakters als Begleiterscheinung aphasicher Storungen. *Zeitschrift fur gesamte Neurologie und Psychiatrie, 45,* 230–241.

Piechowski-Jozwiak, B., & Bogousslavsky, J. (2012). Neurobehavioral syndromes. In M. Paciaroni, G. Agnelli, V. Caso, & J. Bogousslavsky (Eds.), *Manifestations of stroke.* Basel: Karger. (*Frontiers of Neurology and Neuroscience, 30,* 57–60). doi:10.1159/000333410.

Pierce, K., Muller, R. A., Ambrose, J., Allen, G., & Courchesne, E. (2001). Face processing occurs outside the fusiform 'face area' in autism: Evidence from functional MRI. *Brain, 124,* 2059–2073.

Pierrot-Deseilligny, C., Rivaud, S., Gaymard, B., & Agid, Y. (1991). Cortical control of reflexive visually-guided saccades. *Brain, 114,* 1473–1485.

Pigott, S., & Milner, B. (1993). Memory for different aspects of complex visual scenes after unilateral temporal- or frontal-lobe resection. *Neuropsychologia, 31,* 1–15.

Pinel, P., Dehaene, S., Riviere, D., & LeBihan, D. (2001). Modulation of parietal activation by semantic distance in a number comparison task. *Neuroimage, 14,* 1013–1026. doi:10.1006/nimg.2001.0913.

Pissiota, A., Frans, O., Michelgård, A., Appel, L., Långström, B., Flaten, M. A., & Fredrikson, M. (2003). Amygdala and anterior cingulate cortex activation during affective startle modulation: A PET study of fear. *The European Journal of Neuroscience, 18*(5), 1325–1331.

Plutchik, R. (1993). Emotions and psychopathology. In M. Lewis & J. M. Haviland (Eds.), *Handbook of emotions* (pp. 53–66). New York: Guilford.

Poeck, K. (1985). Pathophysiology of emotional disorders associated with brain damage. In P. J. Vinken & G. W. Bruyn (Eds.), *Handbook of clinical neurology* (Vol. 3, pp. 342–367). Amsterdam: Elsevier.

Polk, T. A., Stallcup, M., Aguirre, G. K., Alsop, D. C., D'Esposito, M., Detre, J. A., & Farah, M. J. (2002). Neural specialization for letter recognition. *Journal of Cognitive Neuroscience, 14,* 145–159.

Pollack, J. G., & Hickey, T. L. (1979). The distribution of retino-collicular axon terminals in rhesus monkey. *Journal of Comparative Neurology, 185,* 587–602.

Pollak, C., Thorpy, M. J., & Yager, J. (2010). *The encyclopedia of sleep and sleep disorders* (3rd ed.). New York: Facts on File Publishing.

Pollmann, S. (2010). A unified structural-attentional framework for dichotic listening. In K. Hugdahl & R. Westerhausen (Eds.), *The two halves of the brain: Information processing in the cerebral hemispheres* (pp. 444–468). Cambridge: MIT Press.

Poppelreuter, W. (1917). *Die Psychischen Schaedungen durch Kopfschuss in Kriege 1914–1916.* Leipzig: Voss.

Poppen, J. L. (1939). Ligation of the left anterior cerebral artery: Its hazards and the means of avoidance of its complications. *Archives of Neurology and Psychiatry (Chic.), 41,* 495–503.

Porter, R., & Lemon, R. (1993). *Corticospinal function and voluntary movement.* New York: Oxford University Press.

Powell, W. R., & Schirillo, J. A. (2009). Asymmetrical facial expressions in portraits and hemispheric laterality: A literature review. *Laterality: Asymmetries of Body, Brain and Cognition, 14,* 545–572.

Powell, T. P., Cowan, W. M., & Raisman, G. (1965). The central olfactory connexions. *Journal of Anatomy, 99,* 791–813.

Powers, R. E., Ashford, J. W., & Peschin, S. (2008). *Memory matters.* Alzheimer's Foundation of America, AFA.

Pribram, K. H. (1991). *Brain and perception: Holonomy and structure in figural processing.* New Jersey: Lawrence Erlbaum Associates.

Price, J. L. (1973). An autoradiographic study of complementary laminar patterns of termination of afferent fibers to the olfactory cortex. *Journal of Comparative Neurology, 150,* 87–108.

Prigatano, G. P. (2010). *The study of anosognosia.* New York: Oxford University Press.

Prigatano, G. P., & Schacter, D. L. (1991). *Awareness of deficit after brain injury: Clinical and theoretical issues* (pp. 53–55). Oxford: Oxford University Press. ISBN 0-19-505941-7.

Prinz, P. N., & Raskind, M. (1978). Aging and sleep disorders. In R. Williams & I. Karacan (Eds.), *Sleep disorders: Diagnosis and treatment* (pp. 303–321). New York: Wiley.

Prkachin, K. M., & Silverman, B. E. (2002). Hostility and facial expression in young men and women: Is social regulation more important than negative affect? *Health Psychology, 21*(1), 33–39.

Proulx, M. J. (2010). Synthetic synaesthesia and sensory substitution. *Consciousness and Cognition, 19*(1), 501–503. doi:10.1016/j.concog.2009.12.005.

Proulx, M. J. (February 2013). Blindness: Remapping the brain and the restoration of vision. *Psychological Science Agenda.*

Psychological Assessment Resources. (2003). Computerised Wisconsin Card Sort Task Version 4 (WCST). Psychological Assessment Resources.

Puce, A., Allison, T., Gore, J. C., & McCarthy, G. (1995). Face-sensitive regions in human extrastriate cortex studied by functional MRI. *Journal of Neurophysiology, 74,* 1192–1199.

Pujol, J., Lopez, A., Deus, J., Cardoner, N., Vallejo, J., Capdevila, A., et al. (2002). Anatomical variability of the anterior cingulate gyrus and basic dimensions of human personality. *Neuro-Image, 15,* 847–855.

Quaranta, A., Siniscalchi, M., & Vallortigara, G. (2007). Asymmetric tail-wagging responses by dogs to different emotive stimuli. *Current Biology, 17,* 199–201.

Quiroga, R. Q., Reddy, L., Kreiman, G., Koch, C., & Fried, I. (2005). Invariant visual representation by single neurons in the human brain. *Nature, 435*(7045), 1102–1107.

Quiroga, R. Q., Fried, I., & Koch, C. (2013). Brain cells for grandmother. *Scientific American, 308*(2), 30–35.

Radiopaedia.org. (2011). Radiopaedia.org, the wiki-based collaborative radiology resource. Creative commons. Radiopaedia, 5 July 2008. Web. 07 Oct. 2011. http://radiopaedia.org/images/4975.

Radley, J. J., Williams, B., & Sawchenko, P. E. (2008). Noradrenergic innervation of the dorsal medial prefrontal cortex modulates hypothalamo-pituitary-adrenal responses to acute emotional stress. *The Journal of Neuroscience, 28*(22), 5806–5816.

Rafal, R., Smith, J., Krantz, J., Cohen, A., & Brennan, C. (1990). Extrageniculate vision in hemianopic humans: Saccade inhibition by signals in the blind field. *Science, 250,* 118–121.

Raichle, M. E. (1998). Behind the scenes of functional brain imaging: A historical and physiological perspective. *Proceedings of the National Academy of Sciences of the United States of America, 95*(3), 765–772.

Raikkonen, K., Matthews, K. A., & Salomon, K. (2003). Hostility predicts metabolic syndrome risk factors in children and adolescents. *Health Psychology, 22*(3), 279–286.

Rains, G. D., & Milner, B. (1994). Right-hippocampal contralateral-hand effect in the recall of spatial location in the tactual modality. *Neuropsychologia, 32,* 1233–1242.

Rajput, A. H., Pahwa, R., Pahwa, P., & Rajput, A. (1993). Prognostic significance of the onset mode in parkinsonism. *Neurology, 43,* 829–830.

Raleigh, M. J., McGuire, M. T., Brammer, G. L., Pollack, D. B., & Yuwiler, A. (1991). Serotonergic mechanisms promote dominance acquisition in adult male vervet monkeys. *Brain Research, 559*(2), 181–190.

Ralph, M. R., Foster, R. G., Davis, F. C., & Menaker, M. (1990). Transplanted suprachiasmatic nucleus determines circadian period. *Science, 247*(4945), 975–978.

Ramirez-Rodriguez, G., Ortiz-Lopez, L., Dominguez-Alonso, A., Benitez-King, G. A., & Kempermann, G. (2011). Chronic treatment with melatonin stimulates dendrite maturation and complexity in adult hippocampal neurogenesis of mice. *Journal of Pineal Research, 50,* 29–37.

Randolph, C. (1998). *Repeatable battery for the assessment of neuropsychological status (RBANS).* San Antonio: Psychological Corporation.

Rapp, P. R., & Amaral, D. G. (1989). Evidence for task-dependent memory dysfunction in the aged monkey. *Journal of Neuroscience, 9,* 3568–3576.

Rastad, C., Ulfberg, J., & Lindberg, P. (2011). Improvement in fatigue, sleepiness, and health-related quality of life with bright light treatment in persons with seasonal affective disorder and subsyndromal SAD. *Depression Research Treatment,* 543906.

Rastatter, M. P., & McGuire, R. A. (1990). Some effects of advanced aging on the visual-language processing capacity of the left and right hemispheres: Evidence from unilateral tachistoscopic viewing. *Journal of Speech & Hearing Research, 33,* 134–140.

Raver, C. (2004). Placing emotional self-regulation in sociocultural and socioeconomic contexts. *Child Development, 75,* 346–353.

Raz, N., Gunning, F. M., Head, D., Dupuis, J. H., McQuain, J., Briggs, S. D., Loken, W. J., Thornton, A. E., & Acker, J. D. (1997). Selective aging of the human cerebral cortex observed in vivo: Differential vulnerability of the prefrontal gray matter. *Cerebral Cortex, 7*(3), 268–282. doi:10.1093/cercor/7.3.268.

Raz, N., Lindenberger, U., Rodrigue, K. M., Kennedy, K. M., Head, D., Williamson, A., & Acker, J. D. (2005). Regional brain changes in aging healthy adults: General trends, individual differences and modifiers. *Cerebral Cortex, 15,* 1676–1689.

Reiss, A. J., Jr., & Roth, J. A. (Eds.). (1994). *Understanding and preventing violence* (Vol. 3). Washington, DC: National Academy Press.

Reitan, R. (1958). Validity of the trail making test as an indicator of organic brain damage. *Perceptual and Motor Skills, 8,* 271–276. doi:10.2466/pms.1958.8.3.271.

Renoux, G., Biziere, K., Renoux, M., Guillaumin, J., & Degenne, D. (1983). A balanced brain asymmetry modulates T cell-mediated events. *Journal of Neuroimmunology, 5,* 227–238.

Reuter-Lorenz, P. A. (2013). Aging and cognitive neuroimaging: A fertile union. *Perspectives on Psychological Science, 8,* 68–71.

Reuter-Lorenz, P. A., & Cappell, K. A. (2008). Neurocognitive aging and the compensation hypothesis. *Current Directions in Psychological Science, 17,* 177–182. doi:10.1111/j.1467-8721.2008.00570.x.

Reuter-Lorenz, P., Stanczak, L., & Miller, A. (1999). Neural recruitment and cognitive aging: Two hemispheres are better than one, especially as you age. *Psychological Science of Journal, 10,* 494–500.

Rey, A. (1964). *L'examen clinique en psychologie.* Paris: Presses Universitaires de France.

Reynolds, C. R. (2002). *Comprehensive trail making test.* Austin: Pro-Ed. doi:10.1177/0734282905282415.

Rhodes, R. D., Hu, S. R., & Harrison, D. W. (2012). Diminished right frontal capacity in high-hostiles: Facial dystonia and cardiovascular reactivity. Manuscript submitted for publication.

Richter, C. P., & Hines, M. (1938). Increased spontaneous activity produced in monkeys by brain lesions. *Brain, 61,* 1–16.

Ridderinkhof, K. R., Ullsperger, M., Crone, E. A., & Nieuwenhuis, S. (2004). The role of the medial frontal cortex in cognitive control. *Science, 306,* 443. doi:10.1126/science.1100301.

Riddoch, G. (1935). Visual disorientation in homonymous half-fields. *Brain, 58,* 376–382.

Riddoch, M. J., Edwards, M. G., Humphreys, G. W., West, R., & Heafield, T. (1998). Visual affordances direct action: Neuropsychological evidence from manual interference. *Cognitive Neuropsychology, 15,* 645–683. doi:10.1080/026432998381041.

Riege, W. H., Metter, E. J., & Williams, M. V. (1980). Age and hemispheric asymmetry in nonverbal tactile memory. *Neuropsychologia, 18,* 707–710.

Riemann, D., Spiegelhalder, K., Nissen, C., Hirscher, V., Baglioni, C., & Feige, B. (2012). REM sleep instability-A new pathway for insomnia? *Pharmacopsychiatry, 45*(5), 167.

Riggs, L. A., & Ratliff, F. (1952). The effects of counteracting the normal movements of the eye. *Journal of the Optical Society of America, 42,* 872–873.

Rilling, J. K., & Insel, T. R. (1999). Differential expansion of neural projection systems in primate brain evolution. *Neuroreport, 10,* 1453–1459.

Ringo, J. L., Doty, R. W., Demeter, S., & Simard, P. Y. (1994). Time is of the essence: A conjecture that hemispheric specialization arises from interhemispheric conduction delay. *Cerebral Cortex, 4,* 331–343.

Rixt, F., Riemersma-van der, L., Swaab, D. F., Twisk, J., Hol, E. M., Hoogendijk, W. J. G., & Van Someren, E. J. (2008). Effect of bright light and melatonin on cognitive and noncognitive function in elderly residents of group care facilities: A randomized controlled trial. *JAMA, 299*(22), 2642–2655. doi:10.1001/jama.299.22.2642.

Rizzolatti, G., Fadiga, L., Gallese, V., & Fogassi, L. (1996). Premotor cortex and the recognition of motor actions. *Cognitive Brain Research, 3*(2), 131–141.

Robins, N. (1967). "Electric Taste" after section of the chorda tympani. *Nature, 214,* 1113–1114.

Robinson, R. G., & Downhill, J. E. (1995). Lateralization of psychopathology in response to focal brain injury. In R. J. Davidson & K. Hugdahl (Eds.), *Brain asymmetry.* Cambridge: MIT Press.

Robinson, R. G., Kubos, K. L., Starr, L. B., Rao, K., & Price, T. R. (1984). Mood disorders in stroke patients. Importance of location of the lesion. *Brain, 107,* 81–93.

Rodgers, A. B. (September 2008). Alzheimer's disease: Unraveling the mystery. National Institute on Aging. National Institutes of Health. U.S. Department of Health and Human Services. NIH Publication Number: 08-3782.

Rogers, L. J. (1982). Light experience and asymmetry of brain function in chickens. *Nature, 297,* 223–225. doi:10.1038/297223a0.

Rogers, L. J., & Anson, J. M. (1979). Lateralisation of function in the chicken fore-brain. *Pharmacology Biochemistry Behavior, 10,* 679–686.

Rogers, L. J., Zucca, P., & Vallortigara, G. (2004). Advantages of having a lateralized brain. *Proceedings of the Royal Society B: Biological Sciences, 271*(Suppl.6), 420–422.

Röhl, M., & Uppenkamp, S. (2012). Neural coding of sound intensity and loudness in the human auditory system. *JARO-Journal of the Association for Research in Otolaryngology, 13*, 369–379.

Roitman, J. D., Brannon, E. M., & Platt, M. L. (2012). Representation of numerosity in posterior parietal cortex. *Frontiers in Integrative Neuroscience, 6*, 25. (Published online 2012 May 31. PMCID: PMC3364489). doi:10.3389/fnint.2012.00025.

Roizen, J. (1997). Epidemiological issues in alcohol-related violence. In M. Galanter (Ed.), *Recent developments in alcoholism* (Vol. 13, pp. 7–40). New York: Plenum Press.

Roland, P. E. (1976). Astereognosis: Tactile discrimination after localized hemispheric lesions in man. *Archives of Neurology, 33*(8), 543–550.

Rombaux, P., Huart, C., De Volder, A. G., Cuevas, I., Renier, L., Duprez, T., & Grandin, C. (2010). Increased olfactory bulb volume and olfactory function in early blind subjects. *Neuroreport, 21*(17), 1069–1073. doi:10.1097/WNR.0b013e32833fcb8a.

Rosenthal, N. E., Sack, D. A., Gillin, C., Lewy A. J., Goodwin, F. K., & Davenport, Y., et al. (1984). Seasonal affective disorder: A description of the syndrome and preliminary findings with light therapy. *Archives of General Psychiatry, 41*, 72–80.

Roskes, M., Sligte, D., Shalvi, S., & De Dreu, C. K. W. (2011). The right side? Under time pressure, approach motivation leads to right-oriented bias. *Psychological Science*, Online First, November, 1–5.

Ross, E. D. (1981). The aprosodia: Functional-anatomic organization of the affective components of language in the right hemisphere. *Archives of Neurology, 38*, 561–569.

Ross, E. D., & Mesulum, M. M. (1979). Dominant language functions of the right hemisphere? Prosody and emotional gesturing. *Archives of Neurology, 36*, 144–148.

Ross, E. D., & Monnot, M. (2008). Neurology of affective prosody and its functional-anatomic organization in right hemisphere. *Brain & Language, 104*(1), 51–74.

Ross, E. D., & Rush, A. J. (1981). Neuroanatomical correlates of depression in brain damaged patients. *Archives of General Psychiatry, 38*, 1344–1354.

Rossi, G. F., & Rosadini, G. R. (1967). Experimental analyses of cerebral dominance in man. In D. H. Millikan & F. L. Darley (Eds.), *Brain mechanisms underlying speech and language*. New York: Grune & Stratton.

Rossion, B., Dricot, L., Devolder, A., Bodart, J.-M., Crommelinck, M., de Gelder, B., & Zoontjes, A. (2000). Hemispheric asymmetries for whole-based and part-based face processing in the human fusiform gyrus. *Journal of Cognitive Neuroscience, 12*, 793–802.

Rossit, S., Fraser, J. A., Teasell, R., Malhotra, P. A., & Goodale, M. A. (2011). Impaired delayed but preserved immediate grasping in a neglect patient with parieto-occipital lesions. *Neuropsychologia, 49*, 2498–2504.

Rostowski, J., & Rostowska, T. (2012). Neuropsychological context of marital functioning. In Dr. T. Heinbockel (Ed.), Neuroscience. ISBN: 978-953-51-0617-3. In Tech Available from: http://www.intechopen.com/books/neuroscience/neuropsychological-aspects-of-marital-functioning-reviewarticles.

Roth, H. L., Bauer, R. M., Crucian, G. P., & Heilman, K. M. (2013). Frontal-executive constructional apraxia: When delayed recall is better than copying. *Neurocase*, (ahead-of-print), 1–13.

Rothi, L. G., Ochipa, C., & Heilman, K. M. (1997). A cognitive neuropsychological model of limb praxis and apraxia. In L. G. Rothi & K. M. Heilman (Eds.), *Apraxia: The neuropsychology of action*. Hove: Psychology Press.

Rothwell, J. C., Thompson, P. D., Day, B. L., Boyd, S., & Marsden, C. D. (1991). Stimulation of the human motor cortex through the scalp. *Experimental Physiology, 76*, 159–200.

Rubia, K., Lee, F., Cleare, A. J., Tunstall, N., Fu, C. H. Y., Brammer, M., & McGuire, P. (2005). Tryptophan depletion reduces right inferior prefrontal activation during response inhibition in fast, event-related fMRI. *Psychopharmacology, 179*, 791–803.

Ruch, T. C., & Shenkin, H. A. (1943). The relation of area 13 on orbital surface of frontal lobes to hyperactivity and hyperphagia in monkeys. *Journal of Neurophysiology, 6*, 349–360.

Ruff, R. M. (1996). *Ruff figural fluency test.* Odessa: Psychological Assessment Resources, Inc.

Ruff, R. M., Evans, R., & Marshall, L. F. (1986). Impaired verbal and figural fluency after head injury. *Archives of Clinical Neuropsychology, 1,* 87–101.

Ruff, R. M., Allen, C. C., Farrow, C. E., Niemann, H., & Wylie, T. (1994). Figural fluency: Differential impairment in patients with left versus right frontal lobe lesions. *Archives of Clinical Neuropsychology, 9,* 41–55.

Rutherford, W. (1886). The sense and hearing. *Journal of Anatomy and Physiology, London, 21,* 166–168.

Sackeim, H. A., & Gur, R. C. (1978). Lateral asymmetry in intensity of emotional expression. *Neuropsychologia, 16,* 473–481.

Sacks, O. (1985). *The man who mistook his wife for a hat, and other clinical tales.* Summit Books. ISBN 0671554719.

Sadato, N. (2005). How the blind "see" Braille: Lessons from functional magnetic resonance imaging. *Neuroscientist, 11,* 577–582.

Sadato, N., Pascuel-Leone, A., Grafman, J., Ibanez, V., Deiber, M. P., Dold, G., & Hallett, M. (1996). Activation of the primary visual cortex by Braille reading in blind subjects. *Nature, 380,* 526–528.

Saint-Cyr, J. A., Taylor, A. E., & Lang, A. E. (1988). Procedural learning and neostriatal dysfunction in man. *Brain, 111*(4), 941–959.

Saint-Hilaire, J. M., Gilbert, M., Bouvier, G., & Barbeau, A. (1981). Epilepsy with aggressive behavior. Two cases with depth-electrodes recordings (author's transl). *Revue Neurologique, 137,* 161–179.

Saj, A., Honoré, J., Bernard-Demanze, L., Devèze, A., Magnan, J., & Borel, L. (2012). Where is straight ahead to a patient with unilateral vestibular loss? *Cortex* (online 18 June 2012). doi:org/10.1016/j.cortex.2012.05.019.

Sakai, K., & Passingham, R. E. (2001). Prefrontal interactions reflect future task operation. *Nature: Neuroscience, 6,* 75–81.

Sakai, K., & Passingham, R. E. (2006). Prefrontal set activity predicts rule-specific neural processing during subsequent cognitive performance. *Journal of Neuroscience, 26,* 1211–1218.

Sakzewski, L., Ziviani, J., Abbott, D. F., Macdonell, R. A., Jackson, G. D., & Boyd, R. N. (2011). Equivalent retention of gains at 1 year after training with constraintinduced or bimanual therapy in children with unilateral cerebral palsy. *Neurorehabilitation and Neural Repair, 25,* 664–671.

Salgado-Delgado, R., Osorio, A., Saderi, N., & Escobar, C. (2011). Disruption of circadian rhythms: A crucial factor in the etiology of depression. *Depression Research Treatment, 2011,* 1–9.

Sambeth, A., Ruohio, K., Alku, P., Fellman, V., & Huotilainen, M. (2008). Sleeping newborns extract prosody from continuous speech. *Clinical Neurophysiology, 119,* 332–341.

Samuels, E. R., & Szabadi, E. (2008). Functional neuroanatomy of the noradrenergic locus coeruleus: Its roles in the regulation of arousal and autonomic function part I: Principles of functional organization. *Current Neuropharmacology, 6*(3), 235–253.

Sanchez-Navarro, J. P., Martinez-Selva, M., & Roman, F. (2005). Emotional response in patients with frontal brain damage: Effects of affective and information content. *Behavioral Neuroscience, 119,* 87–97.

Sanford, F. (1915). Contact electrification and the electric current: Historical notes and comments on the various theories. *Scientific American,* Supplement No. 2081, 322–323.

Sartory, G., Cwik, J., Knuppertz, H., Schürholt, B., Lebens, M., Seitz, R. J., & Schulze, R. (2013). In search of the trauma memory: A meta-analysis of functional neuroimaging studies of symptom provocation in post-traumatic stress disorder (PTSD). *PloS One, 8*(3), e58150.

Satz, P. (1993). Brain reserve capacity on symptom onset after brain injury: A formulation and review of evidence for threshold theory. *Neuropsychology, 7*(3), 273–295. doi:10.1037/0894-4105.7.3.273.

Sauseng, P., et al. (2009). Brain oscillatory substrates of visual short term memory capacity. *Current Biology, 19,* 1846–1852.

Savage-Rumbaugh, E. S., Murphy, J., Sevcik, R. A., Brakke, K. E., Williams, S. L., & Rumbaugh, D. M. (1993). Language comprehension in ape and child. *Monographs of the Society for Research in Child Development, 58*, 1–222.

Savic, I., Berglund, H., & Lindström, P. (2005). Brain response to putative pheromones in homosexual men. *Proceedings of the National Academy of Sciences of the United States of America, 102*(20), 7356–7361.

Savic, I., Berglund, H., & Lindström, P. (2006). Brain response to putative pheromones in lesbian women. *Proceedings of the National Academy of Sciences of the United States of America, 103*(21), 8269–8274.

Savoiardo, M. (1986). The vascular territories of the carotid and vertebrobasilar systems. Diagrams based on CT studies of infarcts. *Italian Journal of Neurological Sciences, 7*(4), 405–409.

Scepkowski, L. A., & Cronin-Golumb, A. (2003). The alien hand: Cases, categorizations and anatomical correlates. *Behavioral and Cognitive Neuroscience Reviews, 2*, 261–277. doi:10.1177/1534582303260119.

Schachter, S. (1968). Obesity and eating: Internal and external cues differentially affect the eating behavior of obese and normal subjects. *Science, 161*, 751–756.

Schaefer, M., Heinze, H. J., & Galazky, I. (2010). Alien hand syndrome: Neural correlates of movements without conscious will. *PloS One, 5*, 15010. doi:10.1371/journal.pone.0015010.

Schaefer, M., Heinze, H. J., & Rotte, M. (2012). Close to you: Embodied simulation for peripersonal space in primary somatosensory cortex. *PloS One, 7*(8), e42308.

Schaffer, C. E., Davidson, R. J., & Saron, C. (1983). Frontal and parietal electroencephalogram asymmetry in depressed and nondepressed subjects. *Biological Psychiatry, 18*, 753–762.

Scheer, F. A., van Doornen, L. J., & Buijs, R. M. (1999). Light and diurnal cycle affect human heart rate: Possible role for the circadian pacemaker. *Journal of Biological Rhythms, 14*(3), 202–212.

Schenkenberg, T., Bradford, D. C., & Ajax, E. T. (1980). Line bisection and unilateral visual neglect in patients with neurologic impairment. *Neurology, 30*, 509.

Schiff, B. B., & MacDonald, B. (1990). Facial asymmetries in the spontaneous response to positive and negative emotional arousal. *Neuropsychologia, 28*, 777–785.

Schildkraut, J. (1995). The catecholamine hypothesis of affective disorders: A review of supporting evidence. *The Journal of Neuropsychiatry and Clinical Neurosciences, 7*(4), 524.

Schmahmann, J. D. (2004). Disorders of the cerebellum: Ataxia, dysmetria of thought, and the cerebellar cognitive affective syndrome. *Journal of Neuropsychiatry and Clinical Neurosciences, 16*(3), 367–378.

Schmeichel, B. J., & Demaree, H. A. (2010). Working memory capacity and spontaneous emotion regulation: High capacity predicts self-enhancement in response to negative feedback. *Emotion, 10*(5), 739.

Schneider-Garces, N. J., Gordon, B. A., Brumback-Peltz, C. R., Shin, E., Lee, Y., Sutton, B. P., Fabiani, M., et al. (2010). Span, CRUNCH, and beyond: Working memory capacity and the aging brain. *Journal of Cognitive Neuroscience, 22*, 655–669.

Schock, L., Bhavsar, S., Demenescu, L. R., Sturm, W., & Mathiak, K. (2013). Does valence in the visual domain influence the spatial attention after auditory deviants? Exploratory data. *Frontiers in Behavioral Neuroscience, 7*, 6.

Scholz, J., & Woolf, C. J. (2002). Can we conquer pain? *Nature Neuroscience, 5 Suppl*, 1062–1067.

Schönwiesner, M., Novitski, N., Pakarinen, S., Carlson, S., Tervaniemi, M., & Näätänen, R. (2007). Heschl's gyrus, posterior superior temporal gyrus, and mid-ventrolateral prefrontal cortex have different roles in the detection of acoustic changes. *Journal of Neurophysiology, 97*(3), 2075–2082.

Schulte-Rüther, M., Markowitsch, H. J., Shah, N. J., Fink, G. R., & Piefke, M. (2008). Gender differences in brain networks supporting empathy. *NeuroImage, 42*, 393–403.

Schutter, D., Putman, P., Hermans, E., & van Honk, J. (2001). Parietal electroencephalogram beta asymmetry and selective attention to angry facial expressions in healthy human subjects. *Neuroscience Letters, 314*(1–2), 13–16.

Schwartze, M., Keller, P. E., Patel, A. D., & Katz, S. A. (2011). The impact of basal ganglia lesions on sensorimotor synchronization, spontaneous motor tempo, and the detection of tempo changes. *Behavioural Brain Research, 216,* 685–691.

Schwarzlose, R. F., Baker, C. I., & Kanwisher, N. (2005). Separate face and body selectivity on the fusiform gyrus. *Journal of Neuroscience, 25,* 11055–11059.

Science Codex. (2013). Chair of DSM-IV task force: Think twice before using the DSM-V, May 20, 2013 (online).

Scott, W. E., Jr. (1966). Activation theory and task design. *Organizational Behavior and Human Performance, 1,* 3–30.

Scott, K. (7 February 2000). Voluntary amputee ran disability site. *The Guardian.*

Scott, J. G., & Schoenberg, M. R. (2011). Frontal lobe/Executive functioning. In M. R. Schoenberg & J. G. Scott (Eds.), *The little black book of neuropsychology* (pp. 219–248). US: Springer. doi:10.1007/978-0-387-76978-3_10.

Sedda, A., & Scarpina, F. (2012). Dorsal and ventral streams across sensory modalities. *Neuroscience Bulletin, 28*(3), 291–300.

Seeley, W. W., Menon, V., Schatzberg, A. F., Keller, J., Glover, G. H., Kenna, H., et al. (2007). Dissociable intrinsic connectivity networks for salience processing and executive control. *The Journal of Neuroscience, 27,* 2349–2356.

Segalowitz, S. J., Unsal, A., & Dywan, J. (1992). CNV evidence for the distinctiveness of frontal and posterior neural processes in a traumatic brain injured population. *Journal of Clinical of Experimental Neuropsychology, 14,* 545–565.

Sela, L., Sacher, Y., Serfaty, C., Yeshurun, Y., Soroker, N., et al. (2009). Spared and impaired olfactory abilities after thalamic lesions. *The Journal of Neuroscience, 29,* 12059–12069.

Selye, H. H. B. (1956). *The stress of life.* New York: McGraw-Hill.

Semenza, C. (1988). Impairment in localization of body parts following brain damage. *Cortex, 24,* 443–449.

Serfaty, M., Kennell-Webb, S., Warner, J., Blizard, R., & Raven, P. (2002). Double blind randomised placebo controlled trial of low dose melatonin for sleep disorders in dementia. *International Journal of Geriatric Psychiatry, 17,* 1120–1127.

Se'vigny, M.-C., Everett, J., & Grondin, S. (2003). Depression, attention, and time estimation. *Brain and Cognition, 53,* 351–353.

Shaffer, R., & Henry, N. (2 April 1981). Hinckley pursued actress for months, letter shows. *Washington Post,* at A1.

Shallice, T., & Cooper, R. P. (2011). *The organization of mind.* Oxford: Oxford University Press.

Shamay-Tsoory, S. G., Tomer, R., Aharon-Peretz, J. (2005). The neuroanatomical basis of understanding sarcasm and its relationship to social cognition. *Neuropsychology, 19*(3), 288–300.

Shapiro, D. M., & Harrison, D. W. (1990). Alternate forms of the AAVLT: A procedure and test of form equivalency. *Archives of Clinical Neuropsychology, 5,* 405–510.

Shapiro, D., Harrison, D. W., Crews, W. D., Jr., & Everhart, D. E. (1996). Age differences in hemispheric activation to sensory condition. *International Journal of Neuroscience, 87,* 249–256.

Sharma, A., & Dorman, M. (2012). Central auditory system development and plasticity after cochlear implantation. In *Auditory prostheses* (pp. 233–255). New York: Springer Publishing.

Shaw, G. (2012). Heading off migraine: What's the evidence for non-pharmaceutical approaches? *Neurology Now, 8*(3), 23–30.

Sheline, Y. I., Barch, D. M., Donnelly, J. M., Ollinger, J. M., Snyder, A. Z., & Mintun, M. A. (2001). Increased amygdala response to masked emotional faces in depressed subjects resolves with antidepressant treatment, An FMRI study. *Biological Psychiatry, 50,* 651–658.

Shenal, B. V., & Harrison, D. W. (2004). Dynamic lateralization: Hostility, cardiovascular regulation, and tachistoscopic regulation. *International Journal of Neuroscience, 114,* 335–348.

Shenal, B. V., Rhodes, R. D., Moore, T. M., Higgins, D. A., & Harrison, D. W. (2001). Quantitative electroencephalography (QEEG) and neuropsychological syndrome analysis. *Neuropsychology Review, 11*(1), 31–44.

Shenal, B., Harrison, D. W., & Demaree, H. A. (2003). The neuropsychology of depression: A literature review and preliminary model. *Neuropsychology Review, 13*(1), 33–42.

Shi, F., Liu, B., Zhou, Y., Yu, C., & Jiang, T. (2009). Hippocampal volume and asymmetry in mild cognitive impairment and Alzheimer's disease: Meta-analyses of MRI studies. *Hippocampus, 19*(11), 1055–1064.

Shima, K., & Tanji, J. (1998). Both supplementary and presupplementary motor areas are crucial for the temporal organization of multiple movements. *Journal of Neurophysiology, 80*(6), 3247–3260.

Shimojima, Y., Tujii, T., Yanagisawa, A., Tajino, K., Kanda, H., & Yamagami, M. (2003). Hostility disturbs learning. *Perceptual and Motor Skills, 96*(2), 616–622.

Shin, Y. W., Kim, D. J., Ha, T. H., Park, H. J., Moon, W. J., et al. (2005). Sex differences in the human corpus callosum: Diffusion tensor imaging study. *Neuroreport, 16,* 795–798.

Shinohara, Y., Hosoya, A., Yamasaki, N., Ahmed, H., Hattori, S., Eguchi, M., Yamaguchi, S., Miyakawa, T., Hirase, H., & Shigemoto, R. (2012). Right-hemispheric dominance of spatial memory in split-brain mice. *Hippocampus, 22,* 117–121. doi:10.1002/hipo.20886.

Shorter, E. (1997). *A history of psychiatry* (pp. 227–228). Hoboken: Wiley.

Shulman, L. M., Taback, R. L. T., Bean, J., & Weiner, W. J. (2001). Comorbidity of the nonmotor symptoms of Parkinson's disease. *Movement Disorders, 16,* 507–510.

Shultz, R. T. (2005). Developmental deficits in social perception in autism: The role of the amygdale and fusiform face area. *International Journal of Developmental Neuroscience, 23,* 125–141.

Shultz, R. T., Gautheir, I., Klin, A., Fulbright, R. K., Anderson, A. W., Volkmar, F., Skudlarski, P., Lacadie, C., Cohen, D. J., & Gore, J. C. (2000). Abnormal ventral temporal cortical activity during face discrimination among individuals with autism and Asperger syndrome. *Archives of General Psychiatry, 57,* 331–340.

Siddiqui, F., & D'Ambrosio, C. (2012). Chapter 9: Sleep disorders in older patients. *Aging and Lung Disease, 173–188,* Pearson.

Siegel, J. M. (1979). Behavioral functions of the reticular formation. *Brain Research Reviews, 1,* 69–105.

Siegel, J. M. (1985). Anger and cardiovascular risk in adolescents. *Health Psychology, 3,* 293–313.

Siemer, M. (2005). Mood-congruent cognitions constitute mood experience. *Emotion, 5,* 296–308.

Sigelman, C. K., & Rider, E. A. (2011). *Lifespan human development.* Belmont: Wadsworth Cengage Publishers.

Silberman, E. K., & Weingartner, H. (1986). Hemispheric lateralization of functions related to emotion. *Brain and Cognition, 5,* 322–353.

Silver, J., & Arciniegas, D. B. (2007). Pharmacotherapy of neurospsychiatric disturbances. In N. D. Zasler, D. I. Katz, & R. D. Zafonte (Eds.), *Brain injury medicine: Principles and practice* (pp. 963–993). New York: Demos.

Silveri, M. M., Sneider, J. T., Crowley, D. J., Covell, M. J., Acharya, D., Rosso, I. M., & Jensen, J. E. (2013). Frontal lobe γ-aminobutyric acid levels during adolescence: Associations with impulsivity and response inhibition. *Biological Psychiatry, 74*(4), 296–304.

Simon, O., Mangin, J. F., Cohen, L., Le Bihan, D., & Dehaene, S. (2002). Topographical layout of hand, eye, calculation, and language-related areas in the human parietal lobe. *Neuron, 33,* 475–487. doi:10.1016/j.neuroimage.2004.09.023.

Sinclair, D. C. (1955). Cutaneous sensation and the doctrine of specific energy. *Brain, 78*(4), 584–614.

Singer, C., et al. (2003). A multicenter, placebo-controlled trial of melatonin for sleep disturbance in Alzheimer's disease. *Sleep, 26,* 893–901.

Sirigu, A., Grafman, J., Bressler, K., & Sunderland, T. (1991). Multiple representations contribute to body knowledge processing. Evidence from a case of autotopagnosia. *Brain, 114,* 629–642.

Sirois, B. C., & Burg, M. M. (2003). Negative emotion and coronary heart disease. A review. *Behavior Modification, 27*(1), 83–102.

Smith, S. W. (1997). *The scientist and engineer's guide to digital signal processing.* San Diego: California Technical Publishing.

Smith, T. W., & Pope, M. K. (1990). Cynical hostility as a health risk: Current status and future directions. *Journal of Social Behavior and Personality, 5,* 77–88.

Smith, T. W., Galzer, K., Ruiz, J. M., & Gallo, L. C. (2004). Hostility, anger, and aggressiveness and coronary heart disease: An interpersonal perspective on emotion and health. *Journal of Personality, 72,* 1217–1270.

Smith, G. S., Kramer, E., Hermann, C., Ma, Y., Dhawan, V., Chaly, T., et al. (2009). Serotonin modulation of cerebral glucose metabolism in depressed older adults. *Biological Psychiatry, 66,* 259–266.

Smith, E., Kellis, S., House, P., & Greger, B. (2013). Decoding stimulus identity from multi-unit activity and local field potentials along the ventral auditory stream in the awake primate: implications for cortical neural prostheses. *Journal of Neural Engineering, 10*(1), 016010.

Smithuis, R. (2008). Brain ischemia—Vascular territories. The Radiology Assistant, published online, http://www.radiologyassistant.nl/en/484b8328cb6b2.

Smolders, K. C. H. J., de Kort, Y. A. W., & Cluitmans, P. J. M. (2012). A higher illuminance induces alertness even during office hours: Findings on subjective measures, task performance and heart rate measures. *Physiology & Behavior, 107*(1), 7–16.

Snow, V., Beck, D., Budnitz, T., Miller, D. C., Potter, J., Wears, R. L., Williams, M. V., et al. (2009). Transitions of care consensus policy statement: American college of physicians, society of general internal medicine, society of hospital medicine, american geriatrics society, american college of emergency physicians, and society for academic emergency medicine. *Journal of Hospital Medicine, 4*(6), 364–370.

Snyder, K. A., & Harrison, D. W. (1997). The affective auditory verbal learning test. *Archives of Clinical Neuropsychology, 12,* 477–482.

Snyder, K. A., Harrison, D. W., & Gorman, W. J. (1996). Auditory affect perception in a dichotic listening paradigm as a function of verbal fluency classification. *International Journal of Neuroscience, 84,* 65–74.

Snyder, K. A., Harrison, D. W., & Shenal, B. (1998). The affective auditory verbal learning test: Peripheral arousal correlates. *Archives of Clinical Neuropsychology, 13,* 251–258.

Soares, P. P., da Nobrega, A. C., Ushizima, M. R., & Irigoyen, M. C. (2004). Cholinergic stimulation with pyridostigmine increases heart rate variability and baroreflex sensitivity in rats. *Autonomic Neuroscience, 113,* 24–31.

Sobel, N., Prabhakaran, V., Hartley, C. A., Desmond, J. E., Glover, G. H., Sullivan, E. V., & Gabrieli, J. D. E. (1999). Blind smell: Brain activation induced by an undetected air-borne chemical. *Brain, 122,* 209–217.

Sokolov, E. N. (1960). Neuronal models and the orienting reflex. In M. A. B. Brazier (Ed.), *The central nervous system and behavior.* New York: Josiah Macey, Jr., Foundation.

Somerville, L. H., Jones, R. M., Ruberry, E. J., Dyke, J. P., Glover, G., & Casey, B. J. (2013). The medial prefrontal cortex and the emergence of self-conscious emotion in adolescence. *Psychological Science, 24*(8), 1554–1562.

Sommer, I. E., Aleman, A., Bouma, A., & Kahn, R. S. (2004). Do women really have more bilateral language representation than men? A meta-analysis of functional imaging studies. *Brain, 127,* 1845–1852.

Sotres-Bayon, F., Bush, D. E. A., & LeDoux, J. E. (2004). Interactions in fear extinction emotional perseveration: An update on prefrontal-amygdala interactions in fear extinction. *Learning & Memory, 11,* 525–535.

Sovrano, V. A. (2007). A note on asymmetric use of the forelimbs during feeding in the European green toad (Bufo viridis). *Laterality, 12,* 458–463.

Sparks, D. L. (1986). Translation of sensory signals into commands for control of saccadic eye movements: Role of the primate superior colliculus. *Physiological Reviews, 66,* 118–171.

Spenser, S. S., Theodore, W. H., & Berkovic, S. F. (1995). Clinical applications: MRI, SPECT and PET. *Magnetic Resonance Imaging, 13,* 1119–1124.

Sperry, R. W. (1966). Brain bisection and mechanisms of consciousness. In J. C. Eccles (Ed.), *Brain and conscious experience* (pp. 298–313). Heidelberg: Springer-Verlag.

Sperry, R. (1982). Some effects of disconnecting the cerebral hemispheres. *Science, 217,* 1223–1226.

Spielberger, C. D., Johnson, E. H., Russell, S. F., Crane, R., Jacobs, G. A., & Worden, T. J. (1985). The experience and expression of anger: Construction and validation of and anger expression scale. In M. A. Chesney & R. H. Rosenman (Eds.), *Anger and hostility in cardiovascular and behavioral disorders* (pp. 5–30). Washington, DC: Hemisphere.

Spreen, F. O., & Benton, A. L. (1977). *Manual of instructions for the neurosensory center comprehensive examination for aphasia.* Victoria: University of Victoria.

Springer, S. P., & Deutsch, G. (1998). *Left brain, right brain: Perspective from cognitive neuroscience.* New York: W.H. Freeman & Company.

Squire, L. R. (1992). Memory and the hippocampus: A synthesis from findings with rats, monkeys, and humans. *Psychological Review, 99,* 195–231.

Squire, L. R. (2004). Memory systems of the brain: A brief history and current perspective. *Neurobiology of Learning and Memory, 82,* 171–177.

Stahl, S. M., Pradko, J. F., Haight, B. R., Modell, J. G., Rockett, C. B., & Learned-Coughlin, S. (2004). A review of the neuropharmacology of Bupropion, a dual norepinephrine and dopamine reuptake inhibitor. *Primary Care Companion Journal of Clinical Psychiatry, 6*(4), 159–166.

Stanley, M., & Mann, J. J. (1983). Increased serotonin-2 binding sites in frontal cortex of suicide victims. *The Lancet, 321*(8318), 214–216.

Stark, M., Coslett, H. B., & Saffran, E. M. (1996). Impairment of an egocentric map of locations: Implications for perception and action. *Cognitive Neuropsychology, 13,* 481–523. doi:10.1080/026432996381908.

Starkstein, S. E., & Tranel, D. (2012). Neurological and psychiatric aspects of emotion. In Aminoff, Boller & Swaab (Eds.), *Neurobiology of psychiatric disorders E-Book: Handbook of clinical neurology* (Vol. 106, p. 53–74).

Starkstein, S. E., Jorge, R. E., & Robinson, R. G. (2010). The frequency, clinical correlates, and mechanisms of anosognosia after stroke. *The Canadian Journal of Psychiatry, 55,* 355–361.

Starzl, T. E., & Magoun, H. W. (1951). Organization of the diffuse thalamic projection system. *Journal of Neurophysiology, 14*(2), 133.

Steeves, J. K. E., Humphrey, G. K., Culham, J. C., Menon, R. A., et al. (2004). Behavioral and neuroimaging evidence for a contribution of color and texture information to scene classification in a patient with visual form agnosia. *Journal of Cognitive Neuroscience, 16,* 955–965.

Stein, P., Savli, M., Wadsak, W., Mitterhauser, M., Fink, M., Spindelegger, C., Mien, L. K., Moser, U., Dudczak, R., Kletter, K., Kasper, S., & Lanzenberger, R. (2008). The serotonin-1A receptor distribution in healthy men and women measured by PET and [carbonyl-11C]WAY-100635. *European Journal Nuclear Medicine Molecular Imaging, 35*(12), 2159–2168.

Stephan, E., et al. (2003). Functional neuroimaging of gastric distention. *Journal of Gastrointestinal Surgery, 7,* 740–749.

Stephen, J. C., & Tabor, L. (1981). *Experimental Aging Research, 7*(2), 147–158.

Stephens, S. D. (1970). Personality and the slope of loudness function. *The Quarterly Journal of Experimental Psychology, 22,* 9–13.

Stern, K., & McClintock, M. K. (1998). Regulation of ovulation by human pheromones. *Nature, 392,* 177–179.

Stevens, S. S., & Davis, H. (1938). *Hearing: Its psychology and physiology.* New York: Wiley.

Stickgold, R., Hobson, J. A., Fosse, R., & Fosse, M. (2001). Sleep, learning, and dreams: Off-line memory reprocessing. *Science, 294,* 1052–1057.

Strauss, E., Sherman, E. M. S., & Spreen, O. (2006). *A compendium of neuropsychological tests: Administration, norms and commentary* (3rd ed.). Oxford University Press.

Stuss, D. T., & Benson, D. F. (1986). *The frontal lobes.* New York: Raven Press.

Stuss, D. T., & Knight, R. T. (Eds.). (2013). *Principles of frontal lobe function* (2nd ed.). New York: Oxford University Press.

Stuss, D. T., Floden, D., Alexander, M. P., Levine, B., & Katz, D. (2001). Stroop performance in focal lesion patients: Dissociation of processes and frontal lobe lesion location. *Neuropsychologia, 39,* 771–786.

Subramanyam, B., Woolf, T., & Castagnoli, N. (1991). Studies on the in vitro conversion of haloperidol to a potentially neurotoxic pyridinium metabolite. *Chemical Research in Toxicology, 4*(1), 123–128.

Sugiura, M., Sassa, Y., Jeong, H., Miura, N., Akitsuki, Y., Horie, K., et al. (2006). Multiple brain networks for visual self recognition with different sensitivity for motion and body part. *Neuroimage, 32,* 1905–1917.

Suls, J., & Bunde, J. (2005). Anger, anxiety, and depression as risk factors for cardiovascular disease: The problems and implications of overlapping affective dispositions. *Psychological Bulletin, 131,* 260–300.

Sun, T., Patoine, C., Abu-Khalil, A., Visvader, J., Sum, E., Cherry, T. J., Orkin, S. H., Geschwind, D. H., & Walsh, C. (2005). Early asymmetry of gene transcription in embryonic human left and right cerebral cortex. *Science, 308,* 1794–1798.

Sutton, S. K., & Davidson, R. J. (1997). Prefrontal brain asymmetry: A biological substrate of the behavioral approach and inhibition systems. *Psychological Science, 8*(3), 204–210.

Suzuki, M., & Gottlieb, J. (2013). Distinct neural mechanisms of distractor suppression in the frontal and parietal lobe. *Nature Neuroscience, 16*(1), 98–104. doi:10.1038/nn.3282.

Swaab, D. F., Fliers, E., & Partiman, T. S. (1985). The suprachiasmatic nucleus of the human brain in relation to sex, age and senile dementia. *Brain Research, 342*(1), 37–44.

Swann, N. C., Cai, W., Conner, C. R., Pieters, T. A., Claffey, M. P., George J. S., Aron, A. R., & Tandon, N. (2012a). Roles for the pre-supplementary motor area and the right inferior frontal gyrus in stopping action: electrophysiological responses and functional and structural connectivity. *Neuroimage, 59,* 2860–2870. doi:10.1016/j.neuroimage.2011.09.049.

Swann, N., Poizner, H., Houser, M., Gould, S., Greenhouse, I., Cai, W., Strunk, J., George, J., & Aron, A. R. (2012b). Deep brain stimulation of the subthalamic nucleus alters the cortical profile of response inhibition in the beta frequency band: A scalp EEG study in Parkinson's disease. *Journal of Neuroscience, 31,* 5721–5729.

Syka, J. (2002). Plastic changes in the central auditory system after hearing loss, restoration of function, and during learning. *Physiology Review, 82,* 601–636.

Symington, S. H., Paul, L. K., Symington, M. F., Ono, M., & Brown, W. S. (2010). Social cognition in individuals with agenesis of the corpus callosum. *Social Neuroscience, 5*(3), 296–308. doi:10.1080/17470910903462419.

Symonds, L. L., Gordon, N. S., Bixby, J. C., & Mande, M. M. (2006). Right-lateralized pain processing in the human cortex: An FMRI study. *Journal of Neurophysiology, 95*(6), 3823–3830. Epub 2006 Mar 22.

Tajudeen, B. A., Waltzman, S. B., Jethanamest, D., & Svirsky, M. A. (2010). Speech perception in congenitally deaf children receiving cochlear implants in the first year of life. *Otology & Neurotology: Official Publication of the American Otological Society, American Neurotology Society [and] European Academy of Otology and Neurotology, 31*(8), 1254.

Takahashi, N., Kawamura, M., Shiota, J., et al. (1997). Pure topographic disorientation due to right retrosplenial lesion. *Neurology, 49,* 464–469.

Talavage, T. M., Sereno, M. I., Melcher, J. R., Ledden, P. J., Rosen, B. R., & Dale, A. M. (2004). Tonotopic organization in human auditory cortex revealed by progressions of frequency sensitivity. *Journal of Neurophysiology, 91,* 1282–1296.

Talavera, K., Yasumatsu, K., Voets, T., Droogmans, G., Shigemura, N., Ninomiya, Y, & Nilius, B., et al. (2005). Heat activation of TRPM5 underlies thermal sensitivity of sweet taste. *Nature, 438*(7070), 1022–1025.

Tanielian, T. L., & Jaycox, L. H. (2008). Invisible wounds of war: Psychological and cognitive injuries, their consequences, and services to assist recovery. RAND Report.SantaMonica:RAND Corporation.

Tatsch, K., Schwarz, J., Mozley, P. D., Linke, R., Pogarell, O., Oertel, W. H., et al. (1997). Relationship between clinical features of Parkinson's disease and presynaptic dopamine transporter binding assessed with [123I]IPT and single-photon emission tomography. *European Journal of Nuclear Medicine, 24,* 415–421.

Taub, E., & Morris, D. M. (2001). Constraint-induced movement therapy to enhance recovery after stroke. *Current Atherosclerosis Reports, 3,* 279–286.

Taub, E., Uswatte, G., & Elbert, T. (2002). New treatments in neurorehabilitation founded on basic research. *Nature Reviews Neuroscience, 3*(3), 228–236.

Teffer, K., & Semendeferi, K. (2012). Human prefrontal cortex evolution, development, and pathology: Chapter 9. In M. A. Hoffman & D. Falk (Eds.), *Progress in brain research* (Vol. 195). Amsterdam: Elsevier B. V.

Terman, M., & Terman, J. S. (2005). Light therapy for seasonal and non-seasonal depression: Efficacy, protocol, safety, and side effects. *CNS Spectrums, 10,* 647–663.

Teshiba, T. M., Ling, J., Ruhl, D. A., Bedrick, B. S., Peña, A., & Mayer, A. R. (2012). Evoked and intrinsic asymmetries during auditory attention: Implications for the contralateral and neglect models of functioning. *Cerebral Cortex* (first published online February 27 2012). doi:10.1093/cercor/bhs039.

Teuber, H. L. (1955). Physiological psychology. *Annual Review of Psychology, 9,* 267–296.

Teuber, H. L. (1964). The riddle of frontal lobe function in man. In J. M. Warren & K. Akert (Eds.), *The frontal-granular cortex and behavior* (pp. 410–44). New York: McGraw-Hill.

Thatcher, R. W., Moore, N., John, E. R., Duffy, F., Hughes, J. R., & Krieger, M. (1999). QEEG and traumatic brain injury: Rebuttal of the American Academy of Neurology 1997 report by the EEG and Clinical Neuroscience Society. *Clinical Electroencephalography, 30*(3), 94–98.

Thayer, J. F., & Brosschot, J. F. (2005). Psychosomatics and psychopathology: Looking up and down from the brain. *Psychoendocrinology, 30,* 1050–1058.

Thibodeau, R., Jorgensen, R. S., & Kim, S. (2006). Depression, anxiety, and resting frontal EEG asymmetry: A meta-analytic review. *Journal of Abnormal Psychology, 115,* 715–729.

Thomas, O., Avidan, G., Humphreys, K., et al. (2009). Reduced structural connectivity in ventral visual cortex in congenital prosopagnosia. *Nature Neuroscience, 12,* 29–31.

Thomas, N. A., Loetscher, T., Clode, D., & Nicholls, M. E. R. (2012). Right-wing politicians prefer the emotional left. *PLoS One, 7*(5), 36552. doi:10.1371/journal.pone.0036552.

Thompson-Schill, S. L., Bedny, M., & Goldberg, R. F. (2005). The frontal lobes and the regulation of mental activity. *Current Opinion in Neurobiology, 15*(2), 219–224.

Thurstone, L. L. (1938). *Primary mental abilities.* Chicago: University of Chicago Press.

Thurstone, L. L., & Thurstone, T. (1949). *Examiner manual for the SRA primary mental abilities test.* Chicago: Science Research Associates.

Tiwari, D., & Amar, K. (2008). A case of corticobasal degeneration presenting with alien limb syndrome. *Age Ageing, 37,* 600–601. doi:10.1093/ageing/afn103.

Toglia, M. P., & Battig, W. F. (1978). *Handbook of word norms.* Hillsdale: Lawrence Erlbaum.

Tomlin, D., Kayali, M. A., King-Casas, B., Anen, C., Camerer, C. F., Quartz, S. R., & Montague, P. R. (2006). Agent-specific responses in the cingulate cortex during economic exchanges. *Science, 312,* 1047–1050.

Tompkins, C. A. (2012). Rehabilitation for cognitive-communication disorders in right hemisphere brain damage. *Archives of Physical Medicine and Rehabilitation, 93*(1), 61–69.

Tonkonogy, J. M., & Puente, A. E. (2009). *Localization of clinical syndromes in neuropsychology and neuroscience.* New York: Springer Publishing Company.

Tootell, R. B. H., Reppas, J. B., Dale, A. M., Look, R. B., Sereno, M. I., Malach, R., Rosen, B. R., et al. (1995). Visual motion aftereffect in human cortical area MT revealed by functional magnetic resonance imaging. *Nature, 375,* 139–141.

Toplak, M. E., Dockstader, C., & Tannock, R. (2006). Temporal information processing in ADHD: Findings to-date and new methods. *Journal of Neuroscience Methods, 15,* 15–29.

Tramo, M. J. (2001). Biology and music. Music of the hemispheres. *Science, 291*(5501), 54–56.

Tranel, D., & Damasio, H. (1994). Neuroanatomical correlates of electrodermal skin conductance responses. *Psychophysiology, 31,* 427–438.

Tranel, D., Vianna, E., Manzel, K., Damasio, H., & Grabowski, T. (2009). Neuroanatomical correlates of the Benton facial recognition test and judgment of line orientation test. *Journal of Clinical & Experimental Neuropsychology, 31*(2), 219–233.

Troost, B. T. (1980). Dizziness and vertigo in vertebrobasilar disease. Part II. Central causes and vertebrobasilar disease. *Stroke, 11*(4), 413–415.

Tsang, A. H., Husse, J., & Oster, H. (2013). Sleep, energy homeostasis and metabolic syndrome alterations. *Chronobiology and Obesity,* 89–109. New York: Springer.

Tsuda, M., Beggs, S., Salter, M. W., & Inoue, K. (2012). Microglia and intractable chronic pain. *Glia, 61*(1), 55–61.

Tucker, D. M. (1981). Lateral brain function, emotion, and conceptualization. *Psychological Bulletin, 89*(1), 19–46.

Tucker, D. M., & Derryberry, D. (1992). Motivated attention: Anxiety and the frontal executive functions. *Neuropsychiatry, Neuropsychology, and Behavioral Neurology, 5,* 233–252.

Tucker, D. M., & Williamson, P. A. (1984). Asymmetric neural control systems in human self regulation. *Psychological Review, 91,* 185–215.

Tucker, D. M., Watson, R. T., & Heilman, K. M. (1977). Discrimination and evocation of affectively intoned speech in patients with right parietal disease. *Neurology, 27,* 947–950.

Turk-Browne, N. B., Norman-Haignere, S. V., & McCarthy, G. (2010). Face-specific resting functional connectivity between the fusiform gyrus and posterior superior temporal sulcus. *Frontiers in Human Neuroscience, 4,* 176.

Turkeltaub, P. E., Eden, G. F., Jones, K. M., & Zeffiro, T. A. (2002). Meta-analysis of the functional neuroanatomy of single-word reading: Method and validation. *Neuroimage, 16,* 765–780.

Turriziani, P., Carlesimo, G. A., Perri, R., Tomaiuolo, F., & Caltagirone, C. (2003). *Journal of Neurology, Neurosurgery and Psychiatry, 74,* 61–69. doi:10.1136/jnnp.74.1.61.

Turton, A., & Lemon, R. N. (1999). The contribution of fast corticospinal input to the voluntary activation of proximal muscles in normal subjects and in stroke patients. *Experimental Brain Research, 129,* 559–572.

Tuunainen, A., Kripke, D. F., & Endo, T. (2004). Light therapy for non-seasonal depression. *Cochrane Database of Systematic Reviewa, 2,* CD004050.

Twain, M. (n.d.). BrainyQuote.com. BrainyQuote.com Website: http://brainyquote.com/citation/quotes/m/marktwain103892.html#uQHG4URKZTL47ZG4.99. Accessed 3 Jan 2012.

Tzourio-Mazoyer, N., De Schonen, S., Crivello, F., Reutter, B., Aujard, Y., & Mazoyer, B. (2002). Neural correlates of woman face processing by 2-month-old infants. *Neuroimage, 15,* 454–461.

Uddin, L. Q., Kaplan, J. T., Molnar-Szakacs, I., Zaidel, E., & Iacoboni, M. (2005). Self-face recognition activates a frontoparietal "mirror" network in the right hemisphere: An event-related fMRI study. *Neuroimage, 25,* 926–935.

Unsworth, N., Spillers, G. J., & Brewer, G. A. (2012). Working memory capacity and retrieval limitations from long-term memory: An examination of differences in accessibility. *The Quarterly Journal of Experimental Psychology (Hove), 65,* 2397–2410 (Epub ahead of print).

Urgesi, C., Berlucchi, G., & Aglioti, S. M. (2004). Magnetic stimulation of extrastriate body area impairs visual processing of nonfacial body parts. *Current Biology, 13,* 2130–2134.

Vahidnia, A., & Van Der Wolff, D. (2007). Arsenic neurotoxicity: A review. *Human and Experimental Toxicology, 26*(10), 823–832.

Vallortigara, G. (2006). The evolutionary psychology of left and right: costs and benefits of lateralization. *Developmental Psychobiology, 48,* 418–427.

Vallortigara, G., Rogers, L. J., Bisazza, A., Lippolis, G., & Robins, A. (1998). Complementary right and left hemifield use for predatory and agonistic behaviour in toads. *NeuroReport, 9,* 3341–3344.

Van Buren, J. M. (1963). Trans-synaptic retrograde degeneration in the visual system of primates. *Journal of Neurology, Neurosurgery & Psychiatry, 26,* 402–409.

Vandenbroucke, S., Crombez, G., Van Ryckeghem, D., Harrar, V., Goubert, L., Spence, C., Wouter, D., & van Damme, S. (2012). Observing social stimuli influences detection of subtle somatic sensations differently for pain synaesthetes and controls. *Seeing and Perceiving, 25*(1), 19–19.

Van Essen, D. C., Anderson, C. H., & Felleman, D. J. (1992). Information processing in the primate visual system: An integrated systems perspective. *Science, 255*(5043), 419–423.

Van Honk, J., Peper, J. S., & Schutter, D. J. (2005). Testosterone reduces unconscious fear but not consciously experienced anxiety: Implications for the disorders of fear and anxiety. *Biological Psychiatry, 58,* 218–225.

Van Honk, J., Harmon-Jones, E., Morgan, B. E., & Schutter, D. J. (2010). Socially explosive minds: The triple imbalance hypothesis of reactive aggression. *Journal of Personality, 78,* 67–94.

Van Someren, E. J. W., Riemersma, R. F., & Swaab, D. F. (2002). Functional plasticity of the circadian timing system in old age: Light exposure. *Progress in Brain Research, 138,* 205–231.

Van Wagenen, W. P., & Herren, R. Y. (1940). Surgical division of commisural pathways in the corpus callosum: Relation to spread of an epileptic attack. *Archives of Neurology & Psychiatry, 44,* 740–759.

Van Zomeren, A. H., & van den Burg, W. (1985). Residual complaints of patients two years after severe head injury. *Journal of Neurology, Neurosurgery & Psychiatry, 48*(1), 21–28.

Vas, S., Kátai, Z., Kostyalik, D., Pap, D., Molnár, E., Petschner, P., Bagdy, G., et al. (2013). Differential adaptation of REM sleep latency, intermediate stage and theta power effects of escitalopram after chronic treatment. *Journal of Neural Transmission, 120*(1), 169–176.

Verdon, V., Schwartz, S., Lovblad, K.-O., Hauert, C.-A., & Vuilleumier, P. (2010). Neuroanatomy of hemispatial neglect and its functional components: A study using voxel-based lesion-symptom mapping. *Brain, 133,* 880–894.

Victor, T. A., Furey, M. L., Fromm, S. J., Öhman, A., & Drevets, W. C. (2010). Relationship between amygdala responses to masked faces and mood state and treatment in major depressive disorder. *Archives of General Psychiatry, 67,* 1128–1138.

Video. (1993). The Lynchburg story.

Vigneau, M., Beaucousin, V., Herve, P. Y., Duffau, H., Crivello, F., Houde, O., Mazoyer, B., & Tzourio-Mazoyer, N. (2006). Meta-analyzing left hemisphere language areas: Phonology, semantics, and sentence processing. *Neuroimage, 30,* 1414–1432.

Vilensky, J. A. (2012). The neglected cranial nerve: Nervus terminalis (cranial nerve N). *Clinical Anatomy.* doi:10.1002/ca.22130.

Vingerhoets, G., Acke, F., Alderweireldt, A. S., Nys, J., Vandemaele, P., & Achten, E. (2012). Cerebral lateralization of praxis in right-and left-handedness: Same pattern, different strength. *Human Brain Mapping, 33*(4), 763–777.

Vocat, R., Staub, F., Stroppini, T., & Vuilleumier, P. (2010). Anosognosia for hemiplegia: A clinical-anatomical prospective study. *Brain, 133,* 3578–3597. doi:10.1093/brain/awq297.

Vogel, A. C., Petersen, S. E., & Schlaggar, B. L. (2012). Matching is not naming: A direct comparison of lexical manipulations in explicit and implicit reading tasks. *Human Brain Mapping.* doi:10.1002/hbm.22077.

Von Bekesy G. (1960). *Experiments in hearing* (p. 745). New York: McGraw-Hill.

Von Frey, M. (1896). Untersuchung über die Sinnesfunctionen der menschlichen Haut. *Abhandlungen der mathematisch-physischen Klasse der Königichen Sächsischen Gesellschaft der Wissenschaften, 49,* 169–266.

Vuga, M., Fox, N. A., Cohn, J. F., George, C. J., Levenstein, R. M., & Kovacs, M. (2006). Long-term stability of frontal electroencephalographic asymmetry in adults with a history of depression and controls. *International Journal of Psychophysiology, 59,* 107–115.

Wacker, J., Heldmann, M., & Stemmler, G. (2003). Separating emotion and motivational direction in fear and anger: Effects on frontal asymmetry. *Emotion, 3,* 167–193.

Wacker, J., Mueller, E. M., Pizzagalli, D. A., Hennig, J., & Stemmler, G. (2013). Dopamine-D2-receptor blockade reverses the association between trait approach motivation and frontal asymmetry in an approach-motivation context. *Psychological Science.* doi:10.1177/0956797612458935.

Wade, N. J. (1994). A selective history of visual motion aftereffects. *Perception, 23,* 1111–1134.

Wade, N. J. (2003). The search for a sixth sense: The cases for vestibular, muscle, and temperature senses. *Journal of the History of the Neurosciences, 12,* 175–202.

Wager, T. D., Phan, K. L., Liberzon, I., & Taylor, S. F. (2003). Valence, gender, and lateralization of functional brain anatomy in emotion: A meta-analysis of findings from neuroimaging. *NeuroImage, 19,* 513–531.

Wagner, D. D., & Heatherton, T. F. (2010). Giving in to temptation: The emerging cognitive neuroscience of self-regulatory failure. In K. D. Vohs & R. F. Baumeister (Eds.), *Handbook of self-regulatio: Research, theory, and applications* (2nd ed., pp. 41–63). New York: Guilford Press.

Wagner, D. D., Boswell, R. G., Kelley, W. M., & Heatherton, T. F. (2012). Inducing negative affect increases the reward value of appetizing foods in dieters. *Journal of Cognitive Neuroscience, 24*(7), 1625–1633. Posted Online 29 May 2012. doi:10.1162/jocn_a_00238.

Waldstein, S. R., Kop, W. J., Schmidt, L. A., Haufler, A. J., Krantz, D. S., & Fox, N. A. (2000). Frontal electrocortical and cardiovascular reactivity during happiness and anger. *Biological Psychology, 55*, 2–23.

Walker, P. (7 July 2009). Three jailed for arson attack over Muhammad bride novel. *The Guardian News.* http://www.guardian.co.uk/uk/2009/jul/07/muslims-jailed-arson-book-protest. Accessed 3 Aug 2011.

Walker, R., Husain, M., Hodgson, T. L., Harrison, J., & Kennard, C. (1998). Saccadic eye movement and working memory deficits following damage to human prefrontal cortex. *Neuropsychologia, 26*, 1141–1159.

Wall, P. D. (1970). The sensory and motor role of impulses travelling in the dorsal columns towards cerebral cortex. *Brain, 93*(3), 505–524.

Wallentin, M. (2009). Putative sex differences in verbal abilities and language cortex: A critical review. *Brain Language, 108*, 175–183.

Walsh, K. W. (1978). *Neuropsychology: A clinical approach.* Edinburgh: Churchill Livingstone.

Walters, R. P., & Harrison, D. W. (2013a). Frontal regulation of blood glucose levels: Support for the limited capacity model in hostile men. Manuscript submitted for publication.

Walters, R. P., & Harrison, D. W. (2013b). Hostility and the metabolic syndrome: A neuropsychological perspective. Manuscript submitted for publication.

Walters, R. P., Harrison, D. W., Williamson, J., & Foster, P. (2006). Lateralized visual hallucinations: An analysis of affective valence. *Applied Neuropsychology, 13*(3), 160–165.

Wan, H., Sengupta, M., Velkoff, V. A., & DeBarros, K. A. (2005). *65 + in the United States. 2005 U.S. Census Bureau, current population reports* (pp. 23–209). Washington, DC: U.S. Government Printing Office.

Wang, X. J. (2001). Synaptic reverberation underlying mnemonic persistent activity. *Trends in Neurosciences, 24*(8), 455–463.

Wansink, B., Payne, C. R., & Chandon, P. (2007). Internal and external cues of meal cessation: The French paradox redux? *Obesity, 15*, 2920–2924.

Watson, R. T., Heilman, K. M., Miller, B. D., & King, F. A. (1974). Neglect after mesencephalic reticular formation lesions. *Nerulogy, 24*, 294–298.

Watson, R. T., Valenstein, E., & Heilman, K. M. (1981). Thalamic neglect. Possible role of the medial thalamus and nucleus reticularis in behavior. *Archives of Neurology, 38*, 501–506.

Watson, R. T., Valenstein, E., Day, A., & Heilman, K. M. (1994). Posterior neocortical systems subserving awareness and neglect. *Archives of Neurology, 51*, 1014–1021.

Waxman, S. G. (2010). *Clinical neuroanatomy* (26th ed.). New York: McGraw Hill.

Weber, E. H. (1834). *De tactu [Concerning touch].* New York: Academic Press.

Wechsler, D. (1997). *Wechsler memory scale* (3th ed.). San Antonio: The Psychological Corporation.

Wechsler, D. (2009). *Wechsler memory scale* (4th ed.). San Antonio: Pearson.

Wechsler, D., Holdnack, J. A., & Drozdick, L. W. (2009). *Wechsler memory scale: Technical and interpretive manual* (4th ed.). San Antonio: Pearson.

Wedekind, C., & Füri, S. (1997). Body odour preferences in men and women: Do they aim for specific MHC combinations or simply heterozygosity? *Proceedings of the Royal Society B: Biological Sciences, 264*, 1471–1479.

Wedekind, C., Seebeck, T., Bettens, F., & Paepke, A. J. (1995). MHC-dependent mate preferences in humans. *Proceedings of the Royal Society B: Biological Sciences, 260*(1359), 245–249.

Weinstein, E. A. (1981). *Woodrow Wilson: A medical and psychological biography* (p. 399). Princeton: Princeton University Press.

Weissman, D. H., & Woldorff, M. G. (2005). Hemispheric asymmetries for different components of global/local attention occur in distinct temporo-parietal loci. *Cerebral Cortex, 15*, 870–876.

Weissman, M. M., Wickramaratne, P., Nomura, Y., Warner, V., Verdeli, H., Pilowsky, D. J., Grillon, C., & Bruder, G. E. (2005). Families at high and low risk for depression: A 3-generation study. *Archives of General Psychiatry, 62*, 29–36.

Weisz, J., Emri, M., Fent, J., Lengyel, Z., Marian, T., Horvath, G., Bogner, P., Tron, L., & Adam, G. (2001). Right prefrontal activation produced by arterial baroreceptor stimulation: A PET study. *Neuroreport, 12*(15), 3233–3238.

Welberg, L. (2013). Reward: Serotonin promotes patience. *Nature Reviews Neuroscience, 13,* 603.

Wenzel, B. M. (1948). Techniques in olfactometry: A critical review of the last one hundred years. *Psychological bulletin, 45*(3), 231.

Wernicke, C. (1874). *Der Aphasische Symptomencomplex.* Breslau: Cohn and Weigert.

Wernicke C. (1995). The aphasia symptom-complex: A psychological study on an anatomical basis (1875). In P. Eling (Ed.), *Reader in the history of aphasia: From (franz gall to)* (pp. 69–89). Amsterdam: John Benjamins Pub Co.

Wertham, F. (1929). A new sign of cerebellar diseases. *Journal of Nervous and Mental Disorder, 69,* 486–493.

Wertheimer, M. (1923). Laws of organization in perceptual forms. First published as Untersuchungen zur Lehre von der Gestalt II. In *Psycologische Forschung* (Vol. 4, pp. 301–350). (Trans: W. Ellis (1938). *A source book of Gestalt psychology* (pp. 71–88). London: Routledge & Kegan Paul.

Westerhausen, R., & Hugdahl, K. (2010). Cognitive control of auditory laterality. In K. Hugdahl & R. Westerhausen (Eds), *The two halves of the brain: Information processing in the cerebral hemispheres.* Cambridge: MIT Press.

White, D. J., Congedo, M., Ciorciari, J., & Silberstein, R. B. (2012). Brain oscillatory activity during spatial navigation: Theta and gamma activity link medial temporal and parietal regions. *Journal of Cognitive Neuroscience, 24*(3), 686–697. (Posted Online January 30, 2012). doi:10.1162/jocn_a_000987.

Whitehouse, P. J., Price, D. L., Struble, R. G., Clark, A. W., Coyle, J. T., & Delon, M. R. (1982). Alzheimer's disease and senile dementia: Loss of neurons in the basal forebrain. *Science, 215,* 1237–1239.

Wiedemann, G., Pauli, P., Dengler, W., Lutzenberger, W., Birbaumer, N., & Buchkremer, G. (1999). Frontal brain asymmetry as a biological substrate of emotions in patients with panic disorders. *Archives of General Psychiatry, 56,* 78–84.

Wiederhold, M. L. (1970). Variations in the effects of electric stimulation of the crossed olivocochlear bundle on cat single auditory-nerve-fiber responses to tone bursts. *Journal of Acoustical Society of America, 48,* 966–977.

Wikimedia Foundation, Inc. (2011). Wikipedia, the free encyclopedia. Creative Commons: <div xml ns:cc=http://creativecommons.org/ns#.

Wilde, E. A., Biigler, E. D., Haider, J. M., Chu, Z., Levin, H. S., & Li, X. (2006). Vulnerability of the anterior commissure in moderate to severe pediatric traumatic brain injury. *Journal of Child Neurology, 21*(9), 769–776.

Wilkins, C. H., Sheline, Y. I., Roe, C. M., Birge, S. J., & Morris, J. C. (2006). Vitamin D deficiency is associated with low mood and worse cognitive performance in older adults. *American Journal of Geriatric Psychiatry, 14,* 1032–1040.

Williamson, J. B., & Harrison, D. B. (2003). Functional cerebral asymmetry in hostility: A dual task approach with fluency and cardiovascular regulation. *Brain and Cognition, 52,* 167–174.

Williamson, J. B., Shenal, B. N., Rhodes, R., Foster, P., & Harrison, D. W. (2000). Quantitative EEG diagnostic confirmation of expressive aprosodia. *Archives of Clinical Neuropsychology, 15,* 787.

Williamson, J. B., Harrison, D. W., & Walters, R. (2013). The influence of lateralized stressors on cardiovascular regulation and dichotic listening in hostile men. Manuscript submitted for publication.

Willis, W. D., & Coggeshall, R. E. (2004). *Sensory mechanisms of the spinal cord: Primary afferent neurons and the spinal dorsal horn* (Vol. 1). New York: Springer.

Wilson, S. A. K. (1923). Some problems in neurology, II: Pathological laughing and crying. *Journal of Neurology & Psychopathology, 4,* 299–333.

Wilson, S., & Argyropoulos, S. (2005). Antidepressants and sleep: A qualitative review of the literature. *Drugs, 65*(7), 927–947.

Wilson, E. R. H., Wisely, J. A., Wearden, A. J., Dunn, K., Edwards, J., & Tarrier, N. (2011). Do illness perceptions and mood predict healing time for burn wounds? A prospective, preliminary study. *Journal of Psychosomatic Research, 71*(5), 364–366.

Winn, P. (Ed.). (2001). *Dictionary of biological psychology*. New York: Routledge (Taylor & Francis Group Publishers).

Wise, R. A. (1996). Addictive drugs and brain stimulation reward. *Annual Reviews, 19,* 319–340.

Wise, R. A., et al. (1989). Brain dopamine and reward. *Annual Reviews, 40,* 191–225.

Witt, K., van Dorn, R., & Fazel, S. (2013). Risk factors for violence in psychosis: Systematic review and meta-regression analysis of 110 studies. *PloS One, 8*(2), e55942.

Wittenberg, G. F., & Schaechter, J. D. (2009). The neural basis of constraint-induced movement therapy. *Current Opinion in Neurology, 22*(6), 582.

Wittling, W. (1990). Psychophysiological correlates of human brain asymmetry: Blood pressure changes during lateralized presentation of an emotionally laden film. *Neuropsychologia, 28,* 457–470.

Wittling, W. (1995). Brain asymmetry in the control of autonomic-physiologic activity. In R. J. Davidson & K. Hugdahl (Eds.), *Brain asymmetry* (pp. 305–357). Cambridge: MIT Press.

Wittling, W., & Genzel, S. (1995). Brain asymmetries in cerebral regulation of cortisol secretion. *Homeostasis, 36,* 1–5.

Wodarz, R. (1980). Watershed infarctions and computed tomography. A topographical study in cases with stenosis or occlusion of the carotid artery. *Neuroradiology, 19,* 245–248.

Wolpert, D. M., Goodbody, S. J., & Husain, M. (1998). Maintaining internal representations: The role of the superior parietal lobe. *Nature Neuroscience, 1,* 529–533.

Wood, R. L. (1991). Critical analysis of the concept of sensory stimulation for patients in vegetative states. *Brain Injury, 5*(4), 401–409.

Wood, A. G., Saling, M. M., Abbott, D. F., & Jackson, G. D. (2001). A neurocognitive account of frontal lobe involvement in orthographic lexical retrieval: An fMRI study. *NeuroImage, 14,* 162–169.

Woods, A. J., Philbeck, J. W., & Wirtz, P. (2013). Hyper-arousal decreases human visual thresholds. *PloS One, 8*(4), e61415.

Woodworth, R. S., & Sherrington, C. S. (1904). A pseudaffective reflex and its spinal path. *Journal of Physiology, 31,* 234.

Woolf, N. J., Eckenstein, F., & Butcher, L. (1984). Cholinergic systems in the rat brain: I. Projections to the limbic telencephalon. *Brain Research Bulletin, 13,* 751–784.

Woollett, K., & Maquire, E. A. (2011). Acquiring "the Knowledge" of London's layout drives structural brain changes. *Current Biology,* published online 08 December.

Woolsey, C. N. (1933). Postural relations of the frontal and motor cortex of the dog. *Brain, 56,* 24–370.

Woolsey, C. N. (1952). Patterns of localization in sensory and motor areas of the cerebral cortex. In *The biology of mental health and disease* (pp. 193–206). New York: Hoeber. (399–441).

Woolsey, C. N., Marshall, W. H., & Bard, P. (1942). Representation of cutaneous tactile sensibility in the cerebral cortex of the monkey as indicated by evoked potentials. *Bulletin of the Johns Hopkins Hospital, 71,* 399–441.

Woolsey, C. N., Erickson, T. C., & Gilson, W. E. (1979). Localization in somatic sensory and motor areas of human cerebral cortex as determined by direct recording of evoked potentials and electrical stimulation. *Journal of Neurosurgery, 51,* 476–506.

World Health Organization. (2012). *Visual impairment and blindness*. World Health Organization.

Wurtman, R. J., Axelrod, J., & Phillips, L. S. (1963). Melatonin synthesis in the pineal gland: Control by light. *Science, 142*(3595), 1071–1073.

Wurtz, R. H., & Goldberg, M. E. (1989). *The neurobiology of saccadic eye movements*. Amsterdam: Elsevier.

Wyatt, T. D. (2003). *Pheromones and animal behaviour: Communication by smell and taste*. Cambridge: Cambridge University Press.

Wyke, M. (1966). Postural arm drift associated with brain lesions in man: An experimental analysis. *Archives of Neurology, 15,* 329–334.

Wysocki, C. J., Louie, J., Leyden, J. J., Blank, D., Gill, M., Smith, L., & Preti, G. (2009). Cross-adaptation of a model human stress-related odour with fragrance chemicals and ethyl esters of axillary odorants: Gender-specific effects. *Flavour and Fragrance Journal, 24*, 209–218.

Yamuy, J., Mancillas, J. R., Morales, F. R., & Chase, M. H. (1993). *C-fos* expression in the pons and medulla of the cat during carbachol-induced active sleep. *Journal of Neuroscience, 13*, 2703–2718.

Yarbus, A. L. (1961). Eye movements during examination of complicated objects. *Biofizika, 6*, 52–56.

Yeates, K. O., Bigler, E. D., Gerhardt, C. A., Rubin, K. H., Stancin, T., Taylor, H. G., & Vannatta, K. (2012). Theoretical approaches to understanding social function in childhood brain insults: Toward the integration of social neuroscience and developmental psychology. In V. Anderson & M. H. Beauchamp (Eds.), *Developmental social neuroscience and childhood brain insult: Theory and practice.* New York: The Guilford Press.

Yeomans, J. S., & Frankland, P. W. (1995). The acoustic startle reflex: Neurons and connections. *Brain Research Reviews, 21*, 301–314.

Yerkes, R. M., & Dodson, J. D. (1908). The relation of strength of stimulus to rapidity of habit-formation. *Journal of Comparative Neurology and Psychology, 18*, 459–482.

Yoon, B. W., Morillo, C. A., Cechetto, D. F., & Hachinski, V. (1997). Cerebral hemispheric lateralisation of cardiac autonomic control. *Archives of Neurology, 54*, 741–744.

Young, A. A. (2012). Brainstem sensing of meal-related signals in energy homeostasis. *Neuropharmacology, 63*(1), 31–45, ISSN 0028-3908, doi:10.1016/j.neuropharm.2012.03.019. http://www.sciencedirect.com/science/article/pii/S0028390812001189.

Zakzanis, K. K., Mraz, R., & Graham, S. J. (2005). An fMRI study of the trail making test. *Neuropsychologia, 43*, 1878–1886.

Zamrini, E. Y., Meador, K. J., Loring, D. W., Nichols, F. T., Lee, G. P., Figueroa, R. E., et al. (1990). Unilateral cerebral inactivation produces differential left/right heart rate responses. *Neurology, 40*, 1408–1411.

Zangwill, O. L. (1960). *Cerebral dominance and its relation to psychological functions.* Edinburgh: Oliver & Boyd.

Zarling, E. J. (1984). Abdominal epilepsy: An unusual cause of recurrent abdominal pain. *American Journal of Gastroenterology, 79*(9), 687–688.

Zatorre, R. J., & Belin, P. (2001). Spectral and temporal processing in human auditory cortex. *Cerebral Cortex, 11*, 946–953.

Zatorre, R. J., Evans, A. C., Meyer, E., & Gjedde, A. (1992). Lateralization of phonetic and pitch discrimination in speech processing. *Science, 256*, 846–849.

Zatorre, R. J., Evans, A. C., & Meyer, E. (1994). Neural mechanisms underlying melodic perception and memory for pitch. *Journal of Neuroscience, 14*, 1908–1919.

Zernecke, R., Haegler, K., Kleemann, A. M., Albrecht, J., Frank, T., Linn, J., & Wiesmann, M. (2011). Effects of male anxiety chemosignals on the evaluation of happy facial expressions. *Journal of Psychophysiology, 25*, 116–123.

Zgalhardic, D. J., Foldi, N. S., & Borod, J. C. (2004). Cognitive and behavioral dysfunction in Parkinson's disease: Neurochemical and clinicopathological contributions. *Journal of Neural Transmission, 111*, 1287–1301.

Zicafoose, B. F., Davis, R. A., Oliveira, H. E., Franchina, J. J., Harrison, D. W., & Shapiro, D. M. (1988a). Effects of intrainstitutional relocation on physiological, verbal, and motor performance in elderly patients. Second annual Helen Yura Research Symposium, Old Dominion University.

Zicafoose, B. F., Davis, R. A., Oliveira, H. E., Mays, M. J., Franchina, J. J., Harrison, D. W., & Shapiro, D. M. (1988b). Effects of intrainstitutional relocation on hospitalized elderly patients. Third Annual Conference on Gerontological Nursing, Clemson.

Zorzi, M., Priftis, K., & Umilta, C. (2002). Brain damage—Neglect disrupts the mental number line. *Nature, 417*, 138–139.

Index

CPSIA information can be obtained
at www.ICGtesting.com
Printed in the USA
BVOW07s0348170118
505408BV00025B/114/P